Lecture Notes in Artificial Intelligence 2159

Subseries of Lecture Notes in Computer Science
Edited by J. G. Carbonell and J. Siekmann

Lecture Notes in Computer Science
Edited by G. Goos, J. Hartmanis, and J. van Leeuwen

Springer
Berlin
Heidelberg
New York
Barcelona
Hong Kong
London
Milan
Paris
Tokyo

Jozef Kelemen Petr Sosík (Eds.)

Advances in Artificial Life

6th European Conference, ECAL 2001
Prague, Czech Republic, September 10-14, 2001
Proceedings

Series Editors

Jaime G. Carbonell,Carnegie Mellon University, Pittsburgh, PA, USA
Jörg Siekmann, University of Saarland, Saarbrücken, Germany

Volume Editors

Jozef Kelemen
Petr Sosík
Silesian University, Faculty of Philosophy and Science
Institute of Computer Science
74601 Opava, Czech Republic
E-mail: {petr.sosik/jozef.kelemen}@fpf.slu.cz

Cataloging-in-Publication Data applied for

Die Deutsche Bibliothek - CIP-Einheitsaufnahme

Advances in artificial life : 6th European conference ; proceedings / ECAL 2001, Prague, Czech Republic, September 10 - 14, 2001. Jozef Kelemen ; Petr Sosík (ed.). - Berlin ; Heidelberg ; New York ; Barcelona ; Hong Kong ; London ; Milan ; Paris ; Tokyo : Springer, 2001
(Lecture notes in computer science ; Vol. 2159 : Lecture notes in artificial intelligence)
ISBN 3-540-42567-5

CR Subject Classification (1998): I.2, J.3, F.1.1-2, G.2, H.5, I.5, J.4, J.6

ISBN 3-540-42567-5 Springer-Verlag Berlin Heidelberg New York

Springer-Verlag Berlin Heidelberg New York
a member of BertelsmannSpringer Science+Business Media GmbH

http://www.springer.de

Printed in Germany

Typesetting: Camera-ready by author, data conversion by PTP Berlin, Stefan Sossna
Printed on acid-free paper SPIN 10840282 06/3142 5 4 3 2 1 0

Preface

Why is the question of the difference between living and non-living matter intellectually so attractive to the man of the West? Where are our dreams about our own ability to understand this difference and to overcome it using the firmly established technologies rooted? Where are, for instance, the cultural roots of the enterprises covered nowadays by the discipline of Artificial Life? Contemplating such questions, one of us has recognized [6] the existence of the eternal dream of the man of the West expressed, for example, in the Old Testament as follows: ... the Lord God formed the man from the dust of the ground and breathed into his nostrils the breath of life, and the man became a living being (Genesis, 2.7). This is the dream about the workmanlike act of the creation of Adam from clay, about the creation of life from something non-living, and the confidence in the magic power of technologies. How has this dream developed and been converted into a reality, and how does it determine our present-day activities in science and technology? What is this confidence rooted in? Then God said: "Let us make man in our image... " (Genesis, 1.26). Man believes in his own ability to repeat the Creator's acts, to change ideas into real things, because he believes he is godlike. This confidence is – using the trendy Dawkins' term – perhaps the most important cultural meme of the West.

In Prague there is a myth from the Middle Ages about this dream and confidence. According to it [8,10,11], a famous Prague Rabbi, Judah Loew ben Bezalel, who lived at the end of the 15th and at the beginning of the 16th century (a real person buried in Prague's Old Town Jewish cemetery), once constructed a creature of human shape – the Prague Golem. The construction of Golem had two main phases. First, the rabbi and his collaborators formed an earthen sculpture of a man-like figure. Second, he found the appropriate text, wrote it down on a slip of paper and put it into Golem's mouth. So long as this seal remained in Golem's mouth, the Golem could work and do his master's bidding, and perform all kinds of chores for him, helping him and the Jews of Prague in many ways. The Golem was alive (if we can call such a state alive).

The development of the old idea of Golem has two basic sources: well-developed practical skills in pottery, perhaps the highest technology of the age of the origination of the Old Testament, and the magical exegesis of the Sefer Yezirah (Book of Creation) with the idea of the creative power of the letters (symbols) [3]. Our highest technology of today is the information technology based on computers. Observing the cognitive robotics projects of the end of the 20th and the beginning of the 21st century, we may see some similarities with the work of the rabbi from five centuries ago. The power of the right interplay of silicon-based computer hardware with the software in the form of strings of well-combined symbols breathes life into our artificial agents in many surprising ways. The construction of hardware, the writing and implementing of software in order to have autonomous systems accomplishing specific tasks are, in a certain sense, the same activities as those related with the construction of the Golem.

From the perspective of computer science, Golem is nothing more than a finite-state automaton (or maybe an instance of a Turing Machine). Its abilities are limited to the performance of tasks according to the instructions from their programmer similarly to those of the present-day non-autonomous machines. Golem's ability to interact with its environment is considerably limited (according to the tradition, Golem could not speak). It has nothing like beliefs, intentions, culture, internal imaginary worlds, etc. These insufficiencies limit Golem's cognitive abilities as well as the cognitive abilities of our robots today.

In 1915, centuries after Rabbi Loew's entrance into the depository of the stories about Golems, an extremely sensitive Prague writer, Franz Kafka, published a short story entitled Die Verwandlung (The Metamorphosis) in the journal Die weißen Blätter [7]. From our point of view, this is a typical Kafka-style story about the importance of the environment for body and mind. One morning the hero of the story, Gregor Samsa, woke up as an insect and as a result of this metamorphosis, his whole life changed dramatically. His new body perceived the unchanged environment in ways that were completely different from those of his original human body. Moreover, Samsa is, because of the change of his body, perceived in completely different ways by his relatives, friends, and colleagues. His unchanged mind – full of unchanged knowledge, reasoning, and emotions – found the new ways of existence and social functioning of the new body only very laboriously and under very dramatic circumstances. Kafka's artistic feeling and reflection of the problems inside the triangle of mind, body, and environment, described in The Metamorphosis, convert nowadays into fundamental scientific problems related to the construction of intelligent alive agents. However, there is also another perspective from which we can observe Samsa in his changed form but unchanged environment. It is the perspective of Samsa's original environment (i.e. his family, his colleagues, etc.). From this perspective, we can see the importance of social interactions for the formation of the individuality of a subject. We realize that the individuality of the subject emerges, at least to a certain extent, from its interactions with other individuals. This experience is another important motivation for the present-day efforts in creating and studying collective intelligence and societies of robots.

Here are a few notes on the early history of robots. Approximately in the same period in which Franz Kafka wrote the Metamorphosis another recognized Prague writer Karel Čapek wrote his play Rossum's Universal Robots [2]. It is well known that the word robot is of Slavonic origin and that it was used for the first time in this play. However, it was invented during the summer of 1919 in the Slovak spa Trenčianské Teplice by Karel's brother Josef, who chose the word instead of Karel's original idea to use a word of Latin origin – the word labor. Čapek's play was performed for the first time at the Prague National Theater in 1921. At he beginning of the story, Čapek's robots are devoid of any emotions. Then, step by step, as the level of their mutual interaction develops, they turn into beings with a complex system of motivations for doing things, they turn more and more into emotional beings engaged in intensive communication with other robots as well as with humans, beings capable of complex feelings such as compassion, love, faith, enmity, and anxiety. In this famous play, artificially fabricated bodies have their own beliefs, intentions, culture, internal imaginary worlds, in other words, their own emerged minds.

From the position of computer science, robots, at least in R. U. R., strictly overcome the traditional computational barrier given by the well-known Church-Turing Thesis, as well as the famous undecidability theorem by Gödel. Perhaps just because they interact very intensively with other robots, they understand the goals of the humans, the history of mankind, and they share and modify human culture.

In any case, our present-day research tries to show that massive parallel interactions of a suitable computational agent with not only its actual environment but also with the "cultural history" of this environment could exceed the computational power of the Turing Machine. Contemporary research contributes significantly to our ability to construct artificial agents which will be much more alive than our clever machines today. At the end of Čapek's story, robots become the origin of a new race, thus replacing the human race on the planet. The only living man on Earth, an old scholar called Alquist, recognizes two of the robots as Primus and Helena, the first couple of robots falling in love – Adam and Eve of the generation of beings that may once replace the humans. Years later, Prague's underground philosophy will deal with this idea.

Sometime in the late 1970s of the 20th century, a Prague philosopher and writer Zbyněk Fišer, alias Egon Bondy, wrote (and published first as a "samizdat" in 1983) an essay entitled Juliiny otázky (Julia's Questions) [1]. From the philosophical positions, the essay deals with the question of the evolution of mankind, with its future forms. More precisely, Bondy looks for an answer to the provoking questions like: Is man the final stage (something like a goal) of the (biological) evolution? And if not – this view seems to be quite acceptable for him – what will come in the future? What will replace the homo sapiens species in the line of development? Bondy states that the emancipation from the biological background, its disposal or defeat does not and cannot mean, of course, some immaterial existence of intelligence. A more acceptable opinion for him is that it means the achievement of a state in the development of ontological reality which will be able to produce a more adequate organization of matter. Bondy is not able to call this level of development any other way than some artificial form of existence. Artificial in the sense that this existence is not biogenous, but created by human intellect. Now, we can pose another provoking question: Is the enterprise of the Artificial Life a kind of prologue for creating such organization of matter?

Čapek's and Kafka's visions reflected the turbulent social situation of capitalism in the starting decades of the 20th century. The most general idea of computation we have at hand at the beginning of the 21st century was initiated in the same period. It is the idea of computing machine, the idea of different kinds of computationally equivalent formal models of computation, inspired, for example, by a mechanical activity of a worker (a human being performing computation) considered by Emil Post, and by a mechanical typing machine, the original inspiration of Alan Turing (for more about the stories see [5]). I am sure that this concept is – at least to some extent and in a certain sense – also a reflection of some of the specifics of the very basic thinking paradigm dictated by the actual historical state of the development of human society which produced it.

Fišer's alias Bondy's contemplation about the future of mankind is the result of another thinking paradigm – that of dialectic materialism. However, this Weltanschauung has been formed by the experiences of a philosopher living in the conditions of the so-called real socialism in Czechoslovakia in the 1960s and 1970s of the 20th century. While the social structure based on production and profit was, at the beginning of the 20th century, oriented mainly towards the mechanization and automation of work (in factories as well as in abstract computing devices), and, at least in a certain specific meaning, towards dealing with inorganic matter, the social structure based on dialectic materialism was oriented mainly towards changing the life-style of human beings ("Create a new man for the new future!" was a favored slogan of those times, at least in Czechoslovakia). There is an interesting parallel between the paradigm of mastering inorganic matter and the paradigm of mastering information and organic matter, isn't there?

Seen from the perspective of the science and technology of the late 20th and early 21st centuries, Artificial Life is maybe a contemporary reflection of the above mentioned ancient dream and confidence. It is a field whose fundamental starting point – the concept of life as such – arises from our actual level of knowledge and culture. It seems that this concept necessarily contains an ingredient of something secret. Understanding the principles of the functioning of some entity, we tend to view it as a kind of mechanism rather than a living body. As Rodney Brooks claims in one of his provoking considerations, he sees beetles as machines. Many more examples of this paradigmatic movement could be found, but probably the most apparent one is the starting technological explosion in the field of genetic engineering. On the other hand, as our machines become gradually very complex systems with sometimes hardly predictable behavior, they seem to exhibit some attributes of living matter. And maybe our recent artificial animal-like toys, such as Tamagotchi or cybernetic dogs, are only vanguards of the future invasion of artificial creatures adopting more and more life-like features.

Hence it is no surprise that Artificial Life is now a field – as has been written by Dario Floreano and his co-editors in the preface to the proceedings of the last European Conference on Artificial Life of the 20th century [4] – where biological and artificial sciences meet and blend together, where the dynamics of biological life are reproduced in the memory of computers, where machines evolve, behave, and communicate like living organisms, where complex lifelike entities are synthesized from electronic chromosomes and artificial chemistries.

The adventure of Artificial Life reflects this important paradigmatic change from the inorganic into the organic in the study of computation. It reflects it as a shift from the basic inspiration by mechanical events, like the movement of a reading/typing head one step to the left or to the right, towards phenomena such as massive parallelism, situatedness, embodiment, communication, gene recombination, DNA replication, mutation, evolution, etc. All of these phenomena – and many more – are dealt with in the contributions to the present volume which includes the texts of 5 invited lectures, 54 short communications, and 25 poster presentations selected by the program committee of 6th European Conference on Artificial Life held in Prague, September 10–14, 2001. For better orientation, the submitted contributions and poster texts are divided – more or less accurately according their subjects – into sections: Sec. 2 contains contributions on agents in environments, Sec. 3 on artificial chemistry, Sec. 4 on cellular and neuronal systems, Sec. 5 on collaborative systems, Sec. 6 on evolution, Sec. 7 on robotics, Sec. 8 on vision, visualization, and communication, and Sec. 9 on miscellaneous topics.

References

1. Bondy, E.: *Filosofické eseje, sv. 2. (Philosophical Essays, vol. 2).* Dharma Gaia, Praha, 1993; in Czech
2. Čapek, K.: *R. U. R.* London, 1923 (translated by P. Selver)
3. *Encyclopaedia Judaica* (CD-ROM Edition). Keter Publishing, Jerusalem, 1997
4. Floreano, D., Nicoud, J.-D., Mondada, F.: Preface. In: *Advances in Artificial Life* (D. Floreano et al., eds.). Springer-Verlag, Berlin, 1999, pp. v–vii
5. Hodges, A.: *Alan Turing – The Enigma of Intelligence.* Unwin Paperbacks, London, I983
6. Kelemen, J.: *Můj Golem – eseje o cestě Adama ke kyborgovi (My Golem - Essays on Adam's Route to Cyborg).* Votobia, Olomouc, 2001; in Czech
7. Kafka, F.: *The Metamorphosis.* Vitalis, Prague, 1999
8. Petiška, E.: *Golem.* Martin, Prague, 1991
9. *The Holy Bible – New International Edition.* Zondervan Bible Publishers, Grand Rapids, Mich., 1985
10. *The Prague Golem – Jewish Stories of the Ghetto.* Vitalis, Prague, 2000
11. *The Universal Jewish Encyclopedia in Ten Volumes*, vol. 5. Universal Jewish Encyclopedia Co., New York, 1941

Acknowledgements

The 6th European Conference on Artificial Life in Prague, the proceedings of which you have in your hands, is the result of a collective effort of many colleagues and institutions. Let us express our personal indebtedness and gratitude to all of them, namely:

to all contributors for their high quality contributions and for careful preparation of their texts;

to all our respected colleagues in the Program Committee for their cooperation in assuring a high-level scientific meeting. It was an honor for us to work with them;

to all referees who cooperated with the Program Committee during the evaluation process;

to Action M Agency, and especially to Milena Zeithamlová and Lucie Váchová, for making the local arrangements for the ECAL 2001;

to Jiří Ivánek, vice-rector of the Prague University of Economics, his staff, and the management of the Prague University of Economics, for providing an appropriate place and conditions for ECAL 2001;

to Roman Neruda and Pavel Krušina from the Institute of Computer Science of the Academy of Science of the Czech Republic in Prague, for their work with the ECAL 2001 homepage;

to CTU – EU Center of Excellence MIRACLE for financial support of the ECAL 2001;

to Günter Bachelier for donation of computer graphics to the author of the ECAL 2001 best student contribution;

to Róbert Kelemen for the logo of ECAL 2001 and for designing the ECAL 2001 poster;

to Roman Petrla, technical assistant of Petr Sosík;

to Hana Černínová from the Institute of Computer Science of the Silesian University in Opava, for her active assistance in managing the ECAL 2001 program matters and for technical help in editing these proceedings;

to Springer-Verlag, and especially to Alfred Hofmann, for publishing this volume and for perfect cooperation.

July 2001 Jozef Kelemen and Petr Sosík

Organization

ECAL 2001 was organized by the Institute of Computer Science of the Silesian University in Opava, the University of Economics in Prague, the Academy of Sciences of the Czech Republic, and the Czech Technical University in Prague, in cooperation with the Action M Agency.

Program Committee

Mark A. Bedau	Reed College, Portland, OR
Seth Bullock	University of Leeds, UK
Thomas Christaller	NRCI, Sankt Augustin, Germany
Július Csontó	Technical University, Košice, Slovakia
Peter Dittrich	University of Dortmund, Germany
J. Doyne Farmer	Santa Fe Institute, NM
Dario Floreano	EPFL, Lausanne, Switzerland
Inman Harvey	University of Sussex, Brighton, UK
Jozef Kelemen	**Program Chair**, Silesian University, Opava, Czech Republic
Hiroaki Kitano	Sony Computer Science Laboratory, Japan
Jean-Arcady Meyer	École Normale Supérieure, Paris, France
Alvaro Moreno	University País Vasco, San Sebastián, Spain
Chrystopher Nehaniv	University of Hertfordshire, UK
Jason Noble	University of Leeds, UK
Gheorghe Păun	Romanian Academy, Bucharest, Romania
Luc Steels	VUB, Brussels, Belgium
Lynn A. Stein	F. W. Olin College of Engineering, Needham, MA
Charles E. Taylor	UCLA, Los Angeles, CA

Organizing Committee

Jiří Ivánek	University of Economics, Prague
Jozef Kelemen	Silesian University, Opava
Jiří Lažanský	Czech Technical University, Prague
Roman Neruda	Academy of Sciences of the Czech Republic, Prague
Milena Zeithamlová	Action M Agency, Prague

Referees

Andrew Adamatzky
Chris Adami

Mark A. Bedau
Seth Bullock
Trevor Collins
Julius Csontó
Kyran Dale
Ezequiel Di Paolo
Peter Dittrich
Keith Downing
Marc Ebner
Anders Erikson
Arantza Etxeberria
J. Doyne Farmer
Jeff Fletcher
Dario Floreano
John Hallam
Inman Harvey
Takashi Hashimoto
Peter T. Hraber
Phil Husbands
Takashi Ikegami
Jozef Kelemen
Alica Kelemenová
Jan T. Kim
Dušan Krokavec
Vladimír Kvasnička
Marián Mach
Carlo C. Maley
Claudio Mattiussi
Chris Melhuish
Jean-Arcady Meyer
Alvaro Moreno
Chrystopher Nehaniv
Jason Noble
Ludo Pagie
Gheorghe Păun
John Pepper
Daniel Polani
Diego Rasskin-Gutman
Erik Rauch
Thomas S. Ray
Christopher Ronnewinkel
Hiroki Sayama
Anil Seth
Cosma Shalizi
Peter Sinčák
Andre Skusa
Petr Sosík
Pietro Speroni di Fenizio
Hideaki Suzuki
Yasuhiro Suzuki
Charles E. Taylor
Christy Warrender
Richard Watson
Kyubum Wee
Ron Weiss
Claus O. Wilke
Jens Ziegler

Table of Contents

Invited Lectures

Agents in Environments

Artificial Chemistry

Cellular and Neuronal Systems

Collaborative Systems

Evolution

Robotics

Vision, Visualisation, Language, and Communication

Miscellaneous

Computing in Nonlinear Media: Make Waves, Study Collisions

Andrew Adamatzky

University of the West of England, Bristol, UK
andrew.adamatzky@uwe.ac.uk
http://www.ias.uwe.ac.uk/~a-adamat

Abstract. Over the last decade there has been growing interest in new computing algorithms, architectures and materials. Computation based on wave dynamics and reaction-diffusion processes in chemical, physical and biological systems is one of the new approaches being followed. In this talk I will provide a brief account of the subject. Nonlinear media exhibit a variety of spatio-temporal phenomena. Circular waves, spiral waves, and self-localized mobile excitations are the most familiar examples. How to use these phenomena to perform useful computations? I will show that diverse problems are solved in active nonlinear media, where data and results are given by spatial defects and information processing is implemented via spreading and interaction of phase or diffusive waves. Amusing examples from various fields of science will illustrate vitality of the approach: thin layer chemical reactors, cellular automata machines, diffusive ant families, molecular arrays, and pools of doxastic entities.

1 Introduction

Non-linear media reveal variety of wave phenomena in their space-time dynamic: from diffusive to phase waves and from target waves to mobile self-localizations. How can we make use of these phenomena in a design of advanced computing architectures or algorithms? Take, for example, a chemical medium and change local concentrations of reagents in few sites. Diffusive or phase waves are generated and spread. The waves interact one with another. They form dynamic or stationary, precipitate, structure as a result of their interaction. The medium's micro-volumes update their states simultaneously, molecules also diffuse and react in parallel. That is why the medium can be thought of as a massive parallel processor. In the processor data and results of a computation are symbolized by concentration profiles of reagents while the computation is achieved via spreading and interaction of the waves.

A field of reaction-diffusion and excitable computing is rapidly expanding: it is already took up advanced materials and smart matter, computational complexity and theory of computation, social dynamic and robotics applications. It would be unfair to promise a coherent picture of the evolving area. Therefore in this lecture[1] I make known rather personal views on a progress of non-linear me-

[1] The lecture is partly supported by The Royal Society.

J. Kelemen and P. Sosík (Eds.): ECAL 2001, LNAI 2159, pp. 1–10, 2001.

dia computers and outline some perspectives of future development. A detailed treatment of the subject could be found in the book [1].

2 Reaction-diffusion and excitation

The majority of physical, chemical, biological and social processes can be described in terms of propagating fronts of various sorts. Typical examples include dynamic of excitation in heart and neural tissue, calcium waves in cell cytoplasm, spreading of genes in population dynamic, and forest fire. A nonlinear chemical medium is bistable, i.e. each micro-volume has at least two steady stable states between which the micro-volume switches. In the chemical media fronts of diffusing reactants propagate with constant velocity and wave form. The reagents of the wave front convert reagents ahead of the front into product left behind [8].

In excitable chemical media wave propagation occurs because of coupling between diffusion and autocatalytic reactions. When autocatalytic species are produced in one micro-volume of the medium they diffuse to the neighbouring micro-volumes and they trigger an autocatalytic reaction there. That is why an excitable medium responds to perturbations, which exceed the excitation threshold, by generating excitation waves [8], [1].

Excitation waves are not mechanical waves. They do not conserve energy (however conserve waveform and amplitude), do not interfere and generally do not reflect [12]. Because of these properties the excitation waves play an essential role of information transmission in active non-linear media processors.

3 What is a reaction-diffusion processor?

A reaction-diffusion processor is merely a container, or a reactor, filled with chemicals, or reagents; the reactor transforms input data to output data in a sensible and controllable way via spatial spreading and interaction of the reactants.

A medium of the reactor is at rest at the beginning of a computation. Data are transformed into the spatial configuration of a geometrical object. This configuration is cast onto the medium. After projection, the elements of the data configuration form an ornament of local disturbances in the medium's characteristics, e.g. drops of a reagent or an excitation pattern. The local disturbances generate waves. The waves spread in the medium. Eventually the waves, originating form different data sources, meet each other. They somehow interact and produce either a concentration profile of a precipitate or a stationary excitation structure. This emerging pattern represents the result of the computation.

Essentially, we transform our problems to make them solvable by nonlinear media instead of trying to modify the media to solve the problems. As, back to 1876, Lord Kelvin wrote [13]: *It may be possible to conceive that nature generates a computable function ... directly and not necessarily by approximation as in the traditional approach.* Hopefully, most problems with natural spatial parallelism can be computation efficiently solved in nonlinear active media [1].

4 Specialized processors

Generally, a specialized processor is a computing device designed to solve the only one particular problem, possibly with different data sets.

Let me give you a clue as to how a reaction-diffusion medium, being the specialized processor, can solve a 'real' problem. We could start with problems that have an underlying spatial structure, and thus are open for intuitive solutions. Two problems can be straightforwardly solved in a reaction-diffusion or an excitable medium: a Voronoi diagram and a skeleton. A Voronoi diagram problem is to construct a polygon around each point of a given set in such a manner that any point inside the polygon lies closely to the point that generated the polygon than to any other point of the given set. A skeleton of a planar contour is a set of centers of bitangent circles which lie entirely inside the contour.

To construct these structures we employ distance-to-time transformations. They are implemented as a diffusive or phase waves spread. For example, in the Voronoi diagram of a planar set, waves are generated at given sites of data space. The waves spread out. They interact with each other. They form a stationary structure as a result of this interaction (Figure 1). The situation of a skeleton

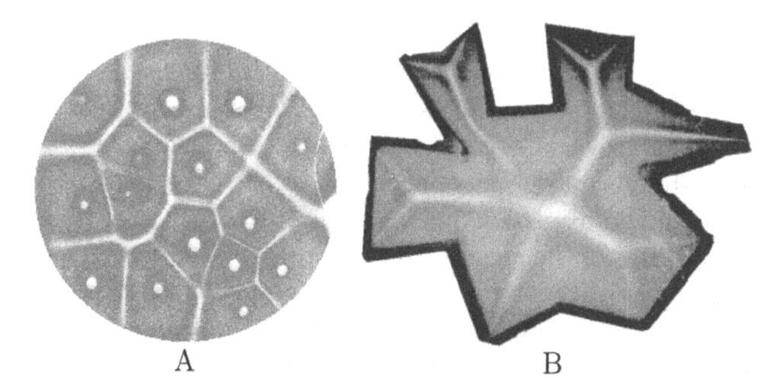

Fig. 1. A Voronoi diagram (A) and a skeleton (B) computed in the real chemical processor. **A.** White discs represent points of given planar set, light segments represent edges of Voronoi cells, dark zones are the sites with precipitate. From [1]. **B.** The initial planar shape is shown in dark colour. The precipitate is grey. Segments and curves of the skeleton are uncoloured. From [1].

is essentially the same. The data shape is projected onto the chemical medium. The medium's sites corresponding to sites with given edge shapes get a specific reagent. The reagent diffuses inwards the shape. A precipitate is formed almost everywhere except sites where two or more waves meet each other. Therefore uncoloured sites of the medium represent edges of the skeleton (Figure 1).

Fig. 2. Reparation of a broken contour in Belousov-Zhabotinsky medium [15]. Published with kind permission of Nicholas Rambidi.

I would say that operations of image processing are more typical problems solved in non-linear media. Take a task of a broken contour reparation in a light-sensitive Belousov-Zhabotinsky medium [15]. A broken contour is projected onto the excitable medium. We want to restore the contour's continuity. Sites belonging to the contour are excited. Excitation waves propagate in the medium (Figure 2). Thus the broken edges of the contour, represented by the excitation waves, expand and finally merge (Figure 2). As a result, we have an inflated but restored shape [15]. Many more examples can be found in [1].

5 Universal processors

In previous section we tackled specialized processors, those build for particular problems. The specialized processors are sometimes inconvenient. Thus, for example, one could not use a reaction-diffusion processor that constructs the skeleton of a planar shape to do arithmetical calculations. A device is called computation universal if it computes any logical function. Are reaction-diffusion and excitable media universal computers?

To prove a medium's universality one have to represent quanta of information, routes of information transmission and logic gates, where quanta information are processed, in the states of the given system. Here I give an overview of two types of computation universality: architecture-based and collision-based universality.

Architecture-based computation assumes that a Boolean circuit is embedded into a system in such manner that all elements of the circuit are represented by the systems stationary states. The architecture is static. Mass transfer or kinetic-based, logical gates [6] and wave gates [17] give us examples of architecture-based universality. When designing a mass transfer gate one assemble the gate of three chemical reactors, connected by tubes. Two reactors represent input variables and the third reactor is an output. In the setup of [6] binary states of two input reactors and one output reactor are represented by levels of acidity in the reactors. Types of logical operations, realizable by the gate, are determined by flow rates between the reactors (Figure 3, A).

The example of a wave gate deals with explicit physical construction of elementary logical gates [17]. It employs particulars of wave interaction in an excitable medium, namely Belousov-Zhabotinsky reaction, to perform a calculation of logical functions at junctions of excitation corridors. Boolean states are

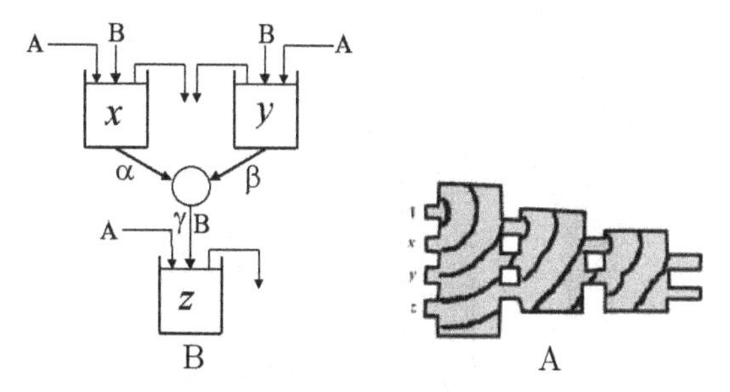

Fig. 3. Two examples of structure-based chemical gates. **A.** Mass transfer based chemical device for computation of basic logical gates. x, y and z are acidity levels in the reactors; A and B are feeding tubes for supply of A and B reagent solutions with an indicator; α and β are coupling coefficients; γ is a flow rate. Modified from [6]. **B.** A wave realization of $x \vee y \vee z$ gate. Waves in output chambers are asynchronous in the case of (`False`, `False`, `False`,) input. If one of the inputs bears a `Truth` value, represented by an excitation wave, the output chambers exhibit synchronous waves, that is represent `Truth`. Modified from [16].

represented by presence or absence of a phase wave at a specified location. The computation of a logical function is based on a particular wave-nucleation size critical for successful generation of a wave [17], [16] and phenomenology of waves interaction (Figure 3, B).

A dynamic, or collision-based, computation employs finite patterns, mobile self-localized excitations, traveling in space and executing computation when they collide with each other. There are no predetermined wires. `Truth` values of logical variables are represented by either absence or presence of the traveling information quanta or by various states of the quanta. A typical interaction gate has two input wires, by 'wire' we mean the trajectory of the mobile self-localization, and, usually, three output wires. Two output wires represent the localizations' trajectories when they continue their motion undisturbed. The third output wire represents the trajectory of a new localization, formed as a result of the collision of two incoming localizations.

Nature gives us various types of nonlinear media that exhibit self-localized mobile patterns in their evolution and therefore are potential candidates for the role of universal computers. The localizations include breathers, solitons, light bullets, excitons, defects, localizations in granular materials, and worms in liquid crystals. Complete catalogues of logic gates, implemented using each of one of these particular species can be found in [1].

Below I discuss two examples of collision-based logic gates in cellular automata models of two-dimensional molecular arrays.

The first example deals with breathers, traveling defects found in numerical experiments on space-time collective dynamic of nonlinear discrete systems. If we take a Hamiltonian with two variables, describing transverse displacement of two bases belonging to base pairs of a DNA molecule, we find that intrinsic local modes are spontaneously emerge. The modes persist because of localiza-

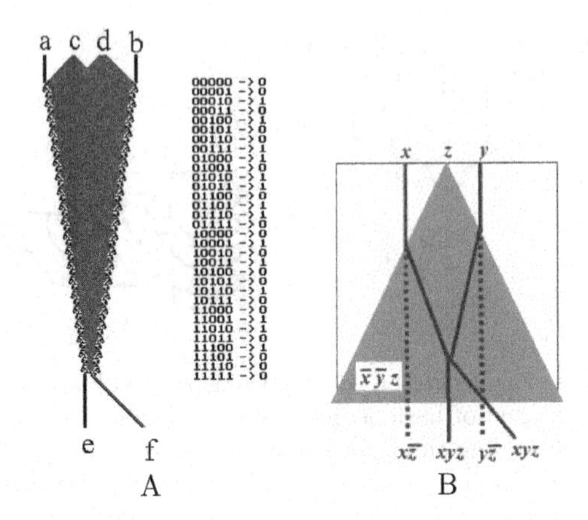

Fig. 4. Interaction of standing localizations with non-localized waves. (**A**) Space-time evolution of one-dimensional cellular automaton, which demonstrates interaction of localizations and non-localized excitations. Time arrow looks down. Rules are shown on the right. Initial positions of two localizations and two sources of excitation are marked by letters **a** and **b** (localizations), and **c** and **d** (sources of spreading excitation). The interaction results in generation of one standing localization (**e**) and one mobile localization (**f**). (**B**) Possible gate built in the result of the interaction.

tion of thermal fluctuations and exchange of energy [9], [10]. When two or more breathers collide, they reflect, slow down, pass undisturbed or form new localization. Thus, enormous variety of logic gates can be implemented. One of examples is shown in Figure 4; dozens of nonetheless impressive instances can be found in [1].

We are discussing light-sensitive mono-molecular aggregates in the second example. If we optically excite an electronic system in the aggregate the competitive mechanisms of delocalization and localization are activated. A localization of excitation emerges due to interplay between these two mechanisms. Take for example a Scheibe aggregate. This is a two-dimensional mono-molecular array, where oxacyanine dye donor molecules are mixed with thiacyanine acceptor molecules [14], [5].

What happens when a photon hits into the aggregate? A domain of excitation is generated. The domain occupies certain number of dye molecules oscillating in phase at the frequency of the fluorescence. The domain moves unchanged across the mono-molecular arrays with the supersonic speed (see e.g. [5] and [4]). A molecule in Scheibe aggregate contacts exactly six neighbouring molecules. Therefore it would be reasonable to represent the aggregate by a two-dimensional hexagonal array, cells of which obey simple rules of finite automata. One example of collision implementation of a logical gate is shown in Figure 5. The truth value of a variable x is defined by presence of a localizaton built of two excited and two refractory cell states. The truth value of a variable y is given by presence of a localization formed of three excited and three refractory sites. The

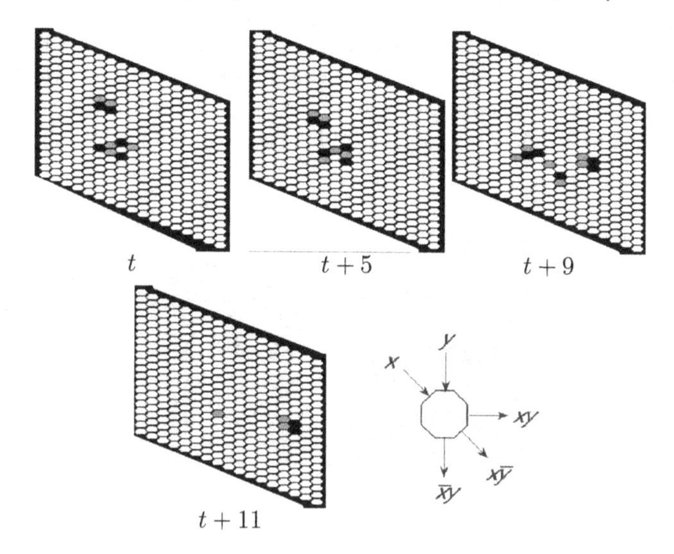

Fig. 5. A collision between two mobile excitations in a cellular automata model of a Scheibe aggregate, and a logical gate realized as a result of the collision.

localizations move toward each other along intersecting trajectories. Eventually they collide. As a result of the collision and subsequent local perturbations, one mobile localization if formed. This localization represents value of xy. It travels along a trajectory perpendicular to trajectory of the signal y.

Quite similar traveling localization could be observed in microtubular arrays, which undergo a continuous process of assembling and disassembling. When a new tubulin subunit is added to an existing microtubule, GTP is hydrolized to GDP and some energy is released. This energy is utilized in changing conformation of the molecule and orientations of electric dipoles. Thus, a defects of antialigned dipoles emerge [7]. The dynamic of defects may be as well used for collision-based computing [1].

6 Discovery of computing abilities

Which problems can be solved in what types of nonlinear media? Should we fabricate these media from scratch or could we instead search for already existing species in nature? To answer the questions I will show which parameters of behaviour of medium's local elements are essential in morphological, dynamic and computational classification of excitable lattices.

A threshold of excitation seems to be a most suitable parameter in the case of excitable media. Unfortunately, the threshold poorly characterizes the space-time excitation dynamic. To discover richer regimes of the excitation dynamic and thereby finding new computing abilities, I enhance the excitation threshold to an excitation interval. In a lattice with local transition rules and 8-cell neighbourhood a cell is excited if the number of excited neighbours belongs to a certain interval $[\theta_1, \theta_2]$, $1 \leq \theta_1 \leq \theta_2 \leq 8$. Using excitation intervals we are able to classify the lattice excitation regimes and to select those suitable for wave-based computing (Figure 6). As we see in the table (Figure 6), an evolution of excitable

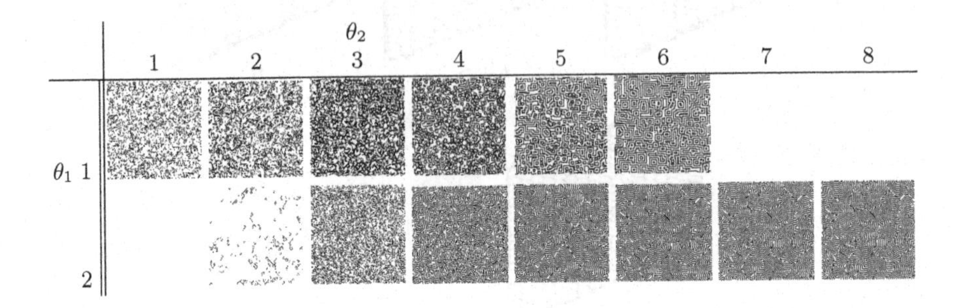

Fig. 6. Morphology of interval excitation.

lattice from sense-less quasi-chaotic activity patterns to formation of proto-waves and then proper labyrinthine-like structures is guided mainly by upper boundary of the excitation interval $[\theta_1, \theta_2]$. When the boundary θ_2 is increased computation capabilities of the medium are changed from image processing (contouring, segmentation, filtration) to universal computation (via colliding localizations) to solution of computational geometry problems (shortest path, spanning tree) ([1]).

7 Molecules with personalities

What if molecules behave humanly? What if humans act like molecules? Chemical media are not the only ones described by reaction-diffusion equations. Thus, not rarely, even crowds, especially in extreme situations of panic or excitement, act according to laws of physics.

What if we think on crowds as drops of reagents? They possibly diffuse. But will they interact? What patterns of precipitates will they form? When fronts of combatants collide and engage into severe fighting they leave dead bodies behind (Figure 7, A). Could we construct a Voronoi diagram with a help of fighting crowds? Certainly. Let us consider ant-inspired mobile finite automata, inhabiting a two-dimensional lattice. Each automaton not only moves, it also picks up and drops pebbles at the lattice nodes. Ideally, we wish to identify those nodes of the lattice which are the edges of the Voronoi cells of the labeled nodes, or data points. Every automaton is given a colour corresponding to one of the labeled nodes. Automata start their walks at the nodes of the given set, which is a seed for the lattice subdivision. Automata carry pebbles. When automata of different colors meet at a node they become agitated and therefore drop their

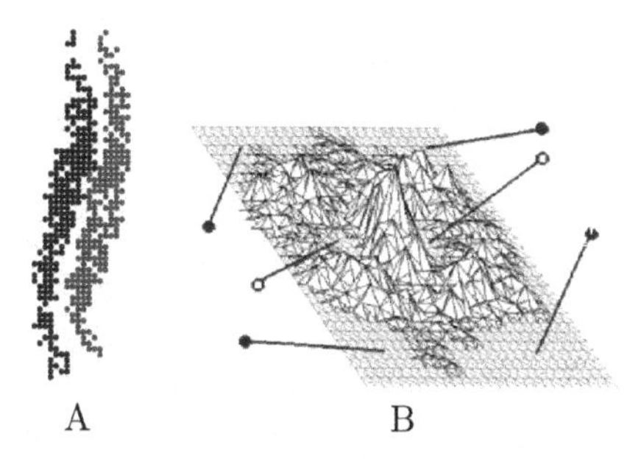

Fig. 7. Computation with spreading and reacting crowds. **A.** Collision of two fronts of virtual combatants. Simulated in ISAAC software [11]. **B.** Pebble-built Voronoi diagram of five lattice nodes, constructed in a lattice swarm model [1]. Data nodes are marked by pins; closed Voronoi cells are indicated by pins with hollow heads.

pebbles. When this process has run its course, the number of pebbles of each color at each node of the lattice indicate membership degree of this node in the set of edge nodes of the Voronoi diagram (Figure 7, B).

What if we use doxastic or affective components instead of chemicals? Are their any particulars in development of cognitive or emotional reactors? We can extract ten abstract reagents which constitute 'material' base of a chemistry of mind. They are knowledge, ignorance, delusion, doubt, and misbelief, all are derivative of belief; and happiness, sadness, anger, confusion and anxiety, as primitive types of emotions.

Analyzing binary interactions of doxastic and affective reagents one can derive several realistic scenarios [2], [3]. Some sets of reactions exhibit quite dull behaviour, others are more interesting. Thus, for example, the following set of reactions show rich oscillating behaviour:

$$\texttt{Anger} + \texttt{Happiness} \longrightarrow \texttt{Confusion} + \texttt{Happiness} + \texttt{Anger}$$
$$\texttt{Anger} + \texttt{Confusion} \longrightarrow \texttt{Confusion}$$
$$\texttt{Happiness} + \texttt{Confusion} \longrightarrow \texttt{Anger} + \texttt{Happiness}.$$

Two types of reactor vessels are worth to investigate: stirred reactors and thin layer, or non-stirred, reactors. Depending on a particular architecture of the para-chemical reactions and relations between doxastic and affective components the system's behaviour exhibits spreading of phase and diffusive waves, formation of domain walls, pseudo-chaotic behaviour and generation of ordered patterns. A simple case of one-dimensional reaction and diffusion is shown in Figure 8.

Finally, I portray a consciousness mind as a chemical reactor, where basic types of emotions and beliefs are symbolized by chemical species, molecules or reagents. The approach is analogical to Minski's many-agents mind, Dennet's space-time consciousness, and Albert's many-minds-in-one-head theory.

Fig. 8. Diffusion of a doubt (δ) in a one-dimensional layer of knowledge (shown by dots). The doubt reacts with the knowledge and a misbelief is produced. Ignorance (ι) and delusion (ϵ) emerge as byproducts of the diffusion and reaction. Certain norms are applied, see [2] for details.

References

1. Adamatzky A. *Computation in Nonlinear Media and Automata Collectives* (Institute of Physics Publishing, 2001).
2. Adamatzky A. Space-time dynamic of normalized doxatons: automata models of pathological collective mentality *Chaos, Solitons and Fractals* **12** (2001) 1629–1656.
3. Adamatzky A. Pathology of collective doxa. Automata model. *Appl. Math. Comput.*.
4. Bang O., Christiansen P.L., If F., Rasmussen K. Ø. and Gaididei Y. Temperature effects in a nonlinear model of monolayer Scheibe aggregates *Phys. Rev.* **E 49** (1994) 4627 – 4635.
5. Bartnik E.A. and Tuszynski J.A. Theoretical models of energy transfer in two-dimensional molecular assemblies *Phys. Rev.* **E 48** (1993) 1516 – 1528.
6. Blittersdorf R., Müller M. and Schneider F. W. Chemical visualization of Boolean functions: A simple chemical computer *J. Chem. Educat.* **72** (1995) 760 - 763.
7. Brown J.A. and Tuszynski J.A. Dipole interactions in axonal microtubules as a mechanism of signal propagation *Phys. Rev. E* **56** (1997) 5834 – 5839.
8. Epstein I.R. and Showalter K. Nonlinear chemical dynamics: oscillations, patterns and chaos *J. Phys. Chem.* **100** (1996) 13132–13147.
9. Forinash K., Peyrard M. and Malomed B. Interaction of discrete breathers with impurity modes *Phys. Review E* **49** (1994) 3400 – 3411.
10. Forinash K., Cretegny T. and Peyrard M. Local modes and localizations in a multicomponent lattices *Phys. Rev.* **55** (1997) 4740 – 4756.
11. Ilachinski A. Irreducible semi-autonomous adaptive combat (ISAAC): An artificial-life approach to land warfare *CNA Research Memo. 91-61.10* (1997).
12. Krinski V.I. Auotwaves: results, problems, outlooks In: *Self-Organization: Autowaves and Structures Far From Equilibrium* (Springer–Verlag, 1984) 9–19.
13. Thomson W. (Lord Kelvin) On an instrument for calculating integral of the product of two given functions *Proc. Roy. Soc. London* **24** (1876) 266–275.
14. Moebius D. and Kuhn H. Energy transfer in monolayers with cyanine dye Scheibe aggregates *J. Appl. Phys.* **64** (1979) 5138-5141.
15. Rambidi N.G., Maximychev A.V. and Usatov A.V. Molecular neural network devices based on non-linear dynamic media *BioSystems* **33** (1994) 125–137.
16. Steinbock O., Kettunen P. and Showalter K. Chemical wave logic gates *J. Phys. Chem.* **100** (1996) 49, 18970–18975.
17. Tóth A. and Showalter K. Logic gates in excitable media *J. Chem. Phys.* **103** (1995) 2058 – 2066.

Ant Algorithms Solve Difficult Optimization Problems

Marco Dorigo

IRIDIA
Université Libre de Bruxelles
50 Avenue F. Roosevelt
B-1050 Brussels, Belgium
mdorigo@ulb.ac.be

Abstract. The *ant algorithms* research field builds on the idea that the study of the behavior of ant colonies or other social insects is interesting for computer scientists, because it provides models of distributed organization that can be used as a source of inspiration for the design of optimization and distributed control algorithms. In this paper we overview this growing research field, giving particular attention to ant colony optimization, the currently most successful example of ant algorithms, as well as to some other promising directions such as ant algorithms inspired by labor division and brood sorting.

1 Introduction

Models based on self-organization have recently been introduced by ethologists to study collective behavior in social insects [2,3,5,14]. While the main motivation for the development of these models was to understand how complex behavior at the colony level emerges out of interactions among individual insects, computer scientists have recently started to exploit these models as an inspiration for the design of useful optimization and distributed control algorithms. For example, a model of cooperative foraging in ants has been transformed into a set of optimization algorithms, now known as ant colony optimization (ACO) [22,24], capable of tackling very hard computational problems, such as the traveling salesman [28,20,29,26,60], the quadratic assignment problem [47,46,35,61], the sequential ordering problem [33], the shortest common supersequence problem [49], various scheduling problems [57,12,48], and many others (see Table 1). More recently, ACO has also been successfully applied to distributed control problems such as adaptive routing in communications networks [54,17]. Another model, initially introduced to explain brood sorting in ants, was used by computer scientists to devise a distributed algorithm for data clustering [43,39]. And a model of flexible task allocation in wasps has become a distributed algorithm for dynamic scheduling or resource allocation in a factory or a computer network [9]. This line of research, termed *ant algorithms* [21,23,27] or *swarm intelligence* [2,4] has also met the interest of roboticists for the design of distributed algorithms for the control of swarms of robots.

J. Kelemen and P. Sosík (Eds.): ECAL 2001, LNAI 2159, pp. 11–22, 2001.

In the following we will overview the main results obtained in the field of ant algorithms. In Section 2 we introduce the ACO metaheuristic, currently the most successful example of ant algorithms, using the classical traveling salesman problem as an example, and we overview the results obtained on a number of problems. In Section 3 we briefly present other types of ant algorithms that, although still in the phase of exploratory research, look promising and maybe one day will be as successful as ACO has shown to be.

2 From Real Ants to Ant Colony Optimization

An important insight of early research on ants' behavior was that in many ant species the visual perceptive faculty is very rudimentarily developed (there are even ant species which are completely blind) and that most communication among individuals, or between individuals and their environment, is based on the use of chemicals, called *pheromones*, produced by the ants. Particularly important for the social life of some ant species is the *trail pheromone*, a pheromone that individuals deposit while walking in search for food. By sensing pheromone trails, foragers can follow the path to food discovered by other ants. This collective pheromone-laying/pheromone-following behavior whereby an ant is influenced by a chemical trail left by other ants was the inspiring source of ant colony optimization, as explained in the following by means of the double bridge experiment.

2.1 The Double Bridge Experiment

The double bridge experiment [36] is an important experiment in the field of ant algorithms. In fact, it gave the initial inspiration [20,28] to all the research work that led to the definition of the ACO metaheuristic. In the double bridge experiment, see Figure 1, an ant nest is connected to a food source via two paths of different length. At start time all ants are in the nest and they are left free to move. The experimental apparatus is built in such a way that the only way for the ants to reach the food is by using one of the two bridge branches. In the initial phase the ants move randomly and they choose between the shorter and the longer branch with equal probability. While walking ants deposit on the ground a pheromone trail; when choosing their way, ants choose with higher probability those directions marked by a stronger pheromone concentration. As those ants choosing the shorter branch will also be the first to find the food and to go back to the nest, the pheromone trail on the shorter branch will grow faster, increasing this way the probability that it will be used by forthcoming ants. This auto-catalytic (positive feedback) process is at the heart of the auto-organizing behavior that very quickly leads all the ants to choose the shortest branch. A similar mechanism can be used by opportunely defined *artificial ants* to find minimum cost paths on graphs, as explained in the following.

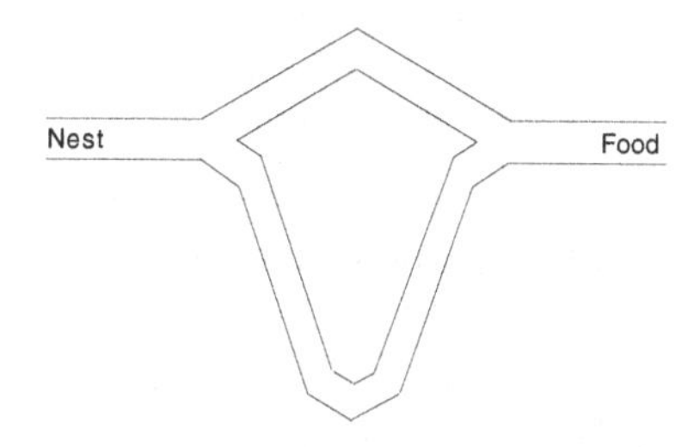

Fig. 1. Experimental setup for the double bridge experiment. Modified from [14,36].

2.2 Artificial Ants for the Traveling Salesman Problem

To illustrate how artificial ants can solve optimization problems, we consider an easy-to-formulate but hard-to-solve combinatorial problem: the well-known traveling salesman problem (TSP), a problem known to be $\mathcal{NP}$-hard.

Consider a set of cities and a set of weighted edges (the edges represent direct connections between pairs of cities and the weights may represent, for example, the distance between the connected cities) such that the induced graph is connected. The TSP is easily described as follows: find the shortest tour that visits each city in the set once and only once. The (constructive and stochastic) algorithm followed by the artificial ants exploits virtual pheromone concentrations associated to the edges connecting pairs of cities: artificial ants build tours, selecting which next city to hop to among a list of non-visited cities depending on city distance and on pheromone concentration. The shorter tours are reinforced by increasing the pheromone concentrations. Then evaporation, which consists in decreasing the pheromone values, is applied to all edges. After a number of iterations, very good solutions are discovered. In addition to finding a very good solution, the algorithm maintains a pool of alternative portions of solutions: this feature may become particularly interesting when the problem is dynamically changing (as most real-world problems are), since the algorithm can focus the search toward this pool of alternative portions of solutions.

The traveling salesman problem obviously lends itself to an ant-based description. But many other optimization problems can be solved with the same approach, because they can be formulated as minimum cost path problems on graphs: the ant-based approach has been shown to be extremely efficient on structured (real-world) instances of the quadratic assignment problem, on the sequential ordering problem, the vehicle routing problem, the shortest common supersequence problem, and many others (see Table 1).

2.3 The Ant Colony Optimization Metaheuristic

Although the algorithms developed for the above-mentioned applications differ in many details among themselves, still their artificial ants share the basic behavior

```
procedure ACO metaheuristic
   ScheduleActivities {possibly in parallel}
      ManageAntsActivity()
      EvaporatePheromone()
      DaemonActions() {Optional}
   end ScheduleActivities
end ACO metaheuristic
```

Fig. 2. The ACO metaheuristic in pseudo-code. Comments are enclosed in braces. The **ScheduleActivities** construct may be executed sequentially, as typically happens in combinatorial optimization problems, or in parallel, as done for example in routing applications. The procedure DaemonActions() is optional and refers to centralized actions executed by a daemon possessing global knowledge.

and cooperation mechanisms as the TSP artificial ants explained above (see also [20,28,29]). This fact was recently captured in the definition of a common framework, called ACO metaheuristic [22,24]. Informally, the ACO metaheuristic (see also Figure 2) can be defined as follows (the reader interested in a more formal definition should refer to [22]).

A colony of (artificial) ants concurrently and asynchronously build solutions to a given discrete optimization problem by moving on the problem's graph representation, where each feasible path encodes a solution of the problem. They move by applying a stochastic local decision rule that exploits pheromone trail values. By moving, ants incrementally build solutions to the optimization problem. Once an ant has built a solution, or while the solution is being built, the ant evaluates the (partial) solution and deposits pheromone on the graph components it used. This pheromone information directs the search of the ants in the future.

Besides ants' activity, an ACO algorithm includes two additional procedures: *pheromone trail evaporation* and *daemon actions* (the last component being optional). Pheromone evaporation is the process by means of which the pheromone trail intensity on the components decreases over time. From a practical point of view, pheromone evaporation is needed to avoid a too rapid convergence of the algorithm towards a sub-optimal region. It implements a useful form of *forgetting*, favoring the exploration of new areas of the search space. Daemon actions can be used to implement centralized actions which cannot be performed by single ants. Examples are the activation of a local optimization procedure, or the collection of global information that can be used to decide whether it is useful or not to deposit additional pheromone to bias the search process from a non-local perspective. As a practical example, the daemon can choose to deposit extra pheromone on the components used by the ant that built the best solution.

It is interesting to note that the ACO approach has a feature which makes it particularly appealing: it is explicitly formulated in terms of computational

Table 1. Some of the current applications of ACO algorithms. Applications are listed by class of problems and in chronological order.

Problem name	*Authors*	*Algorithm name*	*Year*	*Main references*
Traveling salesman	Dorigo, Maniezzo & Colorni	AS	1991	[20,28,29]
	Gambardella & Dorigo	Ant-Q	1995	[30]
	Dorigo & Gambardella	ACS & ACS-3-opt	1996	[25,26,31]
	Stützle & Hoos	$\mathcal{MMAS}$	1997	[60,58,61]
	Bullnheimer, Hartl & Strauss	AS_{rank}	1997	[8]
Quadratic assignment	Maniezzo, Colorni & Dorigo	AS-QAP	1994	[47]
	Gambardella, Taillard & Dorigo	HAS-QAP[a]	1997	[35]
	Stützle & Hoos	$\mathcal{MMAS}$-QAP	1997	[56,61]
	Maniezzo	ANTS-QAP	1998	[44]
	Maniezzo & Colorni	AS-QAP[b]	1999	[46]
Scheduling problems	Colorni, Dorigo & Maniezzo	AS-JSP	1994	[10]
	Stützle	AS-FSP	1997	[57]
	Bauer et *al.*	ACS-SMTTP	1999	[1]
	den Besten, Stützle & Dorigo	ACS-SMTWTP	1999	[12]
	Merkle, Middendorf & Schmeck	ACO-RCPS	2000	[48]
Vehicle routing	Bullnheimer, Hartl & Strauss	AS-VRP	1997	[6,7]
	Gambardella, Taillard & Agazzi	HAS-VRP	1999	[34]
Connection-oriented network routing	Schoonderwoerd *et al.*	ABC	1996	[54,53]
	Di Caro & Dorigo	AntNet-FS	1998	[18]
Connection-less network routing	Di Caro & Dorigo	AntNet & AntNet-FA	1997	[16,17,19]
Sequential ordering	Gambardella & Dorigo	HAS-SOP	1997	[32,33]
Graph coloring	Costa & Hertz	ANTCOL	1997	[11]
Shortest common supersequence	Michel & Middendorf	AS-SCS	1998	[49,50]
Frequency assignment	Maniezzo & Carbonaro	ANTS-FAP	1998	[45]
Generalized assignment	Ramalhinho Lourenço & Serra	MMAS-GAP	1998	[52]
Multiple knapsack	Leguizamón & Michalewicz	AS-MKP	1999	[41]
Optical networks routing	Navarro Varela & Sinclair	ACO-VWP	1999	[51]
Redundancy allocation	Liang & Smith	ACO-RAP	1999	[42]
Constraint satisfaction	Solnon	Ant-P-solver	2000	[55]

[a] HAS-QAP is an ant algorithm which does not follow all the aspects of the ACO metaheuristic.
[b] This is a variant of the original AS-QAP.

agents. While it may in principle be possible to get rid of the agents to focus on the core optimizing mechanism (reinforcement and evaporation), the agent-based formulation may prove to be a useful aid for designing problem-solving systems. Routing in communications networks is a very good example of this aspect. Routing is the mechanism that directs messages in a communications network from their source nodes to their destination nodes through a sequence of intermediate nodes or switching stations. Each switching station has a routing table that tells messages or portions of messages called packets where to go given their destinations. Because of the highly dynamic nature of communications networks due to the time-varying stochastic changes in network load, as well as to unpredictable failures of network components, areas of the network may become congested and new routes have to be discovered dynamically. In the ant-based approach [17,54] ant-like agents reinforce routing table entries depending on their experience in the network: for example, if an agent has been delayed a long time because it went through a highly congested area of the network, it will only weakly, or not at all, reinforce routing table entries that send packets to that area of the network. A forgetting (or evaporation) mechanism is also applied regularly to refresh the system (to avoid obsolete solutions being maintained). AntNet [17], an ACO algorithm designed for routing in packet-switched networks, was shown to outperform (in realistically simulated conditions) all routing algorithms in widespread use, especially, but not only, in strongly variable traffic conditions. Implicitly maintaining a pool of alternative partial routes is the way the system copes with changing conditions and allows it to be flexible and robust.

3 Other Applications Inspired by Social Insects

As we said, ant colony optimization algorithms are only one, although the most successful, example of ant algorithms. Wagner et al. have proposed two algorithms also inspired by the foraging behavior of ant colonies for the exploration of a graph. Other researchers have taken inspiration from other social insect behaviors, such as division of labor, brood sorting and cemetery organization, to propose new types of distributed, multi-agent algorithms, as explained in the following.

3.1 Foraging and Graph Exploration

Taking inspiration from the pheromone-laying/pheromone-following behavior of ant colonies, Wagner et al. [62,63] have proposed two algorithms for exploring a graph called respectively *Edge Ant Walk* [62] and *Vertex Ant Walk* [63] in which one or more artificial ants walk along the edges of the graph, lay a pheromone trail on the visited edges (respectively nodes) and use the pheromone trails deposited by previous ants to direct their exploration. Although the general idea behind the algorithm is similar to the one that inspired ant colony optimization, their goal and implementation are very different. In the work of Wagner et al., pheromone trail is used as a kind of distributed memory that directs the ants towards unexplored areas of the search space. In fact, their goal is to cover the graph, that is to visit all the nodes, without knowing the graph topology. They were able to prove a number of theoretical results, for example concerning the time complexity for covering a generic graph. Also, they recently extended their algorithms [64] so that they can be applied to dynamically changing graphs. A possible and promising application of this work is to Internet search, where the problem is to track down the hundreds of thousands of pages added every day [40] (as well as the ones that disappear).

3.2 Division of Labor and Dynamic Task Allocation in Robots

Division of labor is an important and widespread feature of colonial life in many species of social insects. In ant colonies, for example, workers and soldier ants are typically concerned with nest maintenance and nest defense, respectively. Still, individuals of one class can, if necessary, perform activities typical of the other class. This self-organizing, adaptive aspect of labor division can be explained by a simple behavioral threshold model: although each ant is equipped with a complete set of behavioral responses, different ants have different threshold for different behaviors. For example, a soldier generally has a low threshold for defense related activities and a high threshold for nest maintenance duties, while for workers it is just the opposite. An high demand of workers, due for example to the sudden need for extra brood care, can lead to the involvement of soldiers in this, for them atypical, activity. Interestingly, ants in a same class have similar, but not identical, threshold levels for the same activity. This differentiation determines a continuum in the space of behavioral responses so that the ant colony as a whole can adapt to continuously changing environmental conditions. Krieger and Billeter [38] used the threshold model of ants to define a distributed labor division system for a group of robots that have the goal of collecting items dispersed in an arena. Their experiments have shown that robots governed by threshold-based behavior activation can accomplish the task, and that the system as a whole presents inherent fault-tolerance and graceful degradation of performance: the faulty behavior of one or more robots causes the automatic adaptation in the behavior of the other robots with only minor decreases in overall performance. Most important, this result was accomplished without designing any explicit mechanism of communication among robots, and without the faulty situations being explicitly included in the robots control programs. Although the experiments were run with small toy robots in university lab experimental conditions, it seems clear that the approach has great potentialities for the control of fleets of industrial robots in unstructured environments.

3.3 Brood Sorting, Cemetery Organization, and Data Clustering

Another typical activity that can be observed in ant colonies is clustering of objects. For example, ants cluster corpses of dead ants into cemeteries, or food items into the nest. The clustering activity is performed in a completely distributed way, that is, without any central control mechanism. The way ants accomplish this can be explained by a simple model, similar to the one used for labor division. In the model, ants are endowed with "pick up" and "drop" behaviors and these behaviors are activated with probabilities that are function of both a threshold and of environmental conditions [15]. The environmental condition is in this case given by the density of items in the neighborhood of the ants location. While moving around, an ant carrying an item has a high probability to drop it in zones where a high density of the same item is found, but a low probability to drop it in zones where there is a low density. On the contrary, when an unloaded ant meets an item, it will pick it up with high probability if the zone in which it is located has a low density of that same item, with low probability otherwise. If more kinds of items are present and ants use different thresholds for different items, these are sorted into different clusters. This simple mechanism has been put to work for data visualization and clustering. Kuntz et al. [39], building on an algorithm first proposed by Lumer and Faieta [43], consider the following problem. Given a set of n-dimensional data represented as points in a n-dimensional space, and a metric d which measures the distance between pairs of data items, project the points on a plane so that points in the plane belong to a same cluster if and only if the corresponding data items are similar in the n-dimensional space under the metric d. To let artificial ants solve this problem, the initial projection of data items on the plane is done randomly. Then artificial ants move randomly on the plane and pick up or drop projected data items using rules equivalent to those of the threshold model explained above. The results obtained are qualitatively comparable to those obtained by more classic techniques such as spectral decomposition or stress minimization, but at a much lower computational cost. Moreover, the technique can be easily extended to other difficult problems such as multidimensional scaling (i.e., the problem of transforming a squared matrix of distances among pairs of points into the coordinate of the original points), or data sorting.

4 Conclusions

The researchers' interest in ant algorithms has recently greatly increased, due to both the charm of the ant colony metaphor and the excitement caused by some very promising results obtained in practical applications. Although the approach is very promising, a more systematic comparison with other heuristics is required. We also need to better understand why ant algorithms work so well on certain types of problems, and to clearly identify the problem characteristics which make a problem susceptible of being successfully tackled by an ant algorithm. Finally, results on the theoretical properties of these algorithms are most of the times missing. A notable exception concerns some important instances of ACO algorithms, for which convergence to the optimal solution has recently been proved [37,59].

Acknowledgments. This work was partially supported by the "Metaheuristics Network", a Research Training Network funded by the Improving Human Potential programme of the CEC, grant HPRN-CT-1999-00106. The information provided is the sole responsibility of the authors and does not reflect the Community's opinion. The Community is not responsible for any use that might be made of data appearing in this publication. Marco Dorigo acknowledges support from the Belgian FNRS, of which he is a Senior Research Associate.

References

1. A. Bauer, B. Bullnheimer, R. F. Hartl, and C. Strauss. An ant colony optimization approach for the single machine total tardiness problem. In *Proceedings of the 1999 Congress on Evolutionary Computation (CEC'99)*, pages 1445–1450. IEEE Press, Piscataway, NJ, 1999.
2. E. Bonabeau, M. Dorigo, and G. Theraulaz. *Swarm Intelligence: From Natural to Artificial Systems.* Oxford University Press, New York, NJ, 1999.

3. E. Bonabeau, M. Dorigo, and G. Theraulaz. Inspiration for optimization from social insect behavior. *Nature*, 406:39–42, 2000.
4. E. Bonabeau and G. Theraulaz. Swarm smarts. *Scientific American*, 282(3):54–61, 2000.
5. E. Bonabeau, G. Theraulaz, J.-L. Deneubourg, S. Aron, and S. Camazine. Self-organization in social insects. *Tree*, 12(5):188–193, 1997.
6. B. Bullnheimer, R. F. Hartl, and C. Strauss. Applying the Ant System to the vehicle routing problem. In S. Voß, S. Martello, I. H. Osman, and C. Roucairol, editors, *Meta-Heuristics: Advances and Trends in Local Search Paradigms for Optimization*, pages 285–296. Kluwer Academic Publishers, Dordrecht, 1999.
7. B. Bullnheimer, R. F. Hartl, and C. Strauss. An improved ant system algorithm for the vehicle routing problem. *Annals of Operations Research*, 89:319–328, 1999.
8. B. Bullnheimer, R. F. Hartl, and C. Strauss. A new rank-based version of the Ant System: A computational study. *Central European Journal for Operations Research and Economics*, 7(1):25–38, 1999.
9. M. Campos, E. Bonabeau, G. Theraulaz, and J.-L. Deneubourg. Dynamic scheduling and division of labor in social insects. *Adaptive Behavior*, 8(3):83–96, 2000.
10. A. Colorni, M. Dorigo, V. Maniezzo, and M. Trubian. Ant System for job-shop scheduling. *JORBEL – Belgian Journal of Operations Research, Statistics and Computer Science*, 34(1):39–53, 1994.
11. D. Costa and A. Hertz. Ants can colour graphs. *Journal of the Operational Research Society*, 48:295–305, 1997.
12. M. L. den Besten, T. Stützle, and M. Dorigo. Ant colony optimization for the total weighted tardiness problem. In M. Schoenauer, K. Deb, G. Rudolph, X. Yao, E. Lutton, J. J. Merelo, and H.-S. Schwefel, editors, *Proceedings of PPSN-VI, Sixth International Conference on Parallel Problem Solving from Nature*, volume 1917 of *Lecture Notes in Computer Science*, pages 611–620. Springer Verlag, Berlin, Germany, 2000.
13. M. L. den Besten, T. Stützle, and M. Dorigo. Design of iterated local search algorithms: An example application to the single machine total weighted tardiness problem. In E. J. W. Boers, J. Gottlieb, P. L. Lanzi, R. E. Smith, S. Cagnoni, E. Hart, G. R. Raidl, and H. Tijink, editors, *Proceedings of EvoStim'01*, Lecture Notes in Computer Science, pages 441–452. Springer Verlag, Berlin, Germany, 2001.
14. J.-L. Deneubourg, S. Aron, S. Goss, and J.-M. Pasteels. The self-organizing exploratory pattern of the Argentine ant. *Journal of Insect Behavior*, 3:159–168, 1990.
15. J.-L. Deneubourg, S. Goss, N. Franks, A. Sendova-Franks, C. Detrain, and L. Chrétien. The dynamics of collective sorting: Robot-like ants and ant-like robots. In J.-A. Meyer and S. W. Wilson, editors, *Proceedings of the First International Conference on Simulation of Adaptive Behavior: From Animals to Animats*, pages 356–363. MIT Press, Cambridge, MA, 1991.
16. G. Di Caro and M. Dorigo. AntNet: A mobile agents approach to adaptive routing. Technical Report IRIDIA/97-12, IRIDIA, Université Libre de Bruxelles, Belgium, 1997.
17. G. Di Caro and M. Dorigo. AntNet: Distributed stigmergetic control for communications networks. *Journal of Artificial Intelligence Research*, 9:317–365, 1998.
18. G. Di Caro and M. Dorigo. Extending AntNet for best-effort Quality-of-Service routing. Unpublished presentation at *ANTS'98 – From Ant Colonies to Artificial Ants: First International Workshop on Ant Colony Optimization* http://iridia.ulb.ac.be/ants98/ants98.html, October 15–16 1998.

19. G. Di Caro and M. Dorigo. Two ant colony algorithms for best-effort routing in datagram networks. In Y. Pan, S. G. Akl, and K. Li, editors, *Proceedings of the Tenth IASTED International Conference on Parallel and Distributed Computing and Systems (PDCS'98)*, pages 541–546. IASTED/ACTA Press, Anheim, 1998.
20. M. Dorigo. *Optimization, Learning and Natural Algorithms* (in Italian). PhD thesis, Dipartimento di Elettronica, Politecnico di Milano, Italy, 1992. 140 pages.
21. M. Dorigo, E. Bonabeau, and G. Theraulaz. Ant algorithms and stigmergy. *Future Generation Computer Systems*, 16(8):851–871, 2000.
22. M. Dorigo and G. Di Caro. The Ant Colony Optimization meta-heuristic. In D. Corne, M. Dorigo, and F. Glover, editors, *New Ideas in Optimization*, pages 11–32. McGraw Hill, London, UK, 1999.
23. M. Dorigo, G. Di Caro, and T. Stützle (Editors). Special issue on "Ant Algorithms". *Future Generation Computer Systems*, 16(8), 2000. 104 pages.
24. M. Dorigo, G. Di Caro, and L. M. Gambardella. Ant algorithms for discrete optimization. *Artificial Life*, 5(2):137–172, 1999.
25. M. Dorigo and L. M. Gambardella. Ant colonies for the traveling salesman problem. *BioSystems*, 43:73–81, 1997.
26. M. Dorigo and L. M. Gambardella. Ant Colony System: A cooperative learning approach to the traveling salesman problem. *IEEE Transactions on Evolutionary Computation*, 1(1):53–66, 1997.
27. M. Dorigo, L. M. Gambardella, M. Middendorf, and T. Stützle (Editors). Special issue on "Ant Algorithms and Swarm Intelligence". *IEEE Transactions on Evolutionary Computation*, 2002.
28. M. Dorigo, V. Maniezzo, and A. Colorni. Positive feedback as a search strategy. Technical Report 91-016, Dipartimento di Elettronica, Politecnico di Milano, Italy, 1991.
29. M. Dorigo, V. Maniezzo, and A. Colorni. The Ant System: Optimization by a colony of cooperating agents. *IEEE Transactions on Systems, Man, and Cybernetics – Part B*, 26(1):29–41, 1996.
30. L. M. Gambardella and M. Dorigo. Ant-Q: A reinforcement learning approach to the traveling salesman problem. In A. Prieditis and S. Russell, editors, *Proceedings of the Twelfth International Conference on Machine Learning (ML-95)*, pages 252–260. Morgan Kaufmann Publishers, Palo Alto, CA, 1995.
31. L. M. Gambardella and M. Dorigo. Solving symmetric and asymmetric TSPs by ant colonies. In *Proceedings of the 1996 IEEE International Conference on Evolutionary Computation (ICEC'96)*, pages 622–627. IEEE Press, Piscataway, NJ, 1996.
32. L. M. Gambardella and M. Dorigo. HAS-SOP: An hybrid Ant System for the sequential ordering problem. Technical Report IDSIA-11-97, IDSIA, Lugano, Switzerland, 1997.
33. L. M. Gambardella and M. Dorigo. Ant Colony System hybridized with a new local search for the sequential ordering problem. *INFORMS Journal on Computing*, 12(3):237–255, 2000.
34. L. M. Gambardella, È. D. Taillard, and G. Agazzi. MACS-VRPTW: A multiple ant colony system for vehicle routing problems with time windows. In D. Corne, M. Dorigo, and F. Glover, editors, *New Ideas in Optimization*, pages 63–76. McGraw Hill, London, UK, 1999.
35. L. M. Gambardella, È. D. Taillard, and M. Dorigo. Ant colonies for the quadratic assignment problem. *Journal of the Operational Research Society*, 50(2):167–176, 1999.

36. S. Goss, S. Aron, J. L. Deneubourg, and J. M. Pasteels. Self-organized shortcuts in the Argentine ant. *Naturwissenschaften*, 76:579–581, 1989.
37. W. J. Gutjahr. A graph-based Ant System and its convergence. *Future Generation Computer Systems*, 16(8):873–888, 2000.
38. M. J. B. Krieger and J.-B. Billeter. The call of duty: Self-organised task allocation in a population of up to twelve mobile robots. *Robotics and Autonomous Systems*, 30:65–84, 2000.
39. P. Kuntz, D. Snyers, and P. Layzell. A stochastic heuristic for visualizing graph clusters in a bi-dimensional space prior to partitioning. *Journal of Heuristics*, 5:327–351, 1999.
40. S. Lawrence and C. L. Giles. Searching the world wide web. *Science*, 280:98–100, 1998.
41. G. Leguizamón and Z. Michalewicz. A new version of Ant System for subset problems. In *Proceedings of the 1999 Congress on Evolutionary Computation (CEC'99)*, pages 1459–1464. IEEE Press, Piscataway, NJ, 1999.
42. Y.-C. Liang and A. E. Smith. An Ant System approach to redundancy allocation. In *Proceedings of the 1999 Congress on Evolutionary Computation*, pages 1478–1484. IEEE Press, Piscataway, NJ, 1999.
43. E. Lumer and B. Faieta. Diversity and adaptation in populations of clustering ants. In J.-A. Meyer and S. W. Wilson, editors, *Proceedings of the Third International Conference on Simulation of Adaptive Behavior: From Animals to Animats 3*, pages 501–508. MIT Press, Cambridge, MA, 1994.
44. V. Maniezzo. Exact and approximate nondeterministic tree-search procedures for the quadratic assignment problem. *INFORMS Journal on Computing*, 11(4):358–369, 1999.
45. V. Maniezzo and A. Carbonaro. An ANTS heuristic for the frequency assignment problem. *Future Generation Computer Systems*, 16(8):927–935, 2000.
46. V. Maniezzo and A. Colorni. The Ant System applied to the quadratic assignment problem. *IEEE Transactions on Data and Knowledge Engineering*, 11(5):769–778, 1999.
47. V. Maniezzo, A. Colorni, and M. Dorigo. The Ant System applied to the quadratic assignment problem. Technical Report IRIDIA/94-28, IRIDIA, Université Libre de Bruxelles, Belgium, 1994.
48. D. Merkle, M. Middendorf, and H. Schmeck. Ant colony optimization for resource-constrained project scheduling. In *Proceedings of the Genetic and Evolutionary Computation Conference (GECCO-2000)*, pages 893–900. Morgan Kaufmann Publishers, San Francisco, CA, 2000.
49. R. Michel and M. Middendorf. An island model based Ant System with lookahead for the shortest supersequence problem. In A. E. Eiben, T. Bäck, M. Schoenauer, and H.-P. Schwefel, editors, *Proceedings of PPSN-V, Fifth International Conference on Parallel Problem Solving from Nature*, volume 1498 of *Lecture Notes in Computer Science*, pages 692–701. Springer Verlag, Berlin, Germany, 1998.
50. R. Michel and M. Middendorf. An ACO algorithm for the shortest supersequence problem. In D. Corne, M. Dorigo, and F. Glover, editors, *New Ideas in Optimization*, pages 51–61. McGraw Hill, London, UK, 1999.
51. G. Navarro Varela and M. C. Sinclair. Ant colony optimisation for virtual-wavelength-path routing and wavelength allocation. In *Proceedings of the 1999 Congress on Evolutionary Computation (CEC'99)*, pages 1809–1816. IEEE Press, Piscataway, NJ, 1999.

52. H. Ramalhinho Lourenço and D. Serra. Adaptive approach heuristics for the generalized assignment problem. Technical Report Technical Report Economic Working Papers Series No.304, Universitat Pompeu Fabra, Dept. of Economics and Management, Barcelona, Spain, 1998.
53. R. Schoonderwoerd, O. Holland, and J. Bruten. Ant-like agents for load balancing in telecommunications networks. In *Proceedings of the First International Conference on Autonomous Agents*, pages 209–216. ACM Press, 1997.
54. R. Schoonderwoerd, O. Holland, J. Bruten, and L. Rothkrantz. Ant-based load balancing in telecommunications networks. *Adaptive Behavior*, 5(2):169–207, 1996.
55. C. Solnon. Solving permutation constraint satisfaction problems with artificial ants. In W. Horn, editor, *Proceedings of the 14th European Conference on Artificial Intelligence*, pages 118–122. IOS Press, Amsterdam, The Netherlands, 2000.
56. T. Stützle. $\mathcal{MAX}$–$\mathcal{MIN}$ Ant System for the quadratic assignment problem. Technical Report AIDA-97-4, FG Intellektik, FB Informatik, TU Darmstadt, July 1997.
57. T. Stützle. An ant approach to the flow shop problem. In *Proceedings of the 6th European Congress on Intelligent Techniques & Soft Computing (EUFIT'98)*, volume 3, pages 1560–1564. Verlag Mainz, Wissenschaftsverlag, Aachen, 1998.
58. T. Stützle. *Local Search Algorithms for Combinatorial Problems: Analysis, Improvements, and New Applications.* Infix, Sankt Augustin, Germany, 1999.
59. T. Stützle and M. Dorigo. A short convergence proof for a class of aco algorithms. Technical Report IRIDIA/2000-35, IRIDIA, Université Libre de Bruxelles, Belgium, 2000.
60. T. Stützle and H. H. Hoos. The $\mathcal{MAX}$–$\mathcal{MIN}$ Ant System and local search for the traveling salesman problem. In T. Bäck, Z. Michalewicz, and X. Yao, editors, *Proceedings of the 1997 IEEE International Conference on Evolutionary Computation (ICEC'97)*, pages 309–314. IEEE Press, Piscataway, NJ, 1997.
61. T. Stützle and H. H. Hoos. $\mathcal{MAX}$–$\mathcal{MIN}$ Ant System. *Future Generation Computer Systems*, 16(8):889–914, 2000.
62. I. A. Wagner, M. Lindenbaum, and A. M. Bruckstein. Smell as a computational resource – a lesson we can learn from the ant. In *Proceedings of the Fourth Israeli Symposium on Theory of Computing and Systems (ISTCS-99)*, pages 219–230, 1996.
63. I. A. Wagner, M. Lindenbaum, and A. M. Bruckstein. Efficient graph search by a smell-oriented vertex process. *Annals of Mathematics and Artificial Intelligence*, 24:211–223, 1998.
64. I. A. Wagner, M. Lindenbaum, and A. M. Bruckstein. ANTS: Agents, networks, trees and subgraphs. *Future Generation Computer Systems*, 16(8):915–926, 2000.

The Shifting Network: Volume Signalling in Real and Robot Nervous Systems

Phil Husbands, Andy Philippides, Tom Smith, and Michael O'Shea

CCNR, University of Sussex, Brighton BN1 9QH, UK
philh@cogs.susx.ac.uk

Abstract. This paper presents recent work in computational modelling of diffusing gaseous neuromodulators in biological nervous systems. It goes on to describe work in adaptive autonomous systems directly inspired by this: an exploration of the use of virtual diffusing modulators in robot nervous systems built from non-standard artificial neural networks. These virtual chemicals act over space and time modulating a variety of node and connection properties in the networks. A wide variety of rich dynamics are possible in such systems; in the work described here, evolutionary robotics techniques have been used to harness the dynamics to produce autonomous behaviour in mobile robots. Detailed comparative analyses of evolutionary searches, and search spaces, for robot controllers with and without the virtual gases are introduced. The virtual diffusing modulators are found to provide significant advantages.

1 Introduction

This paper describes some of the main thrusts of an ongoing interdisciplinary study of diffusing neuromodulators in real and artificial systems. After explaining the motivations and biological background of the project, the key results from recent detailed computational models of nitric oxide (NO) diffusion from neural sources are discussed. This leads to a description of work on more abstract artificial neural systems heavily inspired by the biology. These so-called GasNets are used as artificial nervous systems for mobile autonomous robots. Detailed comparative studies of evolutionary robotics experiments involving GasNets and non-GasNets are introduced. These include investigations into the formal evolvability of such systems. The paper closes with a sketch of current and future directions of the project.

2 Biological Background and Motivation

As the Brain Sciences have advanced is has become more and more clear that nervous systems are electrochemical devices of enormous complexity and subtlety [8,15]. While the transmission of electrical signals across neuronal networks is regarded as a fundamental aspect of the operation of nervous systems, neurochemistry adds many dimensions to the picture. Cells can respond to chemicals that they themselves synthesize (autocrine signaling), to chemicals that diffuse from very nearby sources

J. Kelemen and P. Sosík (Eds.): ECAL 2001, LNAI 2159, pp. 23-36, 2001.

(paracrine signaling) or to chemicals that diffuse over greater distances or are carried by blood and tissue fluids [15,16]. The responses that these chemicals elicit are legion and can vary according to the internal and environmental states of the cells involved. Important classes of chemicals involved in the functioning of nervous systems include: neurotransmitters, receptors, neuromodulators and second messengers. Traditionally, chemical signaling between nerve cells was thought to be mediated solely by messenger molecules or neurotransmitters which are released by neurons at synapses [16] and flow from the presynaptic to postsynaptic neuron. Because most neurotransmitters are relatively large and polar molecules (amino acids, amines and peptides), they cannot diffuse through cell membranes and do not spread far from the release site. They are also rapidly inactivated by various reactions. Together these features confine the spread of such neurotransmitters to be very close to the points of release and ensure that the transmitter action is transient. In other words, chemical synaptic transmission of the classical kind operates essentially two-dimensionally (one in space and one in time). This conventional interpretation is coupled to the idea that neurotransmitters cause either an increase or a decrease in the electrical excitability of the target neuron. According to a traditional view of neurotransmission therefore, chemical information transfer is limited to the points of connection between neurons and neurotransmitters can simply be regarded as either excitatory or inhibitory.

In recent years a number of important discoveries have necessitated a fundamental revision of this model. It is now clear that many neurotransmitters, perhaps the majority, cannot be simply classified as excitatory or inhibitory [8]. These messenger molecules are best regarded as 'modulatory' because among other things they regulate or modulate the actions of conventional transmitters. Modulatory neurotransmitters are also 'indirect' because they cause medium- and long-term changes in the properties of neurons by influencing the rate of synthesis of so called 'second messenger' molecules. By altering the properties of proteins and even by changing the pattern of gene expression, these second messengers cause complex cascades of events resulting in fundamental changes in the properties of neurons. In this way modulatory transmitters greatly expand the diversity and the duration of actions mediated by the chemicals released by neurons. The action of neurotransmitters also depends on the receptors they bind to. Although most receptors are highly selective, responding to a single transmitter only, most transmitters can bind to a variety of receptors, with different consequences for different transmitter receptor pairings, even in the same cell [16]. There are a great variety of receptors on different types of cells suggesting the possibility of a combinatorially explosive range of pairings and effects. However, when coupled with this expanded picture of the nervous system, it is the recent discovery that the gas nitric oxide is a modulatory neurotransmitter that has opened entirely unexpected dimensions in our thinking about neuronal chemical signaling [5,6,10]. Because NO is a very small and nonpolar molecule it diffuses isotropically within the brain regardless of intervening cellular structures [25]. NO therefore violates some of the key tenets of point-to-point chemical transmission and is the first known member of an entirely new class of transmitter, the gaseous diffusable modulators (carbon monoxide is another example). NO is generated in the brain by specialised neurons that contain the neuronal isoform of the calcium activated enzyme, nitric oxide synthase or nNOS [1]. NO synthesis is triggered when the calcium concentration in nNOS-containing neurons is elevated, either by electrical activity or by the action of other modulatory neurotransmitters. The existence of a freely diffusing modulatory transmitter suggests a radically different form of signalling in which the transmitter acts four-dimensionally in space and time, affecting volumes of the brain containing many neurons and synapses [1]. NO cannot be contained by biological membranes, hence its release must be coupled directly to its synthesis. Because the synthetic enzyme nNOS can be distributed throughout the neuron, NO can be generated and released by the whole neuron. NO is therefore best regarded as a 'non-synaptic' transmitter whose actions moreover cannot be confined to neighbouring neurons [9,18].

The emerging picture of nervous systems sketched above -- as being highly dynamical, with many codependent processes acting on each other over space and time -- is thoroughly at odds with simplistic connectionist models of neural information processing. Importantly, the discovery of diffusible modulators shows that neurons can interact and alter one another's properties even though they are not synaptically connected. Indeed, all this starts to suggest that rather than thinking in terms of fixed neural circuits, a picture involving shifting networks – continually functionally and structurally reconfiguring – may be more appropriate. Of course many Alife practitioners reject the simple information processing models, but even so, by far the most popular kind of system used as artificial nervous systems are networks of nodes connected by virtual 'wires' along which inhibitory or excitatory 'electrical' signals flow. Although few would claim these are adequate models of the brain, their origins are in principles abstracted from the neuroscience of several decades ago. Although there are many possible levels of abstraction, new styles of artificial nervous systems directly inspired by contemporary understandings of brains as electrochemical machines may be a very fruitful avenue in our quest to develop artificial systems capable of more interesting and useful adaptive behaviours than we can currently manage. At the same time, the study of such systems should bring us deeper understandings of the principles underlying the functioning of real brains.

Given the limitations of current technology, if implemented versions of such systems are to act in real time as sensorimotor control systems for autonomous agents, they must necessarily abstract away much of the detailed complexity of real nervous systems. However, we believe that enough will be left behind to make this a worthwhile endeavour. In tandem with this kind of biologically inspired investigation, there is another far more detailed, more direct, form of modelling that can also be very useful. It is, as yet, very difficult to gather detailed empirical findings on such phenomena as the diffusion dynamics of NO in different parts of the nervous system, because the necessary experimental apparatus has not yet been developed. However, it is possible to build detailed computational models that capture certain salient features of these phenomena in an accurate way. These models are computationally expensive and do not run in real time, but the data they produce can make important contributions to our understanding of the biological processes.

This paper discusses examples of both kinds of work; two aspects of our ongoing investigation of the role of diffusing neuromodulators.

3 Modelling NO Diffusion in Real Neuronal Networks

In the previous section the role of NO in neuronal volume signalling was sketched. NO spreads in three dimensions away from the site of synthesis regardless of intervening cellular or membrane structures [25]. Another very important feature of NO signalling follows from the fact that nitric oxide synthase is soluble and thus highly likely to be distributed throughout a neuron's cytoplasm. This means that the whole surface of the neuron is a potential release site for NO, in marked contrast to conventional transmitter release. These properties suggest that the 3D structure of the NO source, and of any NO sinks, will have a profound influence on the dynamics of NO spread. Hence an accurate structure-based model of neuronal NO diffusion is an indispensable tool in gaining deeper insights into the signalling capacity of the molecule.

Figure 1 shows the results generated by the first accurate model of NO diffusion from continuous biologically realistic structures [19]. The source is an irregular neuron-like structure where the main cell body is a hollow sphere (NO is synthesized in the cell walls but not in the interior of the sphere). A sink has been placed just to the right of the cell body. Diffusion was modelled using accurate difference equation methods on a fine grid [19]. The figure shows the evolution of NO concentration during and after a 100ms burst of synthesis. Two very interesting observation are that the concentration remains high near the centre of the cell body long after synthesis has finished and that there is a significant delay between the start of synthesis and a rise in concentration for points distant from the main cell body. These observations follow from a 'reservoir effect' where NO diffuses into the centre of the hollow structure and is then 'trapped' there by a pressure gradient resulting in a slow time-delayed release [19]. Such a phenomenon, with its possible functional implications, would not have been observed in a less accurate point-source type model [25].

NO is also synthesized in another kind of irregular structure – namely a mesh of fine neuronal fibres in the mammalian cerebral cortex [20]. This mesh, or plexus, arises from a small population of neurons. As one of the biological affects of NO is to dilate the walls of blood vessels, the plexus might mediate the link between increased neural activity and increased blood supply to the same volume of the cortex. However, the vast majority of fibres in the plexus have been shown to be too small to generate above (biological) threshold concentrations of NO. This situation is again ripe for investigation with computational models. Using the same techniques as for the study illustrated in Figure 1, Philippides et al. have modelled the diffusion of NO from plexus structures [20]. Figure 2 shows results from a model investigating the volume over which NO concentrations are above threshold for sources made from regular arrays of very fine tubular structures. We see that once the density of fibres rises above a certain limit, the concerted effect of several very fine sources is to raise concentrations to significant levels. Further computational studies have shown how a random mesh of fine (rather than course) fibres is an ideal structure to ensure a uniform concentration over the plexus [20]. This is exactly the kind of structure found in the cortex, hence these models may point towards an important mechanism for allowing highly targeted NO 'clouds' in the brain.

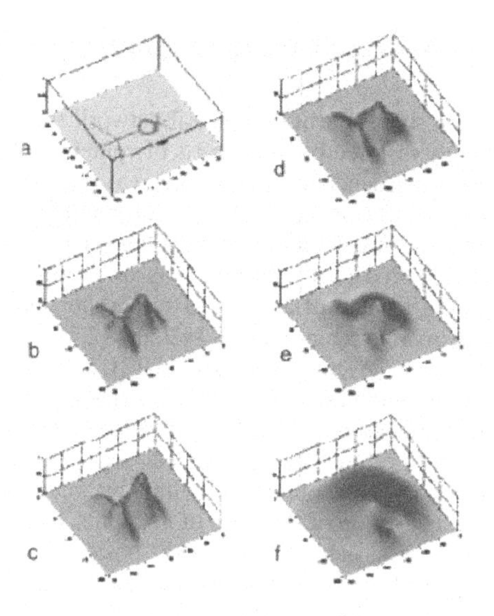

a) position of neuron and sink, b) NO concentration at t=50ms during synthesis, c) concentration at t=100ms: the end of synthesis, d) conc. at t=150ms, e) conc. at t=250ms, f) conc. at 750ms.

Fig. 1. Diffusion of NO from an irregular neuron being influenced by a nearby sink. NO concentration is shown at several time intervals following the initiation of a 100ms burst of synthesis. A 2D slice through the structure is illustrated here. See text for further details.

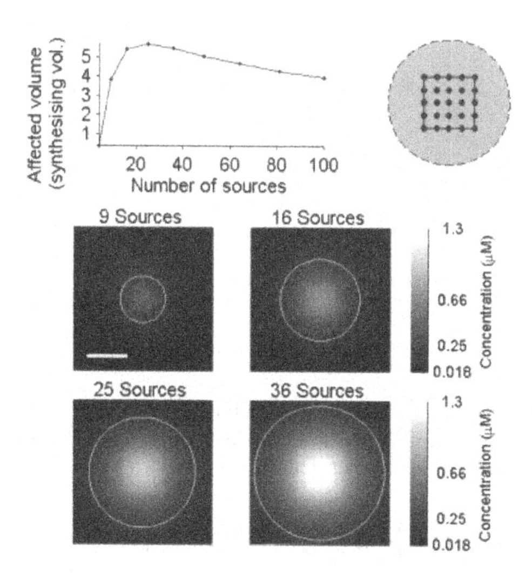

Fig. 2. Different numbers of very fine tubular NO sources arranged in regular arrays affect different volumes of tissue.

4 GasNets: From Neuroscience to Engineering

This section describes one style of artificial neural network from a class of networks whose operation is strongly inspired by those parts of contemporary neuroscience that emphasize the complex electrochemical nature of real nervous systems. So-called GasNets incorporate virtual diffusing gaseous neuromodulators and are used as artificial nervous systems for mobile autonomous robots. They are being investigated as potentially useful engineering tools and as a way of gaining helpful insights into biological systems. While a number of authors have incorporated global analogues of chemical signalling systems into agent control systems [2,7], as far as we are aware this work, which dates back to several years ago [11,13], is the first to concentrate on local processes, with virtual modulators diffusing over space and time.

The basic GasNet networks used in many recent experiments [13] are discrete time step dynamical systems built from units connected together by links that can be excitatory (with a weight of +1) or inhibitory (with a weight of -1). The output, O_i^t, of node *i* at time step *t* is a function of the sum of its inputs, as described by Equation 1. In addition to this underlying network in which positive and negative 'signals' flow between units, an abstract process loosely analogous to the diffusion of gaseous modulators is at play. Some units can emit virtual 'gases' which diffuse and are capable of modulating the behaviour of other units by changing their transfer functions in ways described in detail later. This form of modulation allows a kind of plasticity in the network in which the intrinsic properties of units are changing as the network operates. The networks function in a 2D plane; their geometric layout is a crucial element in the way in which the 'gases' diffuse and affect the properties of network nodes, as illustrated in Figure 3. This aspect of the networks is described in more detail later.

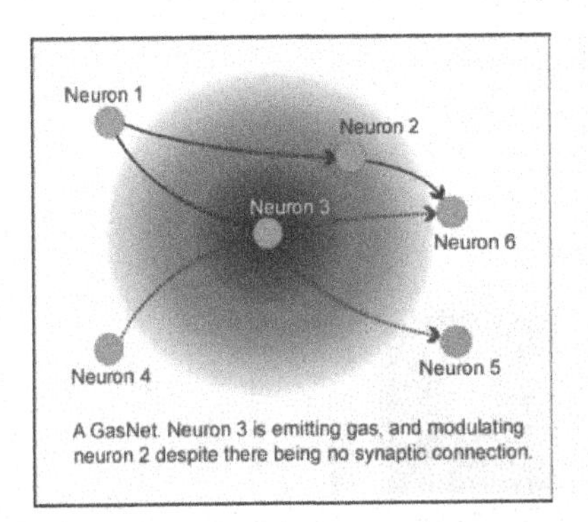

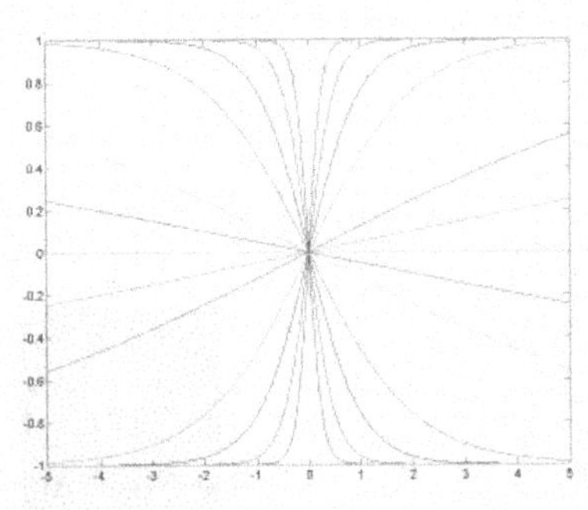

Fig. 3. GasNet operation depends on the geometric layout of the nodes in a 2D plane. The righthand side of the diagram shows how the shape of the tanh transfer function depends on the gain parameter k_i^t, see text for further details.

$$O_i^t = \tanh[k_i^t(\sum_{j \in C_i} w_{ji} O_j^{t-1} + I_i^t) + b_i] \qquad (1)$$

Where C_i is the set of nodes connected to node *i*, I_i^t is the external (sensory) input to node *i* and b_i is a genetically set bias. Each node has a genetically set default transfer function gain k_i^0. The right hand side of Figure 3 shows the shape of the function tanh(kx) over the range [-5,5] for a discrete set of values of k between –4 and 4. It is this gain parameter that is modulated by the diffusing virtual gases in the networks. This means that while the gases are active the shapes of the node transfer functions are being altered from time step to time step. The mechanism for this is explained in the next section.

4.1 Diffusion and Modulation

The virtual diffusion process is simple in order to be computationally fast so that GasNets can be used to control robots in real time. For mathematical convenience there are two gases, one whose modulatory effect is to increase the transfer function gain parameter and one whose modulatory effect is to decrease it. It is genetically determined whether or not any given node will emit one of two 'gases' (gas 1 and gas 2), and under what circumstances emission will occur (either when the 'electrical' activation of the node exceeds a threshold, or the concentration of one of the gases (genetically determined) in the vicinity of the node exceeds a threshold). The electrical threshold used in the experiments described later was 0.5, the gas concentration threshold 0.1. Allowing these two highly biologically inspired possibilities [6,10] is important – it provides a mechanism for rich interaction between two processes, the 'electrical' and the 'chemical'. A very abstract model of gas diffusion is used. For an emitting node, the concentration of gas at distance *d* from the node is given by Equation 2. Here, r_i is the genetically determined radius of influence of the ith node, so that concentration falls to zero for $d>r_i$. This is loosely analogous to the length constant of the natural diffusion of NO, related to its rate of decay through chemical interaction. $T_i(t)$ is a linear function that models the build up and decay of concentration after the node has started/stopped emitting. The slope of this function is individually genetically determined for each emitting node, C_0 is a global constant. For full details see [13].

$$C_i(d,t) = C_0 e^{-2d/r_i} \times T_i(t) \tag{2}$$

At each time step the gain parameter, k_i^t , for the node transfer function at each node (see Equation 1), is changed (or *modulated*) by the presence of gases at the site of the node. Gas 1 increases the value of k_i^t in a concentration dependent way, while gas 2 decreases its value. Concentration contributions from nodes within range of any given site are simply added together. The modulatory effects of the two gases are then summed to calculate the value of k_i^t at each time step. Each node has its own default rest value for the gain parameter, the virtual gasses continually increase or decrease this value. Referring to the right-hand side of Figure 3, this modulation can potentially have drastic effects on a nodes's transfer function, dramatically increasing or decreasing, or even flipping the sign of, its slope. This means that the networks are usually in flux, with rich dynamical possibilities.

Since there were no pre-existing principles for the exact operation and design of such networks, it was decided to allow most of their detailed properties to be genetically specified, giving the possibility of highly non-uniform dynamically complex networks. Hence, in most experiments to date nearly everything is up for grabs: the number of nodes in a network; the way they are connected; the position of the nodes on the 2D plane; the individual properties of each node controlling when (if at all) they emit a gas; which gas is emitted and how strongly; how and if nodes are connected to sensors or motors, as well as various properties of the sensors and motors themselves [13]. About 20 variables per node are needed to describe all this. Our experience has been that a well setup evolutionary search algorithm is a good tool for exploring the space of such systems [12], looking for interesting and useful examples that deepen our understanding of autonomous adaptive systems or provide practical engineering advantages such as robustness and reliability [12,17]. The next section gives an example of using GasNets in such an evolutionary robotics setting.

5 Experimental Comparison

Various forms of GasNet have been used as robot controllers for a variety of tasks and robots [11,13]. A very large number of runs of one particular experimental setup have been carried out, giving us enough data to be able to make statistically significant claims. In this series of experiments GasNets were evolved to control a robot engaged in a visually guided behaviour involving shape discrimination. A simple robot with a fixed CCD camera as its main sensor moved in an arena as illustrated in Figure 4. Two light coloured shapes, a rectangle and a triangle, were placed against a darker background on one of the walls. The task was to reliably move to the triangle, while ignoring the rectangle, from a random initial orientation and position under highly variable lighting conditions. The relative positioning of the shapes, in terms of which was on the left and which on the right, was made randomly.

As well as network size and topology, and all the parameters controlling virtual gas diffusion and modulation, the robot visual morphology, i.e. the way in which the camera image was sampled, was also under unconstrained genetic control. This was achieved by allowing the evolutionary search algorithm to specify the number and position of *single* pixels from the camera image to use as visual inputs. The grey scale intensity value of these pixels (normalised into the range [0.0,1.0]) were fed into the network, one for each genetically specified visual input node in the net. This is illustrated in the bottom left quadrant of Figure 4. Note that this means that the evolved control systems were operating with extremely minimal vision systems, just a few single pixel values. Given the very noisy lighting conditions and the minimal visual input, the shape discrimination task becomes non-trivial.

All the evolutionary runs were carried out using a Jakobi minimal simulation of the robotic setup. The methodology behind these ultra-lean ultra-fast simulations was developed by Jakobi [14] to address one potential problem with evolutionary approaches to exploring classes of robotic control systems: the time taken to evaluate behaviours over many generations. Through a careful use of noise and important decisions about what not to model, a minimal simulation will run very fast but behaviours evolved in them will transfer to the real robots. For full details of the minimal simulation used for the triangle rectangle task see [14]. In the experiment described here, all successful evolved controllers crossed the reality gap: they generated the same behaviours on the real robot as in simulation. Success was defined as being able to move to the triangle and stay there 30 times in direct succession from random starting positions and orientations, under very noisy lighting conditions and irrespective of the relative positioning of the shapes on the same wall. The great advantage of using minimal simulations in this work is that we were able to perform many complete evolutionary runs and hence derive meaningful statistics.

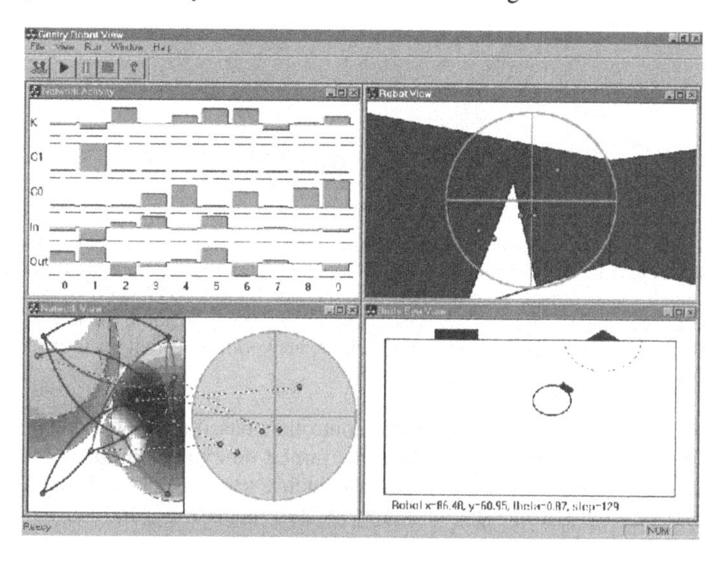

Fig. 4. The visualisation tool used with the minimal simulation of the shape discrimination task. The top right quadrant shows the view through the robot's camera, the bottom right gives a bird's eye view of the robot moving in the arena. The left-hand side of the screen illustrates the structure (including visual morphology) and functioning of the GasNet controlling the evolved robot. The shading in the network representation at extreme bottom left shows the gas concentrations in the network plane at the instant the snapshot was taken. The darker the shading the higher the concentration. See text for further details.

The initial set of evolutionary GasNet experiments with this task resulted in highly robust controllers emerging about 10 times faster than in earlier runs with conventional connectionist networks [13]. Subsequent comparative runs have concentrated on identifying whether or not the virtual gas modulation mechanism was at the root of this speed up. The key result is illustrated in Figure 5. In all experiments the genotypes were strings of integers encoding the various properties of the controllers and coupled visual morphologies, a geographically distributed genetic algorithm was used with a population of 100 [4], a number of different mutation operators were used in tandem, including node addition and deletion operators. The fitness function was based on a weighted average of final distances to the triangle over a set of evaluations from different initial conditions and different relative positioning of the shapes. Poor scores were weighted more heavily than good scores, encouraging robustness by requiring uniformly high scores across the whole evaluation set.

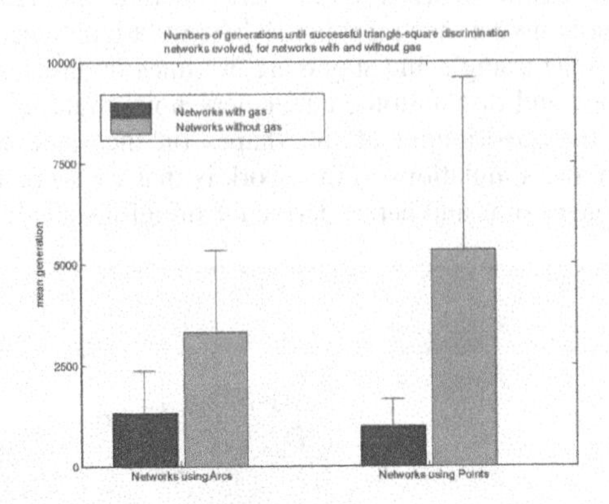

Fig. 5. The average number of generations needed to find controllers giving perfectly successful behaviour on the triangle rectangle problem. The dark columns are for networks with the gas mechanism turned on. The light columns are for networks with the gas mechanism turned off. The figure illustrates two sets of 20 runs in each condition; the difference between the left and right-hand sets is the way in which the network connectivity was encoded. See text for further details.

It can clearly be seen in Figure 5 that controllers based on networks with the virtual gas diffusion and modulation mechanisms turned on evolve significantly faster than those that are identical in every respect (including genotypes and all the evolutionary machinery) except that the gas mechanisms are rendered inoperative. This result has been repeated under various different encoding schemes and for a wide range of mutation rates [23]. The clear implication is that GasNets are more evolvable – their search space is more amenable to the form of evolutionary search used – than the various other forms of network explored. Obviously this could be a potentially very useful property and it is looked at in more detail in the next section.

Nearly all the successful GasNet controllers that were examined in detail exhibited surprisingly simple structures (a typical example is shown in Figure 6) relying on a very small number of visual inputs, although their internal dynamics, supported by interwoven 'chemical' and 'electrical' processes, were often intricate [13]. A number of interesting sub-networks, such as oscillators making use of spatial aspects of the modulation and diffusion processes [13], were independently evolved in several runs, suggesting that they are easily found building blocks that the evolutionary process can make good use of.

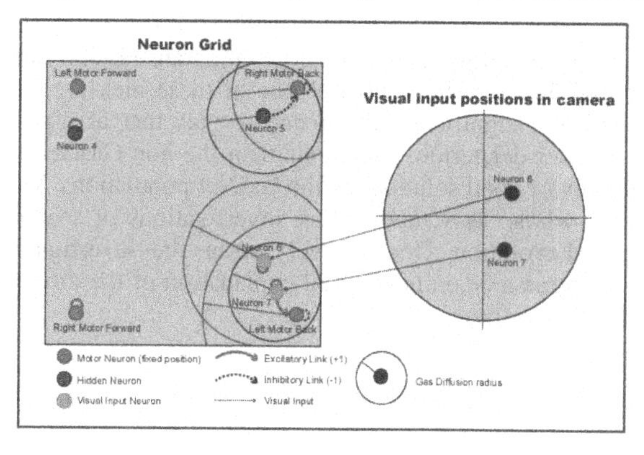

Fig. 6. A typical evolved GasNet controller for the triangle rectangle task illustrating the kind of structural simplicity often found in highly robust solutions. See text for further details.

6 Evolvability and Search Space Properties

The key result illustrated by Figure 5, that, for a particular evolutionary search algorithm, it is easier to find GasNet controllers for the triangle-rectangle task than non-GasNet controllers, tells us that there must be differences in the properties of the two search spaces. Understanding more about what this difference is may help us gain some valuable insights into the dynamics of artificial evolution and the nature of complex search spaces, as well as understanding more about the potential of GasNets. Smith et al. have published a number of papers working towards this goal [21,22,23]. The earliest studies in this series applied a whole range of standard search space 'complexity' and 'smoothness' metrics to the two spaces. These all failed to predict any difference between the spaces [22]. However, the research revealed a number of likely reasons for this: the spaces both appeared to be highly anisotropic, there is strong evidence for large neutral networks permeating the spaces, and a very large percentage of genotypes have negligible fitness. These and other properties combine to make the standard metrics useless for describing the pertinent differences between the spaces. Another (probably interrelated) reason is illustrated by the left-hand graphs in Figure 7 which show the median number of generations needed to reach a given fitness level for GasNets and non-GasNets. There is no difference in the two graphs for fitnesses of less than about 0.5. Fitnesses greater than this value are extremely unlikely to be found in random samples on which most of the basic metrics are based. Hence the focus of the work has shifted to the notion of evolvability and ways to measure it [23]. Evolvability is related to the ability of an individual or population to generate fit variants [24]. A useful measure of this is this transmission function which gives the distribution of all possible offspring for a given individual or population [3]. In the work described here, a variety of mutation operators are used, but no crossover operator is employed. This has allowed transmission functions to be approximated through massive sampling of mutated populations saved from many evolutionary runs, so that many levels of initial fitness are represented. Smith has devised a number of evolvability metrics based on the transmission function [23]. The right-hand graphs in Figure 7 show plots of one of these measure (explained in the caption). The small, but significant, difference reveals that at higher fitnesses the GasNet space has fewer deleterious mutations than the non-GasNet space. This will lead to a larger variety of good solutions in the GasNet populations. While this is very likely not to be the whole story, and further investigations of search dynamics are underway, it must aid evolution. This discovery means the investigations are starting to bear fruit and we hope to soon have a full explanation of the differences in search performance on the two spaces.

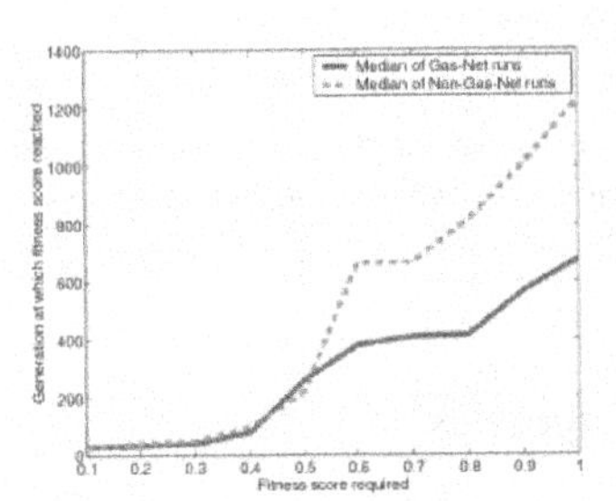

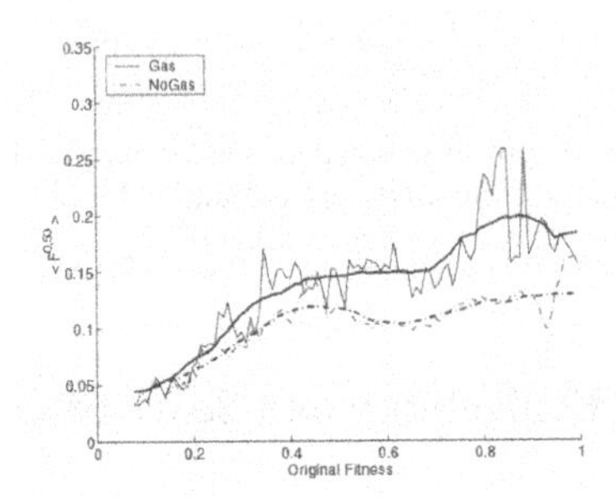

Fig. 7. The left-hand graphs show the median number of generations needed to reach a given fitness score for a large set of evolutionary runs. The right-hand graphs show a small, but significant, difference in the expected fitness of the bottom 50% of mutations applied to solutions of a given initial fitness. See text for further details.

7 Future Directions

There are many extensions to all aspects of the work described in this paper, some planned and some already underway. A number of these will be briefly discussed here.

As far as the computational modelling of the volume signalling roles of NO is concerned, obvious candidates for further work are: modelling diffusion from larger more complex structures and groups of structures, and introducing functional roles for the gas in detailed simulations of the behaviour of small neuronal networks. Both of these present non-trivial technical challenges and would require significant computing resources. However, given how much has been learnt from the studies carried out to date, as outlined in Section 3, it is important that such work is carried out.

The details of the 'electrical' and 'chemical' aspects of GasNets are, to some extent, rather arbitrary. There are numerous other forms that the modulations, node internal dynamics and virtual diffusion processes could take. Many of these are very worthwhile investigating in order to gain a deeper understanding of a whole class of systems. Such insights should usefully inform development in autonomous adaptive systems as well as systems level neuroscience. A number of interesting alternative modulation schemes being investigated include:

- Site specific modulations. The modulation depends on the presence of a 'receptor'. Virtual gases can trigger a range of modulations in a single network, depending on which receptors are present.

- Modulation of other adaptive processes (such as Hebbian synaptic changes). This could add a very powerful dimension to the evolution of adaptive mechanisms as advocated by Floreano et al. [17].

- Modulations at many different time scales, including permanent changes, are common in biology and are likely to play an important role in artificial systems.

Of course investigations into autonomous adaptive systems cannot focus solely on specific behaviour generating mechanisms. A bigger picture, involving overall architectures, body and sensor morphologies, developmental processes and a host of other issues surrounding embodied behaviour in the world, must be borne in mind.

The search space analyses outlined in the last section are part of an on-going investigation and in the future we wish to incorporate analyses of robot behaviours and their underlying controllers into the story.

8 Conclusions

The sciences of the real and the artificial have much to offer each other. This seems especially true in the study of adaptive behaviour and the mechanisms underlying it. This paper has outlined a multi-stranded interdisciplinary project that has striven to exploit the synergies at the interface between neuroscience and contemporary AI/Alife, and in so doing has advocated a shift towards rich electrochemical models and analogies. While the kind of creative exploratory work favoured by many Alife researchers are regarded as very important, and some of the work described here falls into that category, the authors stress the need for strong theoretical underpinnings and detailed analysis wherever possible. Without these, significant progress is unlikely.

References

1. Bredt DS and Snyder SH (1990) Isolation of nitric oxide synthetase, a calmodulin-requiring enzyme. *Proc Natl Acad Sci USA* **87**: 682-685.
2. Brooks, R.A. (1994) Coherent Behavior from Many Adaptive Processes. In: D. Cliff and P. Husbands and J.-A. Meyer and S.W. Wilson (Eds.), *From Animals to Animats 3: Proceedings of The Third International Conference on Simulation of Adaptive Behavior*, 22--29, MIT Press/Bradford Books, Cambridge, MA.

3. Cavalli-Sforza, L. and Feldman, M. (1976). Evolution of continuous variation: Direct approaches through joint distribution of genotypes and phenotypes. *Proc. Nat. Academy of Sciences*, USA, **73**:1689-1692.
4. Collins, R. and Jefferson, D. (1991) Selection in massively parallel genetic algorithms. In: R. K. Belew and L. B. Booker (Eds), Proceedings of the Fourth Intl. Conf. on Genetic Algorithms, ICGA-91, 249--256, Morgan Kaufmann.
5. Gally JA, Montague PR, Reeke Jnr GN and Edelman GM (1990) The NO hypothesis: possible effects of a short-lived, rapidly diffusible signal in the development and function of the nervous system. *Proc Natl Acad Sci USA*, **87**:3547-3551.
6. Garthwaite J, Charles SL and Chess-Williams R (1988) Endothelium-derived relaxing factor release on activation of NMDA receptors suggests role as intracellular messenger in the brain. *Nature* **336**: 385-388.
7. Grand, S. Creatures: An exercise in Creation, *IEEE Intelligent Systems magazine*, July/August 1997.
8. Hall ZW (1992) An Introduction to Molecular Neurobiology. Sinauer Associates Inc, Sunderland, Massachusetts.
9. Hartell NA (1996) Strong activation of parallel fibres produces localized calcium transients and a form of LTD that spreads to distant synapses. ***Neurons*** **16**: 601-610.
10. Holscher, C. (1997) Nitric oxide, the enigmatic neuronal messenger: its role in synaptic plasticity. *Trends Neurosci.* **20**: 298-303.
11. Husbands, P. (1998) Evolving Robot Behaviours with Diffusing Gas Networks, In: P. Husbands and J.-A. Meyer (1998), 71-86.
12. P. Husbands and J.-A. Meyer (Eds) (1998) *EvoRobot98: Proceedings of 1st European Workshop on Evolutionary Robotics*, Springer-Verlag LNCS 1468.
13. P. Husbands and T. Smith and N. Jakobi and M. O'Shea. Better Living through Chemistry: Evolving GasNets for Robot Control, *Connection Science*, **10(3&4)**, 185-210, 1998.
14. Jakobi, N. (1998) Evolutionary Robotics and the Radical Envelope of Noise Hypothesis, *Adaptive Behavior*, **6(2)**: 325-368.
15. Kandel, E. (1976) *The cellular basis of behavior*. Freeman.
16. Katz B (1969) *The release of neural transmitter substances*. Liverpool University Press.
17. Nolfi, S. and Floreano, D. (2000). Evolutionary Robotics: The biology, intelligence and technology of self-organizing machines. MIT Press.
18. Park J-H, Straub V and O'Shea M (1998) Anterograde signaling by Nitric Oxide: characterization and in vitro reconstitution of an identified nitrergic synapse. *J Neurosci* **18.**
19. Philippedes, A. and P. Husbands and M. O'Shea. Four Dimensional Neuronal Signaling by Nitric Oxide: A Computational Analysis. *Journal of Neuroscience* **20(3)**: 1199--1207, 2000.
20. A. Philippedes and P. Husbands and T. Lovick and M. O'Shea (2001). Targeted gas clouds in the brain. (submitted)
21. T. Smith and P. Husbands and M. O'Shea (2001). Neutral Networks and Evovability with Complex genotype-Phenotype Mapping. *Proc. ECAL'01*. LNCS, Springer.
22. T. Smith and P. Husbands and M. O'Shea (2001). Not Measuring Evovability: Initial Investigations of an Evolutionary Robotics Search Space. In *Proc. CEC'01*, IEEE Press.
23. T. Smith and P. Husbands and M. O'Shea (2001). Evolvability, Neutrality and Search Difficulty. (submitted)
24. Wagner, G. and Altenberg, L. (1996). Complex adaptations and the evolution of evolvability. *Evolution*, **50(3)**:967-976.
25. Wood J and Garthwaite J (1994) Model of the diffusional spread of nitric oxide - implications for neural nitric oxide signaling and its pharmacological properties. *Neuropharmacology* **33**: 1235-1244.

A Study of Replicators and Hypercycles by Typogenetics

V. Kvasnička, J. Pospíchal, and T. Kaláb*

Department of Mathematics, Slovak Technical University, 812 37 Bratislava, Slovakia
{kvasnic,pospich}@cvt.stuba.sk, kalab@eset.sk

Abstract. A Typogenetics is a formal system designed to study origins of life from a "premordial soup" of DNA molecules, enzymes and other building materials. It was introduced by Hofstadter in his book *Dialogues with Gödel, Escher, Bach: An Eternal Golden Braid* [17]. Autoreplicating molecules and systems of mutually replicating and catalyzing molecules (hypercycles) were modeled in the present paper in a very simplified way. Abstracted molecules in a form of strands are used in a model of a vessel, where "chemical reactions" occur. The approach is very similar to evolutionary algorithms. While a small hypercycle of two molecules mutually supporting their reproduction can be created without extreme difficulties, it is nearly impossible to create a hypercycle involving more than 4 autoreplicators at once. This paper demonstrates, that larger hypercycles can be created by an optimization and inclusion of new molecules into a smaller hypercycle. Such a sequential construction of hypercycles can substantially reduce the combinatorial complexity in comparison with a simultaneous optimization of single components of a large hypercycle.

1 Introduction

Studies of origins of life belong to crucial topics of artificial life. Unfortunately, a more realistic study using quantum or computational chemistry methods is not feasible, since the number of molecules required for such a task is too huge. It is therefore necessary to use simplified models of molecules and reactions. One of the most famous models in this field is the Typogenetics, a formal system initially devised by Douglas Hofstadter in his famous book *Dialogues with Gödel, Escher, Bach: An Eternal Golden Braid* [17] (cf. refs. [23,30,31]). Typogenetics replaces DNA molecules by strings (called the strands), defining a DNA molecule as a sequence of four kinds of letters, when each letter codes a basis (chemical group forming building block of DNA). The really significant simplification consists in definition of parts of DNA sequences which code elementary operations acting in turn on the DNA sequence itself. In reality a part of the DNA substrings code the instructions prescribing a production of enzymes. Each enzyme codes a different type of operation

* A part of this work was done by T. Kaláb in the framework of his M.Sc. Thesis at the Institute of Informatics, Fac. of Mathematics, Physics and Informatics, Comenius University, Bratislava, Slovakia.

J. Kelemen and P. Sosík (Eds.): ECAL 2001, LNAI 2159, pp. 37-54, 2001.

acting on DNA. These enzymes then use the DNA which coded them as a plan to create a copy of this DNA or a copy of another molecule (another strand in the Typogenetics formalism). While in the real life these substrings used for definition of enzymes are very large, in Typogenetics these codes of "enzymes" or "elementary operations" are smaller by orders of magnitude. Typogenetics is using the sequence of elementary operations coded by a strand to transform this strand (parent) onto another strand (offspring). Typogenetics was discussed by Hofstadter in connection with his attempt to explain or classify a „tangled hierarchy" of DNA considered as replicative systems.

Typogenetics as presented by Hofstadter [17] was not formulated in a very precise and exact way, many concepts and notions were presented only in a "fuzzy" verbal form and the reader was left to an improvisation and an ad-hoc additional specification of many notions of Typogenetics. Morris [23] was the first one who seriously attempted to formulate the Typogenetics in a precise manner and presented many illustrative examples and explanations that substantially facilitated an understanding of Typogenetics. Almost ten years ago Varetto [30] has published an article where he demonstrated that Typogenetics is a proper formal environment for a systematic constructive enumeration of strands that are capable of an autoreplication. Recently, Varetto [31] published another paper where Typogenetics was applied to a generation of the so-called tanglecycles that are simplified versions of hypercycles [8,9] of Eigen and Schuster.

The purpose of the present paper is to introduce a simplified version of Typogenetics that will be still capable to form a proper environment for Artificial Life studies of autoreplicators and hypercycles, both entities that belong to basic concepts of modern efforts [1-7,10,12-14,16,18-21,23-26,28,29] to simulate life *in-silico*. Simplification of our version of Typogenetics consists mainly in trimming of an instruction set, where all instructions that introduce or delete bases in strands were omitted. It is demonstrated that a construction of autoreplicators and hypercycles belongs to very complicated combinatorial problems and therefore an effort to generate them by a systematic constructive enumeration is hopeless. This is the main reason why we turned our attention to evolutionary methods of spontaneous emergence of autoreplicators and hypercycles. One of objectives of the present paper is to demonstrate an effectiveness of a simple version of evolutionary algorithm to create autoreplicators and hypercycles in a way closely related to Darwinian evolution.

The paper is organized as follows: Basic principles of a simplified version of Typogenetics are described in Section 2. Strands are determined as strings composed of four symbols A, C, G, and T. Then a DNA is specified as a double strand composed of a strand and its complementary strand. An expression of strands by enzymes is discussed in Section 3. A simple way how to assign an enzyme to an arbitrary strand is demonstrated. The enzyme is composed of a sequence of elementary instructions and the so-called binding site. In our simplified Typogenetics we retain only those instructions that do not change the length of strands, which excludes for example instructions for insertion or deletion of bases. An action of enzyme upon the strand is strongly deterministic, it is applied to the binding site which first appears when going on the strand from the left to the right. Section 4 is devoted to a specification of autoreplicators. These entities are determined as double strands with such a property that each of both autoreplicator strands is replicated by application of an enzyme. Firstly the double strands are separated. Then each strand produces an enzyme, which is in turn applied to the same strand and produces its complementary DNA copy. The enzyme is produced from the code by a prescription "start from the left, translate a couple of entries into an instruction and move to the right", creating from neighboring couples of strand entries a sequence of "instructions". This sequence of "instructions", which is a sort of metacode of an enzyme, is in this formalism equated with an enzyme. Instructions of such an enzyme usually do not make the copy of its "parental" strand by a straightforward "start from the left, copy and move to the right". They work more like a Turing machine on a tape (a metaphor from computer science), where the instructions can move the enzyme to the left or to the right on the strand. Such a copying process can create the copy e.g. with starting from the middle and jumping back and forth to the left and right, adding entries to the copy of a strand from both sides in turn. The copy can even be created in nonadjacent parts with the conjunctive entries copied at the end. This specification of autoreplicators represents, in fact, a hard constraint, so that their construction is nontrivial combinatorial problem. Fortunately, it can be effectively solved by making use of evolutionary methods. Hypercycles composed of double strands are studied in Section 5. The notion of hypercycles [8,9] is a generalization of autoreplicators such that a hypercycle is a cyclic kinetic structure, where a replication of its *i*th constituent is catalyzed by an enzyme produced by the previous (*i*-1)-th constituent. Hypercycles are considered in recent efforts of artificial life [27] as a proper formal tool suitable for specific explanation of a phenomenon called the increase of complexity. We show that evolutionary algorithms are capable of inducing an emergence of hypercycles from a population initialized by random strands. More complicated hypercycles (composed of three or four replicators) represent for evolutionary algorithms very hard combinatorial problems. This is the main reason why we turned our attention to a sequential step-by-step method of their construction, where a given hypercycle is evolutionary enlarged to a larger hypercycle by adding one additional replicator.

2 Basic Principles of Typogenetics

Let us consider a set B={A, C, G, T}, where elements – bases are called the adenine, cytosine, guanine, and thymine, respectively. These four elements are further classified as purines (A and G) and pyrimidines (C and T), i.e. $B=B_{pur}\cup B_{pyr}$. Moreover, bases are alternatively called *complementary*, A is complementary to T and C is complementary to G. In other words, we say that A is paired with T and C is paired with G (and vice versa). In Typogenetics the basic entity is the so-called *strand S* determined as a string composed of four different bases

$$S = X_1X_2...X_n \in B^* = B \cup B^2 \cup B^3 \cup ... \quad (1)$$

This strand is composed of n bases $X_1, X_2, \ldots, X_n \in B$, where the natural number n specifies its *length*, $|S|=n$. Examples of strands are CCA, TTGGACTTG, ..., their lengths are 3 and 9, respectively.

A *complementary strand* $\bar{S}$ with respect to the strand $S = X_1X_2...X_n$ is determined by $\bar{S} = \bar{X}_1\bar{X}_2...\bar{X}_n$, where $\bar{A} = T, \bar{C} = G, \bar{G} = C$, and $\bar{T} = A$ are complementary bases. For illustration let us consider the above mentioned two strands S_1=CCA, S_2=TTGGACTTG, their complementary forms are $\bar{S}_1 = \text{GGT}$ and $\bar{S}_2 = \text{AACCTGAAC}$, respectively. A *double strand* called the *DNA* is specified by two strands S (lower strand) and R (upper strand) that are complementary, $R = \bar{S}$ (see Fig. 1)

$$A = \begin{pmatrix} R \\ S \end{pmatrix} = \begin{pmatrix} \bar{S} \\ S \end{pmatrix} \tag{2}$$

where a base X_i is paired with Y_i of $R = Y_1Y_2...Y_n \in B$ (and vice versa), for i=1,2,…,n.

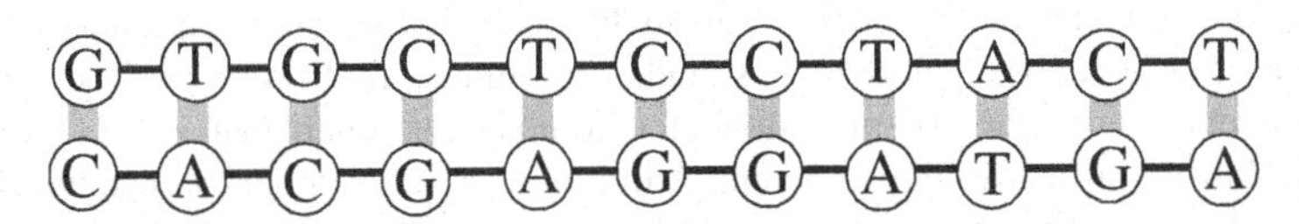

Fig. 1. An illustrative example of DNA composed of two complementary strands S (lower strand) and R (upper strand). Shaded areas represent interactions between complementary bases (realized by two or three hydrogen bonds).

In order to simplify our forthcoming considerations, let us enlarge the set B by a symbol that corresponds to an empty position on the strands, this additional symbol is represented by a hash symbol (#). If a strand contains at least one hash, then the strand is called the *quasistrand* (e.g. S=AACG##CCT). Formally, the above notion is simply realized when the set B is enlarged to $\tilde{B} = B \cup \{\#\}$, then quasistrands are determined as strings of

$$\tilde{B}^* = \left(B \cup \{\#\}\right)^* = \{\text{A}, \text{AA}, ..., \text{A\#C}, ..., \text{CCGT}, ..., \text{GGT\#\#T}, ...\} \tag{3}$$

A *distance* between two (quasi)strands $S = X_1X_2...X_n \in \tilde{B}$ and $R = Y_1Y_2...Y_n \in \tilde{B}$ (of the same length) is determined as follows

$$d(S,R) = 1 - \frac{1}{n}\sum_{i=1}^{n} \delta(X_i, Y_i) \tag{4a}$$

$$\delta(X,Y) = \begin{cases} 1 & (\text{if } X = Y \text{ and } X,Y \in B) \\ 0 & (\text{otherwise}) \end{cases} \tag{4b}$$

where a unit increment corresponds to a situation when both bases are identical and not equal to hash symbols. For illustration, let us consider two strands S =CGTT###AAT and R =TGT###AAAG , according to (4a-b), the distance between them is

$$d(S,R)=1-\frac{1}{10}(0+1+1+0+0+0+0+1+1+0)=1-\frac{4}{10}=\frac{3}{5}$$

A zero distance between two strands S and R means that they are identical and do not contain hash symbols.

3 An Expression of Strands by Enzymes

The purpose of this Section is to specify one of the most important concepts of Typogenetics, an expression of a strand by a sequence of instructions, that is called euphemistically the *enzyme*. Let us consider a set

$$B^2=\{\text{AA},\text{AC},\text{AG},...,\text{TT}\} \qquad \textbf{(5)}$$

composed of sixteen base pairs (doublets). Each strand $S=X_1X_2...X_n\in B^*$ can be expressed by making use of doublets of (5) as follows

$$\begin{aligned} S&=D_1D_2...D_p && (\text{for } n=2p) \\ S&=D_1D_2...D_pX_{2p+1} && (\text{for } n=2p+1) \end{aligned} \qquad \textbf{(6)}$$

where the first (second) possibility is applicable if the length of S is even (odd). Let us consider two mappings

$$instruction: B^2\rightarrow\{mvr,mvl,cop,off,rpy,...\} \qquad \textbf{(7a)}$$

$$inclination: B^2\rightarrow\{s,l,r\} \qquad \textbf{(7b)}$$

where the first mapping *instruction* assigns to each strand a sequence of instructions that will be sequentially performed over the strand when an enzyme (specified by the strand and the second mapping *inclination*) is applied. Details of these mappings will be specified later.

If doublets of a strand are mapped by (7a-b) (see Table 1), we arrive at the so-called *primary structure of the enzyme* that is specified by a sequence of instructions

$$instruction(S)=instr(D_1)-instr(D_2)-...-instr(D_p) \qquad \textbf{(8a)}$$

A *tertiary structure* (2D) of the enzyme is determined by the mapping *inclination*, it offers the following sequence of inclinations assigned to doublets (see Table 1)

$$inclination(S)=inclin(D_1)-inclin(D_2)-...-inclin(D_p) \qquad \textbf{(8b)}$$

Table 1. Specification of mappings *instruction* and *inclination*.

No.	Doublet	Instruct.	Inclin.	No.	Doublet	Instruct.	Inclin.
1	*AA*	***mvr***	***l***	9	*GA*	***rpy***	***s***
2	*AC*	***mvl***	***s***	10	*GC*	***rpu***	***r***
3	*AG*	***mvr***	***s***	11	*GG*	***lpy***	***r***
4	*AT*	***mvl***	***r***	12	*GT*	***lpu***	***l***
5	*CA*	***mvr***	***s***	13	*TA*	***rpy***	***r***
6	*CC*	***mvl***	***s***	14	*TC*	***rpu***	***l***
7	*CG*	***cop***	***r***	15	*TG*	***lpy***	***l***
8	*CT*	***off***	***l***	16	*TT*	***lpu***	***l***

Table 2. Description of single instructions from Table 1

No.	Instruction	Description
1	***cop***	Enzyme turns on copy mode, until turned off, enzyme produces complementary bases
2	***off***	Enzyme turns off copy mode
3	***mvr***	Enzyme moves one base to the right
4	***mvl***	Enzyme moves one base to the left
5	***rpy***	Enzyme finds nearest pyrimidine to the right
6	***rpu***	Enzyme finds nearest purine to the right
7	***lpy***	Enzyme finds nearest pyrimidine to the left
8	***lpu***	Enzyme finds nearest purine to the left

Both sequences (8a-b) that are assigned to a strand specify a transformation of the original (parent) strand onto a derived (offspring) strand. Loosely speaking, this transformation is considered as an application of the corresponding enzyme specified by sequences (8a-b), where the enzyme is visualized as a robot arm operating on the given strand, carrying out the commands that are coded by a sequence (8a), which is unambiguously determined by mapping (7a) based on the strand doublets (see also Table 1). Single instructions are specified by Table 2.

What remains to be determined is a starting position on the strand, where a sequence of enzyme actions is initialized. Such a position is called the *binding site* and it is represented by a base. An application of enzyme is then started on the first base (going from the left to the right) on the strand. If the strand does not contain such a base, than we say that the given enzyme is inapplicable to the strand. The binding site X is specified by the sequence of inclinations (8b) such that going successively from left to right, we construct recurrently a sequence of arrows oriented to the right, left, up, or down. This process is initialized by the first position such that it is automatically set to arrow $\Rightarrow$, see Fig. 2, so that the first inclination is not enacted. When the sequence of inclinations is constructed or analyzed, we get the direction of the last arrow. The binding site is unambiguously determined by the first inclination symbol and by the last arrow (see Table 3)

$$X = f\left(\text{first inclination symbol, last arrow}\right) \tag{9}$$

This formula simply determines the binding site on the strand, e.g. according to Table 3, a sequence of arrows presented by diagram E in Fig. 2 determines the binding site X=A. It means that a corresponding enzyme is initially applied to a base A (going first from the left on a strand). Many different enzymes can have the same binding site. Formally, the whole procedure of construction of an enzyme assigned to a strand S is expressed by

$$enzyme(S) = \left(instruction(S), X\right) \tag{10}$$

where its first component corresponds to an instruction sequence (8a) and the second component specifies a binding site. This relatively complicated way of determination of the binding site was introduced by Morris [23]. Original Hofstadter's approach [17] is much simpler, the binding site is specified only by the last arrow in the 2D enzyme structure, i.e. the type of the last arrow directly specifies a binding site.

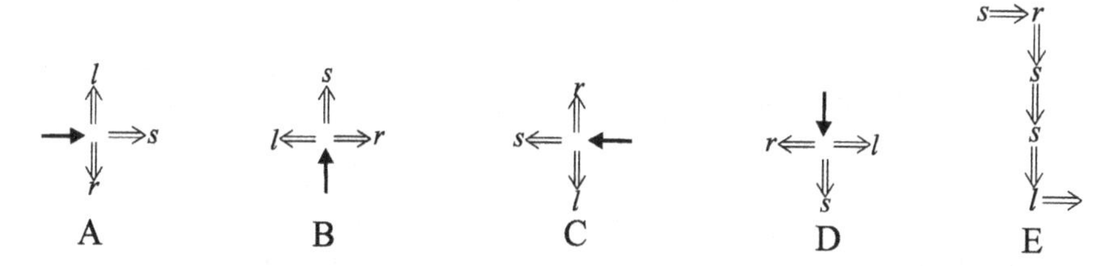

Fig. 2. An outline of four different cases of local properties of inclinations (A to D), where an initial arrow is specified by a black bold arrow. For instance, the first diagram A represents three possible folds (directions of double arrow) created from the bold arrow (oriented from the left to the right) if inclinations ***s***, ***l***, and ***r*** is applied. Diagram E corresponds to a 2D structure produced by an inclination sequence ***s-r-s-s-l***.

Table 3. Different possibilities for binding site determination

No.	1st inclin.	Last arrow	Binding	No.	1st inclin.	Last arrow	Binding
1	*s*	⇒	A	7	***l***	⇐	G
2	*s*	⇑	C	8	***l***	⇑	T
3	*s*	⇓	G	9	***r***	⇑	A
4	*s*	⇐	T	10	***r***	⇐	C
5	***l***	⇓	A	11	***r***	⇒	G
6	***l***	⇒	C	12	***r***	⇓	T

****s***(straight), ***l***(left), and ***r***(right).

For a given strand S and its *enzyme*(S) we may introduce the so-called replication process consisting in an application of the *enzyme*(S) to the strand S. This replication process is formally composed of the following two steps:

Step 1. Construction of an enzyme composed of a sequence of instructions (amino acids)

$$instruction(S) = instr(D_1) - instr(D_2) - ... - instr(D_p) \quad \textbf{(11a)}$$

and a binding site X, i.e.

$$enzyme(S) = (instruction(S), X) \quad \textbf{(11b)}$$

Step 2. Enzyme *enzyme*(S) is applied to the strand S so that its application is initialized at the base X incoming first from the left and then instructions are step-by-step performed over the strand.

This simple process of transformation of the (*parent*) strand S onto another quasistrand (in general, it may contain also hash symbols) R is called the *replication*

$$replication(S) = R \quad \textbf{(11c)}$$

A strand R (*offspring*) is created in the course of replication as a result of the replication process if in some replication stage the enzyme was switched to *on* mode. In general, this strand R may be composed of a number of empty hash symbols that appear in the resulting strand when its length is smaller than a length of the parent strand S. If the result of replication is not a continuous strand, a strand R is defined as the first continuous part of the result of replication. Diagrammatic representation of the above two-step transformation (replication) is outlined in Fig. 3.

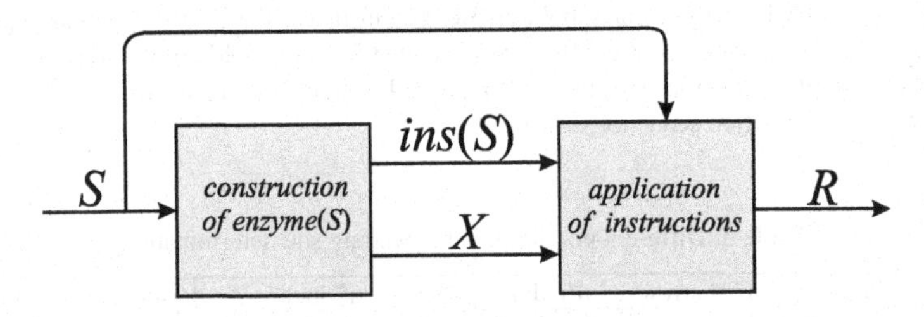

Fig. 3. A schematic outline of a replication process of a strand S. At the first stage an enzyme *enzyme*(S) is constructed, then, at the second stage this enzyme is applied to the strand S by a set of instructions *ins(S)*, starting at a binding site X. Loosely speaking, we may say, that each strand contains a necessary information for its replication process.

Finally, we will discuss how to apply an enzyme *enzyme*(S) to a strand S. Let us postulate that the enzyme is specified by $enzyme(S) = (instruction(S), X)$, where X specifies a binding site on the strand S. Two different situation should be distinguished:

(1) If $X \notin S$, then the enzyme is inapplicable to the strand.

(2) If $X \in S$, then the enzyme is applied to the first appearance (from the left) of the base X. Enzyme instructions (amino acids) are sequentially step-by-step applied to the strand S.

4 Autoreplicators

One of the central notions of artificial (or algorithmic) chemistry [1-7,10,12-14,16,18-21,23-26,28,29] are *autoreplicators*, initially introduced in the beginning of seventies by Eigen and Schuster [8,9] as hypothetical biomacromolecules that are endowed with standard "mass-law" kinetics and that are capable of autoreplication catalyzed by themselves. These authors demonstrated that in this "abiotic" level there is already possible to observe phenomena closely resembling Darwinian evolution based on the surviving of best fitted individuals (i.e. best adapted biomacromolecules).

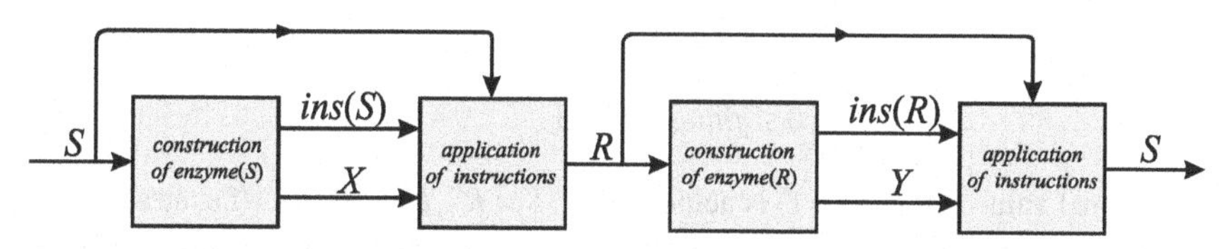

Fig. 4. A schematic outline of an autoreplication process of a strand S, it may be considered as a double application of a scheme presented in Fig. 3. If the strand S is an autoreplicator, then an output from two replications is again the same strand S.

A double strand

$$A = \begin{pmatrix} R \\ S \end{pmatrix} \quad \textbf{(12)}$$

is called the *autoreplicator* if the replication process applied to both its parts results in

$$replication(S) = R \text{ and } replication(R) = S \quad \textbf{(13a)}$$

in a composed form

$$replication(replication(S)) = S \quad \textbf{(13b)}$$

i.e. the strand S is replicated to R, and the strand R is replicated to S. In typogenetic environment we manipulate always with single strands, the above presented definition should be considered as a two-step process: in the first step the strand S is replicated to an offspring R, and then R is replicated to the next offspring identical with the parent strand S, see Fig. 4. The requirement, that each of autoreplicator's complementary strands is replicated exactly onto the other one, is very restrictive.

4.1 An Evolutionary Construction of Autoreplicators

For an application of evolutionary methods [11,15,22] to construction of autoreplicators we need a quantitative measure of a fact whether a strand is autoreplicator or not. We introduce the so-called *fitness* of strands that achieves the maximal value if the strand is an autoreplicator. Let us have a strand *S*, its fitness will reflect its ability to be an autoreplicator. In particular, let $R = \overline{S}$ be a complementary strand to the original strand *S*, applying to these two strands independent replication processes we get

$$replication(S) = R' \text{ and } replication(R) = S' \tag{14}$$

Then a fitness of *S* is determined as follows

$$fitness(S) = \frac{1}{2}\left(2 - d\left(S, \overline{R}'\right) - d\left(R, \overline{S}'\right)\right) \tag{15a}$$

with values ranged by

$$0 \leq fitness(S) \leq 1 \tag{15b}$$

Its maximal value ($fitness_{max}$=1) is achieved for $S = \overline{R}'$ and $R = \overline{S}'$ (i.e. strands *S*, *R'* and *R*, *S'* are complementary). This means that the maximal fitness value is achieved for strands that are autoreplicators.

A mutation represents a very important innovation method in evolutionary algorithms. In particular, going from one evolutionary epoch to the next epoch, individuals of a population are reproduced with small random errors. If this reproduction process were always without spontaneously appearing errors, than the evolution would not contain "variations" that are a necessary presumption of the Darwinian evolution.

Let us consider a strand $S = X_1X_2...X_n$, this strand is transformed onto another strand $T = Y_1Y_2...Y_n$, (where *Y*'s are bases or empty symbols) applying a stochastic mutation operator O_{mut}

$$T = O_{mut}(S) \tag{16}$$

This operator is realized in such a way that going successively from the left to the right each element (base) with a small probability P_{mut} is either changed (mutated) to another base, or deleted from the strand, or enlarged from the right by a new randomly selected base

$$O_{mut}(\text{AACGTTA}) = \begin{cases} \underline{\text{T}}\text{ACGTTA} & (\text{mutation}) \\ \text{AA}_{\text{,}}\text{GTTA} & (\text{deletion}) \\ \text{AACG}\underline{\text{A}}\text{TTA} & (\text{insertion}) \\ \text{AACGTTA} & (\text{exact copy}) \end{cases} \tag{17}$$

where the first three particular cases (mutation, deletion, and insertion) are realized with the same probability.

Evolution of strands towards an emergence of autoreplicators in a population (composed of strands that are considered as objects of Darwinian evolution) is simulated by a simple evolutionary algorithm. Individuals in a population are represented by single strands, and in a reproduction process a mutation operator is applied on a randomly selected parent – strand, creating one offspring. A strand is quasirandomly selected to a reproduction process, a probability of this selection is proportional to the strand fitness. The reproduction process consists in simple copy process, where a strand is simply reproduced with a possibility of appearance of stochastic mutations (specified by the probability P_{mut}). If a new population composed of offspring of reproduction process has the same number of individuals as the original population, then the old population is updated by the new population.

4.2 Results of Computer Simulations of Emergence of Autoreplicators

The above formal definition of the autoreplicator is relatively complicated, it requires two-step process to verify whether a strand is an autoreplicator. Varetto [31] studied a systematic constructive way for the construction of autoreplicators, which is applicable for shorter strands or for strands with the same repeated "motif". In order to demonstrate full capacity of Typogenetics for AL studies, a simple evolutionary algorithm is applied to achieve an evolutionary spontaneous emergence of autoreplicators. The basic parameters of the algorithm were set as follows: size of population N=1000, minimal and maximal lengths of strands l_{min}=15 and l_{max}=30. Probability P_{mut} was set variable during the course of evolution, at the beginning of evolution its value is maximal $P_{mut}^{(max)}$ and then it decreases to a minimal value $P_{mut}^{(min)}$. A current value of probability for an evolutionary epoch t is determined by

$$P_{mut} = P_{mut}^{(max)} - \left(P_{mut}^{(max)} - P_{mut}^{(min)}\right)\frac{t}{t_{max}} \tag{18}$$

where t_{max} is the length of evolution (maximal number of epochs). In our calculations we set $P_{mut}^{(max)} = 0.01$ and $P_{mut}^{(min)} = 0.001$.

A proper method how to visualize evolution is a plot of distance between the temporarily best strand $S_{best}^{(t)}$ (specified for the evolutionary epoch t) and the best strand resulting from the whole evolution $S_{best}^{(all)}$. Since the strands $S_{best}^{(t)}$ and $S_{best}^{(all)}$ may be, in general, of different length, the distance specified in Section 2 (see eqs. (4a) and (4b)) is not applicable for this consideration. It means that we have to determine a notion of distance in a more general way than that one mentioned in Section 2. Let us consider two strands $S=X_1X_2...X_n$ and $R=Y_1Y_2...Y_m$, their lengths are $|S|=n$ and $|R|=m$, respectively. Let $p=\min\{m,n\}$ be a minimal distance of strands S and R, then an alternative distance between them is determined by

$$D(S,R) = |S| + |R| - 2\sum_{i=1}^{p} \delta(X_i, Y_i) \quad \textbf{(19)}$$

where δ is an analogue of Kronecker's delta already defined by (4b). A positive value of this new distance reflects a measure of difference between strands S and R, its vanishing value corresponds to a fact that both strands are identical. A plot of $D\left(S_{best}^{(t)}, S_{best}^{(all)}\right)$ visualizes a way of approaching of temporarily best strands through the evolution to the final and resulting best strand that may be considered as a result of the evolutionary emergence of autoreplicators.

Different plots are shown in Fig. 5. The first diagram A corresponds to plots of maximal and mean fitness and a frequency of appearance of temporarily best strand. At the beginning of evolution there appeared a mixture of different strands. As the population was more evolved (say starting from 500 epochs), where a final solution (an autoreplicator) was already created, its fraction of appearance almost monotonously increases to unit value. The second diagram B corresponds to a plot of a distance D between temporarily best strand and the best final strand (autoreplicator) produced by the evolution of population. We see that the distance decreases with small fluctuations so that starting from the half of evolution this distance is vanishing, i.e. the correct strand (or strands) has emerged from the evolution.

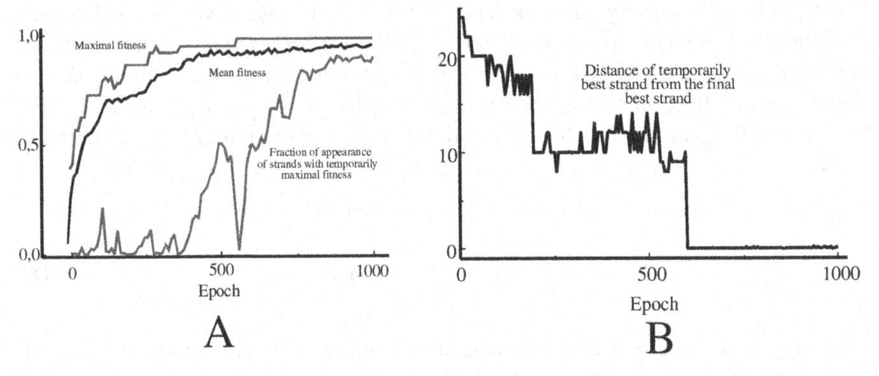

Fig. 5. Two different plots that characterize evolutionary emergence of autoreplicators. Diagram A shows plots of maximal fitness, mean fitness, and a frequency of appearance of temporarily best strand. Diagram B shows plot of distance $D\left(S_{best}^{(t)}, S_{best}^{(all)}\right)$, where $S_{best}^{(t)}$ is a temporarily best strand (for an epoch t) and $S_{best}^{(all)}$ is a best strand (an autoreplicator) produced by the evolution of population. The distance D is determined by (19). The displayed plot indicates that distance D "monotonously" decreases (with some small fluctuations due to a random genetic drift in the population) to zero value, which indicates a spontaneous emergence of an autoreplicator ($S_{best}^{(all)}$) at the end of evolution.

The following set of observation from our numerical results can be formulated (cf. ref. [13]):

(1) *There do not exist dramatic changes in the composition of best strands throughout the whole population period. Rather, we see that evolution of autoreplicators is very opportunistic, it contains only small changes in compositions of strands such that whole evolution is inherently directed to an emergence of autoreplicators.*

(2) *Moreover, there exist long evolutionary periods in which the maximal fitness is kept fixed and small changes appeared in composition of strands. Such evolutionary periods are called the neutral periods, in which evolution "gather" an information for changes that lead to a substantial increase of quality of strands towards their ability to be autoreplicators.*

5 Hypercycles

According to Eigen and Schuster [8,9], a *hypercycle* is a kinetic composition of replicators, where a replication of A_i is catalyzed by enzymes produced by the previous replicator A_{i-1}

$$A_i + E_{i-1} \rightarrow 2A_i + E_{i-1} \text{ (for } i=1,2,...,n)$$

where E_{i-1} is an enzyme produced by the previous replicator A_{i-1}, and $A_0=A_n$, $E_0=E_n$.

Hypercycles may be considered as multilevel hierarchical catalytic kinetic systems. They represent an important concept of the current mental image of an abiotic period of molecular evolution. Autoreplicators, which emerged in the first stage of this evolution, may be integrated into higher level kinetic systems that represent units relatively independent from other autoreplicators or hypercycles. Moreover, hypercycles represent an uncomplicated example of an increase of complexity [27], with a well described mathematical model and a simple computer implementation [9].

Let us consider a sequence of replicators S_1, S_2, ..., S_n, that are mutually related in such a way that a replication of S_i is catalyzed by an enzyme *enzyme*(S_{i-1}) produced by the previous strand S_{i-1} (a previous strand with respect to S_1 is a strand S_n). Applying a metaphor of chemical reactions, a hypercycle can be represented as a sequence of the following reactions

$$S_i \xrightarrow{enzyme(S_{i-1})} S_i + \overline{S}_i \quad and \quad \overline{S}_i \xrightarrow{enzyme(\overline{S}_{i-1})} \overline{S}_i + S_i \qquad \textbf{(20)}$$

for $i=1,2,\ldots,n$. We see that their precise determination is highly restrictive and may give rise to very serious doubts whether hypercycles can exist and be constructed (e.g. within Typogenetics).

Recently, Varetto [31] has introduced the so-called tanglecycles as an alternative to our hypercycles that were specified in a way *closely related to their original meaning* proposed by Eigen and Schuster [8,9]. In particular, in a specification of tanglecycles there is suppressed an autoreplication character of strands, Varetto only required that there exists a replication of a strand S_i onto another strand S_{i+1} and this process is catalyzed by an enzyme of S_{i-1} strand (he does not specify properties of complementary strands taking part in the tanglecycle)

$$S_i \xrightarrow{enzyme(S_{i-1})} S_i + S_{i+1} \quad (i = 1,2,...,n) \tag{21}$$

where $S_0=S_n$ and $S_{n+1}=S_1$. The main difference between hypercycles and tanglecycles consists in the fact that the strands in hypercycles, unlike the tanglecycles, are coupled only through enzymatic catalysis, while in tanglecycles inside a "replication" of S_i there is created the forthcoming strand S_{i+1}. The present version of our Typogenetics machinery is not applicable to a study of tanglecycles, since a replication product S_{i+1} of a strands S_i (see eq. (21)) should be a complementary strand to S_i, i.e. we could not expect that by applying a sequence of reactions (21) for $n \geq 3$ we get at its end a product identical with the initial strand S_1.

5.1 An Evolutionary Construction of Hypercycles

A population P is composed of hypercycles (or hopefully future hypercycles) that are composed of the same number n of strands. Each hypercycle of the population is evaluated by a fitness that reflects an ability of all its components to autoreplicate themselves. Hypercycles are selected quasirandomly (with a probability proportional to their fitness) to a simple reproduction process with a possibility of stochastic mutations (controlled by a probability P_{mut}). The design of the evolutionary algorithm is the same as for the evolution of single autoreplicators.

The fitness of a hypercycle is determined as follows: Let us consider a hypercycle and its complementary form

$$\boldsymbol{x} = (S_1, S_2, ..., S_n) \quad \text{and} \quad \overline{\boldsymbol{x}} = (\overline{S}_1, \overline{S}_2, ..., \overline{S}_n) \tag{22a}$$

$$S_i \xrightarrow{e(S_{i-1})} S_i + R_i \quad \text{and} \quad \overline{S}_i \xrightarrow{e(\overline{S}_{i-1})} \overline{S}_i + R_i' \tag{22b}$$

Each i-th component (S_i and $\overline{S}_i$) is evaluated by a "local" fitness

$$fitness_i = \frac{1}{2}\left(2 - d(S_i, \overline{R}_i) - d(\overline{S}_i, \overline{R}_i')\right) \tag{23}$$

A fitness of the hypercycle $\boldsymbol{x}$ is determined as a *minimum* of local fitness of its constituents

$$fitness(\boldsymbol{x}) = \min_i fitness_i \tag{24}$$

Loosely speaking, a fitness of a hypercycle is determined by a local fitness of its weakest replicator (a chain is as strong as its weakest link). A Darwinian evolution of strands towards an emergence of hypercycles in a population is simulated by a simple evolutionary algorithm used for an evolution of autoreplicators.

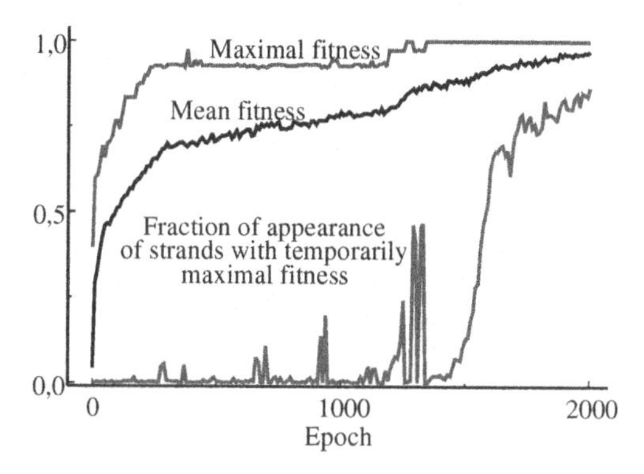

Fig. 6. Three different plots that characterize evolutionary emergence of hypercycles composed of two replicators (2-hypercycle).

The basic parameters of the present evolutionary algorithm that was used for an emergence of hypercycles are set as follows: size of population N=2000, mutation probability P_{mut}=0.001, minimal and maximal lengths of strands l_{min}=15 and l_{max}=30, and the evolution of population is watched two thousands epochs (i.e. t_{max}=2000). The evolutionary approach produced 2-hypercycles; if the same approach was used for higher hypercycles, then we never succeeded in their emergence. Main conclusions from computer simulations of evolutionary emergence of hypercycles are following:

(1) *An evolutionary emergence of hypercycles composed of more than two autoreplicators is a very improbable evolutionary event. In other words, it represents for evolutionary algorithms a very difficult combinatorial tasks.*

(2) *More complex hypercycles may be evolutionary constructed from simpler hypercycles such that they are enlarged by another autoreplicator with evolutionarily optimized composition.*

5.2 Creation of Larger Hypercycles from Simpler Hypercycles

The main conclusion of our simulations, outlined in the previous subsection, is that an evolutionary construction of hypercycles composed of more than two replicators belongs to very hard combinatorial tasks. This is the main reason why we turn our attention to another evolutionary possibility of their construction. A simple way of enlargement of a smaller hypercycle onto a bigger one such that a replicator is incorporated may be formulated in a form of an evolutionary algorithm. Let us consider a hypercycle $\boldsymbol{x} = (S_1, S_2, ..., S_n)$ composed of n replicators with their complementary strands, and let it be enlarged by a replicator denoted by S_{n+1}. We postulate that this new strands S_{n+1} is incorporated into the hypercycle in two stages, such that (1) its replication is catalyzed by an enzyme $enzyme(S_n)$ and (2) its $enzyme(S_{n+1})$ catalyzes a replication of the strand S_1

$$S_{n+1} \xrightarrow{e(S_n)} S_{n+1} + R_{n+1} \text{ and } \overline{S}_{n+1} \xrightarrow{e(\overline{S}_n)} \overline{S}_{n+1} + R'_{n+1} \tag{25a}$$

$$S_1 \xrightarrow{e(S_{n+1})} S_1 + R_1 \text{ and } \overline{S}_1 \xrightarrow{e(\overline{S}_{n+1})} \overline{S}_1 + R'_1 \tag{25b}$$

A fitness of the new strand S_{n+1} is determined as follows:

$$fitness(S_{n+1}) = \frac{1}{4}\left(4 - d\left(S_{n+1}, \overline{R}_{n+1}\right) - d\left(\overline{S}_{n+1}, \overline{R}'_{n+1}\right) - d\left(S_1, \overline{R}_1\right) - d\left(\overline{S}_1, \overline{R}'_1\right)\right) \tag{26}$$

with values ranged by $0 \leq fitness(S_{n+1}) \leq 1$. Its maximal value corresponds to a situation where the strand S_{n+1} is exactly replicated to a complementary strand $\overline{S}_{n+1}$ (catalyzed by $enzyme(S_n)$) and the strand S_1 is exactly replicated to $\overline{S}_1$ (catalyzed by $enzyme(S_{n+1})$), and similarly for the replication of complementary strands $\overline{S}_{n+1}$ and $\overline{S}_1$.

The above approach to a construction of larger hypercycles from smaller ones can be simply implemented within an evolutionary algorithm. The basic advantage of the suggested method is its capability to overcome a combinatorial complexity that has severely plagued standard evolutionary approach discussed in the previous subsection 4.1. This standard approach is now modified in such a way that from a previous calculation we know an n-hypercycle. Its enlargement by a new strand S_{n+1} is evolutionary optimized (according to fitness (26)) while the original n-hypercycle is kept fixed through the whole enlargement evolution. We present here only a simple illustrative example of 3-hypercycle that was constructed by an evolutionary enlargement process of the already known 2-hypercycle. This 3-hypercycle is composed of three DNAs A_1, A_2, and A_3

$$A_1 = \begin{pmatrix} \text{CGATGCAAAGAAAGAAAAAC} \\ \text{GCTACGTTTCTTTCTTTTTG} \end{pmatrix}, \quad A_2 = \begin{pmatrix} \text{GCCTCTCTTTTTCGGAGAGA} \\ \text{CGGAGAGAAAAAGCCTCTCT} \end{pmatrix}$$

$$A_3 = \begin{pmatrix} \text{CGGCAGAGAAAAAAAAGAGC} \\ \text{GCCGTCTCTTTTTTTTCTCG} \end{pmatrix}$$

These double strands form a 3-hypercycle with components specified by the following reactions:

$$S_1 \xrightarrow{enzyme(S_3)} S_1 + \overline{S}_1, \quad \overline{S}_1 \xrightarrow{enzyme(\overline{S}_3)} \overline{S}_1 + S_1$$

$$S_2 \xrightarrow{enzyme(S_1)} S_2 + \overline{S}_2, \quad \overline{S}_2 \xrightarrow{enzyme(\overline{S}_1)} \overline{S}_2 + S_2$$

$$S_3 \xrightarrow{enzyme(S_2)} S_3 + \overline{S}_3, \quad \overline{S}_3 \xrightarrow{enzyme(\overline{S}_2)} \overline{S}_3 + S_3$$

6 Summary

It seems, according to our results, that Typogenetics offers new analogies and formal tools for computer scientists active in artificial life. A central "dogma" of the typogenetics is that strands have twofold role: *First* they are replicators, and *second*, they code an information about the way of their replication. Formally, Typogenetics may be considered as a molecular automaton that on its input reads strands and on its output it replicates strands. To make such an automaton more interesting, we may endow strands with additional properties enabling them to behave in some specific manner. In the present simple approach, strands have infinite resources for their replications. If we introduce a limited space of resources, then we get an additional selective pressure (a struggle for raw materials) with respect to a selection entirely based on strand fitness that reflect their capability of replication. An introduction of a "geographical" distributions of strands in a population might be very important, which was already clear from the evolutionary construction of hypercycles. In that case the population could not be considered as a homogeneous one in a well-stirred vessel. A replication function of strands usually requires only a fraction of the enzyme that is coded in the strand; it is then possible, in general, to code additional strand or enzyme properties that may give rise to an emergence of new properties and hierarchically organized structures. Summarizing, Typogenetics represents a very rich and flexible formal tool, closely related to basic concepts of molecular biology, that opens new possibilities and horizons for artificial life activities and efforts.

Acknowledgments. This work was supported by the grants # 1/7336/20 and # 1/8107/01 of the Scientific Grant Agency of Slovak Republic.

References

1. Adami, C.: Introduction to Artificial Life. Springer-Verlag, New York (1998)
2. Banzhaf, W.: Self-replicating sequences of binary numbers – I. Foundations, Self-replicating sequences of binary numbers – II. Strings of length N=4. Biological Cybernetics 69 (1993) 269-281
3. Banzhaf, W.: Self-organization in a system of binary strings. In: Brooks, R., Maes, P. (eds.): Artificial Life IV. MIT Press, Cambridge, MA (1994) 109-119
4. Banatre, J.-P., Le Metayer, D.: The Gamma Model and its Discipline of Programming. Sci. of Progr. 15 (1990) 55-77
5. Berry, G., Boudol, G.: The Chemical Abstract Machine. Theoret. Comp. Sci. 96 (1992) 217-248
6. Dittrich, P., Banzhaf, W.: Self-Evolution in a Constructive Binary String System. Artificial Life 4 (1998) 203-220
7. Dittrich, P.: Artificial Chemistries, a tutorial held at European Conference on Artificial Life, September 13-17, 1999, Lausanne, Switzerland (lecture is available at the address: http://ls11-www.cs.uni-dortmund.de/achem).
8. Eigen, M.: Self-organization of matter and the evolution of biological macromolecules. Naturwissenschaften 58 (1971) 465-523
9. Eigen, M., Schuster, P.: The Hypercycle: A Principle of Natural Self-Organization. Springer-Verlag, Berlin (1979).

10. Farmer, J. D., Kauffman, S. A., Packard, N. H.: Autocatalytic replication of polymers. Physica D 22 (1982) 50-67
11. Fogel, D. B.: Evolutionary Computation. Towards a New Philosophy of Machine <Intelligence. IEEE Press, Piscataway, NJ (1995)
12. Fontana, W.: Algorithmic chemistry. In: Langton, C. G., Taylor, C., Farmer, J.D., Rasmussen S. (eds.): Artificial Life II. Addison-Wesley, Reading, MA (1992) 159-210
13. Fontana, W., Schuster, P.: Continuity in evolution. On the nature of transitions. Science 280 (1998) 1451-1455
14. Fontana, W., Wagner, G., Buss, L.: Beyond digital naturalism. Artificial Life 1 (1994) 211-227
15. Goldberg, D. E.: Genetic Algorithms in Search, Optimization, and Machine Learning. Addison-Wesley, Reading, MA (1989)
16. Haken, H.: Pattern Formation and Pattern Recognition - An Attempt to a Synthesis. In: Haken, H. (ed.): Pattern Formation by Dynamical Systems and Pattern Recognition. Springer Verlag, Heidelberg (1979)
17. Hofstadter, D.: Dialogues with Gödel, Escher, Bach: An Eternal Golden Braid. Basic Books, Inc, New York (1979) chapters XVI and XVII
18. Hosokawa, K., Shimoyama, I., Miura, H.: Dynamics of Self-Assembling Systems – Analogy with Chemical Kinetics. In: Brooks, R., Maes, P. (eds.): Artificial Life IV. MIT Press, Cambridge, MA (1994) 172-180
19. Ikegami, T., Hashimoto, T.: Active mutation in self-reproducing networks of machines and tapes. Artificial Life 2 (1995) 305-318
20. Kauffman, S. A.: Autocatalytic sets of proteins. Journal of Theoretical Biology 119 (1986) 1-24
21. Kvasnicka, V., Pospichal, J.: Autoreplicators and Hypercycles in Typogenetics. THEOCHEM (in press). Available at the address http://math.chtf.stuba.sk/pub/vlado/Artificial_Chemistry/Typogenetics_paper_THEOCHEM_final.pdf. An extended version of this paper is available at the address http://math.chtf.stuba.sk/pub/vlado/Artificial_Chemistry/Typogenetics_paper_ECAL_old.pdf
22. Mitchell, M.: An Introduction to Genetic Algorithms. MIT Press, Cambridge, MA (1996)
23. Morris, H. C.: Typogenetics: A logic for artificial life. In: Langton, C. G. (ed.): Artificial Life. Addison-Wesley, Reading, MA (1989) 369-395
24. Pargellis, A. N.: The spontaneous generation of digital "life". Physica D, 91 (1996) 86-96
25. Rasmussen, S., Knudsen, C., Feldberg, R., Hindsholm, M.: The Coreworld: emergence and evolution of cooperative structures in a computational chemistry. Physica D 42 (1990) 111-134
26. Ray, T. S.: An approach to the synthesis of life. In: Langton, C. G., Taylor, C., Farmer, J.D., Rasmussen, S. (eds.): Artificial Life II. Addison-Wesley, Reading, MA (1992) 371-408
27. Schuster, P.: How does complexity arise in evolution. Complexity 2 (1996) 22-30.
28. Stadler, P. F., Fontana, W., Miller, J. H.: Random catalytic reaction networks. Physica D 63 (1993) 378-392
29. Suzuki, Y., Tanaka, H.: Symbolic chemical system based on abstract rewriting and its behavior pattern. Artificial Life and Robotics 1 (1997) 211-219
30. Varetto, L.: Typogenetics: An artificial genetic system. Journal of Theoretical Biology, 160 (1993) 185-205
31. Varetto, L.: Studying artificial life with a molecular automaton. Journal of Theoretical Biology 193 (1998) 257-285

Emergence of a Super-Turing Computational Potential in Artificial Living Systems*
(Extended Abstract)

Jiří Wiedermann[1] and Jan van Leeuwen[2]

[1] Institute of Computer Science, Academy of Sciences of the Czech Republic,
Pod Vodárenskou věží 2, 182 07 Prague 8, Czech Republic
jiri.wiedermann@cs.cas.cz

[2] Department of Computer Science, Utrecht University,
Padualaan 14, 3584 CH Utrecht, the Netherlands
jan@cs.uu.nl

Abstract. The computational potential of artificial living systems can be studied without knowing the algorithms that govern the behavior of such systems. What is needed is a formal model that neither overestimates nor underestimates their true computational power. Our basic model of a single organism will be the so-called cognitive automaton. It may be any device whose computational power is equivalent to a finite state automaton but which may work under a different scenario than standard automata. In the simplest case such a scenario involves a potentially infinite, unpredictable interaction of the model with an active or passive environment to which the model reacts by learning and adjusting its behaviour or even by purposefully modifying the environment in which it operates. One can also model the evolution of the respective systems caused by their architectural changes. An interesting example is offered by communities of cognitive automata. All the respective computational systems show the emergence of a computational power that is not present at the individual level. In all but trivial cases the resulting systems possess a super-Turing computing power. That is, the respective models cannot be simulated by a standard Turing machine and in principle they may solve non-computable tasks. The main tool for deriving the results is non-uniform computational complexity theory.

1 Introduction

A tantalizing question in computational mind modeling is the following: if it is true that the mind can be modeled by computational means, how can we explain the fact that mathematicians are often able to prove "mechanically unprovable" theorems, i.e. theorems whose truth or falsity cannot be algorithmically proved within a given formal system (e.g. corresponding to a computer simulating the

* This research was partially supported by GA ČR grant No. 201/00/1489 and by EC Contract IST-1999-14186 (Project ALCOM-FT).

J. Kelemen and P. Sosík (Eds.): ECAL 2001, LNAI 2159, pp. 55–65, 2001.

mind) due to Gödel's incompleteness theorem. In an extensive discussion of the respective problems R. Penrose [6] conjectures that there must be some so far unknown faculty of the brain that gives it a non-computable, non-algorithmic, "super-Turing" power in some cases.

In this paper we offer a plausible explanation of this phenomenon in the realm of artificial living systems. We will show that under a certain not-commonly considered computational scenario the ability to surpass the computational limits of a single Turing machine can emerge in non-uniformly evolving families or communities of far simpler computational devices than Turing machines, viz. finite automata.

The plan of the approach is as follows. First, in Section 2 we introduce our basic tool for modeling a single living organism — an interactive cognitive automaton seen as a finite discrete-state computational device. Then, in Section 3 we model the evolution of such devices by means of potentially infinite sequences of cognitive automata of increasing size and show that the resulting "families" possess the super-Turing computing power. For a formal treatment of the respective issues we use basic notions from non-uniform complexity theory. Next, in Section 4 we show that so-called active cognitive automata that can move in an interactive environment and modify it at will, gain the computing power equivalent to that of the standard (or interactive) Turing machine. Finally, in Section 5 we consider evolving communities of communicating active cognitive automata and for such communities we will show the emergence of super-Turing computing power. The merits of the respective results from the viewpoint of computational cognition will be discussed in Section 6.

All the above mentioned results are based on (old and new) results from non-uniform computational complexity theory. The paper opens a new application area for this theory by interpreting its results in terms of cognitive and evolutionary systems. Doing so sheds new light on the computational potential of the respective systems.

Proofs are omitted in this extended abstract.

2 Cognitive Automata

When modeling living organisms in order to study their computing potential it is important to keep in mind that the computational power of a model can be studied without actually knowing the concrete algorithms that are used by the model in concrete situations. What we have to know is the set of the elementary actions of the given model (its "instruction set") and the scenario of its interaction with its environment (what data can appear at its input, whether and how this data depends on previous outputs from the system, whether the system can "off-load" its data to the environment, and so on).

The second fact that we have to take into account is the crucial difference between the requirements put upon a model in case one merely wants to simulate the behaviour or actions of some (living) system, and those in case we also want to investigate its computational potential. In the former case the choice of a more powerful model than is necessary is acceptable since this can simplify the task of simulation. In the latter case the same choice would lead to the overestimation of the computing potential of the system at hand. Thus, in the latter case the model must neither be too powerful nor too weak: it must exactly capture the computing power of the modeled system.

Fortunately, also in such a case we are in a much better situation than it might appear. Despite their unprecedented complexity (when measured in terms of complexity of human artifacts) it is commonly believed that each living organism can enter into only a finite — albeit in most cases astronomic — number of distinguishable internal configurations. From the space limitations put upon this paper we cannot afford to give exhaustive arguments in favor of this fact. Instead, *we take it as our fundamental assumption that a living organism interacting with its environment can in principle be modeled by a finite discrete-state machine.* In the sequel we will call any finite discrete-state machine used in the above mentioned modeling context a *cognitive automaton.* A finite automaton presents a paradigmatic example of such a device. Other examples of cognitive automata are combinatorial circuits [1], discrete neural (cf. [5], [17]) or neuroidal [11] nets, neuromata [7] and various other computational models of the brain (cf. [17], [18]).

The next thing that our model has to capture is the fundamental difference between the standard scenario of computations by a finite automaton or Turing machine, and that of "computations" by a cognitive automaton. In the former case we assume a finite input that is known, fixed prior to the start of the computation. After starting the computation neither additional inputs nor changes in already existing, but not-yet-read inputs are allowed. In the next run of the machine, with new input data, both the finite automaton and the Turing machine must start again from the same initial configuration as in the case of previous inputs. There is no way to transfer information from past runs to the present one. Under this computational scenario, the respective machines are prevented from learning from their past experience.

Contrary to this, living organisms interacting with their environment process their inputs as delivered by their sensory systems without interruption. The inputs "appear", in an on-line manner, unexpectedly and as a rule they must be processed in real-time (this seems to be a necessary condition for the emergence of at least a rudimentary form of consciousness — cf. [20]). In principle the respective computations never terminate and are practically limited only by the lifespan of the organisms at hand. Moreover, the inputs stream into their cognitive systems in parallel via numerous channels and the systems process them also in a parallel manner. In most cases, once read, the original input is no longer available. The number of input channels depends on the size (or complexity) of the system at hand. In addition, especially the inputs into more complex systems that modify their environment or communicate with other systems may depend on the previous actions of a system or the reactions of other systems. Thus the systems gain a potential ability to learn from their own mistakes or experience. The respective computational scenario is a scenario of perpetual interactive learning.

Formally, any cognitive automaton realizes a translation $\Phi_{m,n}$ that transforms infinite input sequences of Boolean m-tuples into similar sequences of n-tuples, for certain $m, n \geq 1$. Depending on the type of formal device used in place of the cognitive automaton, this device can read and produce the tuples in parallel via many input/output ports (as in the case of neural nets), or these tuples are "packed" into symbols of some finite alphabet and are read or produced in this packed form in one step via a single port (as in the case of finite automata, or interactive Turing machines from Section 4).

From [21] and [7] the next theorem follows:

Theorem 1. *Let $\Phi_{m,n} : \{\{0,1\}^m\}^\omega \to \{\{0,1\}^n\}^\omega$, with $m, n \geq 1$, be a translation acting on infinite streams. Then for any $m, n \geq 1$ the following are equivalent:*

- *$\Phi_{m,n}$ is realized by a discrete neural net;*
- *$\Phi_{m,n}$ is realized by a neuroidal net;*
- *$\Phi_{m,n}$ is realized by a neuromaton;*
- *$\Phi_{m,n}$ is realized by a finite (Mealy) automaton.*

Any translation $\Phi_{m,n}$ that is realized by one of the above mentioned devices will be called a *regular translation.*

Although the learning potential of finite automata is not quite obvious, it can be easily observed e.g. in the case of neuroids [11]. Namely, they can be seen as "programmable neurons" since this ability was their primary design goal. The basic set of elementary operations of a cognitive automaton in order to obtain the potential for the development of cognitive abilities via learning is proposed in [11], [18] and [20]. Nevertheless, as mentioned above, the exact form of learning algorithm is unimportant for determining the computational power of the respective devices.

3 Families of Cognitive Automata

In order to be able to process more complicated translations than the regular ones and also to reveal a dependence of computational efficiency on the size of the underlying devices we will consider families of cognitive automata.

Let $\mathbf{F}_p = \{\mathcal{N}_1, \mathcal{N}_2, \ldots \mathcal{N}_i, \ldots \mid size(\mathcal{N}_i) \leq p(i)\}$ be an infinite *family of cognitive automata* of increasing size bounded by function p. Such a family is also called a non-uniform family since in general there need not exist an algorithmic way to compute the description of $\mathcal{N}_i$, given i. Thus there need not be a 'uniform' way to describe the members of the family. Intuitively, the only way to describe the family is to enumerate all its members.

Definition 1. *Let $\mathbf{F}_p$ be a family of cognitive automata, and let $\Phi_{m,n}$ be the translation realized by automaton $\mathcal{N}_m \in \mathbf{F}_p$ for some $n \geq 1$. Then the* non-uniform translation *$\Psi(\mathbf{F}_p)$ realized by $\mathbf{F}_p$ is the set of translations realized by the members of $\mathbf{F}_p$, i.e. $\Phi(\mathbf{F}_p) = \{\Phi_{m,n} | m \geq 1\}$.*

Note that in the above definition, for any given $m \geq 1$, an (infinite) input sequence of m-tuples is processed by the automaton $\mathcal{N}_m$ that is "specialized" in the processing of m-tuples. The class of translations realized by families of neuromata (i.e. of discrete neural nets reading their inputs via a single input port [7]) of polynomially bounded size will be denoted as POLY-NA. In addition to this class we will also consider the classes LOG-NA (the translations realized by neuromata of logarithmic size), POLY-NN (the translations realized by standard recurrent, or cyclic, discrete neural nets of polynomial size, reading their inputs in parallel via m ports), and POLY-FA (the translations realized by finite automata with a polynomial number of states).

Our main tool for characterizing the computational efficiency of the families of cognitive automata will be interactive Turing machines with advice. Non-interactive versions of such machines were introduced by Karp and Lipton [3] who established the foundation of non-uniform complexity theory. Machines with advice are akin to oracle machines as already introduced by Turing [9]. Effectively, an oracle allows inserting outside information into the computation. This information may depend on the concrete input and is given for free to the respective oracle machine whenever the oracle is queried. Advice functions are a special kind of oracle, where the queries can depend only on the size of the input to a machine. Intuitively, the information delivered by an oracle makes sense only for the given input; the information offered by an advice can be used for all inputs of the same size and this is the only oracle information given for all these inputs. With the help of advice a machine may easily gain a super-Turing computing power, because there is no requirement of "computability" on advice (cf. [1], [3], or [13]). Advice is not necessarily more restrictive than the use of an arbitrary oracle, because one can combine all oracle-values ever queried in computations on inputs of size n into one advice value $f(n)$. This is why one usually imposes size-bounds on advice.

Definition 2. *An advice function is a function* $f : \mathbf{Z}^+ \rightarrow \Sigma^*$. *An advice is called* $S(n)$*-bounded if for all* n, *the length of* $f(n)$ *is bounded by* $S(n)$.

Technically, a Turing machine with an advice function f operates on its input of size n in much the same way as a standard Turing machines does. However, such machine can also call its advice by entering into a special query state. After doing so, the value of $f(n)$ will appear at the special read-only advice tape. From this moment onward the machine can also use the contents of this tape in its computation.

It is intriguing to consider the effect of providing cognitive automata with some means to query oracles. It is reasonable to assume that a $p(n)$-size bounded cognitive automaton can only issue oracle queries of size $p(n)$ in its computation on inputs of size n.

Theorem 2. *Let* $\mathbf{F}_p$ *be a family of cognitive automata that use some Turing machine as oracle. Then* $\Psi(\mathbf{F}_p)$ *can be realized by a Turing machine using* $O(p)$*-bounded advice.*

A converse statement also holds but its formulation is beyond the scope of the present paper.

For the classes of translations realized by (interactive) Turing machines with advice we introduce a notation similar to that used in the theory of non-uniform complexity classes (cf. [1]).

Definition 3. *Let* $\mathbf{x} = \{x_t\}_{t \geq 0}$, *let* f *be an advice function and let* $\mathbf{x} \diamond \mathbf{f} = \{\langle x_t, f(t) \rangle\}_{t \geq 0}$, *where the broken brackets denote the concatenation of two strings surrounded by the brackets. Then the class* $\mathcal{C}/\mathcal{F}$ *of translations consists of the translations* Φ *for which there exists a* $\Phi_1 \in \mathcal{C}$ *and a* $f \in \mathcal{F}$ *such that for all* $\mathbf{x}$: $\Phi(\mathbf{x}) = \Phi_1(\mathbf{x} \diamond \mathbf{f})$.

Thus, a translation Φ belongs to $\mathcal{C}/\mathcal{F}$ iff Φ is realized by a Turing machine from complexity class $\mathcal{C}$ with advice function $f \in \mathcal{F}$. Common choices considered for $\mathcal{C}$ that we will use are: *LOGSPACE* ('deterministic logarithmic space'), *PSPACE* ('polynomial space'), etc. Common choices for $\mathcal{F}$ are *log*, the class of logarithmically bounded advice functions, and *poly*, the class of polynomially bounded advice functions.

In non-uniform computational complexity theory and in the theory of neurocomputing (cf. [5]) the following assertion is proved.

Theorem 3. *For the classes of non-uniform translations the following equalities hold:*

- *POLY-NN=POLY-NA=PSPACE/poly*;
- *LOGSPACE-NN=LOGSPACE-NA=LOGSPACE/log*;
- *POLY-FA=LOGSPACE/poly*.

For more information about the complexity of non-uniform computing we refer to [1].

4 From Cognitive Automata to Cognitive Turing Machines

Consider now a cognitive automaton enhanced by an apparatus that enables it to move around in its living environment and to mark the environment in a way that can later be recognized again by the automaton at hand. The resulting device is called an *active cognitive automaton*. It can store and retrieve information in/from its environment and thus it bears a similarity with robotic cognitive systems. Models of finite automata which can read inputs from a two-way input tape and in addition can also mark input tape cells have been studied for years in automata theory (cf. [16]). The respective machines are provably computationally more powerful than non-marking automata. When we allow a finite set of marks that can be placed to or removed from a potentially infinite environment, one obtains an interactive Turing machine [12], [13], [15]. Interactive Turing machines are a similar extension of standard Turing machines as was the extension of finite automata towards active cognitive automata. Thus, an interactive Turing machine is a Turing machine that translates infinite streams of input symbols into similar streams of output symbols under an interactive scenario as described in Section 2. Several further conditions may be imposed on the way the machine interacts, e.g. to model the bounded delay property that cognitive systems often display in their respond behaviour.

Theorem 4. *The computational power of active cognitive automata is equivalent to the computational power of interactive Turing machines.*

Turing [8] saw "his" machine (i.e., the Turing machine) as a formalized model of a "computer", which in his days meant "a person who calculates". Such a person computes with the help of a finite table (that corresponds to a "program") that is held in a person's head, and further using a (squared) paper, pencil and a rubber. In accordance with Turing's own belief generations of researchers working in artificial intelligence and philosophers of mind have believed that the Turing machine as a whole corresponds to the model of the above mentioned human computer. Our previous short discussion suggests that within the model of a computing person one has to distinguish among three components: the machine's finite control (its "program"), its "sensors, effectors and motoric unit" (movable read/write head) and its environment (the tape). Hodges, Turing's biographer, writes in [2]: *" Turing's model is that of a human mind at work"*. This is only partially correct: in Turing's model, merely the machine's finite control corresponds to the mind of the modeled calculating person.

5 Communities of Active Cognitive Automata

Ultimately, active cognitive automata are of interest only in large conglomerates, interacting like "agents" of individually limited powers. A community of active cognitive automata (or shortly: a community of agents) is a time-varying set of devices consisting at each moment of time of a finite set of active cognitive automata of the same type sharing the same environment. Each automaton makes use of a piece of its immediate environment as its private external memory giving it the computing power of an interactive Turing machine (as stated in Theorem 4). Each automaton has its own input and output port. The ports of all automata altogether present the input and output ports of the community. The number of these ports varies along with the cardinality of the community. Within their set the automata are identified by a unique name (or address).

The agents (automata) can communicate by sending their outputs as inputs to other automata identified by their addresses, or by writing down a message into their environment which can be read by other automata. One can see it also as if the agents move in their environment and encounter each other randomly, unpredictably, or intentionally, and exchange messages. Who encounters whom, who will send a message, as well as the delivery time of each message is also unpredictable. The idea is to capture in the model any reasonable message delivery mechanism among agents — be it the Internet, snail mail, spoken language in direct contact, via mobile phones, etc. Moreover, the agents are mortal: they emerge and vanish also unpredictably.

The description of a community of agents is given at each time by the list of names of all living agents at that time, the list of all transient messages at that time, including the respective senders and addressees, the time of the expedition of each message, its addressee, and the message delivery times. Note that in general most of the required parameters needed in the instantaneous description of a community at a given time are non-computable (since according to our description of community functioning they are unpredictable). Nevertheless, at each moment in time they can be given by a finite table. Therefore the description of the whole community at any time is always finite.

In [13], [15] the following result has been proved for the case of "real" agents communicating via the Internet.

Theorem 5. *The computational power of communities of agents is equivalent to the power of interactive Turing machines with an advice function whose size grows linearly with the processing time.*

Note that similar to the case of infinite families of cognitive automata, communities of active cognitive automata have a super-Turing computing power. The source of this power is given by the potentially unlimited cardinality of the community and by the non-computable characteristics of the community size in the unpredictable interaction among community members and those of their existence span (leading to non-uniformity of the resulting system).

6 Afterthoughts

The results in the previous theorems have interesting interpretations in the world of cognitive automata and computational cognition.

Theorem 1 describes the equivalence among the basic types of cognitive automata. In complexity theory numerous other models of non-uniform computation are known — such as combinatorial or threshold circuits and other types of neural nets, especially the biologically motivated ones (cf. [4]). Nonetheless, the computational equivalence of the respective models indicates that computational cognition is a rather robust phenomenon that can in principle be realized by various computational models which are equivalent to finite automata. We said "in principle" because in practice much will depend upon the efficiency of such models. For instance, in [20] a principle of consciousness emergence in cognitive systems is sketched. In the simplest case consciousness takes the role of a control mechanism that, based on feed-back information from a system's sensors, verifies the correct realization of motoric actions to which orders have been issued. If these actions are not performed in accordance with these orders, the consciousness will realize it and take care about the appropriate remedy. In order to fulfill this role consciousness must operate in real time w.r.t. the speed of the system. The system must react fast enough to be able to recognize the erroneous realization of its orders and take the appropriate measures in time that still give opportunity for the realization of rescue actions. In practice such requirements disqualify "slow" systems and support the specialized, fast or "economical" solutions. It is known that there are cognitive tasks that can be realized by a single biological neuron (over n inputs) whereas the equivalent neural nets requires a quadratic number of standard neurons [4].

Theorem 2 suggests that the evolution of families of cognitive automata that use "recursively enumerable" information from an external source and which might lead to the emergence of a super-Turing computing power, can be simulated using Turing machine models with a very limited, global learning facility.

Theorem 3 illustrates the various degrees of efficiency of certain classes of cognitive automata. For instance, the family of neural nets of polynomial size has the power of PSPACE/*poly* whereas families finite automata of the same size only have the power of LOGSPACE/*poly*. It also demonstrates the emergence of super-Turing power in the course of non-uniform evolution within families.

Theorem 4 points to a jump in the computational power of sufficiently developed active cognitive automata equipped by sensors that can scan the environment and by effectors by which the automata can modify their environment. The corresponding individual active cognitive automata (or cognitive robots) gain the power of interactive Turing machines. In other words, the original finite-state computing device capable of reaching but a bounded number of configurations will turn into a device which can reach a potentially unbounded number of configurations. This is a nice argument that qualitatively illustrates e.g. the revolutionary contribution of the development of the script to the development of human civilization.

The study of the computational power of interactive Turing machines was initiated and studied in [12]. Roughly speaking, the respective theory leads to a generalization of standard computability theory to the case of infinite computations. The results from [12] indicate that by merely adding interactive properties and allowing endless computations, one does not break the Turing computational barrier. The resulting devices are not more powerful than classical Turing machines. They simply compute something different than the latter machines. Thus the new quality is only brought into computing by letting non-predictability enter into the game (cf. [15]).

Theorem 5 asserts that a community of active cognitive automata has a much greater power than the sum of the powers of the individual automata. Here we see the emerging non-recursive computing power of unpredictable external information entering into a system. Do the results of Theorem 5 really mean that the corresponding systems can solve undecidable problems? Well, they can, but only under certain assumptions. In order for these systems to simulate a Turing machine with advice they need a cooperating environment. Its role is to deliver the same information as is offered by the advice. Thus the respective results are of a non-constructive nature: both the advice and the corresponding inputs from the environment exist in principle but there is no algorithmic way to obtain them. In practice the assumptions of the existence of external inputs suitable for the solution of a concrete undecidable problem, such as the halting problem, are nor fulfilled. Hence, without such "right" inputs no community of cognitive automata will solve an undecidable problem. On the other hand, no Turing machine without an advice could simulate e.g. the (existing) Internet — simply because the Internet develops in a completely unpredictable, non-algorithmic way. One can say that the current Internet realizes a concrete non-algorithmic translation that, however, emerges somehow "all by itself", by the joint interplay of all users who operate and upgrade the Internet in a completely unpredictable manner. All users jointly play the role of an "advice" — nonetheless the respective advice keeps emerging on-line, incrementally, is "blind", possessing as a whole no purposeful intention. The same holds for the development of a human society — it also evolves in a non-algorithmic way and therefore cannot be modeled by a single computer (without advice).

What remains to be done is answering Penrose's question from the introduction of this paper. To see the idea, consider the information computed and stored in a long run by the community of cognitive robots. By virtue of Theorem 5 this is non-computable information. Now, each member of this community has access to this information which effectively plays the role of an advice and thanks to this, in principle each member of the community gains a super-Turing computing power.

7 Conclusion

The previous results can be seen as applications of computability theory to artificial life systems. The main result explaining the emergence of the super-Turing computing potential within the respective systems certainly justifies the approach and points to the increasing role that computer science will play in problems related to understanding the nature of the emergence of life in general and intelligence in particular (cf. [19]). The above results point to quite realistic instances where the classical paradigm of a standard Turing machine as the generic model of all computers which is able to capture all computations, is clearly insufficient. It appears that the time has come to reconsider this paradigm and replace it by its extended version — viz. interactive Turing machines with advice. For a more extended discussion of the related issues, see [13].

References

1. Balcázar, J. L. — Díaz, J. — Gabarró, J.: Structural Complexity I. Second Edition, Springer, 1995, 208 p.
2. Hodges, A.: Turing — A Natural Philosopher. Phoenix, 1997, 58 p.
3. Karp, R.M. — Lipton, R. J.: Some connections between non-uniform and uniform complexity classes, in *Proc. 12th Annual ACM Symposium on the Theory of Computing* (STOC'80), 1980, pp. 302-309.
4. Maass, W. — Bishop,C., editors: Pulsed Neural Networks. MIT-Press (Cambridge), 1998.

5. Orponen, P.: An overview of the computational power of recurrent neural networks. Proc. of the Finnish AI Conference (Espoo, Finland, August 2000), Vol. 3: "AI of Tomorrow", 89-96. Finnish AI Society, Vaasa, 2000.
6. Penrose, R.: The Emperor's New Mind. Concerning Computers, Mind and the Laws of Physics. Oxford University Press, New York, 1989.
7. Šíma, J. — Wiedermann, J.: Theory of Neuromata. *Journal of the ACM,* Vol. 45, No. 1, 1998, pp. 155–178.
8. Turing, A. M.: On computable numbers, with an application to the Entscheidungsproblem, Proc. London Math. Soc., 42-2 (1936) 230-265; A correction, ibid., 43 (1937), pp. 544-546.
9. Turing, A.M: Systems of logic based on ordinals, Proc. London Math. Soc. Series 2, 45 (1939), pp. 161-228.
10. Turing, A. M.: Computing machinery and intelligence, Mind 59 (1950) 433-460.
11. Valiant, L.G.: Circuits of the Mind. Oxford University Press, New York, Oxford, 1994, 237 p., ISBN 0–19–508936–X.
12. van Leeuwen, J. — Wiedermann, J.: On algorithms and interaction, in: M. Nielsen and B. Rovan (Eds), Mathematical Foundations of Computer Science 2000, 25th Int. Symposium (MFCS'2000), Lecture Notes in Computer Science Vol. 1893, Springer-Verlag, Berlin, 2000, pp. 99-112.
13. van Leeuwen,J. — Wiedermann, J.: The Turing machine paradigm in contemporary computing, in: B. Enquist and W. Schmidt (Eds), Mathematics Unlimited - 2001 and Beyond, Springer-Verlag, 2001, pp. 1139-1155.
14. van Leeuwen, J. — Wiedermann, J: A computational Model of Interaction in Embedded Systems. Technical Report TR CS-02-2001, Dept. of Computer Science, University of Urecht, 2001
15. van Leeuwen, J, — Wiedermann, J.: Breaking the Turing Barrier: the Case of the Internet. Manuscript in preparation, February, 2001.
16. Wagner,K. — Wechsung, G.: Computational Complexity, VEB Deutscher Verlag der Wissenschaften, Berlin 1986, 551 p.
17. Wiedermann, J.: Toward Computational Models of the Brain: Getting Started. *Neural Network World,* Vol. 7, No. 1, 1997, pp. 89-120.
18. Wiedermann, J.: Towards Algorithmic Explanation of Mind Evolution and Functioning (Invited Talk). In: L. Brim, J. Gruska and J. Zlatuška (Eds.), Mathematical Foundations of Computer Science, Proc. of the 23-rd International Symposium (MFCS'98), Lecture Notes in Computer Science Vol. 1450, Springer Verlag, Berlin, 1998, pp. 152–166.
19. Wiedermann, J.: Simulated Cognition: A Gauntlet Thrown to Computer Science. *ACM Computing Surveys*, Vol. 31, Issue 3es, paper No. 16, 1999.
20. J. Wiedermann: Intelligence as a Large-Scale Learning Phenomenon. Technical Report ICS AV CR No. 792, 1999, 17 p.
21. Wiedermann, J.: The computational limits to the cognitive power of neuroidal tabula rasa, in: O. Watanabe and T. Yokomori (Eds.), *Algorithmic Learning Theory,* Proc. 10th International Conference (ALT'99), Lecture Notes in Artific. Intelligence, Vol. 1720, Springer Verlag, Berlin, 1999, pp. 63-76.

Eco-Grammars to Model Biological Systems: Adding Probabilities to Agents

Sergio O. Anchorena and Blanca Cases

Dpt. of Computer Languages and Systems.
University of the Basque Country
Apdo. 649, 20080, San Sebastián, Spain
{ylbanoss, jipcagub}@sc.ehu.es

Abstract. The aim of this paper is to define probabilistic Eco-grammar systems and some of their applications in the field of evolutionary biology. A probabilistic Eco-grammar system is composed of agents that select the rules of internal growth as well as the action rules according to distributions of probability. The environment is $0L$ probabilistic. Our interest is centered in the study of the normalized populations of symbols obtained along a derivation. In this paper we show that the probabilistic approach applied to Eco-grammar systems allows to model the evolutionary stable strategies of Maynard Smith.

1 Preliminaries: Eco-Grammar Systems

Eco-grammar systems (EG) are a model of computation with universal computational capability, proposed as a formal framework to model ecosystems in [3,4, 10,12]. EG systems are composed of mutually interdependent Eco-agents, acting on a common environmental string (and possibly other Eco-agents) according to action rules. This model is defined by Csuhaj-Varjú, Kelemen, Kelemenová and Paŭn [4].

The states of the Eco-agents, and of the environmental string, are described by strings of symbols. An alphabet V is any non empty and finite set of symbols. V^* denotes the set of all the strings over V. The empty string is $\varepsilon \in V^*$. Eco-agents act synchronously by discrete steps, applying rewriting rules over a string of symbols (representing the environment or other Eco-agents). The environment,is a $0L$ [12] system that rewrites in parallel the symbols that are not handled by the Eco-agents.

Definition 1. *A $0L$ scheme is a pair $E = (V_E, P_E)$ such that $V_E = \{a_1, ..., a_m\}$ is an alphabet and P_E is a set of parallel rewriting rules of the form $a_i \to \alpha_i$, with $a_i \in V_E$, $\alpha_i \in V_E^*$. If $x = a_{i_1}...a_{i_k} \in V_E^k$, we say that $x\prime$ is derived in one step from x in E, denoted $x = a_{i_1}...a_{i_k} \Rightarrow_E \alpha_{i_1}...\alpha_{i_k} = x\prime$ iff $\forall j\,(1 \le j \le k)\ a_{i_j} \to \alpha_{i_j} \in P_E$. We assume that P_E is complete, that is, if there exist a symbol $a_i \in V_E$ such that there are not rules in P_E with left hand side a_i, we apply by default the identity rule $a_i \to a_i$. If $x \in V_E^*$ then $P_E(x) = \{x\prime :\ x \Rightarrow_E x\prime\}$.*

J. Kelemen and P. Sosík (Eds.): ECAL 2001, LNAI 2159, pp. 66–75, 2001.

Definition 2. *An Eco-grammar system [4] is a tuple* $\Sigma = (E, A_1, \ldots, A_n, w_{E_0})$ *where* $E = (V_E, P_E)$ *is a* $0L$ *environment, the tuples* $A_i = (V_i, P_i, R_i, \varphi_i, \psi_i, w_{i_0})$ *are the Eco-agents and* $w_{E_0} \in V_E^*$ *is the environmental axiom.*

Definition 3. *An Eco-agent is defined as the tuple* $A_i = (V_i, P_i, R_i, \varphi_i, \psi_i, w_{i_0})$ *where:*

- V_i *is the alphabet of the Eco-agent.*
- P_i *is a finite set of* $0L$ *rules called developmental rules of the Eco-agent, defined over the alphabet* V_i.
- R_i *is a set of sequential rewriting rules, called action rules, of the form* $a \rightarrow \beta$ *with* $a \in V_E^+$, $\beta \in V_E^*$.
- $\varphi_i : V_E^* \rightarrow 2^{P_i}$, *the sensor function, is a computable function that selects from the environmental word* w_E, *the active subset* $P_{i_j} = \varphi_i(w_E) \subseteq P_i$ *of internal developmental rules* P_i. *The subset* P_{i_j} *is completed to be a* $0L$ *system.*
- $\psi_i : V_i^* \rightarrow 2^{R_i}$, *is the action function, a computable function that selects the active subset of the (sequential rewriting) set of rules of* R_i, *that change the state of the environment depending on the internal state of the Eco-agent.*
- $w_{i_0} \in V_E^+$, *is the axiom of the Eco-agent.*

The action of an Eco-agent at step of derivation k is manifested in the application of a rewriting rule to a symbol of the environmental word or the internal state of a different Eco-agent. An Eco-Grammar system combines the synchronous action of all the Eco-agents, each one applying a rewriting rule in sequential mode to the selected sub-string, with the action of the environment, that applies a $0L$ scheme to the remaining symbols.

Definition 4. *Let* $\Sigma = (E, A_1, \ldots, A_n, w_{E_0})$ *be an Eco-Grammar system with environment* $E = (V_E, P_E)$. *A configuration is a word*

$$\sigma = (w_E, w_1, ..., w_n) \in (V_E \cup \{",", "(", ")"\})^*$$

where $w_E \in V_E^*$ *is the environmental word, and each* $w_i \in V_i^*$ *is the internal word of Eco-agent* A_i. *The set of delimiters is disjoint to the alphabets of Eco-agents, and* P_E *is completed to include delimiters.*

A step of derivation in Σ *is represented as*

$$\sigma = (w_E, w_1, ..., w_n) \Rightarrow_\Sigma (w_E\prime, w_1\prime, ..., w_n\prime) = \sigma\prime$$

iff:

- $w_E \in V_E^*$ *is the state of the environment and each* $w_i \in V_i^*$ *is the internal state of Eco-agent* A_i.
- $w_i\prime\prime \in \varphi_i(w_E)(w_i)$ *is the internal word of* A_i *after the application of the internal* $0L$ *scheme* $\varphi_i(w_E) \subseteq P_i$ *selected by the sensor* φ_i.
- $\sigma = z_1 a_{i_1} z_2 a_{i_2}, \ldots, z_n a_{i_n} z_{n+1}$ *and there exists an ordination of the Eco-agents* $A_{i_1} ... A_{i_n}$ *such that the rule* $a_{i_j} \rightarrow \beta_{i_j} \in \psi_{i_j}(w_{i_j}\prime\prime)$ *is selected by the action function* ψ_{i_j} *from the intermediate internal state* $w_{i_j}\prime\prime$, *producing the configuration* $\sigma\prime = z_1\prime\beta_{i_1} z_2\prime\beta_{i_2}, \ldots, z_n\prime\beta_{i_n} z_{n+1}\prime$.

– *The sub-words non expanded by the Eco-agents are expanded by the environment, that is,* $\forall j\,(1 \leq j \leq n+1) z_j \Rightarrow_E z\prime_j$ *is a* 0*L step of derivation.*

The Eco-agent A_i interacts with A_j if there exists the possibility of modifying A_j: $V_i \cap V_j \neq \emptyset$. To simplify the definitions, we deal with non interactive Eco-agents in this work. That is, the alphabet of all the Eco-agents is the environmental alphabet V_E.

2 Probabilistic Eco-Grammar Systems with Reactive-Adaptive Eco-Agents

An Eco-Grammar system with probabilistic Eco-agents assigns to the developmental and action rules of the Eco-agents a probability of selection, according to the state of the environment and to the internal state of the Eco-agent. Following [2], we deal with the number of occurrences (population or absolute frequency) of a symbol $a_i \in V_E$ in a word w, denoted $|w|_{a_i}$.

If $R = \{item_i, \ldots, item_{\#R}\}$ is any non empty and ordered finite set with cardinal $\#R$, a distribution of probability π over R is a vector $\pi = (\pi_1, \ldots, \pi_{\#R})$ such that $\pi_i \in [0,1]$ is the probability of selecting $item_i$ and $\sum_{i=1}^{\#R} \pi_i = 1$. We consider that any finite set of rules is ordered.

Definition 5. *A probabilistic* 0*L scheme [1] is a tuple* $E = (V_E, P_E, q)$ *where the pair* (V_E, P_E) *is a* 0*L system and* $q = (q_1, ..., q_m)$ *is a vector of distributions of probability. Each* q_i *represents a distribution of probability* q_i *over the subset of rules with left hand side* a_i*, that is,* $\{a_i \to \alpha \in P_E\}$.

Definition 6. *A probabilistic EG system is a tuple* $\Sigma = (E, A_1, \ldots, A_n, w_{E_0})$. *The environment* $E = (V_E, P_E, q)$ *is a* 0*L probabilistic scheme. Each* A_i *is a probabilistic Eco-agent and* w_{E_0} *is the environmental axiom.*

A probabilistic Eco-agent is such that the sensor function and the action function perform a probabilistic selection of the rules of internal growth and action, respectively. We can study Probabilistic *EG* systems from the points of view of the sensor function and of the action function. Consider the second case. The interior of the Eco-agent is irrelevant. The Eco-agent simply reacts selecting the action rules with probabilities that vary in function of the state of the environment. We call those systems probabilistic Reactive *EG* systems, or *REG* systems.

In probabilistic *REG* systems, the definition of Eco-agents A_i is notably simplified reducing them to a set of action rules R_i and to a distribution of probability $\psi_i(w_E)$ over R_i, depending on the state of the environment w_E.

If we permit in the scheme of a reactive Eco-agent the change over time k of the distribution of probability $\psi_i^k(w_E)$ assigned to the rules R_i, the Eco-agent is Reactive-Adaptive.

Definition 7. *A Probabilistic EG system with reactive-adaptive Eco-agents is represented by the tuple* $\Sigma = (E, A_1, \ldots, A_n, w_{E_0})$, *where each Eco-agent* $A_i = (R_i, \psi_i^0, \rho_i)$ *is such that:*

- R_i *is a set of action rules and* ψ_i^0 *is the reaction function at step 0 of derivation.* $\forall k \in \mathbb{N}$, ψ_i^k *denotes the reaction function at time step k.*
- $\psi_i^k(w_{E_k}) = (\psi_{i_1}^k(w_{E_k}), ..., \psi_{i_{\#R_i}}^k(w_{E_k}))$ *is a distribution of probability that selects the rule* $r_{i_j} \in R_i$ *with probability* $\psi_{i_j}^k(w_{E_k})$.
- ρ_i *is a learning rule [7]. If* $w_{E_k} \in V_E^*$ *is the environmental word ,* $\psi_i^{k+1} = \rho_i(\psi_i^k, w_{E_k})$, *that is, the reaction function changes at step* $k+1$ *according to its own definition and to the environmental string at step k.*

We denote the class probabilistic EG's with reactive-adaptive Eco-agents by $AREG$. A hierarchy of Probabilistic $AREG$ systems is organized depending on the following parameters: the reaction scope, that determines the situation of the Eco-agent for observation and action, and the presence or not of a learning rule.

The absence of a learning rule is reduced to $\rho_i = I$, such that $\forall k \; \psi_i^{k+1} = I(\psi_i^k, w_{E_k}) = \psi_i^k = \psi_i^0 = \psi_i$. That is, the reaction function is non time dependent. The learning rule depends always on the whole environmental word, to reflect the action of the environment over the organism independently of the scope of reaction of the Eco-agent.

Respecting to the scope of reaction, at the step k of derivation, $AREG$ Eco-agents observe the whole environmental word w_{E_k} and after they select the action rule to be applied anywhere if possible. We say that the Eco-agents are locally reactive, or AR_LEG, when the mechanism of derivation allows firstly the selection of a symbol $a \in V_E$ in the environmental string $w_{E_k} = xay$, and secondly, in function of the selected symbol $a \in V_E$ they get a distribution of probability to select the action rule. The Eco-agent is simple if he is non-sensitive, that is, the reaction function does not depend on what is observed. The class of non adaptive Simple Eco-agents is denoted by SEG.

Scope of reaction

	The string	**The symbol**	**None**
	$w_{E_k} \in V_E^*$	$a \in V_E$, $w_{E_k} = xay$	
I	REG $\psi_i^{k+1}(w_{E_k}) = \psi_i(w_{E_k})$	R_LEG $\psi_i^{k+1}(a) = \psi_i(a)$	SEG $\psi_i^{k+1} = \psi_i$
ρ	$AREG$ $\psi_i^{k+1}(w_{E_k}) = \rho(\psi_i^k, w_{E_k})(w_{E_k})$	AR_LEG $\psi_i^{k+1}(a) = \rho(\psi_i^k, w_{E_k})(a)$	$ASEG$ $\psi_i^{k+1} = \rho(\psi_i^k, w_{E_k})$

3 Evolutionary Stable Strategies: Basic Concepts

Evolutionary games are based on the concept of Evolutionary Stable Strategy, ESS, created by Maynard Smith [7,8], who, by means of this concept and the application of a modified version of the theory of games, wanted to build a mathematical model of the combats between animals, looking for the strategies of behaviour that would result favored.

If $\bar{e} = (e_1, \ldots, e_k)$ is a non zero vector such that $e_i \in \mathbb{R}$ represents the population of a specie S_i, the normalized population is $\bar{p} = \frac{1}{\sum_{i=1}^{k} e_i} \times \bar{e}$. If $\bar{e} = \bar{0}$ is the null vector, the normalized population is $\bar{p} = \bar{0}$.

An evolutionary stable strategy is defined as having the following property: if the majority of the members of a high population adopts the strategy, not any mutant strategy could intrude the population. In other words, a strategy is stable in the evolutionary sense when there is not a mutant strategy giving a superior Darwinian efficacy to the individuals.

There exists a mathematical definition of an evolutionary stable strategy. Consider the matrix (table 3) of payments in a game. I is a ESS, if given any mutant strategy J one of the following conditions does hold [7,8]:

E	I	J
I	E(I,I)	E(I,J)
J	E(J,I)	E(J,J)

condition 1: $E(I,I) > E(J,I)$
condition 2: $E(I,I) = E(J,I)$ and $E(I,J) > E(J,J)$

Pure strategies and mixed strategies: A pure strategy is such that the player selects a tactic inside of a set of possible tactics for the encounters during the game, and maintains it during the game. Following Maynard Smith [7] a mixed strategy is one that recommends to follow a different tactic in the game according to a specific distribution of probabilities.

Assume that the members of the population combat in random encounters and, consequently, each individual reproduces those individuals in the class itself (individuals that use the same strategy) according to accumulated payment (the pay is translated as Darwinian efficacy). If there exists a evolutionary stable strategy, then the population should evolve toward it.

4 The Game of the Falcon and the Dove

The game of the "falcon" and the "dove", designed by Maynard Smith [7,14], consists in a simple a model that aims to explain the combats in the interior of a population in which there are only two possible tactics: either the role of "falcon" or the role of "dove". The tactic of the falcon consists in fighting until the end, intensifying the combat until, the animal wins (the adversary retreats or is seriously damaged), or gets gravely hurt. The tactic of the dove consists in avoiding hard fights. If the adversary initiates an advance, the dove goes away before getting damaged. The assignment of scores represents the expected pays after each possible encounter, presented in table 4.

Serious damage: -C
Victory: +V
Run away: 0
Long encounter: -L

E	FALCON (F)	DOVE (D)
FALCON (F)	$\frac{1}{2} \times (V - C)$	+V
DOVE (D)	0	$\frac{1}{2} \times (V - L)$

It is easy to see that pure strategies, "play always the role of falcon" or "play always the role of dove" do not fill the conditions to be evolutionary stable strategies.

The model predicts that, if an evolutionary stable strategy were exist, it should be a mixed strategy [8], that consist in playing the role of falcon with probability p and the role of dove with probability $1-p$. This two cases should be equivalent [9]: a)there are sub-populations of pure strategies, being the proportion of falcons p and b)each individual show mixed behaviour playing the role of $falcon$ with probability p.

We present a numerical example. The initial biological efficacy for all the individuals is 1. The table 4 of the game is obtained adding the payment of the combat to the initial efficacy.

C=9/2
V=5/2
L=1/2

E	FALCON (F)	DOVE (D)
FALCON (F)	$1+\frac{1}{2}\times(V-C)=0$	$1+V=7/2$
DOVE (D)	$1+0=1$	$1+\frac{1}{2}\times(V-L)=2$

It is easy to see that the only evolutionary stable strategy is to play the role of falcon with probability $p^*=3/5$ and to play the role of dove with probability $1-p^*=2/5$.

5 Modeling the Dynamics of Evolutionary Stable Strategies

The static analysis of the ESS inherent to the theory of games can be transformed in dynamic using: a)locally reactive and non adaptive R_LEG Eco-agents that follow pure strategies and b) Locally reactive and adaptive AR_LEG Eco-agents that follow mixed strategies.

In both cases, the alphabet is $V=\{F,D\}$ where symbol F represents a falcon and D a dove in the environmental string. The environment is the erasing $0L$ system: $E=(V,P_E,(q_1,q_2)=$ with $P_E=\{F\rightarrow\varepsilon,D\rightarrow\varepsilon\}$ and $q_1(F\rightarrow\varepsilon)=1$, $q_2(D\rightarrow\varepsilon)=1$. That is, the environment has a regulatory function of the amount of symbols in the environmental string.

We assign to the Eco-agents rewriting rules $F\rightarrow\alpha$ and/or $D\rightarrow\beta$. The right-hand sides α, β are strings in $\{F,D\}^*$ where the occurrences of F and D have a numerical relation extracted from the table 4 of payments of the game.

The Eco-agents represent a sample of the population. An Eco-agent is a pure falcon, A_F, a pure dove A_D or an adaptive mixed individual A_M that, at step k of derivation, plays the role of a falcon with probability p_k and the role of a dove with probability $1-p_k$. The Eco-agents are locally reactive, they randomly select a position in the environmental string and apply the rule according to the symbol that they find. The encounters between individuals are produced through the symbols that the Eco-agents find in the environment: An Eco-agent A_F that selects symbol D represents the encounter of a falcon (Eco-agent) with a dove (environmental symbol).

In table 5 we represent the three types of Eco-agents. In table 5 we set some notation common for the two models (k is the step of derivation).

		F	D
	Rules	$\psi_F(F)$	$\psi_F(D)$
A_F	$F \to \varepsilon$	1	0
	$D \to F$	0	0.05
	$D \to F^2$	0	0.15
	$D \to F^3$	0	0.30
	$D \to F^4$	0	0.30
	$D \to F^5$	0	0.15
	$D \to F^6$	0	0.05
	Mean	$F \to \varepsilon$	$D \to F^{3,5}$
	Rules	$\psi_D(F)$	$\psi_D(D)$
A_D	$F \to D$	1	0
	$D \to D^2$	0	1
	Mean	$F \to D$	$D \to D^2$

		F	D
	Rules	ψ_F^0	ψ_D^0
A_M	$F \to \varepsilon$	p	0
	$D \to F$	0	0.05p
	$D \to F^2$	0	0.15p
	$D \to F^3$	0	0.30p
	$D \to F^4$	0	0.30p
	$D \to F^5$	0	0.15p
	$D \to F^6$	0	0.05p
	$F \to D$	1-p	0
	$D \to D^2$	0	1-p

Environmental string at step k, w_{E_k} has symbols:	F	D
Populations of symbols (frequency):	F_k	D_k
Normalized populations (relative frequency):	$f_k = \frac{F_k}{F_k+D_k}$	$d_k = 1 - f_k$

5.1 Sub-populations of Pure Strategies: A Model Using Non Adaptive Simple Reactive Eco-Agents $R_L EG$

We build a probabilistic $R_L EG$ system: $\Sigma = (E, A_1, \ldots, A_n, w_{E_0})$ composed of pure falcons A_F and pure doves. Notice that the values of the evolutionary game of falcons and doves given in table 4 are exactly the values for the right hand sides of the average rules in table 5. For example, to kill a dove allows a falcon to reproduce in a mean of $7/2 = 3.5$. We can represent this fact as the mean rule $D \to F^{3.5}$.

Table 5 together with table 5.1 show the definitions for the analysis of the model:

n Eco-agents of type:	A_F	A_D
Frequencies of the Eco-agents:	n_F	n_D
Probabilities of the Eco-agents:	$p = \frac{n_F}{n_F+n_D}$	$1-p$

The average number of falcons at the step $k+1$ is

$$F_{k+1} = n \times p \times 3.5 \times d_k \qquad D_{k+1} = n \times (1-p) \times (f_k + 2 \times d_k) \tag{1}$$

Thus, from 1 the equations that govern the normalized populations are:

$$f_{k+1} = p \times 3.5 \times d_k \qquad d_{k+1} = (1-p) \times (f_k + 2 \times d_k) \tag{2}$$

The model should represent adequately the encounter of the population with itself. If the n Eco-agents are considered a sample of the population of falcons and doves living in the environmental string, we would demand the proportion of falcons p of the sample to be identical to the proportion of falcons of the environment, $p = f_k$ and $1 - p = d_k$.

$$1 = f_{k+1} + d_{k+1} = (1-p) \times (2,5 \times p + 2) \tag{3}$$

From 3 we get:

$$p = f_{k+1} = f_{k+1}/(f_{k+1} + d_{k+1}) = 3,5 \times p/(2,5 \times p + 2) \tag{4}$$

This equation has a single solution $p* = 3/5$, that corresponds to the evolutionary stable strategy: If the proportion of falcons in the string w_{E_0} is $f_0 = 3/5$, and the proportion of Eco-agents that adopt the pure strategy of falcon is identical, $p* = p = f_0 = 3/5$, we can ensure that the population will evolve in a way such that $\forall\, k \in \mathbb{N}\ f_k = p* = 3/5$.

5.2 Individuals Following Mixed Strategies: A Model Using Simple Adaptive-Reactive Eco-Agents AR_LEG

An Eco-agent that follows a mixed strategy with probability $p_k = p_{F_k}$ of playing the role of falcon and probability $p_{D_k} = (1 - p_k)$ of playing the role of dove at the step k of derivation (the initial probability is p_0), is an Eco-agent adaptive and locally reactive described by: $A_M = (\psi_M, R_M, \rho)$.

The set of rules $R_M = R_F \cup R_D$ includes the rules of pure falcons and pure doves. As can be observed in table 5 $\psi_F^0 = p \times \psi_F(F) + (1-p) \times \psi_D(F)$ and $\psi_D^0 = p \times \psi_D(F) + (1-p) \times \psi_D(D)$ is the initial distributions of probabilities of the mixed Eco-agent, where $p = p_0$ is the initial probability of playing the role of falcon and psi_F, psi_D are the reaction functions of the pure Eco-agents defined in the previous model (trivially extended to the whole set of rules R_M).

ψ_M is the function of local reaction, depending on time k. The learning rule is ρ="copy the normalized population", such that:

$$\psi^{k+1} = \rho(\psi^k, w_{E_k}) = f_k \times \psi_F^0 + d_k \times \psi_D^0$$

The equations that describe the value of F_k and D_k for the offspring $k+1$ will be:

$$F_{k+1} = 3.5 \times n \times p_k \times (1 - p_k)$$
$$D_{k+1} = n \times (1 \times p_k(1 - p_k)) + 2 \times (1 - p_k)^2$$

Consequently, in terms of normalized populations:

$$f_{k+1} = 3,5 \times f_k \times (1 - f_k) \tag{5}$$
$$d_{k+1} = (1 - f_k) \times (f_k + 2 \times (1 - f_k)) \tag{6}$$

The normalized population becomes stable when $f_{k+1} = f_k = p_{k+1} = p_k$. On the other hand:

$$1 = f_{k+1} + d_{k+1} = (1 - p) \times (2,5 \times p + 2) \quad (7)$$

From 5 and 7, the state of equilibrium, when $p_{k+1} = p_k = p*$, is given for $p* = \frac{3,5p*}{(2,5p*+2)}$, that solved results $p* = 3/5$, which is the value of p for the evolutionary stable strategy.

Given that at each offspring the increment of a perturbation is no more that a geometrical progression, where the derivative of $p*$ is the rate of growth, we obtain the condition that defines the stability [13] calculating $\lambda = |\frac{\partial f\mu(p)}{\partial p}|_{p*}$, where $f\mu(p) = |\frac{3.5\times p}{2.5\times p+2}|$ and the $\lambda = |\frac{7}{(2.5\times p+2)^2}|_{p*}$

If $|\lambda|_{p*} < 1$ the equilibrium is stable, because the effect of a perturbation tends to disappear.

If $|\lambda|_{p*} > 1$ the equilibrium is unstable, and the effect of a perturbation tends to increase.

If $|\lambda|_{p*} = 1$ the equilibrium is indifferent, and the effect of a perturbation tends to be maintained.

In our case, we will analyze the value λ in $p* = 3/5$.

$|\lambda|_{p*} = \frac{7}{[(2.5\times 3/5)+2)]^2} = 0.57 < 1$, and hence the equilibrium is stable.

Conclusions

The main contribution of this paper consists in the definition of probabilistic EG systems and their classification according to the perceptive and cognitive capabilities of the Eco-agents from the point of view of the action. We distinguish REG systems, whose Eco-agents are reactive, R_LEG, with Eco-agents locally reactive, and simple, SEG. We define adaptive systems as those that have a learning rule, and the classes $AREG$, AR_LEG and $ASEG$. Along the work we pretended to show as convenient would result the introduction of probabilities in the Eco-agents to model life like behaviours. Probabilistic Eco-agents are used to model the evolutionary game representing numbers in the payment matrix.

The main advantage is that the static approach to the problem of the evolutionary stable strategies becomes here in the description of a symbolic dynamic, performed by Eco-agents, in which we study the normalized population of symbols that are obtained in average. In this wise, we have identified the parameters that give rise to a stable normalized population of symbols, identifying it with an ESS. We got a model of ESS: a) By means of R_LEG, systems, maintaining a proportion of $3/5 \times F$, $2/5 \times D$ in the initial string in the composition of pure falcons and doves. b) Using AR_LEG systems, with adaptive Eco-agents, that perform mixed strategies. The results are in case b) identical to case a). Consequently, we dispose of a formal tool to model complex dynamics, of great interest in the field of biological research, whose computational capabilities are not yet totally investigated.

Acknowledgements. This paper has been sponsored by the Spanish Interdepartmental Commission of Science and Technology (CICYT), project number TEL1999-0181.

References

1. Alfonseca, M., Ortega, A. and Suarez, A. Celular Automatas and probabilistic L Systems. In [6]
2. Cases, B. y Alfonseca, M. Eco-grammar Systems applied to the analysis of the dinamycs of populations in artificial Worlds, in [5].
3. Dassow, J., Paŭn, G. and Rozenbberg, R. (1997) Grammar Systems, in [11]
4. Csuhaj-Varjú, E.; Kelemen, J.; Kelemenová, A.; Paŭn,G.: "Eco-Grammar Systems: A Grammatical Framework for Life-Like Interactions", Artificial Life vol. 3 #1 (winter 1997), p. 1-38.
5. Freund, R.; Kelemenov á (ed.) (2000), A. Grammar Systems 2000, Silesian University, Czech Republic.
6. Martín Vide, C. and Mitrana V. (Eds) Words, sequence, grammars, languages: where biology, computer science, linguistics and mathematics meet. Vol II. Springer London (2000).
7. Maynard Smith, J.: La evolución del comportamiento, en Evolución, Libros de Investigación y Ciencia, Ed. Prensa Científica, Barcelona,(1978) pp 116-126.
8. Maynard Smith, J. Evolution and Theory of Games. Cambridge University Press, Cambridge(1982).
9. Montero, F. & Morán, F. Biofísica: Procesos de Autoorganización en Biología. Eudema, Madrid(1992).
10. Paŭn, G. (Ed.) (1995) Artificial Life: Gramatical Models. Black Sea University, Bucharest, Romania.
11. Rozenberg, G. y A. Salomaa (Eds) (1997) Handbook of Formal Languages. Vol. 2: Linear modeling: Background and application, Springer, Berlin (1997).
12. Salomaa, A. (1995) Developmental Models for Artificial Life: Basics of L Systems. In [10]
13. Soé, R.; Manrubia, S. (1996) Orden y caos en sistemas complejos. Edicions UPC, Barcelona.
14. Weibull, J. (1995) Evolutionary Game Theory. The MIT Press, Cambridge MA & London.

Dynamics of the Environment for Adaptation in Static Resource Models

Mark A. Bedau

Reed College, 3203 SE Woodstock Blvd., Portland OR 97202, USA
mab@reed.edu
http://www.reed.edu/~mab

Abstract. We measure the environment that is relevant to a population's adaptation as the information-theoretic uncertainty of the distribution of local environmental states that the adapting population experiences. Then we observe the dynamics of this quantity in simple models of sensory-motor evolution, in which an evolving population of agents live, reproduce, and die in a two-dimensional world while competing for resources. Although the distribution of resources is static, the agents' evolution creates a dynamic environment for adaptation.

1 The Environment for Adaptation

The process of adaptation is driven by the environment, but it also shapes the environment. If we focus on the environment for adaptation, that is, the aspect of the environment that is relevant to adaptive evolution, we would tend to agree with Lewontin [7] that organisms construct their environment:

> The environment is not a structure imposed on living beings from the outside but is in fact a creation of those beings. The environment is not an autonomous process but a reflection of the biology of the species.

But to think clearly about how organisms and their environment affect one another, we must first clarify what the environment for adaptation is. How can it be measured? How is it affected by the behavior of the evolving population? This paper takes a first step at answering these questions.

Part of an answer to these questions involves distinguishing two kinds of local environmental states: those that exist in the environment and so could conceivably be relevant to the agents' adaptation, and those that the agents experience and so that affect their adaptation. To appreciate how evolution is both cause and effect of the environment, one must focus on the latter. In this paper we propose how to measure the environment for adaptation, and we observe its dynamics in evolving systems with fixed resource distributions—what we call *static resource models*. We see that even simple systems with static resource distributions have a dynamic environment for adaptation.

Classifying and quantifying the environment for adaptation is the subject of a variety of recent work in artificial life. The relevance of this work includes both

J. Kelemen and P. Sosík (Eds.): ECAL 2001, LNAI 2159, pp. 76–85, 2001.

its implications for theoretical insights into the evolutionary process and its practical consequences for engineering and controlling evolving systems. Wilson [14] and Littman [8] provide an abstract categorization of environments, with Wilson focusing on the degree of non-determinism in an environment and Littman characterizing the simplest agent that could optimally exploit an environment. But neither method provides a dynamic quantitative measure of the environment the population experiences. Other recent work experimentally investigates agents adapting in different environments. Todd and Wilson [12] introduce an experimental framework for investigating how adaptation varies in response to different kinds of environments, and Todd et al. [13] demonstrate different adaptations in different kinds of environments. In neither case, though, is environmental structure actually classified or measured. The present project grows out of previous work on how adaptability depends on quantitative measures of environmental structure [2,4,5]. What is novel here is the focus on the environment actually experienced, which allows the measures to be dynamic where the previous ones were static. While Lewontin [7] verbally stresses the centrality of the environment for adaptation, being able to quantify and observe it provides the empirical footing needed to respond to skeptics [6].

2 Adaptation in Packard's Static Resource Models

All of our empirical observations are from computer simulations of a certain model—originated by Norman Packard [9,3]—that is designed to be a very simple model of the evolution of sensory-motor strategies. Packard's model consists of agents sensing the resources in their local environment, moving as a function of what they sense, ingesting the resources they find, and reproducing or dying as a function of their internal resource levels. The model's spatial structure is a grid of sites with periodic boundary conditions, i.e., a toroidal lattice. Resources are immediately replenished at a site whenever they are consumed. The agents constantly extract resources and expend them by living and reproducing. Agents ingest all of the resources (if any) found at their current location and store them internally. Agents expend resources at each time step by "paying" (constant) "existence taxes" and "movement taxes" (variable, proportional to distance moved). If an agent's internal resource supply drops to zero it dies.

Each agent moves each time step as dictated by its genetically encoded sensory-motor map: a table of behavior rules of the form IF (environment j sensed) THEN (do behavior k). An agent receives sensory information about the existence of resources (but not the other agents) in the von Neumann neighborhood of five sites centered on its present location in the lattice. Thus, each sensory state j corresponds to one of $2^5 = 32$ different detectable local environments. Each behavior k is a jump vector between one and fifteen sites in any one of the eight compass directions, or it is a random walk to the first unoccupied site, so an agent's genome continually instructs it to move somewhere. Thus, an agent's genotype, i.e., its sensory-motor map, is just a lookup table of sensory-motor rules. But the space in which adaptation occurs is fairly large, consisting of $121^{32} \approx 10^{66}$ distinct possible genotypes. In a resource field missing some of von Neumann neighborhoods, the number of *effectively* different genotypes will smaller (about 10^{29} for the resource fields studied here).

An agent reproduces (asexually, without recombination) if its resource reservoir exceeds a certain threshold. The parent produces one child, which starts life with half of its parent's resource supply. The child also inherits its parent's sensory-motor map, except that point mutations can replace the behaviors linked to some sensory states with randomly chosen behaviors (there is roughly one mutation per reproduction event, on average). A time step in the simulation cycles through the entire population and has each agent, in turn, complete the following sequence of events: sense its present von Neumann neighborhood, move to the new location dictated by its sensory-motor map (if that site is already occupied, it randomly walks to the first unoccupied site), consume any resources found at its new location, expend resources to cover existence and movement taxes, and then, if its resource reservoir is high enough or empty, either reproduce or die. A given simulation starts with randomly distributed agents containing randomly chosen sensory-motor strategies. The model contains no *a priori* fitness function, as Packard [9] has emphasized. Agents with maladaptive strategies find few resources and thus to die, taking their sensory-motor genes with them; by contrast, agents with adaptive strategies tend to find sufficient resources to reproduce, spreading their strategies (with mutations) through the population.

Here we restrict resources to blocks: a square grid of lattice sites which all have the same fixed quantity of resources. In a resource distribution formed of blocks, every site that is not inside a block is part of a resource-free desert. We focus on two distributions, one with many small blocks randomly sprinkled across space, the other with one large block. If the resource distribution contains just one block, then all evolution concerns strategies for exploiting the resources on the block. But if the resource distribution contains many blocks scattered in space, then each block is an evolutionary island supporting its own evolutionary development and subject to migration from nearby islands. So a behavioral innovation that originates on one island can hop from island to island and eventually colonize the entire archipelago. Resource blocks come in different sizes (widths). If the block is small enough (width ≤ 3) then the agent's sensory information always unambiguously indicates the agent's exact location on the block (NW corner, middle, N edge, etc.). As the width increases above 3, so does the ambiguity of the agents' sensory information [4]. Agents cannot always tell exactly where they are, for the sites inside the block all look alike, as do the sites at its edge. Here we studied two size blocks: one (3×3) that is unambiguous, and another (30×30) with rampant ambiguity.

Agents on a given 3×3 block usually all follow the same strategy, for agents following heterogeneous strategies tend to collide and be bumped into the resource desert. The strategy observed are cycles jumping through a sequence of sites on the block. The simplest cycles (period 2) consist of jumping back and forth between two sites. Since a 3×3 block contains 9 distinct sites, it can support at most a period 9 strategy. A period n strategy has room for at most

$n-1$ agents (one agent in the cycle must move first and the space to which it is jumping must be unoccupied). Thus, longer period strategies can support larger populations because they can exploit more of the energetic resources on a block All agents reproduce at the same rate, so a block with a larger population will produce offspring at a higher rate. Thus, blocks with larger period strategies will exert greater migration pressure and thus will have a selective advantage. So evolution in an archipelago of tiny 3×3 resource islands will exhibit one main kind of adaptive event: lengthening the period of an existing strategy. Agents on one large 30×30 block tend to exhibit a fluidly changing ecosystem of coexisting strategies. These strategies fall into two main categories: Edge strategies and the Random strategy. The Random strategy relies on one central gene: if inside the middle of a block (i.e., if sensing the neighborhood with resources at all five sites), do a random walk to the first unoccupied site. (Recall that one of the possible genetically encoded behaviors k is a random walk.) Although unsuccessful on a 3×3 block, a 30×30 block is large enough for the Random strategy to succeed. An Edge strategy consists of moving in a straight line when inside the block and jumping back into the middle of the block when detecting the edge. A proliferating Edge strategy will fill a region of the block with a perpetually rolling population. 30×30 blocks have three main kinds of adaptive events: lengthening an Edge strategy's jump back, discovering a new Edge strategy compatible with the strategy ecology, and discovering the Random strategy.

3 A Measure of the Environment for Adaptation

We are interested in how best to measure the environment that is relevant to a population's adaptation. For the sake of concreteness, we will develop and apply our methods in the context of the static resource models described above. The quantities we define and observe fit within a family of measures that has been developed and studied in similar contexts previously [2,4,5].

The basic idea behind our measure of the environment for adaptation is quite simple. The first step is to define a partition of relevant local environments. Of all the environmental states that exist or could be defined in an evolving system, first focus on those that are *local* to the individual agents in the system in the sense that different agents can be in different local environmental states at the same time. Next limit attention to just those that are *relevant* to the survival or reproducibility of the agents, i.e., those that can be expected to be relevant to the population's adaptation. This includes only those states that can affect or be affected by the agents. Finally, consider a *partition* of local relevant environments—a set of states that is disjoint and exhaustive in the sense that at each time each individual agent is unambiguously in exactly one of the states. For example, in Packard's model studied here, a natural partition of local relevant environments is the set of different von Neumann neighborhoods that the agents can distinguish (the presence or absence of resources in each of the five detectable neighboring sites). At each time each agent is in exactly one of these local states. The states affect the agents by being their sensory input, and the agents affect the states through their behavior [2]. The states are relevant to the agents' adaptation, for the agents' adaptive task is to evolve a genome that associates each those states with an appropriate behavioral rule in a sensory-motor strategy and natural selection shapes only the rules for those von Neumann neighborhoods that are actually experienced by the population.

During the course of evolution, the agents experience states in the partition of local relevant environments with different frequencies. The frequency (probability) distribution of the states is a straightforward reflection of the environment that is currently relevant for the population's adaptation. That is, let $\{e_i\}$ be the partition of local relevant environments, and let $P^E(e_i)$ be the frequency with which agents in the population experience the ith local state e_i (during some time window) in the global environment E. Then the probability distribution P^E reflects the nature of the environment that is *currently relevant* to the population's adaptation. It is important to note that the distribution P^E is defined relative to the partition $\{e_i\}$. Different partitions will yield different probability distributions, so the usefulness and interest in the distribution P^E hinges on the choice of partition $\{e_i\}$.[1] In Packard's model where $\{e_i\}$ is the von Neumann neighborhoods, P^E shows the frequencies with which the population visits different neighborhoods.

The final step in our measure of the environment for adaptation is to quantify the variety in P^E, and information-theoretic uncertainty (Shannon entropy) [11] is the natural tool to use:

$$H(P^E) = -\sum_i P^E(e_i) \log_2 P^E(e_i). \tag{1}$$

$H(P^E)$ reflects two aspects of P^E: its width (number of different neighborhood indices i) and flatness (constancy of the value of $P^E(e_i)$ for different i). The wider and flatter P^E is, the more uncertainty there is about which neighborhood the agents will experience and the higher $H(P^E)$ will be.

Even in static resource models, P^E and so $H(P^E)$ can change, for P^E depends on the adapting population's behavior. As the population's behavior evolves, the population may experience different local environments, so P^E and thus $H(P^E)$ may be dynamic. The dynamics of $H(P^E)$ simultaneously reflects both cause and effect of the adapting population. That is, $H(P^E)$ measures both the environment *created by* the population's *past* adaptation and the environment *relevant for* the population's *future* adaptation, as reflected in the local environmental states the population experiences. The population's adaptation to its environment changes the frequency with which it experiences different local environments, and this creates a self-organizing dynamic in $H(P^E)$. (Parisi,

[1] Since the distribution P^E pools information about the local relevant environments experienced by all the agents, it in effect *averages* this information across the population. If different subsets of agents experience substantially different subsets of local environments (e.g., if they exist in different niches), the averaging in P^E will obscure this difference. To counteract this, one could collect different probability distributions for different subsets of agents.

Nolfi, and Cecconi [10] and Bedau [2] study this issue.) The dynamics of $H(P^E)$ show how adaptation changes the environment for subsequent adaptation.

To get a feel for how P^E and $H(P^E)$ reflect the environment for adaptation, assume the environment consists of an archipelago of randomly scattered 3×3 resource blocks in Packard's model. Since prolonged stay in the resource desert is lethal, the population will overwhelmingly experience some subset of the nine neighborhoods on the block. For example, if the populations on the different resource islands are all exhibiting the same period two cyclic strategy, then the bulk in $P^{3\times3}$ is evenly divided between the two neighborhoods visited (agents bumped off out of the cycle will visit a few other local environments before perishing). As evolution increases the length of the strategy followed by the agents, the population will encountered new environmental states and $P^{3\times3}$ will stay flat but become wider. Thus the environment for adaptation, $H(P^{3\times3})$, will tend to increase. The magnitude of this trend is bounded by the number of different resource sites on the tiny block; with nine distinct resource sites, $H(P^{3\times3}) \leq \log_2 9 \approx 3.17$.

Now consider an environment containing only one large 30×30 block. Whether the block is populated by an ecology of coexisting Edge strategies or dominated by the Random strategy, most of the population will be inside the block, so the bulk of $P^{30\times30}$ will reflect that one neighborhood with most of the rest shared among various edge neighborhoods (plus a small fraction for the agents bumped into the desert). The dynamics of $H(P^{30\times30})$ can be varied. Lengthening the inside jump of an Edge strategy increases the relative frequency with which the population encounters the neighborhood inside the block, and this makes $P^{30\times30}$ less flat and thus shrinks $H(P^{30\times30})$. Discovering a new Edge strategy raises the number of sites experienced in the ecology of strategies, so $P^{30\times30}$ becomes wider and $H(P^{30\times30})$ rises. Discovering the Random strategy typically increases the frequency with which the population encounters the inside the block, so $P^{30\times30}$ narrows and $H(P^{30\times30})$ falls. Thus, the environment for adaptation on a 30×30 block should both rise and fall, but the eventual evolution of the Random strategy will ultimately drive it to a low value.

4 Observations of the Environment for Adaptation

We observed $H(P^E)$ in 100 3×3 blocks scattered randomly in space, and in one 30×30 block. Both resource distributions pump resources into the environment at the same rate, so they can support comparable maximum population sizes. In this resource-driven and space-limited model, population size is a good reflection of the dynamics of adaptation (fitness). Except for varying the resource distribution, all the parameters in the models observed were the same.

Evolution on an archipelago of 3×3 blocks always exhibits a pattern of period-lengthening adaptations. The population starts with a small period cycle on one of the blocks. This strategy migrates across the archipelago, until it has colonized virtually all the blocks. Eventually the appropriate mutations will create a longer-period cycle on one of the blocks and the agent with this longer cycle

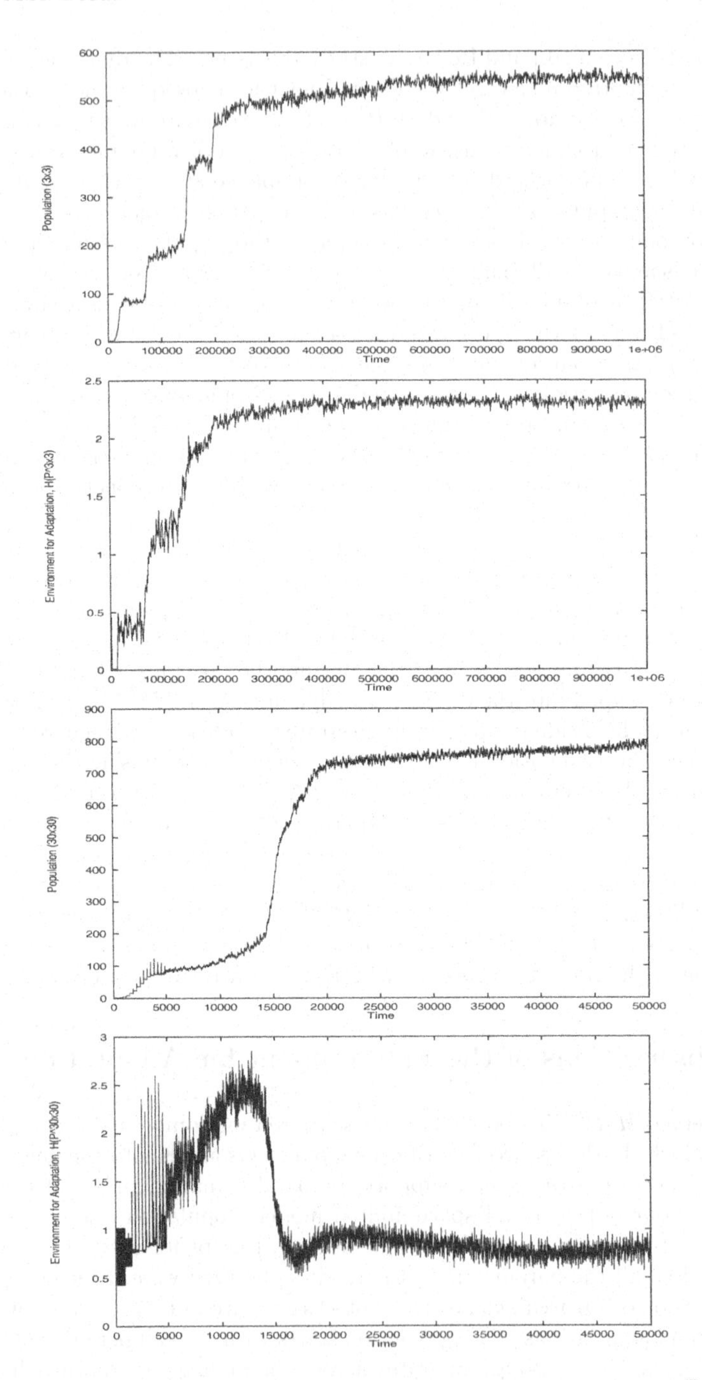

Fig. 1. Dynamics of population size and environment for adaptation, $H(P^E)$, in two typical simulations. The top two graphs are from an archipelago of 100 3×3 blocks. The bottom two graphs are from one 30×30 block.

will sometimes supplant the agents on that block with the shorter cycle, so the block is now populated by a tiny subpopulation with the longer cycle. Children from this subpopulation migrate to other blocks and supplant the shorter-cycle agents there, until they eventually colonize the entire archipelago. This sequence is repeated again and again with ever longer-period cycles, until the cycle length gets close to its maximum possible value (nine).

The top two graphs in Figure 1 show time series of population level and the environment for adaptation, $H(P^E)$, in a typical simulation in an archipelago of 3×3 blocks. The course of evolution has four epochs. Epoch I starts with a period-2 cyclic strategy on a single block. As this strategy reproduces and starts to migrate go other blocks, the population rises, until about all the blocks are colonized. Epoch II starts at about time 75,000 when a period-lengthening mutation enables a period-3 cycle to become established on one of the blocks. The 3-cycle then invades the rest of the archipelago. Epoch III starts at around time 150,000 when a coordinated combination of mutations create a 5-cycle on one block, which then invades the rest of the archipelago. Another period-lengthening mutation on one block initiates Epoch IV with a 6-cycle at around time 200,000. This strategy then colonizes the rest of the archipelago. These period-lengthening adaptations each cause a rise in $H(P^E)$, as the more complicated strategies make the experienced environment more complex.

Evolution on the 30×30 block tends to exhibit its own distinctive pattern of adaptations. One or two edge strategies in the initial random seed population will live on the block, growing to fill the space available in the niche on the block used by those strategies. From time to time new strategies arise by mutation, but most die off quickly. Some find an unoccupied niche on the block and coexist with their ancestral strategies. Others compete for niche space with ancestral strategies and supplant their ancestors. In this way, the block supports a changing ecosystem of subpopulations surviving through different strategies in different regions on the block. Eventually, though, the Random strategy always gets a foothold in the population and drives all Edge strategies extinct.

The bottom two graphs in Figure 1 show data from a typical illustration of a co-evolving succession of Edge strategies ending with domination by the Random strategy. The course of evolution again has four epochs. In Epoch I, one strategy (call it B11) jumps off the bottom edge a half-dozen steps into the block. The population of agents following this strategy grows until at about time 4000 it fills the space along the entire bottom edge with a population size pushing 80. Except for momentary spikes (caused when a crop of children get bumped off the block, thus briefly experiencing the desert neighborhood before dying), $H(P^E)$ is basically stable during this epoch. The entire population experiences two neighborhoods: the inside of the block and the bottom inside edge. Epoch II starts around time 6000 when a second strategy (R11) starts to coexist with B11 by occupying the block's right edge. This strategy lives only right along the edge, so only a handful of agents can occupy the niche created by this strategy. $H(P^E)$ shows a slight rise a subpopulation experiences two new neighborhoods (right inside edge and bottom right corner). Epoch III starts around time 9000 when a third strategy (L11) shares the block with B11 and R11 by exploiting the hitherto unused left edge with a small jump into the block. L11 grows to fill its niche along the left edge, with a subpopulation of a couple dozen. $H(P^E)$ rises again during this epoch as a new neighborhood (inside left edge) gets experienced more and more frequently. Epoch IV starts around 13000 when the Random strategy arises inside the block and it quickly sweeps across the block, knocking B11, R11, and L11 out of their niches and filling virtually all the space on the block. $H(P^E)$ drops dramatically because the neighborhoods experienced by the population shows much less variety. As the population grows, more agents experience an edge or the desert, so $H(P^E)$ rises slightly.

5 Conclusions

Two main conclusions follow from this work, one methodological and the other substantive. The methodological conclusion is that we now have a dynamic and quantitative reflection of the environment that the evolving population actually experiences—the environment for adaptation. This measure of the environment for adaptation is quite general and feasible, and it can straightforwardly be applied to many other evolving systems. Its drawback is that it depends on an *a priori* partition of the environmental states relevant to the population's adaptive evolution. Such a partition is easy in the simple systems studied here but more complex systems can present difficulties, especially if qualitatively new kinds of environmental states emerge unpredictably in the course of evolution.

The substantive conclusion is that the environment for adaptation is *dynamic* even in simple static resource models like those studied here. Sometimes a population evolves a more complicated behavioral strategy which enables it to experience and exploit more of the environment's complexity. Subsequent evolution then takes place in the context of this more complicated environment for adaptation. In other cases the population benefits by evolving a simpler behavioral strategy and thus simplifying its environment. An agent's environment for adaptation includes the rest of the population with which the agent interacts. In static resource models, agent–agent interactions like collisions can be a significant part of the adaptive landscape, so the shape of the landscape depends on the collection of behavioral strategies in the population. Space is itself a resource [1], and available space is dynamic since it depends on the changing spatial location of the population with which agents interact. The dynamics of the environment for adaptation strongly depend on the actual course that evolution happens to take. Repeating a simulation with exactly the same resource distribution (data not shown here) can yield qualitatively different environmental dynamics, which shows the "fluidity" of these environmental dynamics. These conclusions pertain in the first instance just to the particular systems studied here, but they are likely to hold quite generally.

Acknowledgements. For helpful discussion or comments on the manuscript, thanks to Peter Godfrey-Smith, Norman Packard, Peter Todd, Marty Zwick, and two anonymous reviewers. For assistance with this research, thanks to Matt Giger and John Hopson. Thanks to the Santa Fe Institute for hospitality and computational resources that supported some of this work.

References

1. Abrams, P. (1992). Resource. In E. F. Keller and E. A. Lloyd, eds., *Keywords in Evolutionary Biology* (pp. 282–285). Cambridge, Mass: Harvard University Press.
2. Bedau, M. A. (1994). The Evolution of Sensorimotor Functionality. In P. Gaussier and J. -D. Nicoud, eds., *From Perception to Action* (pp. 134-145). New York: IEEE Press.
3. Bedau, M. A., Packard, N. H. (1992). Measurement of Evolutionary Activity, Teleology, and Life. In C. G. Langton, C. E. Taylor, J. D. Farmer, S. Rasmussen, eds., *Artificial Life II* (pp. 4431–461). Redwood City, Calif.: Addison-Wesley.
4. Fletcher, J. A., Zwick, M., Bedau, M. A. (1996). Dependence of Adaptability on Environment Structure in a Simple Evolutionary Model. *Adaptive Behavior* **4**, 283–315.
5. Fletcher, J. A., Bedau, M. A., Zwick, M. (1998). Effect of Environmental Structure on Evolutionary Adaptation. In C. Adami, R. K. Belew, H. Kitano, and C. E. Taylor, eds., *Artificial Life VI* (pp. 189–198). Cambridge, MA: MIT Press.
6. Godfrey-Smith, Peter. (1996). *Complexity and the Function of Mind in Nature.* Cambridge: Cambridge University Press.
7. Lewontin, R. C. (1983). The Organism as the Subject and Object of Evolution. *Scientia* 118, 63–82. Reprinted in R. Levins and R. Lewontin, *The Dialectical Biologist* (pp. 85–106). Cambridge, MA: Harvard University Press.
8. Littman, M. L. (1993). An Optimization-Based Categorization of Reinforcement Learning Environments. In J. A. Meyer, H. L. Roiblat, and S. W. Wilson (Eds.), *From Animals to Animats 2* (pp. 262–270). Cambridge, MA: Bradford/MIT Press.
9. Packard, N. H. (1989). Intrinsic Adaptation in a Simple Model for Evolution. In C. G. Langton, ed., *Artificial Life* (pp. 141–155). Redwood City, Calif.: Addison-Wesley.
10. Parisi, D., Nolfi, S., Cecconi, F. (1992) Learning, Behavior, and Evolution. In F. Varela and P. Bourgine, eds., *Towards a Practice of Autonomous Systems* (pp. 207–216). Cambridge, Mass.: Bradford/MIT Press.
11. Shannon, C. E., Weaver, W. (1949). *The Mathematical Theory of Communication.* Urbana, Ill.: Univ. of Illinois Press.
12. Todd, P. M., Wilson, S. W. (1993). Environment Structure and Adaptive Behavior From the Ground Up. In J. A. Meyer, H. L. Roiblat, and S. W. Wilson (Eds.), *From Animals to Animats 2* (pp. 11–20). Cambridge, MA: Bradford/MIT Press.
13. Todd, P. M., Wilson, S. W., Somayaji, A. B., Yanco, H. A. (1994). The Blind Breeding the Blind: Adaptive Behavior Without Looking. In D. Cliff, P. Husbands, J. -A. Meyer, and S. W. Wilson (Eds.), *From Animals to Animats 3* (pp. 228–237). Cambridge, MA: Bradford/MIT Press.
14. Wilson, S. W. (1991). The Animat Path to AI. In J. -A. Meyer and S. W. Wilson (Eds.), *From Animals to Animats* (pp. 15–21). Cambridge, MA: Bradford/MIT Press.

Adaptive Behavior through a Darwinist Machine

F. Bellas, A. Lamas, and R.J. Duro

Grupo de Sistemas Autónomos Universidade da Coruña, Spain
gsa@cdf.udc.es

Abstract. In this paper we propose a mechanism based on Darwinist principles applied on-line that provides organisms with the capability of adapting through the use of their interaction with their surroundings in order to improve the level of satisfaction obtained. The mechanism involves a two level concurrent operation of evolutionary processes. The processing carried out in the first, or unconscious level, leads to a current, or conscious model of the world and the organism, which is employed for evaluating candidate strategies, using as fitness the predicted motivation satisfaction.

1 Introduction

For real autonomy and adaptability to emerge, artificial agents must be endowed with cognitive mechanisms whose operation and adaptability to changing environments, requirements and tasks are not predetermined by the designer. This allows the artificial agent to organize its own knowledge from its own perceptions and needs.

In this paper we are going to consider a model, whose foundations were laid in [6], based on some not too well known theories within the field of cognitive science that relate the brain or neural structure with its operation through evolutionary processes in somatic time. These theories are: the Theory of Evolutionary Learning Circuits (TELC) [1]; the Theory of Selective Stabilization of Synapses (TSSS) [2]; the Theory of Selective Stabilization of Pre-Representations (TSSP) [3] and the Theory of Neuronal Group Selection (TNGS) or "Neural Darwinism" [4]. For an excellent review see [5]. Each theory has its own features but they all lead to the same concept of cognitive structure and their main ideas are that the brain adapts its neural connections in real time through evolutionary or selectionist processes and this somatic selection determines the connection between brain structure and function. As a reference of work in a similar direction we must cite Nordin et al. [7].

2 Cognitive Mechanism

The final objective of any cognitive mechanism is to link what the organism perceives and what it has done to what it must do in order to satisfy its motivations. In this line, a cognitive model must be given by three basic elements: *strategies, world models* and *internal models*. A *strategy* is just a sequence of actions applied to the effectors. An ideal *strategy* would depend on the current perceptions, internal state and motivations. On the other hand, a *world model* is a function that relates the sensory inputs of the agent in the instant of time t with the sensory inputs in t+1 after applying a *strategy*. A *world model* permits evaluating the consequences of actions on the environment as perceived by the agent (sensed values in $t+1$). Finally, an *internal model* is a function that relates the sensorial inputs t+1 with the internal state according to the motivation for the agent.

J. Kelemen and P. Sosík (Eds.): ECAL 2001, LNAI 2159, pp. 86-89, 2001.

The problem is how to connect these three elements into an on line mechanism that permits the agent to generate *original solutions* making use of *previous experience* and minimizes *designer intervention* so that the motivations are fulfilled.

The way we have solved the problem is shown in Fig. 1. The final objective of the mechanism is to obtain the strategy the agent must execute in the real world to satisfy its motivations. This strategy is the *Current Strategy* and is applied to the *Environment* through the *Effectors* providing new *Sensing* values for the agent. Thus, after each iteration, we have a new action-perception pair obtained from the real world. These action-perception pairs are stored in the *Short-Term Memory* and are used as the fitness function for the evolution of the world and internal models.

We have three evolutionary processes in our mechanism (Fig. 1). A *World Model Evolver* manages world models through evolution. This structure evolves a population of world models and each moment of time selects the best one according to the information it has available about the real world. The *Internal Model Evolver* manages the internal model base in a similar manner. Finally, a *Strategy Evolver* that evolves strategies and when the agent needs one it provides it.

Each one of these evolutionary processes starts from the population stored in a memory (*World Model Memory, Internal Model Memory and Strategy Memory*). The contents of these memories are random for the first iteration of the mechanism. The individual with the highest fitness value after evolution according to the information of the short-term memory is selected as the current model. In the case of strategies these are tested during their evolution by implementing a virtual environment using the current models, that is, the current world model provides the sensory inputs in instant t+1 for the strategy we want to test. The current internal model uses these predicted values as input to predict the internal state resulting from the strategy. The strategy with the highest fitness value is selected as the *Current Strategy* and is applied to the *Environment*. This basic cycle is repeated and, as time progresses, the models become better adapted to the real world and the selected strategies improve.

The controllers these processes handle are Artificial Neural Networks, whose parameters constitute the "chromosomes" of the genetic algorithms, although any other computational structure could be used.

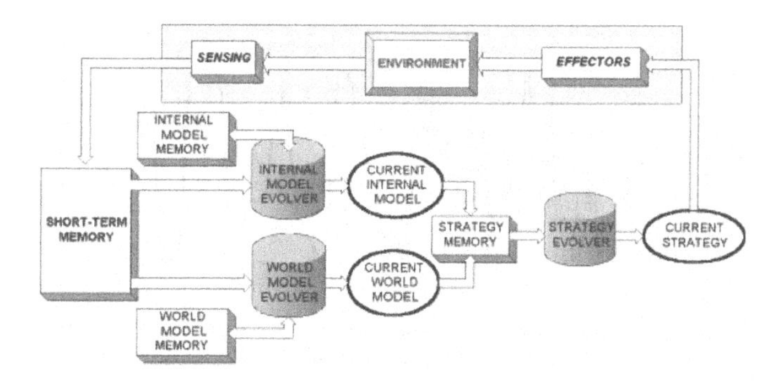

Fig. 1. Block Diagram of the cognitive mechanism where the shadowed area represents the external operation of the mechanism and the remaining area represents the internal operation.

3 Application Examples

We have considered two examples. The first one is a three legged robot with two light sensors in a world with a light that moves up and down in front of it. To be satisfied, the robot must follow the light as closely as possible until its "light reservoir" is filled. The world model and internal model memories are made up of populations of artificial neural networks. The robot must learn to reach the light but initially the models are not adequate as they have not undergone enough evolution. Consequently, the strategies applied are wrong and the robot doesn't follow the light (area A of Fig. 2). As time progresses, the models improve (area B of Fig. 2) until a point is reached where the robot follows the light closely. Once the models are good enough, the robot follows the light perfectly. In order to test the adaptability of the mechanism at this point we change the motivation and now the light hurts and the robot must escape from it. This is shown in the areas C and D of Fig. 2. It is important to see how a change of motivation implies an immediate change in the selected strategies.

As a second example, closer to reality, we have implemented the mechanism on a Hermes II hexapod robot. For the sake of processing power, we ran the mechanism on a 3-D simulation and then transferred it to the real robot. In this case, we have two different tasks, consisting in the robot learning to walk and reaching a tree like object.

First, the robot must learn to walk and the world models have one input corresponding to the distance to the tree (using a virtual sensor from the IR information) and 6 inputs corresponding to the relative phases of each leg. The output was the predicted distance. This way, the mechanism finds that the best gait (combination of phases) in order to advance quickly to the object is similar to a tripod gait. On the left area of figure 3 we represent the simulator and the range of the IR sensors as the robot walks, on the right we see the real robot. As initially the robot had no clue on how to walk or what to do to fulfill its motivations, it started moving its legs randomly producing all kinds of weird gaits until it managed to obtain some gaits that produced motion and thus provided information for the world models and internal models to improve.

Once the robot has obtained an appropriate gait, we change the problem and now the robot must turn, controlling the amplitude of leg motion, with the objective of

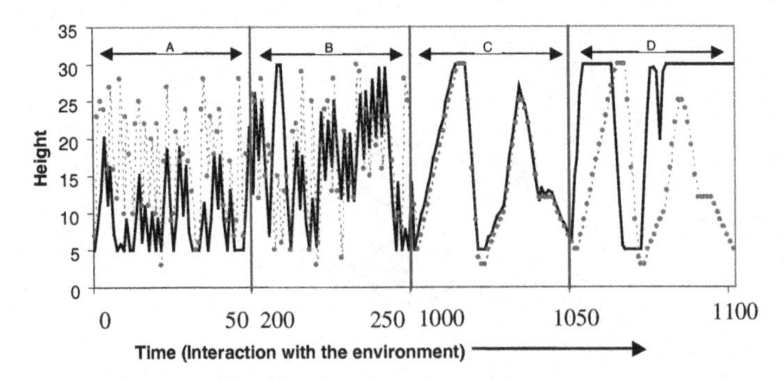

Fig. 2. In the first stages of the mechanism (area A) the robot (dotted line) doesn't follow the light (continuous line), but a few iterations later (area B) the behavior improves a little. In area C the robot follows the light perfectly, and in area D the goal is changed to avoiding the light.

Fig. 3. The Hermes robot develops a gait and strategy to reach the tree-like object.

reaching the tree. The world model now has two sensory inputs (angle and distance to the tree), one amplitude input and two outputs (the predicted perception). The results are good after 200 iterations and the robot reaches the tree (left side of figure 3).

4 Conclusions

An evolutionary learning mechanism for artificial organisms to autonomously learn to interact with their environment without their behavior being completely predetermined has been presented. This structure permits artificial organisms to think things out before acting and provides a framework for learning to take place. It is important to note that during this thinking process, the agent is able to generate original solutions (strategies) the organism has never used before and it is able to include previous experience through evolutionary processes that allow new solutions to arise and, at the same time, permit meaningful combination of previously acquired experience in the form of world or internal models or successful strategies.

Acknowledgements. This work was supported by the FEDER through project PROTECAS N. 1FD1997-1262 MAR. and by the MCYT of Spain through project TIC2000-0739C0404.

References

1. Conrad, M.: Evolutionary Learning Circuits. J. Theor. Biol. 46, 1974, pp. 167.
2. Changeux, J., and Danchin, A.: Selective Stabilization of Developing Synapsis as a Mechanism for Specification of Neural Networks. Nature 264, 1976, pp. 705-712.
3. Changeux, J., Heidmann, T., and Patte, P.: Learning by Selection. The Biology of Learning. P. Marler and H.S. Terrace (Eds.), Springer-Verlag. 1984, pp. 115-133,
4. Edelman, G.M.: Neural Darwinism. The Theory of Neuronal Group Selection. Basic Books 1987.
5. Weiss, G.: Neural Networks and Evolutionary Computation. Part II: Hybrid Approaches on the Neurosciences. Proceedings of the First IEEE Conference on Evolutionary Computation, Vol. 1, 1994, pp. 273-277.
6. F. Prieto, R. J. Duro, J. Santos: Modelo de Mecanismo Mental Darwinista para robots móviles autónomos. Proceedings of SAAEI'96. Zaragoza, Spain, 1996
7. Nordin, P., Banzhaf, W. and Brameier, M.: Evolution of a World Model for a Miniature Robot Using Genetic Programming. Robotics and Autonomous Systems, Vol. 25, 1998, pp. 105-116.

Fault-Tolerant Structures: Towards Robust Self-Replication in a Probabilistic Environment

Daniel C. Bünzli and Mathieu S. Capcarrere[1]

Logic Systems Laboratory
Lausanne Swiss Federal Institute of Technology
CH-1015 INN-Ecublens
{Buenzli.Daniel,Capcarrere.Mathieu}@epfl.ch
http://lslwww.epfl.ch

Abstract. Self-replicating structures in cellular automata have been extensively studied in the past as models of Artificial Life. However, CAs, unlike the biological cellular model, are very brittle: any faulty cell usually leads to the complete destruction of any emerging structures. In this paper, we propose a method, inspired by error-correcting-code theory, to develop fault-resistant rules at, almost, no extra cost. We then propose fault-tolerant substructures necessary to future fault-tolerant self-replicating structures.

1 Introduction

Self-replicating structures in cellular automata have generated a large quantity of papers [10]. In the Artificial Life field, most of these have been devoted to the study of the self-replicating process in itself, as a model of one of life primary property. The underlying motivation of such researches could be expressed as a question: What are the minimal information processing principles involved in self-replication ?

However, if life was an awe inspiring model, simplification was a prime constraint in all these studies, not only for obvious practical reasons, but also as a guide to extract the quintessential ideas behind self-replication. We believe that an always present quality in natural systems was omitted more for the former reasons than the latter principle: *robustness*.

Cellular automata are discrete time and space grid in which all cells update synchronously according to a deterministic rule. No faults ever occurs in the transition rule nor on the cell current state. Nevertheless if we are to look at real biology, it appears at all levels that either insensitiveness to faults or self-repair is a prime condition of survival, the environment being extremely noisy and uncertain. Fault resistant cellular automata designed in the past[2,3,4] are based on hierarchical constructions, i.e., an alteration of the inner architecture of the CA. In this paper, we rather propose to develop fault-tolerant transition rules, keeping the classical CA unchanged. Thereby we create structures that

J. Kelemen and P. Sosík (Eds.): ECAL 2001, LNAI 2159, pp. 90–99, 2001.

are fault-tolerant in their functioning itself, at a minimal cost in terms of space, time or memory.

In section 2, we will accurately define what we mean by a faulty environment, and evaluate its impact on classical self-replicating structures, such as Langton's loop. In the following section, after a short presentation of the error correcting code theory, we will use it to propose a general model of fault-resistant rules. Finally, in section 4, we first present general practical constraints, and then propose two essential fault-resistant substructures for self-replication: a data pipe and a signal duplicator, showing the specificities of such realizations situations compared to the theoretical model.

2 A Non-deterministic Cellular Automata Space

The idea of fault-tolerance is rather vague and our intention in this section is *not* to state any definitive definition. Rather, we propose here, first of all, to delimit the scope of validity of our work, and, secondly, to briefly argue of its validity as a practical environment. Besides, we will show that this noisy environment is more than enough to irrevocably perturbate and thus destroy any of the most famous self-replicating structures.

2.1 Definition of the Faulty Environment

Formally cellular automata are d-dimensional discrete space, in which each point, called a cell, can take a finite number of state q. Each cell updates its own states according to the states of a fixed neighborhood of size n following a deterministic rule. A CA is called uniform when this rule is the same for every cell. Hence a CA may be defined by its transition rule S:

$$S : q^n \rightarrow q$$

In CAs, faults may basically occur at two levels : the synchronization (time) and the cell transition rule (space). We do not cater for the former problem in this article, though we argue in a paper to come [1], that a simple time-stamping method exploiting its inherent parallelism correct efficiently all synchronization faults. As for the latter problem, which could be further divided into reading and writing errors, we model it as a non-deterministic rule. More precisely, there is a probability of faults p_f, such that any given cell will follow the transition rule with probability $(1 - p_f)$, and take any state with probability p_f. This model, though apparently catering only for writing errors, do simulate perfectly reading errors. One may object that we did not mention the major fault problems when a cell simply stops functioning at all. This is not our purpose to treat that kind of 'hardware' problem here, but we may say that our model could be implemented over an embryonics tissue[6] which deals with this kind of malfunction. Besides, one may note that such a permanent failure is not, in itself, fatal to our system, but only weakens all intersecting neighborhoods.

2.2 Classical Self-Replicating Structures Behavior

In this section, we show the brittleness of some famous and less famous classical self-relicating structures. We first consider the Langton's loop[5], we then move on to the more complex Tempesti's loop[11] to finish with one of the simplest self-replicating loop that was described by Chou and Reggia[8]. We do not consider the seminal work of Von Neumann[7] for obvious practical reasons. However one may note that the very accurate nature of that latter work leads to a "natural" brittleness.

Applying this model to the landmark self-replicating structures of Langton demonstrates a high sensitivity to failure. Following our model of noisy environment described above, a probability of fault p_f as low as 0.0001 shows a complete degradation of the self-replication function. Tempesti's structure shows exactly the same type of failure for an even lower p_f. The main cause of this total fault intolerance is due to what we may call irreducible garbage remains. Both these loops have as default rule, no change. When confronted to an unexpected neighborhood, i.e., when the fault only generates surprise and not confusion, the cell remains in its previous state. Then garbage that appears in the void (in the middle of a quiescent zone) is irreducible and remains forever, or at least, and that is the source of this higher-than-expected sensitivity to faults, until the loop meets it for its inevitable loss.

The Chou and Reggia loop[8] (p.286, fig. 1g), begins to show significant degradation with a probability p_f of 0.001. This relative resistance is only due to its small size (3x3), and any fault affecting the loop almost inevitably leads to its complete destruction. One may note that this loop suffers, as the preceding examples, from *irreducible remains.*

It is clear that these rules where not designed to be fault-tolerant and thus their size is their only means of defense. Effectively, their probability of failure is directly proportional to their size, and may be approximated by $(1-(1-p_f)^{size})$. Of course this probability to be accurate should be augmented by the number of neighboring cell, and, as we saw, by the probability of encountering *irreducible remains.*

We now present how to design fault-tolerant structures.

3 Fault-Resistant Rules

In a noisy environment, as defined earlier, a cell cannot trust entirely its neighborhood to define its future state. In other works [4,2,3], where the aim was to design generic faultless cellular automata, a hierarchical construction was used, e.g., a meta-cell of dimension 2 creates a fault-tolerant cell, to the cost of a lot more complex architecture and greater computation time. However our approach here is *not* to create a fault-tolerant CA architecture, but rather to design the rules of a classical CAs so that the emergent, global behavior is preserved. That latter choice allows both a reduced computation time and lesser memory space than the former, the cost being a resulting CA specific to one task. In this section we first briefly present some element of error correcting code theory, on which we base our design of fault-tolerant rule presented in section 3.2.

3.1 Error-Correcting Code

The idea behind the error correcting code theory is to code a message so that when sent over a noisy transmission channel, at reception, one can detect and correct potential errors through decoding, and thus reconstruct the original message.

Formally, let Σ be an alphabet of q symbols, let a code $\mathcal{C}$ be a collection of sequence c_i of exactly n symbols of Σ, and let $d(c_i, c_j)$ be the Hamming distance between c_i and c_j. We call $d(\mathcal{C})$ the minimal distance of a code $\mathcal{C}$ which is defined as $\min_{i \neq j} d(c_i, c_j), c_i, c_j \in \mathcal{C}$. Then the idea of error-correcting code, introduced by Shannon[9], is to decode, on reception, the word received by the closest word belonging to $\mathcal{C}$. Using this simple strategy with a code $\mathcal{C}$ of minimal distance d, allows the correction of up to $(d-1)/2$ errors.

In this theory, one may see that the number of word in $\mathcal{C}$, M, and the minimal distance d, plays against one another for a fixed n and q. To avoid a waste of memory space, in our fault-tolerant rules developed in the next subsection, it would be useful to maximize M for a given d. However it is not always possible to determine it *a priori*, but for $d = 2e+1$, so that we can correct e errors, this maximum M, $A_q(n,d)$, is bounded. The lower and upper bounds known as Gilbert-Varshamov[12] bounds, are:

$$\frac{q^n}{\sum_{i=0}^{d-1} \binom{n}{i} (q-1)^i} \leq A_q(n,d) \leq \frac{q^n}{\sum_{k=0}^{e} \binom{n}{k} (q-1)^k} \tag{1}$$

3.2 Theoretical Aspect of Fault-Resistant Rules

A rule r in a 2-dimensional cellular automata is the mapping between a neighborhood, v, and a future state, c^+, that, if we use Moore neighborhood, may be written as $(v_1, ..., v_9 \rightarrow c^+)$. In the noisy environment, as defined in section 2.1, the values $v_1, ..., v_9$ may be corrupted, thereby inducing in most cases, a corrupted value of c^+, propagating in time the original error. The idea behind this paper is to see c^+ as the message to be transmitted over a noisy channel, and $v_1, ..., v_9$ as the encoding of the message. Then it becomes clear that if this encoding does respect the error correcting code theory constraints exposed above then, even if the neighborhood values are corrupted, it will still be decoded into the correct c^+ future state and the original errors will not last further than one time step.

Of course, the constraints will depend on the number of errors, e, we want to be able to correct for a given neighborhood. We define $(v_1, ..., v_9 \rightarrow c^+, e)$ to be the transition rule r resisting e errors. In consequence, we also define $V(r)$ to be the set of rules co-defined by r, $V(r) = ((\overline{x} \rightarrow c^+) \mid d(\overline{x}, (v_1, ..., v_9)) \leq e)$. Then it appears clearly that if each rule of the co-defined set of rules is at a minimal hamming distance of 1 of any other rule of the co-defined set of rules, that is, if each main rule is at minimal hamming distance of $2e+1$ of any other main rule, then, reinterpreting the result of Shannon above, we can guarantee that the CA thus defined will be able to function correctly providing that at most e errors happen at any one time step in each neighborhood. As one may note, the size of $V(r)$ is rather large. More precisely, for a neighborhood of size n, and an alphabet of q symbols we have:

$$|V(r)| = \sum_{k=0}^{e} \binom{n}{k} (q-1)^k \qquad (2)$$

A *conflict* appears when the intersection between every co-defined set of rules is not empty. For instance, for a von Neumann neighborhood, $r_1 = (03234 \rightarrow 1, 1)$ and $r_2 = (03034 \rightarrow 2, 1)$ are in major conflict as the neighborhood (03234) lies both in $V(r_1)$ and $V(r_2)$. To avoid wasting too many possible different neighborhoods, we distinguish *major* and *minor* conflicts. A major conflict appears when the future state of the conflicting rules are different. However minor conflicts, when future states of the conflicting rules are identical, are reasonable as it does not prevent error correction. Thus, we have the following reinterpreted Shannon's theorem:

Theorem 1. *Let $\mathcal{R}$ be all the transition rules of a CA C, then if we have:*

$$\forall_{i \neq j} (r_i \rightarrow c_i^+), (r_j \rightarrow c_j^+) \in \mathcal{R}, \quad d(r_i, r_j) \geq 2e+1 \textbf{ or } c_i^+ = c_j^+$$

we can guarantee that the CA C will correct up to e errors occurring at any one time step in any neighborhoods.

We have now defined theoretically how to conceive a fault-tolerant structures on a classical CA. In the following section, we study the practical application of the above defined constraints to peculiar substructures of interest for self-replication. On this more practical aspect of the question, it would be interesting to know how many different rules are available for use. If we do not take into account the distinction of minor and major conflict, and as we need all the rules to be at a hamming distance of $d = 2e+1$ to guarantee distinct co-defined set of rules, the number of rules available is given by the bounds of Gilbert-Varshamov (eq. 1). Nevertheless these bounds are rather large, and do not include the large advantages brought by the minor conflict exception. In any case the waste of possible rule is usually quite important self-replicating structures. If we look at classical structures, such as Langton's loop, the number of used rule compared to the number of rules available (namely q^n), is well within the Gilbert-Varshamov bounds for a minimal number of errors $e = 1$. On the other hand, even for $e = 1$, if the probability of failure of a cell is p_f, and the size of the neighborhood is n, then the probability of failure of our structure, whatever the size of this structure, is reduced, roughly, to $1 - ((1-p_f)^9) - 9 * p_f(1-p_f)^8$.

4 Fault-Resistant Structures

In this section, we propose to apply practically the theoretical ideas put forward in the previous section. We will only consider rules that can correct up to 1 error in their neighborhood, as it seems a reasonable compromise between the fault-tolerance capabilities and the implied "waste" of rules. In the same way of thinking, we will only consider two-dimensional Moore neighborhood, as it provides a wealth of rules compared to the von Neumann neighborhood at no real additional cost in terms of computational time, memory space or physical complexity.

The aim of this research is to develop a fault-tolerant self-replicating structure. Firstly, we will study some common problems and desirable properties to construct such structures. Secondly, in section 4.2, we show how to practically construct a fault-tolerant data pipe, a structure essential to convey information. Finally, in section 4.3, we propose a fault-tolerant signal duplicator, central to the whole idea of self-replication.

4.1 Properties, Methods, and Problems

Our CA's structure are quite classically defined as an "active" structure in a pool, possibly infinite, of quiescent cells (henceforth represented by the '.' symbol). Hence the first necessary rule is $r_q = (......... \rightarrow ., 1)$. Its co-defined set of rules $V(r_q)$ defines all the rules with one active cell. It has the advantage to cover what we called in section 2.2, the irreducible remains. Effectively, any fault creating a single active cell in a quiescent neighborhood will be reverted to the quiescent state at the next time step. This eliminates the main problem encountered by the non fault-tolerant structures. Although the encounter of our fault-tolerant rules with the remains would not be as deadly as for the classical loops, this prevents accumulation of errors in the same neighborhood over time. Nevertheless, it is important to note that, consequently to the definition of r_q, we cannot use any rules with less than 3 active cells in its neighborhood. Its co-defined set of rules would otherwise intersects with $V(r_q)$ and create a major conflict unless, of course its future state is '.'. This last constraint which may be seen as a disadvantage in terms of the global size of the structure, is not much of a problem on the point of view of error-correction as the fault-tolerance capabilities of our structures only depend on the size of the neighborhood and not of the global structure.

That latter implied property brings us to a more general remark about the method to use when creating fault-tolerant structures. If we take figure 1, one would be tempted not to define the rule $r_1 = (........M \rightarrow ., 1)$ arguing that it is covered by $V(r_q)$. However, this would be mistake as $V(r_1) \neq V(r_q)$, we would then lose our guarantee of error correction. Hence, minor conflicts are handy as they allow us to define rules such as r_1 and r_q simultaneously, nevertheless they do not make the definition of r_1 unnecessary if we are insure the fault-tolerance property.

Besides these kind of tricks one should use, we are currently developing a more formal method to 'automatically' create fault-tolerant rules. At the moment, the semi-algorithm goes as follows: When going from the structure from one time step to the next one, we define the rules (with $e = 1$) for each cell in the structure and its neighborhood, going from left to right and top to bottom. After each rule definition, we test for major conflicts. This way to proceed prevents a lot of useless work, but does not discard the design by hand of the initial structure and its successors.

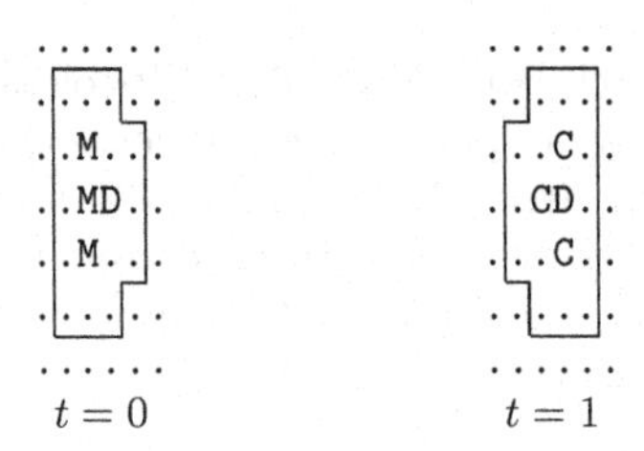

Fig. 1. Structure and its neighborhood

4.2 Safe Data Pipe

Now that we have seen the theoretical aspects and some general practical constraints, we are able to define a fault tolerant data pipe. These kind information transmission structures are essential for self-replicating loops. A typical example of these is given in figure 2.

0+0000	00+000	000+00
t	$t+1$	$t+2$

Fig. 2. Usual data pipe

One sees immediately the problem with that kind of structures. It implies the two following rules, $r_1 =$ (0..0...+. $\rightarrow$ +, 1) and $r_2 =$ (+..0...0. $\rightarrow$0, 1), which obviously provoke a major conflict. In fact, we remember that, for $e = 1$, each rule must be at a minimal hamming distance of 3. This constraint deals also with the quiescent rule constraint.

We can see in figure 3 an example of a fault-tolerant data pipe for $e = 1$ addressing that constraint. It is able to transmit trains of data. As one may see two types of data are transmitted in this example :L which uses 3 states to move itself, and + which uses 4. As we will explain later, the 4 states are needed as it is the *head* of the data train.

```
.........        .........        .........
000{00[00        000}00]00        0000{00[0
00{L0[+00        000}L0]D0        000{L0[+0
000{00[00        000}00]00        0000{00[0
.........        .........        .........
  t = 0            t = 1            t = 2
```

Fig. 3. Fault-tolerant data pipe

We will now have a more detailed look at this fault-tolerant data pipe:

- at $t = 0$, the upper and lower “wings” are necessary to maintain the '0' state situated at the 'east' of the head of the signal, the '+'state. The state '0' is designed to conduct the head.
- at $t = 1$, we have a transitory structure. This transition is necessary to maintain the integrity of the pipe. The upper and lower “wings” not moved yet would create a conflict if they were still in the state '['. We would end up with the conflicting transition rules (0..00D[[.$\rightarrow$[, 1) and (0..000+[. $\rightarrow$0, 1). This transitional step is there to create a diversity in the neighborhoods, thereby suppressing all conflicts.
- at $t = 2$, we get back to the original the structure, the train of signal (that one may define as L+) has moved forward.

As we noted earlier, a supplementary state was required for the head of the signal, the data '+', than for the rest of the signal 'L'. It can be clearly seen why if we imagine the situation without that supplementary state (see figure 4).

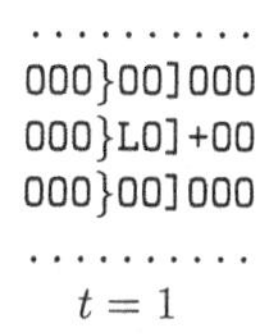

Fig. 4. Problem with the head signal

In this situation we have to define the rule (0000000+0 $\rightarrow$ 0, 1), which then conflict with the rule (000000[+[$\rightarrow$1, 1) defined at the preceding time step. The transitory state 'D' solves that conflict. The neighborhood of L, and any other following signal being more diverse, that supplementary transitory state is useless.

4.3 Signal Duplication

Now that we have, through the detailed definition of a fault-tolerant data pipe, seen the practical aspects of making fault-tolerant structures, we propose in this last section to quickly view a fault-tolerant signal duplicator. This peculiar substructure is always present in self-replicating structures. Actually, its the core of self-replication. The data pipe is there to maintain the structure, the duplicator is there to start a new construction, the replicate. We can check that this rather more complex structures only requires the use of two more states, C and K, besides the state needed by the data pipe described above.

```
....000.....   ....000.....   ....000.....   ....000.....
....000.....   ....000.....   ....000.....   ....000.....
....000.....   ....000.....   ....000.....   ....000.....
....000.....   ....000.....   ....000.....   ....000.....
....000.....   ....000.....   ....000.....   ....000.....
0]00CCC00000   00[0CCC00000   00]0CCC00000   000[CCC00000
0]D00C000000   0[+00C000000   00]D0C000000   00[+0C000000
0]0000000000   00[000000000   00]000000000   000[00000000
............   ............   ............   ............
   t = 0          t = 1          t = 2          t = 3

....000.....   ....000.....   ....000.....   ....000.....
....000.....   ....000.....   ....000.....   ....000.....
....000.....   ....000.....   ....000.....   ....000.....
....000.....   ....000.....   ....000.....   ....000.....
....000.....   ....000.....   ....000.....   ....000.....
000]CCC00000   0000+CC00000   0000DCC00000   0000DDC00000
000]DC000000   000[+C000000   0000]+000000   0000[D000000
000]00000000   0000[0000000   0000]0000000   00000[000000
............   ............   ............   ............
   t = 4          t = 5          t = 6          t = 7

....000.....   ....000.....   ....000.....   ....000.....
....000.....   ....000.....   ....000.....   ....000.....
....000.....   ....000.....   ....000.....   ....000.....
....000.....   ....000.....   ....000.....   ....000.....
....000.....   ....000.....   ....0D0.....   ....[+[.....
0000+KC00000   0000K+K00000   0000[]]00000   0000C[C[0000
00000KD00000   00000[+00000   00000C]D0000   00000C[+0000
00000]000000   000000[00000   000000]00000   0000000[0000
............   ............   ............   ............
   t = 8          t = 9          t = 10         t = 11

....000.....   ....000.....   ....000.....   ....000.....   ....000.....
....000.....   ....000.....   ....000.....   ....000.....   ....0D0.....
....000.....   ....000.....   ....0D0.....   ....[+[.....   ....]]].....
....0D0.....   ....[+[.....   ....]]].....   ....0[0.....   ....000.....
....]]].....   ....0[0.....   ....000.....   ....000.....   ....000.....
0000CCC]0000   0000CCC0[000   0000CCC0]000   0000CCC00[00   0000CCC00]00
00000C0]D000   00000C0[+000   00000C00]D00   00000C00[+00   00000C000]D0
0000000]0000   00000000[000   00000000]000   000000000[00   000000000]00
............   ............   ............   ............   ............
   t = 12         t = 13         t = 14         t = 15         t = 16
```

Fig. 5. Duplication of a signal

5 Concluding Remarks

We have shown that it is possible to create fault-tolerant structures by acting on the rules only. While not totally free of costs in terms of memory or computational time, the advantage of this method, in these terms, compared to classical, i.e., hierarchical, fault-tolerant CAs is evident. In terms of error correction capabilities, the rule-based method renders the probability of failure proportionate to the, relatively very small, size of the neighborhood. However, as we have seen in section 4.1, the possibility of algorithmically, and thus automatically, develop fault-resistant rules is still embryonic. Further work will pursue in this direction, thereby eliminating the main drawback of our method. Effectively, at the moment we only have task specific hand-made rules.

The two substructures presented are at the core of self-replicating loops, and we should be able in the near future to present a complete self-replicating structure resistant to a noisy environment. We believe that this will be attained with no more complexity, in terms of the number of state or the CA architecture, than most classical loops presented to date.

References

1. Mathieu S. Capcarrere. Asynchronous cellular automata: Efficient designed and evolved solutions for the synchronization and density problems. *Submitted*, 2001.
2. Peter Gács. Self-correcting two-dimensional arrays. In Silvio Micali, editor, *Randomness in computation*, volume 5 of *Advances in Computing Research*, pages 223–326, Greenwich, Conn, 1989. JAI Press.
3. Peter Gács. Reliable cellular automata with self-organization. In *Proceedings of the 38th IEEE Symposium on the Foundation of Computer Science*, pages 90–99, 1997.
4. Masaretu Harao and Shoici Noguchi. Fault tolerant cellular automata. *Journal of computer and system sciences*, 11:171–185, 1975.
5. Christopher G. Langton. Self-reproduction in cellular automata. *Physica D*, 10:135–144, 1984.
6. Daniel Mange, Moshe Sipper, Andre Stauffer, and Gianluca Tempesti. Towards robust integrated circuits: The embryonics approach. *Proceedings of the IEEE*, 88(4):516–541, april 2000.
7. John Von Neumann. *Theory of Self-Reproducing Automata.* University of Illinois Press, Illinois, 1966. Edited and completed by A. W. Burks.
8. James A. Reggia, Hui-Sien Chou, and Jason D. Lohn. Self-replicating structures : Evolution, emergence and computation. *Artificial Life*, 4:283–302, 1998.
9. C. E. Shannon. *The mathematical theory of communication.* University of Illinois Press, 1949.
10. Moshe Sipper. Fifty years of research on self-replication: An overview. *Artificial Life*, 4:237–257, 1998.
11. Gianluca Tempesti. A new self-reproducing cellular automaton capable of construction and computation. In F. Morán, A. Moreno, J. J. Merelo, and P. Chacon, editors, *Advances in Artificial Life: Proc. 3rd Eur. Conf. on Artificial Life (ECAL95)*, volume 929 of *LNAI*, pages 555–563. Springer-Verlag, 1995.
12. R. R. Varshamov. Estimate of the number of signals in error correcting codes. *Dokl. Akad. Nauk. SSSR*, 117:739–741, 1957.

Survival of the Unfittest? – The Seceder Model and its Fitness Landscape

Peter Dittrich[1] and Wolfgang Banzhaf[1]

Informatik XI
University of Dortmund
D-44221 Dortmund, Germany
{dittrich, banzhaf}@LS11.cs.uni-dortmund.de
http://ls11-www.cs.uni-dortmund.de/

Abstract. The seceder model is an extremely simple individual based model which shows how the local tendency to be different gives rise to the formation of hierarchically structured groups, called the seceder effect. The model consists of a population of simple entities which reproduce and die. In a single reproduction event three individuals are chosen randomly and the individual which possesses the largest distance to their mean is reproduced by creating a mutated copy (offspring). The offspring replaces a randomly chosen individual of the population. In this contribution we investigate the effective fitness landscape of the seceder model. Fitness is measured as reproductive success. The investigation of the fitness landscape revealed an on the first view counterintuitive phenomena: The individuals of the basic seceder model are always located in the worst regions of the fitness landscape where the replication rate is relatively low.

1 Introduction

The question of how groups emerge spontaneously from local interactions of individuals is investigated in many different disciplines such as biology, physics, sociology, or computer science. In evolutionary biology the question how evolutionary branching and speciation take place is approached by developing formal models which demonstrate the formation of groups [3, 6, 11, 14, 19]. These models are individual-based in contrast to macro-evolution models which assume a species or group as a given elementary unit [2, 16]. There has also been an increasing interest from statistical physics to deal with simple evolutionary models [2, 3, 6].

The diffusion and separation of individuals in genotype or trait space is either achieved by drift in a neutral fitness landscape [11] or by introducing an explicit fitness function [3, 6, 7, 13] which causes disruptive selection. Sometimes additional explicit functions are introduced to model strength of competition between individuals and ecological interactions [3]. Such functions are also needed to model the benefit of communication among groups on several levels [6].

J. Kelemen and P. Sosík (Eds.): ECAL 2001, LNAI 2159, pp. 100–109, 2001.

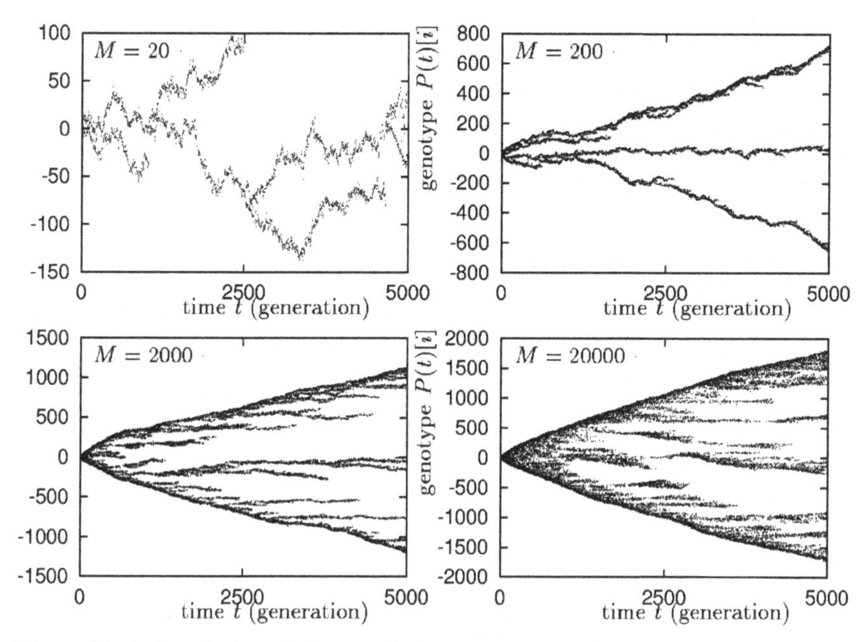

Fig. 1. Typical evolution of the population structure of the seceder model for population size $M = 20, 200, 2000, 20000$ and $\sigma = 1$. The population is initialized at $t = 0$ with $P(0)[i] = 0$. A small dot represents an individual i where its genotype $P(t)[i]$ specifies its ordinate value (vertical position).

The seceder model [5] is a microscopic model of an evolving population where the fitness landscape depends on the current population structure, like in [3, 11, 8]. The proposed mechanism is simple compared to other individual-based models [1, 6] for the formation of species or hierarchical organizations. But despite of its simplicity it shows comparably complex behavior. The seceder model does not require global energy functions [3, 6], spatially separated populations [12, 14], or sexual recombination [3, 11, 13].

The question of how microlevel actions explain macro-level regularities is also a central question in sociology [17]. Here, the seceder model may be a contribution as a social mechanism [10] for explaining how individual imitative behavior for the purpose of being different counter-intuitively can lead to the formation of groups on the macro-level. It should also be noted that the mechanism of the seceder model can be used to build practical applications in computer science. For example, it can be applied as a diversity maintenance method for evolutionary optimization algorithms where the reduction of diversity often causes a premature convergence and thus a bad performance of the optimization algorithm [18].

2 The Seceder Model

In following the basic seceder model is defined. In the seceder model an *individual* is represented by a real number. The *population* of size M is represented by an array $P = \{P[1], \ldots, P[M]\}$ of individuals $P[i] \in \mathcal{R}$. The population evolves over time according to the following algorithm:

Algorithm 1 (basic seceder model)

<u>**while**</u> $\neg$*terminate*() <u>**do**</u>

$s_1 := P[randomInt(1, M)]$	choose three individuals randomly
$s_2 := P[randomInt(1, M)]$	
$s_3 := P[randomInt(1, M)]$	
$\mu := f_{sel}(s_1, s_2, s_3)$	select individual with largest distance to others
$\lambda := \mu + N(0, \sigma)$	create offspring by adding a random number
$P[randomInt(1, M)] := \lambda$	replace randomly chosen individual
$t := t + 1/M$	increment time counter

<u>**od**</u>

The selection function

$$f_{sel}(g_1, g_2, g_3) = \begin{cases} g_1 & \text{if } F_1 \geq F_2 \wedge F_1 \geq F_3, \\ g_2 & \text{if } F_2 \geq F_1 \wedge F_2 \geq F_3, \\ g_3 & \text{otherwise,} \end{cases} \quad \text{where } F_i = |g_i - \frac{1}{3}(g_1 + g_2 + g_3)|, \tag{1}$$

returns the argument which possesses the largest distance to the mean of the three arguments. One iteration of the above algorithm is called a *step* and M iterations are called a *generation* which is used to measure time. The *distance* between two individuals is measured by the Euclidean distance (see definition of F_i in Eq. 1). *Mutation* is performed by adding a normally distributed random number with mean 0 and variance $\sigma = 1$ denoted by $N(0, \sigma)$. The population is usually initialized with one genotype, $P[i] = 0$ at $t = 0$. The procedure $randomInt(a, b)$ returns a uniformly distributed random number out of $\{a, a+1, \ldots, b\}$. The algorithm implies that the population size is constant and that an individual may have an arbitrary number of offsprings or may have no offspring at all. We write $P(t)$ for the population at time t, and $P(t)[i]$ for the i-th individual of population $P(t)$. Figure 1 shows how the population evolves over time in typical simulations for four different population sizes.

3 The Effective Fitness Landscape of the Basic Seceder Model

There are many ways to measure fitness [15]. Here we measure fitness as reproductive success ([9], Chapter 13, p. 366). Formally: Given a population

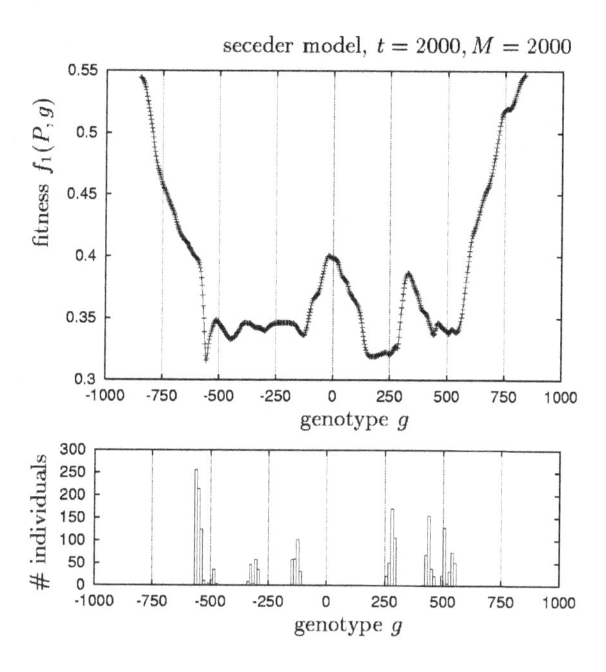

Fig. 2. Typical fitness landscape of the seceder model with population size $M = 2000$ after $t = 2000$ generations. High fitness value correspond to high replication rate.The lower graph shows the distribution of genotypes in the population.

$P = \{P[1], \ldots, P[M]\}$, the fitness of genotype $g \in \mathcal{R}$ is defined as:

$$f_1(P, g) = \frac{1}{M^2} \sum_{i,j=1}^{M} \begin{cases} 1 & \text{if} \quad g = f_{sel}(g, P[i], P[j]), \\ 0 & \text{otherwise.} \end{cases} \tag{2}$$

This fitness $f_1(P, g)$ measures the probability that a genotype g would be reproduced if it is chosen as the first individual s_1 in Alg. 1.

Figure 2 and 3 show typical fitness landscapes appearing in the seceder model for population size $M = 2000$ and an initial population with $P(0)[i] = 0$. The distribution of the population in genotype space is also plotted. We can see that (maybe surprisingly) the population is located in regions of low fitness (equal to low replication rate). In Fig. 2 the system is shown at an early stage ($t = 2000$) where the population diameter is still relatively small. The effective fitness landscape is more rugged than at a later point in time shown in Fig. 3. There, at $t = 20000$, three main arms coexist. Their individuals reside in the lowest regions of the fitness landscape. Some individuals of the outer arms are located on the increasing steep outer slope of the fitness landscape. Thus they tend to reproduce more likely and cause the two outer arms to depart from each

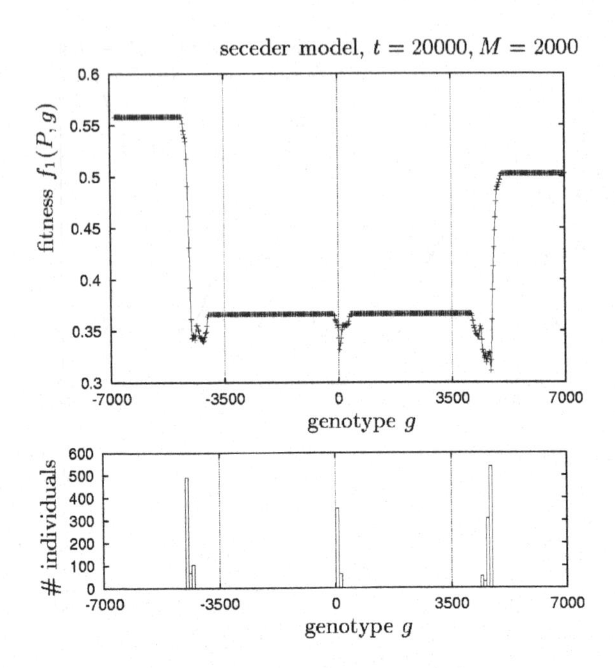

Fig. 3. Typical fitness landscape of the seceder model with population size $M = 2000$ after $t = 20000$ generations. The lower graph shows the distribution of genotypes in the population.

other. This effect will become more clear in the following when we investigate the seceder model with additional selection pressure.

4 The Seceder Model with Additional Selection Pressure

We will now extent the seceder model by an additional selection pressure by introducing an explicit "fitness" function $f : \mathcal{R} \rightarrow \mathcal{R}$ [4]. The additional selection pressure is added by modifying the death rate of a genotype g according to f. A high value $f(g)$ corresponds to high death rate and thus to low fitness[1]. This is achieved by the following algorithm:

Algorithm 2 (seceder model with additional selection pressure)
Same as Alg. 1, but we replace the insertation of the offspring (namely the line "$P[randomInt(1, M)] := \lambda$") by the following algorithm which is in fact a tournament:

[1] Note that in general in biological populations high death rate does not necessarily imply low fitness.

$i := randomInt(1, M)$ choose two candidates for replacement
$j := randomInt(1, M)$
$k := selectProportional(f(P[i]), f(P[j]))$ select one of them based on f
$P[k] := \lambda$ replace it by the offspring

The non-deterministic function $selectProportional(F_1, \ldots, F_n)$ returns an index $i \in \{1, \ldots, n\}$ where the probability that i is returned is $F_i / \sum_{j=1}^{n} F_j$. We will investigate the behavior of the model for the following function:

$$f(x) = |x|^{\alpha}. \tag{3}$$

With $\alpha = 0$ we obtain the basic seceder model. For $alpha > 0$ the death rate increases with distance to the origin in genotype space. There is one global minimum of the death rate. Note that even $f(g)$ can be zero (here, for $g = 0$), the death rate never becomes zero when Alg. 15 is applied. Furthermore note that the death rate is not only a function of genotype. The actual (absolute) death rate of a genotype depends on the other genotypes present in the population because tournament selection is applied.

Figure 4 shows how the additional selection pressure influences the time evolution of the population structure. Typical simulations are shown for population size $M = 1000$ and selection pressure $\alpha = 0, 0.01, 2, 4, 9, 15$. If the selection pressure is strong ($\alpha > 9$ for $M = 1000$) the time evolution is quite complex. New groups [5] emerge, move through genotype space, split up, and die. There is an ongoing change of the population structure. A simple stationary state never appears.

5 The Effective Fitness Landscape of Seceder Model with Additional Selection Pressure

In order to derive the effective fitness landscape for the seceder model with additional selection pressure we assume that the population size M is large. This allows to separately handle the "seceder" reproduction part (part A) and the additional selection part (part B) because if the population size is large it is unlikely that an individual is chosen in part A and part B during one step of Alg. 15. So in principal we can take the fitness landscape f_1 from the basic seceder model and subtract the effect of additional selection part, denoted by the function h:

$$f_2(P, g) = \underbrace{3 \frac{1}{M^2} \sum_{i,j=1}^{M} \begin{cases} 1 & \text{if} \quad g = f_{sel}(g, P[i], P[j]) \\ 0 & \text{otherwise} \end{cases}}_{f_1(P,g)} - \underbrace{2 \frac{1}{M} \sum_{i=1}^{M} \frac{f(g)}{f(g) + f(P[i])}}_{h(P,g)}. \tag{4}$$

Figure 6 shows three plots of fitness landscapes for three different situations that appear in simulations with Alg. 15, $M = 1000, \alpha = 9$. These situations are

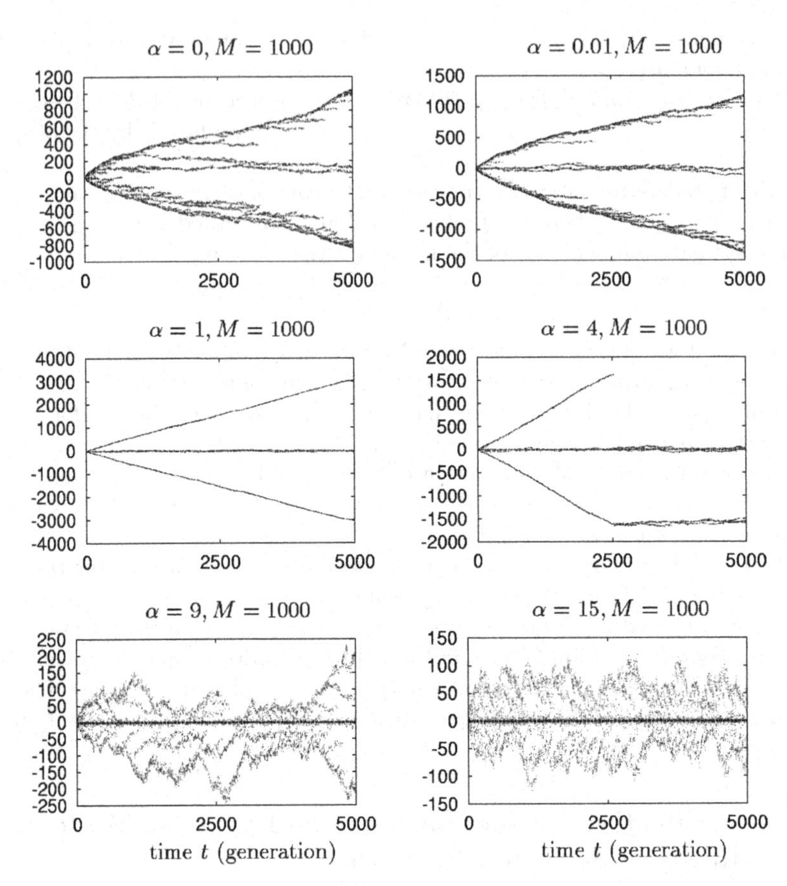

Fig. 4. Typical simulations of the seceder model with additional selection pressure. The strength of the additional selection pressure (death rate) increases with α. Initialization: $P(0)[i] = 0$. Parameters: $f(x) = |x|$, $M = 1000$, $\sigma = 1$. High value of $f(x)$ corresponds to high death rate of an individual with genotype $x \in S$.

take from a run shown in Fig. 5. They are typical for a (a) shrinking, (b) steady, and (c) increasing population diameter.

Many individuals reside in regions of low fitness. But some groups are not in local minima of the fitness landscape. Some groups are even completely located on a slope of the landscape. These are the groups that are moving through genotype space (compare Fig. 6 with Fig. 5). By looking at the gradient of the fitness landscape we can also predict the movement of a group. But this prediction is only possible for a short amount of time in the future because change in the population causes also change of the fitness landscape.

Looking at the rightmost group of individuals in Fig. 6 (c). This group moves away from the origin. We may imagin for a short while that the group moves up the hill of the fitness landscape. But the change of the population causes

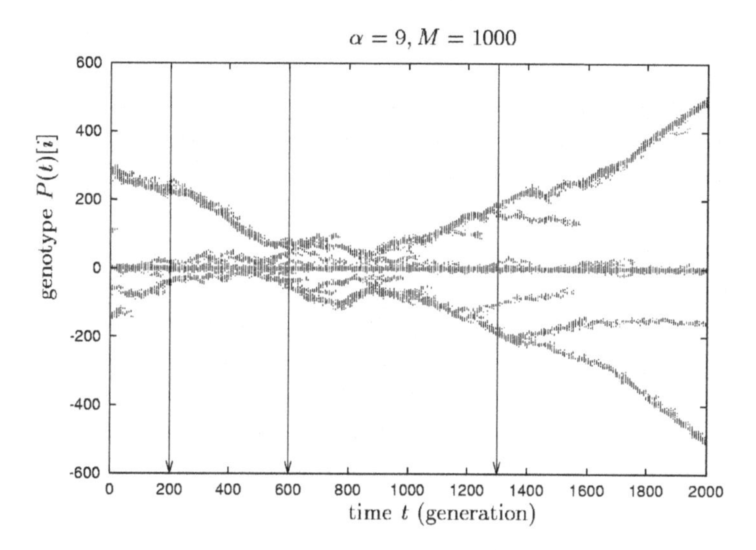

Fig. 5. The fitness landscape of the plotted run is displayed for $t = 200$, $t = 600$, and $t = 13000$ in Fig. 6.

this fitness hill to flatten and finally to get a shape similar to the big rightmost fitness hill in Fig. 6 (a). If that happens the group will move back towards the origin or even may die out if its effective fitness becomes too low, as can be seen in Fig. 4 for $\alpha = 9$.

6 Summary and Conclusion

We have defined an effective fitness landscape for the seceder model and its variant, the seceder model with additional selection pressure. The fitness landscape has been found to be useful to get a deeper insight into the dynamic behavior of the model. The effective fitness allows short time prediction of the movement of a main arm (group of similar individuals). But should neither be regarded as a *cause* for the long-term dynamics, nor as a cause for the population structure.

A naive model of evolution draws a picture of "species" or entities moving in a fitness landscape. During this movement (caused by reproduction and mutation) entities increase their fitness, climb up the hills of the fitness landscape, and get "caught" in local optima. Surprisingly, in the seceder model it is exactly the other way round. The entities reside in regions of low fitness, mostly in local *minima* of the fitness landscape.

Obviously the seceder model alone is also a naive model of evolution. Real biological evolution is a process which can only be understood as an interplay of many mechanisms. One such mechanism is probably the seceder effect investigated in that contribution.

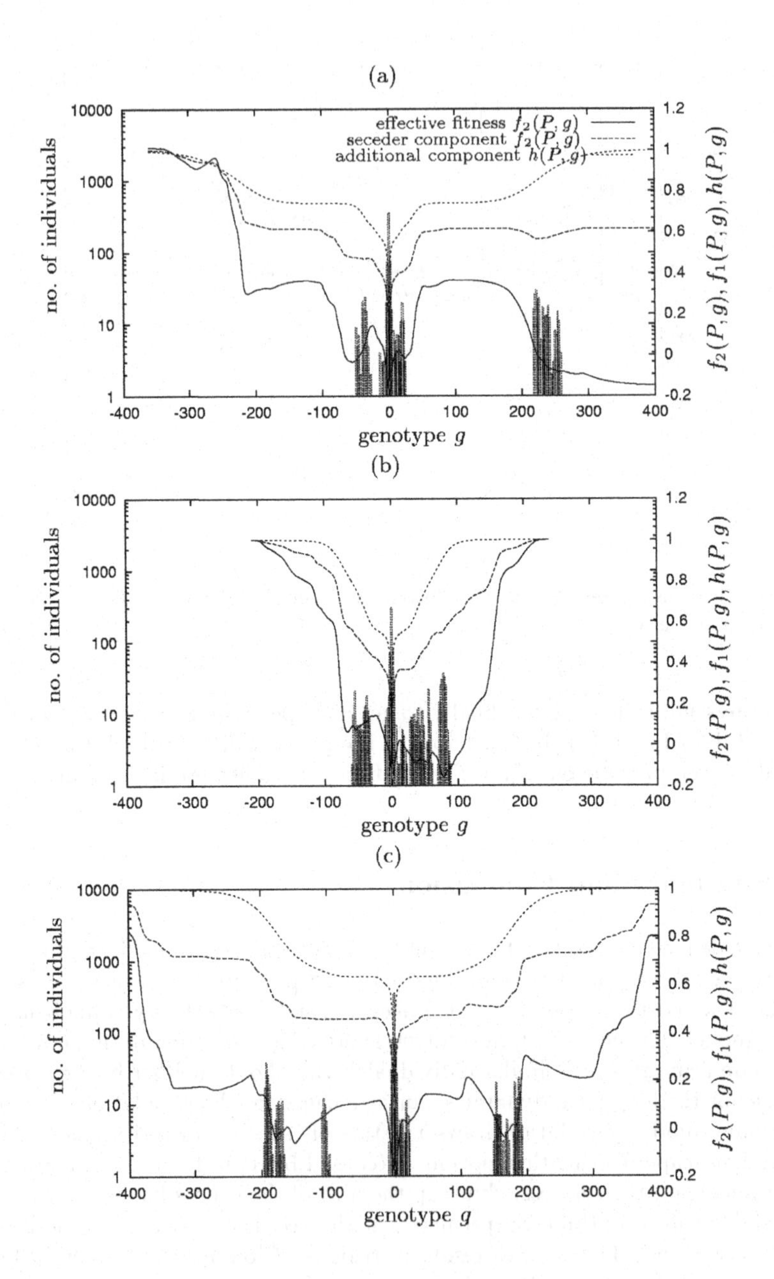

Fig. 6. Typical fitness landscape of the seceder model with additional fitness pressure. Population size $M = 1000$ after $t = 5000$ generations. The distribution of genotypes in the population is now plotted inside the graph. $f(x) = |x|^\alpha, \alpha = 9$.

How can we transfer our findings to the real-world? Can we expect the seceder effect to occur in natural, technical or social systems? Probably not in its pure form as exhibited by the basic seceder model. But the effect could very likely be observed overlayed by other fitness influencing components as modeled by the extended seceder model (Alg. 15). Applying the model to real-world systems in different domains like those mentioned above, would be an interesting prospect for future research.

References

1. F. Bagnoli and M. Bezzi Species formation in simple ecosystems. *Int. J. Mod. Phys. C*, 9(4):555–571, June 1998.
2. P. Bak and K. Sneppen Punctuated equilibrium and criticality in a simple model of evolution. *Phys. Rev. Lett.*, 71:4083–6, 1993.
3. U. Dieckmann and M. Doebeli On the origin of species by sympatric speciation. *Nature*, 400(6742):354–357, 1999.
4. P. Dittrich. The seceder effect in bounded space. *InterJournal*, 2000. presented at International Conference on Complex Systems, 21-26 May, 2000, Nashua, NH, InterJournal status: submitted, manuscript number: 363.
5. P. Dittrich, F. Liljeros, A. Soulier, and W. Banzhaf. Spontanous group formation in the seceder model. *Phys. Rev. Lett.*, 84:3205–8, 2000.
6. B. Drossel Simple model for the formation of a complex organism. *Phys. Rev. Lett.*, 82(25):5144–7, 1999.
7. M. Ebner, R. A. Watson, and J. Alexander. Co-evolutionary dynamics on a deformable landscape. In *Proceedings of the 2000 Congress on Evolutionary Computation (CEC'2000)*, volume 2, pages 1284–1291, San Diego Marriott Hotel, La Jolla, CA, 2000. IEEE.
8. M. Eigen and P. Schuster. The hypercycle: a principle of natural self-organisation, part A. *Naturwissenschaften*, 64(11):541–565, 1977.
9. Douglas J. Futuyma *Evolutionary Biology.* Sinauer, Sunderland, MA, 1997.
10. P. Hedström and R. Swedberg, editors. *Social Mechanisms : An Analytical Approach to Social Theory*, Cambridge, MA, 1998. Cambridge Univ. Pr.
11. P.G. Higgs and B. Derrida Stochastic models for species formation in evolving populations. *J. Phys. A*, 24(17):985–991, Sep. 1991.
12. K. Johst, M. Doebeli, and R. Brandl Evolution of complex dynamics in spatially structured populations. *Proc. R. Soc. Lond. Ser. B*, 266(1424):1147–1154, 1999.
13. A. S. Kondrashov and F. A. Kondrashov Interactions among quantitative traits in the course of sympatric speciation. *Nature*, 400(6742):351–354, 1999.
14. F. Manzo and L. Peliti Geographic speciation in the Derrida-Higgs model of species formation. *J. Phys. A*, 27(21):7079–7086, Nov. 1994.
15. R. E. Michod *Darwinian Dynamics – Evolutionary Transitions in Fitness and Individuality.* Princeton University Press, Princeton, NJ, 1999.
16. M. E. J. Newman Self-organized criticality, evolution, and the fossil extinction record. *Proc. R. Soc. London Ser. B*, 263:1605–10, 1996.
17. T. Thomas Schelling. *Micro Motives and Macrobehavior.* Norton, New York, 1978.
18. N. Shamir, D. Saad, and E. Marom Preserving the diversity of a genetically evolving population of nets using the functional behavior of neurons. *Complex Systems*, 7(5):327–46, 1993.
19. G. F. Turner and M. T. Burrows A model of sympatric speciation by sexual selection. *Proc. R. Soc. Lond. Ser. B*, 260(1359):287–292, 1995.

Evolving Multi-agent Networks in Structured Environments

T. Glotzmann, H. Lange, M. Hauhs, and A. Lamm

BITÖK, University of Bayreuth

Abstract. A crucial feature of evolving natural systems is parallelism. The simultaneous and distributed application of rules (governed by e.g. biochemistry) is generally considered as *the* preposition to build up complex structures. In this article, the potential of agent-based modelling equipped with a concurrent rewriting rule system for artificial evolution is investigated. The task given to the system (pattern construction drawing from a small pool of symbols) is sequential in character, but has to be solved by a strictly parallel rule system. This requires special care in setting up the environment, and it turns out that this is accomplished only by virtue of a hierarchy of levels and modularisation. The proposed three level hierarchy resembles stages of natural evolution in which the emergence of stabilizing mechanisms and cooperative behaviour can be studied. A few preliminary simlation runs are shown and discussed.

1 Introduction

Agent based modelling has become one of the core areas of AL research. A natural framework for agent-based programming are object-oriented high level languages. However, truly *autonomous* objects are not provided by these, as they are usually transformed into serialized machine code by the compiler. An extension towards truly parallel systems is the concept of *concurrent (multiset) rewriting systems* (Boudol and Berry 1992, Banatre and Le Metayer 1995, Meseguer 1996) on the theoretical side and multiprocessing (including multithread) techniques on the practical side. Concurrency of and communication among agents are two complementary aspects of natural systems capable of decision making, or generally of systems capable of interaction (Milner 1989). Thus, we expect that for AL research this architecture is also advantageous.

Any AL investigation requires a decision or hypothesis about which aspects of living systems are given a priori to or in the system and which aspects can be expected to result from exposing such systems to structured environments. Here we opted for a rather complex concurrency and interaction structure in which the population of agents is embedded. At the same time, relatively simple features and restrictions are imposed on the agents themselves. The environment is also simple - a small number of resources is provided without spatial distribution.

The theoretical motivation for this approach is taken from formal models of interaction (Wegner and Goldin 2000). We speculate that persistence of life (and also biological memory) is first of all conditioned by the form of interaction of agents with the abiotic world and that the type of interaction can be studied in an abstract manner by AL simulations. A related approach with a different notion of environment are the so-called ecosystem grammars (Csuhaj-Varjú et al. 1997).

J. Kelemen and P. Sosík (Eds.): ECAL 2001, LNAI 2159, pp. 110–119, 2001.

2 Building Blocks of the Multiagent Simulations

2.1 Concurrent Rewriting as Chemical Dynamics

Concurrent rewriting systems operate on terms that do not necessarily have normal forms by rules that are performed in parallel (Meseguer 1996, 1998). If the rule system performs circular transformations on the elements of the set of terms, a set may indefinitely change. Furthermore concurrent rewriting rules operate on *multisets* rather than on single elements as Lindenmayer do rules (Prusinkiwiecz and Lindenmayer 1990). Thus, a rule generally specifies the transformation of a multiset into another multiset. Rules can make use of variables and wildcards in its templates as exemplified in file systems and databases. Wildcards can only occur on the left side of a rule for pattern matching. Variables can occur on both sides and are then reproduced by the rule. In this way, a rule matches on a class of multisets rather than on a single multiset. In special cases, the multiset on either side of the rule can be empty (consuming or generating rules) or contain one element only.

The syntax of rules in our implementation of an environment, addressed as *Pool* in the following, is described as follows: Let Σ be a set of symbols, X a set of variables and $T = (\Sigma \cup X)^*$ the set of templates made of symbols and variables. Then the rule system R is described as

$$R = \{ r: T^* \to T^* \mid \forall (r: \alpha \to \beta) : X_\beta \subseteq X_\alpha, \alpha, \beta \in T^* \}$$

where X_α is the set of variables occurring on the left-hand side and X_β the set of variables occurring on the right-hand side of a rule. $X_\beta \subseteq X_\alpha$ makes sure that all variables occurring on the right-hand side of a rule are already defined on the left-hand side. T^* is seen as the set of all multisets over T.

For our purpose, we consider concurrent rewriting adapted for the simulation of chemical dynamics (Suzuki et al. 1997, Speroni et al. 2001). Consisting of symbols of an alphabet, i.e. atoms, elements can be considered as molecules and rules as chemical reaction equations, if an appropriate conservation law holds (i.e. the total sum of symbols is equal on both sides of the rule). For example the rule

$$x\text{–}CH_2OH \quad O_2 \to x\text{–}COOH \quad H_2O$$

represents the oxidation of any alcohol to an organic acid where x is a variable (thermodynamical restrictions can also be encoded in such rules).

2.2 The World Where Agents Live

The environment in which our agents live consists of a network of *places*. Each place is considered as point-like, i.e. is devoid of any structure. Everything contained in a given place, elements as well as agents, is visible and accessible for all agents there. A network topology is induced by the interconnection of places. Thus the *Pool*-world is discrete and distances are measured by counting the places on the shortest path between two locations. Agents and their offspring can move from place to place within this network. From an outside perspective the dynamics of agents is concurrent in a physical and technical sense (realised as different threads or different CPUs).

2.3 The Agent's Architecture

The agent's interior is very similar to the outside world: it consists of a pool containing elements and a system of rewrite-rules applying to these elements; such a single rule is called a micro-agent. In order to shuffle elements from the world outside into the agent and vice versa interface-rules are provided. They work just like normal rules but the left side of their formal notation takes elements from one (e.g. the external) pool and their right side writes products into the other (e.g. the internal) pool. In parallel to the world outside an agent can be structured hierarchically into subagents (which are similarly structured as the agents) resting inside an agent's pool which can consist of sub-subagents again, and so on (Fig. 1), resembling a russian doll. In this way, rule systems are grouped into functional units.

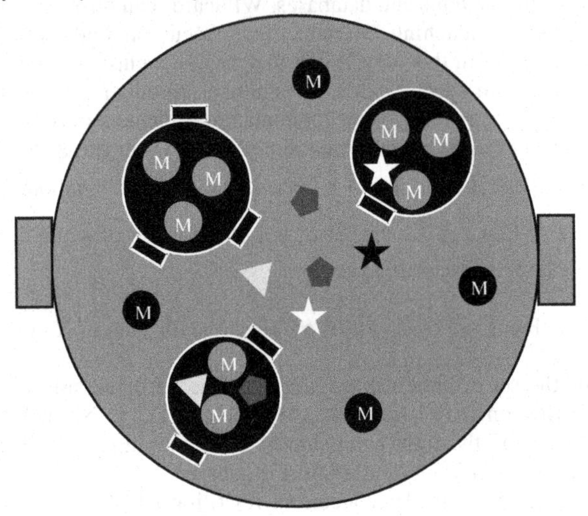

Fig. 1. An agent consisting of a pool with several elements (in the center), three subagents and some microagents. □ :interfaces, ○: agent/subagents, M: micro-agents (rules)

3 Evolution of String-Generating Agents in *Pool*

In string-generating systems, as discussed here, rules perform mass-preserving operations like simulated reactions in chemical models. An exception to this restriction is the use of *markers*, which are special symbols, that can be created and/or destroyed without such restrictions. They are normally taken from a different alphabet than the string-symbols and provide a coordination of rules.

Seen as a simplified chemical system, agents synthesizing and transforming strings out of letters and string fragments shall be subject to evolution driven by genetic operations and place specific selection. Based on the architecture of chapter 2, we implemented an evolutionary system of agents in Java. The program, i.e. the rule-set, of an agent constitutes the genome and is represented as XML code. Parsing such a genome, an agent can be formed and started. The *Pool* system is strictly individual-based. All the agents act asynchronously and are started as independent threads (Oaks and Wong 1999). In contrast to the majority of artificial evolutionary systems, this system is not generation-based (where reproduction and scheduling is provided externally). Here actions, offspring creation and death is controlled by each agent's clock. Lifetime and offspring number of an agent depend on its fitness, which in turn is computed by a supervisor object linked to each agent and evaluating its output.

3.1 Description and Realisation of *Pool*

In *Pool* we study the potential of concurrent rewriting systems to simulate the appearance of collective behavior. To this end we introduce instantaneous fitness measures that evaluate each agent's performance with respect to a given task at the respective places. We investigated the example of a string recognising and synthesizing system. Given a pool of letters, the optimal agent produces the word EVOLUTION as output after digesting appropriate letters from the *Pool*. Fitness is calculated by estimating the number of mutation steps the agent will still need to synthesize the desired string. In *Pool*, a full set of genetic operations is implemented acting on rules (microagents) or subagents. Rules can be mutated by insertion or deletion, rules as well as subagents may be duplicated, and crossover can take place.

3.2 Fitness Calculation and Selection

Providing the evolving system with an external fitness functional is an often unwanted but at the same time often chosen way to let the evolution proceed. Ideally, an Artificial Life system and its individuals should have an implicitly defined fitness, as the prescription of a given task to be solved may exclude the possibility of open-ended evolution (Bedau et al. 1998).

In *Pool*, the number of offspring of an agent is determined by its fitness. Furthermore, the life span an agent can have is inherited from its parent. Finally, a place can have limited capacity for agents, i.e. when a new agent is created at an already occupied place, the agent either is discarded or drives another agent off the place. In the latter case, selection is performed by eliminating agents with lesser fitness than the new one has (resp. its parent). Additionally, in *Pool* an agent related closer to the new one will be pushed off more probable than others. This should conserve high genetic differentiation.

4 The Evolution of EVOLUTION in Pool

In the EVOLUTION example, the task of agents is to synthesize the string "EVOLUTION" from a couple of letters taken from a pool by concurrently applying rewrite-rules. It is important to note that this task is not only a matter of permutation but also of the correct *sequential* composition of rules which in principle act parallel. So for instance rule-application on elements which are pre-products of those the rule should apply to would be disastrous in many cases!

The implementation of this example was realized in two variants. We will show that the straightforward fully flexible single level model does not lead to successful task solving within feasible CPU times, and that a hierarchy of levels is required much more efficient.

4.1 A Single Level Model

In a first approach, agents at all places were allowed to make use of all features of the rewrite-system. Besides ordinary symbols, rules could contain variables for pattern matching as described in 2.1 as well as marker symbols. Markers are necessary for copying elements (they are non-conservative) or for ordering rule application. Furthermore, each individual was equipped with a set of different output interfaces, which contained variables too, leading to highly non-deterministic output behaviour in addition to the non-deterministic behaviour of the inner rule system. Here, fitness calculation faced two major observational problems:

What is the agent's spectrum of possible behaviours, when the observer only has access to its *actual* behaviour within a *finite* run? What is the adequate duration of a run or number of runs to be able to classify the agent's behaviour? It is known from theoretical computer science that this problem is theoretically undecidable. A pragmatic solution is needed that balances the trade-off between allocating CPU time to the testing phase (continuing fitness testing) and the model revision phase (reproduction). Obviously this solution will depend on the structure of the environment.

The use of variable-containing templates for pattern matching and construction in combination with parallelism and non-determinism leads to random behaviour. How should an observer decide whether a product of an agent came forth due to the agent's skills or is due to accidental features of that run?

In our experiment we investigated different stochastic strategies to measure the reliability of outputs, i.e. to distinguish systematic from random fitness scores. However, in a multitude of runs, which were frequently varied, the "gamblers" always were the winners and evolution stopped in a stable state with agents working completely at random. This may be partially attributable to the simplicity of our example (the odds of random gamblers decay multiplicatively with string length), but also points to a fundamental feature of natural evolution not present in our example: *modularization of tasks* and the *emergence of hierarchies*. We thus propose the use of a hierarchical system in the next section, in which the overall task is broken into subtasks that may occur sequentially in evolving systems. The spontaneous appearance of hierarchical structures within the system itself is well beyond the scope of this preliminary investigation and may well be one of the truly hard problems of agent-based simulations.

4.2 The Evolutionary Hierarchy

In a second approach, we split the above task of synthesizing the string "EVOLUTION" from single elements into modules that were assigned to different levels. These levels stand for various "epochs" of evolution. Transitions among levels were automated and hierarchical in the sense that once an agent has solved the task of his level it is transferred/restarted at the next higher level in the hierarchy.

We simplified the agent's architecture relative to the one used in the first approach: An agent has one interface only for each letter it is able to read, it may have only two kinds of rules in its inside (i.e. joiners that link two strings together and splitters that split), and may produce only a single output string repeatedly. The genome contains the patterns of the rules ensuring that rules only match to really existing strings or strings which can be synthesized by rules of the same set. The rules no longer make use of variables nor markers. Here we will describe a world that was divided into *three* different levels representing different evolutionary epochs: a pattern recognition level, an optimisation and a synthesis level.

The Pattern Recognition Level. At the first (basic) level all agents live in one place in a world of unlimited resources. There is no direct competition among agents. The task consists of synthesizing and recognizing three letter strings (the patterns) only. The interpretation of this world is that agents need to discover the synthesis of this three-letter-string (syllables) in order to attain maximal fitness.

By means of rule application, agents take letters and synthesize strings out of them. At the end of their lifetime, they place a number of offspring according to their fitness and disappear, resembling r-strategies in the language of population biology (May 1974). Agents undergo mutations until they discover how to synthesize the sub-strings (syllables) of the target-word. In the EVOLUTION example three types of specialists evolve ("EVO", "LUT", "ION"). Selection follows the principle of capacity-limitation in this place (cf. 3.2).

The moment an individual has achieved to fulfil its task, its offspring are no longer placed in the same level but in one out of three places of the next hierarchical level (one for each discovered syllable).

The Optimisation Level. The first level describes the situation of a world with abundant resources that need only to be discovered by agents. Once an agent has access to this resource it engages in exponential growth until an (external) limitation becomes effective. The automatic "transfer" to the next level skips this phase of exponential growth and assumes that successful agents from the first level would soon find themselves exposed to different selection pressures.

This second level represents a finite world in which each substring (syllable) population becomes rewarded for its efficiency in its own place. The aim is to minimise string generation times by eliminating redundant rules which don't contribute to the string synthesis. Except fitness calculation (see chapter 3.2), all other conditions are the same as in the first level. The only fitness criterion (besides not changing the output previously learned) is the minimisation of the number of input-interfaces. The moment an agent accomplishes to read only those letters it needs for synthesis, it qualifies for the next level.

The Synthesis Level. At the next stage of evolution, the success of the evolved agents may change the total abundance of primary resources; the world is not only finite, it may also become instationary. That is why we introduce a feedback link from the highest to the lowest level. Such a feedback between agent success (fitness) and resource availability transcends the simple fitness measures that were used before, as formerly fit agents may drive the whole system into instability.

At this level the task is posed such that agents must interact/combine in order to produce a secondary resource from the primary ones (letters). Thus the successful output of the composite word "EVOLUTION" is not interpreted as pattern recognition (see basic level), but as a (true) synthesizing task. This scenario implies to keep all module-building agents in the race by posing a synthesis task that needs all three of them. The product may be considered as a secondary resource and the corresponding novel pattern recognition task is started in the first level again.

In the upper level, all agents operate on the same pool of limited resources competing directly (and interactively) with each other. This place has got unlimited capacity for the number of agents, but their population is limited by the finite pool of joint resources (i.e. letters). If an agent doesn't achieve to produce an output during its life span, it may not have offspring and goes extinct. Letters consumed by an agent stay in it until it produces an output or dies. In this case the output string and also the strings left in a dead agent are decomposed into single letters refilling the pool for all other agents.

Experience shows that mutations in this level are often lethal, thus destabilizing the system. Agents at this level therefore no longer undergo mutation, but may use crossover. Before they can do so, their rule set has to be duplicated by a genetic operator. This takes place in form of building two subagents out of an agent or subagent and concatenating the strings each of them produces. An offspring of an EVO-producing agent for instance would produce EVOEVO after duplication. Thus the agents' architecture is organized as a binary tree resulting from repeated duplication. By means of crossover, partial trees are exchanged leading to the production of strings consisting of different syllables.

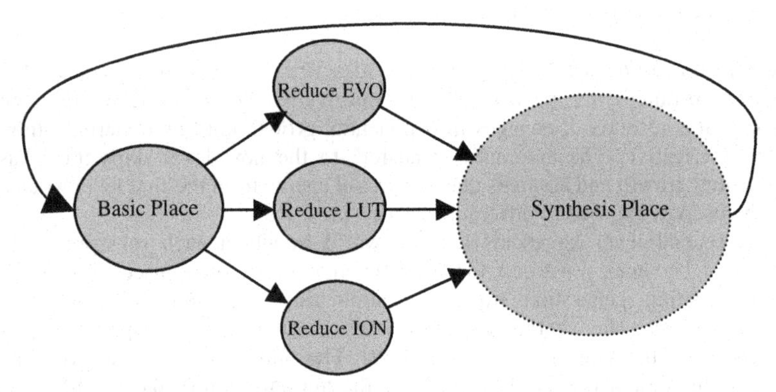

Fig. 2. The hierarchy of *Pool*. The three levels are shown from left to right. The target string EVOLUTION is synthesized in the Synthesis Place at level 3. After that, a new pattern recognition task is posed in the basic place that corresponds to decomposition of the products within the highest place (arrowed loop).

Similar to the first level, an agent's fitness is calculated depending on the distance of the resulting to the target-string, but the agent is no longer rewarded by driving someone off the place, but by getting a longer life span. Thus, a fitter agent can survive periods where no letters are available because a multitude of agents has consumed some but not all of the letters they need (deadlock). Agents with lesser fitness will starve in this case, releasing their content of letters into the world which can be snatched by the longer living ones. This scenario resembles a K-strategy in terms of population biology.

Outputs are broken into single letters, but with one important exception: the target-string. Thus the moment the target-string is „generated" by an agent, it will accumulate and single letters will become less abundant, causing starvation of the whole population. The target string is considered as a secondary resource for which a corresponding pattern recognition task is introduced at the basic level. The interpretation of this world is that successful interaction among efficient agents may produce new resources (that need to be recognized). A successful recognition of this properly encoded pattern by specialized agents of the basic level results in decomposition of the corresponding composite word in the highest level once this agent arrives there. If this happens before the population of the highest level has died out, the latter is rescued by breaking that string and will recover. Otherwise the whole population will die out. Finally, we extended the set of target-strings to all syllable-level permutations of EVOLUTION. In this way, different results can trigger different code-strings to be discovered in the basic place. Fig. 2 visualizes the hierarchical structure of places and the tasks performed at each level.

5 First Results

We discuss sample simulation runs from the hierarchical setting of ch. 4.2. Runs of the EVOLUTION example always follow the same scheme in the first two levels: After some time, one of the specialists producing EVO, LUT or ION has evolved and moves to the corresponding code reduction place. Typically after the occurrence of some hundred agents, all three kinds of specialists will have evolved (specifying time in terms of generations doesn't make sense in an asynchronous system).

Code reduction in the second level typically converges after the life cycles of a few hundred individuals. Then the first of the specialists with minimal required input moves to the synthesis level. There it will multiply itself and duplicate its output string until the occurrence of a string with three repetitions of the syllable it is specialized for. This is the state of maximal fitness with one kind of syllable only. After some additional time, the other specialists will appear in the place as well.

As long as no crossover takes place, the agents only compete for resources with individuals of the same kind. After crossover, the resulting agents compete with two or all three kinds of specialists as well as with each other.

Now further progression separates into two possible scenarios: Once one or more strings with three syllables occur, they accumulate while binding further resources. If the appropriate destructor starting in the basic pool doesn't evolve quickly enough, the population in the synthesis place starves and vanishes except for a few individuals arriving from the reduction places. Fig. 3 shows an example population which died out at time 550 due to the lack of a destructor for LUT-ION-EVO. The system will stay in a quasi steady state of arriving and starving until the adequate destructors are found in terms of agents which can produce the syllable encoding the accumulated string. These strings will be dissolved into their letters and a new evolution will start in the third level until another undissolvable string occurs (Fig. 3).

The second scenario is that the desired destructor appears early enough. Then the population will recover and the agents producing the concerned string will survive with high probability. A difficult but achievable task now is to produce a different target string (Fig. 4). This event leads to a new critical state where the whole population is endangered.

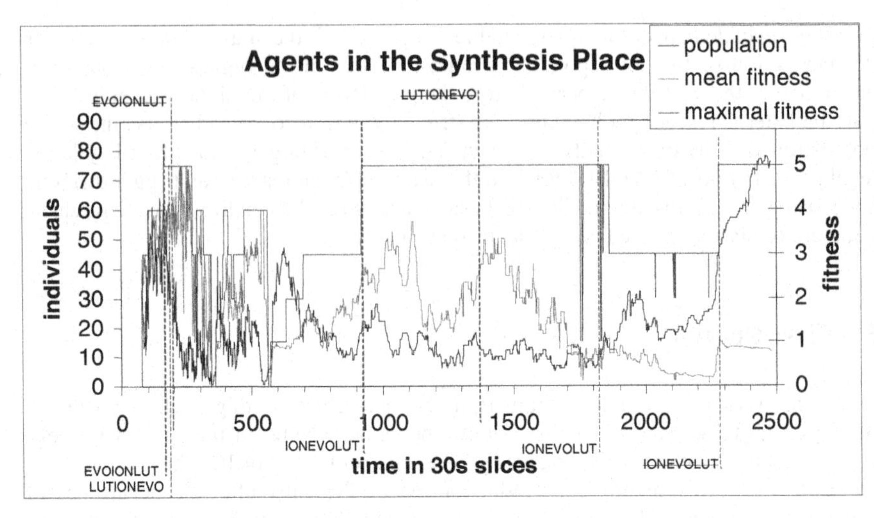

Fig. 3. Population dynamics at the synthesis (highest) level. Shown is the number of viable agents, each of which is executed as a thread. The first appearance of the target strings (e.g. "EVOIONLUT" after 180 time slices) is marked with the corresponding string, the extinction of the last agents capable of producing a specific string by the strings stroked through. Whenever a target string appears the maximal fitness within the population jumps to its theoretical maximum. The relationship between max and mean fitness indicates the diversity of the respective population.

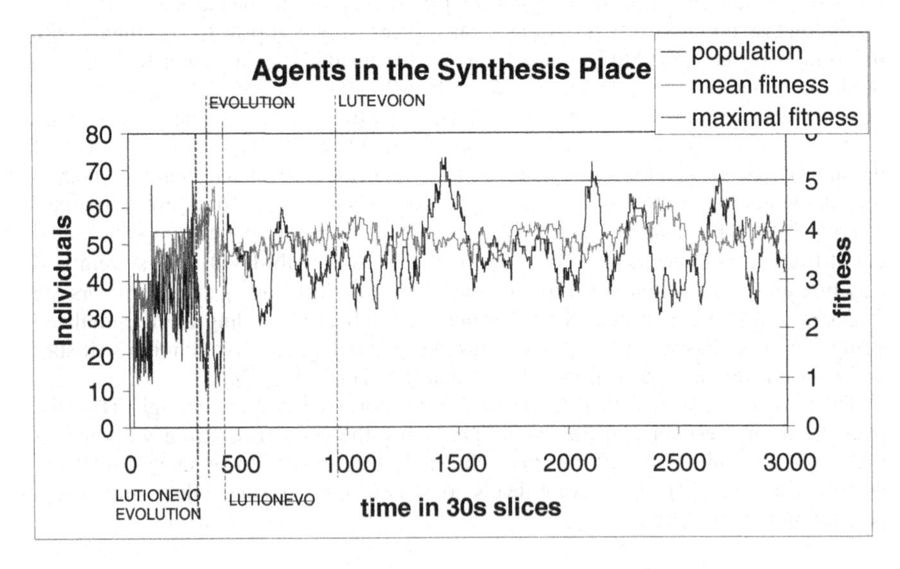

Fig. 4. Same as Fig. 3 for another example in which the target string EVOLUTION has been produced. In this case, LUT-ION-EVO and EVO-LUT-ION could soon be decomposed, so the population didn't die out.

6 Conclusions

In order to simulate some basic aspects of natural evolving systems, we investigated a multiagent simulation setting representing a concurrent rewriting rule system. Although working strictly parallel, the agents are provided with a sequential task, the synthesis of a target string from elementary symbols. A single level realization did not reliably succeed in this task. Structuring the world can lead to non-trivial evolution much more effective. In this case, the population often stabilizes in some kind of "modus vivendi", a state where the population can survive with no need of further evolution. The way we decided to go is to locally separate evolutionary steps. This is of course an additional input of information into the system, but very crucial to prevent equilibrium at an early stage.

In *Pool* as well as in nature, crises or catastrophic events force evolution to bring forth new phenomena. In a limited world those are eventually caused by the organisms themselves.

7 References

J.-P. Banatre and D. Le Metayer (1995): Gamma and the chemical reaction model. In: Proceedings of the Coordination'95 workshop. IC Press, London.

Boudol, G. and Berry, G. (1992): The chemical abstract machine. Theoretical Computer Science 96, 217-248.

Bedau, M. A., Snyder, E. and Packard, N.H. (1998): *A Classification of Long-Term Evolutionary Dynamics*. In: Adami, C., Belew, R.K., Kitano, H. and Taylor, C.E. (eds.), Artificial Life VI. MIT Press, Cambridge

Csuhaj-Varjú, E., Kelemen, J., Kelemenová, A. and Paun, G. (1997): Eco-Grammar Systems: A Grammatical Framework for Studying Lifelike Interactions. Artificial Life 3, 1-28.

Goldberg, D.E. (1989): Genetic Algorithms in Search, Optimization, and Machine Learning, Addison-Wesley: Reading.

May, R.M. (1974): Stability and complexity in model ecosystems. Princeton Univ. Press.

Milner, R. (1989): *Communication and Concurrency*. Prentice Hall, Hertfordshire.

Meseguer, José (1996): Rewriting Logic as a Semantic Framework for Concurrency: a Progress Report, SRI International.

Meseguer, José (1998): *Research Directions in Rewriting Logic*, In: U. Berger and H. Schwichtenberg, editors, Computational Logic, NATO Advanced Study, Institute, Marktoberdorf, Germany, July 29 - August 6, 1997. Springer-Verlag.

Oaks, S., Wong, H. (1999): *Java Threads*, O'Reilly, Beijing.

Prusinkiewicz, P. and Lindenmayer, A. (1990): The Algorithmic Beauty of Plants. Springer, New York.

Speroni, P., Dittrich, P. and Banzhaf, W. (2001): Towards a Theory of Organizations. In: Lange, H. (ed.): Proceedings of the 4th German Workshop on Artificial Life. Bayreuther Forum Ökologie (in press).

Suzuki, Y., Tsumoto, S. and Tanaka, H. (1997): Analysis of Cycles in Symbolic Chemical System based on Abstract Rewriting System on Multisets. Artificial Life V, 522-528. MIT press.

Wegner, P. and Goldin, D. (2000): *Coinductive Models of Finite Computing Agents*. Electronic Notes in Theoretical Computer Science 19, 21.

Suicide as an Evolutionarily Stable Strategy

Steven Mascaro, Kevin B. Korb, and Ann E. Nicholson

School of Computer Science and Software Engineering
Monash University, VIC 3800, Australia,
{stevenm,korb,annn}@csse.monash.edu.au

Abstract. Ethics has traditionally been the domain of philosophers, pursuing their investigations *a priori*, since social experimentation is not an option. After Axelrod's work, artificial life (ALife) methods have been applied to social simulation. Here we use an ALife simulation to pursue experiments with ethics. We use a utilitarian model for assessing what is ethical, as it offers a computationally clear means of measuring ethical value, based on the utility of outcomes. We investigate the particular action of altruistic suicide fostering the survival of others, demonstrating that suicide can be an evolutionarily stable strategy (ESS).

Keywords: Altruism, artificial life, ethics, social simulation, evolutionarily stable strategies.

1 Introduction

Since its inception, research into ethics has been the domain of philosophers, pursuing their investigations for the most part *a priori*, without benefit of experimental interventions in society. Since experimental manipulations of society just to investigate the ethical value or disvalue of some behaviour would themselves almost certainly be unethical, this is hardly surprising. Nevertheless, there is an important, and ancient, philosophical tradition which would support at least the relevance of such experimentation, namely utilitarianism. The absolutist ethics of most religions dictate that the ethical virtue, if any, of an act inheres in the act itself, and so ethical experimentation would not only be unethical it would also be pointless: the experimental outcomes come after the act in question, and so are irrelevant. Utilitarianism, by contrast, asserts that the moral worth of an action is measured by the utility of its outcomes: an act is good if it leads to states of high positive utility and bad if it leads to states of high negative utility. Adopting such a view, we can pursue an experimental program to examine the ethical value or disvalue of such actions as euthanasia, abortion, rape and racial discrimination. Of course, experimentation in human society would be unlikely to be tolerated; however, the recent emergence of artificial life (ALife) simulation methods allow us to pursue such a program ethically.[1]

[1] Unless, of course, one is inclined to take *very* seriously Chris Langton's claim that ALife forms are truly alive. But we think such claims are (so far) unjustified.

J. Kelemen and P. Sosík (Eds.): ECAL 2001, LNAI 2159, pp. 120–132, 2001.

The aim of our project, then, is to explore the consequences of actions construed as ethical or unethical under various circumstances in an ALife simulation environment. Since our ALife populations *evolve,* our project is equally aimed at exploring the evolvability of ethical and unethical actions, and also at whether various social actions (e.g., altruistic actions) present initially are lost due to selection pressure or contrariwise maintained in a stable state at some level above the random mutation rate — i.e., are evolutionarily stable strategies (ESS).

In this paper we describe both our simulation environment and our first trial experiment, the evolutionary and ethical characteristics of suicide fostering the survival of others, i.e., altruistic suicide. This experiment illustrates suicide not only as altruistic, but also as an evolutionarily stable strategy: the population retains altruistic suicide at a rate much larger than can be explained by the (re-) introduction of behaviours by random mutation.

2 Background

The ideas behind this research project stem in part from the pioneering work of Axelrod [1] on the evolution of behaviour in the iterated prisoner's dilemma. Since then, a large variety of "social simulation" studies have been pursued. Researchers are using ALife to investigate, for example, social communication networks [3], economics [4,6] and disease transmission [3]. The extension of social simulation techniques to the study of specifically ethical questions seems a very natural one. Such an extension does require the introduction of some way of measuring the ethical value of the actions under study, in order to interpret the results. The more orthodox, absolutist ethical systems are not very promising for this purpose; it is at best unclear how to rank the ethical value of actions all of which violate some qualitative principles of behaviour. Fortunately, utilitarianism offers a direct and computationally clear means of measuring the ethical value of actions. Utilitarianism is an important theory in the western philosophical tradition, commonly associated with Bentham [2], but originating at least with Plato's *Republic.* Such a use of utilitarianism is in principle a direct extension of the approach to artificial intelligence that views its subjects as decision-theoretic agents, as advocated, for example, by [9].

Given that agents have utilities — numeric values assigned to the situations which the agents are in (the experiences they have) which reflect their preferences[2] — utilitarianism asserts (in one of its forms) that the ethical action in any situation is the one that maximizes the total utility over all the agents in a population. The value of an action a for an individual following this approach is

$$v(a) = \sum_i u_i(o) \quad (1)$$

[2] See [10] for the very general conditions in which numeric utility functions may be constructed from preference orderings.

where $v(a)$ is the moral value of action a and $u_i(o)$ is the utility of outcome state o (resulting from action a) for individual i.[3]

Since the moral worth of an action can only be judged properly according to the known and knowable circumstances surrounding the act (i.e., the agent cannot be expected to be prescient of the outcomes), the *actual* outcomes cannot truly be used to assess moral worth when deciding what action to take. Instead, the outcomes reasonably anticipated must serve; in other words, the moral worth of an action is more properly asserted to be the sum of its *expected* utilities over all agents in the environment:

$$v(a) = \sum_j \sum_i u_i(o_j)p(o_j|a) \tag{2}$$

where o_j is an outcome state and $p(o_j|a)$ is its probability given the action a.

A standard approximation, sanctioned by the Law of Large Numbers, is the substitution of long-term frequencies for probabilities. In an evolutionary setting it is fair to say that the repeated exposure of a type of action to various kinds of environments will produce a large number of tests of the properties of the action. In consequence, the outcomes of the action will occur with a certain frequency, translatable to probabilities that can be substituted into Equation 2[4] — hence the tendency of that type of action to induce overall positive or negative utilities may be assessed relative to particular environments. By examining the tendency of an action to produce either positive or negative average utilities under various circumstances, an evolutionary experimental ethics may find answers to moral questions which no amount of purely analytic ratiocination may yield.

3 Evolutionary Ethics

"Evolutionary ethics" as a philosophical thesis avers that ethical values are to be found in what promotes or impedes survival and reproduction. Here we introduce quite a different meaning: evolutionary ethics as the empirical exploration of utilitarian value in an evolving population. In any particular environment, some actions will be evolvable without being ethical (e.g., rape in many environments), and others will be ethical without being evolvable (e.g., mass altruistic suicide of a kin group). Thus, we can imagine a 3-dimensional lattice for classifying behaviours according to their evolvability and ethical value relative to the environment. Table 1 shows a slice of this lattice, with ethical value on one axis and evolvability on the other, the slice capturing a hypothetical "World X". For example, according to this table, World X is one where altruistic suicide is both ethical and evolvable, whereas rape could evolve but would be unethical.

[3] The outcomes are, of course, future outcomes dependent upon the action selected. Therefore, it would be appropriate to take into account the time discount factor for valuing future events. For simplicity, we ignore that issue here.

[4] Though this equation is never explicitly calculated in our simulations, it is implicit in our ethical calculations.

Table 1. Actions mapped by ethical value and evolvability in some possible world X

World X	ETHICAL	UNETHICAL
EVOLVABLE	Altruistic Suicide	Rape
UNEVOLVABLE	Mass Suicide	(Widespread) Incest

By a behaviour being evolvable we mean that the behaviour is an ESS — an *evolutionarily stable strategy*, as defined by John Maynard Smith: "An ESS is a strategy such that, if all the members of a population adopt it, then no mutant strategy could invade the population under the influence of natural selection" [8]. A prototype of an ESS is TIT-FOR-TAT in the iterated prisoner's dilemma, which resists invasion by either the SUCKER or DEFECTOR strategies [1]. In our experiments, strategies are always mixed (i.e., probabilistic); therefore, the criterion for an ESS we apply is to test whether the behaviour resists disappearing in the presence of alternative behaviours. In particular, we consider here whether the probability of altruistic suicide drops to the background mutation rate (below which it cannot reliably fall) when the option of not committing suicide is available (as it always is).

For many behaviours it is possible to find environments (worlds) where they are variously ethical or unethical, and others again for which they are variously evolvable or unevolvable. So what is of interest is not simply inventing worlds where behaviours often thought to be unethical turn out to be ethical; rather, it is investigating the boundaries separating those environments in which a behaviour is ethical vs. unethical, and again evolvable vs. unevolvable, and determining what factors explain the differences, that is of real interest. If the demarcating factors correspond to factors in our world, there is a possibility that evolutionary ethical simulations will shed light on issues of importance to us.

4 The Ethics Simulation Environment

Our ethics simulation environment shares with genetic algorithms the concepts of a population composed of agents who have genotypes which guide their behaviour [7]. Genetic operators during reproduction mix the genes of reproducers (crossover) and modify the genetic determinants of behaviour randomly in mutations. As is standard in ALife (but not genetic algorithms), the environment in which the agents act include not only externally determined features (such as food), but also the agents themselves and their behaviour. Agents may eat, attempt to mate, move around their two-dimensional world, and die from poor health, old age, or suicide.

There are two units of time in the simulation, one for the purposes of simulation and the other for statistics collection. Every agent gets a chance to perform one action during a *cycle*; the order in which agents perform actions is randomly selected for each cycle. The second time unit is the *epoch*, at the end of which assorted statistics are collected; the epoch extends over a fixed number of cycles, specifically five in this study.

The Board. The board consists of a two-dimensional 6x6 grid of cells, each capable of holding an unlimited number of items, such as food or agents. Attempts to walk past a board boundary will result in a failed action.

Food. Agents can eat food that is distributed uniformly around the board. The food contains nutrition that the agents can absorb to increase their health, which enables them to survive and reproduce. Food is generated each cycle and has a short lifetime relative to agents. In this study, the food is distributed across cycles according to either a sinusoidal or constant distribution function. Sinusoidal distribution provides for periods of high and low availability of food, mimicking seasonal fluctuations or, perhaps, periodic droughts.

Agents. Each agent has a chromosome (which determines its behavioural tendencies), a location and orientation, an age, a level of health and a randomly selected age of "natural" death (assuming it does not die of "non-natural" causes, such as lack of food, in the meantime). The chromosomes have "heads" and "bodies". The head contains general information for the agent, including mutation rates for other values in the chromosome. During mating, these mutation rates are themselves modified in offspring by global meta-mutation rates. For the initial cycle, 1000 agents are generated with random properties (including chromosomes), except for health. This is set high enough for all agents to immediately commence mating.

The chromosome body is made up of a list of production rules (condition-action pairs). The number varies from individual to individual. The rules are examined during each simulation cycle, with the first rule having a condition that matches the agent's environment observation used to select the agent's action. The action part of a rule consists of a probability vector across action types; i.e., the action taken is determined probabilistically, in a way that depends upon the environmental conditions. If no rule is matched during a cycle, a default action probability vector (in the chromosome's head) is used.

Rules are conditioned on the agent's observations of the environment. The conditions are patterns that are matched against the environment, allowing for values that are greater than, less than or equal to some value specified in the condition, or possibly accepting any environmental value for a particular measurement. If all conditions in a rule are met, then its action probability vector is used to select the agent's action. The actions currently available to agents are: moving forward, turning, resting, eating, mating and suiciding. An agent will rest for one cycle when its attempt to perform another action fails, such as attempting to eat when there is no food available.

Health and Utility. An agent's initial health is given to it by its parents; that is, each parent donates a portion of its own health to give its offspring a start in life, currently 50 health units from each parent. Subsequently, an agent's health depends upon events. Eating nutritious food results in a direct positive health effect; movement costs a unit of health. Each action also has a utility associated with it, given the circumstances. Thus, successful mating has a positive utility, as does eating nutritious food. Utilities flow on to health, and so are simultaneously indirect health determiners. Utility and health units are kept on the same scale, so that one negative utile (utility unit) decrements health one unit.

Observations. An agent makes four observations of its environment and itself. The health and age of the agent make up the internal observations; the external observations consist of the number of agents on the board and the amount of food available.

Moving and Eating. An agent can rotate, move to a neighbouring cell, or consume food in its own or a neighbouring cell.

Reproduction. Since parents must donate 50 units of health to their children, there are two preconditions to successful mating: (1) there must be a mate available in the current or a neighbouring cell; and (2) both potential parents must have sufficient health available. This required health level is actually set above 100, at 200 units, implying some "maturation" period for infants to reach reproductive health and not allowing weak individuals to reproduce. If mating is successful, then an offspring is produced whose chromosome is a combination of its parents' chromosomes.

The offspring's chromosome head is taken randomly from one parent, with the mutation rates in the head mutated according to the meta-mutation rates. The chromosomal body is a uniform crossover of the parent bodies, subject to mutation. Since the parents' body lengths can be different, this uniform crossover is somewhat different from the usual genetic algorithm crossover. A locus, as if for a single point crossover, is uniformly randomly selected from each parent chromosome. The offspring chromosome contains as many rules as does the first parent from its first rule to its locus plus as many rules as the second parent from its locus plus one to the end of its chromosome. This allows the complexity (chromosomal length) of offspring to itself evolve. The actual rules composing the child chromosome are taken with equal probability from either parent at the corresponding site, unless one parent's chromosome is too short, in which case the remaining rules are taken from the longer parent.

Suicide. Suicide causes an agent's health to drop to 0.

Assigned Values for Utilities and Health Effects. The actions (for this simulation) have at most two outcomes, resulting from success or failure — both having an associated utility and health effect. The assigned values are illustrated in Table 2 (with health effects in parentheses). In addition to these is a health effect of -2 for living one cycle.

The values in the table have been chosen to correspond with values that might be expected in analogous natural systems. What is important is not the exact

Table 2. The utilities and health effects (in parentheses) associated with the outcomes of actions. A dash indicates that the outcome for the action in that row is not possible

	SUCCESS	FAILURE
WALK	5 (-10)	0 (0)
TURN	1 (-2)	-
EAT	10 (70)	0 (0)
MATE	25 (-50)	0 (0)
SUICIDE	0 (-all health)	-

numbers, but rather the approximate ratios. The assumption underlying the assignment of utilities is of the moderately content agent (note that all utilities are positive). Therefore suicide would always result in lower accumulated utility for the life of the agent (provided that not every action attempted fails).

The ethical investigations of the simulation will obviously be dependent upon the utilities chosen here. Certainly, one could choose any combination of utilities that suits. However, we believe that the utilities are sufficiently simple and provide for scenarios which are of genuine interest. Also note that we have run simulations with randomly differing but similar utilities that have produced equivalent results, so the results do not rely on having carefully tweaked utility values.

5 Experimental Results

The behaviour we have chosen to investigate in the first instance is altruistic suicide. In these ethical simulations, an altruistic act may be considered to be any act which reduces the expected fitness of the individual, but is beneficial to the expected fitness of the remaining population. An alternative interpretation, justified on utilitarian grounds, is that an altruistic act is one which reduces one's own expected future utility while enhancing the global expected future utility of the remaining population. Here we employ the latter interpretation. We assess the altruism of an action by running the simulation multiple times, recording the population's utility with and without the action being available. In this experiment, the expected future utility is (as stated) always non-negative. Hence, suicide is (almost) always detrimental to the agent. It is altruistic, therefore, whenever it enhances the utility of the population remaining.

5.1 The Evolutionary Stability of Suicide

Figure 1a shows the proportion of suicides to all actions performed each epoch over an average of 39 simulations when a constant amount of food is generated per cycle (dashed line) and again when food is generated over time according to a sine wave with an extreme amplitude (solid line) over an average of 44 simulations. Figure 1b shows the average population sizes in both of these simulation types. In all simulations the proportion of suicides performed by the agents is

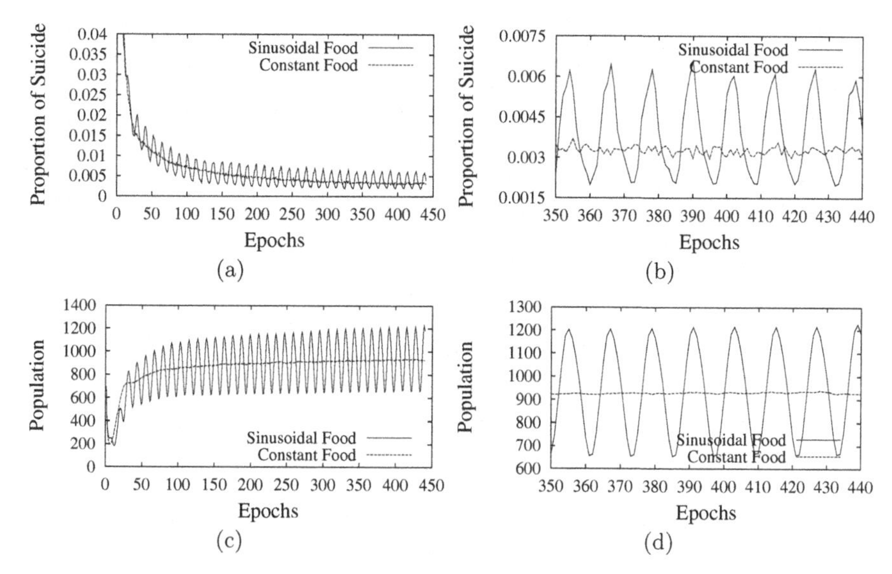

Fig. 1. (a) Proportion of suicides to all actions in simulations with sinusoidal (solid) and constant (dashed) food; (b) an enlargement of the last 90 epochs in (a); (c) Average populations for sinusoidal (solid) and constant (dashed) food simulations; (d) an enlargement of the last 90 epochs in (c)

very high in the beginning, because the initial agents are generated randomly; suicide begins as approximately equiprobable amongst all of the actions available. In both simulations, these proportions both drop off sharply to begin with, as would be expected. Soon, however, the suicides in the sinusoidal simulations become clearly cyclical, with peaks in suicide corresponding with peaks in population (visible in Figures 1(c) and (d)). Recalling that the graph shows suicide as a proportion of all actions (and hence filters the effects of population size and number of actions performed), the presence of this cyclicality indicates that the agents have evolved this behaviour in response to the seasonal nature of the environment. Contrasting this to the constant food simulation, we find no cyclicality. We confirmed this by performing a significance test on the difference between the peaks and troughs of the suicides of the sinusoidal simulations ($t(22) = 18.5$, $p < 0.01$).[5]

The graphs in Fig. 1 demonstrate that suicide, given cyclical food distribution, is evolutionarily stable. It is clearly above the background mutation rate. We can establish this even though we did not measure the background mutation rate directly (since the mutations of the action probabilities are governed by

[5] In a variation of the suicide experiment, we modelled yeast-like suicide, whereby yeast cells suicide and are cannibalised by other yeast cells. With this modification, the frequency of suicide was higher; in particular, troughs were higher than in the cyclical suicide simulation.

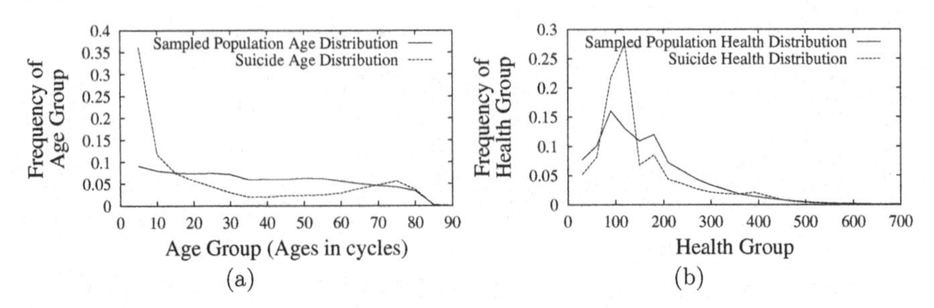

Fig. 2. (a) The distribution of ages and (b) the distribution of health amongst a sample of 2000 agents in the general population and amongst 2000 of those that suicide over approximately the last 100 epochs

multiple mutation rates), by comparison of the suicide rate with and without cyclical food distribution (which itself cannot be *less* than the background mutation rate, though it may be higher). Note also that the background mutation rate is not cyclical, and so can not be the cause of the observed cycles.

5.2 Explaining Suicide

The evolution of altruistic behaviour has long been of interest to biologists and sociobiologists (cf. Wilson's *Sociobiology* 1975). One accepted explanation of the possibility of such evolution is that it occurs via kin selection: the altruistic act reduces the probability of the actor reproducing, but enhances the probability of its kin reproducing, thus tending to preserve or increase the presence of the gene(s) supporting the altruistic action [5]. Actions studied by biologists have been either less dramatic than suicide (e.g., donating food) or have been directed at close relatives (e.g., sacrificing one's life for one's children). Nevertheless, kin selection seems to be the only possible explanation of our demonstration, under appropriate conditions, of suicide being an ESS.

Who is suiciding? The graphs in Fig. 2 contrast the distributions of age and health of the general population with those of the agents that perform suicide. There is a marked difference between these, with age presenting the greatest contrast. Somewhat surprisingly, younger agents are performing the bulk of suicides; more intuitive is the smaller peak in suicides amongst old agents. We expected that agents with lower health would have occasion to suicide altruistically, and Fig. 2b supports that. We speculated, therefore, that youth suicide in the simulation is a side-effect of the correlation between youth and low health.

The benefits of suicide when the agent is either elderly or low in health are quite evident. By performing the sacrifice at a late age, the consequences to an agents fitness will not be great, since they will have few reproductive opportunities remaining. An elderly agent may benefit its own genetic endowment (through its children and other kin) more by withdrawing from the environment and releasing resources than by attempting to reproduce. Such freed resources

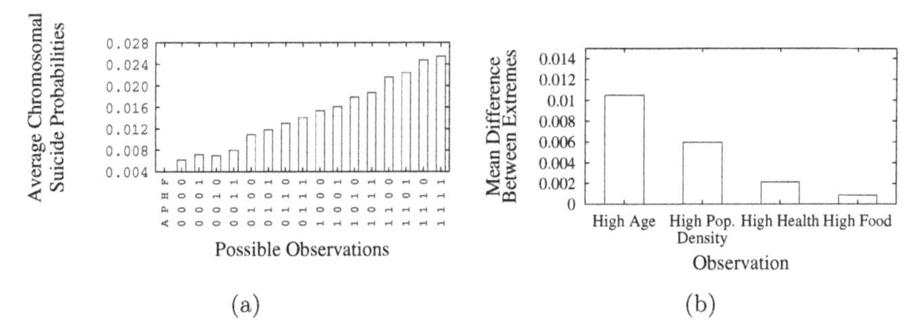

Fig. 3. Suicide with sinusoidal food distribution. (a) Average chromosomal probabilities of suicide for the last 50 epochs. On the horizontal axis, A=Age, P=Population, H=Health, F=Food; 0 indicates a low observation (the low observations are Health: 0 health units; Age: 0 cycles; Population: 0 agents; Food: 0 pieces of food), while 1 indicates a high observation (the high observations are Health: 600 health units; Age: 80 cycles; Population: 1600 agents; Food: 300 pieces of food). (b) Mean differences of suicide probabilities given observation extremes (e.g., high vs. low population)

include food and also mating opportunities. By disengaging in competition for these resources, nearby relatives will gain some advantage in reproduction, which in turn will promote genes that favour such altruistic behaviour.

The case for suicide amongst agents with low health is similar. Such an agent would not be able to mate prior to rebuilding its health. If resources are scarce, or about to become scarce, altruistic suicide favouring nearby kin would again be selected for. Of course, the negative effects of suicide would be at a minimum when an agent has both low health and old age.

Under what conditions is suicide more likely? Figure 3a shows how suicide probabilities have evolved in the chromosomes of agents given a cyclical food pattern — that is, the genetic basis of suicide in the simulation. All the possible extreme environments were examined to find the probabilities for suicide the agents have in such conditions, making sixteen different environments. These environments are listed along the horizontal axis, while the vertical axis indicates the normalized probabilities of suicide under those conditions.

There is a clear correlation between age and suicide. This supports our view that young age suicides are prominent because of the strong correlation between youth and low health in our simulation — that is, the chromosomal probabilities show that it is not young age as such that is causing suicides. Also, high populations lead to greater suicide rates, as indicated in Fig. 3b.

5.3 Kin Selection

One would expect the genetic relatedness between agents to be correlated to the distance between them. We measure the genetic diversity (the inverse of genetic

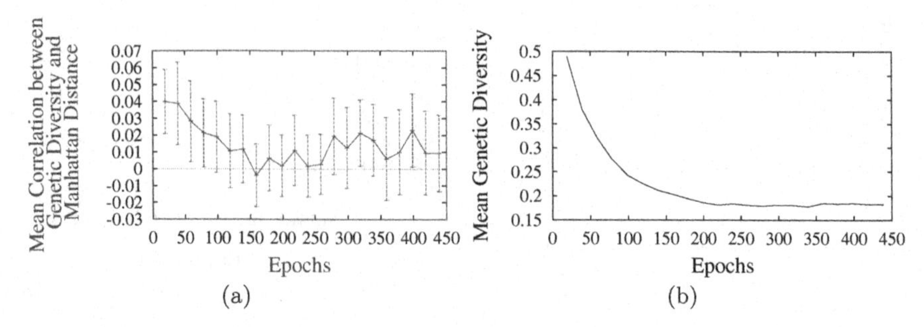

Fig. 4. (a) The mean correlation between genetic diversity and Manhattan distance over time, with 99% confidence intervals. (b) Mean genetic diversity over time. 30 pairs of agents are sampled per epoch over periods of 20 epochs

relatedness) for two agents a and b stochastically as:

$$\mathrm{D}(a,b) = \sqrt{\sum_{i=0}^{n} \frac{||apv(a, obs_i) - apv(b, obs_i)||}{n}} \tag{3}$$

where n is the number of environments (observation vectors) randomly sampled, obs_i is that vector of observations, apv is a function from an agent and an observation vector to the agent's action probability vector.[6] The difference between action vectors for the two agents is calculated for n observation vectors, and the root-mean-square of these differences gives the final value. This is a measure of the average difference between the probability vectors of the two agents under various environments; the environments chosen are weighted according to the environments actually encountered by the agents during the simulation.

We investigated whether genetic diversity is related to the Manhattan distance between agents. The Manhattan distance is the sum of the magnitudes of the x- and y-translations needed to move from one agent to the other.[7] We found that genetic diversity is low at zero distances (i.e., in the agent's own square), but greater at other distances, allowing kin selection to operate. Although there is only a small linear correlation of 0.039 between distance and genetic diversity (measured during the early epochs 0 – 20, when the population genetics is undergoing the greatest change), this was statistically significant ($p<0.001$).

Figure 4a shows the correlations between genetic diversity and the Manhattan distance (averaged over 44 simulations). The graph shows the correlation diminishing from an initial value of 0.039 to zero, after about 150 epochs; after that they appear to drift randomly. This is what we would expect given an initial period of rapid evolution, in which conditional suiciding behaviour is taken up by the wider population due to selection pressure, followed by an equilibrium

[6] The double bars represent the magnitude of a vector. Hence, in the equation, we are taking the magnitude of the difference vector.

[7] We used this for simplicity, despite the fact that it is not the optimal distance measure, since the agents may move diagonally.

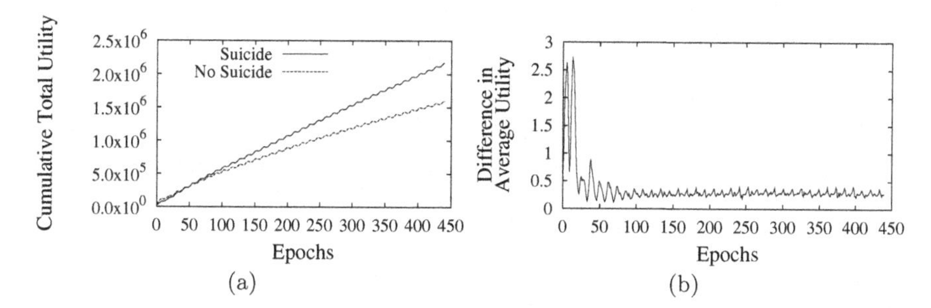

Fig. 5. Comparison of (a) cumulative total utilities and (b) difference in average utility (suicide minus no suicide)

state in which the population is homogeneous. A homogenous population implies less variation in the values of genetic diversity (since they are uniformly low), in turn implying lower correlation values. Figure 4b shows, in support of this theory, that genetic diversity itself declines rapidly in the early stages of the simulation, with genetic homogeneity achieved around epoch 150.

5.4 The Ethical Value of Suicide

To judge the ethical value of suicide we can compare the total cumulative utilities of simulations which are identical except in allowing or disallowing suicide. Figure 5a is the outcome for cumulative total utility given sinusoidal food supply.[8] (The ripples show the deleterious effect of drought on the utilities of the agents.) Whereas the total utility in the simulations with suicide is initially lower, the situation soon reverses, with the cumulative total utilities of the two simulation types subsequently diverging. Figure 5b shows that the average utility with suicide is greater at every epoch, hence the lower total utility in the early epochs of the suiciding simulations is entirely due to smaller population size. Despite smaller early populations, it does not take long for the ethical virtue of assisting others to survive droughts to assert itself in higher cumulative utility. It is worth noting what is sometimes forgotten in discussions of utilitarian ethics, that the proper assessment of the ethical value of the action requires us to consider not just the immediate but also the future utilities of agents: the cumulative graphs of both average and total utilities settle into a roughly linear form only after about 80 epochs. After about 150 epochs, total utility tells the ethical story: in this world (that is, under the circumstances of this simulation), altruistic suicide is quite often the ethical option.

[8] The averages of 44 simulations with suicide and of 31 simulations without suicide were used.

6 Conclusion

We have demonstrated that suicide can be both an evolutionarily stable strategy and altruistic. More generally and importantly, we have illustrated the use of ALife simulations for testing applied ethical propositions of general interest. We believe this creates an new opportunity to discover empirical facts about important ethical problems computationally.

References

[1] R. Axelrod. *The Evolution of Cooperation.* Basic Books, New York, 1984.
[2] J. Bentham. *Introduction to the Principles of Morals and Legislation.* 1789.
[3] J.M. Epstein and R. Axtell. *Growing Artificial Societies.* MIT Press, 1996.
[4] N. Gilbert and R. Conte. *Artificial Societies: The Computer Simulation of Social Life.* London: UCL, 1995.
[5] W.D. Hamilton. The genetical evolution of social behavior. *Journal of Theoretical Biology,* 7,Papers I&II:1–16 & 17–52, 1964.
[6] J. Holland and J. Miller. Artificial adaptive agents in economic theory. *American Economic Review,* 81:365–370, 1991.
[7] John Holland. *Adaptation in Natural and Artificial Systems.* University of Michigan Press, 1975.
[8] J. Maynard Smith. *Evolution and the Theory of Games.* Cambridge University Press, 1982.
[9] Stuart Russell and Peter Norvig. *Artificial intelligence: A Modern Approach.* Prentice-Hall, 1995.
[10] J. von Neumann and O.Morgenstern. *Theory of Games and Economic Behavior.* Princeton University Press, 1947.
[11] E.O. Wilson. *Sociobiology: The New Synthesis.* Harvard, 1975.

Eden: An Evolutionary Sonic Ecosystem

Jon McCormack

School of Computer Science and Software Engineering
Monash University, Clayton Campus Victoria 3800, Australia
jonmc@csse.monash.edu.au

Abstract. This paper describes an Artificial Life system for music composition. An evolving ecology of sonic entities populate a virtual world and compete for limited resources. Part of their genetic representation permits the creatures to make and listen to sounds. Complex musical and sonic relationships can develop as the creatures use sound to aid in their survival and mating prospects.

1 Introduction

«Man is nature creatively looking back on itself»
Friedrich Von Schlegel [1]

Music, like all complex creative endeavors, has drawn from a vast range of human experiences in search of expression. A great source of this expression is often nature itself. Over all other art forms, music seems open to 'the purest expression of order and proportion, unencumbered as it is by material media' [2], so it therefor seems natural for composers to look to artificial life and artificial nature as a source of creative inspiration.

In many ways, artificial life adopts a process-based methodology, shifting the emphasis from material to mechanisms. Formal process mechanisms have existed in music for some time and had a profound impact in music of the twentieth century [3].

This paper describes a novel music composition system that draws from artificial life techniques in its methodology.

1.1 Simulation and Composition

In discussing computer music in this paper, it is important to differentiate between music *simulation* and music *composition*. With simulation, the primary goal is for the computer to simulate an existing composer, genre, or playing style. For example, Johnson-Laird developed a system to improvise performance similar to that of jazz musicians [4]. Ebcioglu devised an expert system to generate chorales in the style of J.S. Bach [5]. As a rather crude generalization, music simulation is a decomposition problem; original compositional techniques, unique to the computer, fair better when they draw from generative or evolutionary methodologies (like Artificial Life).

Of course, it is possible to use evolutionary techniques to evolve works that mimic a particular style, and this has been the focus of much research into evolutionary composition. In their survey paper of evolutionary music composition systems [6], Todd and Werner make the important observation that in evolutionary systems it is necessary to determine which individuals are more 'fit' than others (in the case of a

J. Kelemen and P. Sosík (Eds.): ECAL 2001, LNAI 2159, pp. 133-142, 2001.

compositional system, which individuals are better 'composers') and hence should survive and have offspring. Todd and Werner see the fitness evaluation as an integral part of the evolutionary system – to be performed by a critic of some sort, be it human, rule-based, learning-based or even co-evolved. It could be argued that this interpretation while legitimate from an evolutionary computing perspective, limits the potential of computer-based composition. This is because it assumes that (i) critics know at every intermediate step that one approach will be musically 'better' than another, and (ii) that the creative process can be adequately expressed using some formalized system [7]. That is, it assumes that musical criticism is understanding rather than interpretation. It is suggested that the role of the non-human 'critic', requires considerable domain-specific knowledge, and that such knowledge can be difficult to quantify and encode [8], [9].

1.2 Evolutionary Systems

In natural selection, organisms adapt to their environment. It is possible to view a creature's morphology as the result of adaptation to its particular environment (physical, temporal and social). For the system described in this paper, the 'critic' in the terminology of Todd and Werner is the environment itself. The only domain-specific knowledge is a simplified physical system, which (superficially) mimics the dynamics of sound in the real world. The human 'composer' acts in the mode of meta-creator, designing *environments* and observing the results (visually and sonically) as creatures within the world adapt to that environment.

In the natural world, evolution is not only about survival, however. In sexual species, sexual selection plays an important role [10]. In particular, mating calls represent one of the earliest forms of communication [11] and have roots in the origins of music itself. The artificial evolution of mating calls has been studied by Werner and Dyer [12]. Many famous composers have drawn from birdsong and mating calls as compositional material. Janequin, Beethoven, Messiaen, and Rautavaara for example, have all made extensive use of mating calls in their compositions.

1.3 Related Work

Much of the system described in this paper draws its inspiration and methodology from John Holland's *Echo* [13], particularly in the use of rule-based methods for the internal decision-making system of creatures. Many others have used evolutionary systems as a basis for composition, but in the main for compositional simulation [9], [6], rather than as a new form of creative tool for the composer.

A more closely related system to the one described here would be that of Dahlstedt and Nordahl [14]. Their *Living Melodies* system uses a genetic programming framework to evolve an ecosystem of musical creatures that communicate using sound. *Living Melodies* differs from *Eden* in that it assumes all creatures have an innate 'listening pleasure' that encourages them to make noise to increase their survival prospects. *Eden* contains no such inducement, beyond the fact that some sonic communication strategies that creatures discover should offer a survival or mating advantage. Hence, only some instances of evolution in *Eden* result in the use of sonic communication, whereas in *Living Melodies*, *every* instance evolves sonic communication.

2 The *Eden* System

Like many Alife worlds [15], [16], [17], [14], the *Eden* world operates over a two-dimensional rectangular lattice of *cells* that develop globally at discrete time steps. Each cell can contain multiple elements, the principle types being *rock*, *bio-mass* (food) and *evolving creatures* (who are appetizing and carnivorous). Rocks and bio-mass do not undergo evolution, however the bio-mass operates under a feedback control model similar to that of Lovelock's *Daisyworld* [18]. The intensity of radiant energy falling on the bio-mass is seasonally adjusted, giving rise to cycles of growth and decay.

Rocks are placed in the world according to a simple diffusing model. If a rock is placed in a cell, then no further growth is possible in that cell. Rocks do not grow or change, but they do provide places of refuge allowing creatures to hide from predators. Each of the three major entities in the world has a distinctive 'color', allowing those entities with appropriate sensors to use color information to distinguish between the various types of matter in the world.

2.1 Performance System

The evolving creatures use a rule-based performance system, similar to that described by Holland [13]. The overall structure of this system is shown in Figure 1. Creatures have a series of environmental sensors that detect the physical qualities of the surrounding environment. This sensory information is passed as *messages* to a rule-based system that performs internal processing on a list of current active messages at each time step. As a result of this processing, the performance system may decide to output a message indicating that an action should be taken.

A creature's sensors and actions are drawn from a finite set of possibilities and do not evolve. Evolution of the performance system (how the sensory information is turned into actions) does undergo an evolutionary process.

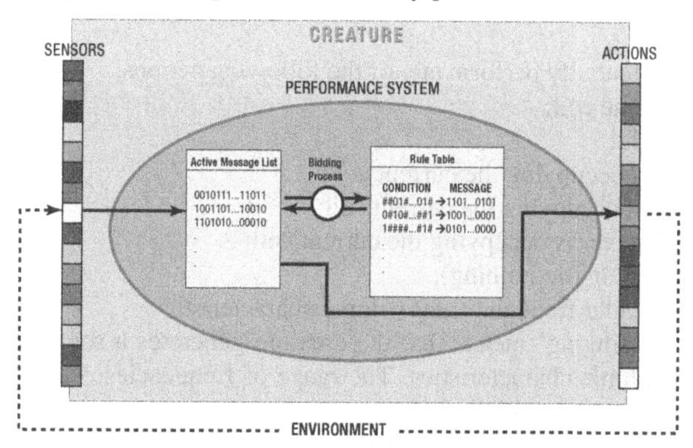

Fig. 1. The principle components of a creature: a series of sensors that detect environmental and local stimuli; a rule-based performance system that evolves; and a set of actions that the creature can perform in the world

2.2 Sensors

A creature may 'sense' the following information from the environment:

- *Color* of the contents of the cell the creature is currently occupying, and the color of the cells in the front, left and right directions of the creature;
- *Nutritional value* of other entities in the current cell (creatures are carnivorous and may kill and eat each other). Rocks have no nutritional value;
- *Sound* information – details of the sound arriving at the current cell;
- *Pain* level – an introspection as to damage the creature is suffering. Creatures will 'feel' pain if for example they are being hit by another creature or are very hungry;
- *Energy* level – an overall measure of how healthy the creature is.

The primary goal of the system is for the creatures to develop interesting sonic behaviour, hence a great deal of bandwidth in the sensors is devoted to sensing sound. Sound sensors detect sound pressure levels across a range of frequency bands, giving the creature the potential ability to distinguish many different types of sound. This ability to 'hear' is complemented by the range of sounds a creature can potentially make, which is detailed in section 2.3.

A physical model determines how the sound is propagated through the environment [19]. Sound arriving at a cell from multiple sources takes into account distance, frequency, and intermediate obstructions. No real attempt is made to simulate the psychoacoustic properties of sound beyond the exponential relationship between pressure levels and perception of intensity. The perceptual mechanism for loudness behaves in an exponential way, as it does for humans,

$$L = 20 \times \log_{10}(P/P_0) \tag{1}$$

Where L is the sound pressure level in decibels (dB), P_0 a reference pressure corresponding roughly to the threshold of hearing in humans [20, page 1055]. Psychoacoustic properties are perceptual, and additional properties would be difficult to integrate into the system at its current level of physical modeling.

2.3 Actions

A creature can potentially perform any of the following actions:

- *Move* forward one cell;
- *Turn* left or right;
- *Eat* whatever is occupying the current cell;
- *Hit* what ever is occupying the current cell;
- *Mate* with whatever is occupying the current cell;
- *Rest* for a time step (do nothing);
- *Sing* with particular frequency and volume characteristics.

The action of 'singing' means that the creature generates a sound with particular frequency and volume characteristics. The range of frequencies possible mirrors that which the creature can potentially hear.

The intent to perform an action does not necessarily mean it can actually be carried out. Action *messages* are sent to the environment, where they are tested for physical possibility. For example, it is not possible to move into a rock. Other actions may be capable of being performed, but of course may be of little use, or even detrimental to the creature's health (e.g. trying to eat when there is noting to eat on the cell).

The environment of *Eden* enforces a simple physical model on the world and its inhabitants. All actions carry an energy penalty; the amount of this penalty determined by the physical effort needed to perform the action. A creature dies if its energy level reaches zero. A creature's internal representation keeps track of its current physical properties such as mass, velocity, color and energy level. Many of these properties may change over its lifetime. For example, as a creature eats bio-mass, its mass increases and physical actions like moving around cost more.

2.4 Rules

Messages passed from the sensor system are processed by an internal rule-based, message passing system, similar to that described by Holland [13]. Since this system is described in detail by Holland, the description here will be minimal, highlighting the differences between Holland's system and that of *Eden*.

Messages are stored in an *active message list*, in order of their arrival into the system. Messages are 32 bit binary strings, in the case of messages that come from the environment the string represents sensory information. A database of *rules* (called a *rule-table*) is maintained as part of each creature's internal representation.

Each rule consists of two components, a *condition* string and an *output message* string. Condition strings are composed from an alphabet of three possible symbols: *0*, *1*, or *#*. To see if a rule should be applied, the condition string undergoes a bitwise check with a message string from the active message table. For each bit, a *0* or *1* in the condition string matches the same symbol in the message string in the same bit position. A # in the condition string matches either a *0* or a *1* in the same bit position. The successful match of a message from the active message table with a rule's condition string results in the rule's output message being placed in the active message table[1].

2.4.1 Credit Assignment

Rules incorporate a *credit assignment* system, whereby each rule is assigned a credit value, representing the 'usefulness' of the rule to the organism. New rules begin with a default credit value, and must *bid* to be used. When more than one rule matches a given input string, the rule with the highest bid wins. In the case of equal highest bids, the winning rule is chosen at random from amongst the highest bidders. Bidding is proportional to a rule's credit value and it's *specificity* (the less # symbols in the condition string the more specific it is – a condition string comprised of only # symbols will match every input string, but always bid 0).

If a rule is the successful bidder on a message from the environment (a sensor message), it pays its bid to the environment. If it is bidding on a message generated from another rule it pays the rule that generated the message. Only the winning rule must pay out its bid – the losing bidders do not loose any credit. Again, the reader is referred to [13] for details.

---

[1] Subject to successful bidding, detailed in the next section.

2.4.2 Credit Payoffs

Each creature's health and energy levels are monitored via a *health index*. If the creature is finding food and not being attacked for instance, the health index will increase. If the creature is running around aimlessly or being attacked, its health index will decrease.

Let $H_t(\lambda)$ represent the health index of a particular creature, λ at the current timestep *t*, and $H_{t_i}(\lambda)$ be the health index at some previous time t_i (where $t - t_i > 0$). For each creature, the cumulative differential of the health index is monitored and when it its magnitude exceeds some constant, α_λ, a *credit payoff* is performed on the active rules (i.e. when the inequality $|H_{t_i}(\lambda) - H_t(\lambda)| \geq \alpha_\lambda$ is satisfied).

For a given credit payoff C_i, all the rules that were successful bidders since the last credit payoff C_{i-1} are kept in a list. In addition, the total credit paid out to the environment since the last credit payoff, E_i, is kept. All active rules are paid out proportionally according to the formula,

$$P_i = \frac{kE_i}{f_{R_i}} C_i \tag{2}$$

Where f_R is the frequency of the particular rule *R* in the list, P_i is the credit value added to the rule's current credit value and *k* is a constant. C_i may be positive (meaning an increase in health) or negative (a decrease in health). Recall that C_i is a differential value, representing the rate of change in health. The number of time steps between successive payoffs will be dependent on how quickly or slowly the creature's health is changing. For example, if a creature is being attacked and losing health quickly, payoffs will be more frequent. The rules involved in letting the creature get hit will also decrease in credit quickly (hopefully soon being outbid by other rules that may prove more successful, if the creature is to survive).

Using this payoff system of equation (2), over time rules that are helpful to the creature's survival gain credit. Using the table-based approach of active rules allows rules that indirectly increase health to receive appropriate credit. For example, while the rule to 'eat when you find food' is a good one, you may need to walk around and look for food to find it first. The rules for walking and turning, although they decrease health in the short term, may result in finding food. This increases health in the longer term. If such rules are helpful in increasing health, their credit will increase.

2.5 Evolution

No artificial life system is complete without evolution[2] and *Eden* is no exception. The genetic algorithm used is based on the *Schemata* approach of Holland [21], [13]. Only the particular implementational aspects of *Eden's* evolutionary components relevant to this discussion are detailed here.

Recall from section 2.3 that mating is a possible action for a creature. The basic units of genetic exchange through mating are the creature's rule-tables. Mating will

[2] 'No artificial life systems of this ilk' may be a better generalization.

succeed only if a creature is over a certain age and is healthy enough. Successful mating, while costing energy, does not adversely affect health. For two creatures that mate, the most successful (highest credit) rules are crossed over in the hope that the resultant rules may also be successful. Additionally, rule mutation and creation are possible. The probability of these events occurring can be controlled interactively by the user at run time.

The observant reader will note that the selection of rules based on their strength during crossover represents a *Lamarckian* evolution [22], since learned behavior is passed from parents to offspring. This is done for efficiency reasons, as it results in the discovery of interesting strategies for survival more quickly than for a Darwinian approach. It is quite possible to bypass the Lamarckian components of the evolutionary system, if required.

3 Implementation

Although the primary goal of *Eden* is the evolution of sonic communication for the purposes of music composition, the program has a visual dimension as well. As explained in section 1.2, the composer has the role of 'meta-creator', and a visual feedback system was considered a reasonable way of facilitating this role. It is also an interesting way to observe, anecdotally at least, the behavior of creatures in the world.

Representation of the entities of *Eden* is done using tiling patterns, loosely based on Islamic ornamental patterns [23]. The resultant images formed by a grid of cells (as shown in figure 2), suggest a continuous mass of substance, as opposed to squares containing individual entities.

3.1 Interaction

While *Eden* is running, the user has control over a number of parameters. These include mutation and random rule generation probability during breeding, selection of coefficients that control bio-mass efficiency and distributions, and physical parameters related to the environment. Real-time statistical and environmental information is also provided. A sample screen shot is shown in Figure 2.

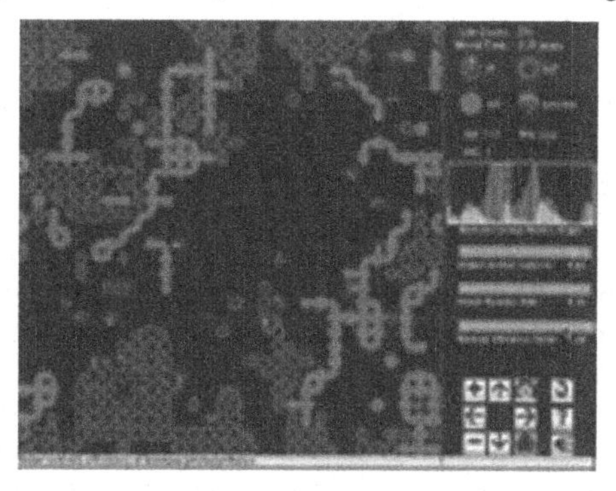

Fig. 2. A screen shot of *Eden* in operation showing the world (left) populated with rocks, bio-mass and evolving creatures. The user interface and statistical controls are on the right side of the image

3.2 Sound

Sound listening and sound generation is split into frequency bands. For simplicity, the current implementation uses three frequency bands. One could imagine the frequency bands as 'low', 'mid', and 'high'. Creatures can generate sound at four possible levels of intensity at each band, giving a total gamut of 64 distinct sounds. This selection is somewhat arbitrary, chosen as a reasonable tradeoff between storage space and sonic diversity. There is no reason why it could not be extended if required.

All 64 sounds are pre-generated, using frequencies within the (human) audio spectrum (20-20,000Hz). The base frequencies chosen to represent 'low', 'mid' and 'high' were 100, 1000 and 10,000Hz. Each sound is of a few milliseconds duration, roughly equal to a single time step. A separate software program pre-generates the sound set based around user-supplied frequencies. Experiments have been performed with different sound sets, and the generation and selection of sound sets could be considered part of the compositional process. It is important to note however, that the creatures do not 'hear' the timberal properties of the sound – they just register as bit patterns on the individual's sensors. The choice of sonic qualities of the sound set is for the benefit of the human listener, as a sonification of the process.

When the program is running, sounds are replayed in real time as the creatures sing them, thus creating the composition. A creature may sing different sounds each time step, thus permitting the generation of more sonically complex sounds that change in tonal characteristics over time. The ability to control levels at different spectra also contributes to the tonal evolution of sounds that the creatures make.

4 Results

As mentioned in section 1.3, there is no 'hardwired' impetus for a creature to actually make or listen to a sound. They will only do so if it increases their chances of survival or mating.

A number of simulations have been run in relatively small worlds (usually a grid size of 50 x 50 cells). Starting with random rules takes a long time to evolve to anything useful, so controls are provided to seed new creatures with some basic 'instinctual' rules to help them survive (e.g. '*if on top of food then eat*'). The user can choose dynamically how much or how little 'instinct' they would like the initial population of creatures to have. If the population falls below a given threshold, the world is automatically populated with a fresh set of creatures.

In many simulation runs sound does not play a major role in survival. While sound can be heard, analyses of the rules show that this is principally due to mutation (for example the 'eat' action mutates into a 'sing' action). Such rules do not survive for long, if they do not provide an increase in the health index. Singing costs energy, so it will only be used if it can provide some survival or mating advantage.

In some simulation runs however, sound does provide interesting survival advantages. For example, offspring of a certain parent used sound to signal an abundance of food, calling its siblings to share in the find. A creature's 'ears' listen in a forward facing direction, over a conical shaped area, so moving forward when hearing a food call is a reasonable strategy. A creature's (fixed) morphology dictates that it can hear at much greater distances than it can see. After many generations of using this method of signaling for food, new behaviors eventually emerged that exploited the food-signaling tendency, calling creatures over and then hitting, killing, and eating them. Why go searching for food when it will come when called...

Creatures also exploit the frequency dependent nature of their 'voices' and 'ears'; often groups (children of a common ancestor) will use only a particular frequency band when communicating. The use of different frequency bands for different tasks (food signals and mating) have also been observed.

More results that are interesting are obtained when some basic 'instinctual' rules for using sound are seeded into new creatures. For example, if the world contains creatures that cry out when being attacked, others soon learn to avoid going near those who are screaming. Behaviors also emerged whereby creatures would use signals to find mates, or to deceive in order to attack and eat another individual.

Creatures also evolve strategies for coping with seasonal change. For example, in the winter months, when food is scarce, some individuals prefer to 'hibernate' resting until food becomes more abundant and easier to find, then bulking up in the summer months in a period of frenetic activity.

Perhaps the most interesting properties that the system exhibits is the 'evolution' of sounds as the generations pass by. Simulations often begin with chaotic cacophony of sound that slowly becomes more 'pure' and sparse as the inhabitants of *Eden* evolve and discover rules that exploit the use of sonic communication. Certainly, the sonic development exhibits characteristics of dynamics and counterpoint, often considered the basis of many good compositions.

5 Conclusion

Clearly there are a number of limitations in the system described. In the physical world, organisms evolved sensory organs because they were useful for survival or mate selection. Whereas in *Eden*, the sensory organs are provided unconditionally and the internal structure of the creature evolves around them.

Secondly, the internal performance system could be improved. While the rule-based system described does permit complex sequences of actions to evolve, the discovery of such actions can be difficult (the longer the dependency of rules on each other, the more difficult they are to discover). It is planned to replace the rule-based system with a non-deterministic finite state automata (NFA) based system, in the hope that this may permit the evolution of a more complex use of communication.

Nonetheless, despite these limitations, as a compositional system, *Eden* is capable of producing compositions that have interesting qualities[3]. The goal of this work is not simulation or mimicry of existing compositional techniques or styles, but to expand the possibilities for composers who want to 'creatively look back on nature' to paraphrase Von Schlegel. Audio samples from some evolved worlds of Eden are available on-line for the listener to judge for themselves at:
http://www.csse.monash.edu.au/~jonmc/projects/eden.html.

Acknowledgements. Eden was developed during a fellowship provided by the Australia Council.

---

[3] Of course, 'interesting' is a subjective term and one always tends to find ones own creations interesting.

References

1. Schlegel, F.V.: *Lucinde and the Fragments*, University of Minnesota Press (1971)
2. Loy, G.: *Composing with Computers – a Survey of Some Compositional Formalisms and Music Programming Languages*, In *Current Directions in Computer Music Research.* M.V. Matthews and J.R. Pierce, (eds.). MIT Press: Cambridge, MA. (1989) 291-396
3. Xenakis, I.: *Formalized Music: Thought and Mathematics in Composition*, Indiana University Press (1971)
4. Johnson-Laird, P.N.: *Human and Machine Thinking*, Lawrence Eribaum Associates (1993)
5. Ebcioglu, K.: *An Expert System for Schenkerian Synthesis of Chorales in the Style of J.S. Bach.* In *1984 International Computer Music Conference.* San Francisco, ICMA, (1984) 135-142
6. Todd, P.M. and G.M. Werner: *Frankensteinian Methods for Evolutionary Music Composition*, In *Musical Networks: Parallel Distributed Perception and Performance.* N. Griffith and P.M. Todd, (eds.). MIT Press/Bradford Books: Cambridge, MA. (1998)
7. Solomonoff, R.J.: *The Discovery of Algorithmic Probability: A Guide for the Programming of True Creativity.*, In *Computational Learning Theory: Eurocolt '95.* P. Vatanyi, (ed.). Springer-Verlag: Berlin. (1995) 1-22
8. Wolpert, D. and W. Mcready: *No Free Lunch Theorms for Search,* Technical Report SFI-TR-95-02-010, Santa Fe Institute. (1995)
9. Wiggins, G., et al.: *Evolutionary Methods for Musical Composition.* in *Proceedings of the CASYS98 Workshop on Anticipation, Music & Cognition.* (1999)
10. Darwin, C.R.: *The Desent of Man, and Selection in Relation to Sex.* Reprinted 1981 by Princeton University Press ed. London, John Murray (1871)
11. Hauser, M.D.: *The Evolution of Communication.* Cambridge, MA, MIT Press (1997)
12. Werner, G.M. and M.G. Dyer: *Evolution of Communication in Artifical Systems*, In *Artificial Life II.* C.G. Langton, (ed.). Addison-Wesley (1991) 659-682
13. Holland, J.H.: *Hidden Order: How Adaption Builds Complexity*, Helix Books (1995)
14. Dahlstedt, P. and M.G. Nordahl: *Living Melodies: Coevolution of Sonic Communication.* in *First Iteration: a conference on generative systems in the electronic arts.* In: A. Dorin and J. McCormack (eds.): Melbourne, Australia, Centre for Electronic Media Art, (1999) 56-66
15. Ulam, S.: *Random Processes and Transformations.* In *Proceedings of the International Congress on Mathematics* (1952) 264-275
16. Conrad, M. and H.H. Pattee: *Evolution Experiments with an Artificial Ecosystem.* Journal of Theoretical Biology, **28** (1970) 393-401
17. Packard, N.H.: *Intrinsic Adaption in a Simple Model for Evolution.* In C.G. Langton (ed.) *Artificial Life, SFI Studies in the Sciences of Complexity.* Los Alamos, NM, Addison-Wesley, (1988) 141-155
18. Watson, A.J. and J.E. Lovelock: *Biological Homeostasis of the Global Environment: The Parable of Daisyworld.* Tellus, **35B** (1983) 284-289
19. Roederer, J.: *Introduction to the Physics and Psychophysics of Music.* Second ed. New York, Springer-Verlag (1975)
20. Roads, C.: *The Computer Music Tutorial.* Cambridge, Mass., MIT Press (1996)
21. Holland, J.H.: *Adaption in Natural and Artificial Systems: An Introductory Analysis with Applications to Biology, Control, and Artificial Intelligence.* Second ed. MIT Press Cambridge, MA (1992)
22. Bowler, P.J.: *Lamarckism*, In *Keywords in Evolutionary Biology.* E.F. Keller and E.A. Lloyd, (eds.). Harvard University Press, Cambridge, MA. (1992) 188-193
23. Grünbaum, B. and G.C. Shephard: *Interlace Patterns in Islamic and Moorish Art.* In *The Visual Mind: Art and Mathematics.* M. Emmer, (ed.). MIT Press: Cambridge, MA. (1993).

New Hybrid Architecture in Artificial Life Simulation

David Kadleček and Pavel Nahodil

Faculty of Electrical Engineering, CTU in Prague, Department of Cybernetics
Technicka 2, 166 27, Praha 6, Czech Republic
kadlecd@rtime.felk.cvut.cz
nahodil@fel.cvut.cz

Abstract. This paper gives concise overview of our approach to the simulation of living forms and shows how it is possible, with help of the AI bottom-up methods, ethological and evolution principles, to simulate agents who have needs and desires, reflexes and instincts, but also have ability to perform deliberative tasks and to accomplish more complex goals. We have implemented a virtual environment – Artificial Life Simulator and our ideas and approaches were tested in practical situations.

1. Introduction

Intelligent life forms are partly bottom-up and partly top-down structures that have been created by an evolutionary process [2]. Our work tries to find a compromise between bottom-up and top-down approaches with by designing hybrid architecture. A system based on this architecture has properties that cannot evolve in time as well us parts that evolve and adapt in time [4]. The most important parts of the system are the Vegetative System Block, the Action Selection Block and the Conception Block. The main aim is to show how to connect several levels of the mind, namely reflexes and instincts, self-preservation and deliberative thinking, into one system. The work shows that the resulting less complex as well as more complex behaviors are influenced by all the levels of the mind. Such a combination gives a powerful and robust system able to survive and evolve in various environments. Agents based on this architecture are emergent, pro-active, social, autonomous and adaptive [1]. All structured information inside the agents and virtual environments as well as the messages flowing between the agents are described with the XML language. The proposed architecture, its possibilities, advantages and disadvantages were tested in a virtual environment implemented in Java. The combination of the Java and XML enables us to create a virtual environment that is very flexible and has high diversity of properties and conditions that influence an agent's behavior.

J. Kelemen and P. Sosík (Eds.): ECAL 2001, LNAI 2159, pp. 143-146, 2001.

2. Hybrid Combination of the Behavioural and Knowledge Based Approach to Create the Artificial Life

The architecture that we propose gives the agent basic instincts and reflexes and motivates him to explore and evolve in time. His behavior is a combination of both emergent and deliberative factors. He has an efficient capacity to make complex decisions and plan his goals in the long term. We have divided agent's control and mind functionality into three main blocks. All these blocks are independent and specialized. We will discuss these three blocks in more detail.

The first block is the **Vegetative System Block**. A set of drives is used to measure events, for unsupervised learning. Stimuli coming from the Perception Layer are usually transformed to drives. These drives can accumulate, be transformed into electronic signals and stimulate or inhibit creation of another drive. Simply said, the vegetative system transforms these drives and chemicals to electrical stimuli and vice versa. Any drive or chemical can change its concentration or transform to another. The chemicals and drives that we are using is a simplified subset of substances found in the existing vegetative systems. For simplicity, we use only a subset of a real vegetative system and its functionality (e.g.: transformations, chemical reactions) is also simplified. Proposed chemicals and drives are listed bellow:
Testosterone, Estrogen, Energy, Water, Cyanide, Progesterone, Histamine, Alcohol, Fear toxin, Wounded, General Cure, Antihistamine, Immunity level, Anabolic steroid, Endorphin, Adrenaline, Life, Injury, Stress, Pain, Hunger, Coldness, Hotness, Tiredness, Sleepiness, Loneliness, Crowdedness, Fear, Boredom, Anger, Sex drive, Reward and *Punishment.*

In addition to the chemicals and drives, we also define a set of chemical devices. These devices enable transformations of the chemicals and drives and can work together in one circuit. These devices are: *Chemo-Emitter, Chemo-Receptor, Chemo-Container* and *Chemo-Pipe*. The second block is the **Action Selection Block** [2], [4]. All tendencies of the system as well as stimuli from another blocks are combined here and "the best action" according to the situation is selected. Proposed architecture enables the combination of reflexes and instincts as well as tendencies from higher cognition parts in one block and to take all of them into account. For the action selection, we are using a mechanism most similar to the mechanism proposed by Tyrell [5]. In addition we have added some improvements [4] and made some changes to the architecture:

- **Pre-computed rate between consummatory acts and appetitive behaviors**

If the appetitive behavior is activated only with the intention to achieve some consummatory act, we must consider all three parts of the flow (Appetitive behavior, Taxi and Consummatory act) as one chain of actions. Each of the three actions would have no utility without other two remaining ones. Therefore, instead of considering these actions as three separated actions, we will consider them as one action with three parameters. Its utility therefore depends on all three parts. It is typical that the appetitive behavior is some times interrupted by an unexpected action. For example if an animal is looking for a water source, it can be interrupted by predators or if the animal perceives some use full subject. The appetitive behavior can move on to another appetitive behavior. The probability that the appetitive behavior will not be finished and therefore the taxi and the consummatory act will not be performed depends on estimated duration of the appetitive behavior and on the environment. We usually don't know how the appetitive behavior will take and therefore we must estimate its duration. An environment with a high density of predators or things of interest means that behavior can be interrupted very often.

In case of Taxi, we usually know how far away the subject of our desire is, when the Appetitive behavior is going to move on to the Taxi. We can therefore predicate how much time it will take if we know the speed of our agent.

The utility of the action is therefore

$U = U_{CA}/(1 + c_e * (d_A + d_T))$,

where c_e *is constant of the environment,* U_{CA} *is the utility of the consummatory act in the case the agent could perform the act immediately (No Appetitive behavior or Taxi is needed before the consummatory act),* d_A *is estimation of the duration of the Appetitive action and* d_T *is Taxi duration.*

- **Parallel execution of low-level actions.**

The ASM, we have proposed, can select more than one action in one step and multiple actions thus run at the same time. If we look at typical elementary actions such as "eat", "step forward", "step right", "sleep" etc., we can see that some actions can be performed simultaneously but other not. For example "eat" and "step forward" can run simultaneously whereas it is very difficult to go forward and backward at the same time. It is similar to sleeping and eating. We must define which actions are compatible and can run parallel in our system. This definition depends on physical properties of our environment and on conception of actions. If we want to use this approach, we must define relationship among actions. These relationships determine, which action can run parallel and which not. We have tested this during simulations and realized that this approach is possible and very effective.

The third block of our architecture is the **Conception Block**. This block contains all higher cognition parts as planning, learning and social behavior. Such behaviors can be achieved using various approaches. They can be coded into the artificial neural networks, Petri nets, scripts written in some pseudo language, or whatever else. The Conception Block can contain specialized sub-blocks. Each sub-block is specialized for a particular activity and built by different techniques.

The main three blocks must be connected together somewhere. They use the same low-level actions. The idea behind this is that all organisms in Nature also use the same muscles to move, use the same parts of body to perform desired actions regardless of whether the action is excited by some reflex or is a part of higher cognition plan. Therefore, these blocks cannot be wholly separated from each other. We have combined the Vegetative System, Action Selection and Conception blocks together. Stimuli from these blocks are combined and the best actions according to our simulated environment are selected. The Vegetative System Block gives stimuli related to drive homeostasis (equilibrium), satisfying internal needs, system preservation etc. The structure of the Action Selection Block gives an agent basic instincts and reflexes, and enables emergent behavior. The Conception Block performs planning, learning and social behavior. Plans driven by this block can be always interrupted in the case that some more emergent action needs to be performed. A more emergent action is some reflex or some internal need that must be satisfied.
The system has a set of fixed parameters and a set of free parameters. The free parameters can change in time and let the system to evolve according to local conditions. We enable such feature using evolutional and genetic principles.

Our agents can be "male" or "female" and as they become adult, they can mate with each other. The parameters of these two adult agents are combined and transferred onto their descendant. This part of the simulations is based on well-known techniques of genetic and evolutionary programming. Some parameters of the described agent's architecture are optional and change during evolution, other are fixed. Instincts and reflexes cannot evolve in time so rapidly and therefore they have more fixed parameters. In addition, the Perception Layer, Actuation Layer and Attention Layer were used.

3. Conclusion

Building on our previous work (in Mobile Robots Group at Dept. of Cybernetics) in robots / agents behaviour design we have shown our approach to a simulation of the artificial life, which is based on hybrid combination of the bottom-up (behaviour based) and top-down (knowledge based) approach [3], [4]. During the simulations we have proved that stimuli coming from reflexes, instincts and the vegetative system can be combined together with deliberative tasks and can work very efficiently together. We have proved that it is possible to execute more than one motor action in a time and that the resulting less complex as well as more complex behavior is influenced by all the levels of the mind.

Acknowledgement. This work was supported by the Ministry of Education of the Czech Republic within the frame of the Prof. Vladimír Mařík's project "Decision Making and Control for Manufacturing" number MSM 212300013. The support of the Universities Development Fund of the Czech Republic, under the Project number FR564 / 2000 to Pavel Nahodil, is also gratefully acknowledged.

References

1. Ferber, J.: Multi Agent Systems. An Introduction to Distributed Artificial Intelligence. Addison-Wesley. (1999).
2. Maes, P.: A Bottom-Up Mechanism for Action Selection in Artificial Creatures. In: Proc. From Animals to Animats Conference '91, (1991), 238 - 248.
3. Svatoš, V., Nahodil, P., Kadleček, D., Kurzveil, J., Maixner V.: Community of Mobile Robots in Education. In: Proc. of IFAC EPAC 2000 Conference: The 21st Century Education in Automation and Control, Skopje, Macedonia, (2000), 53 - 57.
4. Nahodil, P., Svatoš, V.: Sense-Action Encoding in Hybrid Decision System. In Proc. MIC2001, IASTED Int. Conference, Innsbruck, Vol. 1, (2001), 414 - 419.
5. Tyrrell, T.: Computational Mechanisms for Action Selection. Ph.D. Thesis, Edinburgh University. (1999)

"In Silico" Experiments on Heavy Metal Sorption by Algal Biomass

Jana Kadukova[1], Marian Mihalo[2], Martin Palko[2], and Marek Polak[2]

[1] Institute of Metalurgy, University of Technology in Kosice, Letna 9, 043 85 Kosice, Slovakia
kadukova@hfnov.tuke.sk
[2] Department of Cybernetics and Artificial Intelligence, University of Technology in Kosice, Letna 9, 041 20 Kosice, Slovakia
{mihalo, palkom, polak}@alife.tuke.sk

Abstract. A lot of localities in our environment are contaminated by toxic metals. Some of the living organisms are able to clean it. This is the result of their resistance mechanisms which allow organisms to survive and they appear as sorption of the harmful elements from the environment. The "in vitro" experiments are rather expensive and time consuming. The possibility of reducing the costs led to the idea of building a computer model and perform some experiments "in silico". Now a basic model is ready and its functionality is being checked.

1 Introduction

Resistance mechanisms of organisms against toxic metals are based on few different mechanisms provided by algae cells which make them possible to survive in a heavily polluted environment. Due to these mechanisms, populations of some species may be used for sorption of metals from contaminated waters. The experiments take a long time with high money expenses. With the artificial life simulation based on multi-agent simulation system we are preparing a simulator that could at least partially substitute the real experiments with real organisms [1].

2 Essence of Sorption Ability

The researchers found various organisms with developed resistance mechanisms. These mechanisms can be divided into three groups according to their localisation [2]:

1. mechanisms on the cell wall,
2. mechanisms of the cell wall and cytoplasmic membrane,
3. intracellular mechanisms.

Function of the resistance mechanisms mostly appears as a "cleaning power" (or sorption ability) that removes heavy-metal ions from the surrounding accumulating them within the body, binding to the cell wall or creating low-toxic compounds. However, not all of the organisms are suitable to be used for technological purposes

J. Kelemen and P. Sosík (Eds.): ECAL 2001, LNAI 2159, pp. 147–150, 2001.

of this kind. But microorganisms, the algae above all, seem to have the proper nature. Usage of them has a lot of advantages [1]:

- high sorption capacity,
- relatively rapid sorption process,
- minimal nutrient expenses to cultivate the biomass and keep it useable,
- possibility to use dead cells (due to passive mechanisms on the cell wall),
- functioning under normal pressure and temperature,
- heavy-metal concentrates accumulated by the cells are suitable for conventional metal-recovery processes,
- biomass can be used repeatedly.

3 "In Silico" Model

To build a computer model, we chose the specie of *Chlorella kessleri*. It is a unicellular freshwater green alga and its sorption mechanisms have all the advantages mentioned thereinbefore. (You can find more information in [1], [3], [4].)

Considering one alga cell as an autonomous agent we started to create the model adhering the bottom-up approach. Our model consists of two components: model of the environment and model of the cell.

3.1 Model of the Environment

The environment is designed as a 4-layer 2-D discrete grid. The separate layers represent position of cells, concentration of heavy metal, concentration of minerals necessary for cells to survive and concentration of products of the cell metabolism (high amount can cause death of cells).

The first layer is changed, if a cell dies (it is removed) or reproduces (mostly 4 new agents are inserted). Cells can also move depending on the simulated situation – either slow steady flow or test-tube shaking. The content of remaining three layers is affected by the activity of agents and moved according to diffusion principles.

3.2 Model of the Cell

The *Chlorella* cell is represented by an algorithm that provides all necessary vital functions: the initialization by the "birth" of the cell, treating heavy-metal ions, uptake of minerals (feeding), metabolism, excreting metabolism products, reproduction, death.

The metabolism is provided by an amount of internal energy of the cell [2]. Each action (reproduction, some metabolic functions, sorption and higher level of toxic metal in environment) causes the decrement of internal energy. The energy is incremented as a result of photosynthesis and the quantity depends on the count of neighbour cells which shade the incoming light [5], [6].

Behaviour of the model is affected by number of parameters and constants that determine the function of all actions and properties of the environment.

3.3 Implementation

To realise the model and perform the simulation we took the system Swarm. It is a collection of software libraries based on a general purpose programming language providing support for agent-based simulation programming [7].

4 Experiments

The structure of the model developed till now is rather simple and is not dedicated to any certain metal. But we have run many simulations and tried to make the model follow the dynamics of real heavy metal sorption (arsenic, similar aurum and copper above all). Sorrow, we don't have enough results of the "in vitro" experiments. On figures 1 and 2 you can see a comparison of some experiments undertaken "in vitro" and "in silico". The second one does not copy the real dynamics accurately, because copper has slightly different impact on the cell metabolism from aurum.

Some experiments were also carried out to designate the dynamics of the population growth. We intend to use evolutionary algorithms, because we hope they will help us to adjust the model (the model parameters). But more results from real experiments must be ready first.

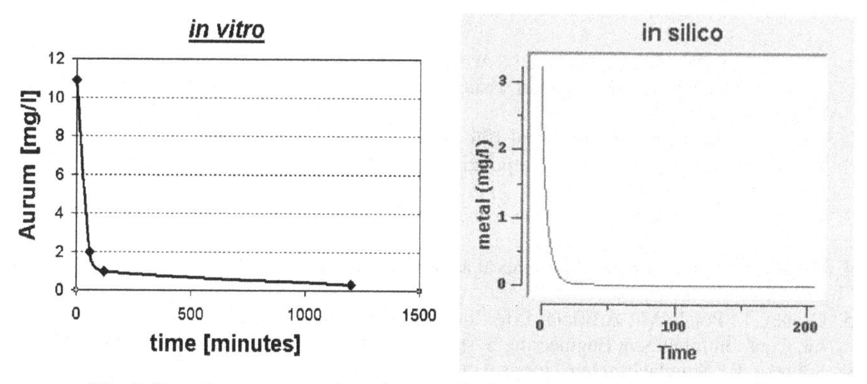

Fig. 1. Experiment on sorption of aurum (1 discrete time step = 10 minutes)

5 Conclusion

We believe, our work, after being finished, will help the biologists with their research of alga *Chlorella kessleri* and its sorption mechanisms. However, it will require a lot of experiments and alternating the model structure, its algorithm and properties.

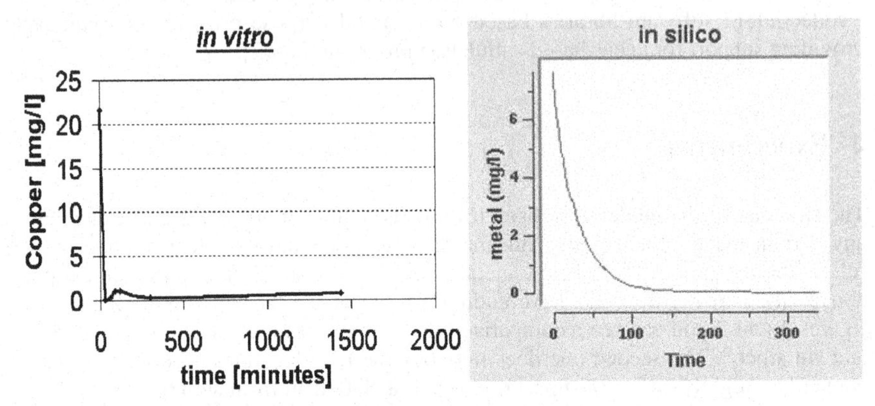

Fig. 2. Experiment on sorption of copper (1 discrete time step = 10 minutes)

References

1. Csonto, J., Polak, M., Kadukova, J.: Artificial Life Simulation of Living Alga Cells and its Sorption Mechanisms. In: Special Issue of International Journal of Medical Informatics. Elsevier, accepted.
2. Polak, M., Kadukova, J.: Artificial Life Simulation of Living Algae Cells. In: Sincak, P., Vascak, J. (eds.): Quo Vadis Computational Intelligence. Physica-Verlag Heidelberg New York (2000) 324-328.
3. Hindak, F.: Studies on the chlorococcal algae (Chlorophyceae), II. VEDA Bratislava (1980) 59-65.
4. Hindak, F.: Studies on the chlorococcal algae (Chlorophyceae), III. VEDA Bratislava (1984) 177-181.
5. Csonto, J., Polak, M.: Artificial Life Simulation of Alga Chlorella kessleri. In: 1999 IEEE Int. Conf. on Intelligent Engineering Systems Proc. ELFA, Ltd. Kosice (1999) 295-298.
6. Sobotka, B.: Simulacia zelenych rias v prostredi Swarm. Thesis, KKUI TU Kosice (2000).
7. Swarm Development Group: Swarm User Guide. Santa Fe Institute (2000). http://www.swarm.org

This work is supported by grant VEGA (1/8135/01) – "Life Simulators and their applications".

Spatially Explicit Models of Forager Interference

Anil K. Seth

CCNR/COGS, University of Sussex, Brighton, BN1 9QH, UK
and
The Neurosciences Insitute, 10640 John Jay Hopkins Drive, San Diego, CA92121
anils@cogs.susx.ac.uk

Abstract. In behavioural ecology the 'interference function' relates the intake rate of a focal forager to the density of conspecifics, and has been (and remains) the subject of extensive theoretical and empirical activity. This paper describes a novel agent-based model of interference characterised by the use of genetic algorithms to evolve foraging behaviours in a spatially explicit environment. This model provides a flexible platform for modelling interference phenomena, and addresses several conceptual and practical problems with orthodox approaches. Its flexibility is demonstrated by an exploration of how, in the context of interference, some instances of individual irrational behaviour can be understood as a consequence of adaptation to a group context.

1 Introduction

When multiple agents forage for limited resources, the intake rate per agent can be considered to be a function of (at least) both resource density and agent density. If the influence of agent density is non-negligible, this function describes an *interference* relationship, where interference is defined as the more-or-less immediately reversible decline in intake rate due to the presence of conspecifics [7]. In behavioural ecology, models of interference are widely used for the prediction of spatial distributions of foragers over environments of varying resource availability, but their potential utility is by no means limited to the biological domain - consider for example a collection of robots foraging for rock samples, or a group of information agents scouring the internet.

This paper first outlines orthodox approaches to modelling interference, identifying some serious conceptual and practical difficulties which have to do with (a) respecting the distinction between *behaviour* (observed agent-environment interactivity) and *mechanism* (agent-side structure subserving this interactivity), and (b) accommodating the role of individual optimality. An alternative modelling strategy is then described which (a) uses genetic algorithms (GAs) to evolve forager behaviour, and (b) restructures the problem of interference at the level of individual perception and action in a spatially explicit environment. Analysing an example of such a model, I argue that it provides a conceptually satisfying and empirically flexible platform for modelling interference phenomena. This flexibility is exemplified by an exploration of an instance of individual irrational behaviour - in the context of interference - that can be understood

J. Kelemen and P. Sosík (Eds.): ECAL 2001, LNAI 2159, pp. 151–160, 2001.

as a consequence of adaptation to group situations, providing insights which - I will argue - would have been difficult to attain using orthodox methods. At a general level, this research is an example of how the techniques of artificial life can contribute to theoretical biology and behavioural ecology, and how the conceptual apparatus of biology can, in turn, enrich the field of artificial life.

1.1 A Brief History of Interference

In a recent survey, Van der Meer & Ens [8] distinguish two approaches to modelling the interference function, 'phenomenological' and 'mechanistic'. The former originates from Hassell & Varley's identification, from empirical data, of a linear relationship between the logarithms of intake rate and agent density, the slope of which was identified with the level of interference [1]. This model and its derivatives, although widely used, have been criticised for (a) lacking any interpretation of interference at the level of underlying behaviour-generating mechanisms [8], and (b) failing to adequately describe subsequent empirical data, some of which describes a non-linear log-log relationship and hence a level of interference that varies with agent density [3].

The mechanistic approach involves the construction of individual-based models in which the actions of individual agents follow pre-specified rules. The way in which the application of these rules interacts with agent density in the determination of intake rate can then be derived. Ruxton et al. [3], for example, divide a population of agents into a number of mutually-exclusive states (prey-handling, searching, fighting, etc.), and sets of 'reactions' are defined between these states; upon encountering a food item, an agent would move from a searching state to a prey-handling or feeding state, and so on. Differential equations are constructed expressing transition rates between these states, and the equilibrium solutions of these equations yield the interference function. The intuition underlying this approach is that individual agents behave like 'aimless billiard balls', or molecules in a chemical reaction, moving randomly and interacting upon collision in ways determined by the reaction matrix. Interference is explicitly associated with particular states, for example fighting, or state transition sequences, such as kleptoparasitism. Significantly, these models do *not* normally involve simulations of agents actually moving around in spatially explicit environments, rather, agents are abstract entities characterised only by being in particular states (drawn from a pre-determined repertoire), with the 'environment' consisting only of the set of probabilities for the various state transitions.

Whilst these models certainly have the potential to address the criticisms of the phenomenological approach mentioned above, a problem remains in that agents cannot normally be guaranteed to behave 'optimally'; that is, the inflexible nature of the state transitions experienced by agents in most mechanistic models provides no guarantee that individual agents will follow sequences of states that maximise their intake. This is a serious difficulty because the use of interference to predict spatial distributions depends on the assumption of individual optimal foraging [8], and indeed it might be reasonably argued that interference *per se* is of little interest outside of a context in which agents attempt to maximise intake. Stillman et al. [6] have recently attempted to address this problem with a mechanistic model in which state transitions are made according to the relative costs and benefits they offer. Their model, however, imputes considerable computational power and statistical agility to the agents; also, the optimal 'rules' themselves, if not the state transitions, must still be pre-specified.

Another problem with most mechanistic models (including that of Stillman et al.) is that the explicit identification of a set of 'behavioural' states (or state transition sequences) with interference, is a phenomenological process in just the same sense as is the identification of interference directly from empirical data. In other words, existing mechanistic models can be criticised on exactly the same grounds that proponents of these models criticised earlier phenomenological models: the components of the models require explanation in terms of behaviour-generating mechanisms. Care must be taken, especially in so-called 'mechanistic' models, not to confuse behavioural and mechanistic levels of description. Interference - a behavioural phenomenon - should be modelled as *arising from* the interaction of agents (each with its own internal behaviour-generating mechanism) in a shared environment, and not introduced as an *a priori* mechanistic component of the model itself.

2 A Spatially Explicit Model of Interference

This section describes a mechanistic model of interference with two novel characteristics. First, the use of GAs to partially specify agent mechanism entails consistency with the assumption of individual optimal foraging, without necessarily requiring that agent-side mechanisms be complex. Second, modelling at the level of individual perception and action in a spatially explicit environment means that interference no longer need be explicitly identified with *a priori* model components. The structure of the model is deliberately simple, and indeed is designed to capture the simplest possible scenario in which interference can be explored in a spatial context. Simple neural network controllers are embedded in equally simple agents that move around a simulated single-patch environment, collecting food items, with artificial evolution specifying network parameters such that agents forage 'optimally'. The observed relationship between agent density and intake rate then yields the interference function.

Each agent possesses 3 sensors, 2 responding to the distance to the nearest food item, and 1 reflecting an internal battery level $\mathcal{B}$ (agents cannot directly sense the presence of conspecifics). The distance sensors range linearly from 100.0 (at the item) to 0.0 ($\geq$ 200.0 arbitrary distance units - DUs - away), and the battery sensor ranges linearly from 0 to 200 (the maximum, and initial, $\mathcal{B}$). If the nearest food item is to the left of the agent, the left sensor responds with 20% greater activation (and *vice versa* if the object is to the right). Internally, each agent comprises a simple feedforward neural network (fig. 1). The input units linearly scale the sensor values to the range [0.0,1.0], and units in subsequent layers transform (sigmoidally) their input sum (plus a threshold value) to outputs in the range [0.0,1.0]. Motor outputs are scaled to [-10.0,10.0] to set the wheel speeds; a value of 10.0 on both wheels translates to a speed of 2.8 DUs per time-step. Real valued genotypes of length 32, with each locus constrained to the range [-1.0,1.0], specify the weight and threshold values.

The environment comprises an infinite plane containing stationary food items (radius 8.0). Both food items and agents (radius 2.5) are always initialised within a 200.0 by 200.0 'patch'. Each time an agent encounters food, the food item disappears to be immediately replaced in another random location in the patch, thus ensuring a constant density of available food (important in view of the distinction between interference and the straightforward depletion of resources). Each encounter with food fully replenishes $\mathcal{B}$, which otherwise depletes at a rate of 1 unit per time-step; if $\mathcal{B}$ reaches 0 the agent 'dies'. Encounters with conspecifics (if any) have no effect. Fig. 1 illustrates a typical environment populated by 3 agents, together with a representation of a single agent.

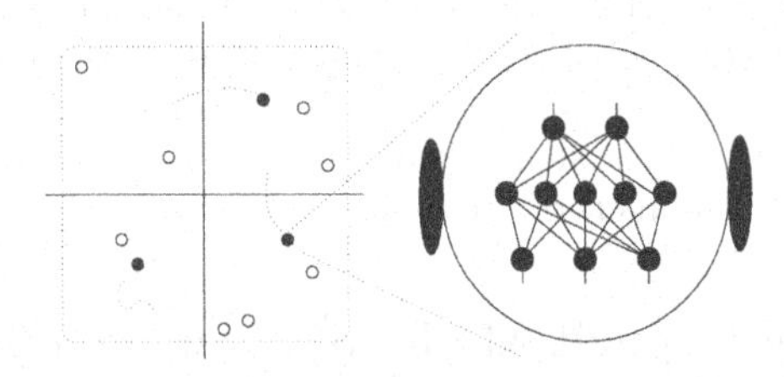

Fig. 1. Depiction of a typical environment populated by 3 agents (filled circles) and 8 food items (clear circles).

Foraging behaviour is evolved over 800 generations of a distributed GA.[1] Genotype fitnesses are averaged over 4 evaluations per generation. Each evaluation begins by decoding the genotype into a single agent, placed at random in the patch along with 8 randomly scattered food items (with a minimum spacing of 25 DUs between objects). Fitness is then assessed using $\mathcal{F} = \sum_{t=1}^{800} \mathcal{B}/200$, where t indexes time-steps; each evaluation lasts for a maximum of 800 time-steps. This fitness function rewards agents that live long and forage efficiently, without specifying the structure of the behaviour through which this is to be achieved. After a sufficient number of generations have elapsed to ensure 'optimal' foraging (800 is quite enough, fitness reliably asymptotes after 400), an interference function can be derived by decoding the fittest genotype into an agent and assessing its intake (averaged over 500 evaluations) both in isolation (as in the GA itself), and also in the presence of n 'clonal' conspecifics *for all* $n \in \{1, 12\}$. This basic model, in which agents are evolved *in isolation*, shall be called 'condition A'.

A 'condition B' is also explored in what follows, in which there are 13 distinct GAs in the evolutionary phase (indexed by n), one for each group size from $n = 1$ to $n = 13$. Consider the case, for example, $n = 8$. In the corresponding GA each genotype is always decoded into 8 clones which co-exist in the environment. The genotype fitness is determined by that of a randomly selected agent clone from

[1] Population size 100, crossover rate 0.5, point mutation rate 0.05.

this group. The testing phase proceeds as before, but of course each group is now derived from the fittest genotype from the corresponding GA. Thus, all groups in condition B comprise of agents that were *evolved in the presence of the same number of conspecifics also present during testing.* The purpose of distinguishing these conditions is to illuminate an instance of suboptimal individual behaviour which can be understood in terms of adaptation to a group context (see [4] for previous work along these lines).

3 Results

Agents evolved in conditions A and B shall from now on be referred to as A-agents and B-agents. The results presented below derive from 8 complete repetitions of both evolution and testing phases of each condition.

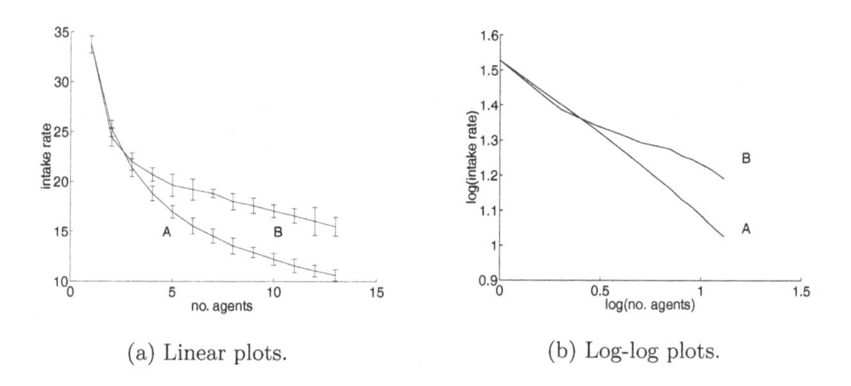

(a) Linear plots.

(b) Log-log plots.

Fig. 2. Interference functions for conditions A and B. Each data point is the average of 8 runs of each condition. In (a) error bars represent standard deviations, and in (b) logarithms are taken to base 10.

Fig. 2 illustrates the interference functions obtained in each condition. In 2(b) there is a strikingly clear linear relationship for A-agents between the logarithms of agent density and intake rate (and a corresponding smooth curve in 2a, just as first identified in the field by Hassell & Varley), with the constant slope of this line indicating a constant level of interference.

For B-agents, by contrast, this linear relationship is no longer apparent. Not only is the observed level of interference generally lower than for A-agents, but this difference becomes more apparent at higher agent densities. Two observations follow immediately from these results. The first is that interference - both constant and density dependent - can indeed be effectively modelled using only the sensorimotor interactions of groups of agents, without recourse to explicit behavioural states and associated transition rules. Secondly, that somehow, B-agents - evolved in the presence of conspecifics - are behaving in ways which allow them to forage more effectively in the presence of conspecifics than those (A-agents) evolved in isolation.

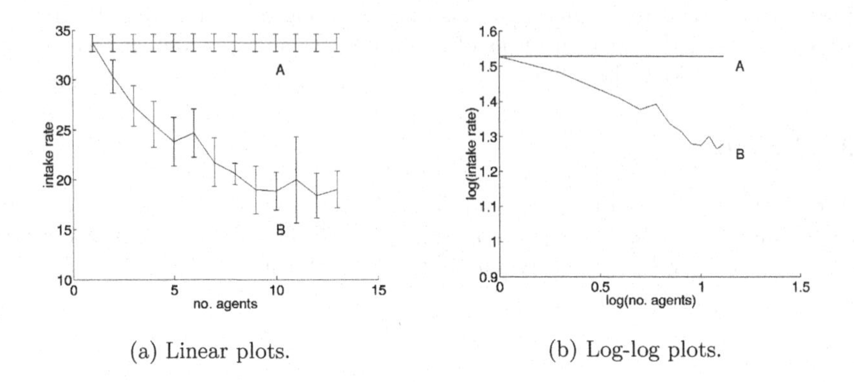

(a) Linear plots.

(b) Log-log plots.

Fig. 3. Intake rates of isolated A- and B-agents. The abscissa of each plot represents the group sizes within which the B-agents were evolved. Each data point is the average of 8 runs of each condition, and error bars represent standard deviations.

Fig. 3 illustrates the foraging performance of *isolated* agents, with the abscissa representing the group sizes (for condition B) within which the agents were evolved (condition A will of course provide only a single point on these graphs, but this is extended into a line in order to aid comparison). It is immediately evident that the larger the group a given B-agent was evolved with, the worse it performs when on its own. The same behavioural strategy that delivers an advantage over A-agents in group situations is a handicap for isolated agents.

So what is going on? Inspection of A-agents suggests that they all follow the entirely intuitive strategy (given their sensorimotor poverty) of making for the nearest food item as rapidly as possible. However, it also seems that groups of agents following this strategy tend to become 'clumped' together. By contrast, groups of B-agents, despite suffering the same sensorimotor poverty as A-agents, appear to distribute themselves more evenly throughout the patch. Could it be that the avoidance of clumping both lessens the toll of interference in group situations and leads to inefficient foraging by isolated agents?

It is possible to be more formal with the notion of 'clumpiness'. Let us define the 'instantaneous clumpiness' of a group of agents as $cl_t \in [0.0, 1.0]$ such that $cl_t = 0.0$ indicates a maximally dispersed group, and $cl_t = 1.0$ a maximally clumped group. Given d_i the distance from each (alive) agent i to the nearest (alive) conspecific, r_a the number of alive agents, and $f()$ a Gaussian function of height 1.0, mean 0.0, and radius 15.0, then:

$$cl_t = \frac{1}{r_a} \sum_{i=1}^{r_a} f(d_i).$$

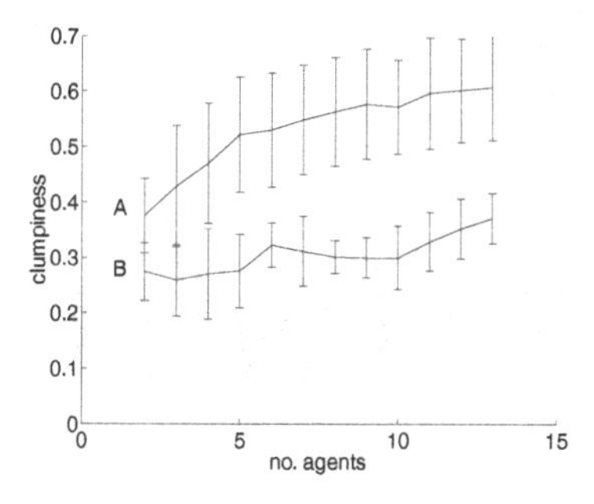

Fig. 4. Average $\overline{cl}$ for groups of 2-13 agents over 8 runs of each condition.

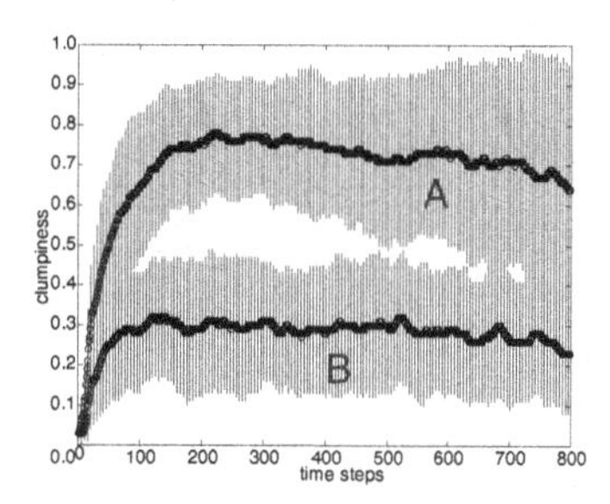

Fig. 5. Average cl_t groups of 8 A- and B-agents. Each line represents the average cl_t of the group at each of 800 time-steps over 500 evaluations, standard deviations are shown.

The 'overall clumpiness' cl of the behaviour of a group of agents is then the average cl_t over all time-steps for which at least one agent was alive (the 'instantaneous clumpiness' cl_t of a single agent 'group' is zero). This metric was used to compare the clumpiness of groups (of evolved agents) of sizes ranging from $n = 1$ to $n = 13$ from each condition. Each group was evaluated 500 times, with a final clumpiness $\overline{cl}$ derived from the average cl. Fig. 4 indicates, as hypothesised, that groups of A-agents present much higher $\overline{cl}$ values than groups of B-agents. An example is illustrated in fig. 5, which contrasts the clumpiness profile (the average of 500 evaluations of 800 time-steps each) of a group of 8 A-agents with a group of 8 B-agents. Both groups, being initially randomly distributed, display an initially low average cl_t, but for the majority of the evaluation the B-agents present much the lower average cl_t.

But how does this difference in propensity to clump come about (remember that the agents cannot see each other), and how could this lead to differences in the level of interference? Further inspection reveals that upon encounter with food, a B-agent will decelerate dramatically for a few time-steps, before accelerating away towards another target (an example is illustrated in fig. 6b). By contrast, A-agents rarely deviate from maximum velocity (fig. 6a). This in itself is enough to ensure that the B-agents become more dispersed. Consider two agents p and q heading towards a single food item from a single starting point. Both agents will remain in close proximity as they approach the item, but only one of them, let us say p, can benefit from it. The item now disappears and both p and q perceive, and begin to move towards, a new nearest item. If they both travel always at maximum speed (A-agents), they will again remain in close proximity as they approach this new item and, again, only one will benefit. However, if p slows down following consumption of the first item, then the situation is different and can be seen to work to the benefit of both agents, especially if they coexist with several others. From the perspective of any individual agent the environment will now be less stable, with the nearest food item suddenly

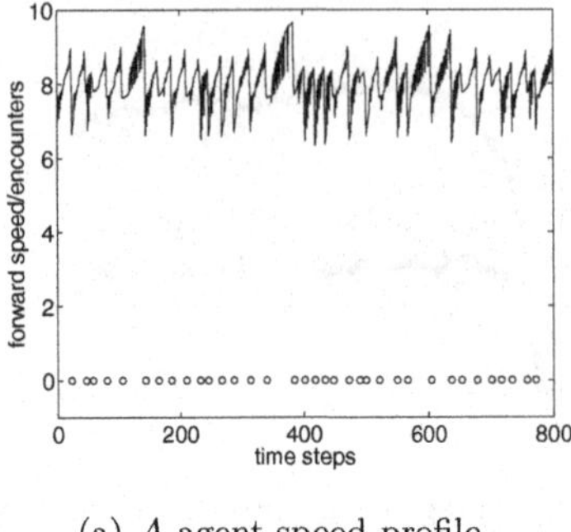

(a) A-agent speed profile.

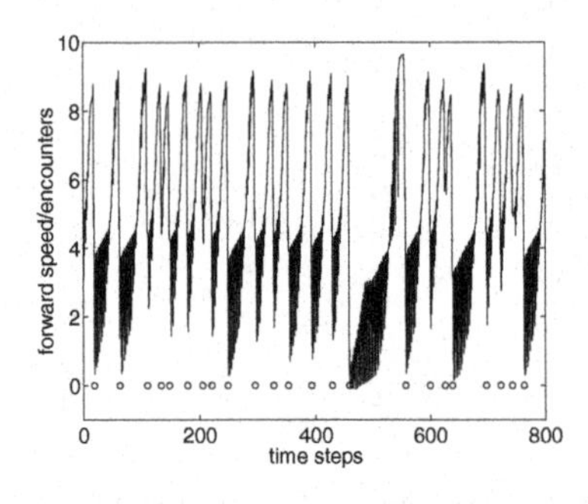

(b) B-agent speed profile.

Fig. 6. Forward speed profiles of (a) an isolated A-agent and (b) an isolated B-agent over one evaluation of 800 time-steps. The circles represent encounters with food items. Similar speed profiles to (a) were observed in 6 out of 8 replications of condition A, and to (b) in all 8 replications of condition B.

changing location as other agents move around and consume. Now, when p decelerates, it becomes possible that after a few time-steps p and q will perceive *different* nearest food items, and so will no longer be in direct competition. And the more dispersed a group is, the more unpredictable the pattern of food depletion will be, and hence the greater individual benefit there is to be gained from being part of a dispersed group. Of course the strategy of slowing down after consumption is clearly *not* the optimal strategy for an *isolated* agent for whom any deviations from maximum speed can only serve to reduce intake rate.

4 Discussion and Summary

The immediate implications of this work for theoretical biology are twofold. First, we have seen how interference - both constant and density dependent - can be modelled without interference itself being explicitly associated with aspects of agent mechanism. Interference in the present model is a consequence of agents foraging 'optimally' in a spatially explicit, limited resource, and shared environment, and, as such, is properly treated as a behavioural phenomenon rather than, improperly, as an *a priori* model component. Second, optimality has been incorporated into a mechanistic approach (answering the call in [8]) without making great demands on the computational agility of the agents themselves (in contrast to Stillman et al. [6]). This has been achieved through the use of GAs in conjunction with simple internal mechanistic substrates and spatially explicit environments, as opposed to the (orthodox) methodology of 'building in' mechanisms of optimality which supervene on pre-determined repertoires of internal behavioural correlates and transitions rules. (A general manifesto for the present methodology appears in [5].)

The clear linear form of the condition A log-log plot (fig. 2b) is particularly striking, both for its congruence with the original observations of Hassell & Varley, and for its contrast with condition B. To understand this contrast, one may consider condition A as a reductionist ideal; the (optimal) behaviour of a single agent is first determined and the behaviour of the group is then derived from the independent (additive, linear) combination of the behaviours of all constituent agents. Condition B, by contrast, does not permit such a conceptual separation of individual and group behaviours. Here, patterns of agent-environment interaction contingent on group behaviour (for example resource instability patterns) can be present *during* optimisation, and hence can influence the trajectory and outcome of the optimisation process, in terms of both individual and group behaviours. Such patterns represent 'historical constraints' insofar as they constitute dynamical invariants which constrain the future dynamics of the system, and which are able to alter their own conditions of realisation.[2] So, whilst condition A treats the process of optimisation and the outcome as entirely separable, condition B allows for historical constraints to arise and persist from their continual interaction. Perhaps, then, the 'ideal' linear interference function observed in condition A is indicative of a rigidly reductionist, ahistorical approach to group foraging, an approach from which dissent would not have been possible under orthodox modelling regimes that necessarily deny the potential for historicity inherent in (the present) condition B. This carries further implications for theoretical biology; it may well be that historical processes of this kind are essential in understanding the dynamics of interference in real situations. Of course it remains an open question whether the departures from 'ideal' interference that have been observed since Hassell and Varley can be attributed to the kind of historical processes elucidated here.

Finally, it has been observed that agents evolved with conspecifics experience lower levels of interference than those evolved in isolation, and conversely that agents evolved with conspecifics do worse, when tested in isolation, than those evolved in isolation. A strategy (deceleration after consumption) was identified, suboptimal for an isolated individual, yet optimal in a group context. In artificial life itself this strategy and its consequences are significant both for the purposes of design and prediction of agent spatial distributions, broadly construed. However, whether or not this strategy can be observed in real organisms is not known, although the fact that most encounters with food items require a certain 'handling time' suggests that, in many real situations, slowing down upon consumption is not something that can be avoided.

The present model deals only with very simple interference situations. It should, however, be easily extensible to cover some of the more complex situations of biological concern, for example the role of individual competitive dif-

[2] As opposed to 'fixed constraints', pertaining to the agent or to the environment, that form part of the *a priori* model structure, or indeed frequency dependent effects in (orthodox) game-theoretic models in which optimal 'outcomes' (now called 'evolutionary stable strategies'), whilst dependent on the contemporaneous distribution of strategies, are still considered independent of the mechanics of the optimisation process. These issues are discussed in detail in [5].

ferences [2]. On the other hand, the implications of the model are by no means restricted to biology; an understanding of the dynamics of interference is of interest any situation in which groups of agents - biological or artificial - pursue a shared goal - cooperatively or competitively - in a limited resource environment. Indeed, the structure and dynamics of the model as it stands (and in particular those interaction patterns that have to do with agent 'clumpiness') may well have closer analogues in artificial situations than in biological contexts. Finally, the origin of the insights described here in the combination of GAs, simple internal mechanisms, and spatially explicit environments, and particularly the accompanying potential for elucidating historical constraints at the level of sensorimotor interaction patterns, suggests that the same insights would have been hard to attain from within the orthodox modelling community described in section 1.1. This may explain, amongst other things, why field biologists have not (to my knowledge) explicitly looked for the kind of suboptimal behaviour described above. Hopefully there is now some motivation for them to do so.

Acknowledgements. Thanks to Phil Husbands, Peter Todd, the members of the CCNR, and my anonymous reviewers for advice, the EPSRC for money (award no. 96308700), and the Sussex University High Performance Computing Initiative for technical support. This work was carried out whilst the author was at Sussex University.

References

1. M.P Hassell and G.C. Varley. New inductive population model for insect parasites and its bearing on biological control. *Nature*, 223:1133–1136, 1969.
2. N. Holmgren. The ideal free distribution of unequal competitors: Predictions from a behaviour-based functional response. *Journal of Animal Ecology*, 64:197–212, 1995.
3. G.D. Ruxton, W.S.C. Gurney, and A.M. Roos. Interference and generation cycles. *Theoretical Population Biology*, 42:235–253, 1992.
4. A.K. Seth. Evolving behavioural choice: An investigation of Herrnstein's matching law. In D. Floreano, J-D. Nicoud, and F. Mondada, editors, *Proceedings of the Fifth European Conference on Artificial Life*, pages 225–236. Springer-Verlag, 1999.
5. A.K. Seth. Unorthodox optimal foraging theory. In J.-A. Meyer, A. Berthoz, D. Floreano, H. Roitblat, and S.W. Wilson, editors, *From animals to animats 6: Proceedings of the Sixth International Conference on the Simulation of Adaptive Behavior*, pages 471–481, Cambridge, MA, 2000. MIT Press.
6. R.A. Stillman, J.D. Goss-Custard, and R.W.G. Caldow. Modelling interference from basic foraging behaviour. *Journal of Animal Ecology*, 66:692–703, 1997.
7. W.J. Sutherland. Aggregation and the 'ideal free' distribution. *Journal of Animal Ecology*, 52:821–828, 1983.
8. J. Van der Meer and B.J. Ens. Models of interference and their consequences for the spatial distribution of ideal and free predators. *Journal of Animal Ecology*, 66:846–858, 1997.

Antigens, Antibodies, and the World Wide Web

Daniel Stow and Chris Roadknight

B54/124, Adastral Park, Martlesham Heath, Ipswich, Suffolk, UK. IP5 7RE
dstow@lineone.net, Christopher.roadknight@bt.com

Abstract. In this report the immune system and its adaptive properties are described and a simple artificial immune system (AIS) model based on the clonal selection theory is presented. This immune system model is demonstrated to be capable of learning the structure of novel antigens, of memory for previously encountered antigens, and of being able to use its memory to respond more efficiently to antigens related to ones it has previously seen (cross-reactivity). The learning, memory and cross-reactivity of the AIS are fruitfully applied to the problem of fuzzy resource identification. Interesting antigen/antibody relationships are also identified.

1 Introduction

Biologically inspired approaches to computing have found a natural application to problems in telecommunications (see [Di Caro & Dorigo 1998; Seth 1998; Roadknight & Marshall 2000] for examples). Modern telecommunication systems can be seen as complex structures built out of simple interacting components, much like biological systems. In this light, the co-operation between the two fields is hardly surprising. A current concern is that the introduction of new Internet services is being held back by human-intensive approaches to network management. It is becoming clear that management of the Internet must be less dependent on manual intervention.

Immune systems have several properties that make them of interest to computer scientists, including diversity, distributedness, adaptability, self-organisation and autonomy [Hunt & Cooke 1996; Somayaji *et. al.* 1998; Hofmeyr & Forrest 1999]. These are properties that they share with Roadknight and Marshall's bacteria.

2 The Artificial Immune System

Immune systems have a variety of appealing properties (condensed from similar lists in [Hunt & Cooke 1996] and [Somayaji *et. al.* 1998]. *Distributedness ,Diversity, Dynamically changing coverage, Imperfect detection, Autonomy, Unsupervised learning,: Content addressable memory, Adaptability.*

As in other machine learning applications of artificial immune systems (e.g. [Hunt & Cooke 1996; Timmis *et. al.* 2000]), the antigens in our model represent the problem that the immune system must learn to solve. In order to show that our model could

J. Kelemen and P. Sosík (Eds.): ECAL 2001, LNAI 2159, pp. 161-165, 2001.

successfully learn about real world data, we developed an encoding scheme to convert URLs, taken from a cache access log file, to 64-bit binary strings. This scheme encodes the first three numbers of the IP address (8 bits each), the number of forward slashes in the URL (3 bits), the number of characters between the first and last forward slash (5 bits), the first six letters of the name of the requested file (5 bits each), and the type of file requested, e.g. GIF, HTML, etc. (2 bits).

Antibodies in our model are 64-bit binary strings, just like antigens. In the biological immune system, molecular binding between antibodies and antigens depends on them having complementary shapes. Similarly, matching between antibodies and antigens in our model depends on the number of complementary bits they have. This was also the matching method used by Hightower *et. al.* [1995]. After counting the number of complementary bits we apply a sigmoid function to give the probability of the antibody binding the antigen. This simulates the idea that antibodies must match antigens over a certain minimum area before they can bind

The genes used to create antibodies are divided into four libraries, each containing a number of 16-bit segments. In our model there are currently 16 segments per library. To create an antibody a segment is picked at random from each library and these segments are joined together. We experimented with two different types of stopping criteria for the learning. One of these was to stop learning when an antibody with a certain minimum score was found. The other involved limiting the number of antigens available for the antibodies to match. Each time an antibody attempted to match an antigen, an antigen had a chance to copy itself. Each time an antigen was matched it was killed. The learning was stopped when there were no antigens left.

Memory in our model is maintained in the simplest way possible – when the system finishes learning about an antigen it finds the highest scoring antibody in the population and makes it a memory antibody. Unlike other antibodies these can live for an unlimited length of time. The number of memory antibodies is limited and when the memory is full the new antibody replaces the oldest.

3 Experimental Results

In our research on our immune system model, we were interested in answering the following question: in what circumstances does it perform best and why? By performance we mean the speed at which it learns to recognise antigens, which will obviously be crucial if it is to be applied in the management of network resources. The first factor that we noticed affecting the performance of the model was the set of antigens itself. Figure 1 shows the speed of learning for antigens encoded from nine different cache log files and four random number seeds. Speed of learning in this case is measured by the average number of generations of clonal selection required to evolve an antibody with a score of at least 95 (out of 100).

The graph shows that the performance is affected far more by the cache log file than by the random number seed. The likely explanation is that different log files have different patterns of repetition of URLs in them. If there is more repetition of the same or similar URLs in the log file, this should be mirrored in the set of antigens. The more repetition there is in the set of antigens, the more useful memory will be.

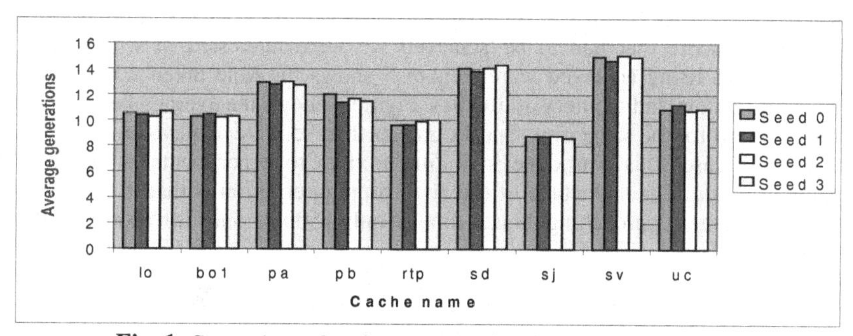

Fig. 1. Comparison of performance on nine different cache log files.

Figure 2 shows the performance on one log file (`lo`) for memory sizes from 0 to 1000 (the maximum possible), and numbers of randomly generated antibodies from 10 to 2000. Here performance is measured by the length of time the system takes to learn 1000 antigens. We discovered, by comparing this graph with the corresponding graph for number of generations, that time is the better measure. With a large number of randomly generated antibodies, the time taken is high but the number of generations is low. This is because the size of the population of antibodies affects the time taken, the more antibodies there are the more time it takes.

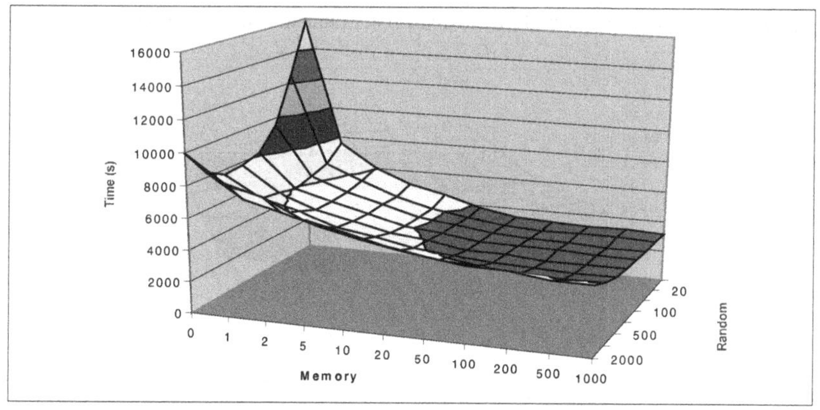

Fig. 2. Performance with different numbers of random and memory antibodies.

The graph shows that the memory size generally has a much larger impact on the speed of learning than the number of random antibodies. Only when there is very little or no memory does the number of random antibodies have much effect. The graph also shows that a memory size of 50 is almost as good as the maximum of 1000, suggesting that repetition in the log files generally happens within a short space of time. If antigens took a long time to be repeated such a short memory would not be very useful. It also shows that the system does not have to remember everything (and thereby use a large amount of space) to be effective. We also experimented with a different stopping criterion for the learning, which involved limiting the amount of antigen available for antibodies to match. The learning was stopped when there was none left. When the initial amount of antigen was 1000 we achieved similar results as shown in Figure 2 for the time taken to learn. However, since we could no longer set the minimum score that had to be achieved we were interested in what the best antibody scores being achieved were. Figure 3 shows the time taken with different numbers of random and memory antibodies. Figure 4 shows the average best antibody scores achieved for the same runs. Scores greater than 95 were achieved in most runs, and in many cases this was done in a similar time to experiments using the old stopping criterion. The advantage of the new stopping criterion is that it can achieve average best scores approaching 98 in a time of only 7500 seconds for 1000 antigens

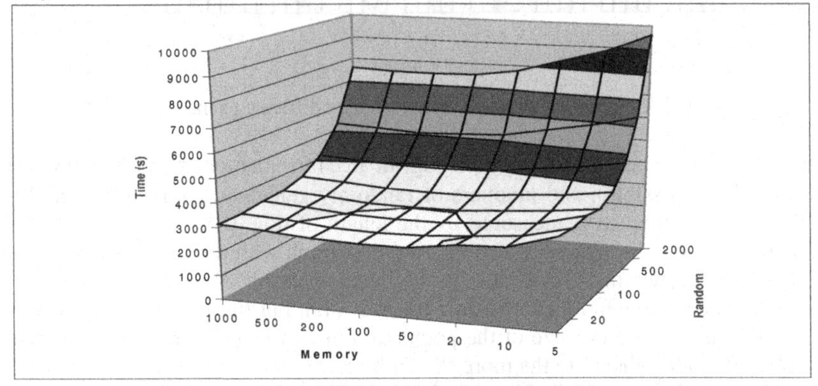

Fig. 3. Performance with different numbers of random and memory antibodies.

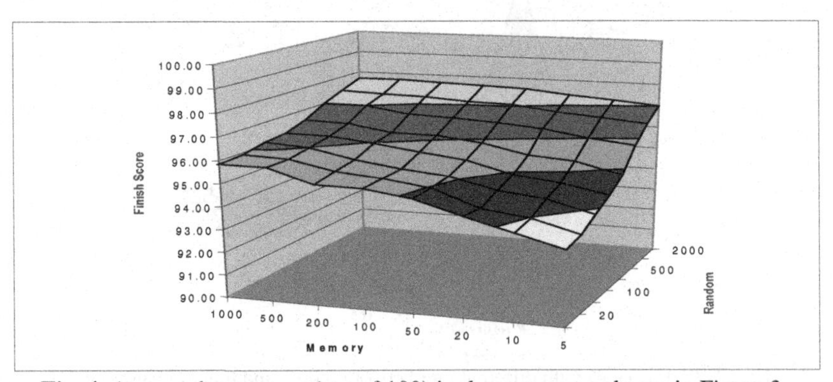

Fig. 4. Average best scores (out of 100) in the same runs shown in Figure 3.

4 Conclusions

We believe that immunological principles can be fruitfully applied to network management, particularly the problem of resource identification. In this work we have presented a simple model of the immune system as a first step towards this. Our immune system model proved to be capable of learning and remembering the structure of novel antigens, encoded from URLs in cache access log files. What is more, the model could achieve this even with limited resources We also suggest that the success of the system is somewhat dependent on patterns in the antigens it must learn about. This may limit the domains to which it can be applied. We will need to consider possible real world applications for the system, which means considering very carefully how antigens are to be encoded so that similar real world situations map to similar antigens, and also how the antibodies can encode real world solutions for these situations.

References

Di Caro, G., & Dorigo, M. (1998). AntNet: Distributed stigmergetic control for communications networks. *Journal of Artificial Intelligence Research*, **9**, pp. 317-365.

Hightower, R., Forrest, S., & Perelson, A.S. (1995). The evolution of emergent organisation in immune system gene libraries. In L.J. Eshelman (ed.), *Proceedings of the Sixth International Conference on Genetic Algorithms*, San Francisco, CA: Morgan Kaufmann, pp. 344-350.

Hofmeyr, S.A., & Forrest, S. (1999). Immunity by design: An artificial immune system. In *Proceedings of the Genetic and Evolutionary Computation Conference (GECCO)*, San Francisco, CA: Morgan Kaufmann, pp. 1289-1296.

Hunt, J.E., & Cooke, D.E. (1996). Learning using an artificial immune system. *Journal of Network and Computer Applications*, **19**, pp. 189-212..

Roadknight, C.M., & Marshall, I.W. (2000). Adaptive management of a future service network using a bacteria inspired genetic algorithm. In *Late Breaking Papers at the 2000 Genetic and Evolutionary Computation Conference*, Las Vegas, Nevada, pp. 331-337.

Seth, A. (1998). A co-evolutionary approach to the call admission problem in telecommunications. In J. Halloran, & F. Retkowsky (eds.), The Tenth White House Papers: Graduate Research in the Cognitive and Computing Sciences at Sussex, Cognitive Science Research Paper 478, School of Cognitive and Computing Sciences, University of Sussex.

Somayaji, A., Hofmeyr, S.A., & Forrest, S. (1998). Principles of a computer immune system. In *1997 New Security Paradigms Workshop*, ACM, pp. 75-82

Timmis, J., Neal, M., & Hunt, J. (2000). An artificial immune system for data analysis. *Biosystems*, **55**, pp. 143-150.

I Like What I Know: How Recognition-Based Decisions Can Structure the Environment

Peter M. Todd[1]* and Simon Kirby[2]

[1] Center for Adaptive Behavior and Cognition
Max Planck Institute for Human Development
Lentzeallee 94, 14195 Berlin, Germany
ptodd@mpib-berlin.mpg.de
http://www-abc.mpib-berlin.mpg.de/users/ptodd

[2] Language Evolution and Computation Research Unit
Department of Theoretical and Applied Linguistics
University of Edinburgh, Scotland, UK
simon@ling.ed.ac.uk
http://www.ling.ed.ac.uk/~simon/

Abstract. Cognitive mechanisms are shaped by evolution to match their environments. But through their use, these mechanisms exert a shaping force on their surroundings as well. Here we explore this cognition-environment interaction by looking at how using a very simple cognitive mechanism, the recognition heuristic for making choices, can result in some objects in the world being much more often recognized and "talked about" than others. An agent-based simulation is used to show what behavioral factors affect the emergence of this environmental structure.

1 Introduction: Cognition Creating Environment Structure

Some things are more famous—more recognized, more talked about—than others. This unsurprising fact about the world can be useful when making decisions: If we have the choice between a recognized and an unrecognized option, we can make a rapid (and often adaptive) decision based on recognition alone. What is more interesting is that some things are *much* more famous, or successful, or noteworthy, than others. The few most productive scientists are cited far more than the majority of their colleagues, and the few most successful bands sell far more records than their competitors. If we plot such quantities as number of citations, or records sold, or city population, or Olympic gold medals won, against the rank-ordered list of authors, bands, cities, or countries, we see a characteristic J-shaped curve (where the "J" is on its side), indicating very few objects with very high values and most objects with rather low values [5]. This striking aspect of environment structure appears to emerge commonly in domains that

* We are indebted to Martin Dieringer for his work programming this model.

J. Kelemen and P. Sosík (Eds.): ECAL 2001, LNAI 2159, pp. 166–175, 2001.

people think about, talk about, and make decisions about. The question we want to explore here is, how does this structure arise?

The obvious answer is that there is some underlying structure in the world that our decision-making and knowledge simply tracks. Some researchers just produce more important work, and some cities are located in more resource-rich locations. But these pre-existing differences do not explain the entirety of this phenomenon in all domains—the action of intelligent agents making choices and communicating information about their world can also shape the distribution of knowledge and choices in the world. This is a specific instance of the more general question of how cognition and environment act to affect each other. Cognitive mechanisms are shaped by their environments, both through evolutionary selection across generations and through learning and development within lifetimes. But by making decisions that guide actions which in turn alter the surrounding world, cognitive mechanisms can also shape their environments in turn. This mutual shaping interaction between cognitive structure and environment structure can result in coevolution between the two over extended periods of time. What will the dynamics of such feedback loops be, and what patterns of structure in both cognition and environment will emerge?

We can observe the outcomes of these interactions in the world around us, for instance in the J-shaped distributions described above. But to understand *how* such interactions work to produce environment structure, we would like to start out with systems that are initially unstructured, particularly lacking pre-existing differences in importance between objects, and then observe how structure can emerge over time. Thus, to get a handle on the questions of environment-behavior coevolution, we can construct simple simulation models of decision-making agents interacting over time with each other and the rest of their environment. We begin here with a model incorporating perhaps the simplest decision heuristic, the *recognition heuristic*. With this model we can study how the use of a simple heuristic shapes the distribution of choices in (and knowledge about) an environment full of different options. In particular, we are interested in whether and in what circumstances the recognition heuristic can itself create a clumpy environment, in which some options are recognized, and chosen, much more often than others (e.g., following a J-shaped distribution). Such environment structure can in turn make the recognition heuristic ecologically rational, that is, able to make beneficial choices. In the rest of this paper, we indicate past work on how cognitive mechanisms can shape their environments and introduce the recognition heuristic as a particularly simple cognitive mechanism to study, describe our model of recognition-based decision making in a group of moving, communicating agents, and discuss some of our initial results on the factors affecting the emergence of environment structure via simple cognition.

2 Background: Behaviors, Environments, and Recognition

Behavior by definition affects the environment (including the behaving organism) in some way, usually in a small way, but sometimes in a manner that has lasting consequences for further behavior. Important theories of how organisms can affect their own selective environments (including their cultures) have been developed, notably dual inheritance (gene/culture) ([2]) and niche construction ([10]) models. These models however typically rely on very simplified behavioral strategies analyzed at a population level. Because agent-based simulation enables the study of population-level effects of decision-making mechanisms operating at an individual level, it allows more explicit models and analyses. This approach has proven particularly fruitful in studying the evolution of language—by simulating populations of agents communicating with an evolving syntactic mechanism, one can gain insights into the way that this mechanism shapes, and is shaped by, the coevolving language being used in the population ([7], [8], [9]).

But language mechanisms are somewhat complex, and we wanted to start here with a simpler cognitive mechanism to explore environment construction. The simplest kind of choice—numerically, at least—is to select one option from two possibilities, according to some criterion on which the two can be compared. The simplest form of this choice occurs when the only information available is whether or not each possibility has ever been encountered before. In this case, the decision maker can do little other than to rely on his or her own partial ignorance, choosing recognized options over unrecognized ones. This kind of "ignorance-based reasoning" is embodied in the recognition heuristic [4]: When choosing between two objects (on some criterion), if one is recognized and the other is not, then select the former. For instance, when deciding which of two cities is larger, say Palo Alto or Weed, California, the recognition heuristic would lead (most of) us to choose Palo Alto (which is the correct choice).

Using the recognition heuristic will be adaptive, yielding good decisions more often than would random choice, in those environments in which exposure to different possibilities is positively correlated with their ranking along the decision criterion being used. Such correlations are likely to be present in species with social information exchange where important environmental objects are communicated and unimportant ones are ignored. For instance, because we talk and hear about large cities more often than small cities (because more happens in them, more people come from them, etc.), using recognition to decide which of two cities is larger will often yield the correct answer (in those cases where one city is recognized and the other is not). Use of the recognition heuristic appears widespread: For instance, Norway rats prefer to eat foods they recognize as having smelled on the breath of another rat over unrecognized foods [3], even when more useful information is available to guide their choice (such as that the rat whose breath they smelled was ill from eating that food—for a model to explain this ignorance of further information, see [11]). Humans make decisions in accordance with the recognition heuristic both in the lab (e.g., choosing recognized cities as larger than unrecognized ones—see [4]) and outside (e.g., choosing stocks to invest in based on recognition of the company, a strategy which can be adaptive in a financial sense as well—see [1]).

3 A General Model of Agents Deciding with Recognition

In the most general setting for a model of agents acting in a world and making decisions on the basis of recognition, we can consider a space of options and a population of agents, each of whom can choose one option at a time. The options are identified just by a unique name, and so the agents can only use name recognition (or lack thereof) to choose between them. We can think of each agent as making a series of choices between two named options presented simultaneously. The environment-structure outcome we are interested in is how the population of agent-choices is distributed over the set of available options at some point in time—more specifically, are choices made by agents evenly distributed over options, or are they clumped or J-distributed?

Even with this very simple setup, we have some variations to consider. We can affect the decisions made in at least four main ways:

- 1. How recognition is determined and remembered over time
- 2. Whether and how recognition information is communicated between agents
- 3. How options are selected to be presented as choice-pairs to an agent
- 4. How choices are made between pairs of presented options

Before we discuss how specific instantiations of these factors affect emergent environment structure, though, we need to make the simulation scenario itself a bit more specific.

4 A More Specific Model: Agents Affecting "City" Sizes

To make a more concrete and intuitively "recognizable" model, we start with a simulation that can be related to the city-size choice task mentioned earlier. Now the options are essentially place-names (locations) in a grid, and agents make choices between these locations and then move to the chosen place. More specifically, agents live on a grid-world (multiple agents can inhabit each location) and have a (recognition) memory of places they have been and have heard about from other agents. Time proceeds in steps, and at each step an agent has a chance of "talking" with other agents about some of the locations it recognizes (and hearing from them about what they recognize in turn). Every so often, each agent is presented with a choice between two locations selected from the grid-world, and the agent must choose one of these locations on the basis of its recognition knowledge, using the recognition heuristic. (Thus if it recognizes one location and not the other, it will choose the former; otherwise the choice will be made at random.) The agent then moves to the chosen location. Even less often, the agent has a chance of forgetting about some of the locations it recognizes. (This can equivalently be thought of as the old agent dying and a new child agent being introduced who first learns some reduced set of what its parent knew.) By instantiating this all in a two-dimensional space, we can monitor how agents are distributed over this space over time, and look for actual population clusters. Given this overview, we can now turn to a consideration of the specific instances possible for the four factors listed in the previous section.

4.1 How Recognition Is Determined and Remembered

An agent could recognize a location after visiting that location, hearing about it from another agent, or seeing it as one of the options presented in its occasional choice-pairs. Here we always incorporate the first source of recognition knowledge, and turn on or off communication to test its effect; we leave out the third indirect form of experience (though even such minimal exposure can have a strong impact in humans, causing what has been called the "overnight fame" effect—see [6]).

It became clear in early runs of our system that individual recognition knowledge must also disappear somehow, or else each agent ends up recognizing nearly all locations and can no longer use the recognition heuristic to choose between them. In this model, recognition memory is reduced at regular (user-specified) intervals as mentioned earlier. Different sets of locations to be forgotten can be selected: a fixed percentage of recognized locations chosen at random; or locations that have not been experienced recently; or the least-often experienced locations. The latter two can also be combined in a decaying memory trace, which we will not elaborate on here—instead, we will focus on the first two forms of forgetting.

4.2 How Recognition Knowledge Is Communicated between Agents

It is possible that agents acting independently on the basis of their own individual experience, choosing to go to locations that they personally recognize by having visited before, could suffice to create emergent environment structure (though we have not seen this happen yet). But it is more likely, and more realistic, that communication between agents will enhance any clustering of choices in the space of options (here, locations)—the "social computation" enabled by a communicating population of simple decision-making agents should lead to greater environmental impact (as has been found in simulations where the interactions of many generations of simple language learners enable syntax to emerge—see [8]). Thus we allow agents to tell each other about locations that they recognize. We must specify who can talk to whom, how often, and about what. In these studies we let agents occupying the same location talk to each other; we set how many other agents a given agent can listen to during each time step, and how many locations each of those other agents can mention to the given agent (who then enters those locations into its memory as "recognized via communication"). Each agent can mention its allotment of locations in the same orders that are possible for forgetting (but reversed): random, newest-experienced first, or most-experienced first. (Note that with communication included this model bears a resemblance to Boyd & Richerson's model of conformist transmission [2], but here knowledge, rather than choices, is communicated, and individual decision making is much more the focus.)

4.3 How Options Are Selected for Presentation as Choice-Pairs

Humans and other animals face choices between recognized and/or unrecognized options all the time—but how are the options we simultaneously face actually determined? At times they may be randomly encountered, or generated in a more systematic way by the environment; in other situations, other individuals may present us with the things to choose between. Similarly, the agents in our model can encounter their two options (at user-specified time intervals) in a few ways. The two locations can be selected at random from the entire grid-world, or in a locally-biased fashion by specifying a distance-factor that determines how rapidly the chance of seeing a given location as an option falls off with distance between the deciding agent and that location (according to $probability = 1/distance^{distancefactor}$). Or the two locations can be chosen from those recognized by other agents in the current location, instantiating another form of local communication, or in proportion to how many agents are at each location in the grid-world, representing "advertising" being sent from cities near and far saying "move here!" (with more advertising coming from larger cities).

4.4 How Choices Are Made between Presented Options

Once a pair of options is presented to an agent, it must use its recognition knowledge to make a choice. This choice can be made using the recognition heuristic as discussed earlier, or it could be made randomly, to provide a basis of comparison with the impact of the recognition heuristic. (Additionally, the "anti-recognition heuristic" could be used, always picking unrecognized over recognized options, which would result in greater individual exploration rather than conformity. This is certainly useful to have in societies and other systems facing an explore/exploit tradeoff, but which we will not explore further here—see [2], [13].) Choices could also be made on the basis of recency or amount of recognition (as for forgetting and communication). Finally, we can allow agents to refuse to make a choice when there is no clear basis for selecting one option over another, for instance, when both are recognized or both unrecognized; such refusal would allow their rate of use of the recognition heuristic to increase, which could enhance its power to shape the environment.

5 Initial Results

Given some combination of the factors just described, we can specify an initial state of the world (how many agents, where they are located, how old they are, what they know if anything at the start—here, we use random positions for some number of agents with random ages and no starting knowledge) and then watch over many timesteps how the environment (distribution of agents and knowledge) changes. (The system is implemented in Java and runs at a good clip in real time for populations of several hundred agents.) There are a few important values to monitor as the simulation progresses: First, we look at the distribution of agents per location, graphed with the locations along the x-axis ordered from left to right by "population size" so that we can see when this distribution becomes more or less J-shaped. Second, we look at the correlation between the recognition for each location (that is, the number of agents that recognize it) and its population size (the number of agents on it)—the higher this correlation is, the more effective and valid is recognition knowledge for choosing the more-populous location to move to (though keep in mind that there is no selective pressure or other advantage for agents to go to more or less crowded locations). And third, we keep track of variables regarding the agents and their knowledge: the mean number of locations recognized, the mean number of decisions made, the number made when recognition distinguished between the choices, and the number of choices to move to the larger of the pair of locations.

We have not yet explored all combinations of the behavioral factors available in this model, but the initial indication is that substantial clustering of knowledge and choices is not so easy to come by. Consider an illustrative example run, in which we used a 10x10 world, 200 agents, a 200-timestep "memoryspan" (meaning that forgetting occurred for each individual every 200 timesteps, plus or minus a 20-timestep standard deviation), a 10-timestep interval between choices and moves (with standard deviation of 4), randomly selected locations as choice options, 50% chance of communicating on each timestep, and 1 location heard from 1 agent every time this communication takes place. The results of this run of 1000 timesteps are shown in Fig. 1. Here we see that the population at this moment is essentially J-distributed, with very few locations with the largest number of agents (10) and most locations with only a few agents (3 or fewer), but this changes from timestep to timestep and can often appear less skewed. The population vs. recognition correlation over time begins low (around 0) in the initial random scattering of individuals, and then as they move around this correlation rises above 0.4 by timestep 300 before falling and rising again (the reason for these apparent cycles is not yet clear). The average number of locations recognized by each individual asymptotes around 38, with 30 of these recognized from hearing about them from other agents, and only 8 coming from "personal" experience (just the proportions one would expect from the rates of choice/movement and communication). Finally, the average number of decisions made so far by each individual (counted since the last time they hit their memoryspan and forgot past experiences) levels off around 9, with half of those made using the recognition heuristic (as would be expected when nearly half of the locations are recognized by each individual), and half again of those being choices to move to the larger of the two locations.

It is this last value—the proportion of recognition-based decisions that result in movement to the higher-population location—that is crucial. Here, with half of those decisions going each way, we see that recognition knowledge has no effect on movement to more populous locations, meaning that the agents will not be able to cluster strongly. This is because the locations that agents recognize are mostly ones they have heard about from others, and hence are usually locations that the others also heard about, meaning that the last time anyone was actually personally *in* the location could have been a long time ago—and any cluster of individuals that were there then could have long since disappeared. Thus, for clusters to build, the information communicated between agents has to be made more timely. There are two main ways to do this: change what the agents remember (and so can talk about at all), and change which locations they choose to talk about at any instant.

When we change both of these factors, making agents preferentially forget the oldest-experienced locations and communicate the most-recently-experienced ones, we do indeed see more environment structure emerge. Now the population vs. recognition correlation averages around .4 (peaking at .6), and a few locations end up with 12 or 13 agents in them. The average recognition per agent falls to 11 locations, because most of what they hear about now they have already heard before. As a consequence, the number of recognition-based decisions also falls, to 2.5 out of 9, but the proportion of these choices to larger-population locations rises slightly, to 60%. This is enough to produce a more-structured environment, but the effect is not as large as might be possible, and our search continues for what factors may be able to increase the environment structure further.

6 Conclusions and Extensions to the Model

In our simple model of the impact of recognition-based decision making on the environmental distribution of choices and agents, it appears that introducing time-dependencies into the memory and communication of agents is sufficient (and perhaps necessary) for allowing population clusters to emerge. This fits in well with the ecological rationality perspective that cognitive structure and environment structure should match to produce adaptive behavior [12]—given that the environment changes over time, agents' knowledge should also change so that old possibly outdated information is no longer relied on.

So far, our search for the factors that most enable environment construction has been carried out by hand, manipulating the factors ourselves and running new simulations. But this process could potentially be sped up by enabling the coevolution of the recognition heuristic and the structure of the environment, allowing agents to inherit the exact strategies they will use—both their decision strategies (e.g., recognition-based or anti-recognition-based) and their communication strategies (e.g., tell everyone everything you've seen, or the most recent locations, or none). One question of interest in such a coevolutionary setting is whether there is selection to "the edge of chaos" (or at least to a cognitively advantageous sweet-spot) such that agents end up recognizing around half of

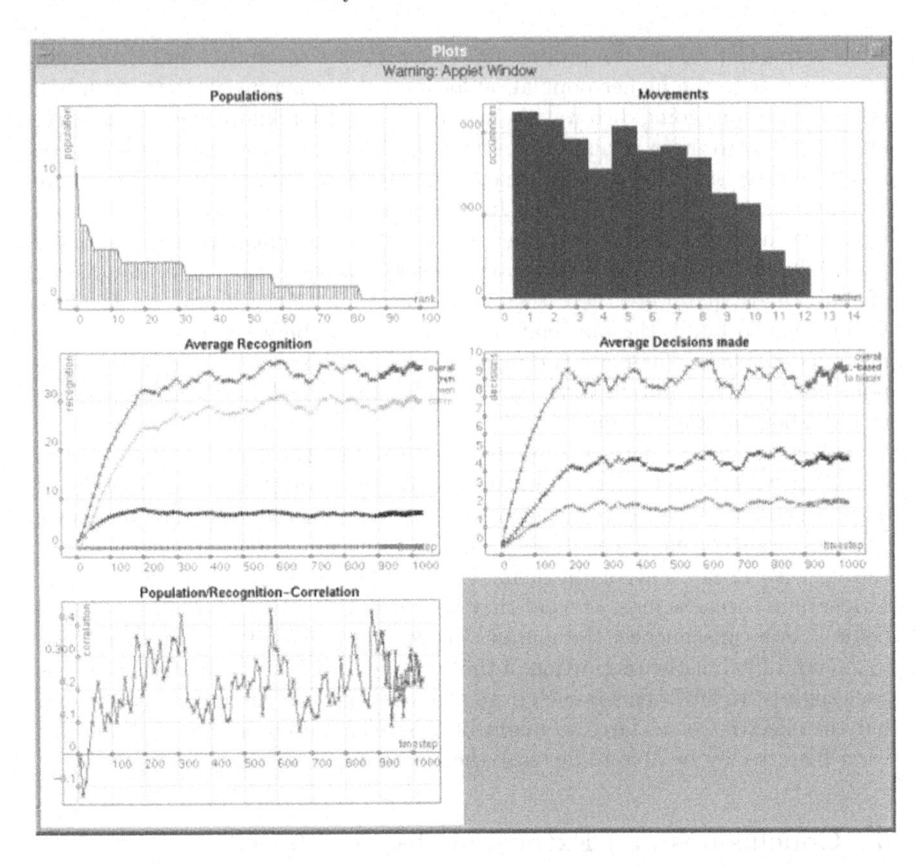

Fig. 1. Statistics for a sample run after 1000 timesteps, showing the distribution of agent population per location against ranked location for this timestep (basically J-shaped), distribution of frequency of movements by their distance, mean recognition knowledge per agent over time (showing from top to bottom total number of locations recognized, number recognized via communication, and number recognized via own experience), mean number of decisions made per agent over time (top to bottom: total, recognition-based, and number that were to the larger-population location), and correlation between population-size and recognition of each location over time.

the population-clusters they could move to and thus can continue to make good recognition-based decisions. Finally, we can consider the ecological rationality of the recognition heuristic in these situations by introducing some fitness differential into the environment—that is, by making some locations "better" or "worse" somehow than others, to see if (and how fast) the population of agents discovers this additional environment structure. If we build in a few "bad" locations where the death rate is higher than elsewhere, for instance, then presumably fewer agents will leave this location and spread the (neutral) word about it, so over time it will be less visited. We could then also give agents the ability to communicate a valence, good or bad, about the locations they have visited (which could be perfectly remembered by audience members, or eventually forgotten, which seems to be the fate of quality tags in other domains), to see how this alters the physical (population-based) and cultural (knowledge-based) environments of the population as a whole. These agents may not know much, but they like what they (and others) know, and that can be enough to change the structure of their world.

References

1. Borges, B., Goldstein, D.G., Ortmann, A., Gigerenzer, G.: Can ignorance beat the stock market? In: Gigerenzer, G., Todd, P.M., ABC Research Group: Simple Heuristics That Make Us Smart. Oxford University Press, New York (1999) 59–72
2. Boyd, R., Richerson, P.J.: Culture and the Evolutionary Process. University of Chicago Press, Chicago (1985)
3. Galef, B.G., Jr.: Social influences on the identification of toxic foods by Norway rats. Animal Learning & Behavior **18** (1987) 199–205.
4. Goldstein, D.G., Gigerenzer, G.: The recognition heuristic: How ignorance makes us smart. In: Gigerenzer, G., Todd, P.M., ABC Research Group: Simple Heuristics That Make Us Smart. Oxford University Press, New York (1999) 37–58
5. Hertwig, R., Hoffrage, U., Martignon, L.: Quick Estimation: Letting the Environment Do the Work. In: Gigerenzer, G., Todd, P.M., ABC Research Group: Simple Heuristics That Make Us Smart. Oxford University Press, New York (1999) 209–234
6. Jacoby, L.L., Kelley, C., Brown, J., Jasechko, J.: Becoming famous overnight: Limits on the ability to avoid unconscious influences of the past. Journal of Personality and Social Psychology **56** (1989) 326–338
7. Kirby, S., Hurford, J.R.: Learning, culture and evolution in the origin of linguistic constraints. In: Husbands, P., Harvey, I. (eds.): Fourth European Conference on Artificial Life. MIT Press/Bradford Books, Cambridge MA (1997) 493–502
8. Kirby, S.: Syntax out of learning: the cultural evolution of structured communication in a population of induction algorithms. In: D. Floreano, J.D. Nicoud, F. Mondada (eds.): Advances in Artificial Life. Lecture notes in computer science, Vol. 1674. Springer-Verlag, Berlin Heidelberg New York (1999)
9. Kirby, S.: Spontaneous evolution of linguistic structure: An iterated learning model of the emergence of regularity and irregularity. IEEE Transactions on Evolutionary Computation (in press).
10. Laland, K.N., Odling-Smee, J., Feldman, M.W.: Niche Construction, Biological Evolution and Cultural Change. Behavioral and Brain Sciences **23** (2000) 131-146
11. Noble, J., Todd, P.M., Tuci, E.: Explaining social learning of food preferences without aversions: An evolutionary simulation model of Norway rats. Proceedings of the Royal Society of London B: Biological Sciences **268** (2001) 141–149.
12. Todd, P.M.: The ecological rationality of mechanisms evolved to make up minds. American Behavioral Scientist **43** (2000) 940–956
13. Wilson, S.W.: Explore/exploit strategies in autonomy. In: In P. Maes, M.J. Mataric, J.-A. Meyer, J. Pollack, S.W. Wilson (eds.): From animals to animats 4: Proceedings of the Fourth International Conference on Simulation of Adaptive Behavior. MIT Press/Bradford Books, Cambridge MA (1996) 325–332

Bio-Language for Computing with Membranes

Angel V. Baranda[1], Juan Castellanos[1],
Fernando Arroyo[2], and Carmen Luengo[2]

[1] Dept. Inteligencia Artificial
Facultad de Informática, Universidad Politécnica de Madrid
Campus de Montegancedo, Boadilla del Monte, 28660 Madrid, Spain
{jcastellanos}@fi.upm.es
http://www.dia.fi.upm.es
[2] Dept. Lenguajes, Proyectos y Sistemas Informáticos
Escuela de Informática, Universidad Politécnica de Madrid
Crta. de Valencia km. 7, 28031 Madrid, Spain
{farroyo, cluengo}@eui.upm.es
http://www.lpsi.eui.upm.es

Abstract. This paper presents a bio-language that integrates data structures and processes for a formal computational model from the field of computing with membranes. A bio-language is a new concept extracted from biological behaviour. In order to define such a bio-language, the static and dynamic structures of a Transition P system are formalised.

1 Introduction

Natural Computing is a new field of research trying to imitate the way nature computes. Of course, nature not only computes at neural or genetic level, but also at the cellular level. Many biological systems are well defined by various types of membranes, starting with cell membranes, going to the skin of organisms. Any biological system can be viewed as a hierarchical construct where the interchange of materials takes place. This process of materials interchange can be interpreted as a computing process [1]. Taking into account this point of view, we can say that membrane computing is a general architecture of "living organisms", in the sense of Artificial Life [2].

Computing with Membranes uses a hierarchical membrane structure in the same sense that it is used in the chemical abstract machine of *Berry and Boudol* [3], and the basic ideas of evolution rules as multisets transformation are inspired in G systems introduced by *Banâtre* [4].

Alive data structures can be defined as data structures containing the code they need to process them. In order to achieve program execution it is needed to add a reactive ingredient, which will start the computational process. Moreover, an alive data structure must be a structure defined recursively, containing active principles or distributed functions which define the code to be executed over the data structure components in a parallel way. This transformation process or bio-process is able to modify the own alive data structure.

J. Kelemen and P. Sosík (Eds.): ECAL 2001, LNAI 2159, pp. 176–185, 2001.

In order to simulate a bio-process, it is necessary to define a global function F which transforms the alive data structure. This simulation process will be an external process, so a compilation process will generate this global function in a transparent manner for users.

A bio-language is a bio-processes programming language. Therefore, different types of biological processes need different types of bio-languages. In this paper, we will describe a formal way to obtain a bio-language for simulating computing with membranes.

Following points will describe Transition P systems as a particular system of Computing with Membranes, passing through static and dynamic structure of P systems until giving a formal definition of a Transition P system. This formal definition will facilitate the bio-language specification.

2 Informal Description of Transition P Systems

Within Natural Computing, P systems is a new computing model based on the way nature organises cellular level in living organisms. Different processes developed at this level can be thought as computations.

Among different variants of P system, we have chosen Transitions P system as objects for our studying. Transition P systems, and many other membrane computing models, have two different components: the static one or super-cell, and the dynamic one compounded by evolution rules.

Describing the super cell concept, we can say that a super cell is a hierarchical structure of biuniquely labelled membranes. A membrane can contain none, one or more than one membrane. The external one is named "skin" and it can never be dissolved. Membranes without inner membranes are called elementary membranes. A membrane structure can be represented in a natural way as a *Venn diagram.*

Membranes define regions, that is the area in-between a membrane and and the membranes directly included in it, if there are any. Regions contain a multiset of objects (symbols from a given alphabet) and a set of evolution rules. In some sense, rules consume objects from its region and send objets to other adjacent regions or to their regions. It can be also defined a priority relationship defining a partial order among rules of regions.

Transition P systems evolve due to evolution rules defined inside their regions. This evolution process is done in a parallel way and in an exhaustive manner, that is, each executable rule is executed in a non-deterministic way compiting with the other executable rules. Moreover, Transition P systems are able to modify their static structure when they are evolving; this is possible because of some evolution rules of regions are able to dissolve the external membrane defining the region [1].

Finally, we can say that Transition P systems compute sending objets to a determined region named i_0, and computation finish when any rule can be executed in the system.

3 Static Structure of Transition P Systems: Algebraic Definition

Once Transition P systems have been described informally, we will present them in a more formal manner.

In order to provide a static description of a Transition P system, it will be needed to define the following elements: Multisets, Evolution Rules, Trees, Regions and finally Transition P systems.

Therefore, it is very important to determine data structures for defining multisets, evolution rules and regions and the operations that can be done over such a data structures [5].

3.1 Multisets

The natural way to represent object inside regions is the multiset. The region contains a collection of different objects, so they can be represented by a determined multiset. Let us to define more precisely a Multiset.

Let U an arbitrary set and let N the natural number set: a multiset M over U is a mapping from U to N.

$$\begin{aligned} M : U \rightarrow N \\ a \rightarrow Ma \end{aligned} \tag{1}$$

An important subset of a multiset is called *Support* of the multiset. This set is formed by those elements of U, which their image is higher than zero.

Support Multiset: $\mathit{Supp}\, M = \{a \in U | Ma > 0\}$.

Let us give some more definition over multisets.

Definitions: Let M, M_1 and M_2 multisets over U, then:

Empty Multiset: Is the 0 map, that is, $\forall a \in U$, $0a = 0$.

Multiset Inclusion: $M_1 \subset M_2$ if and only if $\forall a \in U$, $(M_1\, a) < (M_2\, a)$.

Multiset Union: $\forall a \in U$, $(M_1 \cup M_2)\, a = (M_1\, a) + (M_2\, a)$.

Multiset Difference: $\forall a \in U$, $(M_1 - M_2)\, a = (M_1\, a) - (M_2\, a)$, if $M_1 \subset M_2$.

These definitions will help us to formalise the different components of P systems.

3.2 Evolution Rules

Once have been defined formally multiset, we can define evolution rules in terms of multisets.

Evolution rules are responsible for Transition P system evolution. They make evolve their regions. An evolution rule is formed by one antecedent and by one consequent. Both of them could be represented by multiset over different sets. Moreover, an evolution rule can make disappear the external membrane of its region, which produces changes in the Transition P system static structure. With these considerations, we can define an evolution rule as follows:

Let L a label set, let U an object set, and let $D = \{out, here\} \cup \{in\, j | j \in L\}$ the set of region labels to which a rule can send objects. An evolution rule with label in L and objects in U is a tern (u, v, δ), where u is a multiset over U, v is a multiset over $U \times D$ and $\delta \in \{dissolve, not\ dissolve\}$.

With this evolution rules representation, it can be defined several operations over evolution rules.

Definitions: Let $r = (u, v, \delta)$, $r_1 = (u_1, v_1, \delta_1)$ and $r_2 = (u_2, v_2, \delta_2)$ be evolution rules with labels in L and objects in U. Let $a \in U$ and n a natural number. Then:

Evolution rules add: $r_1 + r_2 = (u_1 \cup u_2, v_1 \cup v_2, d_1 \vee d_2)$.

Product of an evolution rule by a natural number: $n\ r = \sum_{i=1}^{n} r$.

Inclusion of evolution rules: $r_1 \subset_u r_2$ if and only if $u_1 \subset_u u_2$.

Input of an evolution rule: $Input\ r = u$.

Dissolve: $Dissolve\ r = \delta$.

Now it is necessary to define some functions over evolution rules related to the rule consequent. These functions will provide important information over what happens when the rule is applied.

Membrane labels set the rule is sending objects:

$$Ins\ r = \{j \in L | (\_, in_j) \in Supp\ v\}.$$

Evolution rule's outputs:

$$\begin{aligned}(OutputToOut\ r)a &= v(a, out)\\ (OutputToHere\ r)a &= v(a, here)\\ ((OutputToIn\ j)r)a &= v(a, in\ j)\end{aligned}$$

These last functions $OutputToOut$, $OutputToHere$, $OutputToIn$, return the multiset that the rule is sending to his father ($ToOut$), to itself ($ToHere$) and to a determined region ($ToIn \cdots$). They will be very useful in order to define the dynamic structure of a Transition P system.

3.3 Trees and Membrane Structure

A Tree over an arbitrary set U can be defined as a pair where the first component is an element in U and the second one is a set of trees over U.

In this definition the second element of the pair can be the empty set, in which case the element is a leaf.

Over trees can be defined the Multiplicity function that determines the number of occurrences of a given element a in the tree.

Multiplicity: Let $T = (u, S)$ be a tree, where $u \in U$ and S is a set of trees over U, then

$$(Mult\ T)a = \begin{cases} 1 + \sum_{s \in S}(Mult\ s)a \text{ if } a = u \\ \sum_{s \in S}(Mult\ s)a \quad\quad \text{if } a \neq u \end{cases} \tag{2}$$

The Transition P system membrane structure can be defined as a tree in which U is the empty set. In this case, the skin membrane is the roof of the tree and the elementary membranes are the leaves.

3.4 Regions

Membranes define regions, that is, a region is the inter-membrane area. Regions in the membrane structure are uniquely labelled in some set L; contain objects, and evolution rules with a priority relationship defining a partial order among rules of the region.

Therefore, a region can be defined in a more formal manner as:

Let L a label set and U an object set. A region with labels in L and objects in U is a tern $(l, \omega, (R, \rho))$ where $l \in L$, ω is a multiset over U, R is a set of evolution rules with labels in L and objects in U and ρ is a partial order relation over R.

3.5 Transition P System

At this point, all data structures we need to define the static structure of a Transition P system have been presented. So we can define a Transition P system as a tree of regions uniquely labelled. In a much more formal way:

Let L a label set, and let U an object set. A Transition P system with labels in L and objects in U is Π, where Π is a tree of regions with labels in L and objects in U provided $\forall(l \in L), (Mult\ \Pi)(l, \_, \_) < 2$. Moreover, we can say that a Transition P system with labels in L and objects in U is a pair whose first element is a region with labels in L and objects in U and the second one is a set of Transition P systems with labels in L and object in U, and regions are uniquely labelled.

$$\Pi = ((l, \omega, (R, \rho)), \Pi\Pi) \text{ and } \forall l \in L, (Mult\ \Pi)(l, \_, \_) < 2 \tag{3}$$

Where ω is a multiset over U, R is a set of evolution rules over L and U, ρ is a partial order relationship over R and $\Pi\Pi$ is a set of Transition P system with labels in L and objects in U.

This algebraic representation define precisely the static structure of Transition P systems.

We have name it Static Structure because this representation do not take into account evolution in Transition P systems, only represent Transition P systems in a static manner.

4 Dynamic Structure of a Transition P System

With dynamic structure Transition P systems, we are naming everything related to evolution. During evolution, a Transition P system is changing its region contents in several ways. Objects are changed by evolution rules application.

Evolution rules that can be applied in a determined configuration when the P system has evolved cannot be applied anymore, and even, a rule can dissolve the external membrane of the region and then the region disappears from the static structure of the P system sending its objects to its father region and disappearing its evolution rules.

We can find some dependencies of active structures used to define the dynamic structure of a Transition P systems. We need to classify evolution rules in order to know which ones can be applied in a determined Transition P system evolution step. Moreover, if we consider the region content as a multiset over an object set, we can consider evolution rules as a set in order to build a multiset of evolution rules. This new point of view is very useful in order to describe in a simpler manner non-deterministic evolution of Transition P systems.

4.1 Classification of Evolution Rules

First of all, we will define maximal set of a given partially ordered set. This maximal set is formed by those elements that there are not higher elements than they are in the set.

Let $(U, >)$ a partial order over U. Then:

$$((Maximal\,U)\ <) = \{u|(u \in U) \wedge (\neg(\_ < u))\} \tag{4}$$

This definition will be very useful in order to explain the dynamic description of Transition P systems.

Useful Rules: A rule in a rules set is useful over a given labels set L if and only if the $Ins\, r$ is included in L. So we can obtain the Useful rules set by:

Let R an evolution rule set with labels in L and objects in U, let $L' \subseteq L$, then:

$$(Useful\, R)L' = \{r \in R | Ins\, r \subseteq L'\} = \{r|(r \in R) \wedge (Ins\, r \subseteq L')\} \tag{5}$$

Useful Rules in a membrane are those they send objects to their adjacent regions. Therefore we only need to obtain L' as the set of labels from adjacent regions to a given region.

Adjacent regions to a region: Let reg a region of a Transition P system P with labels in L and object in U. Let $(reg, \Pi\Pi)$ the sub-tree of P with root in reg. Let us define adjacent region in P to reg as:

$$ADJACENT_{\Pi}\, reg = \{r|(r, \Pi\Pi') \in \Pi\Pi\} = \{r|(r, \_) \in \Pi\Pi\} \tag{6}$$

Applicable Rules: A rule is applicable over a multiset ω if and only if its antecedent is included in ω. Let R an evolution rules set with labels in L and objects in U, let ω a multiset over U, then:

$$(Applicable\, R)\omega = \{r \in R | Input\, r \subseteq \omega\} = \{r|(r \in R) \wedge (Input\, r \subseteq w)\} \tag{7}$$

Obviously, in a region, only applicable rules can be applied in a determined evolution step in a Transition P system.

Active Rules: A rule r, in a rules set is active over a partial order relationship defined over R if and only if there is not any rule in R higher than r according to the partial order relationship. Notice that such definition is equivalent to the maximal set definition given above.

Let (R, ρ) a partial order relation defined over an evolution rules set R with labels in L and objects in U, then:

$$(Active\ R)\rho = (Maximal\ R)\rho \tag{8}$$

4.2 Classification of Multisets over Evolution Rules

With these definitions, it is possible determine exactly evolution rules that can be applied in order to make evolve the region reg. By definition, they are those rules satisfying following equations:

$$L' = \{l \in L | (l, \_, \_) \in ADJACENT_{\Pi}\ reg\}$$
$$Active(Applicable((Useful\ R)L')\omega)\rho \tag{9}$$

Traditional description of Transition P system dynamic describes evolution in a region in terms of every rule satisfying (9) can be applied in parallel "for all occurrences of rule antecedent multiset, for all regions at the same time"[1]. What this sentence is telling us in terms of operations defined among rules is the following: We must search for a lineal combination of rules satisfying (9) being complete; that is no one more rule satisfying (9) can be added to it.

On the other hand, parallel application of two rules r_1, $r_2 \in R$ at the same evolution step in a region, is equivalent to applying the rule $r_1 + r_2$. In the same sense, we can say that applying n times the same rule in parallel has the same effect that applying the rule defined by nr.

Therefore, parallel rules application inside a region can be substituted by a lineal combination of evolution rules from the region. Moreover, lineal combinations of evolution rules can be represented as multisets over R, the set of evolution rules of the region.

Finally, the region evolve in only one step by application of only one rule formed by a complete lineal combination of evolution rules in R satisfying (9).

Following definition will provide the set of multiset of evolution rules able to make evolve one region in only one evolution step.

Definitions: Let R an evolution rules set with labels in L and objects in U. Let MR, MR_1 and MR_2 multisets over R and let ω a multiset over U.

MultiAdd: This operation performs the transformation of a multiset of evolution rules in a lineal combination of evolution rules. From now on, we will keep in mind this equivalence between lineal combinations of evolution rules and multiset of evolution rules.

$$\oplus MR = \sum_{r \in R} (MRr)r \tag{10}$$

MultiInclusion$_u$: This binary relationship defines a partial order relation over multiset over R. It will be very useful in order to determine which multiset over R will be completes.

$$MR_1 \sqsubseteq_u MR_2 \equiv (\oplus MR_1) \subset_u (\oplus MR_2) \tag{11}$$

MultiApplicable: This is the set of multiset over R (evolution rules set) that can be applied over ω.

$$(MultiApplicable\ R)\omega = \{MR|(MR : R \to N) \wedge ((Input\ (\oplus MR)) \subseteq w)\} \tag{12}$$

MultiComplete:

$$(Multicomplete\ R)\omega = (Maximal((MultiApplicable\ R)\omega)) \sqsubseteq_u \tag{13}$$

The above expression describes the set of complete multisets over $(R, \sqsubseteq_u)$ and ω.

Finally, let P a Transition P system with labels in L and objects in U, let $reg = (l, \omega, (R, \rho))$ one region of P and let $(reg, \Pi\Pi)$ the sub-tree of P which root is reg. The set of complete rules multiset is defined by following equations:

$$LABELS\ = \{l \in L|(l, \_, \_) \in ADJACENT_{\Pi}\ reg\} \tag{14}$$

$$ACTIVES\ = Active(Applicable((Useful\ R)LABELS)\omega)\rho \tag{15}$$

$$COMPLETES\ = (Multicomplete\ ACTIVES)\omega \tag{16}$$

The $LABELS$ set defines the set of adjacent regions'labels to region reg.

The $ACTIVES$ set defines the set of evolution rules that can be applied in order to make the region evolves.

The $COMPLETES$ set defines the set of complete multisets over R (the set of the region evolution rules) capable of making evolve the region in one step by application of one and only one lineal combination of evolution rules. Therefore, there are as many different possibilities for the region evolution as elements $COMPLETES$ has. A non-deterministic election of a multiset in $COMPLETES$ is enough for having non-deterministic evolution in the region.

Once has been formally stated the local non-determinism, it is possible to describe the global non-deterministic parallelism in a Transition P system in terms of regional local non-deterministic parallelism. That is, lets evolve independently every region in the Transition P system applying one of the $COMPLETES$ multiset in each one, and non-deterministic election of a $COMPLETES$ multiset in each region will provide parallel non-deterministic evolution for the Transition P system.

Application of a rule in a region produces changes in the P system. The produced changes can be summarised by:

- Rule antecedent is consumed from region multiset.
- New objects from rule consequent are sent to itself, to its father region or to their ADJACENT regions.

– Membrane is dissolved if the rule dissolves the membrane. In this case, the region and its evolution rules dissapear from the system and the resulting object multiset of the region, after evolution, is sent to its father region.

These changes are produced in parallel for every region in the Transition P system.

5 Programming Transition P System: Bio-Language for Transition P System

Programming Transition P system not is writing a code in a determined programming language. Programming Transition P system is developing a Transition P system architecture -membrane structure- and projecting on it data and operations in order to situate objects and evolution rules inside Transition P system's regions. Code execution is obtained making evolve the P system and results are obtained by examination of a given region in the P system. Of course, code execution is made in a non-deterministic and massive parallel way.

Therefore, and due to P systems particular characteristics, Transition P system language design must be helpful for Transition P system designers. Therefore, we will define our bio-language syntax for defining the static Transition P system structure. We are aware that such structure is changing by evolution. Hence after some evolution steps, it will be useful to check the system evolution. Figure 1 shows the main compilation and decompilation processes during Transition P system evolution.

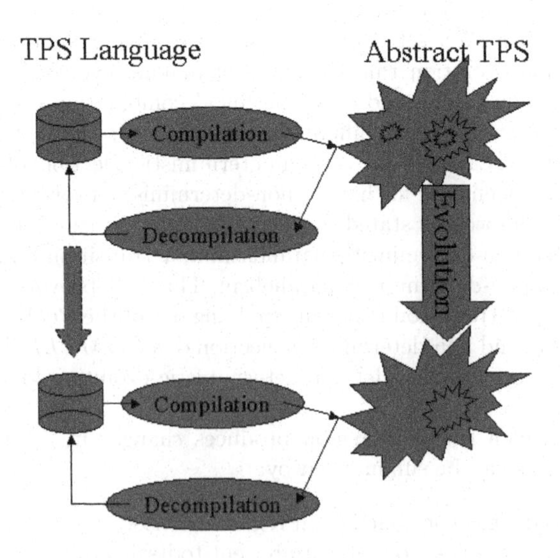

Fig. 1. Compilation and decompilation processes during Transition P system evolution

Now we will describe the syntax of our bio-language by rewriting rules:

- V a set of objects
- Transition P system (TPS): $\Pi \Rightarrow [_l \, Region; \{\Pi\}]_l$
- Region $\Rightarrow$ Objects, Rules, Priorities
- Objects $\Rightarrow \{o^n\}$ where $o \in V, n \in N$
- Rules $\Rightarrow \lambda | rule\{, rule\}$
- Priorities $\Rightarrow \lambda | r_i < r_j \{, r'_i < r'_j\}$ where $i, i', j, j' \in N$
- Addresses $\Rightarrow here|out|l$ where $l \in$ Labels
- ObjTarg $\Rightarrow \{(o, Addresses)^n\}$ where $o \in V$ and $n \Rightarrow 1|2|\cdots$
- Rule $\Rightarrow r_m : Objects \rightarrow ObjTarg \, \delta$ where $\delta \in \{dissolve, \ not \, dissolve\}$

These rewriting rules define the syntax of the TPS bio-language for determining the static structure of a Transition P system at any step of evolution. The internal code execution is embedded in the Transition P system architecture through evolution rules of the Transition P system rules, which define the internal distributed functions of the P system.

6 Conclusions

This paper presents a bio-language for representing an artificial alive data structure in particular for Transition P systems, which can be consider as one example of artificial being. A Transition P system has itself data and code to be executed. The syntax language is not very complicated for helping to develop applications on such a computational paradigm. We hope this approach will be useful for developers of Transition P system architectures.

References

1. G. Paun, Computing with Membranes, Journal of Computer and System Sciences, 61, 1, (2000) 108-143
2. M. Sipper, Studying Artificial Life Using a Simple, General Cellular Model, Artificial Life Journal, 2, 1 (1995), 1-35.
3. G. Berry, G. Boudol, The Chemical Abstract Machine, Theoretical Computer Science, 96 (1992), 217-248.
4. J. P. Banâtre, A. Coutant, D. Le Metayer, A Parallel Machine for Multiset Transformation and its Programming Style, Future Generation Computer Systems, 4 (1988), 133-144.
5. A. Baranda; Castellanos J.; Gonzalo Molina R.; Arroyo, F.; Mingo, L.F., Data Structure for Implementing Transition P System in Silico, Proceedings of the Workshop on Multiset Processing, (2000), 21-34.

Artificial Chemistry: Computational Studies on the Emergence of Self-Reproducing Units

Naoaki Ono and Takashi Ikegami

Institute of Physics, Graduate School of Arts and Sciences,
The University of Tokyo 3-8-1 Komaba Meguro-ku, Tokyo, 153-8902, Japan
{nono,ikeg}@sacral.c.u-tokyo.ac.jp

Abstract. Acquisition of self-maintenance of cell membranes is an essential step to evolve from molecular to cellular reproduction. In this report, we present a model of artificial chemistry that simulates metabolic reactions, diffusion and repulsion of abstract chemicals in a two-dimensional space to realize the organization of proto-cell structures. It demonstrates that proto-cell structures that maintain and reproduce themselves autonomously emerge from a non-organized initial configuration. The results also suggest that a metabolic system that produces membranes can be selected in the chemical evolution of a pre-cellular stage.

1 Introduction

The emergence of cell structures — the leap from pre-cellular to cellular reproduction — is one of the most important transitions in the evolution of life. To understand the minimal properties of living cells, Gánti proposed a model named "chemoton" [6,7] that is composed of three autocatalytic cycles: metabolic, genetic, and membrane subsystems. These cycles maintain themselves co-operatively. Cell membranes are not only useful in keeping molecules from diffusing away, they are essential for the earliest evolution of self-replicating molecules [4,5,12]. It is worth noting that the membranes are maintained dynamically by metabolism inside the cell. In other words, cell is a system that defines itself. Maturana and Varela called this ability to create its borders by itself as "autopoiesis" [22,10,11]. To demonstrate metabolism and self-organization of proto-cell membranes, they proposed an abstract computational model [22] (the original model was re-implemented by McMullin [13]) that is based on artificial chemistry simulated by a two-dimensional cellular automaton. Breyer and others proposed another extension of this model [2] introducing metabolism by catalysts.

One of the advantage of these computational studies of abstract models is that they allows us to simulate an evolution over a long period, which is difficult to simulate experimentally. Computer simulations of abstract models provide a way to bridge the experiments on the organization of proto-cell structures and the theoretical studies on the evolution of primitive organisms [15,16]. Our

J. Kelemen and P. Sosík (Eds.): ECAL 2001, LNAI 2159, pp. 186–195, 2001.

purpose here is to understand how a self-maintaining system emerges and evolves in the pre-cellular stage. In this paper, we will introduce a model of artificial chemistry to demonstrate self-maintenance and self-reproduction of proto-cell structures.

2 Model

Chemicals are represented by abstract particles that react with each other and diffuse in a two-dimensional reaction space that is arranged as a triangular lattice. Note that, unlike usual models of cellular automata, any numbers of particles can be simultaneously placed on a site. A state of a site $\mathbf{x}$ at time step t is represented by a state vector*

$$\mathbf{n}(\mathbf{x},t) = (n_1(\mathbf{x},t), \ldots n_m(\mathbf{x},t) \ldots, n_N(\mathbf{x},t)), \qquad (1)$$

where $n_m(\mathbf{x},t)$ denotes the number of the m-th type of particles on the site $\mathbf{x}$. The possible number of particle types is fixed at N in this model. The total number of particles is also conserved.

2.1 Repulsion Effects

Particles diffuse on lattice sites with random walks that are biased by repulsive forces between particles. First, the repulsion potential of a particle is given by a summation of all repulsion from the particles in the same and the adjacent six sites;

$$\Psi_m(\mathbf{x},t) = \sum_{|\mathbf{x}'-\mathbf{x}|\leq 1} \sum_{m'=1}^{N} n_{m'}(\mathbf{x},t)\psi_{mm'}(\mathbf{x}'-\mathbf{x}), \qquad (2)$$

where $\psi_{mm'}(d\mathbf{x})$ expresses the repulsion on an m-th type particle from an m'-th type particle. The probability with which a particle moves to an adjacent site depends on the gradient of the potential, given by,

$$p_m(\mathbf{x} \to \mathbf{x}', t) = k_{dif}\, f(\Psi_m(\mathbf{x}',t) - \Psi_m(\mathbf{x},t)), \qquad (3)$$

where k_{dif} is the diffusion coefficient which is a common constant for all particles. Namely, particles move to the lower potential sites with this probability, in practice we use the following forms as the function f:

$$f(\Delta E) = \frac{\Delta E}{e^{\beta \Delta E} - 1} \qquad (4)$$

where β denotes the Boltzmann constant which gives the inversed temperature of the system. We assume that the system is in contact with a heat bath so that the temperature is kept constant. The value of β is normalized and fixed to 1 throughout all the simulations below.

Repulsion between two particles depends on both their particle types and their spatial configuration. We define three groups of particles; hydrophilic, hydrophobic, and neutral. Hydrophilic and hydrophobic particles repel each other strongly. A neutral particle interacts with other particles relatively weakly. To make clear the role of membranes, we consider two types of hydrophobic particles, namely the isotropic ($\mathbf{M_i}$) and anisotropic ($\mathbf{M_a}$) types. Isotropic hydrophobic molecules are represented by particles that repel hydrophilic particles regardless of the spatial orientation (see Fig. 1a). When they are mixed with hydrophilic particles, they form spot-shaped clusters, similar to oil on water. On the other hand, repulsion of anisotropic particles depends on their configuration. Figure 1b illustrates the repulsion potential strength around an anisotropic particle. In one direction, the repulsion is stronger than in the other directions. According to its symmetry, it can turn into three distinct orientations ($\mathbf{M_a^0}$, $\mathbf{M_a^1}$, $\mathbf{M_a^2}$). Rotation of particles is executed stochastically. Probabilities of the rotation depend on the difference of the potential as follows,

$$p_{M_a^h \to M_a^{h'}}(\mathbf{x}, t) = k_{rot}\, f(\Psi_{M_a^{h'}}(\mathbf{x}, t) - \Psi_{M_a^h}(\mathbf{x}, t)), \tag{5}$$

where k_{rot} gives the rotation coefficient. Particles rotate to the lower potential directions. However, we introduce repulsion between $\mathbf{M_a}$s in the neighbor sites, so that particles tend to align in the same orientation. According to these effects, the anisotropic particles tend to arrange themselves in the same orientation and forms thin, membrane-shaped cluster.

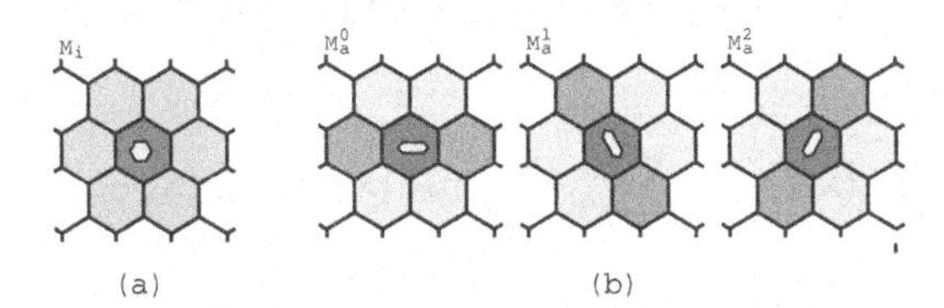

Fig. 1. Repulsion potential around a hydrophobic particle. Depth of grey corresponds to the degree of repulsion on hydrophilic particles in the site from the hydrophobic particle in the centre site. (a) Isotropic repulsion of a particle $\mathbf{M_i}$ (b) Anisotropic repulsion of particles $\mathbf{M_a}$.

2.2 Metabolism

We introduce chemical reactions to consider metabolism of particles. As a minimal model for a metabolic system of a proto-cell, we use five distinct chemical species; **A**, **M**, **X**, **Y** and **W**. Chemical reactions are simulated by stochastic transitions among these particle types. Figure 2 illustrates the assumed reaction paths among them.

A particle **A** represents an "autocatalytic" particle, which catalyses the replication of another **A** particle by consuming a "resource" particle **X**. It can also produce a "membrane" particle **M** by consuming a particle **X**. These particles spontaneously decay into "waste" particles **Y**. A particle **W** represents "water" which does not change into any other chemical. We assume that particles **A** and **W** are hydrophilic, particle **M** is hydrophobic, particles **X** and **Y** are neutral, respectively. We assume that the resource particles have higher chemical potential than other particles so that reactions proceed according to the difference of their chemical potential. Additionally, we introduce an external energy source that recycles **Y** into **X** to maintain a non-equilibrium system.

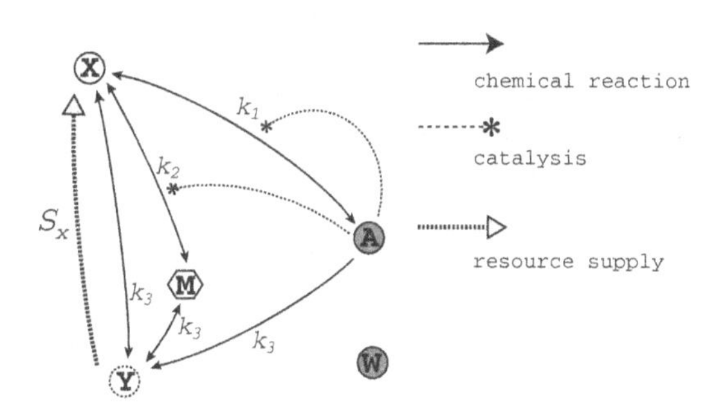

Fig. 2. Reaction paths. Particle **A** catalyses production of particles **A** and **M** from particle **X**. All particles decay into particle **Y**. There is an external energy source that supplies particle **X**.

2.3 Reaction Kinetics

We assign chemical potential (G_m) to each chemical types and calculate the transition probability from the difference in the chemical potential between two types, using the following formula,

$$p_{m \to m'}(\mathbf{x}, t) = k_l \, f(G_{m'} - G_m + \Psi_{m'}(\mathbf{x}, t) - \Psi_m(\mathbf{x}, t)), \quad (6)$$

where k_l gives the rate coefficient of the reaction between m and m'. It is worth noting that the change in repulsion potential of the particles is also taken into account. According to this term, the environment of the particles affects the rate coefficient, for example, it is difficult to synthesize a hydrophobic particle in a hydrophilic environment.

We assume that particle **A** catalyses replication of another **A** using it as a template, as below,

$$X + A_{template} + A_{catalyst} \leftrightarrow A_{copy} + A_{template} + A_{catalyst}. \tag{7}$$

Thus the rate of replication depends on the concentration of both the template and catalyst particles. It should be approximated by a function proportional to the square of the concentration of the autocatalyst as below,

$$k_1(\mathbf{x}, t) = k_0 + C_A \, n_{\mathbf{A}}(\mathbf{x}, t)^2 \tag{8}$$

$$k_2(\mathbf{x}, t) = k_0 + C_M \, n_{\mathbf{A}}(\mathbf{x}, t)^2, \tag{9}$$

where k_1 and k_2 represent the rate coefficients of $X \leftrightarrow A$ and $X \leftrightarrow M$ respectively. C_A and C_M denote the activity coefficients of the catalyst, and $n_{\mathbf{A}}(\mathbf{x}, t)$ gives the number of **A** particles on the site $\mathbf{x}$ at time t. These reaction rates have a constant base rate coefficient k_0. The coefficient of spontaneous decay is given as a common constant (k_3). We assume that an external energy source changes **Y** into **X** with a constant rate coefficient S_X. The rate coefficients between **X** and **Y** are modified as below,

$$\mathbf{X} \xrightarrow{k_3} \mathbf{Y} \tag{10}$$

$$\mathbf{Y} \xrightarrow{k_3 + S_X} \mathbf{X}. \tag{11}$$

The simulations were performed with a Monte-Carlo method. At each time step, we computed the potential and the transition probabilities of diffusion, rotation and reaction for all particle types on all lattice sites, then synchronously update their states according to the probabilities.

3 Results

We mainly focus on simulations starting from a homogeneous configuration[1], where each site has a sufficient number of **X** and **A** to sustain their replication. We compared two cases where the metabolic system produces (i) isotropic hydrophobic particles, and (ii) anisotropic particles.

Initially, the production of **A** and **M** from **X** catalysed by **A** begins in both cases. When a sufficient number of **M** has been produced, a phase separation between **M** and other particles takes place (Fig. 3). In the isotropic case, **M** accumulates into spatial spot patterns. Since the spots are maintained by the metabolic cycle, they finally reach a stable regular pattern similar to Turing-patterns [21,14,9]. On the other hand, in the anisotropic case, **M** forms membrane-shaped clusters. Then, competition for resource particles between the cells with an enclosed membrane begins. Figure 4 shows how cells evolve.

[1] The numbers of particles **A**, **M**, **X**, **Y**, and **W** on each site are (15, 15, 10, 10, 50), respectively.

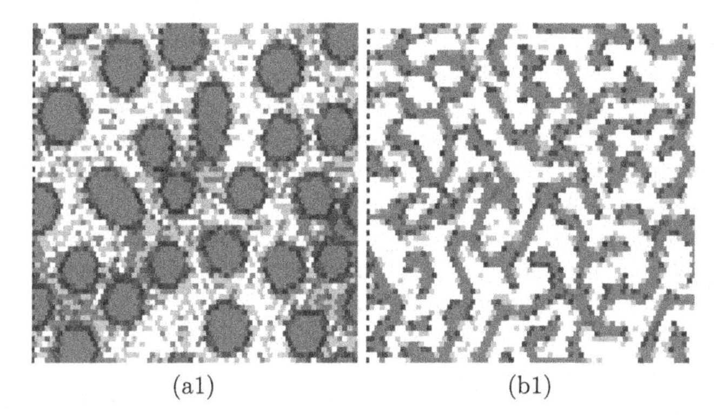

Fig. 3. Phase separation from a homogenous initial configuration (t = 6000). The lattice size is 64 by 64. The boundaries are periodic. Depth of red, yellow and blue represent the density of **M**, **A** and **W**, respectively. (a1) Isotropic particles form spot-shaped clusters. (b1) Anisotropic particles form membrane-like clusters.

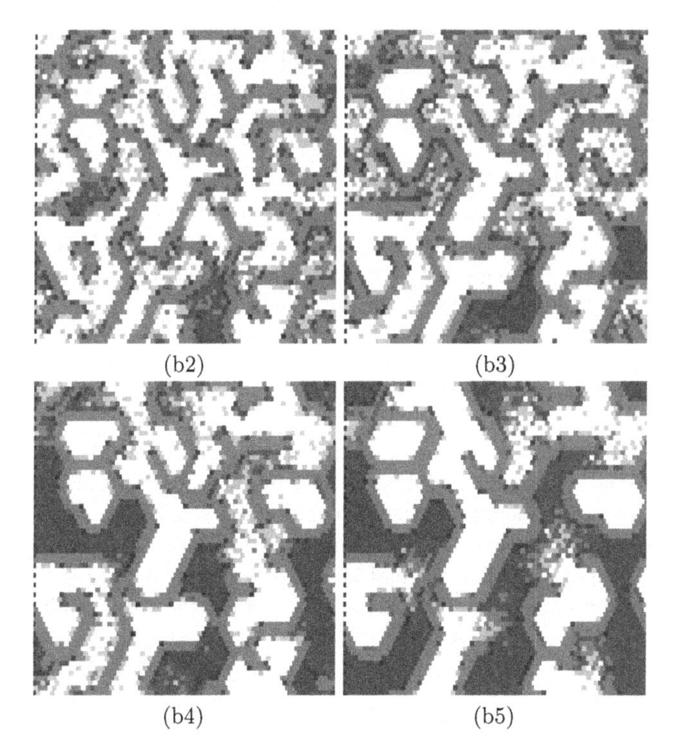

Fig. 4. Emergence of a proto-cell structure after membrane formation (continued from Fig. 3(b1)) (t = 12000 (b2), 18000 (b3), 24000 (b4), 30000 (b5)). In the anisotropic case, osmosis of resource particles across membranes occurs. As a result of competition, some closed cells survives and begin to grow.

The mechanism of this competition is simply osmosis caused by the gradient of **X** between the two sides. Because we assume that **X** and **Y** do not repel **M** very strongly, particles selectively permeate the membranes. If the concentration of **A** is slightly higher on one side than on the other, **X** is consumed more rapidly on this side. According to the gradient, more **X** will come to this side, and more **X** is catalysed into **A**. This positive feedback makes one side rich in **A** and the other side poor. As a result, the winning cell becomes larger by absorbing the others. When the density of **A** is lower than a certain threshold, it cannot sustain metabolism any longer and decay into **Y**. The difference between the two cases becomes clearer when the resource supply is poor. In the isotropic case, metabolism cannot be sustained, so that particles **A** and **M** gradually decay and finally disappear (Fig. 5(a1-3)). On the other hand, in the anisotropic case, a few cells remain and successfully grow (Fig. 5(b1-3)). Of course, if the resource supply is too low, it is impossible to sustain metabolism in either cases. However, there is a certain parameter region where particles producing membranes only can survive.

In the anisotropic case, we notice that once a few cell structures are formed, a reproduction process is launched. With the assimilation of particles by osmosis, a cell grows in size. When the size of a cell reaches a certain threshold, new membrane partitions begin to form within it. Eventually, a cell divides into several daughter cells. This process takes place recursively as long as there exists enough resource particles.

In sum, the emergence of a reproductive cell is as follows: 1) Starting from a random initial state, membrane particles form entangled clusters. 2) When these clusters form a closed barrier, they separate metabolizing particles into cells. 3) Competition occurs among cells. Some cells that outperform others are selected. 4) The surviving cells begin to reproduce themselves.

4 Discussion and Conclusions

We have developed a model of an autopoietic proto-cell structure and have shown that membranes are maintained dynamically by the autocatalytic metabolic system inside, and that the membrane naturally divides and reproduces the cell.

This reproductive feature is indispensable for living cells — we name it "reproductive autopoiesis". The original idea of autopoiesis has stressed the importance of an autonomous boundary, especially in an autocatalytic cell system [11]. We have extended it to the self-reproduction of the autopoietic system. Consequently, a self-maintaining boundary at the same time enables self-reproduction of the self-defining system itself. In other words, an autopoietic system induces only stability (maintenance) as well as instability (reproduction) of the system dynamics.

It is clearly demonstrated in Fig. 5 that a reproductive proto-cell emerges from an unorganized initial configuration. This process depends on osmosis across the membrane, which is driven by non-equilibrium reactions of the metabolic system. We presume that similar results will be observed when this

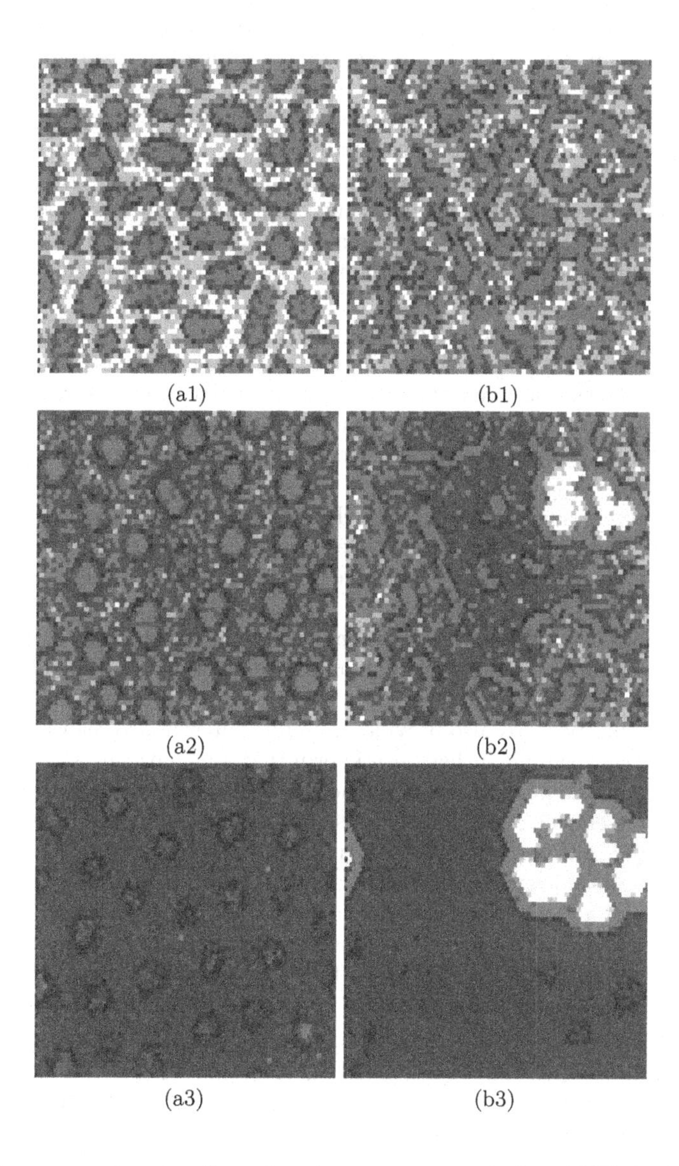

Fig. 5. Emergence of proto-cells under a low resource supply (t = 15000 (a1,b1), 30000 (a2,b2), 45000 (a3,b3)). The rate of resource supply is low so that particles cannot sustain metabolism in the isotropic case; in the anisotropic case, there are a few cells that absorb resource particles from the environment and succeed in metabolizing. As they grow, they reproduce themselves spontaneously.

model is extended to the three-dimensional cases. It would thus explain the transition from molecular to cellular reproduction in the earliest stage of life evolution. This proto-cell structure is a simple example of an autonomous agent that procures an energy resource from outside and metabolizes it to survive. In this model, we have defined the chemical and spatial potential of particles, and can easily define the entropy of the distribution. We are now analysing the flows of energy and entropy around the cells, and how they change through reproduction. Kauffman has discussed how the complexity of life has increased in the course of evolution [8]. Quantitative analysis of the emergence and evolution of proto-cells with respect to the thermodynamic framework should be discussed in future studies.

When a proto-cell structure reproduces, those component chemical particles will be inherited by their daughter cells. It is the emergence of a new heritable system. This transition alters the process of natural selection. A group selection of autocatalytic molecules has been thought to be a prerequisite for evolution of large reaction cycles, because larger autocatalytic cycles are simply weak against the invasion of parasites [19,20,3].

In the present simulations above, cells do not leave but stick together when they divide. This is because membrane formation is simulated by anisotropic clustering of hydrophobic molecules rather than a bilayer membrane of amphiphilic molecules. Our recent model which simulates the dynamics of amphiphilic molecules demonstrates fissions of autopoietic micelles or proto-cells. Various shapes of cells and their dynamics of fission observed in that model is reported elsewhere [17].

The present approach also provides a useful framework to realize the roles of membranes. For example, it has been known that a membrane can catalyse synthesis of membrane molecules synergistically, so that it forms a self-reproducing structure by itself when proper molecules are supplied [1,23,18]. These synergistic effects of membrane particles are well demonstrated by simulating the interaction between metabolic systems and membranes. We have also shown that selective permeability of a membrane leads to the emergence of proto-cell structure. The next steps are to examine the kind of membranes that can promote metabolism most efficiently, and the ways in which they evolve. More studies on membrane shape/function based cellular evolution is required to answer these questions.

Acknowledgement. This work is partially supported by the COE project ("complex systems theory of life") and Grant-in-aid (No. 11837003) from the Ministry of Education, Science, Sports and Culture.

References

[1] Bachmann, P. A., Luisi, P. L. and Lang, J. Autocatalytic Self-Replicationg Micelles as Models for Prebiotic Structures, *Nautre*, **357** (1992), 57–59.
[2] Breyer, J., Ackermann, J. and McCaskill, J. Evolving Reaction-Diffusion Ecosystems with Self-Assembling Structures in Thin Films, *Artificial Life*, **4** (1998), 25–40.

[3] Cronhjort, M. B. and Blomberg, C. Cluster Compartmentalization may Provide Resistance to Parasites for Catalytic Networks, *Physica D*, **101** (1997), 289–298.
[4] Eigen, M. and Schuster, P. *The Hypercycle. A Principle of Natural Self-Organization*, Springer-Verlag, Berlin; New York (1979).
[5] Eigen, M., Schuster, P., Gardiner, W. and Winkler-Oswatitsch, R. The Origin of Genetic Information, *Scientific American*, **244**, 4 (1981), 78–94.
[6] Gánti, T. Organization of Chemical Reactions into Dividing and Metabolizing Units: The Chemotons, *BioSystems*, **7** (1975), 15–21.
[7] Gánti, T. Biogenesis Itself, *J. theol. Biol.*, **187** (1997), 583–593.
[8] Kauffman, S. A. *Investigation*, Oxford University Press (2000).
[9] Lee, K. J., Mccormick, W. D., Pearson, J. E. and Swinney, H. L. Experimental Observation of Self-Replication Spots in a Reaction Diffusion System, *Nature*, **369** (1994), 215–218.
[10] Maturana, H. R. and Varela, F. J. *Autopoiesis and Cognition: the Realization of the Living*, D. Reidel Publishing (1980).
[11] Maturana, H. R. and Varela, F. J. *The Tree of Knowledge*, Shambhala Publications (1987).
[12] Maynard Smith, J. and Szathmáry, E. *The Major Transitions in Evolution*, Oxford: Freeman & Co. (1995).
[13] McMullin, B. and Varela, F. J. Rediscovering Computational Autopoiesis, 4th European Conference on Artificial Life (eds.Husbands, P. and Harvey, I.), Brighton, UK (Jul 1997), MIT press.
[14] Meinhardt, H. *Models of Biological Pattern Formation*, Academic Press, London (1982).
[15] Ono, N. and Ikegami, T. Model of Self-Replicating Cell Capable of Self-Maintenance, Proceedings of the 5th European Conference on Artificial Life (ECAL'99) (eds.Floreano, D., Nicoud, J. D. and Mondada, F.), Lausanne, Switzerland (1999), Springer.
[16] Ono, N. and Ikegami, T. Self-Maintenance and Self-Reproduction in an Abstract Cell Model, *J. theor. Biol.*, **206** (2000), 243–253.
[17] Ono, N. Ph.D. thesis, Artificial Chemistry: Computational Studies on the Emergence of Self-Reproducing Units, *Univ. of Tokyo*, (March 2001).
[18] Segre, D., Ben-Eli, D., Deamer, D. and Lancet, D. The Lipid World, *Orig. Life Evol. Biosph.* (2000).
[19] Szathmáry, E. and Demeter, L. Group Selection of Early Replicators and the Origin of Life, *J. theor. Biol.*, **128** (1987), 463–486.
[20] Szathmáry, E. and Maynard Smith, J. From Replicators to Reproducers: the First Major Transitions Leading to Life, *J. theor. Biol*, **187** (1997), 555–571.
[21] Turing, A. M. The Chemical Basis of Morphogenesis, *Philosophical Transactions of the Royal Society (part B)*, **237** (1953), 37–72, (Reprinted in *Bulletin of Mathematical Biology* (1990) **52**, No.1/2 153-197).
[22] Varela, F. J., Maturana, H. R. and Uribe, R. Autopoiesis: The Organization of Living Systems, its Characterization and a Model., *BioSystems*, **5** (1974), 187–196.
[23] Walde, P., Wich, R., Fresta, M., Mangone, A. and Luisi, P. L. Autopoietic Self-Reproduction of Fatty Acid Vesicles., *Jounal of American Chemical Society*, **116** (1994), 11649–11654.

Stability of Metabolic and Balanced Organisations

Pietro Speroni di Fenizio and Wolfgang Banzhaf

Informatik Centrum Dortmund (ICD), Joseph-von-Fraunhofer-Str. 20,
44227 Dortmund, Germany
pietro.s@altavista.net, banzhaf@cs.uni-dortmund.de

Abstract. We investigate the possible organisations emerging from an artificial chemistry (AC) of colliding molecules in a well stirred reactor. The molecules are generated from 7 basic components (atoms), each with a different behavior. After discovering two main types of organisations (metabolic o. and balanced o.), we deepen our analysis by studying their behavior over time. The phases they pass through and their stability with respect to an external influx of random information are examined. We notice that no organisation seems to be totally stable over time, yet metabolic organisations pass through a growth phase with a much higher stability. Lastly we observe how the different phases are triggered by the presence or absence of particular atoms.

Introduction

The last years have seen a growing interest for systems where different elements interact to generate new elements, often different from the ones previously present. Those systems, also known as Artificial Chemistries, have been used to model biological, chemical , ecological and social systems. The revolutionary importance of these systems, in comparison to previous linear or non-linear ones, lies in the fact that it is impossible to study them with the method of differential equations. This impossibility and its relevance has been clearly noticed by Fontana and Buss in [Fontana and Buss, 1996].

If we consider S to be the space of all possible elements which can be generated by an Artificial Chemistry , and P(S) the space of all possible subsets of S, we can see that some of those subsets hold two key properties, (1) the property of being a self-maintaining set of elements, and (2) the property of being a closed set of elements. A self-maintaining set of elements is a set where each element can be generated by the interaction of the elements of this set, while a closed set is a set whose elements, as they interact, cannot generate anything external to the set itself. A set which is both self-maintaining and closed is called an *organisation*. Organisations are a central concept in the field of Artificial Chemistries since they represent points of no return in the space of the possible states of the system. When a system reaches an organisation, it can only remain similar to itself, or devolve into a sub-organisation.

All this has already been investigated in [Fontana and Buss, 1993] [Ikegami and Hashimoto, 1996][Dittrich and Banzhaf 1998][Speroni at al, 2000]. Here we wish to address the question what happens if new elements which are not part of the already existing organisation are inserted into the system. Clearly, sometimes the system will

J. Kelemen and P. Sosík (Eds.): ECAL 2001, LNAI 2159, pp. 196-205, 2001.

return to the original organisation again. In other cases, it will adapt to a more complex organisation, and sometimes it will pass through a transition phase and reach a different organisation altogether. We are thus studying the stability of the generated organisations respect to an external influx of elements. To study this question we use a system based on combinators [Hindley and Seldin, 1986], [Speroni, 2000].

The System

Our Artificial Chemistry keeps a finite number of basic elements (atoms) which interact to generate combined elements (molecules). The behavior of the system is heavily dependent on the set of atoms which are available at certain moments. Since some organisations tend to consume more and more of the same atoms, they often prepare the right environment for the next organisation to arise. More specifically, in our system we collide algebraic algebraic structures called combinators to generate new combinators. The space of combinators has the following qualities: (1) It is infinite, (2) each combinator directly represents a different operator, (3) every possible operator from the space of strings to the space of strings is present and (4) each combinator (in its normal form) can be expressed using a finite set of basic elements.

Table 1. The seven basic combinators C, R, W, B, I, S, K. s_0 represents the rest of the string. s_0 can be equal to ∅ or to an arbitrary string.

Atom	Applied Atom	Reduced string	Atom	Applied Atom	Reduced string
C	$C\ x_1x_2x_3\ s_0 \rightarrow$	$X_1x_3x_2\ s_0$	I	$I\ x_1\ s_0 \rightarrow$	$x_1\ s_0$
R	$Rx_1x_2\ s_0 \rightarrow$	$X_1\ s_0,\ x_2$	S	$S\ x_1x_2x_3\ s_0 \rightarrow$	$x_1x_3(x_2x_3)\ s_0$
W	$W\ x_1x_2\ s_0 \rightarrow$	$X_1x_2x_2\ s_0$	K	$Kx_1x_2\ s_0 \rightarrow$	$x_1\ s_0$
B	$B\ x_1x_2x_3\ s_0 \rightarrow$	$x_1(x_2x_3)\ s_0$			

Combinators

Each combinator is a string of symbols with balanced parentheses. Each symbol represents an operator that can be applied to the subsequent elements of the string. In our experiments we used 7 basic operators: B, C, K, I, R, S and W. This set of operators contains two disjointed bases of the space of all possible combinators: B, C, K, W, and K, S, I. Each base can generate every possible combinator. Thus, elements of one base might be expressed as combinators of the other base. Some redundancy will therefore be present in the solutions. The seventh combinator, the 'R' has been specifically designed to permit combinators to release independent sub-units (see results section). Each combinator can be seen as a simple Lisp program, which uses the other programs (as data) to generate new programs.

Each set of parentheses '(',')'encloses a sub-combinator. Each sub-combinator is a combinator in its own right and as such part of the longer combinator. Also each sub-combinator can hold other sub-combinators inside, as a set of parentheses can enclose other sets inside '(( ) ( ))'. In Table 1 all operators, from here on called atoms, and their effect on the subsequent elements can be seen. Each element x_i can be either an atom or a sub-combinator. If it is a sub-combinator it will be copied altogether (by the atoms W and S), destroyed altogether (by the atom K) or released altogether (by the atom R).

When we write a combinator, some parentheses can be considered useless. These might be eliminated, in particular, any set of parentheses with only one or no element inside (example, '( )' ,'(K)'). Also parentheses which start a combinator or a sub-combinator can be eliminated. For instance, (S(I)((K)K)) can be written as SI(KK).

Reduction Operation and Normal Form. Let us consider one combinator. By applying its atoms as operators to the rest of the elements, we move from one combinator to another. This operation is called *reduction*. Each combinator codes for an operator from the space of strings with parenthesis in itself. Each operator can be coded in infinite ways with combinators. When we reduce one combinator to another, through a reduction, the coded operator does not change. An example will clarify this.

If 'a', 'b' and 'x' are combinators, with 'a' and 'b' being one the reduced form of the other ('a'→'b') then 'a'*'x'→'b'*'x'. In our model we consider the list of the operators (the molecules) present at a certain time, regardless of the particular coding with which they were created.

When combinators are reduced, they move from being represented by one string to being represented by another. Sometimes a combinator reaches a form from which no reduction operation is applicable. Those strings are called the *normal form* of the combinator. Not all combinators have a normal form, yet an important theorem in combinator theory declares that *if* a combinator possesses a normal form, this form is unique. It is not essential in which order the atoms are reduced, if a normal form exists, this is unique and it is always possible to reach it [Hindley and Seldin, 1986].

In our system we consider only combinators that possess a normal form and we store combinators in their normal form. When two combinators interact they generate a new combinator and the reduction process starts. A necessary condition for the result to be acceptable is to reach a normal form in less than 'Max_Time' steps.

The Artificial Chemistry

Our system contains a few hundreds of elements in a well stirred reactor, from now on called *the soup*. Those elements interact with each other and generate other combinators, possibly different from both reactants. In our system we do not keep the number of elements fixed. Instead, the number of elements released from the reaction process can vary, depending on the elements involved. What is kept constant, though, is the number of available atoms. Each type of atom exists only with a few copies. Beside the soup we keep a registry (called *pool*) of the available atoms which are supposed to float in the soup. Every time a reaction takes place, the generated combinators consume atoms from the pool, while when the reactants are destroyed their atoms are added to the pool. If there are not enough atoms in the pool or any of the elements does not reach a normal form in a predefined number of steps, the reaction is considered *elastic* in this environment, and the original combinators are retained instead.

When two elements of the soup collide their reaction generates a multiset of new elements. In a formal way we can write the reaction in the following form:

$$A*b \rightarrow a(b) \rightarrow c_1 \ldots c_n \quad (1)$$

with $a,b,c_1,\ldots,c_n$ combinators. c_1 is then the direct result of the normalisation process of a*b, while $c_2,\ldots,c_n$ are the other elements that are released in the operation.

We now present a short example for illustration purposes. Let two elements be: WR and (SKI). If we apply the first to the second we reach WR(SKI) which can be reduced to R(SKI)(SKI), then to SKI releasing a separate copy of SKI.

$$WR*SKI \rightarrow WR(SKI) \rightarrow R(SKI)(SKI) \rightarrow SKI, SKI \quad (2)$$

In our system we measure the time in physical and biological generations. If NMolecule are present at a certain time then, after each interaction, the physical clock will be advanced 1/NMolecule generations, while the biological clock will be advanced by the same quantity, only after non elastic interactions.

The Influx. We provide the system with a continuous influx of random information. Each physical generation, with a probability P_{in} depending upon the number of existing molecules, we randomly assemble a new molecule to be inserted into the soup. The molecule is assembled from the elements of the pool, so the total amount of atoms in the soup remains constant. P_{in}(NMolecules) is an exponential function with

$$P_{in}(NMolecules) = 1 \quad \text{for} \quad NMolecule = MinNMolecules \quad (3)$$

and

$$P_{in}(NMolecules) = 0.5 \quad \text{for} \quad NMolecule = HalfProbMolecules \quad (4)$$

The influx function in this contribution is different from the influx function presented in [Speroni, 2000] and this explains some of the differences in the behaviour of the generated organisations. In order to generate a random molecule we pick each atom with a probability of $1/9^{th}$, open a parenthesis with a probability of $1/9^{th}$ and close one with the same probability. If we close a parenthesis that has never been opened the combinator is terminated. If an atom which is not present in the pool is required the combinator is aborted and a new one is tried. After a combinator has been constructed it is reduced to its normal form and the resulting combinator is inserted into the soup provided normalisation is possible.

The K, R Atom and the Outflux. A seemingly small difference from the previous model [Speroni, 2000] is the presence of two molecules which eliminate sub-combinators. Both the K and the R eliminate sub-combinator from a molecule. The first totally eliminates it (re-splitting it into atoms), while the second releases it as a separate component. In the previous model only one was present (called K, but working as R here, releasing the sub-element). In that situation, if a molecule was assembled, the only way to disassemble it was by applying this molecule to other molecules. Yet some molecules just would not normalise into shorter form. And as more and more molecules would be frozen in this state, the system would stop. As a consequence, that system needed a constant outflux.

In the present system, the need to get rid of bulky molecules is not solved for the system. We leave such job as a problem to be solved by the appearing organisations. And (as we shall see) it is a job they happen to solve really well.

Results:

We asked our system few simple questions:
1. What are the possible organisations reachable?
2. Can those organisations be divided into broader categories?
3. Once the system has found an organisation, how stable is such a solution with respect to a random influx of information?
4. Does such stability depend on the type of organisation?

Because of the universal characteristics of the space we are working with, many organisations are possible. Interestingly, many of them also seem to be accessible to the system (an organisation could be present in the space of all the possible ones, but be too complex to be found). We made 150 runs with parameters: operation type A°B→A(B), atoms used all 7. Number of atoms 2000 of each type. MinNMolecules=50, HalfNMolecules 300, no outflux, no mutation, MaxTry=10000. MaxLength=100, MaxDepth=20. Starting number of randomly generated molecules 300. Length of the experiment 10000 generation. We then made 26 esperiments with the same parameters, but length of the experiment 30000.

Each run is then much longer than the runs in the previous work (10000 and 30000 generations, compared to the 1000 of [Speroni, 2000]). The first interesting observation is that *no organisation seems to be totally stable*. Some experiments reached an organisation, kept it for many thousand generations, then unexpectedly switched to a different one. So far, however, we were not able to gather enough data to check whether there is a power law distribution between size and frequency of extinction events of molecules.

Two types of organisation have been observed, *metabolisms*, and *balanced organisations*. Other organisations could be possible as well, but due to the constant influx with which we fed the system, only relatively stable organisations could be noticed in the experiments.

Metabolism

Metabolisms are particular organisations that were recognised and presented for the combinator system in [Speroni, 2000]. Here, we keep the definition of metabolism given in [Bagley and Farmer, 1992]: "A metabolism takes in material from its environment and reassembles it in order to propagate the form of the host organism...".

Those organisations gained their name by their ability to 'use' external elements as inputs to grow, while being unable to increase their size without an external influx of elements. As they grow those metabolism preserve their digital form: elements present in them, and quantitative relations between them. In the earlier work, the influx of elements would stop after 300 elements were present. Here, the influx follows an exponential law and is never totally absent. For this reason the global behavior of the new metabolism is now qualitatively different. Metabolisms in our system have some unique features that makes them easy to recognise. They have, usually, a small number of different elements, sometimes just one or two, and their interaction doesn't lead to the generation of more elements of the same kind.

An example of a simple metabolism would be a single element 'a' which, applied to itself, generates two copies of itself

$$a*a \rightarrow a,a \tag{5}$$

So two elements would be flowing into the reaction and two elements would be flowing out of it. And the elements are the same so the composition of the soup is not changed. Yet if we observe the system after some time we notice an increase of 'a' elements. How can this happen?

The reason lies in the external random elements that we regularly insert. In fact, many times those elements give rise to pathways which end up with the production of more elements of the metabolism. An example will clarify this. Suppose that the molecule S is thrown into the soup.

$$S*a \rightarrow Sa \tag{6}$$

$$Sa*a \rightarrow Saa \tag{7}$$

$$Saa*a \rightarrow aa(aa) \rightarrow a,a,a,a \tag{8}$$

So the whole reaction was

$$S*3a \rightarrow 4a \tag{9}$$

Reactions like this happen very often and the net result is an increase of elements of the metabolic type in the soup. On the other hand, not all molecules inserted generate pathways that lead to the release of a's. Sometimes the molecules just gets bigger and bigger, soon reaching the physical limits imposed on the dimension of a possible molecule (as one of the system parameters). When this happens, the molecule will just remain present in the soup without being able to participate in any further reaction. This second type of reactions is, in general, relatively rarer, and only some atoms gets frozen as long molecules.

The Active Phase and its Stability. A metabolism will, in general, have various possible phases. The first is always the *active phase*. Here the number of molecules of the system tends to increase as new molecules are randomly inserted. Some junk molecules are generated, too, but don't, normally, interfere with the process. During this phase we can notice a constant increase in the number of molecules, a stability in the number of different molecules, a decrease in entropy, nearly no innovativity, and a constant decrease in the free atoms in the pool. The number of atoms will tend to decrease following a precise relation between the various types of atoms. Once one of the atom types is totally consumed this phase ends and the *resting phase* begins.

In an active metabolism no inserted molecule seems to be able to push the system into another organisation. Each randomly inserted molecule is 'digested' before it can harm the organisation. Digested molecules end up either being molecules of the metabolism or junk long molecules, unable to react.

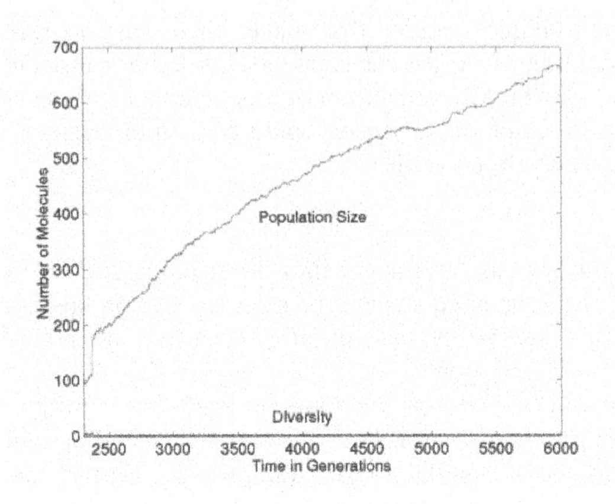

Fig. 1. Example of a metabolism in its active phase.

In some rare case the situation is reverted if a particularly 'harmful' molecule is inserted. Once this happens the metabolism starts loosing elements. In general, elements that revert the metabolism are elements that are able to destroy other molecules without being destroyed (or modified) themselves, for instance of the form 'b' such that b*x→'b'. Often a metabolism (especially in its first active phase) is too simple to deal with those molecules. If this happens the system shrinks in size. Once the number of elements drops to less than 50 molecules (MinNMolecules), many new molecules start to be inserted into the soup. New reactions begin to happen and the system will enter a transition phase. From here it might revert back to the old metabolism (with the dangerous molecule being removed from the scene by the flood of new molecules), to another metabolism, or to a totally different organisation. Even in this latter case we don't have an abrupt transition to a new organisation, but rather first a shift to a different phase. In a futuristic picture were such organisation is man-made, and performing a particular task, it would be possible to invoke some correction operation before the system gets out of control.

Even if no dangerous molecule is inserted, the general behaviour of a metabolism is directly linked to the molecules of the influx. In our standard example each molecule has double the probability to contain a growing atom (S or W) than to contain a shrinking one (K). In another experiment no W were allowed and metabolisms were very rare. Moreover, those metabolisms did not grow monotonically, but instead behaved as in a random walk in the number of molecules.

Rest-Post Metabolic Cloud.. In a limited medium a metabolism cannot grow for ever. Once at least one of the atom types is exhausted, the metabolism passes from the active phase to the resting phase. In this second phase elements are still inserted and the metabolism tries to transform them as it did before. Yet, since there are not enough free atoms in the pool, many transformations stop halfway. Recall that each interaction to be accepted needs to reach a normal form on every molecule generated. Yet in order to 'digest' a molecule many separate reactions are usually necessary, making it possible for the process to stop halfway. If this happen many new molecules are generated, the diversity and entropy increases and many new reactions are tried out.

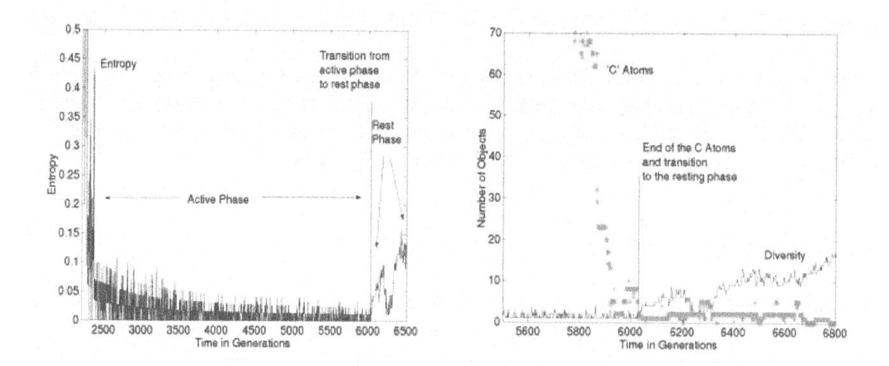

Fig. 2. Transition from active to resting phase. On the left entropy, on the right diversity and Number of C Atoms. Both entropy and diversity increase as one of the atom types is exhausted.

Balanced Organisation

We call *balanced organisations* closed self-maintaining sets which do not need external input to grow. They, too, metabolise external elements, but instead of using their active components to grow (as the metabolism does with the 'S' elements in the

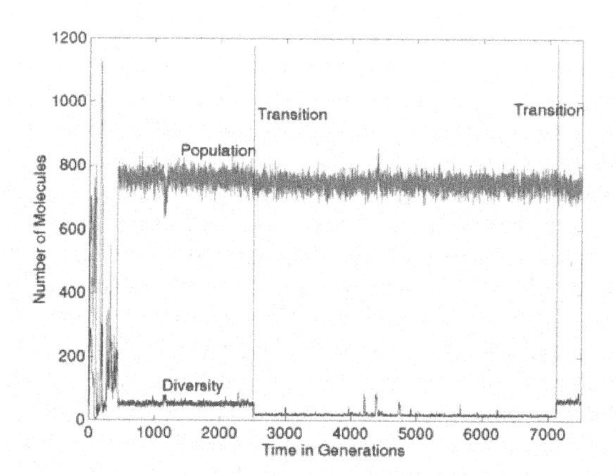

Fig. 3. Transition from a balanced organisation to another.

previous example), they just split them up into their atomic components, then reach back for the elements in the pool to grow. The ability to 'split up' components, unused in metabolic organisations, is new to this version of the system, and presented in 'the K,R atom and the outflux' paragraph. Balancedorganisations often hold two types of molecules, molecules which build and molecules which destroy. The dynamic balance between these two types permits the organistion to build itself continuously and to efficiently destroy the influx elements which could destabilise such balance.

This kind of organisation tends to fluctuate around a particular average value with a corresponding standard deviation. Balanced organisations are not totally stable either, and the system will often percolate to a new organisation being sometimes a balanced organisation, sometimes a metabolism.

The Space of Organisations

A complementary point of view to this dynamic picture is instead to track the movement of the system in the space of all possible organisations. Such a space, spanned by all possible organisations and their relations of intersection $\cap$ and union $\cup$ generates a lattice [Speroni at al, 2000]. Every set of elements 'S' uniquely defines an organisation O_s. Every time we intersect two organisations we uniquely define another organisation, the biggest organisation contained in the intersection. Every time we unite two organisations we, again, uniquely define an organisation, the smallest organisation generated by the union of the two sets. The operation of union is particularly important in our case, since every time we insert a new element into an existing organisation we potentially push it towards a larger organisation. As the system changes in time, its change can be observed as a movement in this lattice of organisations.

Conclusion

As we study specific artificial chemistries more and more properties seem to emerge. We studied a model which used as elements molecules made up of 7 types of basic 'atoms'. We have observed that many organisations are possible, yet none seems totally stable with respect to an external influx of random elements. We discussed how a particular kind of organisation, metabolism, uses such an influx. As long as it is able to transform the elements of the influx, it is able to keep itself relatively stable, while as soon as the influx gets too large or the organisation is unable to transform it any more, it is destabilised. We observed that often the exhaustion of one or more basic atomic types was able to push the system into a different organisation.

The field of artificial chemistries is still a quite young subfield of artificial life. Applications are sporadic at best. Yet the numerous natural examples of systems which contain different interacting elements seem to suggest vast possibilities of application. All too often well planned systems based on few precise components seem unable to cope with the enormous variety of forms that nature provides, and end up being used under very limited work conditions only. The study of Artificial Chemistries, on the other side, of their metabolic organisations and features like stability could permit to organise vast changes in the world that would otherwise be impossible to obtain. If those tools are developed it will be important to slow down to the point of being able to use them with wisdom, being able to use them for entire humanity more than for a selected group of people, for the whole nature more than only for human beings.

Acknowledgements. We wish to thank Peter Dittrich for invaluable comments, and Jens Busch for support and review of an earlier draft of this paper. P.S.d.F acknowledges financial support from BMBF under program 01 IB 801.

References

Bagley J. R. and Farmer J. D. 1992. *Spontaneous Emergence of a Metabolism.* In Proceedings of The Second International Conference on Artificial Life, 94-140. Edited by Langton C. G., Taylor C., Farmer J. D., and Rasmussen S., Redwood City, Addison-Wesley.

Dittrich P. and Banzhaf W. 1998. *Self-Evolution in a constructive Binary String System.* Artificial Life 4(2):203-220.

Fontana W. 1992. *Algorithmic Chemistry.* In Proceedings of The Second International Conference on Artificial Life, 159-209. Edited by Langton C. G., Taylor C., Farmer J. D., and Rasmussen S., Redwood City, Addison-Wesley.

Fontana W. and Buss L. W. 1993. *"The arrival of the Fittest": toward a Theory of Biological Organization.* Bulletin of Mathematical Biology.

Fontana W. and Buss L. W. 1996. *"The barrier of objects: From dynamical systems to bounded organizations"*, In Boundaries and Barriers, J.~Casti and A.~Karlqvist, 56 – 116, Redwood City, Addison-Wesley.

Hindley J. R. and Seldin J. P., 1986. *Introduction to Combinators and λ-Calculus.* Cambridge University Press.

T. Ikegami and T. Hashimoto, Active mutations in self reproducing networks of machine and tapes Artificial Life 2(3):305--318. 1995

Kauffman, S. 1993. *The Origins of Order: Self-Organisation and Selection in Evolution.* Oxford University Press.

Speroni d.F. P. 2000. *A Less Abstract Artificial Chemistry.* In Artificial Life VII: Proceedings of the Seventh International Conference, edited by Bedau M.A., McCaskill J. S., Packard N. H. and Rasmussen S, Cambrdige, MA, MIT Press.

Speroni d.F. P., Dittrich P., Banzhaf W. and Ziegler J. 2000. *Towards a Theory of Organizations.* In Proceedings of the 4th German Workshop on Artificial Life (GWAL 2000).

Suzuki Y., and Tanaka H. 1998. *Order Parameter for Symbolic Chemical System.* In Proceedings of The Sixth International Conference on Artificial Life, 130-139. Edited by Adami C., Belew R., Kitano H., and Taylor C., Cambridge, MA, MIT Press.

Spontaneous Formation of Proto-cells in an Universal Artificial Chemistry on a Planar Graph

Pietro Speroni di Fenizio[1], Peter Dittrich[1], and Wolfgang Banzhaf[1]

Informatik XI
University of Dortmund
D-44221 Dortmund, Germany
{speroni, dittrich, banzhaf}@LS11.cs.uni-dortmund.de
http://ls11-www.cs.uni-dortmund.de/

Abstract. An artificial chemistry is embedded in a triangular planar graph, that allows the molecules to act only locally along the edges. We observe the formation of effectively separated components in the graph structure. Those components are kept separated by elastic reactions from molecules generated inside the component itself. We interpret those components as self-maintaining proto-cells and the elastic nodes as their proto-membrane. The possibility for these cells to be autopoietic is discussed.

1 Introduction

Presently, much research has been directed towards synthesizing life-like structures in digital media. This goal requires a deeper understanding of the term "alive" and its consequences. Some threads of research have focused on reproduction and have generated programs able to reproduce [6,11]. Here we follow a different path by focusing on the ability of any living system to self-repair and to keep its identity as an entity separated from its environment [5,18]. Under those assumption the characteristics of being alive must emerge from the interactions of simpler components. Here we use an artificial chemistry with combinators as molecules. The combinator chemistry is a universal artificial chemistry because it is implicitly defined and it supports universal computation (it can simulate a Turing-machine). In order to permit molecules to generate a "physical" membrane that separates one being from another we embed the artificial chemistry in a graph.

A common feature of artificial chemistries is often the absence of any spatial conditions [4,5]. Rather, molecules are floating around, randomly colliding with each other. Counter-examples exist, however, either in the form of cellular automata which define a space where to insert the elements [1,2,8,9,10], or by giving a position in a Euclidean space to each element and a range of interaction [13,20]. Using a cellular automaton (where each cell can hold one molecule) is simple and elegant, but does not permit an unbounded growth of elements in

J. Kelemen and P. Sosík (Eds.): ECAL 2001, LNAI 2159, pp. 206–215, 2001.

a particular position. The second alternative is more flexible in this regard, yet the number of interactions that the model has to deal with grows as $n!$ (where n is the local number of elements). This is not efficient and in addition difficult to model, since interactions farer away will often be canceled by the effect of nearer ones.

Here we suggest a different approach in which an artificial chemistry (AC) is embedded in a graph, with each molecule being a vertex of the graph and possible interactions being allowed only along the edges of the graph. We suggest here to use a particular graph, namely, a planar triangular graph. A planar triangular graph can be drawn on a sphere without edge crossing, and with each face being triangular. This particular type of graph can be manipulated by adding and deleting nodes with a minimal local rearrangement of the edges. So we can add nodes near a selected node (like in a 2D model). Yet if we circumscribe a set of connected nodes and insert n new nodes, the number of new relations to be taken into account grows only linearly with n.

Like in many artificial chemistries (and in ours in particular) not all molecules can interact with each other. Special conditions may arise under which different molecules that would have destroyed themselves in a well-stirred reactor can instead coexist [1,3]. This gives rise to higher organizational levels [7] by allowing the whole graph to split into "effectively" separated components that temporarily act as different reaction vessels. The separation into different components by membrane-like structures has also been identified as an important aspect of "biological" information processing [12] and chemical evolution [17].

In the following, we present the planar triangular graph as a feasible tool to model spatial structures, apply it to an artificial chemistry based on combinators, and then show how separate components arise naturally. These components can be interpreted as proto-cells. For this purpose we also have to defined what a membrane, a cell, and an autopoietic cell are in our system.

2 Basic Definitions and Operators

Let $G =< V, E >$ be a graph. G is planar if (and only if) it can be embedded in a plane without edges crossing each other. Given a planar graph embedded in the plane and a point x, we define the face (of G) to contain x as the set of all the points in the plane which can be reached by x with a Jordan curve and whose points are disjoint from G ([19], p. 64). A face of G is called triangular if there are exactly three vertices which can be reached from a point inside the face, through a Jordan curve. Or equivalently if the equivalent vertex in the dual graph[1] G' has three edges. A planar graph G is triangular if every face is triangular, or equivalently if the dual of every face in G' has exactly three edges.

Given the application-oriented field we are addressing, we will consider in this paper only graphs with at least five nodes. In order to manipulate a graph we need the following operators to add and delete a node: **Add node:** Given

[1] The dual graph of a planar graph G is constructed by inserting a node for every face of G and connecting nodes whose equivalent faces are neighbors.

any planar triangular graph it is always possible to add a node to a face given as a parameter. The new node is added to the list of nodes, and three edges are added connecting the new node the three nodes of the face. This operation destroys the old face given as a parameter and creates three new faces. **Delete node:** Given any triangular planar graph with at least six nodes, and given any node x it is always possible to delete x and the edges that were connected to x. In this way we create a new planar graph G' with all faces triangular except one. Subsequently, it is possible to add edges connecting the nodes of this non-triangular face to restore the triangular planar graph. We start adding edges from the node which has remained with only two edges, since in a triangular planar graph each node possesses at least three connections

We connect the nodes randomly inside the face, taking care never to connect pairs of nodes which are already connected.

2.1 Artificial Chemistry Based on Combinators

Any artificial chemistry can be embedded in a graph. In our contribution we use an artificial chemistry based on combinators, described in more detail in [14,15,16].

Molecules and Reactions: Briefly we say that there is (1) an (infinite) set of potential molecules S and (2) a reaction mechanism which computes the reaction product for two colliding molecules $x, y \in S$. This reaction mechanism can be written as a function $r : S \times S \to S \times S \times \cdots \times S$. So, there may be an arbitrary number of products. We write the function r as an operator[2] $\oplus$. Molecules are built from a substrate of elements called *atoms*. There are seven types of atoms (I, K, W, R, B, C, S), each with a different function. The total number of atoms in the reactor is kept constant during a run. Free atoms (not bounded in molecules) are separately stored and form a global pool.

Dynamics: At every step we pick two neighboring molecules (x, y) and apply the first x to the second $x \oplus y \longrightarrow x(y)$. As described in [14,15,16], this creates a (multi-)set of new molecules $z_1, \ldots, z_n$. We insert the molecules $z_2, \ldots, z_n$ randomly in the two faces next to the link $x - y$ (Fig. 1). x is replaced with z_1 (the result of the combinator reduction) and y is finally deleted[3]. The process may appear complicated, but has been carefully designed to be coherent for any $n \geq 1$.

3 Results

When observing the behavior of the system, we notice that different molecules tend to cluster in different regions of the graph. At first glance we can immedi-

[2] We use the symbol $\oplus$ instead of the symbol $+$ because differently to chemistry the order of the reactants is important to calculate the product(s).

[3] Note that the operation is slightly different from the operation described in [14] where z_1 replaces x and y would be deleted. This change has been enacted to prevent reactions of the type $x \oplus y \longrightarrow x, y$ to switch the molecules. Thus we permit easy diffusion.

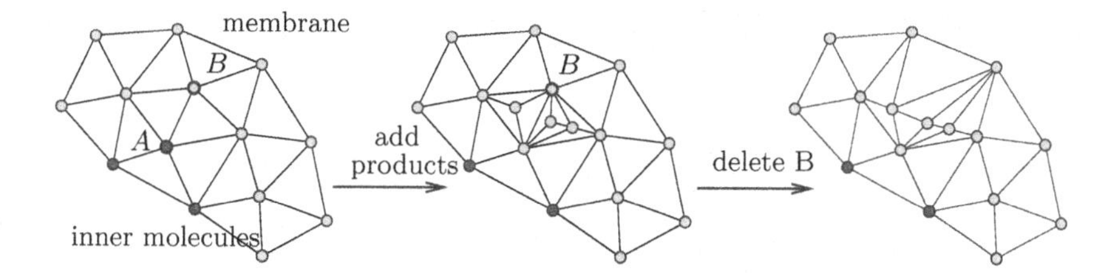

Fig. 1. Illustration of a reaction that creates membrane molecules.

ately notice how the graph helps the system to balance different organizations. Here, an *organization* is defined as a set of molecules that is closed and self-maintaining. This means that every reaction among molecules of the set produce only molecules of that set, and every molecule of the set is produced by at least one reaction among molecules of the set [5]. So, different organizations can be present at the same time (see Fig. 4, below).

Definition and discovery of the membrane: A first theoretical definition of *membrane molecules* is the set of all molecules who cannot interact with the molecule they are linked with. This definition has a number of flaws: real membrane molecules do interact with their neighbors. Their role is to keep their structure while thus separating inside from outside. We suspect that some molecules in the system do act like membrane molecules of biological cells. The second drawback is more technical. We recall that a reaction is possible if the reduction computation of the expression $x(y)$ halts, and if the number of free atoms (in the global pool) is sufficient to support the computation and the results. For this reason some molecules will be unable to interact, regardless of the number of free atoms, while others are reactive only "sometimes". The first type of links will be called "*absolutely*" elastic; while the second type "*effectively*" elastic. In Fig. 4 the absolutely elastic links are removed before drawing the graph to show the formation of cells. This is imprecise, since the membrane could be much thicker, comprising molecules which in theory could interact, but in practice do not interact because the necessary atoms are missing all the time. In our experiments we have discovered clusters of molecules which were unable to interact and divided the graph into separated regions. Molecules which are absolutely unable to interact with themselves obviously form the raw material for membranes. As soon as such molecules cluster, all inner molecules of these clustered non-reactive molecules will build the membrane. For this reason, here, we define as *membrane molecules* those molecules which are absolutely unable to interact with themselves, regardless of their effective position. Thus, molecules which would build a membrane under the right condition.

Definition of cell: As the membrane molecules appear and start to cluster, the graph looses its homogeneity and starts to divide into different regions. If we delete all absolutely elastic links of a graph G, we will obtain a new graph G'. In this new graph any separated component containing more than one node

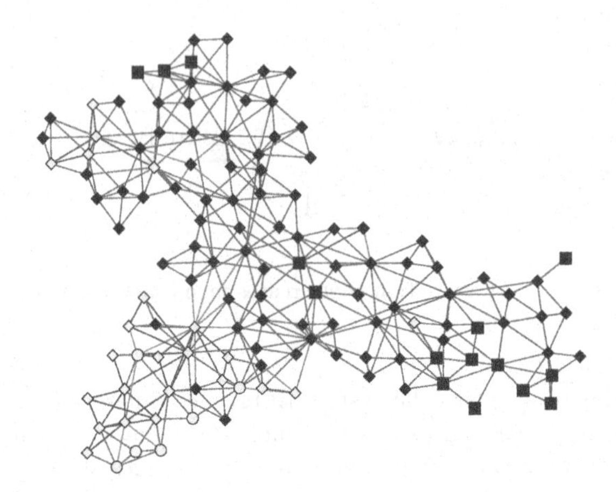

Fig. 2. Example of two cells. Magnification of a sub-graph taken from generation 917. Red edges denote "absolutly" elastic (non-reactive) connections. Membrane molecules: ■(r) and ♦(y_2). Inner molecules: ◊(b) and ○(g). Note that the lower left cell contains a different inner organization (◊and ○) then the small upper left cell (◊only).

will be regarded as a *cell.* Such a definition is purely geometrical, and in no way infers anything about the nature of the cell, or its possible evolution.

Autopoietic cell: Some cells exhibit stronger properties, so that they can be classified as autopoietic cells (example in Fig. 2). In the following we give a tentative definition of what would be an autopoietic sub-system in an artificial chemistry embedded in a graph. We will call I the set of elements inside a cell, and M the set of elements from its membrane. A cell will be called *autopoietic* if: (1) either I is an organization, and I produces M, or (2) I is an organization and $I \cup M$ produces M or (3) $I \cup M$ is an organization.

We will call such an organization the *cell organization.* Autopoiesis also requires that the cell organization is relatively stable with respect to an influx of random noise, e.g., random molecules. So, we require the system to continuously produces membrane molecules in order to maintain the cell as a separate unit. Under the influx of random molecular noise molecules which cannot interact with the organization will also be generated [16]. Being unable to interact, those molecules will naturally connect to the membrane, and as the membrane grows they will be expelled from the cell. In [16] we explain how metabolic organizations under the influx of random noise sometimes produce long unusable molecules. In our system such molecules would end up being part of the membrane. Under the right conditions (excessive number of atoms which build up membrane molecules), such a membrane could also be overproduced, with the effect of splitting the cell into separate cells. Even if all other requirements are satisfied, not every organization producing membrane molecules would form an autopoietic cell. In our system, if membrane molecules are produced too rapidly,

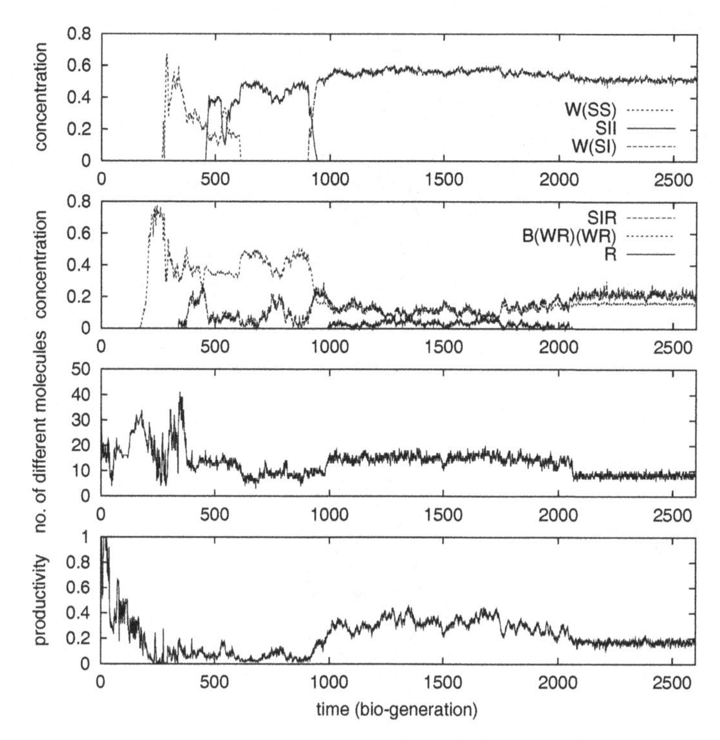

Fig. 3. Time evolution of the concentration of important molecules, number of different molecules (diversity), and productivity (probability that two neighboring molecules can react).

they tend to create big chunks and could not form a smooth membrane, that would otherwise be able to separate the inside from the outside (Fig. 4, middle).

Structure of the process - autopoietic mechanism: We will now discuss in detail a concrete experiment[4] which shows spontaneous formation of autopoietic cells. This will be done by listing its molecules, both the membrane and the inner ones. The main molecules that appeared in this experiment are:

combinator (molecule)	symbol	name	combinator (molecule)	symbol	name
$B(WR)(WR)$	○	g	$W(SS)$	♦	y_1
SIR	◊	b	$S(R(W(SS)))(W(SS))$	●	w
R	□	v_1	SII	■	r
RR	□	v_2	$W(SI)$	♦	y_2

[4] Parameter settings to replicate the experiment: Operation type $x \oplus y \longrightarrow x(y)$. Atoms used: B, C, K, R, I, S, W. Number of atoms: 1000 of each type. The influx of random molecules is exponentially decaying with MinNMolecules = 25, HalfNMolecules = 150. Outflux probability 0.01. Random seed 789. MaxTry = 10000. MaxLength = 100, MaxDepth = 20. Starting number of randomly generated molecules: 150 (initial population size).

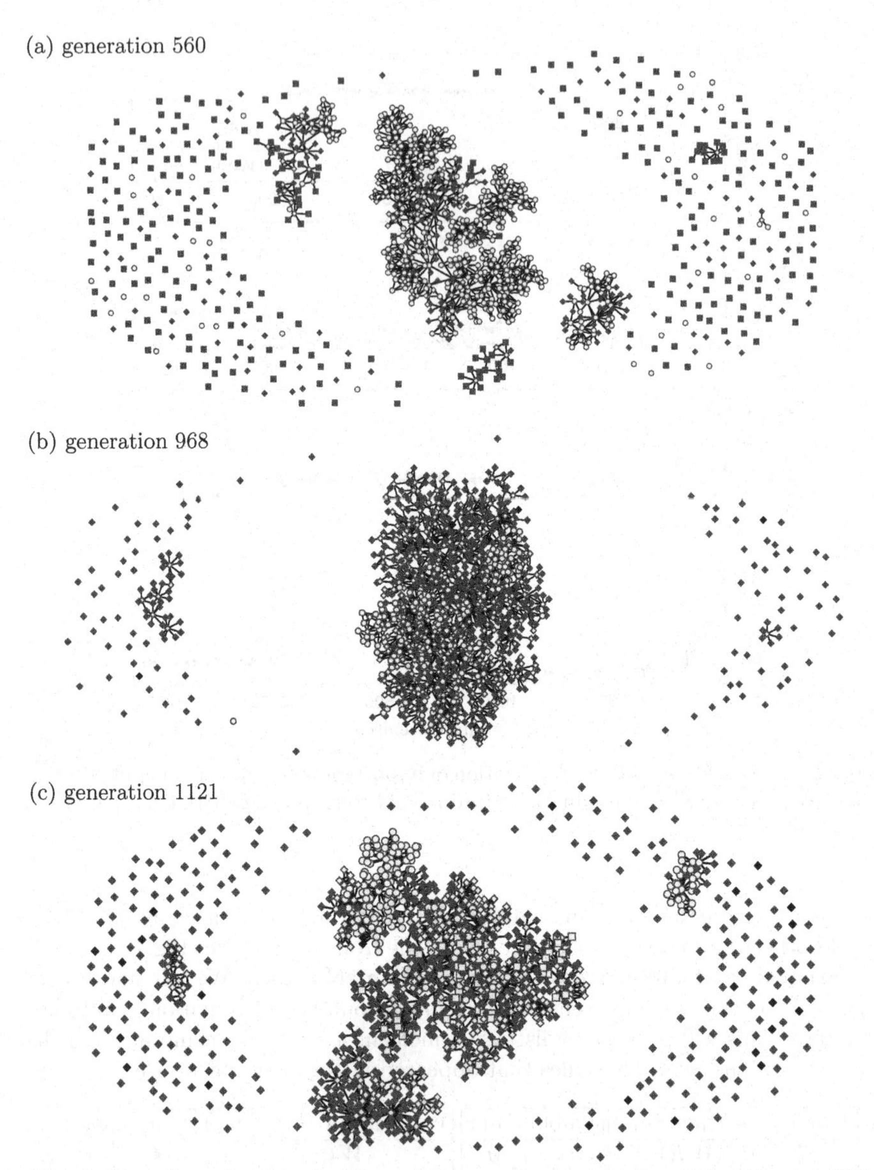

Fig. 4. Two-dimensional embedding of the graph at generation 560, 968, and 1121. Absolutely elastic links are removed before embedding. Note that nearness in the plane does not necessarily infer nearness in the planar graph. The embedding algorithm places nodes without neighbors at the outset of the plane.

The left four being inner molecules; the right four being membrane molecules. Short names are used to ease the following discussion. Two secondary membrane molecules which less frequently appear are:

combinator (molecule)	symbol	name
$S(B(WR)(WR))(B(WR)(WR))$	•	w_2
$S(SIR)(SIR)$	•	w_3

Membrane molecules can never interact with themselves, nor with other membrane molecules. Note that any element of the form $R(x)$ will react with anything else and always creates x by that reaction. Thus $R(x)$ can be seen as an intermediate product. For this reason in our table we will list every molecule $R(R(\ldots(x)\ldots)$ as x. Some of the inner molecules follow general rules. Let $x, y \in S$ be arbitrary molecules these general rules read: $g \oplus y \longrightarrow 4y$; $R(x) \oplus y \longrightarrow x, y$; $RR(x) \oplus y \longrightarrow R(y), x$.

$b = SIR$ acts differently on any molecule, as can be seen in the following table:

$\oplus$	g	b	v_1	v_2	y_1	w	r	y_2
b	$4g$	$2b$	v_2	v_1, v_2	w	$elastic$	$2r$	$3y_2$

What is missing are the reaction products between membrane molecules and inner molecules:

$\oplus$	g	b	v_1	v_2
y_1	$w_2, 3g$	w_3, b	n.h.	n.h.
w	elastic	elastic	n.h.	n.h.
r	$4g$	$2b$	n.h.	n.h.
y_2	$13g$ elastic de facto	$4b$	v_2	$2v_1, v_2$

Some membrane molecules, interacting with inside molecules create intermediate products w_3 and w_2. Both products cannot interact with themselves, nor with membrane molecules. We list their interaction with inside molecules:

$\oplus$	b	w_3
b	$2b$	elastic
w_3	$4b$	elastic

$\oplus$	g	w_2
g	$2g$	$4g$
w_2	$10g$ elastic de facto	elastic

Interactions that do not appear because the corresponding molecules are not present at the same time are marked "n.h." (never happened).

From this tables we can conclude that y_1 cannot be a membrane for b (♦ and ◊, respectively, in Fig. 4 (a)) since in no way can it be generated by their interaction. b instead can use r (■ in Fig. 4) and y_2 (♦ in Fig. 4 (b) and (c)) to create cells as it effectively does (Fig. 2)). g (○) can create cells using any of the above membrane molecules. Yet with y_1 and r (■) it generates them smoothly, ending up with an effective membrane (○ and ■). When y_2 appears with g the formation of membrane cannot be as clearly observed as in the other case, since the membrane ends up generating too big clusters. Either this is an effect of the explosive reaction ($w_2 \oplus g \longrightarrow 10g$), or is an effect of the dynamics. The investigation of the later case is beyond the scope of this paper.

We showed examples from different organizations arising in one run, but where not every membrane molecule was coupled to every organization to generate an effective self-maintaining cell. The other cases seem in all regards autopoietic systems.

Different types of membrane compete: In our experiment we noticed how different types of membrane tend to compete (Fig. 3, upper graph). Since they often use the same type of atoms (S and W) this no surprise. On the other hand, inner molecules can feed upon a bigger variety of atoms, molecules are often divided by the membrane into different cells that do not compete since they feed on different atoms (Fig. 3, second graph from above).

Different runs generate different molecules: Up to now we have spoken about one run only. When more runs are performed, sometimes other cells are generated, based upon different molecules but with similar behavior. Another possible outcome of a run are organizations so different that no membrane is present at all, e.g., where the productivity[5] stabilizes at exactly one.

Since each run generates a totally different experiment, with a completely different behavior, averaging over a number of runs would not give interesting data. For this reason we choose to thoroughly present one run instead. For an analysis of the variability of the possible organization that may appear in different runs see [15].

4 Conclusion

Our results were achieved by permitting any possible molecule to appear in the system (at least in principle). The resulting organizations not only generated a membrane in the "physical" space, but also, through their network of interactions, maintained exactly those molecules which continuously generated the cell.

What we observed is a special case of an autopoietic structure. In a general case the molecules would not be required to remain constant through time. What would remain constant instead is the network of interactions which generate the cell[6]. This is equivalent to saying that a cell can be autopoietic even if it changes its component types, as long as the new components still interact in the same functional way.

In our work the atoms were assumed to have a homogeneous concentration throughout the entire graph. An interesting variation would be to consider them locally in each face. This could give rise to a Darwinian process as cells have to destroy other cell's molecules to free atoms for use in their faces. As we said earlier our results were reached by studying an universal artificial chemistry. Each cell generates its own table of interaction for its molecules. It would be an interesting problem to extract these reaction tables and to use a fast and simple model for studying the evolutionary dynamics of cell assemblies.

[5] Productivity measures the probability that a collision of two molecules is reactive.

[6] If the molecules remain constant this condition is automatically fulfilled

References

1. W. Banzhaf, P. Dittrich, and B. Eller. Selforganization in a system of binary strings with topological interactions. *Physica D*, 125:85–104, 1999.
2. J. Breyer, J. Ackermann, and J. McCaskill Evolving reaction-diffusion ecosystems with self-assembling structure in thin films. *Artificial Life*, 4(1):25–40, 1999.
3. R. Durrett. The importance of being discrete (and spatial). *Theor. Popul. Biol.*, 46:363–394, 1994.
4. J. D. Farmer, S. A. Kauffman, and N. H. Packard Autocatalytic replication of polymers. *Physica D*, 22:50–67, 1986.
5. W. Fontana and L. W. Buss 'The arrival of the fittest': Toward a theory of biological organization. *Bull. Math. Biol.*, 56:1–64, 1994.
6. R. Laing. Some alternative reproductive strategies in artificial molecular machines. *J. Theor. Biol.*, 54:63–84, 1975.
7. B. Mayer and S. Rasmussen. Dynamics and simulation of micellar self-reproduction. *Int. J. Mod. Phys. C*, 11(4):809–826, 2000.
8. B. Mayer and S Rasmussen. The lattice molecular automaton (LMA): A simulation system for constructive molecular dynamics. *Int. J. of Modern Physics C*, 9(1):157–177, 1998.
9. B. McMullin and F. J. Varela Rediscovering computational autopoiesis. In P. Husbands and I. Harvey, editors, *Fourth European Conference on Artificial Life*, pages 38–47, Cambridge, MA, 1997. MIT Press.
10. N. Ono and T. Ikegami. Self-maintenance and self-reproduction in an abstract cell model. *J. Theor. Biol.*, 206(2):243–253, 2000.
11. A. N. Pargellis The spontaneous generation of digital "life". *Physica D*, 91(1-2):86–96, 1996.
12. Gheorghe Paun. Computing with membranes. *J. of Computer and System Sciences*, 61(1):108–143, 2000.
13. M. A. Shackleton and C. S. Winter. A computational architecture based on cellular processing. In M. Holcombe and R. Paton, editors, *Information Processing in Cells and Tissues*, pages 261–272, New York, NY, 1998. Plenum Press.
14. P. Speroni di Fenizio. Building life without cheating. Master's thesis, University of Sussex, Falmer, Brighton, UK, 1999.
15. P. Speroni di Fenizio. A less abstract artficial chemistry. In M. A. Bedau, J. S. McCaskill, N. H. Packard, and S. Rasmussen, editors, *Artificial Life VII*, pages 49–53, Cambridge, MA, 2000. MIT Press.
16. P. Speroni di Fenizio and W. Banzhaf. Metabolic and stable organisations. This volume, 2001.
17. Y. Suzuki and H. Tanaka. Chemical evolution among artificial proto-cells. In M. A. Bedau, J. S. McCaskill, N. H. Packard, and S. Rasmussen, editors, *Artificial Life VII*, pages 54–63, Cambridge, MA, 2000. MIT Press.
18. F. J. Varela, H. R. Maturana, and R. Uribe Autopoiesis: The organization of living systems. *BioSystems*, 5(4):187–196, 1974.
19. R. J. Wilson. *Introduction to Graph Theory*. Oliver and Boyd, Edinburgh, 1972.
20. K.-P. Zauner and M. Conrad. Conformation-driven computing: Simulating the context-conformation-action loop. *Supramolecular Science*, 5(5-6):791–794, 1998.

Understanding the Agent's Brain: A Quantitative Approach

Ranit Aharonov[1], Isaac Meilijson[2], and Eytan Ruppin[3]

[1] Center for Neural Computation
The Hebrew University, Jerusalem, Israel
ranit@alice.nc.huji.ac.il
[2] School of Mathematical Science
Tel-Aviv University, Tel-Aviv, Israel
isaco@math.tau.ac.il
[3] School of Computer Science and School of Medicine
Tel-Aviv University, Tel-Aviv, Israel
ruppin@math.tau.ac.il

Abstract. In recent years there have been many works describing successful autonomous agents controlled by Evolved Artificial Neural Networks. Understanding the structure and function of these neurocontrollers is important both from an engineering perspective and from the standpoint of the theory of Neural Networks. Here, we introduce a novel algorithm, termed PPA (Performance Prediction Algorithm), that quantitatively measures the contributions of elements of a neural system to the tasks it performs. The algorithm identifies the elements which participate in a behavioral task, given data about performance decrease due to knocking out (lesioning) sets of elements. It also allows the accurate prediction of performance due to multi-element lesions. The effectiveness of the new algorithm is demonstrated in two recurrent neural networks with complex interactions among the elements. The generality and scalability of this method make it an important tool for the study and analysis of evolved neurocontrollers.

1 Introduction

The paradigm of Evolved Neurocontrollers has produced promising results in terms of evolving autonomous agents which successfully cope with diverse behavioral tasks [9,7,5,4,3]. A fundamental and challenging question in this field is to understand how the controlling networks perform the tasks successfully. This is important from two complementary views. First, to achieve the goal of developing "smart" controllers, one must gain understanding of the evolved systems. Second, the study of artificial agents can potentially provide insights into the organization of natural systems. For the field of Artificial Life to be useful in this sense, one must develop in-depth understanding of the workings of the artificial agents.

In a previous work we have analyzed evolved neurocontrollers using classical neuroscientific tools, and demonstrated that the control mechanisms evolving

J. Kelemen and P. Sosík (Eds.): ECAL 2001, LNAI 2159, pp. 216–225, 2001.

resemble known findings from neurobiology [1]. In this work we present a new method for analyzing neural systems, both natural and artificial, and demonstrate how it can be used to analyze evolved neurocontrollers. This method deals with the fundamental issue of identifying the significance of elements in the system to the tasks it performs.

Even simple neural systems are capable of performing multiple and unrelated tasks, often in parallel. Each task recruits some elements of the system, and often the same element participates in several tasks. This poses a difficult challenge when one attempts to identify the roles of the network elements, and to assess their contributions to the different tasks. In neuroscience, assessing the importance of single neurons or cortical areas to specific tasks is usually achieved either by assessing the deficit in performance after a lesion of a specific area, or by recording the activity during behavior, assuming that areas which deviate from baseline activity are more important for the task performed. These classical methods suffer from two fundamental flaws: First, they do not take into account the probable case that there are complex interactions among elements in the system. E.g., if two neurons have a high degree of redundancy, lesioning of either one alone will not reveal its influence. Second, they are qualitative measures, lacking quantitative predictions. Moreover, the very nature of the contribution of a neural element is quite elusive and ill defined. In this paper we propose both a rigorous, operative definition for the neuron's contribution and a novel algorithm to measure it. The ability of the new algorithm suggested here, to quantify the contribution of elements to tasks, allows us to understand how the tasks are distributed within the system. Moreover it allows the accurate prediction of performance following damage of any degree to the system.

We first applied the Performance Prediction Algorithm (PPA) to a network hand-crafted to exhibit redundancy, feedback and modulatory effects. Then we used the PPA to study evolved neurocontrollers for behaving autonomous agents [1]. In both cases the algorithm results in measures which are highly consistent with what is qualitatively known a-priori about the systems. The fact that these are recurrent networks suggests that the algorithm can be used in many classes of neural systems which pose a difficult challenge for existing analysis tools. Moreover, the proposed algorithm is scalable and applicable to the analysis of large neural networks. It can thus make a major contribution to studying the organization of tasks in Neural Networks, and specifically in the "brains" of autonomous agents controlled by evolved neurocontrollers.

2 The PPA Algorithm

We consider a network (either natural or artificial) of N neurons performing a set of P different functional tasks. For any given task, we would like to find the *contribution vector* $c = (c_1, ..., c_N)$, where c_i is the contribution of neuron i to the task in question. We suggest a rigorous and operative definition for this contribution vector, and propose an algorithm for its computation.

Suppose a set of neurons in the network is knocked out (lesioned) and the network then performs the specified task. The result of this experiment is described by the pair $< m, p_m >$ where m is an incidence vector of length N, such that $m(i) = 0$ if neuron i was lesioned and 1 if it was intact. p_m is the *performance* of the network divided by the baseline case of a fully intact network.

The underlying idea of our definition is that the contributions of the units are those values which allow the most accurate prediction of performance following damage of any degree to the system. More formally, let the pair $< f, c >$, where f is a smooth monotonous non-decreasing[1] function and c a normalized column vector such that $\sum_{i=1}^{N} |c_i| = 1$, be the pair which minimizes:

$$E = \frac{1}{2^N} \sum_{\{m\}} [f(m \cdot c) - p_m]^2. \tag{1}$$

This c will be taken as the *contribution vector* for the task tested, and the corresponding f will be called its adjoint *performance prediction function.*

Given a configuration m of lesioned and intact neurons, the predicted performance of the network is the sum of the contribution values of the intact neurons $(m \cdot c)$, passed through the performance prediction function f. The contribution vector c accompanied by f is optimal in the sense that this predicted value minimizes the Mean Square Error (MSE) relative to the real performance, over all possible lesion configurations.

The computation of the contribution vector and the prediction function are done separately for each task, by the Performance Prediction Algorithm (**PPA**), using a training set consisting of a subset of all the 2^N possible lesioning configurations. The training error E_t is defined as in equation 1, but only averaging on the configurations present in the training set. The PPA is an EM-style algorithm which iteratively updates f and c until a local minima of the training error is reached. The steps of the PPA are:

- **Step 1:** Choose an initial normalized contribution vector c for the task. If there is no special choice initialize to the normalized effect of lesioning each unit separately.

  Repeat steps 2 and 3 until the error E_t converges or a maximal number of steps has been reached:
- **Step 2: Compute f.** Given the current c, perform isotonic regression [2] on the pairs $< m \cdot c, p_m >$ in the training set. Use a smoothing spline [6] on the result of the regression to obtain the new f.
- **Step 3: Compute c.** Using the current f compute new c values by training a perceptron with input m, weights c and transfer function f. The output of the perceptron is exactly $f(m \cdot c)$, and the target output is p_m. Hence training the perceptron results in finding a new vector c, that given the current function f, minimizes the error E_t on the training set. Finally re-normalize c.

[1] It is assumed that as more important elements are lesioned ($m \cdot c$ decreases), the performance (p_m) decreases, and hence the postulated monotonicity of f.

The output of the algorithm is thus a contribution value for every neuron, accompanied by a function, such that given any configuration of lesioned neurons, one can predict with high confidence the performance of the damaged network. **Thus, the algorithm achieves two important goals:** a) It identifies automatically the neurons or areas which participate in a cognitive or behavioral task. b) The function f predicts the result of multiple lesions, allowing for non linear combinations of the effects of single lesions.

The application of the PPA to all tasks defines a *contribution matrix* C, whose k^{th} column ($k = 1...P$) is the contribution vector computed using the above algorithm for task k, i.e. C_{ik} is the contribution of neuron i to task k.

Introducing the contribution matrix allows us to approach issues relating to the distribution of computation in a network in a quantitative manner. Here we suggest quantitative measures for localization of function and specialization of neurons.

If a task is completely distributed in the network, the contributions of all neurons to that task should be identical (full *equipotentiality* [8]). Thus, we define the *localization* L_k of task k as a deviation from equipotentiality. Formally, L_k is the standard deviation of column k of the contribution matrix divided by the maximal possible standard deviation.

$$L_k = \frac{std(C_{*k})}{\sqrt{(N-1)/N^2}}. \tag{2}$$

Note that L_k is in the range $[0, 1]$ where $L_k = 0$ indicates full distribution and $L_k = 1$ indicates localization of the task to one neuron alone. The degree of localization of function in the whole network, L, is the simple average of L_k over all tasks. Similarly, if neuron i is highly specialized for a certain task, C_{i*} will deviate strongly from a uniform distribution, and thus we define S_i, the *specialization* of neuron i as

$$S_i = \frac{std(|C_{i*}|)}{\sqrt{(P-1)/P^2}}. \tag{3}$$

3 Results

We tested the proposed index on two types of recurrent networks. We chose to study recurrent networks because they pose an especially difficult challenge, as the output units also participate in the computation, and in general complex interactions among elements may arise[2]. We begin with a hand-crafted example containing redundancy, feedback and modulation, and continue with networks that emerge from an evolutionary process. The evolved networks are not hand-crafted but rather their structure emerges as an outcome of the selection pressure to successfully perform the tasks defined. Thus, we have no prior knowledge about their structure, yet they are tractable models to investigate.

[2] In order to single out the role of output units in the computation, lesioning was performed by decoupling their activity from the rest of the network and not by knocking them out completely.

3.1 Hand-Crafted Example

Figure 1 depicts a neural network we designed to include potential pitfalls for analysis procedures aimed at identifying important neurons of the system (see details in the caption). Figure 2(a) shows the contribution values computed by three methods applied to this network. The first estimation was computed as the correlation between the activity of the neuron and the performance of the network[3]. The second estimation was computed as the decrease in performance due to lesioning of single neurons. To allow for comparison between methods these contributions were also normalized such that the sum of their absolute values is 1. Finally, we used the PPA, training on a set of 64 examples. Note that as expected the activity correlation method assigns a high contribution value to neuron 9, even though it actually has no significance in determining the performance. Single lesions fail to detect the significance of neurons involved in redundant interactions (neurons $4-6$). The PPA successfully identifies the underlying importance of all neurons in the network, even the subtle significance of the feedback from neuron 10. We used a small training set (64 out of 2^{10} configurations) containing lesions of either small (up to 20% chance for each neuron to be lesioned) or large (more than 90% chance of lesioning) degree. Convergence was achieved after 10 iterations.

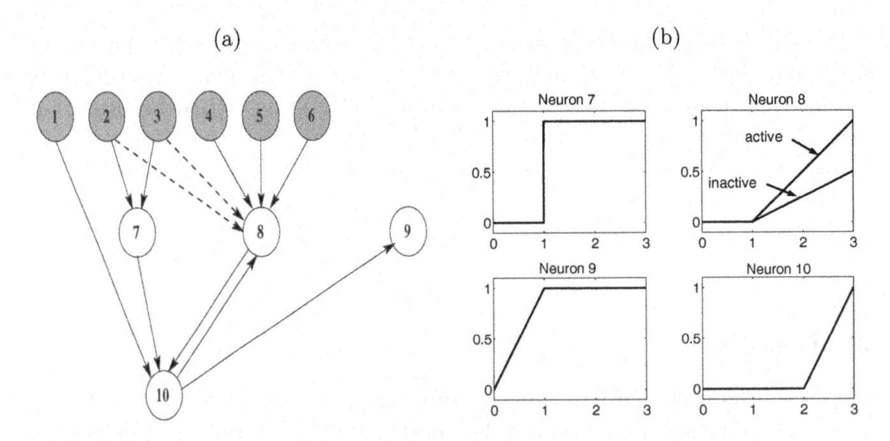

Fig. 1. Hand-crafted neural network: a) Architecture of the network. Solid lines are weights, all of strength 1. Dashed lines indicate modulatory effects. Neurons 1 through 6 are spontaneously active (activity equals 1) under normal conditions. The performance of the network is taken to be the activity of neuron 10. b) The activation functions of the non-spontaneous neurons. The x axis is the input field and the y axis is the resulting activity of the neuron. Neuron 8 has two activation functions. If both neurons 2 and 3 are switched on they activate a modulating effect on neuron 8 which switches its activation function from the inactive case to the active case.

[3] Neuron 10 was omitted in this method of analysis since it is by definition in full correlation with the performance.

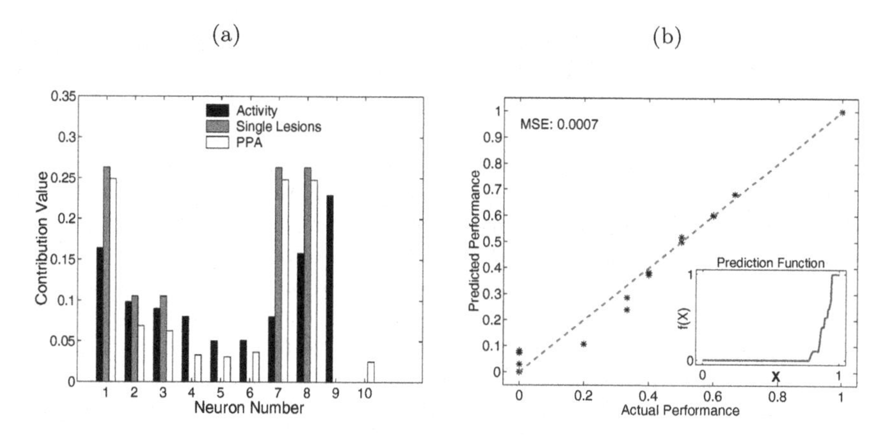

Fig. 2. Results of the PPA: a) Contribution values obtained using three methods: The correlation of activity to performance, single neuron lesions, and the PPA. b) Predicted versus actual performance using c and its adjoint performance prediction function f obtained by the PPA. *Insert:* The shape of f.

As opposed to the two other methods, the PPA not only identifies and quantifies the significance of elements in the network, but also allows for the prediction of performances from multi-element lesions, even if they were absent from the training set. The predicted performance following a given configuration of lesioned neurons is given by $f(m \cdot c)$ as explained in section 2. Figure 2(b) depicts the predicted versus actual performance on a test set containing 230 configurations of varying degrees ($0 - 100\%$ chance of lesioning). The mean square error (MSE) of the performance prediction is only 0.0007. In principle, the other methods do not give the possibility to predict the performance in any straightforward way, as is evident from the non-linear form of the performance prediction error (see insert of figure 2(b)). The shape of the performance prediction function depends on the organization of the network, and can vary widely between different models (results not shown here).

3.2 Evolved Neurocontrollers

Using evolutionary simulations we developed autonomous agents controlled by fully recurrent artificial neural networks. High performance levels were attained by agents performing simple life-like tasks of foraging and navigation. Using various analysis tools we found a prevailing structure of a *command neuron* switching the dynamics of the network between radically different behavioral modes [1]. Although the command neuron mechanism was a robust phenomenon, the evolved networks did differ in the role other neurons performed. When only limited sensory information was available, the command neuron relied on feedback from the motor units. In other cases no such feedback was needed, but other neurons performed some auxiliary computation on the sensory input. We applied the PPA to the evolved neurocontrollers in order to test its capabilities in a system on which we have previously obtained qualitative understanding, yet is still relatively complex.

Figure 3 depicts the contribution values of the neurons of three successful evolved neurocontrollers obtained using the PPA. Figure 3(a) corresponds to a neurocontroller of an agent equipped with a position sensor (see [1] for details), which does not require any feedback from the motor units. As can be seen these motor units indeed receive near zero contribution values. Figures 3(b) and 3(c) correspond to neurocontrollers who strongly relied on motor feedback for their memory mechanism to function properly. The algorithm easily identifies their significance. In all three cases the command neuron receives high values as expected. We also obtained the degree of localization of each network, as explained in section 2. The values are: 0.56, 0.35 and 0.47 for the networks depicted in figures 3(a) 3(b) and 3(c) respectively. These values are in good agreement with the qualitative descriptions of the networks we have obtained using classical neuroscience tools [1].

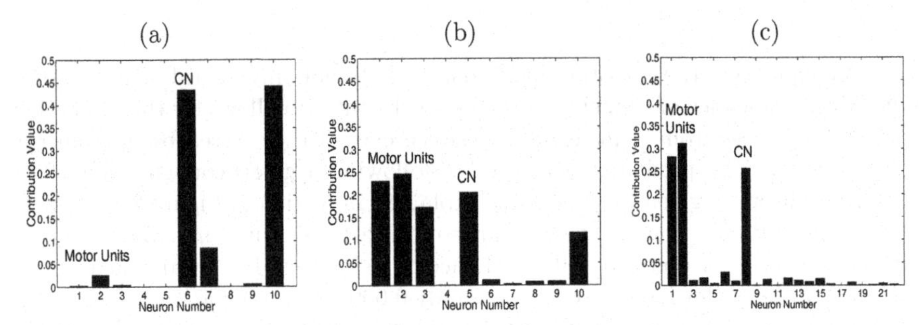

Fig. 3. Contribution values of neurons in three evolved neurocontrollers: Neurons 1-4 are motor neurons. CN is the command neuron that emerged spontaneously in all evolutionary runs.

3.3 Effects of the Training Set

To better understand how the training set effects the results of the PPA we trained the algorithm using different training sets of increasing "lesioning order". Set 1 consisted of all configurations in which a single neuron was lesioned as well as the fully intact and all-lesioned configurations. Set 2 consisted of set 1 augmented by all configurations in which exactly two units are lesioned. Set 3 consisted of set 2 augmented by results from triple lesions. Set 4 consisted of all the 2^N lesioning configurations. As neural network test beds we used the hand crafted example described in section 3.1 and a complex neurocontroller which was unique in demonstrating a relatively large number of significant neurons.

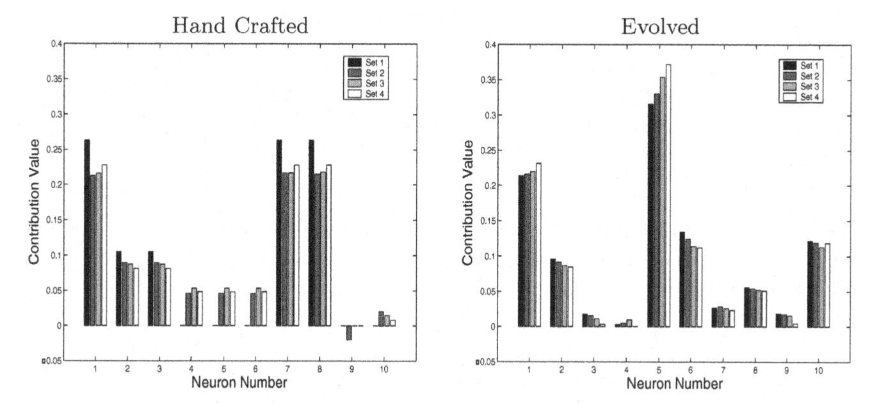

Fig. 4. Contribution values of neurons for varying training sets: The PPA was performed using training sets of increasing lesioning order (see text for details). Left figure is results from the hand-crafted example. Right figure is from the evolved neurocontroller.

Varying the training set has two possible effects on the results of the PPA. First it may shift the values in the contribution vector (c). This is evident in figure 4, especially for the hand-crafted example where "high order interactions" between units exist, e.g. redundancy. Even if the contribution vector remains the same, the fact that the training set consists of more points may also result in a different fit for the performance prediction function (f). In order to study separately the two effects we ran the PPA for each training set. Then we used the resulting c and fitted f using each of the four training sets as fitting sets (analogous to step 2 of the PPA)[4]. Table 1 depicts the performance prediction capabilities as a function of the training set and the fitting set. The prediction capability is measured as the MSE on a chosen test set, here the set of all 2^N lesioning configurations. The diagonal of the table depicts the results of the PPA without any manipulations (the training and fitting sets are identical).

The diagonal of table 1 demonstrates that the prediction performance of the PPA increases with the lesioning order of the training set (this effect is in general true also for random sets of increasing size). But is this due only to a better fit of the function, or do the additional configurations reveal more about the real contributions of the elements? The hand-crafted example clearly demonstrates the latter; the use of higher order fitting sets, although it decreases the MSE does not reach the levels attained by finding c using the PPA with higher order training sets (compare training with set 1, fitting with set 4 vs. training and fitting on set 4). Similar effects, though somewhat weaker, are evident for the evolved neurocontroller studied here. This testifies to the fact that some type of

[4] fitting f with the same training set as the one used for running the algorithm results in the PPA-produced f.

"high order interactions" between the neurons exists within the evolved network. It is interesting to note that many other neurocontrollers do not demonstrate this effect, i.e. the improvement in prediction performance is due to more points for fitting the function, and not to a better estimation of the contributions. This in some sense testifies to their "simplicity".

Table 1. Prediction MSE as a function of training and fitting sets. Each row corresponds to a different training set, and each column to a different fitting set. Sets are ordered by increasing lesioning order as explained in the text.

Hand Crafted				
	set 1	*set* 2	*set* 3	*set* 4
set 1	0.0172	0.0098	0.0062	0.0051
set 2	0.0021	0.0018	0.0007	0.0005
set 3	0.0012	0.001	0.0004	0.0002
set 4	0.0016	0.001	0.0003	0.0002

Evolved				
	set 1	*set* 2	*set* 3	*set* 4
set 1	0.0075	0.0051	0.004	0.0019
set 2	0.0065	0.0044	0.0033	0.0016
set 3	0.0049	0.0034	0.0025	0.0012
set 4	0.0032	0.0024	0.0018	0.0009

4 Discussion

One of the main advantages of using evolutinary algorithms to develop neurocontrollers for autonomous agents is the ability to produce unconstrained solutions, not influenced by our preconseptions of how to solve the task. Moreover, evolutionary techniques permit the development of controllers coping with multiple tasks in parallel. This advantage, however, poses a great challenge for analyzing the resulting networks because they are unconstrained, they multiplex solutions to different tasks, and we have no prior knowledge of their structure-to-function relationships. This work suggests a novel quantitaive method to aid in the analysis of neural networks in general and of evolved neurocontrollers in particular.

We have introduced a novel algorithm termed PPA (Performance Prediction Algorithm) to measure the contribution of neurons to the tasks that a neural network performs. These contributions allowed us to quantitatively define an index of the degree of localization of function in the network, as well as for task-specialization of the neurons. The algorithm uses data from performance measures of the network with different sets of lesioned neurons. We have applied the algorithm to two types of Artificial Recurrent Neural Networks, and demonstrated that it results in agreement with our qualitative a-priori notions and with classical qualitative analysis methods. We have shown that estimation of the importance of system elements, using simple activity measures and single lesions, may be misleading. The new PPA is more robust as it takes into account interactions of higher degrees. Moreover, it serves as a powerful tool for *predicting* damage caused by multiple lesions, a feat that is difficult even when one can accurately estimate the contributions of single elements. The shape of the performance prediction function itself may also reveal important features of the organization of the network, e.g. its robustness to neuronal death.

Application of the PPA to artificial rather than biological "nervous systems" enables one to test many different training sets and study their effect on the resulting contribution values. We have shown that in some cases using training sets including lesions of more neurons allows for better prediction, due to better estimation of the contribution values. This supports the notion that some type of higher order interactions exists among the elements. We are currently working on developing a quantitative measure for this qualitative notion.

As the field of Evolved Neurocontrollers develops, it is expected that many types of interesting and complex controllers will evolve to cope with many different tasks. Being able to analyze the networks, and gain understanding of how they operate will be of great benefit both for developing better controllers and for using the artificial agents as models to help us understand biological agents. The promising results achieved and the potential scalability of the PPA lead us to believe that it will prove extremely useful in obtaining insights into the organization of Evolved Neurocontrollers.

Acknowledgments: We greatly acknowledge the valuable contributions made by Ehud Lehrer, Hanoch Gutfreund and Tuvik Beker.

References

[1] R. Aharonov-Barki, T. Beker, and E. Ruppin. Emergence of memory-driven command neurons in evolved artificial agents. *Neural Computation*, 13(3), 2001.

[2] R. Barlow, D. Bartholemew, J. Bremner, and H. Brunk. *Statistical Inference Under Order Restrictions*. John Wiley, NewYork, 1972.

[3] Dario Floreano and Francesco Mondada. Evolution of homing navigation in a real mobile robot. *IEEE Transactions on Systems, Man, and Cybernetics - Part B*, 26(3):396–407, June 1996.

[4] Dario Floreano and Francesco Mondada. Evolutionary neurocontrollers for autonomous mobile robots. *Neural Networks*, 11(7-8):1461–1478, October-November 1998.

[5] Faustino Gomez and Risto Miikkulainen. Incremental evolution of complex general behavior. *Adaptive Behavior*, 5(3/4):317–342, 1997.

[6] G. D. Knott. Interpolating cubic splines. In National Science Foundation J. C. Cherniavsky, editor, *Progress in Computer Science and Applied Logic*. Birkhauser, 2000.

[7] Jérôme Kodjabachian and Jean-Arcady Meyer. Evolution and development of neural controllers for locomotion, gradient-following and obstacle-avoidance in artificial insects. *IEEE Transactions on Neural Networks*, 9(5):796–812, September 1998.

[8] K. S. Lashley. *Brain Mechanisms in Intelligence*. University of Chicago Press, Chicago, 1929.

[9] Christian Scheier, Rolf Pfeifer, and Yasuo Kunyioshi. Embedded neural networks: Exploiting constraints. *Neural Networks*, (7-8):1551–1569, October-November 1998.

Observations on Complex Multi-state CAs

Eleonora Bilotta[1] and Pietro Pantano[1]

[1] Centro Interdipartimentale della Comunicazione, Università della Calabria,
Arcavacata di Rende,
87036 Rende (CS) , Italia
{bilotta, piepa}@unical.it

Abstract. A genetic algorithm, with a fitness function based on input-entropy, was developed to search for rules with complex behaviour for multi-state CAs. The surprising presence of complex rules founded allows the observation that, in this context too, complexity organises itself through various behaviour of emergent structures, which interact within regular or uniform domains, creating many different configurations, previously noticed for elementary CAs. In this paper observations on the behaviour of multi-state CAs are reported, through the dynamics of the emergent structures or gliders. In particular, it is discovered that the more particles follow local rules, the more complexity develops. This seems to be also validated by input entropy plots, which the particles produce in the dynamics of their related CAs. The paper re-opens the CAs classification problem and gives birth to new insights into some of the mathematical capability of building special configurations, attempting to measure in some way complexity.

1 Introduction

The Cellular Automata (CAs), dynamic structures in which space, time and states of the system are discrete, have long been studied in so much as they are mathematical models able to simulate complex behaviour of some physical and/or biological systems. One of the most interesting characteristics of these systems is the phenomenon by which a high number of elementary units, with the capability of local interaction, obeying the same rule, which is local and deterministic, is the source of numerous emerging configurations and auto-organisation. According to [1] CAs can be classified into 4 principal classes. The first two classes consist of ordered systems. The third class is that of chaotic systems, while the fourth class forms the set of the so-called complex systems, which for several reasons can be considered to be on the border between the class of organised systems and the class of chaotic systems. In complex systems the behaviour of CAs presents interesting characteristics that are not found in chaotic or ordered regimes. Emerging structures, distinct from a background, can be observed, which continuously change over a long transitory, before extinguishing; or the patterns propagate periodically in space-time. These last periodic structures are identified as “gliders” [2] [3] or "particles" [4]. The dynamics of the emergent structures has various consequences, from the extinction to the generation of other chaotic or periodic structures (new gliders), generating continuous transitions from ordered to chaotic behaviour. Some gliders can, in turn, generate other gliders,

J. Kelemen and P. Sosík (Eds.): ECAL 2001, LNAI 2159, pp. 226-235, 2001.

and are known as glider-guns. Other gliders, called glider-eaters, incorporate some pieces of the space-temporal configurations, giving rise, in turn, to important modifications, from a phenomenological point of view. So, a great number of the studies in this field are orientated towards the problems of searching for complex rules and on the problems of classification. Where are these complex rules? The Edge of chaos Hypothesis, formulated by Langton [5], maintains that complex rules of class 4 can be founded in the region between order and chaos [6]. With regards to Langton's suggestion, other researchers have framed theories to confirm or reject the hypothesis of a border between order and chaos [7], [8], [9]. In particular, Mitchell, Crutchfield and collaborators discovered that, whereas their genetic algorithms produced complex rules, it was not deducible that the cellular automata generated could lie near to the separation between simple and chaotic regions.

Recently, Oliveira and colleagues [10] used some parameters of classification, present in literature, to build GAs that carry out computational tasks on CAs.

Nevertheless the majority of the analysis carried out up to now refers to CAs with two states and small neighbourhood. When the number of neighbours and states increase, complex rules become more and more difficult to find. Little attention has been devoted to multi-state CAs. For this reason, utilising the input entropy function as fitness criterion, a genetic algorithm was realised to search for complex rules for multi-state CAs [11], verifying that it was effective as many families of complex rules were found.

The paper is organised as follows. First, after the formal notation used for Cellular Automata and for Genetic Algorithm, a description of domains, particles and interactions for some complex rule for multi-state CAs is presented. Then, some analysis of the growing complexity in the emerging structures for multi-state complex rules is presented. The paper ends by illustrating of how it is possible to compose the particle interactions by means of a law of composition that permits the combining of the CAs space-time behaviour. In the appendix, all the rules cited in the work are noted.

2 Formal Notation

Following Wolfram [1] let a cellular automaton be defined as a discrete dynamic system consisting of a regular lattice of identical cells. The value of the site i in a one-dimensional cellular automaton at time t is indicated by $a_i^{(t)}$. Each value will depend on the values of the sites at the time before t-1. So the rule of the evolution of the site will be given by:

$$a_i^{(t)} = F\left[a_{i-r}^{(t-1)}, a_{i-r+1}^{(t-1)}, \ldots\ldots, a_i^{(t-1)}, \ldots\ldots\ldots\ldots, a_{i+r}^{(t-1)}\right], \quad (1)$$

where r represents the radius of the neighbourhood and $2r+1$ is the number of sites (around the site i) that affects the change of the value of the site itself. The value of the site is supposed to be the whole number included between 0 and k-1.

Following Mitchell [12], a rule can also be written as:

$$G = \langle s_1, s_2, ...s_l \rangle, \quad (2)$$

where every s_i will have a value included between 0 and *k-1*. The number *l* of terms present in the string (2) will be:

$$l = k^{2r+1}. \quad (3)$$

The sequence (2) can be considered as the "genome" of the rule.

This string functions as base structure for the GA we have realised in [11] in order to detect complex rules for multi-state CAs. The fitness function that has been used, is based on Wuensche's input-entropy function [2], which has been rewritten for multi-state CAs in the following way:

$$S^t = -\sum_{i=1}^{k^{2r+1}} \left(\frac{Q_i^t}{n} \times \log\left(\frac{Q_i^t}{n}\right)\right), \quad (4)$$

where *2r+1* is the number of neighbours, *n* is the number of sites, Q_i^t is the frequency of the *i-th* configuration of the neighbourhood at the time *t*.

The use of this genetic algorithm allowed us the obtaining of a great quantity of families of rules with complex behaviour.

One of the first problems to be faced was the analysis of the enormous quantity of rules found by the system and the accompanying difficulty of their classification, particularly given that they present genomes that are not easy to represent (and/or with very wide representations). Other kinds of problems concern the classification of the space-time pattern.

At first, the qualitative data indicates that almost half of the results present the characteristics of complex systems. There are a lot of emerging of configurations that are distinct from a fairly regular or homogeneous background, with periodic configurations, responsible for the phenomenon of the interacting gliders and other chaotic configurations, such that the dynamic of the system continuously varies from order and chaos. In particular, it is observed that:

a. Some systems are characterised by chaotic non-periodic structures, that transform themselves for a long transitory, until they extinguish or merge with other structures;
b. Other systems present the characteristics of interacting gliders, characterised by periodic configurations that immediately emerge and collide, creating still more configurations. The periodic configurations (gliders) separate, sometimes, different homogeneous backgrounds, which create figures whose "borders" interact.

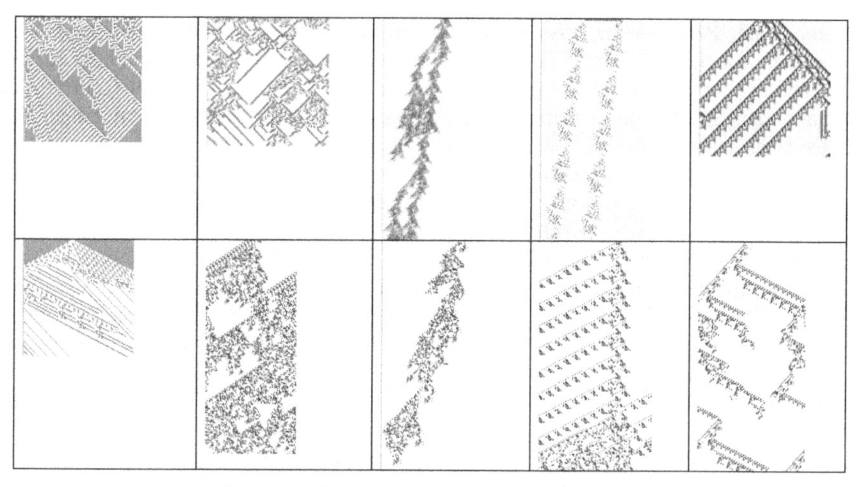

Fig. 1. Various examples of complex systems.

Moreover, between the two typologies exist systems with mixed characteristics, in which periodic and non-periodic structures emerge and interact.

3 Domains, Particles, and Interactions in Some Multi-state CAs

In the following, observations on some complex rules are reported.

In [4] domains, particles and interactions are defined. According to these authors, the study of these features is very important, since a description of higher order of the cellular automaton dynamics is obtained. Instead of describing the CAs by means of the evolution rules or by analysing the behaviour of one single site, this approach takes into consideration the mutual interactions between sites that, by parallel computation, exhibit regular and predictable behaviour.

For example, the rule k3r1A has a periodic domain 2-0 and a regular domain 1200, generated using all three states. As for the regular domains, the rule k4r2A presents a great variety: 110100, 1000, 220000. The characteristic of this last rule is its great regularity that allows the generation, through simple procedures, of various regular domains composing simple or complex gliders.

Various quiescent states can appear for multi-state CAs. For example, for k3r1A, 1 is present as quiescent state, for k3r2A, the quiescent states are 0 and 2.

Besides the regular domains, the complex rules we have been studying present emerging structures of remarkable importance. The majority of these particles present themselves as separators of quiescent states, periodic domains and regular domains, or as separators between these. Very often they appear as connectors between unsynchronised regions. Therefore it is obvious that the greater is the number of regular states and regions, the greater will be the number of particles that will have such characteristics. In this sense, the rule k4r2A presents a great variety of particles of this type. Particles that propagate in a quiescent state are more interesting. For examples, for the rule k3r2-27B a large particle, generated by the code 1111220221, can be seen, which slowly moves towards the right. The period of this particle is greater than 200 temporal steps, while the stationary dimension is greater than 70. Indeed, the particle presents two branches, the first one dies out and the second one reproduces itself and reproduces the entire particle, whose behaviour is in the sequence 21001. The particles with triangular structure produced by k4r2 (we called "fish") are also interesting; such structures spread within a periodical domain 0-3. The particle generated by the code 10, in its smaller dimension, reconnects two unsynchronized regions of this periodical domain.

The rule k4r2A also presents a great variety of particles. On a background 0, there are those generated by the data 20001, 10001, 200010001, which spread with the same speed. Those generated by the data 22, 20022, 3020022, and 31020022, which spread with another speed.

Another important series of very complex emerging particles, in other contexts, are called glider-guns. Particles that generate such types of complex configurations can be also observed in the multi-state CAs. For example, the rule k3r2A has a series of particles generated by the data: 22011, 2200122, and 2200011100102. The code 123021202023100221 gives a configuration of great interest, present in the rule k4r1A, which resides in the quiescent state 3, moves towards the right and generates particles that move towards the left. Such configurations cover the space-time as if they were a regular domain.

Another way of investigating the behaviour of complex CA is to analyse the interaction between particles and other emerging particles, because the quantity of information that an emerging structure brings with it is strictly correlated to its spatial and temporal dimensions [4]. In particular, this informative content is “released” or “transformed” in the interaction with other particles. The temporal scale determines the number of states that the particle can assume over time and that represent its phases.

The investigation of such interaction is also useful for multi-state systems. The emerging behaviour shows two different characteristics: interactions between different particles and synchronization between particles of the same nature. For example, in the rule k4r1A the interaction of two particles may produce: annihilation, absorption of a particle by another (one can absorb the other and vice-versa), birth of another particle, production of a particle that generates a series of equal configurations.

This interaction greatly depends on the incidence angle and is one of the reasons for the unpredictable behaviour of the evolution of the system. It is even more interesting to try to combine two particles that move at the same speed. In this case more complex particles emerge, as for example happens in the k4r2-fish and in the k4r2A. When such constructive combination is possible, the set of particles can also generate extended covering (regular domain) in space-time. For this reason, in some cases it is possible that the regular domains are produced by synchronized particles.

4 The Growing of Complexity

The complex structures described in the previous section, show behaviour of growing complexity. It has been discovered that such complexity can be measured in terms of spatial and temporal dimension of the particles. In some observed cases, the greater the difference between the minimum and maximum dimension, the more complex the emerging particle can be considered. The greater the period, the more complex is the emerging structure. A correlation between this phenomenon, that seems obvious at first sight, and the frequency of consultation of the look-up table by a particle, was found. A direct correlation between the number of used rules and the "visible" complexity of the emerging particles exists. It is finally interesting to see that, in pattern of high complexity, the distribution in the use of rules is peculiar to each structure, as it represents the "signature" of the particle itself.

Therefore, an analysis of the rules used by the emerging structures allows the identification of complexity for the space-time behaviour. We argued that the greater the number of local rules used, the greater is the complexity of the shape of the emerging configuration. Usually with this comes a variation of the frequency of use of the rules expressed in the string that determines the genome.

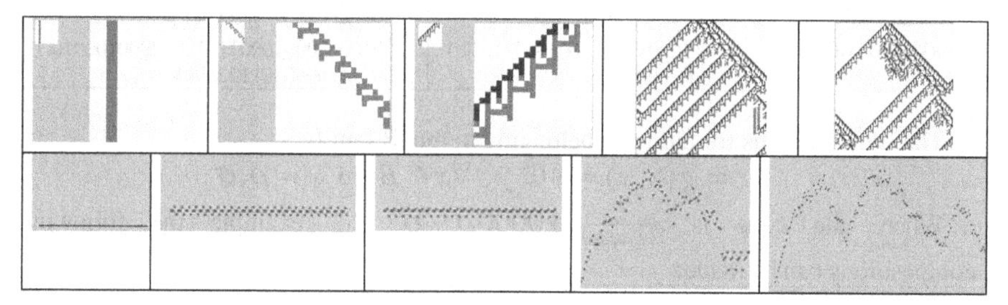

Fig. 2. Various gliders and glider-guns for the rule k4r1A and their related input-entropy plots.

Similar behaviour was observed in the investigation of emerging complex shapes for various multi-state CAs.

5 Interaction between Particles and Rule of Composition of Their Dynamics

Some rules show certain remarkable regular behaviour. For example the rule k4r2A shows various regular domains and various particles. Some particles belong to two principal categories related to the speed they move towards the right:

a) Particles that move with lower speed;
b) Particles that move at higher speed.

There is an enormous quantity of particles belonging to the two families. In reality, very complex particles can be generated by more simple particles through simple rules of composition. A indicates the set of particles belonging to the family a) and B indicates the set of particles belonging to the family b). Greek letters are used to indicate the elementary particles:

A={α – 10001, β – 20001, γ – 310320020001, δ – 22020000001122 0001, ...}
B={ρ – 22, σ – 20022, τ – 3020022, υ - 31020022,
ω – 31020000220000010200000203000 22,
ζ – 220000310200002200000102000000203000 22,
λ – 2200002200002200002200003102000022000001020000002030002 2,.........}

So two applications can be defined:

$$f : AxA \rightarrow A$$
$$g : BxB \rightarrow B$$

They allow the generation of new particles of the family starting from the base particle.

The function f is defined by :

$$z = f(x, y) = x \oplus y,\ \forall x, y \in A$$

Where the code is $\xleftrightarrow[codex]{}000\alpha \vee \beta000\xleftrightarrow[codey]{}$; for example the following compositions can be made:

$\alpha \oplus \alpha$	$\alpha \oplus \beta \oplus \gamma$	$\alpha \oplus \beta \oplus \gamma \oplus \alpha \oplus \delta$
1000100010001	100010002000100031032 0020001	10001000200010003103200200010001000 100022020000001 1220001

The same happens for particles belonging to the B family:

$$z = g(x, y) = x \oplus y,\ \forall x \in B \text{ ed } y = \rho, \sigma$$

Where the code is $\xleftrightarrow[codex]{}0000\rho \vee \sigma$; for example the following compositions can be made:

$\rho \oplus \rho$	$\rho \oplus \sigma$	$\tau \oplus \rho \oplus \sigma$
22000022	22000020022	302002200002200002002 2

Other and more complex laws of composition exist. New gliders belonging to the B family can be created, joining the gliders of the types ρ or σ at the left of the starting particle, followed by four or five zeros. This process produces more complex particles. If we repeat the process many times, we can obtain regular domains in the space-time diagrams.

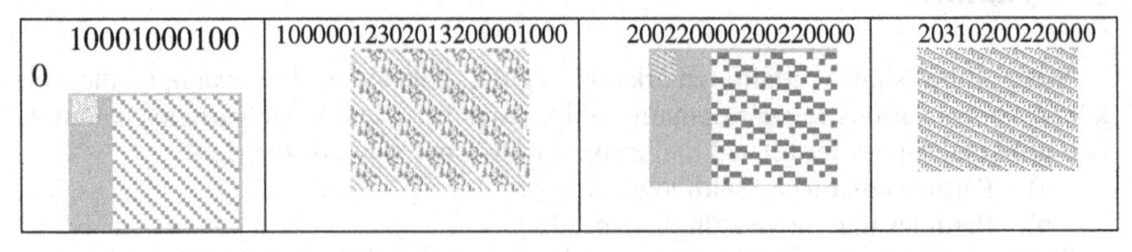

Fig. 3. Examples of regular domains created by glider compositions.

This behaviour has been noted for many multi-state CAs.

6 Conclusions and Open Problems

The aim of this work is to investigate the behaviour of complex rules for multi-state CA, analysing some characteristics of the emerging properties, using the key concepts of domains, particles and interaction between particles. For multi-state CAs too, there exist the following characteristics:

a. Some systems are characterised by chaotic non-periodic structures, which change over a long transitory, until they die out or they merge with other structures;
b. Other systems present characteristics of interactive particles, characterized by periodic configurations which emerge immediately and which collide, so creating still others. The periodic configurations separate different homogeneous/regular backgrounds, which create figures whose "borders" interact.
c. Moreover, between the two typologies there exist systems with mixed characteristics in which periodic and non-periodic structures emerge and interact.
d. Other structures are not classifiable but some base characteristics can be identified, such as the evolutionary period, the direction and speed of the spread of the pattern in space, the coexistence with other patterns (symbiotic process of different type), the exact periods of latency (momentary disappearance), with the exact periods of their reappearance.

From the analysis carried out up to now, it can be said that the particles are more complex the greater is the period and the difference between the maximum and minimum spatial dimensions. Such observation correlates with the analysis of the local rules used. The higher is the number of these particles, the more the particles are complex in the resulting spatial configuration.

It has been shown that it is possible in some way to define some procedures of combination between particles or particles and regular domains. These initial observations, still preliminary, at the moment show that some interesting conclusions about the dynamics of complex systems, already indicated for CAs, are also valid for multi-states CAs, although a comprehensive systematisation and the creation of "catalogues" which provide the complete behavioural repertory of a complex rule in this context, is still a long way off.

Bibliography

1. S. Wolfram, (1984) Universality and Complexity in Cellular Automata, *Physica 10D*, 1-35.
2. H.J. Conway (1982) "What is Life" in "Winning ways for your mathematical plays", Berlekamp, E, J.H. Conway and R.Guy, Vol. 2, chap.25, Academic Press, New York
3. A. Wuensche (1999). Classifying Cellular Automata Automatically: Finding gliders, filtering and relating space-time patterns, attractors basins and the Z parameter, *Complexity,* **4** (3), 47-66.
4. W.Hordijk, C.R.Shalizi, and J.P.Crutchfield (2000) *Upper Bound on the Products of Particle Interactions in Cellular Automata*, Santa Fe Institute, Working Paper, 00-09-052.

5. C.G. Langton (1986) Studying artificial life with cellular automata , *Physica 22 D*, 120-149.
6. C.G. Langton (1990) Computation at the edge of chaos, *Physica 42D*, 12-37.
7. H. Gutowitz , C.G.Langton (1995) *Mean Field Theory of the Edge of Chaos*, in Proceedings of ECAL 3, Springer, Berlin.
8. N.H. Packard (1988) Adaptation toward the edge of chaos. In A.J.Mandell, J.A.S. Kelso and M.F. Shlesinger (eds.), *Dynamics Patterns in Complex Systems*, Singapore, World Scientific
9. M. Mitchell, P.T. Hraber, J. P. Crutchfield (1993) *Revisiting the edge of chaos: evolving cellular automata to perform computations*, Complex Systems 7, 89-130.
10. G.M.B. Oliveira, P.P.B. De Oliveira, N. Omar (2000) Evolving Solutions of the Density Classification Task in 1D Cellular Automata, Guided by Parameters that Estimate their Dynamic Behaviour, , in *Artificial Life Proceedings,* M. A. Bedau, J.S.McCaskill, N.H. Packard, S. Rasmussen (Eds.), The MIT Press, Cambridge, Massachusetts.
11. E. Bilotta , A. Lafusa, P.Pantano (2001) *Searching for CAs complex rules by means of Gas*, submitted for publication.
12. M. Mitchell (1996) *An Introduction to Genetic Algorithms*, 1996, The MIT Press, Cambridge, Massachusetts.

Appendix of the Rules Cited in the Work

k3r1A:001220000100011010200211112

k3r2A:2001010002021001100210010200022202000011100200002102012102110212200021022100201002002101000101001121220100020012100000220000202101200001201000200020022001221110000012200000111000220100102011202000220011110002020201122222000120000102002001120 20

k3r227B:120221211120200100200101022202202220221110110102122121200100122110220121221010011202020212111011012022000010221211011100021020111100220112021020112212021201000010112222012110121010220200111220012210010010010021100021120220011110200012122 0000

k4r1A:331332000301320033122300110222001331212122013033011322022322322

k4r2A:
00000200003331020003010102201201002002100301000000000130013020010030021002133100000020030000000010000020001020200000000010301000202203000311020220233303010000010010002033030023000033231001133012002001301210000230022000000000210020203221002000211033003120022200031200101301310000000330010021000010000011002100200200310033230300300000032010200000111103000010000120220000000000110000020020200230100220201023201000100000310010003203230100100022110112022022002103200203003200021000000000300032012000220300000122003202302023001000130102030222100101200313000220020010010001030100000100022000100030101300002030100230000010000212100000001000002202002000322303000020002213331020301030020331122231220000023300101200223030201330021000310200030200000002000030031203112030002333000010000300002020102020200322000031100200001001303202021000000313000200030000002001032320003102010101000200003103000031000300000101000230000010011200031031000001010000000002013020300023320200120000221001000030121203200010010000000100001303033111110000200200000000 0

k4r2 fish:

000003200231002100000100220310000000000330000000100002313000102000201330000003
100000102001200000010100100200000032002302010003100000300023033303030313000000
3020100200301001300030002000200000203000002031001320031020302000310000122020100
0201030003210130000003002000030000002001020110111203010000022200111103010200030
0023000013030010003100102300203000100020003000301000331000300130000131000000002
0003201120303003000200200012023300323001030100121102002200000302000212000003300
1230100000003020300100231000120010000000200010230001110020301103100030002030000
0220021211010120110002301103200303000112130232200303232000000100030121003133010
0010000033000110302010301020020332100030000033200030230001100320021102113110203
011033011200000003030010010000021003303003030003020200000000000000000033003002
0303210000000000000030000020201000103300100300100021331330201001100001201000002
2200020000212100033002200000230000020200313002021000010031020003013202003102011
0012103002200020203020210022102000000310001000100022000012000110003000213033003

k2r2 B1

3000303003301120000020322230010000033110022002000023021023000232002100303000123
1031330311000022010032113032011000312010001001130100230001200021200313120010301
3010000301233222100012110030203223000000100120030301321000210122000223230000100
0101010200303010030003000000220333001001000010132000301202002130323000300213100
0311013330201002203110311012002001120002120301003130023321232100231100020030232
1030000033030020003120011003322313210200022200102000110103211220101000002300301
3021030031030000330013300310030300002000000021320000000011303100000331233200100
3200113000103001200120010002010320100200131202112000112012210131100000322201100
0300311202222320233101000103200101001312003110012103020200000312000001100203120
0332000030333333001101300102211003202232303013213030202200022033300002232003330
02000020130203300002002110120003201101033000031010000000000031210233010310100022
0001100020320100121222201003012212002010100000230010100220003230322330333000000
2010030020320012330002000011000210022100030000023332213003121203223030301000

k4r2 B3

30001030033011203000203222300100100031100220020010230210230002322202132303000123
1232330311000022000032113032013001312010001001130300230201000021230313120010301
3010000301233202101012110030203223300000100120030311321000210122000223230000100
0100010200303010030103020000220333033001000010132301331201002130323000300213100
0311313300201002202100311010002001121002120301003130023301232130221110020030232
3030030033030022103120011003122313200330022200102000100103211220101000002300301
3021333031030000330013000310030300002020030021320000003013303100002321233200100
3200113200103021230120000002010320100200131202112000112112210131100000322211100
0300311202222320333101000103200101001312303100012103020210000312000001102203120
0330000030333333001101300000211003302232303013213330202203020033300002232303330
02000020130203300002002110220003201101033000033010000000000031210133010310100022
0001100020310100121222201003012212002010100203230010100220011230322330333000000
2010030020320012330002020011300210022100030000123302213003122203223030301000

k4r2 B4

201300201010000203103110233101200000220000300320000000200110012200000000021010100
0000000330100023021000020030101002020000100000301002000030200010320010030001113
0300103022000003013032300003022102001030003010330202100001010000130010232033001
313110201001000000013010003200003003100003000000000000000000002003210030010003000
2220100010102333300210020020330002201320100002000300010030202000000131002022030
222000000002000100301000000000200011310021032002001310100000000203000330200100002
002000002030000000000001132230010002020001020313200202200100300232300001002223000
1000000212100321002011033003033200130002001002001100000202000100200030000310030
30020002102000010200230000000010021000101000010013001000000122300300000023322002
01020010002003201011023002003030021023100031000003100200020000000030031030000000
0002201013003000130000000033203000200000031001300021020000000000002002302020100
0100300100003012010000302030000033000000320003002300000300200001030110001012001
0100000001000001010021031103101213213010300010000000030021232100030000010200

Artificial Neural Networks and Artificial Evolution as Tools to Study Organization of Spatial Behavior in Rats[1]

Eduard Kelemen

Laboratory of Neurophysiology of Memory, Institute of Physiology, Academy of Sciences of the Czech Republic, Videnska 1083, 142 20 Prague, Czech Republic
kelemen@biomed.cas.cz

Abstract. We show that a tiny three layer artificial neural network can solve a version of a spatial task originally designed to test navigation in rats. The network navigates toward a target, which is not perceived directly, its location is defined relative to distant landmarks. Rats can solve a similar task only if they have a functional hippocampus – a brain structure that is thought to be the substrate of a spatial behavior. Our neural network navigates to an "invisible" target using almost optimal trajectories. Activity of neurons of the network is correlated with spatial properties of an environment in a way similar to neurons of hippocampus and related structures of rat brain.

1 Introduction

Place cells (PC) and head direction cells (HDC) are two basic types of neurons with spatially correlated activity that have been identified in rat brain. The properties of these neurons suggest that they are part of a neural navigation system. The firing rate of a place cell increases when the rat is in a particular location of the environment [1]. The discovery of place cells in the rat hippocampus together with studies showing that this structure is necessary for efficient spatial behavior led to the hypothesis, that the hippocampus functions as a cognitive map [2]. The activity of head direction cells is correlated with the azimuth of the rat's head, irrespective of animal's location in the environment. The network of HDCs resembles a compass in its ability to point to the same "absolute" direction regardless of geographic location. HDCs are found in structures functionally related to hippocampus[3].

A number of neural models attempt to mimic the organization and function of PCs and/or HDCs in the effort to understand the spatial abilities of rats [4-9]. In our model we have chosen a different approach. We did not try to create a network organized according to the image of a brain; neither did we intend to model PC or HDC phenomena. Our aim was to make a very simple neural network consisting of 16 neurons that would solve a spatial problem that is often used to test the spatial ability of rodents. We were interested whether phenomena observed in the navigational systems of real animals will emerge in this artificial network.

[1] Work was supported by grant GACR number 309/00/6156

J. Kelemen and P. Sosík (Eds.): ECAL 2001, LNAI 2159, pp. 236-241, 2001.

2 Methods

The problem that our network solves is based on the Morris water maze, which is an experimental task used to study navigation in rats [10]. In this task rats learn a location of a submerged platform in a pool of opaque water. The platform is invisible to the rats, and therefore they have to remember its location using only distant landmarks. The neural network controls the movement of an agent in a simple environment composed of a circular arena with a diameter of 100 cm. The agent's task is to move from any starting location to a hidden circular goal with a diameter of 20 cm. The position of the goal can be recognized only relative to the two light sources, which supply the only sensory input for the network.

The neural network is composed of three layers. In the input layer there are two pairs of sensory neurons. Each pair is sensitive to the intensity of one of the two lights. In each pair of neurons the two sensors are 10 cm from each other. The activity of sensory neuron A_{ax} is computed as follows:

```
Aax = 100 - (distance from neuron to light)
```

In the hidden layer there are 10 neurons, each of them receives inputs from all the sensory neurons. The activity of each of hidden layer neuron A_{bx} is computed as:

$$A_{bx} := \Sigma^{4}_{y=1}(A_{ay} * w_{ay,bx});\ \texttt{if}\ A_{bx} > 100\ \texttt{then}\ A_{bx} = 100;\ \texttt{if}\ A_{bx} < 0\ \texttt{then}\ A_{bx} = 0$$

A_{ay} – activity of input layer neuron, $W_{ay,bx}$ – synaptic weight between sensory neuron a_y and hidden layer neuron b_x. In the third and final layer there are 2 output neurons. Their activity A_{cx} is computed similarly to the activity of the hidden layer neurons:

$$A_{cx} := \Sigma^{10}_{y=1}(A_{by} * w_{by,cx});\ \texttt{if}\ A_{cx} > 100\ \texttt{then}\ A_{cx} = 100;\ \texttt{if}\ A_{cx} < 0\ \texttt{then}\ A_{cx} = 0$$

A_{by} – activity of hidden layer neuron, $W_{by,cx}$ – synaptic weight between hidden layer neuron b_y and output layer neuron c_x. The decision about the direction of further movement of the agent is based on the difference between the activities of the two output neurons.

```
if ( Ac1 - Ac2 ) > 1 then turn 45 degrees left and
move forward; if ( Ac1 - Ac2 ) < -1 then turn 45
degrees right and move forward; if -1 < ( Ac1 - Ac2 ) <
1 then move straight forward.
```

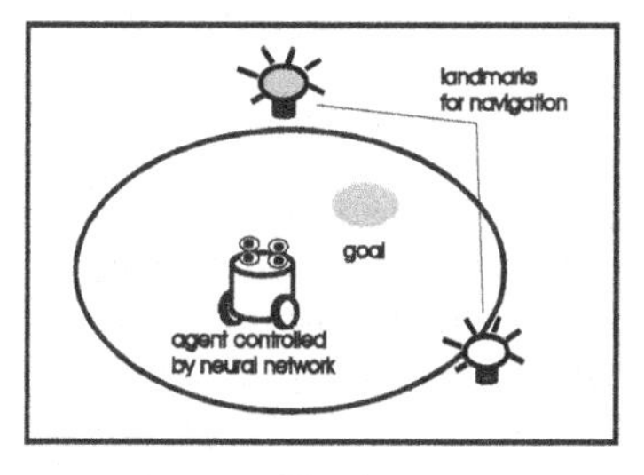

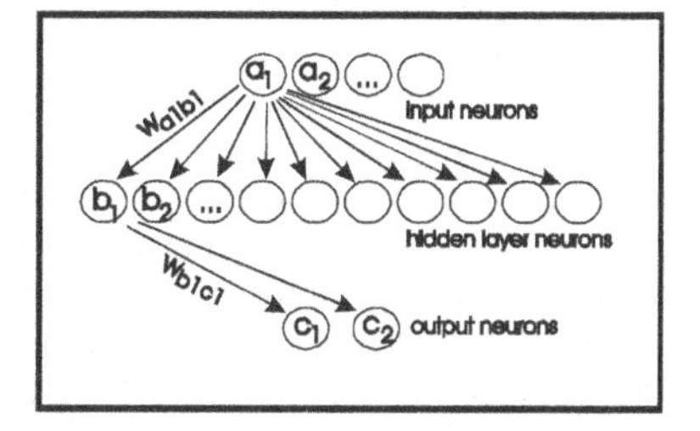

Fig. 1. Scheme of the task and the network architecture.

The synaptic weights of the network are adjusted using artificial evolution. At the beginning 20 sets of synaptic weights are randomly generated. Then each network navigates in the arena for 1000 time steps. The agent starts at one of the pseudo-randomly chosen locations and if it gets to the goal it starts at another location until the time limit is met, or the network crosses a border of the arena. The number of entrances to the target within the time limit is counted for each network and this is used to measure the net's success. The two networks with the best performance are chosen, their synaptic weights are slightly randomly changed to generate the synapses for the 16 networks of the next generation. The other 4 networks of the next generation are randomly generated. This evolution progresses until one of the etworks reaches the target 50 times during the time limit. To reach this criterion network has to generate almost optimal trajectories.

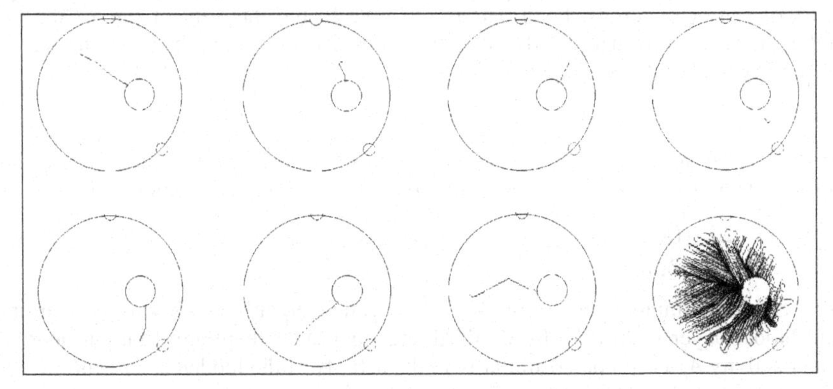

Fig. 2. Figure shows trajectories of the navigating network toward the target from different starting locations. The large circle is the border of the arena, the smaller circle inside the arena shows the position of the hidden target; the two marks on the periphery of the arena are the locations of the light beacons. In the bottom right-hand corner there are multiple paths from different starting points generated by the same network.

3 Results

Six networks were evolved to solve the task. Trajectories generated by these networks are almost optimal and similar to those generated by rats that solve the Morris water maze task (Figure 2).

About 35% of hidden layer neurons are not active at all. Other neurons are active in average in 72% of an environment and their activity is correlated with location in the environment and/or azimuth of movement. Some of these neurons have activity strongly correlated with location and almost not correlated with azimuth similarly to PCs of the rat's hippocampus. On the other hand there are neurons responding to azimuth of movement rather then location. The positional and directionally tuned cells that appeared in the networks do not form distinct categories, but are extreme examples along a spectrum of spatial tuning that is usually controlled to some extent by both the location and direction of the agent (Figure 3). To assess how much activity of neurons varies with direction of movement we computed coefficient of directionality *cd* for each neuron. This coefficient simply tells the average difference between the activity of the neuron when the agent is moving one of 8 directions and mean activity of the neuron:

$$cd = (\Sigma^{N}_{P=1} \Sigma^{8}_{dir=1} (ac(P,dir) - aac(P))) / (N * (maxac-minac) * 8)$$

The surface of arena was divided to pixels (10*10 cm). Value *ac(P,dir)* gives the average activity of the neuron when the agent was on pixel *P* and moving direction *dir*. Similarly an average activity of neuron on pixel *P* across all possible movement directions *aac(P)* was computed. Because the proportion of the environment were neurons were active and absolute activity levels differed in different neurons, the sum in the equation is divided by number of pixels were the neuron was active *N*, and by difference between maximal *maxac* and minimal *minac* activity.

Output neurons of navigating networks were active all over the environment and were stronger correlated with direction of movement than hidden layer neurons (Figure 3).

To analyze whether spatial properties of neurons in our networks evolved as an adaptation to the spatial problems, or whether they are a result of spatially organized neural input, we generated control networks with randomly set synaptic weights. The spatial properties of hidden layer neurons in these randomly generated networks are similar to those of hidden layer neurons of networks that solve a spatial task. In control networks the activity of output neurons is significantly less controlled by orientation than activity of hidden layer neurons, this tendency is just opposite to what we observe in navigating networks (Figure 3).

4 Discussion

We have shown that a three-layer neural network composed of 16 neurons can solve a spatial problem, in which it has to navigate to a hidden target with position defined relative to two distant light sources. The ability of rats to solve a similar task (Morris water task, [10]) depends on the hippocampus. The fact that hundreds of thousands of neurons are needed to solve the task in rat hippocampus while only 16 are needed for the artificial version of the task may be surprising, but there are several important differences between the two situations. Our networks take several hundreds of generations to achieve optimal behavior. The rat learns the location of hidden platform in the water during approximately 40 trials and uses a different learning mechanism. Another important simplification in our model is in the quality of the available sensory information. Our network gets direct information about its distance from the two landmarks, in real animals this information has to be computed from a complex image on the retina.

We observed phenomena similar to those known from navigational system of rats in the firing pattern of neurons of our network. Neurons with activity tuned to location and direction were observed in the hidden layer of the networks that were able to solve the spatial task, as well as in networks with synaptic weights set to random values. This shows that spatially tuned cells emerged as a result of the organization of the sensory inputs, not as a part of a solution to the spatial problem.

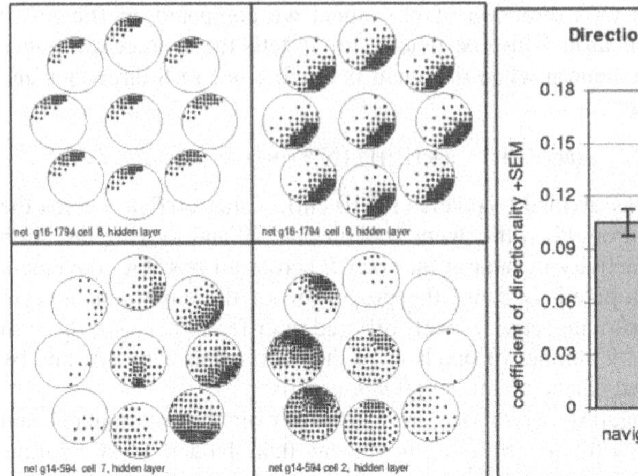

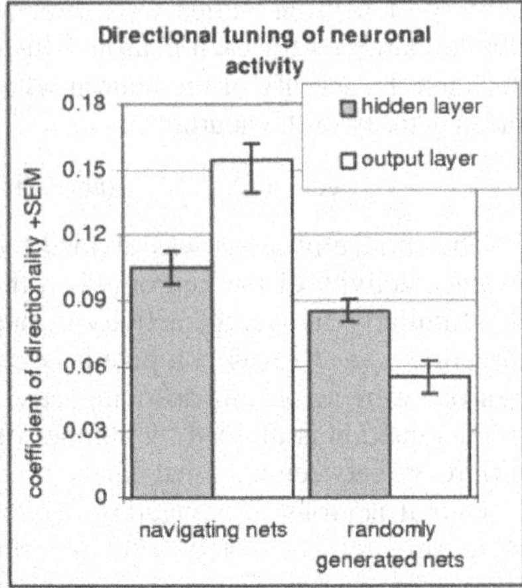

Fig. 3. Left - Each cell's spatial activity is described using 9 circles. The central circle shows average activity of the neuron when network was at different locations of the arena. The higher is neuronal activity at certain place the higher is density of dots. The 8 circles surrounding the central circle show average activity of the neuron when network was at certain location and was moving in one of the eight directions. On the two top figures there are two place cells with location specific activity, their activity does not depend on orientation. In the bottom figures there are neurons with activity correlated with location as well as orientation. Right – Figure shows how much is the neuronal activity influenced by direction of movement in different layers of networks solving a spatial task and networks with randomly generated synaptic weights.

Treves et al. [11] created a similar model of a navigating three-layer neural network that solved a very similar spatial task in an unoptimal way resembling that of hippocampectomized rat. Our network solves the task in an almost optimal way using trajectories very similar to trajectories of a normal rat with functioning hippocampus. The main difference between the two models is in the information perceived by the sensory units. In our case the receptive field of sensors was 360°, so each unit responded to the distance from the landmark irrespective of the agent's orientation. The receptive fields of the sensory units in Treves's et al. experiment were 120° wide. This could explain also the fact that Treves et al. report no pure place cells in the middle layer of their network rather they only found cells sensitive to both location and orientation.

References

1. Muller, R.U., Kubie, J.L., Ranck, J.B., Jr.: Spatial Firing Patterns of Hippocampal Complex-Spike Cells in a Fixed Environment. Journal of Neuroscience 7 (1987) 1935-1950.
2. O'Keefe, J., and Nadel, L. (1978). The Hippocampus as a Cognitive Map. Claderon Press, Oxord.

3. Taube, J.S., Muller, R.U., Ranck J.B., Jr.: Head Direction Cells Recorded from the Postsubicilum in Freely Moving Rats. Journal of Neuroscience 10 (1990) 420-447.
4. Zhang, K.: Representation of Spatial Orientation by the Intrinsic Dynamics of the Head-Direction Cell Ensemble: a Theory. Journal of Neuroscience 16 (1996) 2112-2126.
5. Samsonovich, A., McNaughton, B.L.: Path Integration and Cognitive Mapping in a Continuous Attractor Neural Network Model. Journal of Neuroscience 17 (1997) 5900-5920. 1. Blum, K. I., and L. F. Abbott: A model of spatial map formation in the hippocampus of the rat. Neural Computation 8 (1996) 85-93.
6. Tsodyks, M.: Atractor Neural Network Models of Spatial Maps in Hippocampus. Hippocampus 9 (1999) 481-489.
7. Blum, K.I., Abbott, L.F.: A Model of Spatial Map Formation in the Hippocampus of the Rat. Neural Computation 8 (1996) 85-93.
8. McNaughton B.L., Barnes, C.A., Gerrard, J.L., Gothard, K., Jung, M.V., Knierim, J.J., Kudrimoti, H., Qin, Y., Skaggs, W.E., Suster, M., Weawer, K.L.: Deciphering the Hippocampal Polyglot: the Hippocampus as a Path Integrator System. Journal of Experimental Biology 199 (1996) 173-185
9. Muller, R. U., Stead M.: Hippocampal Place Cells Connected by Hebbian Synapse Can Solve Spatial Problems. Hippocampus 6 (1996) 709-719.
10. Morris, R. G. M., Garrud, P., Rawlins, J.N.P., O´Keefe, J.: Place Navigation Impaired in Rats with Hippocampal Lesions. Nature 297 (1982) 681-683.
11. Treves, A., Miglino, O., Parisi, D.: Rats, Nets, Maps, and the Emergence of Place Cells. Psychobiology 20 (1992) 1-8.

transsys: A Generic Formalism for Modelling Regulatory Networks in Morphogenesis

Jan T. Kim

Institut für Neuro- und Bioinformatik,
Seelandstraße 1a, 23569 Lübeck, Germany
kim@inb.mu-luebeck.de

Abstract. The formal language **transsys** is introduced as a tool for comprehensively representing regulatory gene networks in a way that makes them accessible to ALife modelling. As a first application, Lindenmayer systems are enhanced by integration with **transsys**. The resulting formalism, called **L-transsys**, is used to implement the ABC model of flower morphogenesis. This **transsys** ABC model is extensible and allows dynamical modelling on the molecular and on the morphological level.

1 Introduction

During the last years, regulatory networks consisting of transcription factors and the genes encoding them have received increasing amounts of interest in various biosciences, including bioinformatics and Artificial Life. Regulatory networks are among the key mechanisms of phenotypic realization of genetic information. One of the most prominent phenotypic properties is the morphology of an organism, and consequently, various models and analyses of regulatory networks are concerned with morphogenesis.

Regulatory networks can be modelled by various approaches, e.g. by differential equation systems [10,16] or rule-based systems [7,13]. Furthermore, a multitude of systems for modelling morphogenesis have been developed, including cellular automata [1], Lindenmayer systems (L-systems) [2,5,6,8] and others [3,4,7,15], and such models have been combined with regulatory network models in different ways. These models have led to many new results and insights. However, the development of increasingly detailed and specialized models makes it difficult to integrate the findings obtained with different models into a coherent theoretical framework.

In this paper, the **transsys** modelling framework is introduced which is designed to address this problem. The motivation underlying **transsys** is to provide a framework which (1) is sufficiently generic to integrate several methods for modelling gene regulatory systems, (2) facilitates interfacing with models of growth processes and (3) allows formulation of concise and expressive models to facilitate scientific communication.

J. Kelemen and P. Sosík (Eds.): ECAL 2001, LNAI 2159, pp. 242–251, 2001.

L-systems (see [17] for a general introduction) are powerful systems for modelling morphogenesis. However, even with parametric L-systems, it is not possible to simulate regulatory networks in a concise and comprehensive way because explicit representations for genes and their products are lacking. Therefore, L-systems were considered to be a suitable target of enhancement by transsys. The resulting system allows modelling of wildtype and mutant flower morphogenesis dynamics such that models of single gene mutants can be derived from the wildtype model by single alterations in the transsys model specification. This is demonstrated by implementing the ABC model of flower morphogenesis in *Arabidopsis* [14].

2 The Modelling Framework

Conceptually, regulatory networks consist of two types of node elements, transcription factors and genes that encode transcription factors. Edges in a regulatory network are constituted by activating and repressing effects that factors have on the expression of genes. transsys is a computer language designed to comprehensively represent regulatory networks such that they can be combined with a wide range of dynamical models. A network description is called a *transsys program*. The transsys language is outlined in Sect. 2.1, using the transsys program shown in Fig. 1 as an example. A full specification will be published in a forthcoming technical report[1]. Sections 2.2 and 2.3 describe simulation of transsys network and integration into an L-system framework.

The set of factors in a transsys program is denoted by $\mathcal{F} = \{f_1, f_2, \dots, f_n\}$, the set of genes is denoted by $\mathcal{G} = \{g_1, g_2, \dots, g_m\}$.

The state of a regulatory network in transsys is defined by the concentrations of the factors described in the transsys program. The set of factor concentrations is denoted by $\mathcal{C} = \{c_f : f \in \mathcal{F}\}$. Such a set is also referred to as a *transsys instance*.

Temporal dynamics within a regulatory networks are constituted by changes in concentrations of transcription factors. These changes are computed by numerically integrating a set of differential equations which are specified and parameterized by the transsys program. The change in concentration of factor f is denoted by Δc_f. The amount of factor f synthesized due to expression of gene g is denoted by $\Delta_g c_f$.

2.1 The transsys Language

Transcription Factors. Transcription factors are described by factor blocks, as shown in Fig. 1. A factor is characterized by its name immediately following the factor keyword and its decay rate, specified within the block. The decay rate of a factor f is formally denoted by r_f. In relation to the many biochemical

[1] check http://www.inb.mu-luebeck.de/transsys/ for software and current documents on transsys.

```
transsys example
{
  factor A { decay: 0.05; }
  factor R { decay: 0.05; }

  gene rgene
  {
    promoter
    {
      A: activate(1.0, 10.0);
      R: repress(1.0, 1.0);
    }
    product
    {
      default: R;
    }
  }

  gene agene
  {
    promoter
    {
      constitutive: random(0, 0.1);
      A: activate(0.1, 1.0);
      R: repress(1.0, 1.0);
    }
    product
    {
      default: A;
    }
  }
}
```

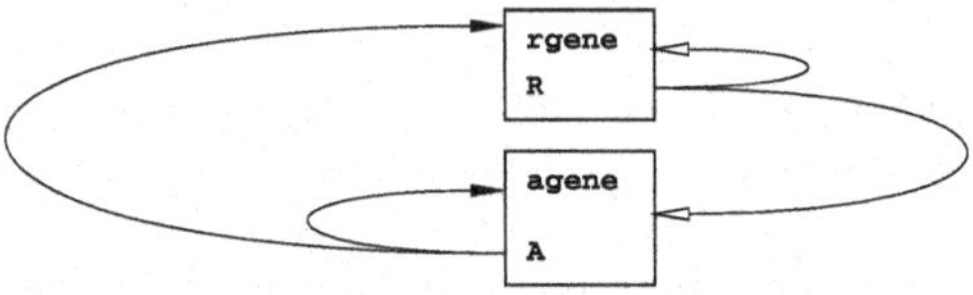

Fig. 1. An example `transsys` program with two factors `A` and `R`, and two genes `agene` and `rgene`. The regulatory interactions expressed by `activate` and `repress` promoter statements can be displayed graphically by arrows with filled or open arrow heads, respectively.

and biological properties of a protein, this is a strongly simplified and abstract representation. It is possible, however, to model effects of protein-protein to some extent, as decay rates are specified by expressions (see below), so they can e.g. be made to depend on activities of other factors.

Genes. Genes are units of genetic information that can be partitioned into a structural part and a regulatory part. The structural part encodes a biological activity which may be realized RNA or by protein molecules. The regulatory part determines the expression pattern of the gene by cis-acting elements which are recognized by transcription factors.

The regulatory and the structural part of a gene are represented in `transsys` by the `promoter` and `product` blocks within a `gene` block (see Fig. 1). The `product` block specifies the name of the factor encoded by the gene. The set of all genes encoding a factor f is denoted by $\mathcal{E}_f$.

The `promoter` block contains a list of statements describing conditions for activation or repression of the gene. A statement models a transcription factor binding site and the effect which binding has on the activation of the gene by describing how to calculate a contribution to activation (see Sect. 2.2). Let a_i denote the contribution of promoter statement i.

There are three types of statements possible in the `promoter` block. `constitutive` is the most generic type, this statement specifies an expression determining an amount of activation:

$$a_i = \text{ result of evaluating expression} \tag{1}$$

`activate` and `repress` statements are both preceded by a factor name f, and both have a list of two expressions as arguments. The arguments determine the specificity, denoted by a_{spec}, and the maximal rate of activation, denoted by a_{max}. The actual amount of activation is calculated according to the Michaelis-Menten-equation:

$$a_i = \frac{a_{\text{max}} c_f}{a_{\text{spec}} + c_f} \tag{2}$$

Repression is calculated by the same formula with the sign reversed:

$$a_i = -\frac{a_{\text{max}} c_f}{a_{\text{spec}} + c_f} \tag{3}$$

Both parameters a_{spec} and a_{max} are specified by expressions, which allows modelling of modulation of activation by protein-protein interactions. The amount of product p synthesized through expression of gene g in a time step is given by

$$\Delta_g c_p = \begin{cases} a_{\text{total}} := \sum_i a_i & \text{if } a_{\text{total}} > 0 \\ 0 & \text{otherwise} \end{cases} \tag{4}$$

Expressions. The form of expressions in `transsys` was designed to match that of expressions in programming languages like C or Java, i.e. the arithmetic and logical operators from these languages were built into `transsys`.

The main difference between expressions in `transsys` and in other languages is the interpretation of identifiers: Identifiers are used exclusively to refer to factor concentrations. Thus, for example, the promoter statement `constitutive: 2 * x` in a gene `g` encoding factor `p` means that the concentration of `f` is increased through expression of `g` by $\Delta_{\text{g}} c_{\text{p}} = 2 \cdot c_{\text{x}}$. This statement could e.g. be used to model a factor `x` in the transcription initiation machinery which determines the basal level of expression.

2.2 Dynamics

Simulation of regulatory network dynamics in `transsys` starts with calculating factor concentration changes according to

$$\Delta c_f(t) = \left(\sum_{g \in \mathcal{E}_f} \Delta_g c_f(t) \right) - r_f(t) c_f(t) \tag{5}$$

where t denotes the time step in which the calculation takes place. Having calculated all concentration changes, $\mathcal{C}(t+1) = \{c_{f_1}(t+1), c_{f_2}(t+1), \ldots, c_{f_n}(t+1)\}$, the set of factor concentrations at time $t+1$, is determined by calculating

$$c_f(t+1) = c_f(t) + \Delta c_f(t) \tag{6}$$

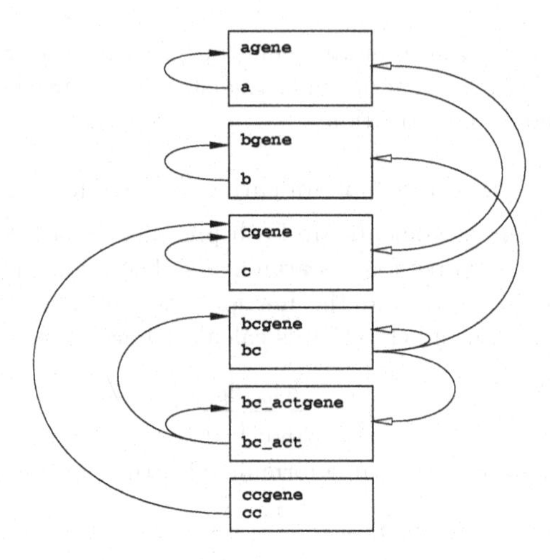

Fig. 2. The `transsys` program implementing the ABC model of flower morphogenesis.

for all factors $f \in \mathcal{F}$. After this derivation of $\mathcal{C}(t+1)$ from $\mathcal{C}(t)$, simulation of growth or diffusion dynamics may be performed, depending on the modelling context, before the next derivation is done.

2.3 `transsys` within an L-System Framework

The formalism designed by combining parametric L-systems with `transsys` is called `L-transsys`. The concept underlying `L-transsys` is a replacement of parameter lists, which are associated with symbols in parametric L-systems, with `transsys` instances.

An `L-transsys` symbol can be associated with one specific `transsys`. In the terms of parametric L-systems, this can approximately be construed as associating the factor concentrations as real-valued parameters with the symbol. However, the factor concentrations in a `transsys` instance are more than an unstructured set of numeric parameters as the instance provides a method for deriving $\mathcal{C}(t+1)$ from $\mathcal{C}(t)$. This allows a straightforward algorithm for interleaved application of `transsys` updates and L-system rule application:

1. Create the axiom (initial string), including initial factor concentrations.
2. For all symbols with a `transsys` instance, derive $\mathcal{C}(t+1)$.
3. From the current string with updated `transsys` instances, derive the next string according to the L-system production rules.
4. Continue with step 2 until desired number of derivations is reached.

3 Results

According to the ABC model of flower morphogenesis [14], flower morphogenesis is controlled by three functions, labelled A, B and C. Expression of only A

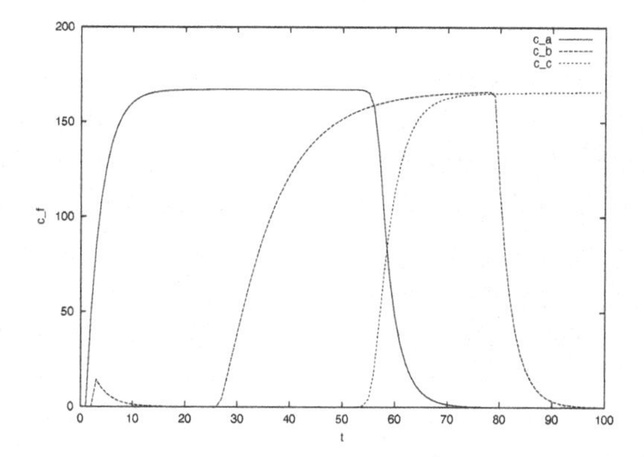

Fig. 3. Temporal dynamics of gene expression in the regulatory network depicted in Fig. 2. The **transsys** instance was initialized with all concentrations set to zero.

Fig. 4. Model of a fully developed wild type flower grown under the control of the ABC **transsys** model depicted in Fig. 2. Sepals are rendered as small green squares, petals as large, white squares, stamens are depicted by blue cylinders and carpels are composed of red boxes. The grey cylinder at the bottom represents the stalk.

function results in formation of sepals, joint expression of A and B functions leads to petal formation, stamens are formed upon joint expression of of B and C, and expression of C alone yields carpels.

The implementation of the ABC model with a **transsys** program is shown in Fig. 2, a listing is provided in Appendix A. In addition to the factors `a`, `b` and `c`, encoded by `agene`, `bgene` and `cgene`, respectively, the model comprises three more factors. Factors `bc` and `bc_act` and the corresponding genes form an oscillating system which is used to control the domain of B function expression. Factor `cc` controls the onset of C function expression. These three **transsys** factors are an *ad hoc* design which is intended to be replaced with more realistic

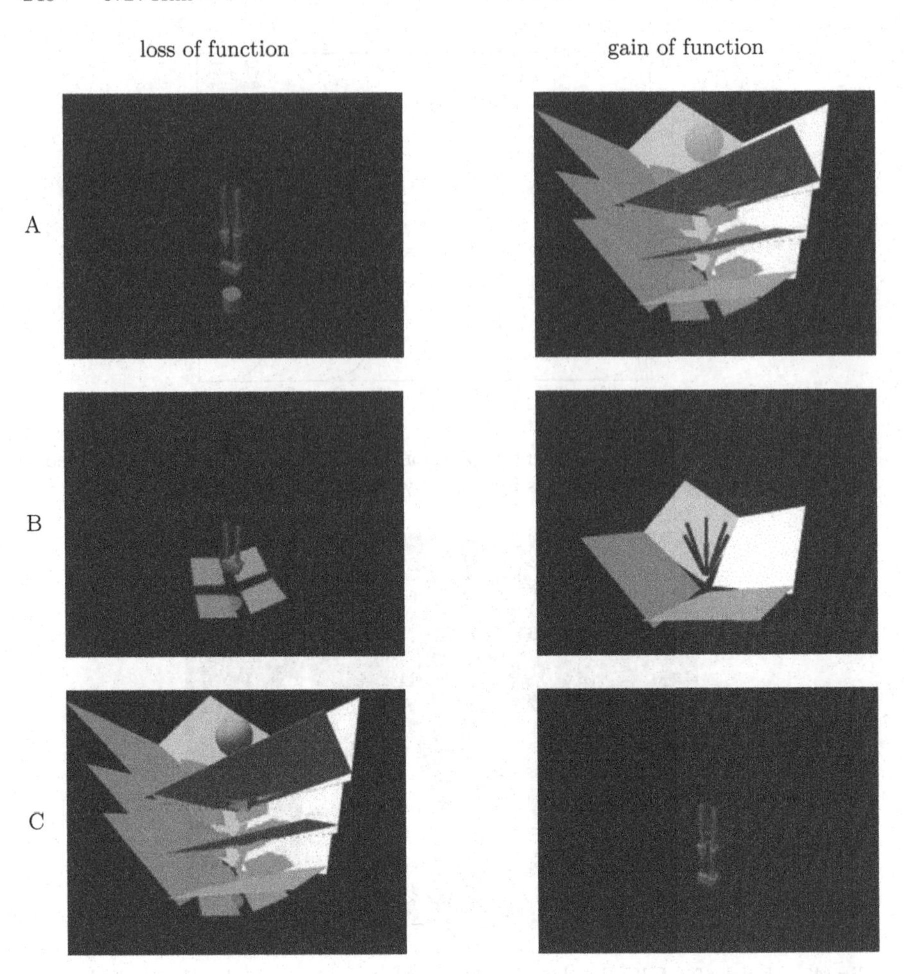

Fig. 5. `L-transsys` models of flowers grown with mutant variants of the ABC network shown in Fig. 2. A sphere represents a meristem (where growth continues), other structures are explained in Fig. 4. Top row: mutants in A function, middle row: mutants in B function, bottom row: mutants in C function. Left column: loss of function mutants, right column: gain of function mutants. Loss of function mutants were modelled by replacing the product of the mutated gene with a factor called `nonfunc`, which does not interact with any promoters. Gain of function mutants were modelled by adding the promoter statement `constitutive: 500;` to the promoter of the mutated gene.

models of molecular factors in the future. Of course, `transsys` can also be used to explore various hypotheses about the regulatory network controlling expession of the A, B and C functions. As shown in Fig. 3, the model satisfactorily matches the dynamics expected in the wildtype situation.

Figure 4 shows the graphical result of incorporating the transsys ABC program into an L-transsys model. The L-system rules in this model implement the growth behaviour outlined above. The morphological whorl structure of an *Arabidopsis* flower is correctly rendered by the L-transsys version of the ABC model.

Loss of function mutants can be modelled in transsys by adding a factor that represents a nonfunctional mutant of a protein and replacing the product of a gene with the nonfunctional factor. Gain of function mutants can be simulated by adding a high level of constitutive expression to the promoter of a gene.

Figure 5 shows the results of simulating both gain and loss of function of A, B and C. All mutations result in phenotype models that are qualitatively correct. The number of whorls is not correctly captured. This is due to the L-system component of the model which uses changes in $c_{\mathtt{a}}$, $c_{\mathtt{b}}$ and $c_{\mathtt{c}}$ to trigger generation of a new whorl, so multiple consecutive whorls with identical organ identities are collapsed into one.

4 Discussion and Outlook

Using the ABC model of flower morphogenesis as an example, it has been demonstrated that concise and expressive models of morphogenetic processes controlled by regulatory networks can be realized with transsys. Individual, localized mutations can be modelled by individual, localized alterations in the L-transsys model specification. This shows that transsys models are not only suitable for representing individual regulatory networks, but beyond that, are also capable of modelling the effects of mutations. Thus, adequation between nature and transsys models extends deeply into the genetic level. In this sense, transsys is a step towards modelling the fundamental principles underlying morphogenetic phenomena, following a traditional ALife modelling spirit of abstracting fundamental dynamical processes underlying life (see [11, preface]).

While the numerical parameters for activation and repression for the ABC model could simply be chosen *ad hoc*, the dependence on such numerical parameterization may become a problem when dealing with larger networks. This problem is currently being addressed by developing a qualitative or a semiquantitative mode of transsys dynamics, and by exploring numerical parameter space with evolutionary approaches, following the approach used in [13].

As concise and comprehensive representations which enable dynamical simulation and incremental modelling, transsys programs provide a suitable format for representing and exchanging regulatory networks. As a start for building a collection of transsys programs, it is planned to implement some networks published in the literature (e.g. [13,18]). Such a study will also serve to assess the level of universality attained with transsys, analogously to the analysis of universality recently conducted for the GROGRA growth grammar system [9]. As an upcoming extension of transsys, integration of diffusion in a cellular lattice framework is currently being developed. This will allow modelling of reaction-diffusion processes, as exemplified by [10,12].

Acknowledgements. I am grateful to Gerhard Buck-Sorlin, Winfried Kurth and Daniel Polani for critical and inspiring discussions regarding various aspects of `transsys`. I acknowledge financial support from the BMBF (Grant 0311378) during the initial phase of `transsys` development.

References

[1] Andrew Adamatzky. Simulation of inflorescence growth in cellular automata. *Chaos, Solitons & Fractals*, 7:1065–1094, 1996.
[2] Gerhard Buck-Sorlin and Konrad Bachmann. Simulating the morphology of barley spike phenotypes using genotype information. *Agronomie*, 20:691–702, 2000.
[3] Kurt Fleischer and Alan Barr. The multiple mechanisms of morphogenesis: A simulation testbed for the study of multicellular development. In Christopher G. Langton, editor, *Artificial Life III*, Santa Fe Institute Studies in the Sciences of Complexity, pages 379–416, Redwood City, CA, 1994. Addison Wesley Longman.
[4] Christian Fournier and Bruno Andrieu. A 3D architectural and process-based model of maize development. *Annals of Botany*, 81:233–250, 1998.
[5] Christian Jacob. Evolution programs evolved. In H.-M. Voigt, W. Ebeling, I. Rechenberg, and H.-P. Schwefel, editors, *PPSN-IV*, pages 42–51, Berlin, Germany, 1996. Springer Verlag.
[6] Jaap A. Kaandorp. Simulation of environmentally induced growth forms in marine sessile organisms. *Fractals*, 1:375–379, 1993.
[7] Jan T. Kim. LindEvol: Artificial models for natural plant evolution. *Künstliche Intelligenz*, pages 26–32, 2000.
[8] Winfried Kurth. Growth grammar interpreter GROGRA 2.4: A software tool for the 3-dimensional interpretation of stochastic, sensitive growth grammars in the context of plant modelling. introduction and reference manual. Technical report, Universität Göttingen, 1994.
`ftp://www.uni-forst.gwdg.de/pub/biom/gro.ps.gz`.
[9] Winfried Kurth. Towards universality of growth grammars: Models of Bell, Pagés, and Takenaka revisited. *Ann. For. Sci.*, 57:543–554, 2000.
[10] Koji Kyoda and Hiroaki Kitano. A model of axis determination for the *Drosophila* wing disc. In Dario Floreano, Jean-Daniel Nicoud, and Francesco Mondada, editors, *Advances in Artificial Life*, Lecture Notes in Artificial Intelligence, pages 473–476, Berlin Heidelberg, 1999. Springer-Verlag.
[11] C.G. Langton, C. Taylor, J.D. Farmer, and S. Rasmussen, editors. *Artificial Life II*, Redwood City, CA, 1992. Addison-Wesley.
[12] H. Meinhardt. Biological pattern formation: New observations provide support for theoretical predictions. *BioEssays*, 16:627–632, 1994.
[13] Luis Mendoza and Elena R. Alvarez-Buylla. Dynamics of the genetic regulatory network for *Arabidopsis thaliana* flower morphogenesis. *J.theor.Biol.*, 193:307–319, 1998.
[14] Elliot M. Meyerowitz. The genetics of flower development. *Scientific American*, 271(5):56–65, 1994.
[15] Karl J. Niklas. Adaptive walks through fitness landscapes for early vascular land plants. *American Journal of Botany*, 84:16–25, 1997.
[16] S.W. Omholt, E. Plathe, L. Oyehaug, and K.F. Xiang. Gene regulatory networks generating the phenomena of additivity, dominance and epistasis. *Genetics*, 155:969–980, 2000.

[17] P. Prusinkiewicz and A. Lindenmayer. *The algorithmic beauty of plants*. Springer-Verlag, New York, 1990.
[18] Günter Theißen and Heinz Saedler. MADS-box genes in plant ontogeny and phylogeny: Haeckel's 'biogenetic law' revisited. *Current Opinion in Genetics and Development*, 5:628–639, 1995.

A The ABC Network in transsys

```
transsys abc
{
  factor a { decay: 0.3; }
  factor b { decay: 0.3; }
  factor c { decay: 0.3; }
  factor bc { decay: 0.6; }
  factor bc_act { decay: 0.1; }
  factor cc { decay: 0.0; }

  gene agene
  {
    promoter
    {
      constitutive: 0.1;
      a: activate(0.000000001, 50.0);
      c: repress(50.0, 100.0);
    }
    product { default: a; }
  }

  gene bgene
  {
    promoter
    {
      constitutive: 0.00000001;
      b: activate(0.0003, 50.0);
      bc: repress(80.0, 5000.0);
    }
    product { default: b; }
  }

  gene cgene
  {
    promoter
    {
      cc: activate(0.5, 4.0);
      c: activate(10.0, 50.0);
      a: repress(1500.0, 20.0);
    }
    product { default: c; }
  }

  gene bcgene
  {
    promoter
    {
      bc_act: activate(1.0, 5.0);
      bc: repress(1.0, 1.0);
    }
    product { default: bc; }
  }

  gene bc_actgene
  {
    promoter
    {
      constitutive: 0.1;
      bc_act: activate(0.05, 3.0);
      bc: repress(0.1, 5.0);
    }
    product { default: bc_act; }
  }

  gene ccgene
  {
    promoter
    {
      constitutive: 0.01;
    }
    product { default: cc; }
  }
}
```

Evolution of Reinforcement Learning in Uncertain Environments: Emergence of Risk-Aversion and Matching

Yael Niv[1], Daphna Joel[1], Isaac Meilijson[2], and Eytan Ruppin[2]

[1] Department of Psychology
Tel-Aviv University, Tel-Aviv 69978, Israel
yaeln@cns.tau.ac.il, djoel@post.tau.ac.il
[2] School of Mathematical Sciences
Tel-Aviv University, Tel-Aviv 69978, Israel
isaco@math.tau.ac.il, ruppin@math.tau.ac.il

Abstract. Reinforcement learning (RL) is a fundamental process by which organisms learn to achieve a goal from interactions with the environment. Using Artificial Life techniques we derive (near-)optimal neuronal learning rules in a simple neural network model of decision-making in simulated bumblebees foraging for nectar. The resulting networks exhibit efficient RL, allowing the bees to respond rapidly to changes in reward contingencies. The evolved synaptic plasticity dynamics give rise to varying exploration/exploitation levels from which emerge the well-documented foraging strategies of risk aversion and probability matching. These are shown to be a direct result of optimal RL, providing a biologically founded, parsimonious and novel explanation for these behaviors. Our results are corroborated by a rigorous mathematical analysis and by experiments in mobile robots.

1 Introduction

Reinforcement learning (RL) is a process by which organisms learn to achieve a goal from their interactions with the environment [15]. In RL, learning is contingent upon a scalar reinforcement signal which only provides evaluative information about how good an action is in a certain situation. Behavioral research indicates that RL is a fundamental means by which experience changes behavior in both vertebrates and invertebrates, as most natural learning processes are conducted in the absence of an explicit supervisory stimulus. A computational understanding of RL is a necessary step towards an understanding of brain functions, and can contribute widely to the design of autonomous artificial learning agents. RL has attracted ample attention in computational neuroscience, yet a fundamental question regarding the underlying mechanism has not been sufficiently addressed, namely, **what are the optimal neuronal learning rules for maximizing reward in reinforcement learning?** In this paper, we use Artificial-life (Alife) techniques to derive the optimal neuronal learning rules

J. Kelemen and P. Sosík (Eds.): ECAL 2001, LNAI 2159, pp. 252–261, 2001.

that give rise to efficient RL in uncertain environments. We further investigate the behavioral strategies which emerge from optimal RL.

RL has been demonstrated and studied extensively in foraging bees. Real [13] showed that when foraging for nectar in a field of blue and yellow artificial flowers, bumblebees exhibit efficient RL, rapidly switching their preference for flower type when reward contingencies were switched between the flowers. The bees also manifested risk averse behavior: in a situation in which blue flowers contained $2\mu l$ sucrose solution, and yellow flowers contained $6\mu l$ sucrose in $\frac{1}{3}$ of the flowers, and zero in the rest, 85% of the bees' visits were to the blue constant-rewarding flowers, although the mean return from both flower types was identical. Such risk-averse behavior has also been demonstrated elsewhere [8], and has traditionally been accounted for by hypothesizing the existence of a subjective non-linear concave "utility function" for nectar [6]. Risk averse behavior is also prominent in humans, and is an important choice strategy, well-studied in economics and game-theory, although its biological basis is not yet firmly established.

A foraging bee deals with a rapidly changing environment - parameters such as the weather, and competition affect the availability of rewards from different kinds of flowers. This implies an "armed-bandit" type scenario, in which the bee collects food and information simultaneously. As a result there exists a tradeoff between exploitation and exploration, as the bee's actions directly effect the "training examples" which it will encounter through the learning process. A notable strategy by which bumblebees (and other animals) optimize choice in such situations is probability matching. When faced with flowers offering similar rewards but with different probabilities, bees match their choice behavior to the reward probabilities of the flowers [9]. This seemingly "irrational" behavior with respect to optimization of reward intake is explained as an Evolutionary Stable Strategy (ESS) for the individual forager when faced with competitors [16], as it produces an Ideal Free Distribution (IFD) in which the average intake of food is the same at all food sources. Using Alife techniques, Seth evolved battery-driven agents competing for two different battery refill sources, and showed that indeed matching behavior emerges only in a multi-agent scenario [14].

In a previous neural network (NN) model, Montague et al. [11] simulated bee foraging in a 3D arena of blue and yellow flowers, based on a neurocontroller modelled after an identified interneuron in the honeybee suboesophogeal ganglion. This neuron's activity represents the reward value of gustatory stimuli, and similar to dopaminergic neurons in the Basal Ganglia, is activated by unpredicted rewards [5]. In their model this neuron is modeled as a linear unit P, which receives visual information regarding changes in the percentages of yellow, blue and neutral colors in the visual field, and computes a prediction error. According to P's output the bee decides whether to continue flying in the same direction, or to change direction randomly. Upon landing, a reward is received according to the subjective utility of the nectar content of the chosen flower [6], and the synaptic weights of the networks are updated according to a special anti-Hebbian-like learning rule. As a result, the values of the weights come to represent the expected rewards from each flower type.

While this model replicates Real's foraging results and provides a basic and simple NN architecture to solve RL tasks, many aspects of the model, first and foremost the handcrafted synaptic learning rule, are arbitrarily specified and their optimality with respect to RL questionable. Towards this end, we use a generalized and parameterized version of this model in order to evolve optimal synaptic learning rules for RL (with respect to maximizing nectar intake) using a genetic algorithm. In contrast to common Alife applications which involve NNs with evolvable synaptic weights or architectures [1,12], we set upon the task of evolving the network's neuronal learning rules. Previous attempts at evolving neuronal learning rules have used heavily constrained network dynamics and very limited sets of learning rules [2,4]. We define a general framework for evolving learning rules, which essentially encompasses **all heterosynaptic Hebbian learning rules**, along with other characteristics of the learning dynamics. Via the genetic algorithm we select bees based solely on their nectar-gathering ability in a changing environment. The uncertainty of the environment ensures that efficient foraging can only be a result of learning throughout lifetime, thus synaptic learning rules are evolved.

In the following section we describe the model and the evolutionary dynamics. Section 3 describes the results of our simulations, and the evolution of RL. In section 4 we analyze the foraging behavior resulting from the learning dynamics, and find that when tested in new environments, our Alife creatures manifest risk aversion and probability matching behaviors. Although this behavior was not selected for, we rigorously prove that these strategies emerge directly from optimal RL. Section 5 describes a minirobot implementation of the evolved RL model, and we conclude with a discussion of the results in section 6.

2 The Model

A simulated bee-agent flies above a 3D patch of 60x60 randomly scattered blue and yellow flowers. A bee's life consists of 100 trials. In each trial the bee starts its descent from a height of 10 units, and advances in steps of 1 unit that can be taken in any downward direction (360° horizontal, 90° vertical). The bee views the world through a cyclopean eye (10° cone view), and in each timestep decides whether to maintain the current heading direction or to reorient randomly, based on the visual input. Upon landing the bee consumes any available nectar in one timestep, and another trial begins. **The evolutionary goal (the fitness criterion) is to maximize nectar intake.**

In the neural network controlling the bee's flight (Fig. 1a), which is an extension of Montague et al's network [11], three modules ("regular", "differential" and "reward") contribute their input via synaptic weights, to a linear neuron P. The regular input module reports the percentage of the bee's field of view filled with yellow $[X_y(t)]$, blue $[X_b(t)]$ and neutral $[X_n(t)]$. The differential input module reports temporal differences of these percentages $[X_i(t) - X_i(t-1)]$. The reward module reports the actual amount of nectar received from a flower $[R(t)]$ in the nectar-consuming timestep[1] and zero during flight. Note that **we do not**

[1] In this timestep it is also assumed that there is no new input $[X_i(t) = 0]$

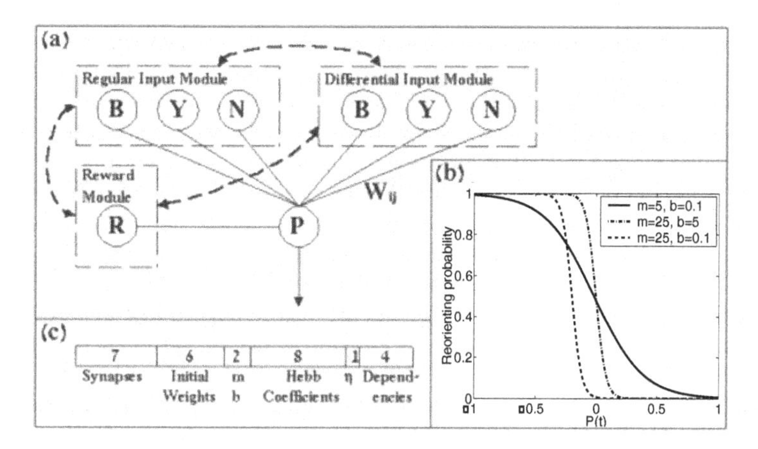

Fig. 1. (a) The bee's neural network controller. (b) The bee's action function. Probability of reorienting direction of flight as a function of $P(t)$ for different values of parameters m, b. **(c) The genome sequence of the simulated bee.**

incorporate any non-linear utility function with respect to the reward. Thus P's continuous-valued output is:

$$P(t) = R(t) + \sum_{i \in \text{regular}} W_i X_i(t) + \sum_{i \in \text{differential}} W_i[X_i(t) - X_i(t-1)]. \quad (1)$$

The bee's action is determined according to the output $P(t)$ using Montague et al's probabilistic action function [6,11] (Fig. 1b):

$$p(\text{change direction}) = 1/[1 + \exp(m \cdot P(t) + b)] \quad (2)$$

During the bee's "lifetime" the synaptic weights of the regular and differential input modules are modified via a heterosynaptic Hebb learning rule of the form:

$$\Delta W_i = \eta(AX_i(t)P(t) + BX_i(t) + CP(t) + D), \quad (3)$$

where η is a global learning rate parameter, $X_i(t)$ and $P(t)$ are the presynaptic and the postsynaptic values respectively, W_i their connection weight, and A-D are real-valued evolvable parameters. In addition, learning in one module can be dependent on another module (dashed arrows in Fig. 1a), such that if module M depends on module N, M's synaptic weights will be updated according to equation (3) only if module N's respective neurons have fired (if it is not dependent, the weights will be updated on every timestep). Thus the bee's "brain" is capable of **a non-trivial axo-axonic gating of synaptic plasticity.**

The simulated bee's genome (Fig. 1c) consists of a string of 28 genes, each representing a parameter governing the network architecture and or its learning dynamics. Seven boolean genes determine whether each synapse exists or not; 6 real-valued genes (range [-1,1]) specify the initial weights of the regular and differential modules (the synaptic weight of the reward module is clamped to

1, effectively scaling the other synapses); and two real-valued genes specify the action-function parameters m (range [5,45]) and b (range [0,5]). Thirteen remaining genes specify the learning dynamics: The regular and differential modules each have a learning rule specified by 4 real-valued genes (parameters A-D of equation (3), range [-1,1]); The global learning rate η is specified by a real valued gene; and four boolean genes specify dependencies of the visual input modules on each of the other two modules.

The optimal gene values were determined using a genetic algorithm. A first generation of bees was produced by randomly generating 100 genome strings. Each bee performed 100 trials independently (no competition) and received a fitness score according to the average amount of nectar gathered per trial. To form the next generation, fifty pairs of parents were chosen (with returns) with a bee's fitness specifying the probability of it being chosen as a parent. Each two parents gave birth to two offsprings, which inherited their parents' genome (with **no Lamarkian inheritence** of learned weights) after performing recombination (genewise, $p = 0.25$) and adding random mutations. Mutations were performed by adding a uniformly distributed value in the range of [-0.1,0.1] to 2% of the real-valued genes, and reversing 0.2% of the boolean genes. One hundred offsprings were created, these once again tested in the flower field. This process continued for a large number of generations.

3 Evolution of Reinforcement Learning

To promote the evolution of efficient learning rules, bees were evolved in an "uncertain" world: In each generation one of the two flower types was randomly assigned as a constant-yielding high-mean flower (containing $0.7\mu l$ nectar), and the other a variable-yielding low-mean flower ($1\mu l$ nectar in $\frac{1}{5}$th of the flowers and zero otherwise). The reward contingencies were switched between the two flower types in a randomly chosen trial during the second or third quarter of each bee's life. Evolutionary runs under this condition typically show one of two types of fitness curves: runs in which reward-dependent choice behavior is successfully evolved are characterized by two distinct evolutionary jumps (Fig. 2a), while unsuccessful runs (which produce behavior that is not dependent on rewards) show only the first jump.

About half of the evolutionary runs were successful. Figure 2b shows the mean value of several of the bees' genes in the last generation of each of five successful runs. The second evolutionary jump characteristic of successful runs is due to the almost simultaneous evolution of 8 genes governing the network structure and learning dependencies. All successful networks have a specific architecture which includes the reward, differential blue and differential yellow synapses, as well as a dependency of the differential module on the reward module, conditioning modification of these synapses on the presence of reward. Thus we find that in our framework, only a network architecture similar to that used by Montague et al. [11] can produce above-random foraging behavior, supporting their choice as an optimal one. However, **our optimized networks utilize a heterosynaptic**

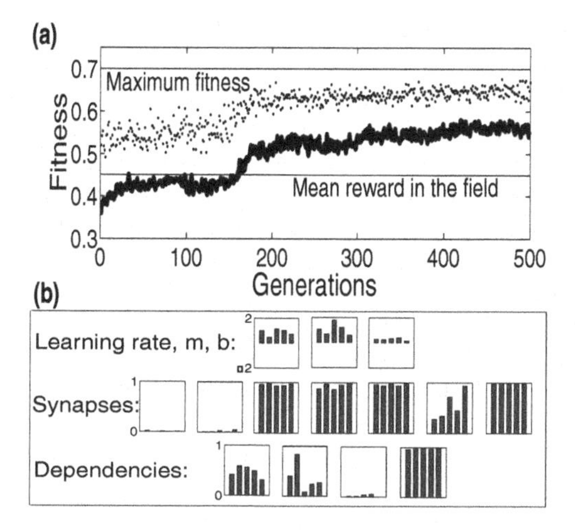

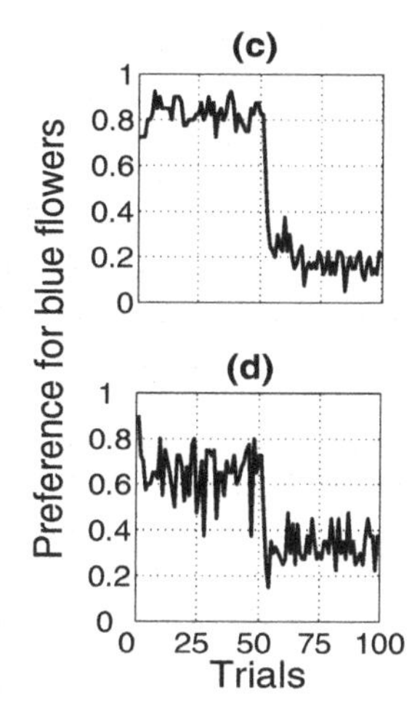

Fig. 2. (a) Typical fitness scores of a successful run of 500 generations. Solid line - mean fitness, dotted line - maximum fitness in each generation. **(b) Mean value of several genes** in the last generation of five successful runs. Each subfigure shows the mean value of one gene in the last generation of five runs.**(c,d) Preference for blue flowers for two different bees** from the last generation of a successful run, averaged over 40 test bouts, each consisting of 100 trials. Blue is the initial constant-rewarding high-mean flower. Reward contingencies are switched at trial 50.

learning rule different from that used by Montague et al., which gives rise to several important behavioral strategies.

Bees from the last generation of a successful run show a marked preference for the high-mean rewarding flower, with a rapid transition of preferences after the reward contingencies are switched between the flower types. An examination of the behavior of the evolved bees, reveals that there are individual differences between the bees in their degree of exploitation of the high-rewarding flowers versus exploration of the other flowers (Fig. 2c,d). This can be explained by an **interesting relationship between the micro-level Hebb rule coefficients and the exploration/exploitation tradeoff characteristic of the macro-level behavior:** In the common case when upon landing the bee sees only one color, the synaptic update rule for the corresponding differential synapse is

$$\Delta W_i(t+1) = \eta[(A-C)\cdot(-1)\cdot[R(t)-W_i(t)] + (D-B)] \tag{4}$$

leading to an effective monosynaptic coefficient of (A-C), and a general weight decay coefficient (D-B). For the other differential synapses, the learning rule is:

$$\Delta W_j(t+1) = \eta(C\cdot[R(t)-W_i(t)] + D). \tag{5}$$

Thus, positive C and D values result in spontaneous strengthening of competing synapses, leading to an exploration-inclined bee. Negative values will result in a declining tendency to visit competing flower types, leading to exploitation-inclined behavior.

4 Emergence of Risk Aversion and Probability Matching

A prominent strategy exhibited by the evolved bees is risk-aversion. Figure 3a shows the choice behavior of previously evolved bees, tested in a new environment where the mean rewards of the two kinds of flowers are identical. Although the situation does not call for any flower preference, the bees prefer the constant-rewarding flower. Furthermore, bees evolved in an environment containing two constant-rewarding flowers yielding different rewards, also exhibit risk-averse behavior when tested in a variable-rewarding flower scenario, thus risk- aversion is not a consequence of evolution in an uncertain environment per se. In contradistinction to the conventional explanations of risk aversion, our model does not include a non-linear utility function. **What hence brings about risk-averse behavior in our model?** Corroborating previous numerical results [10], we prove analytically that this foraging strategy is a direct consequence of Hebbian learning dynamics in an armed-bandit-like RL situation.

The bee's stochastic foraging decisions can be formally modeled as choices between a variable-rewarding (v) and a constant-rewarding (c) flower, based on memory (synaptic weights). We consider the bee's long-term choice dynamics as a sequence of N cycles, each choice of (v) beginning a cycle. The frequency f_v of visits to (v) can be determined by the expected number of visits to (c) in a cycle, and is

$$f_v = \frac{1}{E[1/p_v(W_v, W_c)]} \tag{6}$$

where $p_v(W_v, W_c)$ is the probability of choosing (v) in a trial in which the synaptic weights are W_v and W_c for the variable and the constant flower respectively. We show that if $p_v(W_v, W_c)$ is a positive increasing choice function such that $[1/p_v(W_v, W_c)]$ is convex, the risk order property of $W_v(\eta)$ always implies risk-averse behavior, i.e. **for every learning rate, the frequency of visits to the variable flower (f_v) is less than 50%, further decreasing under higher learning rates.** Our simulations corroborate this analytical result (Fig. 3b).

In essence, due to the learning process in an armed-bandit type situation, the bee makes its decisions based on a biased sampling of the two flowers, and does not compute the long-term mean reward obtained from each flower. This is even more pronounced with high learning rates such as those evolved (~0.8). After landing on an empty flower of the variable- rewarding type and recieving no reward, the bee updates the reward expectation to near zero, and as a result, prefers the constantly rewarding flower, from which it constantly recieves a reward of $\frac{1}{2}\mu l$. As long as the bee chooses the constant-rewarding flower, it will not update the expectation from the variable-rewarding flower, which will remain near zero. Even after an occasional "exploration" trial in which a visit to the variable flower yields a high reward, the preference for this flower will last only until the next unrewarded visit. Note that such abnormally high learning rates were also used in Montague et al.'s [11] model, and have been hypothesized by Real [13].

The simulated bees also demonstrate probability-matching behavior. Figure 3(c,d) shows the previously evolved bees' performance when tested in matching

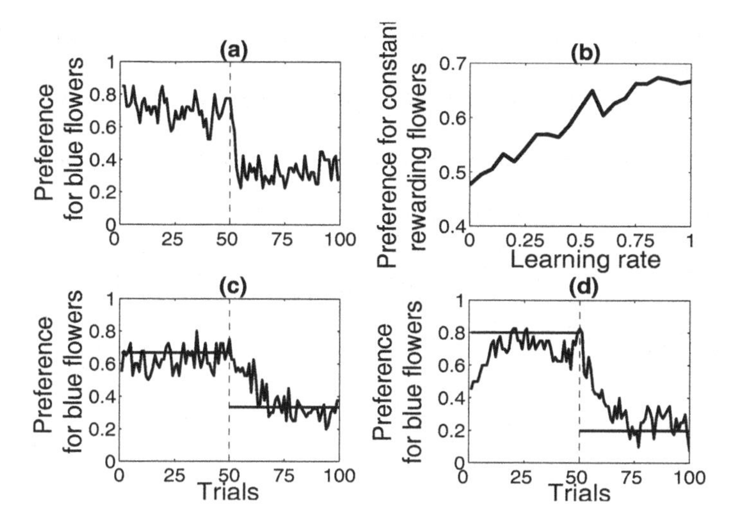

Fig. 3. Preference for blue flowers averaged over 40 previously evolved bees tested in conditions different from those they were evolved in: **(a) Risk aversion** - Although both flower types yield the same mean reward (blue - $\frac{1}{2}\mu l$ nectar, yellow - $1\mu l$ in half the flowers, contingencies switched at trial 50), there is a marked preference for the constant-yielding flower. **(b) Risk aversion is ordered according to learning rate.** Each point represents the percentage of visits to constant-rewarding flowers in 50 test trials averaged over 40 previously evolved bees, with a clamped learning rate. **(c-d) Matching** - All flowers yield $1\mu l$ nectar with reward probabilities for blue and yellow flowers (c) 0.8, 0.4 and (d) 0.8, 0.2 respectively (contingencies switched at trial 50). Horizontal lines - behavior predicted by perfect matching.

experiments in which all flowers yield $1\mu l$ nectar, but with different reward probabilities. In both conditions, the bees show near-matching behavior, preferring the high-probability flower to the low-probability one, by a ratio that closely matches the reward probability ratios. This is again a direct result of the learning dynamics Thus, in contradistinction to previous accounts, matching can be evolved in a non-competitive setting, as a direct consequence of optimal RL.

5 Robot Implementation

In order to assess the robustness of the evolved RL algorithm, we implemented it in a mobile mini-robot by letting the robot's actions be governed by a NN controller similar to that evolved in successful bees, and by having its synaptic learning dynamics follow the previously evolved RL rules. A Khepera mini-robot foraged in a 70X35cm arena whose walls were lined with flowers, viewing the arena via a low-resolution CCD camera (200x200 pixels), moving at a constant velocity and performing turns according to the action function (eq. 2) in order to choose flowers, in a manner completely analogous to that of the simulated bees. All calculations were performed on a Pentium-III 800Mhz computer

(256Mb RAM) in tether mode. Moving with continuous speed and performing all calculations in real-time, the foraging robot exhibited rapid RL and risk-averse behavior (Fig. 4). Thus the algorithms and behaviors evolved in the virtual bees' simulated environment using discrete time-steps hold also in the different and noisy environment of real foraging mini-robots operating in continuous time.

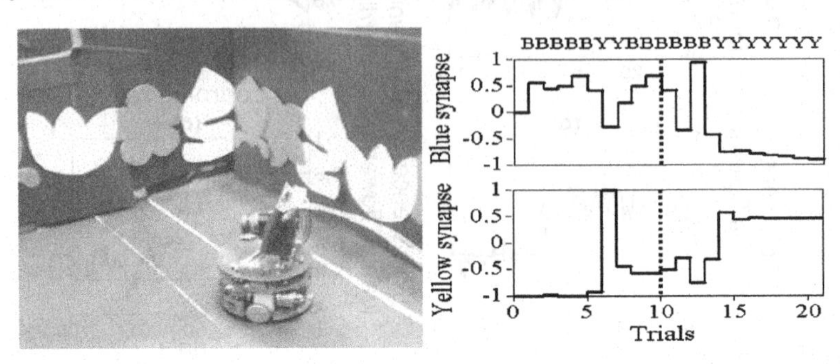

Fig. 4. Synaptic weights of a mobile robot incorporating a NN controller of one of the previously evolved bees, performing 20 foraging trials (blue flowers - $\frac{1}{2}\mu l$ nectar, yellow - $1\mu l$ in half the flowers, contingencies switched after trial 10). **(a)** The foraging robot. **(b)** Blue and yellow differential weights represent the expected rewards from the two flower colors along the trials. **Top:** Flower color chosen in each trial.

6 Discussion

The interplay between learning and evolution has been previously investigated in the field of Alife. Much of this research has been directed to elucidating the relationship between evolving traits (such as synaptic weights) versus learning them [1,7]. A relatively small amount of research has been devoted to the evolution of the learning process itself, most of which was constrained to choosing the appropriate learning rule from a limited set of predefined rules [2,3]. In this work we show for the first time, that optimal learning rules for RL in a general class of armed bandit situations, can be evolved in a general Hebbian learning framework. The evolved heterosynaptic learning rules are by no means trivial, as they include an anti-Hebbian monosynaptic term and employ axo-axonic plasticity modulation. We have no rigorous proof as to their optimality for the given task, but results from multiple evolutionary runs strongly suggest this.

The emergence of complex foraging behaviors as a result of optimal learning per se, demonstrate once again the strength of Alife as a methodology that links together phenomena on the neuronal and the behavioral levels. We show that the fundamental macro-level strategies of risk aversion and matching are a direct result of the micro level synaptic learning dynamics in RL, making additional assumptions conventionally used to explain them redundant. This result is important not only to the fields of Alife and animal learning theories, but also to economics and game theory.

In summary, the significance of this work is two-fold: on one hand we show the strength of simple Alife models in evolving fundamental processes such as reinforcement learning, and on the other we show that optimal reinforcement learning can directly explain complex behaviors such as risk aversion and probability matching, without need for further assumptions.

References

1. D. Ackley and M. Littman. Interactions between learning and evolution. In J.D. Farmer C.G. Langton, C. Taylor and S. Rasmussen, editors, *Artificial Life II*. Addison-Wesley, 1991.
2. D.J. Chalmers. The evolution of learning: An experiment in genetic connectionism. In D.S. Touretzky, J.L. Elman, T.J. Sejnowski, and G.E. Hinton, editors, *Proc. of the 1990 Connectionist Models Summer School*. Mogan Kaufmann, 1990.
3. D. Floreano and F. Mondada. Evolutionary neurocontrollers for autonomous mobile robots. *Neural networks*, 11:1461–1478, 1998.
4. J.F. Fontanari and R. Meir. Evolving a learning algorithm for the binary perceptron. *Network*, 2(4):353–359, November 1991.
5. M. Hammer. The neural basis of associative reward learning in honeybees. *Trends in Neuroscience*, 20(6):245–252, 1997.
6. L.D. Harder and L.A. Real. Why are bumble bees risk averse? *Ecology*, 68(4):1104–1108, 1987.
7. G.E. Hinton and S.J. Nowlan. How learning guides evolution. *Complex Systems*, 1:495–502, 1987.
8. A. Kacelnik and M. Bateson. Risky thoeries - the effect of variance on foraging decisions. *American Zoologist*, 36:402–434, 1996.
9. T. Kaesar, E. Rashkovich, D. Cohen, and A. Shmida. Choice behavior of bees in two-armed bandit situations: Experiments and possible decision rules. *Behavioral Ecology*. Submitted.
10. J. G. March. Learning to be risk averse. *Psychological Review*, 103(2):309–319, 1996.
11. P.R. Montague, P. Dayan, C. Person, and T.J. Sejnowski. Bee foraging in uncertain environments using predictive hebbian learning. *Nature*, 377:725–728, 1995.
12. S. Nolfi, J.L. Elman, and D. Parisi. Learning and evolution in neural networks. *Adaptive Behavior*, 3(1):5–28, 1994.
13. L.A. Real. Animal choice behavior and the evolution of cognitive architecture. *Science*, 253:980–985, August 1991.
14. A.K. Seth. Evolving behavioral choice: An investigation into Herrnstein's matching law. In J. Nicoud D. Floreano and F. Mondada, editors, *Advances in Artificial Life, 5th European Conference, ECAL '99*, pages 225–235. Springer, 1999.
15. R.S. Sutton and A.G. Barto. *Reinforcement learning: An introduction.* MIT Press, 1998.
16. F. Thuijsman, B. Peleg, M. Amitai, and A. Shmida. Automata, matching and foraging behavior of bees. *Journal of Theoretical Biology*, 175:305–316, 1995.

Searching for One-Dimensional Cellular Automata in the Absence of *a priori* Information

Gina M.B. Oliveira[1], Pedro P.B. de Oliveira[1], and N. Omar[2]

[1]Universidade Presbiteriana Mackenzie, R. da Consolação 896, 6° Andar, Consolação
01302-907 São Paulo, SP - Brazil
{gina, pedrob}@mackenzie.com.br

[2]Instituto Tecnológico de Aeronáutica, Praça Marechal E. Gomes 50, Vila das Acácias
12228-901 São José dos Campos, SP - Brazil
omar@comp.ita.cta.br

Abstract. Various investigations have been carried out on the computational power of cellular automata (CA), with concentrated efforts in the study of one-dimensional CA. One of the approaches is the use of genetic algorithms (GA) to look for CA with a predefined computational behavior. We have previously shown a set of parameters that can be effective in helping forecast CA dynamic behavior; here, they are used as an heuristic to guide the GA search, by biasing selection, mutation and crossover, in the context of the Grouping Task (GT) for one-dimensional CA. Since GT is a new task, no *a priori* knowledge about its solutions is presently available; even then, the incorporation of the parameter-based heuristic entails a significant improvement over the results achieved by the plain genetic search.

1 Introduction

Cellular automata (CA) are discrete complex systems which possess both a dynamic and a computational nature. In either case, based only upon their definition, it is not possible to forecast neither their dynamic behavior, nor the computation they will perform when running ([3]). However, various studies in the context of one-dimensional CA have been carried out on defining parameters, directly obtained from their transition rule, which have been to shown to help forecast their dynamic behavior ([2], [7], [8], [14] and [20]).

An important line of investigation is the study of computational ability of CA. One of the novelties of the current trend in such a study is the use of Genetic Algorithms (GA) and other evolutionary techniques to design CA that perform a predefined, specific computation ([1], [6], [10],[11], [12], [13] and [17]).

In tune with that, we carried out a number of experiments with the evolution of one-dimensional CA, to perform the computations featured in the literature, our approach being the use of a set of parameters as an auxiliary metric to guide the GA

J. Kelemen and P. Sosík (Eds.): ECAL 2001, LNAI 2159, pp. 262-271, 2001.

optimization. These parameters, conceived to help forecast the dynamic behavior of one-dimensional CA, were drawn from our previous works ([14] and [15]).

In this work, we emphasized the experiments involving the Grouping Task (GT), an adaptation of Sipper's Ordering Task ([18]). Here, GT is formally defined and results from the task are reported. Details are presented of the modifications made in a GA, so as to incorporate the search heuristic based in the forecast parameters.

In the next section, some concepts about the CA dynamics are reviewed. In Section 3, computational aspects of CA are addressed, and the Grouping Task is defined. The GA environment is described in Section 4, and the results obtained discussed in Section 5. Some final remarks are then made in Section 6.

2 Cellular Automata Dynamics

Basically, a cellular automaton consists of two parts, the *lattice* where states are defined, and the *transition rule* that establishes how the states will change along time ([11]). For one-dimensional CA, the neighborhood size m is usually written as $m = 2r + 1$, where r is called the *radius*. In the case of two-state CA, the transition rule is given by a transition table that lists each possible neighborhood with its output bit, that is, the update value of the center cell of the neighborhood.

Through the analysis of the dynamic behavior exhibited by CA, it was verified they could be grouped into classes. A few rule space classification schemes have been used in the literature; for instance, Wolfram proposed a qualitative behavior classification, which is widely known ([19]). Later on, Li and Packard proposed a series of refinements in the original Wolfram classification ([8] and [9]). The following is one of the versions ([9]) of their classification scheme, which divides the rule space into six classes. **Null (or Homogeneous Fixed Point)**: the limiting configuration is only formed by 0s or 1s. **(Heterogeneous) Fixed Point**: the limiting configuration is invariant (with possibly a spatial shift) by applying the cellular automaton rule, the null configurations being excluded. **Two-Cycle**: the limiting configuration is invariant (with possibly a spatial shift) by applying the rule twice. **Periodic**: the limiting configuration is invariant by applying the automaton rule L times, with the cycle length L either independent or weakly dependent on the system size. **Edge of Chaos, or Complex**: although their limiting dynamics may be periodic, the transition time can be extremely long and they typically increase more than linearly with the system size. **Chaotic**: these rules are characterized by the exponential divergence of its cycle length with the system size, and for the unstability with respect to perturbations.

The dynamics of a cellular automaton is associated with its transition rule. In order to help forecast the dynamic behavior of CA, several parameters have been proposed, directly calculated from its transition table ([2], [7], [8], [15] and [20]). The high cardinality of CA rule spaces make their parameterization a daunting task, which entails the usage of more than a single parameter ([2]).

Before carrying on, it should be remarked that it is not possible to expect the precise forecasting of a generic cellular automaton, from an arbitrary initial configuration.

It has been proved ([3]), that the decision problem associated with the latter generic proposition is undecidable. Hence, all we can expect is really a parameter that can help forecast the dynamic behavior of a cellular automaton; in particular, for present purposes, our analysis is focused on one-dimensional, two-state CA.

From the study of the parameters published in the literature, and also from others investigated by the authors of this paper, we analyzed and selected a set of parameters ([15]). Two of them were chosen from among those published: Sensitivity ([2]) and Z ([20]). In addition to them, we conceived and tested several others, and eventually came up with a set of five parameters. The three new ones in the group were named Absolute Activity, Neighborhood Dominance and Activity Propagation ([13], [14] and [15]). All of them have been normalized between 0 and 1, for one-dimensional CA with any radius, and were formally defined in [15].

3 Computational Tasks in Cellular Automata

Once a computational task is defined, it is far from trivial finding CA that perform it. Manual programming is difficult and costly, and exhaustive search of the rule space becomes impossible, due to its high cardinality. A solution is the use of search and optimization methods, particularly evolutionary computation methods. Our approach relies on a GA to search for CA with a predefined computational behavior, with the central idea that the forecast parameter set we selected is used as an auxiliary metric to guide the processes underlying the GA search ([13]). Packard ([17]) used a GA as a tool to find CA rules with a specific computational behavior. He considered CA rules as individuals in a population and defined their fitness according to their ability to perform the task. In this way, the genotype of the CA was given by its transition rule, and the phenotype by its ability to perform the required task. Crossover among two CA was defined by the creation of two new rules, out of segments of two other rules; mutation was achieved by the random flip of the output bit of one of the transitions of the rule. Here, we rely on a simple GA environment similar to that used in [12], with the appropriate modifications required by the parameter-based heuristic.

The most widely studied CA task is the Density Classification Task (DCT) ([1], [6], [10], [11], [12], [13], [17]). Another is the Synchronization Task (ST) ([4], [12], [16]). In [18] the Ordering Task (OT) was introduced, in the context of non-uniform CA with non-periodic boundary conditions. The goal in this task is to find a CA rule able to sort a random initial configuration formed by 0s and 1s. In other words, the initial configuration is formed by N_0 0s and N_1 1s, randomly distributed among the lattice, and the final configuration is formed by N_0 0s, occupying the leftmost N_0 cells, followed by N_1 1s. Because our approach relies on CA with periodic boundary conditions, OT has to be redefined. As a result, we adapted the Ordering Task for periodic boundary conditions, which yielded what we named the Grouping Task (GT).

Let us consider an arbitrary initial configuration of a lattice with length N, formed by N_0 0s and N_1 1s, randomly distributed. A cellular automaton transition rule is said to be able to solve GT if, after T time steps, the final configuration still has N_0 0s and N_1 1s, but distributed in the following way:

- If $N \neq N_0$ and $N \neq N_1$ (i.e, the state of at least one cell of the initial configuration is different from the others): from the first (leftmost) to the last cell of the final configuration there can only be one transition from 1 to 0 (represented by $\delta_{10} = 1$).
- If $N = N_0$ or $N = N_1$ (i.e, the initial configuration has all the cells in the same state): the first condition - final configuration with the same number of 0s and 1s - is sufficient (in this case, $\delta_{10} = 0$).

T was made equal to $2 \times N$, analogously to what has been imposed on DCT and ST ([11]). Once GT is defined in this way, OT can be viewed as a special case of GT.

4 The GA Environment

The environment built to evolve the GT was similar to those used in DCT and ST experiments ([13] and [16]). In GT experiments, we used radius 2, binary CA, with one-dimensional lattice, using a population of 100 individuals (32 bits each), evolving during 50 generations. Elitism was used at a rate of 20% (that is, the 20 best rules of the population at each generation were always preserved for the next one); the other 80 individuals were obtained through crossover and mutation. Parent selection for the crossover was directly made from the elite without considering each individual fitness. Standard one-point crossover was used at a rate of 80%. Mutation was applied after crossover, in each new generated individual, at a rate of 2% per bit.

For the DCT and ST experiments, each individual evaluation was obtained, at each generation, out of testing the performance of the automaton in 100 different Initial Configurations (ICs), without considering incomplete solutions. However, it was verified that, in the case of GT, under such severe conditions of evaluation, the GA did not converge, finding rules with performance close to 0%. A more flexible evaluation was then used, which accounted also for partial solutions. For each IC tested, the evaluation of the individual/automaton was obtained in the following way:

a. If the IC is correctly grouped in the final configuration (FC), the score is 1.
b. If the IC is not correctly grouped, two variables are obtained over FC: N_{0F}, the number of zeros in FC, and δ_{10F}, the number of 1-to-0 state transitions along FC. Hence, two other variables are calculated: D_0 and D_δ. D_0 is the difference between the actual (N_{0F}) and the correct (N_0) number of zeros in FC: $D_0 = | N_{0F} - N_0 |$. Analogously, D_δ is given by: $D_\delta = | \delta_{10F} - \delta_{10} |$. If either D_0 or D_δ is a high number, it means that the FC is far from the expected configuration, and an upperbound error (ξ_{max}) is defined, as a function of the lattice length N, i.e., $\xi_{max} = \lfloor N/3 \rfloor$.
   b.1. If $D_0 > \xi_{max}$ or $D_\delta > \xi_{max}$, the IC is grouped in a completely wrong way in FC and the score is 0.
   b.2. Otherwise, the IC is partially grouped in FC and the score is given by the expression: $0.5 - (0.5/\xi_{max}) \times \max(D_0 - 1, D_\delta - 1)$.

Summarizing, the score of the individual in relation to each IC can be: 0 (FC is very different from the expected), between 0 and 0.5 (FC is partially grouped), or 1 (FC is successfully grouped). Each individual evaluation was obtained out of testing the performance of the automaton in 100 different ICs. The efficacy of the GA run was measured by testing the performance of the best rule found, at the end of the run in 10^4 different ICs, without considering partial solutions. Initially, the GA search was done without any information about the parameters. The experiments were performed for three lattice sizes (of 9, 19 and 49 cells); the results are presented in Section 5.

Let us explain how the GA environment was modified so as to incorporate the parameter-based heuristic. Basically, the parameter-based heuristic is coded as a function (referred to as *Hp*), which returns a value between 0 and 100 for each transition rule, depending on the rule's parameter values. *Hp* is then used to bias the GA operation, in the following ways: *Selection*: the fitness function of a CA rule is defined by the weighted average of the original fitness function (efficacy in 100 different ICs) and the *Hp* function. *Crossover*: various crossovers are made, at random crossover points, creating twice as much offspring; the two of them with the highest *Hp* value are then considered for mutation. *Mutation*: various mutations are made to a rule, so that the one with the highest *Hp* value is selected for the new population. The heuristic entailed by the parameters may be implemented in various ways. The way used to establish the parameter heuristic in the GT experiments was supplying the value ranges of the parameters in which it would be expected to obtain the CA rules, for some *a priori* knowledge of the problem. In the case of the Grouping Task, as we did not have any information about the parameters, we first ran the GA without the parameter information. Then, the values of 4 of the selected parameters cited in Section 2 were calculated for the best rules found (12 rules with efficacy above 55%) in 40 GA runs; Z parameter was excluded from the original set of 5 parameters because it was shown to be irrelevant for this task. The value ranges found were the following – Sensitivity: [0.30, 0.39]; Absolute Activity: [0.11, 0.19]; Neighborhood Dominance [0.79, 0.86]; and Activity Propagation [0.06, 0.11]. These ranges are interpreted as those with high probability of occurrence of the expected behavior (that is, the capacity to solve the GT). In the case of Sensitivity, for example, that means there is a high probability of the rule being able to perform the GT if its Sensitivity is between 0.30 and 0.39.

The chart depicted in Figure 1 was used to define *Hp*, as a function of each parameter. The interval [MAX_0, MAX_f] represents the one with the highest probability of occurrence of rules able to perform the GT. The evaluation of the rule in respect to any parameter depends on its value, according to the following: if the parameter value is within the interval [MAX_0, MAX_f], the evaluation is 100 (maximum value); otherwise, the value follows the linear functions represented in the chart. For example, in the case of the Sensitivity parameter, evaluation is made by the chart in Figure 2. This evaluation procedure is made for each parameter individually. The total evaluation of the rule, *Hp*, is given by the simple average of the evaluation of the four parameters: Sensitivity, Neighborhood Dominance, Absolute Activity and Activity Propagation. For example, for a rule with Sensitivity = 0.34, Neighborhood Dominance = 0.77, Activity Propagation = 0.45 and Absolute Activity = 0.05, using the ranges above and according to the described evaluation procedure, the evaluation of each one of the parameters will be 100, 98, 75 and 94, respectively. The total evaluation of the rule according to the parameter heuristic (*Hp*) will be 91.75.

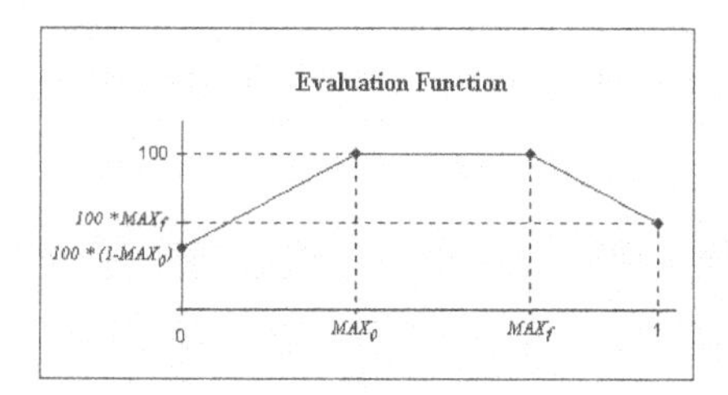

Fig. 1. Evaluation function for each parameter.

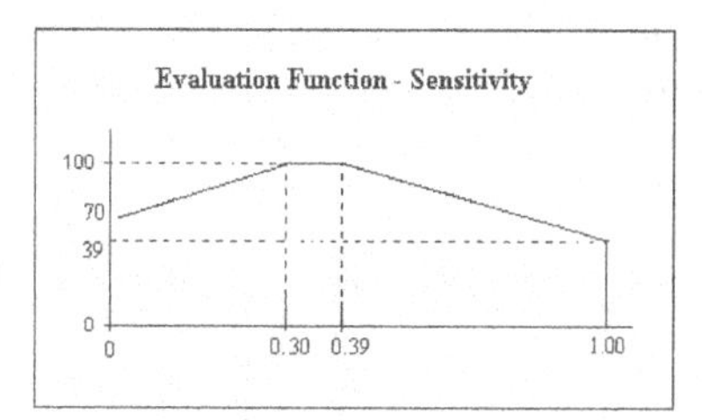

Fig. 2. Evaluation function for the Sensitivity parameter.

Compounded Selection: Selection is achieved through two evaluations: F_o (the original fitness function, that measures the effectiveness of the rule in correctly grouping 100 ICs) and *Hp* (the evaluation function based on the parameters, as described above). The fitness function used (*F*), is a composition of F_o and *Hp*, the parameter-based component being weighted by W_h, and expressed as: $F = F_o + W_h * Hp$.

Biased Crossover: For each pair of rules selected to crossover, *R1* and *R2*, N_{CROSS} crossover points are randomly chosen. The rules *R1* and *R2* are submitted to N_{CROSS} simple crossover and, for each one, two offspring are generated. The forecast parameters of the parameter set are calculated for all offspring and evaluated according to *Hp*. Those with the highest value of *Hp* are selected.

Biased Mutation: Whenever a rule *R* is chosen to be mutated, N_{MUT} different mutations are tested and each resulting rule is evaluated according to the heuristic function *Hp*. The mutated rule with the highest value of *Hp* is selected to be the mutation of *R*.

In all experiments reported in the next section, the following was used: $W_h = 40\%$, $N_{CROSS} = 10$ and $N_{MUT} = 10$.

5 Results

5.1 Lattice with 9 Cells

A series of 40 GA runs was performed without the parameters information, whose results are presented in Table 1, referred to as "Simple 9"; Table 1 also presents the two best rules found. Subsequently, a new experiment was performed, where the ranges of the parameters calculated for the best rules found in *Simple 9*, shown in Section 4, were used as an heuristic. Table 1 shows the efficacy results of this sequence of (also 40) runs, referred to as "Parameter 9", as well as the two best rules found in the experiment. As a result, not only the best efficacy found was higher (80.9%), but also, there was a clear shift in the number of rules found towards the higher efficacy ranges. Clearly, the introduction of the heuristic entailed a substantial improvement in the average efficacy of the rules, as well as in the best rule found.

Table 1. Experimental results: *Simple 9* (S9), *Parameter 9* (P9), *Simple 19* (S19), *Parameter 19* (P19), *Simple 49* (S49), *Parameter 49A* (P49A) and *Parameter 49B* (P49B). Efficacies are given in "%". Best1/Best2 represent the 2 best rules found, and *ER* stands for efficacy ranges.

ER	S9	P9
≤ 30	0	0
(30,40]	2	0
(40,50]	14	0
(50,60]	20	12
(60,70]	4	5
(70,80]	0	21
> 80	0	2
Average	51.8	69.9
Best1	62.8	80.9
Best2	62.3	80.0

ER	S19	P19
≤ 20	14	0
(20,25]	20	3
(25,30]	6	2
(30,35]	0	0
(35,40]	0	0
(40,45]	0	10
> 45	0	25
Average	21.5	43.1
Best1	28.1	48.7
Best2	27.5	47.9

ER	S49	P49A	P49B
≤ 5.0	17	1	0
(5.0,7.5]	12	20	2
(7.5,10.0]	1	4	9
(10.0,12.5]	0	1	1
(12.5,15.0]	0	3	16
> 15.0	0	1	2
Average	4.3	7.8	11.7
Best1	7.5	15.2	16.4
Best2	6.7	13.7	15.8

5.2 Lattice with 19 Cells

After increasing the lattice length to 19 cells, a series of 40 GA runs was performed without the parameters information; the results are presented in Table 1, referred to as "Simple 19", together with the two best rules found. Subsequently, the ranges of the selected parameters presented in Section 4 were used again as an heuristic to guide the GA search and a new experiment, referred to as "Parameter 19", was performed. Table 1 shows the efficacy results of another sequence of 40 runs, and the two best rules found. The best efficacy found was higher (48.7% against 28.1%) and there was again a clear shift in the number of rules found towards the higher efficacy ranges. The final result in the 19-cell lattice (48.7%) is worse than in the one with 9 cells (80.9%), because it is indeed more difficult to group a higher number of cells. But, it is interesting to notice that the introduction of the parameter-based heuristic here entailed a stronger improvement in the average efficacy (100.1%) of the rules than in the experiments with 9 cells (35.1%), as well as in the best rule found (73.3% against 28.8%). A remarkable point to be noticed is the fact that an heuristic derived from CA with 9-cell lattice was used in the search for CA with 19-cell lattice, and even then, a significant improvement was attained. Some experiments with heuristics based on results of *Simple 19* have also been carried out and the results were similar to those obtained in *Parameter 19*.

5.3 Lattice with 49 Cells

The last experiments were performed in 49-cell lattice CA. Initially, a series of 30 GA runs were done without the parameters heuristic, whose results are presented in Table 1, referred to as "Simple 49"; together with the two best rules found. The second experiment used again the ranges of the parameters presented in Section 4, and is referred to as "Parameter 49A"; Table 1 shows the efficacy results of this sequence of (also 30) runs, together with the two best rules found. Although the final result was not so good, there was a strong improvement in the average efficacy (81.3%) with the introduction of the parameter-based heuristic, mainly in the best rule found (102.7%).

In order to try to obtain better results, a new heuristic was built based on the results of the experiments in 49-cell lattice CA. The four selected parameters were calculated for the best rules found in *Parameter 49A* experiment; they are the 5 rules with efficacy above 10%. The ranges found were – Sensitivity: [0.32, 0.40]; Absolute Activity: [0.17, 0.25]; Neighborhood Dominance [0.78, 0.91]; and Activity Propagation [0.06, 0.11] – and they are slightly different from those shown earlier in Section 4. Subsequently, these ranges were used as a new heuristic to guide the GA search and a sequence of 30 runs was performed, referred to as "Parameter 49B". The efficacy results of this experiment and the two best rules found are shown in Table 1. The change of the ranges in the parameter-based heuristic entailed a good improvement in the average efficacy (51.4%) of the rules in relation to the *Parameter 49A* experiment, and a slight increase in the efficacy of the best rule found (7.9%).

6 Final Remarks

Other computational tasks were studied in our approach based on the parameter-based heuristic. In [13], we presented results of experiments involving the DCT, the most widely studied computational task in recent years of CA research ([2], [6], [10], [11], [12] and [17]). In [16], we studied the ST, also investigated by other authors ([4] and [12]). In both cases, it was also verified a significant increase in the GA performance when incorporating the parameter-based heuristic in the search. However, the current experiments involving GT differ from those DCT and ST experiments precisely in the fact that, since it is a new task in the literature, our approach was left without any *a priori* knowledge for usage as heuristic information. All of the heuristics used was based on results of preliminary experiments. In spite of that, the results were extremely encouraging, a clear suggestion that even with no previous knowledge about a task, the heuristic based on the forecast parameters alone, can still guide the GA search towards higher efficiency regions of the search space.

It is important to point out that the heuristics used in *Parameter 19* and *49A* experiments is based on parameters calculated on rules found in experiments involving 9-cell lattice CA. In spite of that, the heuristic improved the efficacy of the rules found in the GA search both in 19- and 49-cell lattices. It should also be observed that, by analyzing the results of the experiments, one can obtain parameter ranges of the best rules, which can then be used to derive a new heuristic. Quite interestingly, this may also improve the efficacy of the new best rules found, in relation to the experiment from where the heuristic is derived. This demonstrates that our approach can be used in an adaptive way, through a continuous refinement of the results.

It is interesting to note that, from the tasks addressed with our approach, GT was the one with the worst final performance: the best rule found for 49cell lattice CA scored 16.4%, while in the DCT and ST, rules have been found with performance higher than 80% for 149-cell lattice CA. However, it should be remarked that, indeed,

GT is inherently much more complicated to accomplish than the other two. In the case of ST, the simplest one among the three, the required final configuration is always the same, regardless of the initial configuration; radius-3 CA rules are known that can solve the task with 100% efficacy, and radius-2 CA rules that can solve it with 94% efficacy ([16]). A for the DCT, the most widely studied task, there are only two required final configurations ([11]) and the CA must converge to one of them depending on the initial configuration; the best rule (radius 3 CA) currently known for the task has about 86% performance ([6]). Back to the GT, notice that there are a number of possible final configurations, in a way that is totally dependent on the initial configuration. Besides, the solution is independent on the "rotation" of the final configuration: for the same initial configuration, various final configurations are accepted. Consequently, GT is a task that requires a more elaborate approach so as to allow the search to be successful. Improvements can be thought of in various aspects of the experiment. The simplest one seems to be to increase the CA radius. Another alternative that deserves investigation is a way to improve the partial evaluations of the solutions. Finally, other techniques of evolutionary computation should be tried, such as coevolution, in a similar way to that already applied with success in the DCT problem in [6].

The incorporation of the heuristic in the GA model was also made in a simple way; for example, the weight of each parameter in the calculation of the *Hp* function is the same. New experiments could be accomplished by looking at ways to evaluate the relative weight of each parameter in the improvement of the obtained results. Furthermore, much more refined techniques could be used to better explore the multi-objective nature of the evaluation function ([5]). Finally, further investigations should be carried out to probe the individual effect of the modifications in the three processes of the GA (selection, mutation and crossover), thus allowing a picture of the individual contributions of each one, for the success of the search. Preliminary analyses indicate that the largest gain is due to the biased crossover and mutation.

Acknowledgments. G.M.B.O. and N.O. are grateful to Univ. Mackenzie and ITA for their support, and CAPES and MACKPESQUISA for funding. P.P.B.O. is grateful to A. Wuensche for various comments, and CNPq (ProTem-CC: CEVAL project) for funding. This work was done while P.P.B.O. was with Universidade do Vale do Paraíba, IP&D, Brazil.

References

1. Andre, D.; Bennett III, F. and Koza, J. (1996). Evolution of Intricate Long-Distance Communication Signals in Cellular Automata Using Programming. In: *Proceedings of Artificial Life V Conference*. Japan, 5.
2. Binder, P.M. (1993). A Phase Diagram for Elementary Cellular Automata. *Complex Systems*, 7:.241–247.
3. Culik II, K.; Hurd, L. P. and Yu, S. (1990). Computation Theoretic Aspects of Cellular Automata. *Physica D*, 45:357-378.

4. Das, R.; Crutchfield, J.; Mitchell, M. and Hanson, J. (1995). Evolving Globally Synchronised Cellular Automata. In: *Proceedings of International Conference on Genetic Algorithms*, San Francisco, 6.
5. Fonseca, C. M. and Fleming, P. J. (1997). Multiobjective Optimization. In Thomas Bäck, David B. Fogel, and Zbigniew Michalewicz, editors, *Handbook of Evolutionary Computation*, volume 1, pages C4.5:1-C4.5:9. Oxford University Press.
6. Juillé, H. and Pollack, J.B. (1998). Coevolving the "Ideal" Trainer: Application to the Discovery of Cellular Automata Rules. In: *Proceedings of Genetic Programming Conference.* Madison, 3.
7. Langton, C.G. (1990). Computation at the Edge of Chaos: Phase Transitions and Emergent Computation. *Physica D*, 42:12-37.
8. Li, W. and Packard, N. (1990). The Structure of Elementary Cellular Automata Rule Space. *Complex Systems*, 4:281-297.
9. Li, W. (1992). Phenomenology of Non-local Cellular Automata. *Journal of Statistical Physics*, 68: 829-882.
10. Mitchell, M.; Hraber, P. and Crutchfield, J. (1993). Revisiting the Edge of Chaos: Evolving Cellular Automata to Perform Computations. *Complex Systems*, 7: 89-130.
11. Mitchell, M. (1996). Computation in Cellular Automata: A Selected Review. In*: Nonstandard Computation.* Weinheim: VCH Verlagsgesellschaft.
12. Mitchell, M.; Crutchfield, J.; and Das, R. (1996). Evolving Cellular Automata with Genetic Algorithms: a Review of Recent Work.. In: *Proceedings of International Conference on Evolutionary Computation and Its Applications*. Moscow, 5.
13. Oliveira, G.M.B., de Oliveira, P.P.B., and Omar, N. (2000). Evolving solutions of the density classification task in 1D cellular automata, guided by parameters that estimate their dynamic behaviour. In: M. Bedau; J. McCaskill; N. Packard and S. Rasmussen, eds. *Artificial Life VII*, 428-436, MIT Press.
14. Oliveira, G.M.B., de Oliveira, P.P.B., and Omar, N. (2001a). Guidelines for Dynamics-Based Parameterization of One-dimensional Cellular Automata Rule Spaces. *Complexity*, 6(2): 63-71, John-Wiley, 2000.
15. Oliveira, G.M.B., de Oliveira, P.P.B., and Omar, N. (2001b). Definition and Applications of a Five-Parameter Characterization of One-Dimensional Cellular Automata Rule Space. *Artificial Life*, under review.
16. Oliveira, G.M.B., de Oliveira, P.P.B., and Omar, N. (2001c). Improving Genetic Search for One-Dimensional Cellular Automata, Using Heuristics Related to Their Dynamic Behavior Forecast. *Proc. of the 2001 IEEE Conference on Evolutionary Computation*, to be held in Seoul, South Korea. To appear.
17. Packard, N. (1988). Adaptation toward the Edge of Chaos. In: *Dynamic Patterns in Complex Systems*. Singapore, 293-301.
18. Sipper, M. (1998). A simple Cellular Automata that solves the density and ordering problems. *International Journal of Modern Physics*, 9 (7).
19. Wolfram, S. (1994). *Cellular Automata and Complexity*. U.S.A.: Addison-Wesley.
20. Wuensche, A. (1999). Classifying Cellular Automata Automatically: Finding gliders, filtering, and relating space-time patterns, attractor basins and the Z parameter. *Complexity*, 4 (3):47-66.

Neutral Networks and Evolvability with Complex Genotype-Phenotype Mapping

Tom Smith[1,2], Phil Husbands[1,3], and Michael O'Shea[1,2]

[1] Centre for Computational Neuroscience and Robotics (CCNR)
[2] School of Biological Sciences
[3] School of Cognitive and Computing Sciences
University of Sussex, Brighton, UK
toms@cogs.susx.ac.uk

Abstract. In this paper, we investigate a neutral epoch during an optimisation run with complex genotype-to-fitness mapping. The behaviour of the search process during neutral epochs is of importance for evolutionary robotics and other artificial-life approaches that evolve problem solutions; recent work has argued that *evolvability* may change during these epochs. We investigate the distribution of offspring fitnesses from the best individuals of each generation in a population-based genetic algorithm, and see no trends towards higher probabilities of producing higher fitness offspring, and no trends towards higher probabilities of not producing lower fitness offspring. A second experiment in which populations from across the neutral epoch are used as initial populations for the genetic algorithm, shows no difference between the populations in the number of generations required to produce high fitness. We conclude that there is no evidence for change in evolvability during the neutral epoch in this optimisation run; the population is not doing anything "useful" during this period.

1 Introduction

Genetic algorithms are classically regarded as performing hill-climbing. Populations of solutions are progressively improved until some acceptable level of fitness is reached, with problems occurring if the system gets stuck in local optima before the required fitness is reached. However, the idea of non-adaptive *neutral* mutation [11, 9, 2, 21] extends this picture to incorporate the idea of connected sets of equal fitness solutions, or *neutral networks*. In this scenario, the population of solutions randomly drifts along a neutral network, occasionally undergoing transitions to higher fitness networks[1], see figure 1. Local optima may not even exist; long periods where the system does not improve in fitness may indicate neutral change rather than simply being stuck in an optimum. Investigation of the population behaviour during these neutral epochs is thus important for artificial life search techniques that evolve problem solutions.

[1] The population may also drop in fitness; work on *error thresholds* looks at the conditions under which this transition to lower fitnesses may occur [16].

J. Kelemen and P. Sosík (Eds.): ECAL 2001, LNAI 2159, pp. 272–281, 2001.

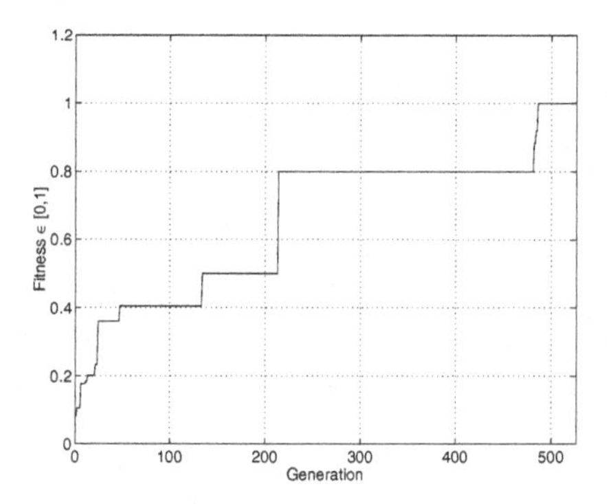

Fig. 1. The fitness of the best individual in a population, over generations. Note the relatively long periods over which fitness does not increase, seen in many optimisation problems. Instead of being stuck in a local optimum, populations may be exploring neutral networks in the space.

In such a space, the emphasis on determining the difficulty of finding good solutions through measures of ruggedness and epistasis may be misplaced. Instead, we will need to look at other properties of the space, and how these properties change during the neutral period. One such property is the *evolvability* of solutions (or populations of solutions), argued in this paper to be equivalent to the capacity of individuals to produce fit offspring. Recent work has shown that this capacity may change during non-adaptive evolutionary phases [25, 24].

In this paper, we investigate the neutral networks and evolvability in a system with an extremely complex genotype-to-fitness mapping. The genotypes code for arbitrarily recurrent neural networks used as robot controllers in a visual shape discrimination. Fitness is evaluated on how close the robot moves to a target shape over an extended period of operation. It is by no means clear that previous work on both neutrality and evolvability [24, 5, 15] will apply to such complex genotype-fitness mapping spaces. The paper extends the analysis presented in previous work [17, 18], here using the idea of the *transmission function* [1] to investigate how evolvability changes during the neutral phase.

The paper is laid out as follows: Section 2 introduces the ideas of evolvability and the transmission function. Section 3 describes the style of neural network used in the work, and the robotics problem on which the evolved network controllers are evaluated. Sections 4 and 5 outline the two experiments carried out, and the results found, and we conclude with discussion.

2 Evolvability and Neutrality

Evolvability is loosely defined as the capacity to evolve, alternatively the ability of an individual or population to generate fit variants [1, 13, 22]. Attempts to rigorously define the concept tend to reflect the background of the researchers involved, although recently there has been more work linking the biological and computer science approaches [3, 20]. Biologists often talk of organisms and structures already pre-adapted to some environment, and their ability to respond to environmental change [12]. Computer scientists tend to talk of evolvability in terms of the ease of finding good solutions in a given search space, closely tied in with work on the properties of search space ruggedness and modality [23, 6, 10, 14, 17].

In this paper, we define evolvability as the ability of individuals to produce fit variants, specifically the ability to both produce fitter variants, and to not produce less fit variants. This definition is intimately tied in with the population transmission function [1, 4]: $T(i \rightarrow j)$, defined as the probability distribution of offspring j obtained through all possible applications of the genetic operators on the parent i (in this work we do not apply recombination, so only a single parent is required). Such a definition encompasses all variation in both the operators and the representation; instead of referring to good and bad genetic operators or good and bad representations, we can talk about the effectiveness of the transmission function. In the remainder of the work, we use the transmission function as short-hand for the distribution of offspring *fitnesses*. Thus the evolvability of an individual or population, i.e. their ability to generate fit variants, is simply a property of the individual or population transmission function.

Researchers have argued that there may be large-scale trends for evolvability to change during evolution [20], and that the capacity can even increase during neutral epochs through the population moving to "flatter" areas of the search space, with fewer deleterious mutations [25, 24]. This effect can occur as the centre of mass of the population moves towards solutions producing a higher number of viable offspring. Evolvability may also change as the population diffuses along neutral networks, thus potentially escaping local optima; adding neutrality may increase evolutionary speed and hence evolvability [15, 5]. This paper investigates these claims in an evolutionary robotics setting with complex genotype-to-fitness mapping. The next section introduces this mapping.

3 An Evolutionary Robotics Search Space

The *GasNet*, introduced by Husbands *et al.* [7, 8], incorporates a mechanism based on the neuron-modulating properties of a diffusing signalling gas into a more standard sigmoid-unit neural network. In previous work the networks have been used in a variety of evolutionary robotics tasks, comparing the speeds of evolution for networks with and without the gas signalling mechanism active [7, 8, 19]. In a variety of robotics tasks, GasNet controllers evolve significantly faster than networks without the gas signalling mechanism. Initial work aimed at identifying the reasons for this faster search has focused on both the underlying search spaces ruggedness and modality [17], and the non-adaptive phases of evolution [18].

3.1 The Task

The evolutionary task is a visual shape discrimination problem; starting from an arbitrary position and orientation in a black-walled arena, the robot must navigate under extremely variable lighting conditions to one shape (a white triangle) while ignoring a second shape (a white square). Both the robot control network (an arbitrarily recurrent neural network incorporating artificial diffusing neuromodulators) and the robot sensor input morphology (the number and position of input pixels used in the visual array) were under evolutionary control. Fitness over a single trial was taken as the fraction of the starting distance moved towards the triangle by the end of the trial period, and the evaluated fitness was returned as the weighted sum of 16 trials of the controller from different initial conditions. For further details of the task, fitness function and genetic algorithm used, see [8, 19].

Success in the task was taken as an evaluated fitness of 1.0 over thirty successive generations of the genetic algorithm. Previous work has shown that controllers incorporating the diffusion mechanism can evolve successful solutions significantly faster controllers without the mechanism enabled [8, 19]. The research presented here is part of an extensive exploration into the reasons for this faster search [17, 18].

3.2 The Solution Representation and Mutation Operator

The neural network robot controllers were encoded as variable length strings of integers, with each integer allowed to lie in the range [0, 99]. Each node in the network was coded for by nineteen parameters, controlling such properties as node connections, sensor input, and node bias. In all experiments, the GA population were initially seeded with networks containing ten neurons. For further details see [8, 19].

Three mutation operators were applied to solutions during evolution. First, each integer in the string had a 3% probability of mutation in a Gaussian distribution around its current value (20% of these mutations completely randomised the integer). Second, there was a 3% chance per *genotype* of adding one neuron to the network, i.e. increasing the genotype length by 19. Third, there was a 3% chance per genotype of deleting one randomly chosen neuron from the network, i.e. decreasing the genotype length by 19.

4 The Experiments

We concentrate on a single evolutionary run, chosen at random from a set of runs used in previous work [8]; figure 2(a) shows the population best and mean fitnesses over generations. In previous analysis of this evolutionary run [18] we have shown that the evolutionary phase lying between generations 100 and 477 is indeed a neutral epoch [21]. First, the variation in multiply evaluated fitnesses for a single genotype can explain the variation in the population best fitness over this period, see figure 2(b). Second, the population is moving significantly through the search space during this phase, and thus not simply stuck in a local optimum.

The first experiment investigates the transmission function, or the probability distribution of offspring fitness (section 2), for the best genotype found at each generation during the evolutionary run. Previous work analysed the

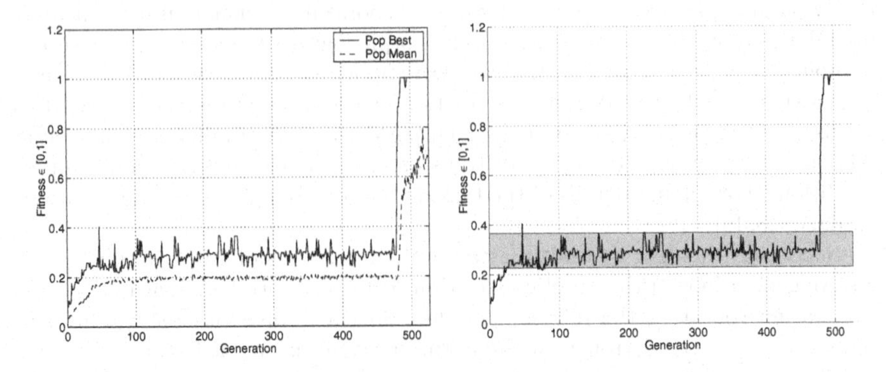

(a) The GA population best and mean fitnesses over generations

(b) Population best fitness, and variation in multiple fitness evaluations of a single genotype (grey band).

Fig. 2. (a) Behaviour of the GA population best and mean fitnesses during the evolutionary run, **(b)** The variation in the population best fitnesses during the neutral epoch (generations 100-477) can be accounted for by the variation in multiple evaluations of a single genotype (shown as the grey band) [18].

probabilities of both the best individual and the population making the transition to a higher fitness [18]. Here we look at the change in the distribution of offspring fitnesses over generations, calculating an approximation to the transmission function through recording the fitnesses of 100,000 applications of the genetic mutation operators to the best individuals of each generation (note that typical genotypes have roughly a few hundred loci, so we are exploring a significant fraction of the neighbouring space through 100,000 offspring). In particular, we are interested in whether the offspring fitness distributions highlight changes in the evolvability of the best individual. The high convergence of the population at each generation implies that changes in the evolvability of the best individual will reflect changes in the population evolvability.

The second experiment empirically tests the predictions made from the evolvability results: does the population evolvability predict the speed of evolution? Five populations from the evolutionary run (populations [100, 200, 300, 400, 477]) were used as initial populations for ten runs each of the evolutionary process (fifty runs in all), to see whether there was a difference in the time taken to reach good solutions. The next section describes the results from the two experiments.

5 Results

5.1 The Transmission Functions

For the best individual of each generation, the transmission functions were approximated through recording fitnesses from 100,000 applications of the mutation operator. Figure 3 shows the fitnesses of the best individual, the best

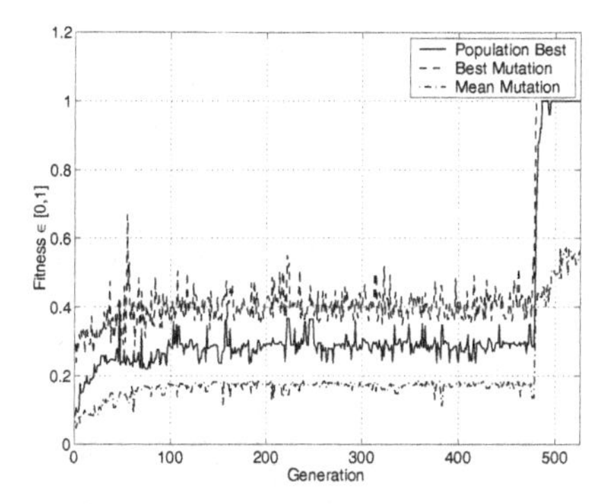

Fig. 3. 100, 000 mutations were applied to each of the best-of-generation individuals (the middle line), to approximate the transmission function distribution of offspring fitnesses. The best and mean mutation fitnesses (the top and bottom lines respectively) show no clear trend during the neutral epoch over generations 100-477.

mutation found and the mean mutation over generations. The graphs closely follow the best individual fitness, rising sharply during the initial hill-climbing period, then staying roughly constant once the neutral epoch is reached around generation 100 (although there is a single high mutation fitness found just after generation 60).

Evolvability is equated with the likelihood of obtaining fit variants, and of not obtaining unfit variants, i.e. the upper and lower tails of the transmission function distribution. Whereas figure 3 tracked only the best and mean fitness of variants over generations, figure 4 shows the percentage of mutations above a fixed fitness (the fitness used here is 0.3 - other fitnesses show similar results), and figure 5 shows the percentage of mutations below two fixed fitnesses (we use 0.1 and 0.2, again other fitnesses show similar results).

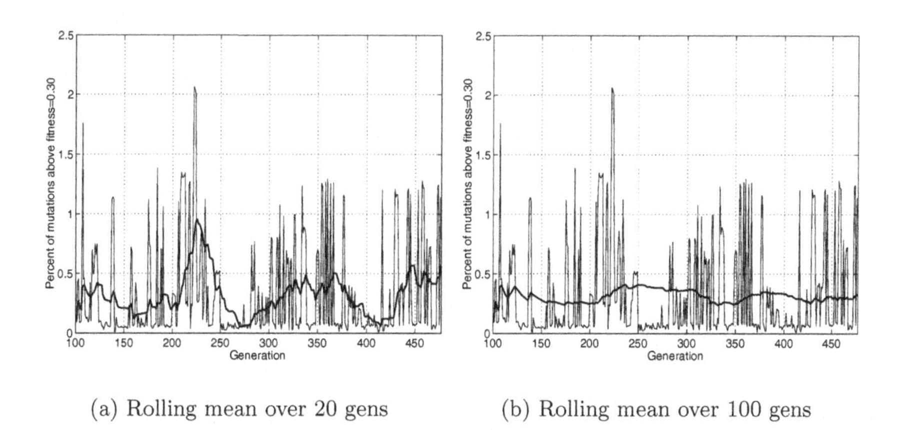

(a) Rolling mean over 20 gens

(b) Rolling mean over 100 gens

Fig. 4. The percentage of mutations above fitness 0.3 from the best-of-generation individuals during the neutral epoch (generations 100-477). **(a)** shows the short-term trends, while **(b)** shows the longer term trend.

We see that the percentage of mutations above a fitness of 0.3 (figure 4(a)) remains extremely low during the neutral epoch, with a maximum of 2%, and the mean remaining generally under 0.5%. Figure 4(a) shows the local trend, the rolling mean over the last 20 generations, which does show signs of movement. However, the longer term trend (figure 4(b) shows the rolling mean over the last 100 generations) shows no such movement.

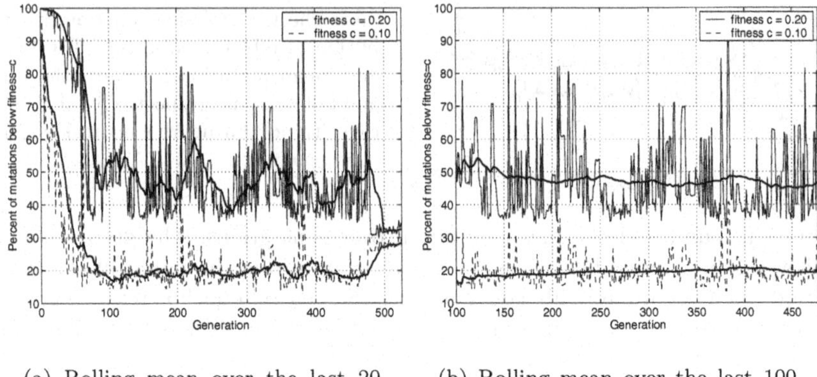

(a) Rolling mean over the last 20 gens for the whole run

(b) Rolling mean over the last 100 gens for the neutral epoch

Fig. 5. The percentage of mutations from the best-of-generation individuals below fitness= [0.2, 0.1]. **(a)** shows the short-term trends over the entire evolutionary run, while **(b)** shows the longer term trend during the neutral epoch (generations 100-477).

The picture for the percentage of mutations below two fixed fitnesses (figure 5) tells a similar story. The number of deleterious mutations falls quickly during the hill-climbing phase of evolution, then stays roughly constant during the neutral epoch; the short-term trends shown in figure 5(a) show the same movement as for the good mutations results, but the long term trend (figure 5(b)) shows no such movement. There is an interesting result from the end of the whole evolutionary run; as fitness dramatically increases at generation 478, the percentage of mutations below fitness 0.1 actually increases. This is discussed further in section 6.

5.2 Repeated Evolution

The results from section 5.1 show that for the search space at hand, there is no long-term trend for change in evolvability during the neutral epoch between generations 100-477. The second experiment tests whether there is any difference in speed of evolution from populations across the epoch. Five populations, from generations [100, 200, 300, 400, 477], were used as the initial populations for the genetic algorithm, and the evolutionary process repeated ten times for each population. Table 1 shows the number of generations required for 100% success on each of the evolutionary runs, while figure 6 shows the median number of generations taken to reach certain fitnesses. Statistical analysis shows no significant differences between the five sets of runs, supporting the hypothesis that there is

	Pop 100	Pop 200	Pop 300	Pop 400	Pop 477
Mean	2,008	2,096	1,901	1,680	3,024
Median	1,522	1,464	932	1,093	1,597
Maximum	4,713	7,696	>10,000	5,707	>10,000
Minimum	353	365	107	353	290

Table 1. Statistics on the number of generations required before the GA reaches 100% success, starting from the 5 populations saved on generations [100, 200, 300, 400, 477]. Note: Runs not reaching success in 10,000 generations were counted as 10,000 for averaging purposes. No significant differences were seen between the populations (Kruskal-Wallis analysis).

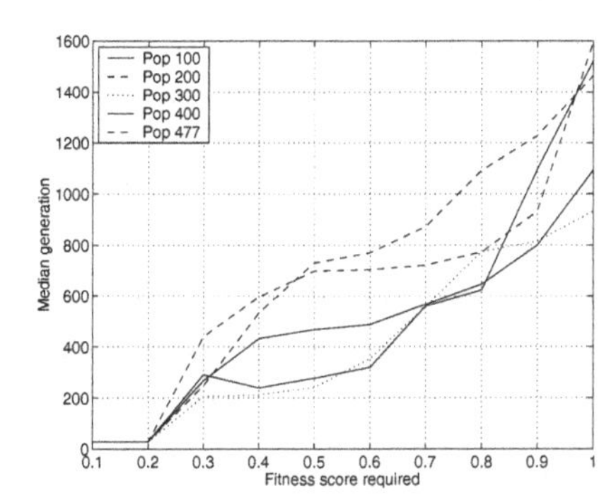

Fig. 6. Median from 10 runs of the number of generations required before the GA reaches a given fitness level, starting from the 5 populations saved on generations [100, 200, 300, 400, 477]. No significant differences were seen between the populations (Kruskal-Wallis analysis).

no change in the evolutionary potential, or evolvability, of the population across the neutral plateau.

6 Discussion

Many complex artificial-life problems such as evolutionary robotics show long periods over which fitness does not markedly increase, classically regarded as points where the population is stuck in local optima. Recent work has shown that the population may be moving along neutral networks in the search space during such periods, and also that the population evolvability may be changing.

In this paper, we have investigated a neutral epoch during an evolutionary process with complex genotype-to-fitness mapping (an evolutionary robotics experiment), and found no evidence for such change in evolvability. The distribution of offspring fitnesses from the best individuals of each generations showed no trend towards a higher probability of producing higher fitness offspring, and no trend towards a higher probability of not producing lower fitness offspring. Further, a second experiment in which populations from across the neutral epoch were used as initial populations for the genetic algorithm, showed no difference between the populations in the number of generations required to produce high fitness. This supports previous work [18], suggesting both that population evolvability does not change across the neutral epoch, and that populations from across the epoch do equally well when straight hill-climbing is used instead of rerunning the GA.

This has implications for artificial-life techniques when evolving solutions to complex problems. In the evolutionary run we have studied here, there is no sense in which the population is doing something "useful" during the neutral epoch - it is not moving to better, i.e. more evolvable, areas of the fitness landscape. Thus the existence of neutral networks in the search space, which allow the evolutionary process to escape from local optima, does not necessarily provide any advantage; in this problem landscape the population does not evolve any faster due to inherent neutrality.

There is no doubt that the presence of neutrality can and does affect population dynamics during evolution, but it may well be that only in a certain class of search spaces does neutrality aid evolution. The use of genetic operators operating on several loci simultaneously, with the ability to alter the genotype length, may render the presence of neutrality less useful than in the fixed length and single-loci mutation genotype-phenotype mappings typically studied in more theoretical work.

Acknowledgements: The authors would like to thank the anonymous ECAL reviewer, Lionel Barnett, Inman Harvey, Andy Philippides, Anil Seth, Lincoln Smith and all the members of the CCNR (http://www.cogs.susx.ac.uk/ccnr/) for constructive discussion. We would also like to thank the Sussex High Performance Computing Initiative (http://www.hpc.sussex.ac.uk/) for computing support. TS is funded by a British Telecom sponsored BBSRC Case award.

References

1. L. Altenberg. The evolution of evolvability in genetic programming. In K.E. Kinnear Jr, editor, *Advances in Genetic Programming*, chapter 3, pages 47–74. MIT Press, 1994.
2. L. Barnett. Ruggedness and neutrality: The NKp family of fitness landscapes. In C. Adami, R.K. Belew, H. Kitano, and C.E. Taylor, editors, *Artificial Life VI: Proceedings of the Sixth International Conference on Artificial Life*. MIT Press / Bradford Books, 1998.
3. C.L. Burch and L. Chao. Evolvability of an RNA virus is determined by its mutational neighbourhood. *Nature*, 406:625–628, August 2000.
4. L.L. Cavalli-Sforza and M.W. Feldman. Evolution of continuous variation: Direct approach through joint distribution of genotypes and phenotypes. *Proceedings of the National Academy of Sciences, USA*, 73:1689–1692, 1976.
5. M. Ebner, P. Langguth, J. Albert, M. Shackleton, and R. Shipman. On neutral networks and evolvability. In *Proceedings of the 2001 Congress on Evolutionary Computation: CEC2001*. IEEE, Korea, 2001. To appear.
6. W. Hordijk. A measure of landscapes. *Evolutionary Computation*, 4(4):335–360, 1996.
7. P. Husbands. Evolving robot behaviours with diffusing gas networks. In P. Husbands and J.-A. Meyer, editors, *Evolutionary Robotics: First European Workshop, EvoRobot98*, pages 71–86. Springer-Verlag, Berlin, April 1998.

8. P. Husbands, T.M.C. Smith, N. Jakobi, and M. O'Shea. Better living through chemistry: Evolving GasNets for robot control. *Connection Science*, 10(3-4):185–210, December 1998.
9. M.A. Huynen, P.F. Stadler, and W. Fontana. Smoothness within ruggedness: The role of neutrality in adaptation. *Proceedings of the National Academy of Sciences, USA*, 93:394–401, 1996.
10. T. Jones and S. Forrest. Fitness distance correlation as a measure of problem difficulty for genetic algorithms. In L.J. Eshelmann, editor, *Proceedings of the Sixth International Conference on Genetic Algorithms (ICGA95)*, pages 184–192. Morgan Kaufmann, CA, 1995.
11. M. Kimura. *The Neutral Theory of Molecular Evolution.* Cambridge University Press, 1983.
12. M. Kirschner and J. Gerhart. Evolvability. *Proceedings of the National Academy of Sciences, USA*, 95:8420–8427, July 1998.
13. P. Marrow. Evolvability: Evolution, computation, biology. In A.S. Wu, editor, *Proceedings of the 1999 Genetic and Evolutionary Computation Conference Workshop Program (GECCO-99 Workshop on Evolvability)*, pages 30–33. 1999.
14. B. Naudts and L. Kallel. A comparison of predictive measures of problem difficulty in evolutionary algorithms. *IEEE Transactions on Evolutionary Computation*, 4(1):1–15, April 2000.
15. M.E.J. Newman and R. Engelhardt. Effects of neutral selection on the evolution of molecular species. *Proceedings of the Royal Society of London, B*, 265:1333, 1998.
16. M. Nowak and P. Schuster. Error thresholds of replication in finite populations - mutation frequencies and the onset of Muller's ratchet. *Journal of Theoretical Biology*, 137:375–395, 1989.
17. T.M.C. Smith, P. Husbands, and M. O'Shea. *Not* measuring evolvability: Initial exploration of an evolutionary robotics search space. In *Proceedings of the 2001 Congress on Evolutionary Computation: CEC2001*. IEEE, Korea, 2001. To appear.
18. T.M.C. Smith, P. Husbands, and M. O'Shea. Neutral networks in an evolutionary robotics search space. In *Proceedings of the 2001 Congress on Evolutionary Computation: CEC2001*. IEEE, Korea, 2001. To appear.
19. T.M.C. Smith and A. Philippides. Nitric oxide signalling in real and artificial neural networks. *BT Technology Journal*, 18(4):140–149, October 2000.
20. P. Turney. Increasing evolvability considered as a large-scale trend in evolution. In A.S. Wu, editor, *Proceedings of the 1999 Genetic and Evolutionary Computation Conference Workshop Program (GECCO-99 Workshop on Evolvability)*, pages 43–46. 1999.
21. E. van Nimwegen, J.P. Crutchfield, and M. Huynen. Neutral evolution of mutational robustness. *Proceedings of the National Academy of Sciences, USA*, 96:9716–9720, 1999.
22. G.P. Wagner and L. Altenberg. Complex adaptations and the evolution of evolvability. *Evolution*, 50(3):967–976, 1996.
23. E. Weinberger. Correlated and uncorrelated fitness landscapes and how to tell the difference. *Biological Cybernetics*, 63:325–336, 1990.
24. C.O. Wilke. Adaptive evolution on neutral networks. *Bulletin of Mathematical Biology*, 2001. Submitted.
25. C.O. Wilke, J.L. Wang, C. Ofria, R.E. Lenski, and C. Adami. Evolution of digital organisms lead to survival of the flattest. *Nature*, 2001. Submitted.

Externally Controllable and Destructible Self-Replicating Loops

André Stauffer and Moshe Sipper*

Logic Systems Laboratory, Swiss Federal Institute of Technology,
CH-1015 Lausanne, Switzerland.

Abstract. Self-replicating loops presented to date are essentially worlds unto themselves, inaccessible to the observer once the replication process is launched. In this paper we present a self-replicating loop which allows for user interaction. Specifically, the user can control the loop's replication and induce its destruction. After presenting the design of this novel loop, we delineate its physical implementation in our electronic wall for bio-inspired applications, the *BioWall.*

1 Introduction: Cellular Automata and Self-Replication

The study of self-replicating machines, initiated by von Neumann over fifty years ago, has produced a plethora of results over the years [1]. Much of this work is motivated by the desire to understand the fundamental information-processing principles and algorithms involved in self-replication, independent of their physical realization [2, 3]. The fabrication of artificial self-replicating machines can have diverse applications, ranging from nanotechnology [4], through space exploration [5], to "intelligent" walls—the latter of which shall be discussed herein.

One of the central models used to study self-replication is that of cellular automata (CA). CAs are dynamical systems in which space and time are discrete. A cellular automaton consists of an array of cells, each of which can be in one of a finite number of possible states, updated synchronously in discrete time steps, according to a local, identical interaction rule. The state of a cell at the next time step is determined by the current states of a surrounding neighborhood of cells. The cellular array (grid) is n-dimensional, where $n = 1, 2, 3$ is used in practice; in this paper we shall concentrate on $n = 2$, i.e., two-dimensional grids.

A major milestone in the history of artificial self-replication is Langton's design of the first self-replicating loop [6]. His 86-cell loop is embedded in a two-dimensional, 8-state, 5-neighbor cellular space; one of the eight states is used for so-called core cells and another state is used to implement a sheath surrounding the replicating structure. Byl [7] proposed a simplified version of Langton's loop, followed by Reggia *et al.* [3] who designed yet simpler loops, the smallest being sheath-less and comprising five cells. More recently, Sayama [8]

* This work was supported in part by the Swiss National Science Foundation under grant 21-54113.98, by the Leenaards Foundation, Lausanne, Switzerland, and by the Villa Reuge, Ste-Croix, Switzerland.

J. Kelemen and P. Sosík (Eds.): ECAL 2001, LNAI 2159, pp. 282–291, 2001.

designed a structurally dissolvable loop, based on Langton's work, which can dissolve its own structure, as well as replicate.

All self-replicating loops presented to date are essentially worlds unto themselves: once the initial loop configuration is embedded within the CA universe (at time step 0), no further user interaction occurs, and the CA chugs along in total oblivion of the observing user.

In this paper we present a self-replicating loop which allows for user interaction. Specifically, the user can control the loop's replication and induce its destruction. These two mechanisms are explained in the next section, followed in Section 3 by their hardware implementation in our electronic wall for bio-inspired applications, the *BioWall.* Finally, we present concluding remarks in Section 4.

2 Loop Design and Operation

Contrary to previous loops, which self-replicate continually, the novel one presented bellow is idle *unless externally activated.*

We consider a two-dimensional, five-neighbor cellular space, with seven basic states per cell (Figure 1; the CA's state-transition rules are given in the Appendix). Our loop, based on the minimal loop of Reggia called UL06W8V [3], comprises four cells. As long as no external input is provided, the loop is inert, continually undergoing a four-time-step cycle (Figure 2).

```
0: empty component
1: building component
2: east-moving signal
3: north-moving signal
4: west-moving signal
5: south-moving signal
6: left-turn signal
```

Fig. 1. The seven basic cellular states used for the idle loop. Replication necessitates additional states discussed ahead.

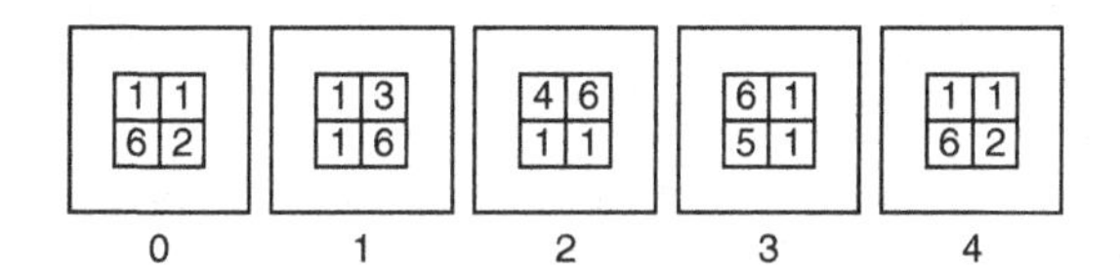

Fig. 2. The four-step idle cycle of the inactive loop.

The user activates the idle loop by providing an external input on one of its four cells (in Section 3 activation occurs by physically touching the BioWall). Activation of the cell leads to the appearance of state 7 instead of state 6 (Figure 3). The activated loop is now ready to self-replicate (if space is available) or self-destruct (if surrounding loops prevent replication).

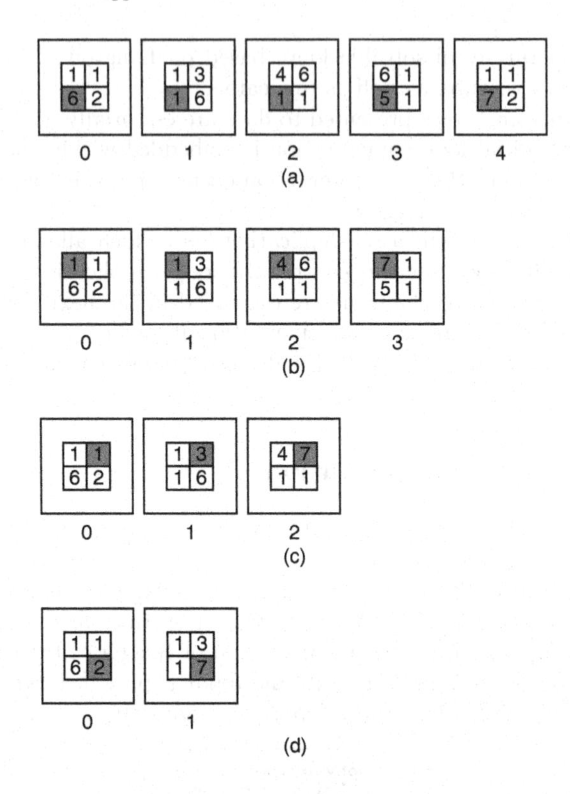

Fig. 3. Activating the idle loop by touching one of its four cells. Depending upon the cell touched, the appearance of state 7 (activated left-turn signal) takes between four (a) to one (d) time steps. In the figures above, the externally activated cell is shadowed.

When surrounded by empty space, the activated loop self-replicates; for example, if the lower-left cell is activated, the loop replicates eastward within eight time steps (Figure 4). Figure 5 presents the four possible eight-step replication processes, induced by activation of the respective cells. Replication requires eight additional cellular states (Figure 6).

When the loop's replication is blocked by another loop, the former initiates an eight-step destruction process (Figure 7); this process can occur in one of four directions, depending on the initially activated cell (Figure 8). While the mother loop can destroy the daughter loop, the contrary is impossible: when the daughter tries to replicate into the mother's space, both loops become idle within four time steps (Figure 9). This deactivation process is shown for all four directions in Figure 10.

The behavior of the loops can be described at a higher level by introducing what we call "macro-notation" (Figure 11). Using these macro symbols we can describe a 12-step process combining activation followed by self-replication (Fig-

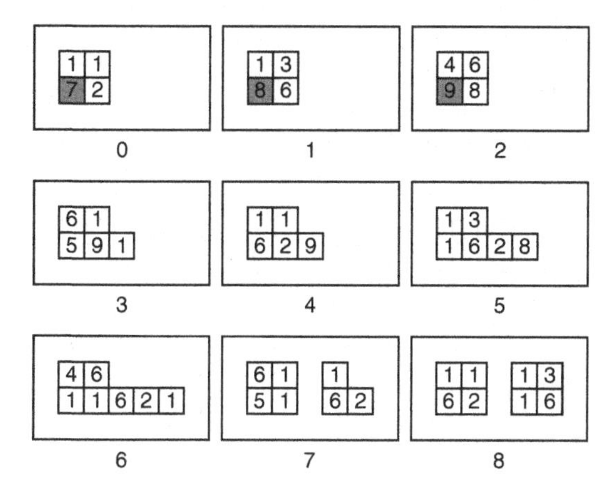

Fig. 4. When the lower-left cell is activated (touched), the loop self-replicates eastward within eight time steps.

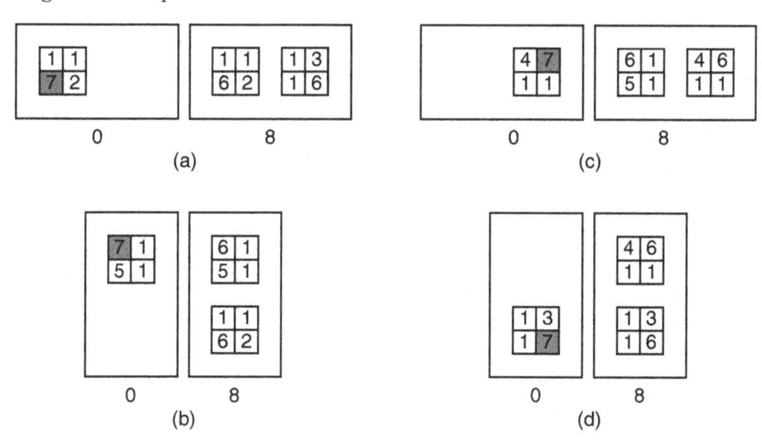

Fig. 5. The four possible activated cells (a-d) and the corresponding replication processes occasioned.

```
 7: activated left-turn signal
 8: first east-branching signal
 9: second east-branching signal
10: first north-branching signal
11: second north-branching signal
12: first west-branching signal
13: second west-branching signal
14: first south-branching signal
15: second south-branching signal
```

Fig. 6. The eight additional cellular states involved in the activation and self-replication process.

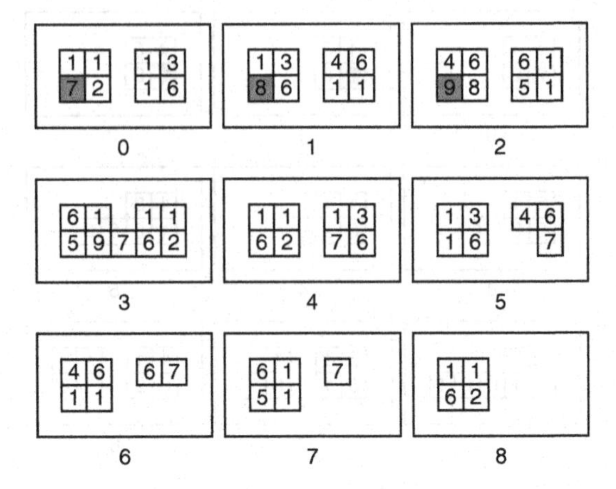

Fig. 7. Activating the lower-left cell results in a destruction process, since the loop is blocked by another loop and thus cannot self-replicate eastward.

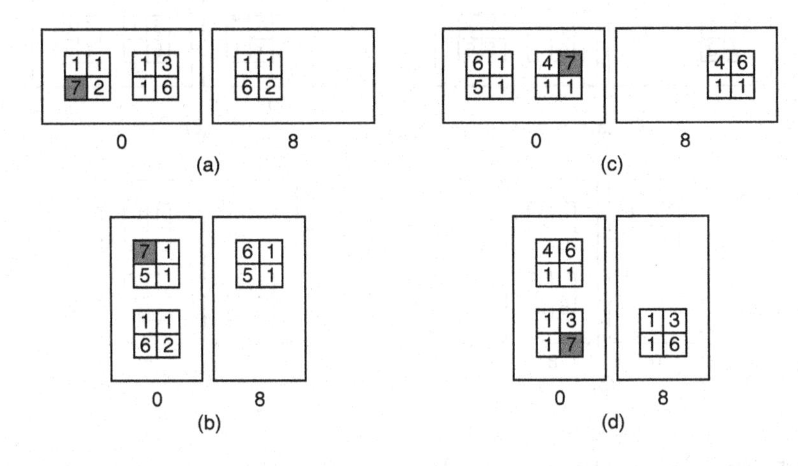

Fig. 8. The four possible activated cells (a-d) and the corresponding destruction processes occasioned.

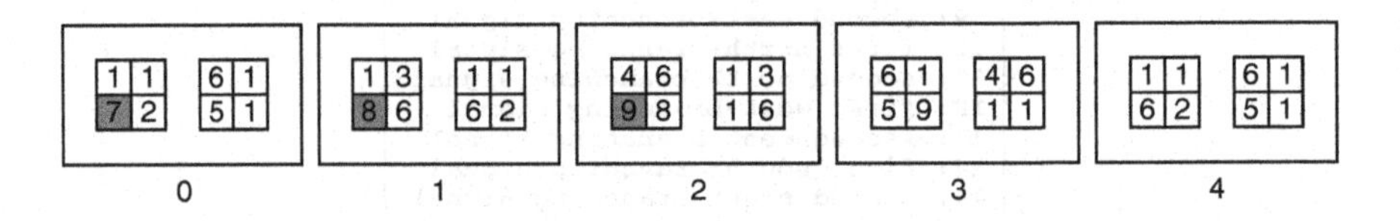

Fig. 9. Daughter deactivation: When the activated daughter loop (left) tries to replicate into the mother loop's (right) space both loops become idle.

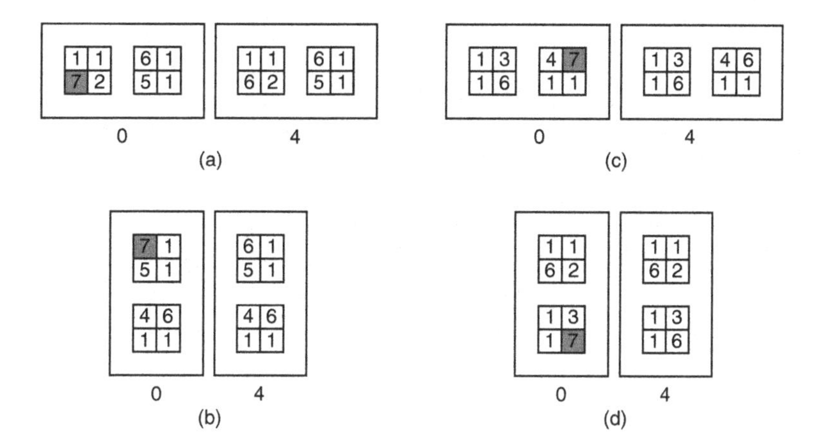

Fig. 10. The four possible activated cells (a-d) and the corresponding deactivation processes occasioned.

ure 12). Three successive 12-step activation-plus-replication processes followed by a single 12-step activation-plus-destruction process are depicted in Figure 13. This kind of macro description allows one to observe and study behavioral patterns at a higher level, similarly to Langton's replicating "colony" [6].

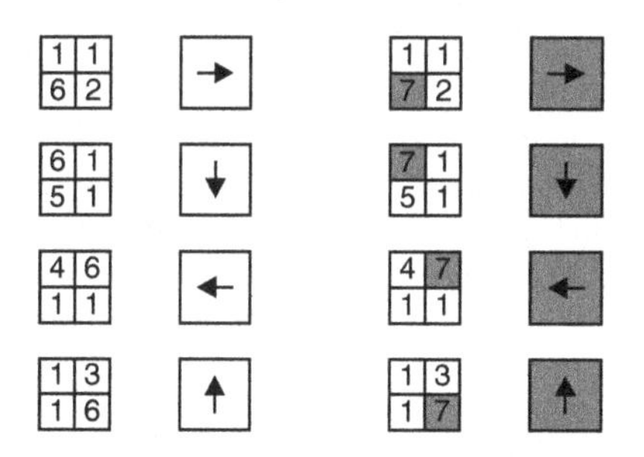

Fig. 11. Macro notation: Idle and activated loops, with their corresponding macro symbols.

3 Physical Implementation: The BioWall

We have physically implemented the five-neighbor CA described above in a two-dimensional electronic wall for bio-inspired applications—the *BioWall*[1]—which

[1] European patent No 01201221.7.

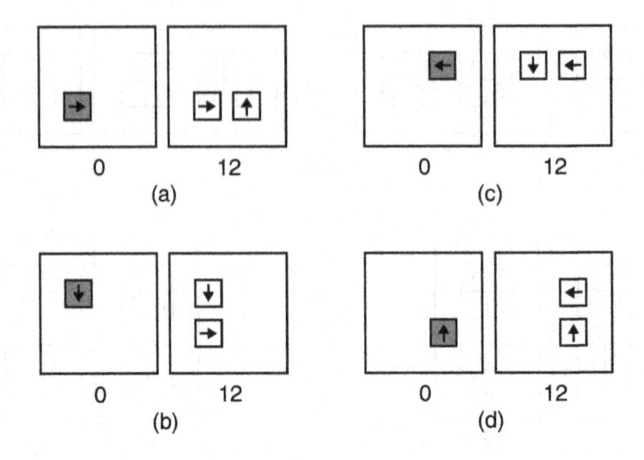

Fig. 12. A macro depiction of the 12-step activation-plus-replication process, shown for the four possible cell activations (a-d).

is an undergoing project in our lab (Figure 14). The BioWall is ultimately intended as a self-repairing display medium, capable of interacting "intelligently" with the user and recovering from faults. In our implementation, each CA cell corresponds to a cell in the wall. The physical cell includes: (1) an input device, (2) a digital circuit, and (3) an output display.

The cell's outer surface consists of a touch-sensitive panel which acts like a digital switch, enabling the user to click on the cell and thereby activate self-replication or destruction (Figure 15).

The cell's internal digital circuit is a field-programmable gate array (FPGA), configured so as to implement: (1) external (touch) input, (2) execution of the CA state-transition rules necessary for loop replication, and (3) control of the output display. This latter is a two color light-emitting diode (LED) display, made up of 128 diodes arranged as an 8×8 dot-matrix, each dot containing a green and a red LED. The display allows the user to view a cell's current state (of the sixteen possible) and whether it is in activated or deactivated mode.

4 Concluding Remarks

We presented an interactive self-replicating loop wherein the user can control the loop's replication and induce its destruction. We then described the loop's physical implementation within our electronic wall for bio-inspired applications, or BioWall. With our current system we can design externally controllable loops of any size—at the price of a larger number of CA state-transition rules.

The ability to interact with a CA universe—a little-studied issue—is of fundamental import where cellular devices are concerned: one must be able to enter input and to view the output if any practical application is envisaged [1]. Our work herein is but a first step in the domain of interactive cellular replicators, an issue which we believe will play an important role in the future of such devices.

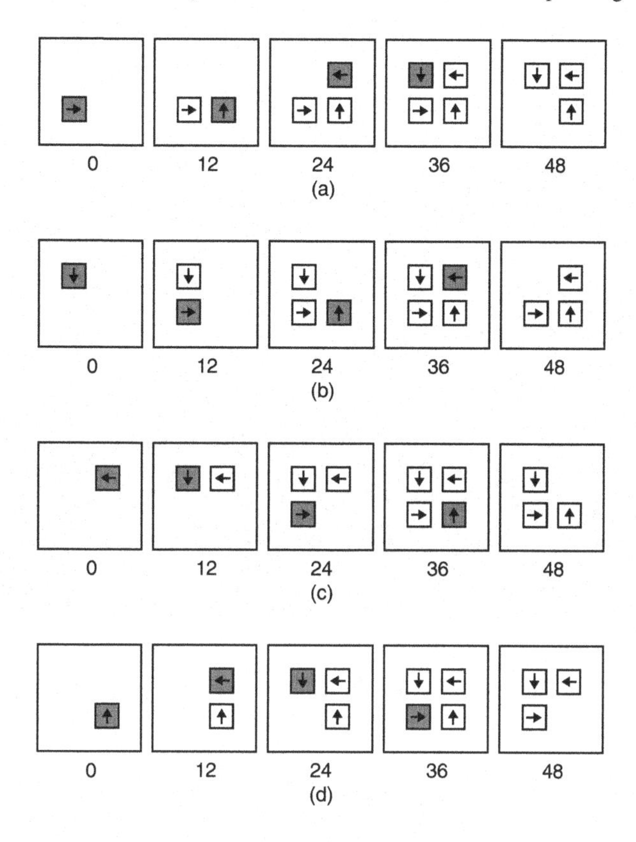

Fig. 13. A macro depiction of three successive activation-plus-replication processes followed by an activation-plus-destruction process, shown for the four possible cell activations of the initial loop (a-d). The loop appearing at time step 36 destroys the original loop after 12 additional steps.

References

1. M. Sipper. Fifty years of research on self-replication: An overview. Artificial Life, 4:237–257, 1998.
2. J. von Neumann. Theory of Self-Reproducing Automata. University of Illinois Press, Illinois, 1966. Edited and completed by A. W. Burks.
3. J. A. Reggia, S. L. Armentrout, H.-H. Chou, and Y. Peng. Simple systems that exhibit self-directed replication. Science, 259:1282–1287, February 1993.
4. K. E. Drexler. Nanosystems: Molecular Machinery, Manufacturing and Computation. John Wiley, New York, 1992.
5. R. A. Freitas and W. P. Gilbreath (Eds.). Advanced automation for space missions: Proceedings of the 1980 NASA/ASEE summer study, Chapter 5: Replicating systems concepts: Self-replicating lunar factory and demonstrations, NASA, Scientific and Technical Information Branch (available from U.S. G.P.O., Publication 2255), Washington D.C., 1980.

Fig. 14. The BioWall used to physically implement our loop (Photograph by A. Badertscher).

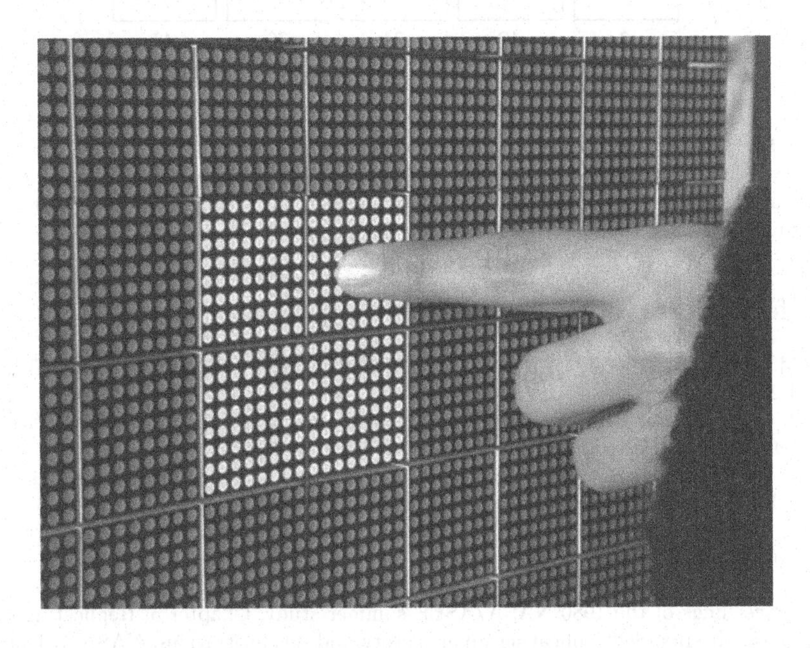

Fig. 15. Touching a cell in order to activate the loop (Photograph by A. Badertscher).

6. C. Langton. Self-reproduction in cellular automata. Physica D, 10:135–144, 1984.
7. J. Byl. Self-reproduction in small cellular automata. Physica D, 34:295–299, 1989.
8. H. Sayama. Introduction of structural dissolution into Langton's self-reproducing loop. In C. Adami, R. K. Belew, H. Kitano, and C. E. Taylor (Eds.), Artificial Life VI: Proceedings of the Sixth International Conference on Artificial Life, pp. 114–122, MIT Press, Boston MA, 1998.

Appendix: Specification of the CA State-Transition Rules

The state-transition rules of the CA implementing the externally controllable and destructible self-replicating loops are given in Figure 16. In this figure, C, N, E, S, and W correspond to the current states of the cell and of its neighbors to the north, east, south, and west, respectively; $C+$ is the state of the cell at the next time step. There are 3 sets of rules in the figure: (1) 108 rules independent of the activation of the cell ($A =$ X), (2) 4 rules when the cell is inactive ($A = 0$), and (3) 4 rules when the cell is activated ($A = 1$).

C,N,E,S,W	->C+	C,N,E,S,W	->C+	C,N,E,S,W	->C+	C,N,E,S,W	->C+
A=X							
0,0,0,2,0	-> 1	1,0,3,8,0	-> 4	5,6,0,1,0	-> 6	7,3,0,0,1	->10
0,0,0,2,1	-> 3	1,4,1,0,0	-> 5	5,6,9,0,0	-> 6	7,0,0,1,4	->12
0,0,3,0,0	-> 1	1,4,10,0,0	-> 5	5,6,11,0,0	-> 6	7,0,1,5,0	->14
0,0,3,1,0	-> 4	1,1,0,0,5	-> 2	5,6,0,14,0	-> 6	8,1,6,0,0	-> 9
0,4,0,0,0	-> 1	1,12,0,0,5	-> 2	5,6,1,15,0	-> 6	8,6,0,0,9	-> 9
0,4,1,0,0	-> 5	2,0,0,0,6	-> 6	5,7,1,0,0	-> 6	8,0,0,0,2	-> 2
0,0,0,0,5	-> 1	2,0,1,0,6	-> 6	6,1,2,0,0	-> 1	9,4,8,0,0	-> 5
0,1,0,0,5	-> 2	2,0,8,0,6	-> 6	6,3,0,0,1	-> 1	9,1,1,0,5	-> 2
0,0,0,0,8	-> 1	2,1,9,0,6	-> 6	6,0,0,1,4	-> 1	9,0,0,0,2	-> 2
0,0,0,0,9	-> 8	2,11,0,0,6	-> 6	6,0,1,5,0	-> 1	10,6,0,0,1	->11
0,0,0,10,0	-> 1	2,13,0,0,6	-> 6	6,3,2,0,1	-> 1	10,0,0,11,6	->11
0,0,0,11,0	->10	2,1,0,0,7	-> 6	6,3,0,1,4	-> 1	10,0,0,3,0	-> 3
0,0,12,0,0	-> 1	3,0,0,6,0	-> 6	6,0,1,5,4	-> 1	11,10,0,0,5	-> 2
0,0,13,0,0	->12	3,1,0,6,0	-> 6	6,1,2,5,0	-> 1	11,1,0,2,1	-> 3
0,14,0,0,0	-> 1	3,10,0,6,0	-> 6	6,12,2,0,0	-> 1	11,0,0,3,0	-> 3
0,15,0,0,0	->14	3,11,0,6,1	-> 6	6,3,0,0,14	-> 1	12,0,0,1,6	->13
1,0,0,0,2	-> 2	3,0,0,6,13	-> 6	6,0,0,8,4	-> 1	12,0,13,6,0	->13
1,0,0,3,0	-> 3	3,0,0,6,15	-> 6	6,0,10,5,0	-> 1	12,0,4,0,0	-> 4
1,0,4,0,0	-> 4	3,0,0,7,1	-> 6	6,3,0,0,8	-> 8	13,0,0,2,12	-> 3
1,5,0,0,0	-> 5	4,0,6,0,0	-> 6	6,0,0,10,4	->10	13,0,3,1,1	-> 4
1,0,0,0,9	-> 9	4,0,6,0,1	-> 6	6,0,12,5,0	->12	13,0,4,0,0	-> 4
1,0,0,11,0	->11	4,0,6,9,0	-> 6	6,14,2,0,0	->14	14,0,1,6,0	->15
1,0,13,0,0	->13	4,0,6,0,12	-> 6	6,0,2,0,1	-> 0	14,15,6,0,0	->15
1,15,0,0,0	->15	4,0,6,1,13	-> 6	6,3,0,1,0	-> 0	14,5,0,0,0	-> 5
1,0,0,2,1	-> 3	4,0,6,15,0	-> 6	6,0,1,0,4	-> 0	15,0,3,14,0	-> 4
1,0,0,2,14	-> 3	4,0,7,1,0	-> 6	6,1,0,5,0	-> 0	15,4,1,1,0	-> 5
1,0,3,1,0	-> 4	5,6,0,0,0	-> 6	7,1,2,0,0	-> 8	15,5,0,0,0	-> 5
A=0							
2,1,0,0,6	-> 6	3,0,0,6,1	-> 6	4,0,6,1,0	-> 6	5,6,1,0,0	-> 6
A=1							
2,1,0,0,6	-> 7	3,0,0,6,1	-> 7	4,0,6,1,0	-> 7	5,6,1,0,0	-> 7

Fig. 16. CA state-transition rules.

The Effect of Neuromodulations on the Adaptability of Evolved Neurocontrollers

Seiji Tokura[1], Akio Ishiguro[2], Hiroki Kawai[2], and Peter Eggenberger[3]

[1] Dept. of Electrical Engineering, Nagoya University,
Nagoya 464-8603, Japan
tokura@cmplx.cse.nagoya-u.ac.jp
[2] Dept. of Computational Science and Engineering, Nagoya University,
Nagoya 464-8603, Japan
{ishiguro, hiroki}@cmplx.cse.nagoya-u.ac.jp
[3] ATR Information Science Division, Kyoto 619-0288, Japan
eggen@isd.atr.co.jp

Abstract. One of the serious drawbacks in Evolutionary Robotics approaches is that evolved agents in simulated environments often show significantly different behavior in real environments due to unforeseen perturbations. This is sometimes referred to as the gap problem. In order to alleviate this problem, we have so far proposed Dynamically–Rearranging Neural Networks(DRNN) by introducing the concept of neuromodulations with a diffusion–reaction mechanism of signaling molecules to so–called neuromodulators. In this study, an analysis of the evolved DRNN and a quantitative comparison with standard neural networks are presented. Through this analysis, we discuss the effect of neuromodulation on the adaptability of the evolved neurocontrollers.

1 Introduction

Recently, the Evolutionary Robotics(ER) approach has been attracting a lot of attention in the field of robotics and artificial life[1]. In contrast to conventional approaches where designers have to construct controllers in a top–down manner, the methods in the ER approach have significant advantages since they can autonomously and efficiently construct controllers by taking *embodiment* (e.g. physical size and shape of robots, sensor/motor properties and disposition, etc.) and the *interaction dynamics* between the robot and its environment into account.

Although the ER approach has the above advantages, there still exist several issues that can not be neglected. One of the most serious issues is known as the *gap* problem: evolved agents in simulated environments often show significantly different behavior in real environments due to unforeseen perturbations. Therefore, it is highly indispensable to develop a method which enables the evolved controllers to adapt not only to specific environments, but also to environmental perturbations.

J. Kelemen and P. Sosík (Eds.): ECAL 2001, LNAI 2159, pp. 292–295, 2001.

In order to alleviate this problem, we have so far proposed *Dynamically–Rearranging Neural Networks*(hereafter, DRNN) by introducing the concept of *neuromodulations*(hereafter, NMs) with a diffusion–reaction mechanism of signaling molecules called *neuromodulators*[2][3]. However, detailed analysis of the evolved DRNN remains uninvestigated.

In this study, we analyze the evolved DRNN and carry out a quantitative comparison of DRNN and standard *monolithic* neural networks(SMNN) in which synaptic weights and neurons' bias are simply the targets to be evolved. We also show how the evolved neuromodulation process works to increase the adaptability by introducing state–transition maps.

2 Basic Concept of DRNN

The basic concept of our proposed DRNN is schematically depicted in Figure 1. As in the figure, unlike conventional neural networks, we assume that each neuron can potentially diffuse its specific(i.e. genetically–determined) type of NMs according to its activity, and each synapse has receptors for the diffused NMs. We also assume that each synapse independently interprets the received NMs, and changes its properties(e.g. synaptic weights). In the figure, the thick and thin lines denote the connections being strengthened and weakened by NMs, respectively. The way of these changes exerted on the synapses is also genetically–determined.

It should be stressed that the *situation evaluation*, *behavior generation* and *regulation mechanisms* can be all embedded within a *monolithic* neural network. Due to this remarkable feature, we expect that whole genetic information can be reduced rather than straightforward approaches.

In summary, contrasting the conventional ER approach that evolves synaptic weights and neuron's bias of neurocontrollers, in this approach we evolve the following mechanisms:

- Diffusion of NMs(when and which types of NMs are diffused from each neuron?)
- Reaction to NMs(how do the receptors on each synapse interpret the received NMs, and modify the synaptic property?)

3 Analysis of the Evolved Agents

3.1 Task: A Peg–Pushing Problem

In this study, we use a *peg–pushing problem* as a practical example. Here the task of the robot is to push the peg toward the light source. The robot used in this experiment is a $Khepera^{TM}$ robot, and is widely used in the ER community. It is equipped with six infra-red sensors which tell the distance to the detected peg in the frontal side, and light direction detectors which view three different angle(left, center and right), and two independently–driven DC motors[3].

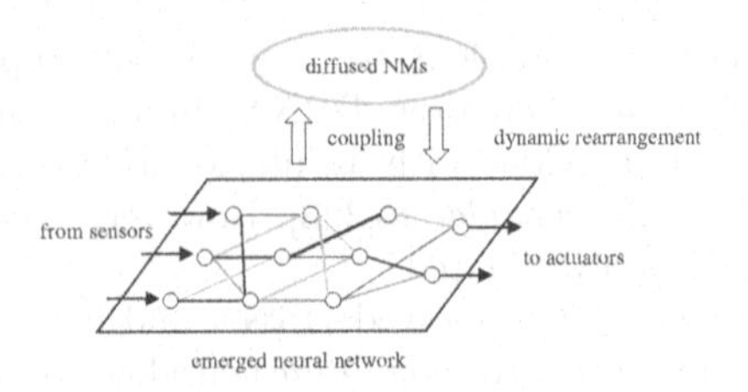

Fig. 1. Basic concept of the DRNN.

3.2 Results

In order to visualize the adaptability of the evolved DRNN and SMNN, here we introduce a technique using a sort of vector map. Due to lack of space, we only show some results. The vertical axis of the vector map(Figure 2[b]) denotes the relative angle to the light source with respect to the heading direction of the robot(see θ_L in the figure [a]). The horizontal axis represents the relative angle to the peg(see θ_P in the figure [a]). Each arrow in the map was drawn in the following way: we first put the robot at a certain state and then recorded the resultant state after one step. For example, the arrow from A to B in figure (b) means that the robot moves to state B when it is put on state A. Note that in this task the robot should reach line C for successful peg–pushing toward the light source. Figure 2(c) represents the resultant vector map in the case of evolving SMNN. It should be noted that the direction of the arrows in some regions are totally inappropriate(i.e. away from line C). Another important thing to be noticed is that there exist some regions containing no arrows(i.e. *empty region*). Due to these circumstances, as in the figure, it is often observed that the robot can not reach line C under the conditions never experienced in the evolutionary process.

Figure 3 shows the obtained vector map by analyzing the evolved DRNN controller. As in the figure, the robot can successfully visit the desired states(i.e. diagonally). Interestingly, as the figure indicates the direction and the length of the vectors are dynamically changed according to the situation in order to eventually reach the diagonal. This is the power of the neuromodulation mechanism. Due to lack of space, we need mention here only that evolving *the way of regulation*(i.e. neuromodulation) instead of simply evolving synaptic weights is much more powerful if one wants to create adaptive neurocontrollers.

4 Conclusions and Further Work

In this study, we analyzed the evolved DRNN and investigated how the neuromodulation process works to increase adaptability against environmental perturbations. We observed that according to the situation the property of the network is dynamically changed in order to generate appropriate behavior. We expect such a *polymorphic* property is essential for adaptive neurocontrollers.

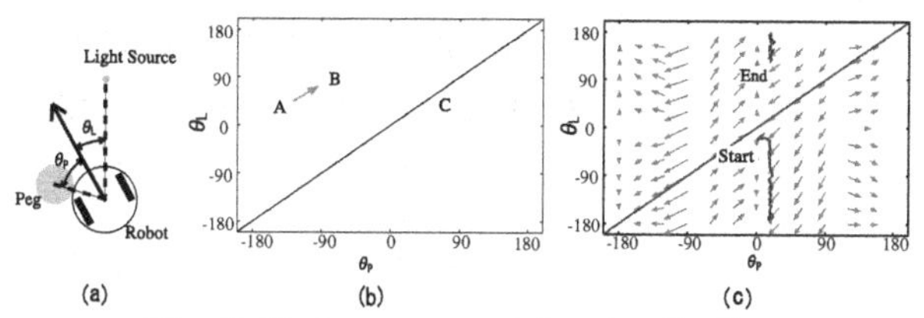

Fig. 2. Vector map used for visualization.

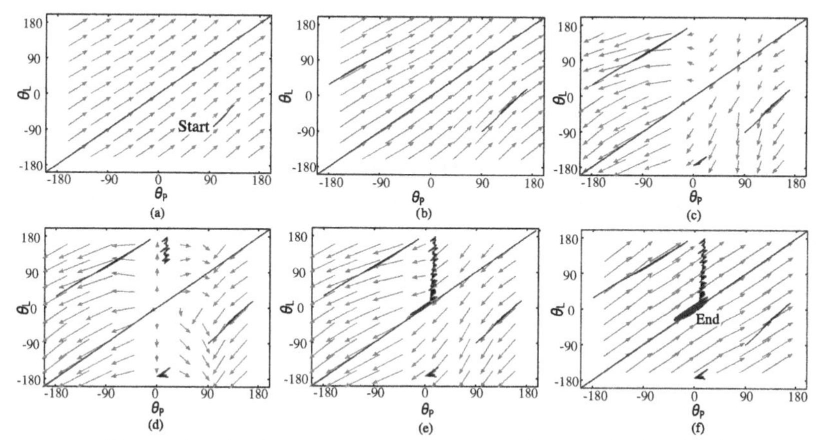

Fig. 3. Obtained vector map in the case of the evolved DRNN.

Currently, we use a simple direct encoding scheme to evolve the DRNN. However, both the network architecture and the way of diffusion and reaction to NMs in the DRNN are closely related not only to the given task but also to the embodiment and the interaction dynamics with the environment. Thus it is preferable to automatically determine these parameters through the evolutionary process. In order to exclude possible presupposition on these parameters we are investigating the concept of the *developmental process*.

References

1. D. Floreano, F. Mondada, "Automatic Creation of an autonomous agent: Genetic evolution of a neural-network driven robot" Proc. of the third International Conference on Simulation of Adaptive Behavior (1994)
2. P. Meyrand, J. Simmers and M. Moulins, "Construction of a pattern-generating circuit with neurons of different networks", *NATURE*, Vol.351, 2 MAY pp.60-63 (1991)
3. P. Eggenberger, A. Ishiguro, S. Tokura, T. Kondo, T. Kawashima and T. Aoki, "Toward Seamless Transfer from Simulated to Real Worlds: A Dynamically-Rearranging Neural Network Approach", *Advances in Robot Learning*(Eds. J. Wyatt and J. Demiris), Lecture Notes in Artificial Intelligence 1812, pp.44-60, Springer (2000)

Testing Collaborative Agents Defined as Stream X-Machines with Distributed Grammars

Tudor Bălănescu[1], Marian Gheorghe[2], Mike Holcombe[2], and Florentin Ipate[1]

[1] Faculty of Sciences, University of Pitesti, Romania
fipate@ifsoft.ro, balanesc@phobos.cs.unibuc.ro
[2] Department of Computer Science, Sheffield University, UK
{m.gheorghe,m.holcombe}@dcs.shef.ac.uk

Abstract. This paper presents a method for testing software collaborative agents. The method is based on the stream X- machine model using as basic processing relations sets of production rules and is built around an existing stream X- machine testing strategy already. The paper investigates different variants of the 'design for test' conditions required by this strategy and their applicability to the case of collaborative agents.

1 Introduction

Although in philosophical literature the notion of an agent as a cognitive subject has been around for a long time, in the last decade this area has also become a major topic of research within artificial intelligence and computer science [18]. As might be expected, within the latter areas, the concept of an agent generally has a more technical meaning, although there is no general consensus on its definition. *Software agents* may be classified according to their mobility or reaction to their environmental stimuli or the attributes they should exhibit. A number of three primary attributes has been identified [22]: *autonomy, learning and cooperation. Collaborative agents* emphasise autonomy - agents can operate on their own - and cooperation - agents have the ability to cooperate with others. Both theoretical aspects pointing out the logic underpinning the collaborative agents as well as the practical issues showing how they are applied to large-scale applications have been investigated [21]. In this context we outline the importance of evaluating collaborative agent systems. The problems related to the verification and validation of these systems to ensure they meet their functional specifications are still outstanding. A specific model of multi agent paradigm called *cooperating distributed grammar systems,* has been devised [8], investigated and developed [20]. In the context of formal software specification, the *X- machine* model introduced by Eilenberg [9] has been reconsidered [12] and proposed as a possible specification language for dynamic systems. Since then a number of further investigations have demonstrated that this idea is of great potential value for both fundamental and practical issues in software engineering [14]. An important aspect approached for *stream X- machines,* a variant of the more general X- machine model, has been that of devising a test method

J. Kelemen and P. Sosík (Eds.): ECAL 2001, LNAI 2159, pp. 296–305, 2001.

that has been proved to find *all faults* of the implementation [15]. Cooperating distributed grammar systems (CDGS) and stream X- machines (SXM) share several common properties as pointed out in [8]:

- the information needed to solve a problem is partitioned into separate and independent knowledge sources (the production rule sets of CDGS and the processing relations/functions of SXM);
- the global database on which various knowledge sources may make changes (the common sentential form of CDGS and the global memory of SXM);
- the control of the opportunistic responses of the knowledge sources to make modifications in the global database (various derivation control strategies of CDGS and the underlying automata of SXM).

It is worth noting that the first and last of these properties emphasise the autonomous behaviour that is expected for the agents, the second issue stresses the cooperation aspects requested for multi-agent systems. The similar properties of both models lead to the definition of the *X- machine model with underlying distributed grammars* (SXMDG) ([4], [10]). A further step towards testing SXMDG has been made by investigating deterministic SXMDG for some specific derivation strategies and regular rules [5]. As testing is crucial for validating an implementation against the specification, this paper will investigate further the testing aspects of SXMDG in the context of building reliable collaborative software agents. The SXMDG model benefits from both components involved and exposes new characteristics of interest: a) the precision and simplicity of the production rules and the derivation strategies of CDGS is used in defining the processing relations of SXM; b) the well established computational power of various CDGS can be transferred to SXMDG ([4], [10]); c) the already proven power of the SXM model in testing the behaviour of an implementation against its specification can be transferred and adapted, as we are going to show in this paper, to SXMDG; d) the property of SXMDG models, inherited from SMSs, of using basic relations/functions that process an input symbol in order to yield a suitable output symbol makes the model suitable for modelling collaborative agents as well as reactive agents [17]; e) various grammar types are used in defining different natural activities that can lead to a promising way of organising system design in a formal manner in the context of SXMDG. The paper presents a method of testing stream X- machines with underlying distributed grammars by adapting the general theory of stream X- machine testing to the characteristics of this new model. In particular, the paper defines a set of 'design for test' conditions which ensure that the behaviour of collaborative agents is under control and observable at any moment. These conditions may sometimes require that extra information is added to the system, but any testing method that can find *all faults* will impose certain restrictions on the behaviour of the system under testing [6]. The paper is structured as follows: section 2 introduces the main concepts concerning cooperating distributed grammar systems, stream X- machines and stream X- machines based on distributed grammars; section 3 introduces the concepts related to stream X- machine testing; the results that refer to the application of the 'design for test' conditions to stream X- machines based on distributed grammars are presented in section 4; conclusions are drawn in the last section.

2 Basic Definitions

Only the basic definitions and notations will be provided in this section. For further details concerning CDGS and SXM the reader may refer to [8] and [14], respectively.

Definition 1. A CDGS *is a system* $\Gamma = (N, A, S, P_1, \ldots P_n)$ *where* N *is the set of* nonterminal symbols, A *is the set of* terminal symbols, S *is the* start symbol, *and* P_i, $1 \leq i \leq n$ *are the* sets of production rules.

Note 1. Let us denote by DM the set of all the derivation modes, i.e.

$$DM = \{t\} \cup \{\leq r, = r, \geq r \mid r \geq 1\}$$

Definition 2. A SXM *is a system* $X = (\Sigma, A, Q, M, \Phi, F, q_0, T, m_0)$ *where* Σ *and* A *are finite sets called the* input set *and* output set *respectively;* Q *is the finite* set of states, M *is, the possibly infinite,* set of memory symbols; $T \subseteq Q$, *is the set of* final states *and* $q_0 \in Q$ *is the* initial state; $m_0 \in M$ *is the* initial memory value; F *is a partial function called the* transition function *defined as* $F : Q \times \Phi \longrightarrow 2^Q$ *and* Φ *is the finite set of* processing relations; *each processing relation is defined as* $\varphi : M \times \Sigma \longrightarrow 2^{A^* \times M}$

Definition 3. A SXMDG *is, for each* $d \in DM$, *a system*

$$X_d = (\Sigma, A, N, Q, M, P, \Phi_d, F, q_0, T, m_0)$$

where $\Sigma, A, Q, M, F, q_0, T, m_0$ *are as in Definition 2 and* N *is a finite set called the set of* nonterminal symbols *such that* $\Sigma \subseteq N \cup A$; $M = (N \cup A)^*$, P *is a set of* n *sets of production rules defined over* $N \cup A$, $P = \{P_1, \ldots P_n\}$ *and* Φ_d *is defined as being* $\Phi_d = \{\varphi_1, \ldots \varphi_n\}$ *with each* φ_i *associated to* $P_i \in P$ *as follows*

$$(x, m') \in \varphi_i(m, \sigma) \text{ if and only if } m\sigma \Longrightarrow^d_{P_i} xm'$$

$\sigma \in \Sigma, x \in A^*, m, m' \in M$ *and* m' *does not begin with a symbol from* A. *As only nonempty output strings will be expected from each* φ_i, *in what follows the set* A^* *will be replaced by* $A^+ = A^* - \{\lambda\}$.

An SXMDG acts as a translator which processes the input strings and works out sequences of symbols according to the rules belonging to the sets P_i.

Note 2. For a language $L \subseteq \Sigma^*$, and an SXMDG X_d we denote by $f_d(L)$ the *language computed* by X_d when L is the input set. For any string $x \in f_d(L)$ there exist: an input string $\sigma_1 \ldots \sigma_p \in L$ where $\sigma_i \in \Sigma$ for all $1 \leq i \leq p$; p output strings $x_1, \ldots x_p$, $x_i \in A^+$ for all $1 \leq i \leq p$ and $x = x_1 \ldots x_p$; p processing relations $\varphi_{j_1}, \ldots \varphi_{j_p}$; p states $q_1, \ldots q_p$ with $q_p \in T$; p memory values $m_1, \ldots m_p$ with $m_p = \lambda$; such that $q_i \in F(q_{i-1}, \varphi_{j_i})$ and $(x_i, m_i) \in \varphi_{j_i}(m_{i-1}, \sigma_i)$ for all $1 \leq i \leq p$.

Example 1. Let us consider the following SXMDG

$$X_{=2} = (\Sigma, A, N, Q, M, P, \Phi_d, F, q_0, T, m_0)$$

where $\Sigma = \{C, D, c\}$, $A = \{a, b, c\}$, $N = \{B, C, B', C', D\}$, $P = \{P_1, P_2, P_3\}$, with $P_1 = \{B \to aB', C \to bC'\}$, $P_2 = \{B' \to aB, B' \to a, C' \to bC, C' \to b\}$, $P_3 = \{D \to aD, D \to a\}$; $\Phi = \{\varphi_1, \varphi_2, \varphi_3\}$, with
$\varphi_1(B, C) = (a, B'bC')$, $\varphi_1(Bb^rCc^s, c) = (a, B'b^{r+1}C'c^{s+1})$, $r, s \geq 1$,
$\varphi_2(B'b^rC'c^{s-1}, c) = \{(a, Bb^{r+1}Cc^s), (ab^{r+1}c^s, \lambda)\}$, $r, s \geq 1$, $\varphi_3(\lambda, D) = (a^2, \lambda)$;
$Q = \{1, 2, 3\}$, $F(1, \varphi_1) = 2$, $F(2, \varphi_2) = 1$, $F(1, \varphi_3) = 3$, $T = \{3\}$, $q_0 = 1$, $m_0 = B$. For the input set $L = \{Cc^{2n-1}D \mid n \geq 1\}$ the language computed by $X_{=2}$ is $f_{=2} = \{a^{2n}b^{2n}c^{2n-1}a^2 \mid n \geq 1\}$.

Indeed, for getting $a^{2n}b^{2n}c^{2n-1}a^2$ we use the input string $Cc^{2n-1}D$, the processing relations $(\varphi_1\varphi_2)^n\varphi_3$ and the states $q_{2k-1} = 2$, $1 \leq k \leq n$; $q_{2k} = 1$, $0 \leq k \leq n$; $q_{2n+1} = 3$; such that for the input strings $\sigma_1 = C; \sigma_i = c$, $2 \leq i \leq 2n$; $\sigma_{2n+1} = D$; the following output symbols $x_i = a$, $1 \leq i \leq 2n-1$; $x_{2n} = ab^{2n}c^{2n-1}$; $x_{2n+1} = a^2$; and memory values $m_0 = B$; $m_{2k-1} = B'b^{2k-1}C'c^{2k-2}$, $1 \leq k \leq n$; $m_{2k} = Bb^{2k}Cc^{2k-1}$, $1 \leq k \leq n-1$; $m_{2n} = m_{2n+1} = \lambda$, are obtained. As it may be seen the SXMDG is powerful enough as it translates a regular language into a non-context-free language by using regular rules.

3 Testing SXM

This section introduces concepts related to the testing method based on SXMs. The idea behind this method is to compare the SXM model of the implementation with the SXM specification of the same system ([15], [14]) and to generate a test set whose successful application will establish that these two machines have identical behaviour. For a SXM $X = (\Sigma, A, Q, M, \Phi, F, q_0, T, m_0)$, the *finite state machine associated* (FSM) with X is defined as $A(X) = (\Phi, Q, F, q_0)$. Note that the set Φ in the definition of $A(X)$ contains in fact labels associated with the partial relations belonging to Φ. For simplicity no distinction is made between the actual relations and the associated labels. The idea of the SXM testing method is to prove that the two stream X- machines (one representing the specification and the other the implementation) have identical input- output behaviour by showing that their associated automata are equivalent. Thus, the generation of the test set used to prove the equivalence of the two X- machines involves two steps: a) the generation of a set of sequences of basic processing relations that tests the equivalence of the two associated automata; b) the translation of the basic processing relations into sequences of inputs. The generation of the set of processing relations (denoted by Z) is based on Chow's W- method [7] that constructs a test set from a FSM specification. In the case of the SXM model, this is the FSM associated with the specification. The generation of Z consists mainly of the generation of two sets. The first set $C \in \Phi^*$, called a *state cover* of $A(X)$, has the property that for any state $q \in Q$ there is $p \in C$ such that p is a path in $A(X)$ from the initial state q_0 to q. The next set $W \in \Phi^*$, called a *characterisation set* of $A(X)$, has the property that if q and q' are two arbitrary distinct states then there is a $p \in W$ such that p distinguishes between q and q'. The string p is said to distinguish between q and q' if p is a path in $A(X)$ that starts in q and p is not a path that starts in q' or vice- versa. A FSM with these properties is called *minimal.* This property does not restrict the generality of our discussion since for any automaton there is a minimal FSM equivalent to it [11]. If the difference between the number of states of the implementation and the number of states of the specification can be estimated and is denoted by k then the set Z we are looking for is: $Z = C(\Phi^{k+1} \cup \Phi^k \ldots \Phi \cup \{\lambda\})W$. Once the set Z has been set up it will be translated into a set of sequences of inputs that can be applied to the two machines. This is the second step of our construction and a *test function* is defined for this purpose. The scope of a test function is to test whether a certain path exists or not in the machines using appropriate input symbols. A test function $test : \Phi^* \longrightarrow \Sigma^*$ is constructed as follows:

- $test(\lambda) = \lambda$
- for any $p_s = \varphi_1 \ldots \varphi_s \in \Phi^*, s \geq 1$, if $p_{s-1} = \varphi_1 \ldots \varphi_{s-1}$ is a path of $A(X)$ that starts in q_0 then $test(p_s)$ is a sequence of inputs that exercise p_s from the initial memory m_0; otherwise $test(p_s) = test(p_{s-1})$.

In other words, for any sequence of basic processing relations p, $test(p)$ exercises the longest prefix of p that is also a path of the machine state plus the function that follows after this prefix, if any. Note that a test function is not uniquely determined, a stream X- machine may have many test functions. Once the set Z and the function $test$ have been determined the test set is $Y = test(Z)$. In order to compute the test function the SXM should be *deterministic* and satisfy some properties called *'design for test'* conditions. A deterministic SXM is one in which for any $\varphi, \varphi' \in \Phi$ and $q \in Q$ if $F(q, \varphi) \neq \emptyset$ and $F(q, \varphi') \neq \emptyset$ then $dom\ \varphi \cap dom\ \varphi' = \emptyset$, where $dom\ \varphi$ denotes the domain of φ. The design for test conditions ensure that the behaviour of the machines will have to be detectable under all conditions. These are *output distinguishability* and *memory completeness* [14] or *weak output distinguishability* and *strong memory completeness* [2]. A SXM is *output-distinguishable* if for any memory value m, input symbol σ and any two processing relations $\varphi, \varphi' \in \Phi$ if there exist an output string $x \in A^+$ and memory values $m_1, m_2 \in M$ with $(x, m_1) \in \varphi(m, \sigma)$ and $(x, m_2) \in \varphi'(m, \sigma)$ then $\varphi = \varphi'$ and $m_1 = m_2$. A SXM is *memory-complete* if for any basic processing relation φ and any memory value m there exists an input σ such that $(m, \sigma) \in dom\ \varphi$. A SXM is *weak output-distinguishable* if for all $\varphi \in \Phi$ there exists an input symbol $\sigma_\varphi \in \Sigma$ such that for all $\varphi' \in \Phi$ it results $\varphi = \varphi'$ and $m_1 = m_1'$ if there exist $m \in M$ and $x \in A^+$ such that $(x, m_1) \in \varphi(m, \sigma_\varphi)$, $(x, m_1') \in \varphi'(m, \sigma_\varphi)$. A SXM is *strong memory-complete* if it is weak output-distinguishable and if for any $\varphi \in \Phi$, $m \in M$ it follows that (m, σ_φ) is in the domain of φ. Furthermore, the (weak) output distinguishability and (strong) memory completeness properties can be relaxed without affecting the validity of the testing method [16].

Definition 4. *Let* $\Sigma_1 \subseteq \Sigma$ *be a subset of the input alphabet of a given SXM* X. *We define by* $V(m_0, \Sigma_1) = \{m_0\} \cup \{m \mid \exists \sigma_1 \ldots \sigma_k \in \Sigma_1^+, \exists \varphi_1 \ldots \varphi_k \in \Phi^+$

and $(x, m) \in (\varphi_1 \ldots \varphi_k)(\sigma_1 \ldots \sigma_k, m_0)$ *for some* $x \in A^+\}$. *Then we define (weak) output distinguishability and (strong) memory completeness with respect to* $V(m_0, \Sigma_1)$.

It is easy to see that a SXM that is (weak) output distinguishable and (strong) memory complete is also (weak) output distinguishable and (strong) memory complete with respect to any $V(m_0, \Sigma_1) \subseteq M$. However, the SXM testing method still remains valid if the usual conditions are replaced by similar conditions with respect to $V(m_0, \Sigma_1)$, where $\Sigma_1 = \Sigma$ in the case of memory-completeness and output-distinguishability and $\Sigma_1 = \Sigma_\Phi = \{\sigma_\varphi \mid \varphi \in \Phi\}$ in the case of strong memory-completeness and weak output-distinguishability [16]. That is, the sufficient conditions in which the testing method is valid are either memory-completeness and ouput-distinguishability with respect to $V(m_0, \Sigma)$ or strong memory-completeness and weak ouput-distinguishability with respect to $V(m_0, \Sigma_\Phi)$. For the example presented the sets C, W are $C = \{\lambda, \varphi_1, \varphi_3\}$, $W = \{\varphi_2, \varphi_3\}$. For $k = 0$, it follows $Z = C(\Phi \cup \{\lambda\})W$, and thus
$Z = \{\lambda, \varphi_1, \varphi_3\}\{\lambda, \varphi_1, \varphi_2, \varphi_3\}\{\varphi_2, \varphi_3\} =$
$\{\varphi_2, \varphi_1\varphi_2, \varphi_2\varphi_2, \varphi_3\varphi_2, \varphi_1\varphi_1\varphi_2, \varphi_1\varphi_2\varphi_2, \varphi_1\varphi_3\varphi_2, \varphi_3\varphi_1\varphi_2, \varphi_3\varphi_2\varphi_2, \varphi_3\varphi_3\varphi_2,$
$\varphi_3, \varphi_1\varphi_3, \varphi_2\varphi_3, \varphi_3\varphi_3, \varphi_1\varphi_1\varphi_3, \varphi_1\varphi_2\varphi_3, \varphi_1\varphi_3\varphi_3, \varphi_3\varphi_1\varphi_3, \varphi_3\varphi_2\varphi_3, \varphi_3\varphi_3\varphi_3\}$
Please note that $X_{=2}$ is deterministic and (weak) output distinguishable but is not (strong) memory-complete. There is a further problem that appears when the application of the test set is evaluated. For any sequence $\sigma_1 \ldots \sigma_n$ in the test set, we have to establish that the outputs produced for any of the inputs $\sigma_1, \ldots, \sigma_n$ to the implementation coincide to those in the specification. This is achieved if either the outputs produced by all prefixes $\sigma_1 \ldots \sigma_i$, $1 \leq i \leq n$, are recorded or the machine meets an additional property, *output-delimited property*, that allows all these intermediary outputs to be deduced from those produced by the overall sequence $\sigma_1 \ldots \sigma_n$. A SXM is *output delimited* if for any two strings of relations $\varphi_1 \ldots \varphi_j$ and $\varphi'_1 \ldots \varphi'_j \in \Phi^*$, any two output strings $x_1 \ldots x_j$ and $y_1 \ldots y_j$ with each $x_i, y_i \in A^+$, from $x_1 \ldots x_j = y_1 \ldots y_j$, $(x_1, m_1) \in \varphi_1(m_0, \sigma_{\varphi_1})$, $(y_1, m'_1) \in \varphi'_1(m_0, \sigma_{\varphi'_1})$, $(x_i, m_i,) \in \varphi_i(m_{i-1}, \sigma_{\varphi_i})$, $(y_i, m'_i,) \in \varphi'_i\ (m'_{i-1}, \sigma_{\varphi'_i})$ for all $2 \leq i \leq, j$ it follows that $x_i = y_i$, for all $1 \leq i \leq j$.

Thus, for an output delimited type an arbitrary path produces the same output as $\varphi_1 \ldots \varphi_j$ on the same input sequence $\sigma_{\varphi_1} \ldots \sigma_{\varphi_j}$ only if it produces the same output as φ_i at each stage $i, 1 \leq i \leq j$. Any type $\Phi = \{\varphi : M \times \Sigma \longleftrightarrow A \times M\}$ containing only relations producing single symbols is output delimited.

4 Testing SXMDG

(Weak) output distinguishability and (strong) memory completeness as well as their more relaxed counterparts are addressed in this section. Output-distinguishability will take place for SXMDG having a LL(1) like property [1]. For a string x, $first_1(x)$ denotes the set $\{a \mid a \in A, \exists r \geq 1, x \Longrightarrow^r ay\}$, $r \in DM$.

Lemma 1. *Any SXMDG X_d, $d \in DM'$ with the property that for any two rules $S \to x_1 \in P_i$, $S \to x_2 \in P_j$, $i \neq j$ it follows $first_1(x_1) \cap first_1(x_2) = \emptyset$, then X_d is* output-distinguishable.

Proof. If two distinct relations φ_i and φ_j, $i \neq j$, are defined for the same memory value and the same input then distinct output values will be generated.

According to lemma 1 the SXMDG $X_{=2}$ defined in example 1 is output-distinguishable. For a given derivation mode $d \in DM$, for each set of productions P_i we denote by $T_i^d = \{B \mid B \in N$ and if $B \Longrightarrow_{P_i}^{d} \alpha$ then $\alpha \in A^+\}$ the set of all nonterminals producing only terminal strings when the derivation mode d is applied. For a nonterminal $B \in N$ and a set of production rules P_i we denote by $L^d(B, P_i) = \{x \mid x \in A^+, B \Longrightarrow_{P_i}^{d} x\}$ the language generated by B applying production rules from P_i and using only once the derivation mode d.

Lemma 2. *Let X_d be an SXMDG and consider $\Sigma_1 = T_1^d \cup \ldots T_n^d$. If $m_0 \in \Sigma$, then $V(m_0, \Sigma_1) = \{m_0, \lambda\}$*

Proof. Direct, from the definition 4.

Lemma 3. *Let us consider an SXMDG X_d having n sets of production rules $P_1, \ldots P_n$ and terminal initial memory value ($m_0 \in A^*$). If there exist n nonterminals $B_1 \in T_1^d, \ldots B_n \in T_n^d$ such that $L^d(B, P_i) \cap L^d(B, P_j) = \emptyset$ for $i \neq j$, then:*

1. *the associated type $\Phi_d = \{\varphi_1, \ldots \varphi_n\}$ is* weak output distinguishable *with respect to the memory $\{m_0, \lambda\}$*
2. *if all P_i contains only regular rules and the derivation mode d is $= r$ then the type Φ is* output delimited.

Proof. 1. It is sufficient to consider $\sigma_{\varphi_i} = B_i$ and to apply Lemma 2.
2. If d is $= r$ then all the output strings have the length k. Such types are output delimited [2].

Example 2. Let us consider
$P_1 = \{S \to aS, S \to aA, A \to bB, B \to c\}$,
$P_2 = \{S \to cS, S \to cB, B \to bA, A \to a, A \to bB, B \to c\}$. For the derivation mode $= 2$ we have: $T_1^{=2} = \{A\}, T_2^{=2} = \{A, B\}$. We can choose $B_1 = A, B_2 = B$. If $m_0 = a$ then according to Lemma 2 we have $V(a, \{A, B\}) = \{a, \lambda\}$. The type $\Phi = \{\varphi_1, \varphi_2\}$ is also output delimited, as stated by Lemma 3.

The memory completeness restriction imposes important hard conditions for the processing relations of SXMDG.

Lemma 4. *If X_d, $d \in DM'$ is a* (strong) memory-complete *SXMDG then for any $P_i \in P$, $B \in N$*

- *there exists a terminal derivation $B \Longrightarrow_{P_i}^{t} x$, $x \in (A \cup N)^*$*

– *there exist* $\sigma \in \Sigma$ *and a derivation* $B\sigma \Longrightarrow^{d}_{P_i} x$, $x \in (A \cup N)^*$, $d \in DM - \{t\}$

Proof. According to (strong) memory-completeness restriction any memory value m and an input σ should have a derivation in any of P_i components starting from $m\sigma$. In particular m may be any nonterminal. In the case $d = t$ each nonterminal should have a terminal derivation and when $d \neq t$ the string $m\sigma$ should start a d derivation in P_i.

Note 3. The processing relation φ_1 defined by the SXMDG $X_{=2}$ presented in example 1, for any input σ and memory $H \in N$ $H \neq B$ and $H \neq C$ does not have (m, σ) in its domain. This result shows that the solution of extending the definition of the processing relations by considering new input symbols is no longer valid in this case [14]. Fortunately, as we have shown in the previous section, a weaker condition is sufficient. This refers only to those memory values that may be computed starting from the initial value m_0 (lemma 3). Furthermore, we will use the variant of design for test conditions that suit best our example, these are strong memory-completeness and weak output distinguishability with respect to some $V(m_0, \Sigma_1)$.

Note 4. These design for test conditions can be met as follows: a) add to the production rule sets a number of extra rules so that the conditions of lemma 3 are satisfied (NB: this may change the resulting language); in the case of example 1 considered the rules $B' \to a \in P_1$, $B \to a \in P_2$ are added; consequently the inputs associated with the processing relations will be the next nonterminals: $\sigma_{\varphi_1} = B$, $\sigma_{\varphi_2} = B'$, $\sigma_{\varphi_3} = D$; b) consider (for testing purposes) $m_0 = \lambda$ and $V(\lambda, \{B, B', D\}) = \{\lambda\}$; then it is easy to see that the SXMDG is strong memor-complete and weak output-distinguishable with respect to $V(\lambda, \{B, B', D\})$.

With these observations in place an algorithm for testing SXMDG in the case $= r$, $r \geq 1$, may be described as follows:

- compute the set Z according to Chow's method as stated in the previous section;
- check whether the conditions described in lemma 3 are fulfilled by the current SXMDG, $X_{=r}$; if these conditions are not satisfied then add the necessary rules according to note 4;
- compute the input test set $test(Z)$.

The set $test(Z) = \{test(y_z) \mid z \in Z\}$, where for $z = \varphi_1 \ldots \varphi_n$, $y_z = \varphi_1 \ldots \varphi_k$, where $k = n$ if z is a path of the SXM and $k = m + 1$, otherwise, where m is such that $\varphi_1 \ldots \varphi_m$ is the longest prefix of z that is also a path of the SXM. Then $test(y_z) = \sigma_{\varphi_1} \ldots \sigma_{\varphi_k}$.

5 Conclusions

In this paper we present a method for testing collaborative software agents modelled by SXMDG systems. The method is a variant of the method originally developed for SXM models. A number of alternative design for test conditions have been considered. The paper shows that the "traditional" design for testing conditions (memory-completeness and output-distinguishability) are not applicable to SXMDS models and that the "usual" recommendation of extending the definition of the processing relations so as to accept new input values is not suitable to this case. On the other hand, the paper considers slightly different design for test conditions (strong memory-completeness and weak output-distinguishability with respect to $V(m_0, \Sigma_1)$) that can be easily imposed on a SXMDS. These conditions seem to be coherent with the behaviour of collaborative agents as their status must be observable and controllable at some discrete moments, defined by the states of the machine. Some problems remain to be further investigated. On the theoretical side, the computational power of the SXMDG having production rules with the property stated in lemma 3 need to be established. From a practical point of view it will be necessary to apply the method developed in this paper to some complex collaborative agent systems in order to prove its usefulness and suitability. A number of approaches dealing with agent models ([13], [17]) outline the parallel behaviour of the agents have been based around communicating stream X- machine concepts [3]. We think that some of the models used in biology or bio-informatics and dealing with communities of agents may be better approached by integrating a membrane like computation [19] with X- machine systems. Some work is in progress and will be reported in a further paper.

References

1. A. Aho, J. Ullman, *The Theory of Parsing, Translation and Compiling,* Vol. I: Parsing, Prentice-Hall, Englewood Cliffs, N.J, 1972.
2. T. Balanescu, Generalized stream X-machines with output delimited type, *Formal Aspects of Computer Science,* accepted, 2001.
3. T. Balanescu, T. Cowling, H. Georgescu, M. Gheorghe, M. Holcombe, C. Vertan, Communicating Stream X- Machines are more than X- Machines, *J. of Universal Computer Sci.,* 5, 9(1999), 492-507.
4. T. Balanescu, H. Georgescu, M. Gheorghe, Stream X- machines with underlying Distributed Grammars, submitted 2001.
5. T. Balanescu, M. Gheorghe, M. Holcombe, A Subclass of Stream X-machines with Underlying Distributed Grammars, *in Proceedings of the International Workshop Grammar Systems 2000,* R. Freund and A. Kelemenova (eds), Bad Ischl, Austria, 2000, 93-111.
6. G. Bernot, M. Gaudel, B. Marre, Software testing based on formal specifications: a theory and a tool, *Software Engineering Journal,* 6(1991), 387-405.
7. T.S. Chow, Testing software design modelled by finite-state machines, *IEEE Transactions on Software Engineering,* 4, 3(1978), 178-187.

8. E. Csuhaj-Varju, J. Dassow, J. Kelemen, Gh. Paun, *Grammar Systems. A Grammatical Approach to Distribution and Cooperation,* Gordon and Breach, London, 1994.
9. S. Eilenberg, *Automata, languages, and machines,* Vol. A, Academic Press, 1974.
10. M. Gheorghe, Generalized Stream X- machines and Cooperating Distributed Grammar Systems, *Formal Aspects of Computer Science,* accepted, 2001.
11. A. Gill, *Introduction to the Theory of Finite-State Machines,* McGraw-Hill, 1962.
12. M. Holcombe, X- machines as a basis for dynamic system specification, *Software Engineering Journal* 3(1988), 69-76.
13. M. Holcombe, Computational models of cells and tissues - machines, agents and fungal infection, presented at UCL, London 15-16 Feb, 2001, submitted to *Briefings in Bioinformatics.*
14. M. Holcombe, F. Ipate, *Correct Systems: Building a Business Process Solution,* Springer Verlag, 1998.
15. F. Ipate, M. Holcombe, An Integration Testing Method That is Proved to Find all Faults, *Intern. J. Computer Math,* Vol 69 (1997), 159-178.
16. F. Ipate, M. Holcombe, Generating test sets from non-deterministic stream X-machines, *Formal Aspects of Computer Science,* accepted, 2001.
17. P. Kefalas, Modelling an Agent Reactive Architecture with X- Machines, *Technical Report,* CITY Liberal Studies, 2000.
18. *Formal Models of Agents,* (J-J. Ch Meyer, P-Y. Schobbens eds), Springer Verlag, LNAI 1760, 1999.
19. Gh. Paun, Computing with membranes, *Journal of Computer and System Sciences,* 61, 1 (2000), 108-143.
20. *Grammatical models of multi-agent systems,* (Gh. Paun, A. Salomaa eds), Topics in Computer Mathematics 8, Gordon and Breach Science, Amsterdam, 1999.
21. A. S. Rao and M. P. Georgeff, BDI Agents: From Theory to Practice, in *Proceedings of the 1st International Conference on Multi-Agent Systems,* San Francisco, USA, 1995, 312-319.
22. M. Wooldridge, N.R. Jennings, Intelligent Agents - Theory and Practice, *Knowledge Engineering Review,* 10, 2(1995), 115-152.

A Three-Dimensional Environment for Self-Reproducing Programs

Marc Ebner

Universität Würzburg, Lehrstuhl für Informatik II,
Am Hubland, 97074 Würzburg, Germany
ebner@informatik.uni-wuerzburg.de
http://www2.informatik.uni-wuerzburg.de/staff/ebner/welcome.html

Abstract. Experimental results with a three-dimensional environment for self-reproducing programs are presented. The environment consists of a cube of virtual CPUs each capable of running a single process. Each process has access to the memory of 7 CPUs, to its own as well as to the memory of 6 neighboring CPUs. Each CPU has a particular orientation which may be changed using special opcodes of the machine language. An additional opcode may be used to move the CPU. We have used a standard machine language with two operands. Constants are coded in a separate section of each command and a special mutation operator is used to ensure strong causality. This type of environment sets itself apart from other types of environments in the use of redundant mappings. Individuals have read as well as write access to neighboring CPUs and reproduce by copying their genetic material. They need to move through space in order to spawn new individuals and avoid overwriting their own offspring. After a short time all CPUs are filled by self-reproducing individuals and competition between individuals sets in which results in an increased rate of speciation.

1 Introduction

Evolution of computer programs, Genetic Programming [17,18,2], is usually a very difficult task because small changes to the program's code are often lethal. Changing a single byte in any large application program is very likely to cause such a severe error that the mutated program no longer performs its intended function. This violates the principle of strong causality [31] which states that small changes to the genotype should have a small effect on the phenotype. In essence, we do not want an entirely random fitness landscape.

Nature's search space is highly redundant [15]. This redundancy is caused in part by a redundancy in the genetic code. In addition, different phenotypes may perform the same function (e.g. different sequences code for the same shape and are able to perform the same enzymatic function because local shape space is only finite). Redundant mappings induce so called neutral networks [14,37]. Neutral networks are genotypes mapping to the same phenotype which are connected via point mutations. It has been argued (see Ebner [9] and Shipman [33])

J. Kelemen and P. Sosík (Eds.): ECAL 2001, LNAI 2159, pp. 306–315, 2001.

that mappings with similar characteristics like nature's search space should also be beneficial for an artificial search technique, i.e. a genetic algorithm [13,11, 20]. Redundant mappings for artificial evolution were investigated by Shipman et al. [34] and Shackleton et al. [32]. We have shown previously that redundant mappings which possess highly intertwined neutral networks increases the evolvability of a population of bit strings [10]. The extent of the neutral networks affects the interconnectivity of the search space and thereby affects evolvability.

In this paper we propose the use of highly redundant mappings for self-reproducing programs. The redundancy in the genotype-phenotype mapping creates the same robustness as it is present in the genetic code. A single exchange of a base-pair in a DNA sequence does not necessarily result in a different sequence of amino acids because the mapping from codon to amino acid is redundant. Therefore, many mutations in our environment are neutral. According to the neutral theory of evolution [16] a large fraction of all mutations in nature are neutral and only a small fraction of the mutations that actually have an effect on the phenotype are beneficial. If the same mechanism is introduced into artificial evolution robustness of programs is increased. A single mutation is no longer lethal for the individual.

We have devised a virtual environment for self-reproducing programs which uses redundant mappings to decode its machine language. Instead of using a highly simplified machine language (such as the Tierran language by Ray [27] which does not use operands), we are using a rather complex machine language were the opcode and two operands are stored in successive memory locations. Constants are also used in our machine language. We integrated them into the instructions and changed the mutation operator instead such that strong causality is preserved. It has been argued that the commands in the Tierra language are like the the amino acids which are used by nature to construct a protein out of the DNA sequence [27]. During a two stage process called transcription and translation, the sequence of base pairs on the DNA are converted into a string of amino acids which fold into a protein. Different protein can fulfill different functions. It is the three-dimensional structure of a protein which is responsible for producing a particular function. In nature, different types of protein may perform the same function if they have the same local structure. Therefore, instead of equating instructions of an assembly language with amino acids we equate instructions with proteins and our processes with cells. Analogies between natural and artificial life are discussed by Davidge [3].

2 The Virtual Environment

Several types of virtual environments for self-replicating programs have already been developed (e.g. Core War [4,5,6,7], Coreworld [24], Luna [23], the Computer Zoo [36], Tierra [26,25,27,28], Network Tierra [29,30], Avida [1] or CoreSys [8]). Spontaneous emergence of self-replication has been investigated by Koza [18] and Pargellis [22,21]. An overview about artificial self-replicating structures is given by Sipper [35]. Our environment consists of a cube of processors. At any

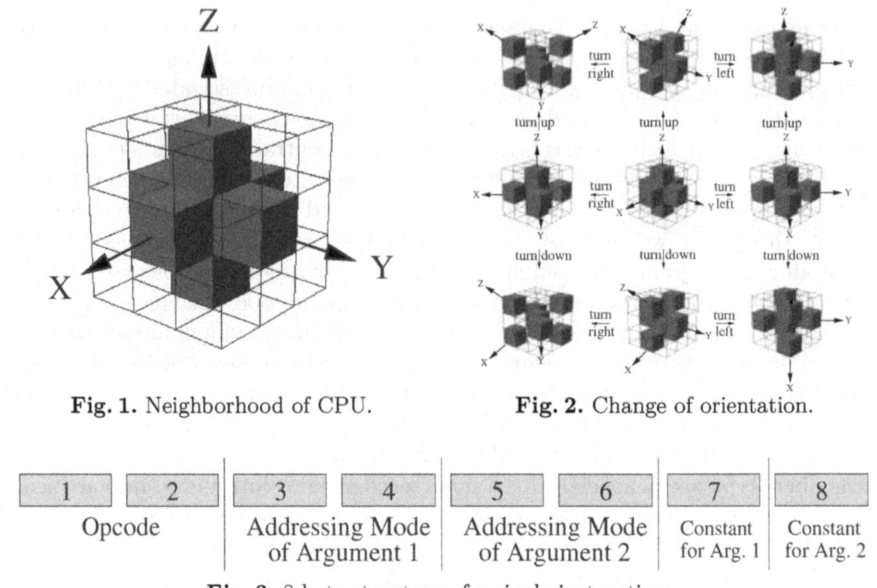

Fig. 1. Neighborhood of CPU.

Fig. 2. Change of orientation.

Fig. 3. 8 byte structure of a single instruction.

point in time at most one process may be running on a CPU. Thus, each CPU can carry at most one individual. The instructions of each individual are stored inside the CPU's memory. Each CPU has 5 registers, a stack with 10 elements, a stack pointer marking the top of the stack, a program counter, a zero flag and a memory of 256 bytes. The registers may be used to hold data as well as to access memory. Each register consist of two parts. The first part specifies a neighboring CPU. The second part specifies an address within the CPU. If an individual tries to access a non-existing address we perform a modulo operation on this register. The neighborhood relation is shown in Figure 1. Each CPU has access to the address space of the CPUs on the left, right, in front of, behind as well as above and below its own CPU. Which CPUs are accessed by a command is determined by the orientation of the CPU. The orientation of each CPU is initially set randomly and may be changed by executing special commands during the run as shown in Figure 2.

We have used a two-address instruction format [19,12]. The list of opcodes is shown in Table 1. The result replaces the value of the second operand. The structure of the machine language is shown in Figure 3. Each command consists of 8 bytes. The first two bytes code for the opcode of the command. Bytes three and four code for the addressing mode of the first operand. Bytes five and six code for the addressing mode of the second operand. Constants are stored in bytes seven and eight. A redundant mapping is used to decode the 2 opcode bytes. Thus, different byte combinations may code for the same opcode. Opcodes are assigned randomly to the different byte combinations. Additional redundant mappings are used to decode the addressing modes of the first and

Table 1. List of opcodes. A "-" in the column for operand 1 or 2 indicates that the opcode does not use this operand. "A" denotes the addressing modes: register indirect, register indirect with auto-increment, register indirect with auto-decrement, and bind. "R" denotes the register addressing mode. "C" denotes the direct addressing mode, and "N" denotes the neighbor addressing mode.

opcode	operand 1	operand 2	name of operation
nop	-	-	no operation
lea	A, N	R	load effective address
move	C, R, A	R, A	move operand 1 to operand 2
clr	-	R, A	set operand to zero
inc	-	R, A	increment operand by one
dec	-	R, A	decrement operand by one
add	C, R, A	R, A	add operand 1 to operand 2
sub	C, R, A	R, A	subtract operand 1 from operand 2
neg	-	R, A	negate operand
not	-	R, A	negate bits
and	C, R, A	R, A	and operand 1 with operand 2
or	C, R, A	R, A	or operand 1 with operand 2
xor	C, R, A	R, A	xor operand 1 with operand 2
shl	-	R, A	shift operand left
shr	-	R, A	shift operand right
cmp	C, R, A	R, A	compare operand 1 with operand 2
beq	-	A	branch on equal (zero flag set)
bne	-	A	branch on not equal (zero flag not set)
jmp	-	A	jump
jsr	-	A	jump to subroutine
rts	-	A	return from subroutine
pop	-	R, A	pop from stack
push	C, R, A	-	push operand to stack
spawn	-	A	spawn new process
kill	-	R, N	kill CPU
turnl	-	-	turn CPU left
turnr	-	-	turn CPU right
turnu	-	-	turn CPU up
turnd	-	-	turn CPU down
fwd	-	-	move CPU forward
label	-	-	marker used for relative addressing

second operand. Because the number of allowed addressing modes can differ for each operator we have used separate mappings to decode the operands for each operator. A standard binary mapping is used to decode the constants.

The following addressing modes are available: direct (e.g. `move #1,r0`), register (e.g. `move r1,r0`), register indirect (e.g. `move (r1),r0`), register indirect with auto-increment (e.g. `move (r1)+,r0`), register indirect with auto-decrement (e.g. `move -(r1),r0`), bind (e.g. `lea <start>,r0`), and neighbor (e.g. `lea [lft],r0`). Depending on the type of command only a subset of all addressing modes can be used. The allowed addressing modes for each command are shown in Table 1. The addressing modes have the usual meaning except for the bind and neighbor addressing modes. Neighbor is used to refer to a neighboring CPU. The bind addressing mode may be used to access memory. This is an addressing mode which is similar to the complementary labels which are used in the Tierran assembly language. In our assembly language, labels may be compiled using the `label` opcode. Each label is assigned a unique constant. This constant is stored in byte 8 of the machine code. If the instruction pointer executes the `label` opcode nothing happens. The bind addressing mode may be used to search for a particular label inside its own or neighboring CPUs. The label with the best match is returned as the resulting address. Special opcodes have been added to change the orientation of a CPU. A CPU may change its orientation by turning to the left (`turnl`), right (`turnr`), up (`turnu`) or down (`turnd`). A process is also able to move through the environment by executing the `fwd` command. This command swaps the specified CPU with the CPU which executed the command. Swapping as opposed to simply copying or copying and erasing the original has been used in order to preserve the contents of the destination CPU.

A command to allocate memory is not included in the instruction set. Other virtual environment for self-reproducing individuals (e.g. Tierra [27] and Avida [1]) include such an instruction. This instruction must be called by each individual which wants to reproduce itself. In our environment there is no need to allocate memory because individuals have read as well as write access to neighboring CPUs. Individuals reproduce themselves by copying their genotype to an adjacent CPU and setting the program counter to the start of the program. Each process is only allowed to execute a limited amount of steps. Thus, it has to reproduce its genetic material before its lifetime is over. In order to avoid periodic effects of the evaluation method we evaluated CPUs at random. On average all CPUs are evaluated for the same number of steps.

3 Reproduction, Variation, and Selection

Individuals reproduce by copying their genetic material to a neighboring CPU and setting the instruction pointer appropriately. In order for evolution to occur, we also need a source of variation and selection. Selection of individuals occurs for three reasons. First, the lifetime of each individual is limited. Each individual may only execute a fixed number of steps before its CPU stops executing instructions. Second, an individual has the possibility of killing another individual and third, space is finite. Several methods to kill another individual exist. An individual may use the kill instruction, reset the instruction pointer using the spawn instruction or simply overwrite the memory of its opponent. Thus, individuals have to protect themselves against other individuals in order to survive. Selection occurs naturally.

In order for evolution to occur, we also need some source of variation. Each memory location is mutated with probability p_{cosmic}. One may equate this type of variation with cosmic rays randomly hitting bytes in memory. If an opcode is to be mutated we simply choose two new bytes for the opcode. The mutations hitting the addressing mode are handled in the same way. Because redundant mappings are used to decode the operands as well as the addressing modes the mutation does not necessarily create a new phenotype. In order to ensure strong causality random mutations of constants are handled differently. If a cosmic mutation hits a constant (bytes 7 or 8 of our machine language) we simply increment or decrement the constant. Thus our environment differs from existing environments in that we use a non-uniform mutation operator. It is highly important that strong causality is preserved. Otherwise we would essentially be performing a random search in the space of self-replicating individuals. Ray has carefully constructed an instruction set with high evolvability by removing all operands [27]. Because we are free to specify the physics of our virtual environment one may as well change the mutation operator which is the option we have chosen for our environment.

Another source of variation results from errors which occur when determining a memory address. The address is either decreased or increased by 1 with probability p_{addr}. Bytes read from memory are decreased or increased by 1 with probability p_{read}. Bytes written to memory are decreased or increased by 1 with probability p_{write}. Thus, an offspring may be different from its parent because of addressing errors, read errors or write errors. In addition, with probability p_{shift} we randomly shift a memory section 8 bytes up or down after a spawn occurred. This causes one command to be removed and one command to be doubled. The command which was doubled is either left untouched, replaced with a NOP or replaced with a random command. The shift operation allows the self-replication programs to grow or shrink in size if necessary.

As the memory fills with self-reproducing individuals, selection sets in. Only those individuals which successfully defend themselves against other individuals will be able to replicate. Short individuals will have an advantage because they are able to reproduce quickly. Another strategy is to develop an elaborate defense mechanism and thereby ensure that the replication will succeed. This might even entail creating signals or tags to make sure that an older individual doesn't destroy its newly created offspring. In principle, it may be possible for the individuals to distinguish their own species from different species by looking at their genotype or inventing special tag bytes. In addition, crossover can develop naturally because at any point in time several individuals may be copying their genetic material into the memory of the same CPU.

4 Experimental Results

Results are described for an environment consisting of a cube of $10 \times 10 \times 10$ CPUs. The probabilities p_{cosmic}, p_{read}, p_{write}, p_{addr}, and p_{shift} have been set to 10^{-6}, 10^{-4}, 10^{-4}, 10^{-3}, and 0.5 respectively. These probabilities were chosen such that the different mutations occur approximately equally often per reproduction. Each CPU has 256 bytes of memory. We filled the memory of all CPUs with NOPs and loaded a manually coded self-reproducing individual into a single CPU. The source code of this individual is shown in Figure 4. The individual first moves one step forward, turns to the left and then up. Next, it determines the beginning and the end of the program and copies its genotype back to front to the CPU on its left side. We performed a number of runs. However, results are only shown for a single run. Thus, the reader may see what actually happens during the run.

Initially, most CPUs are empty and can be used by the individual and its offspring to replicate themselves. Soon, almost all available CPUs are filled. Now individuals have to compete for CPUs. We have analyzed how the original individual speciates. An offspring and its parent are considered to belong to the

```
Addr   Bytes                  Source Code            Comment
---------------------------------------------------------------------
0000 EB6F CB28 EAE2 0000   label  <00>
0008 FB3C 152C 1096 9E2C   fwd                    ; go forward
0010 B609 10CF 621A D479   turnl                  ; turn left
0018 C3C6 CED4 1EB5 5048   turnu                  ; turn up
0020 E5D3 3DC5 EA0D 0083   lea    <00>,R0         ; get beginning
0028 E25E F50F 5332 AA15   lea    <AA>,R1         ; get end
0030 0701 5D91 EF8C 08C9   add    #08,R1          ; add 8 bytes for label
0038 663B E709 F66B 459E   move   R1,R2           ; copy contents of reg.
0040 1B57 A934 F25B B35D   lea    [lft],R2        ; set cpu pointer
0048 87F9 188E 899E CC55   label  <55>
0050 FCC8 4552 E39D 7DC2   move   -(R1),-(R2)     ; copy one byte
0058 B8D6 D016 102A EB77   cmp    R2,R0           ; check if done
0060 51B3 14F3 1302 1455   bne    <55>            ; copy next byte
0068 4322 2624 D15A 3A59   spawn  (R2)            ; spawn new process
0070 DA32 5909 EB54 AB00   jmp    <00>            ; repeat
0078 8090 2B1D EBA6 61AA   label  <AA>
```

Fig. 4. Self-reproducing program. The program copies itself back to front to the CPU on its left side. After all bytes are copied the program spawns a new process. The source code of the program is shown on the right, the assembled bytes on the left. The first column shows the memory addresses.

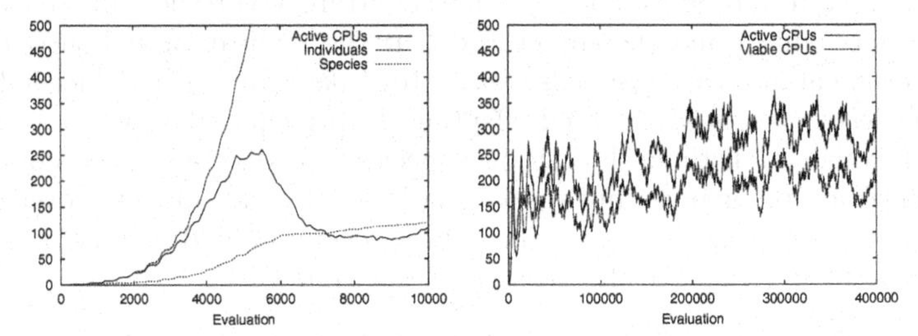

Fig. 5. Relation between speciation and number of active CPUs.

same species if the offspring contains all of the bytes which were executed by its parent. A comparison between speciation and number of active CPUs is shown in Figure 5. The graph on the left shows how the total number of individuals and the total number of species increases over time. The number of active CPUs is also shown. One can clearly see that speciation happens at a faster rate if the number of active CPUs is high. The more active CPUs the higher the speciation rate. Thus, the co-evolutionary environment aids speciation in that due to the presence of other individuals the mutations accumulate in the offspring. All individuals try to replicate themselves, but only if there are a number of active CPUs in the neighborhood, mutations start to accumulate and new species are created. How the number of active CPUs changes during the run can be seen on the graph on the right. The number of individuals which have successfully created an offspring (viable individuals) are also shown in the same graph.

The average size of viable individuals is shown in Figure 6. Also shown in Figure 6 is the size of the three most dominant species, the average gestation

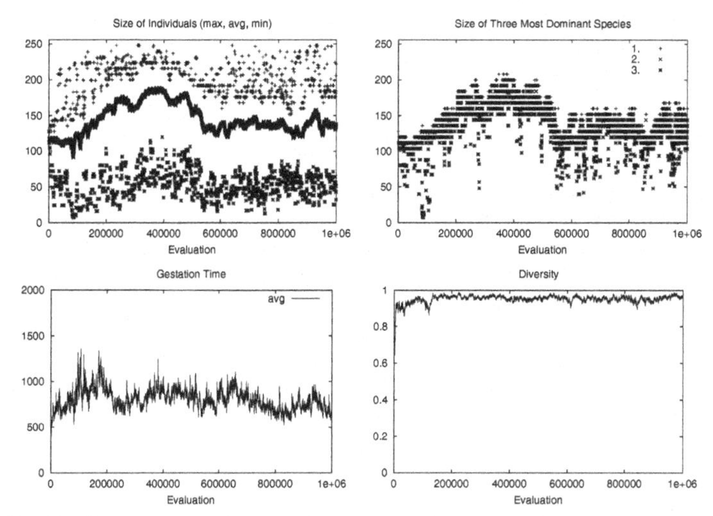

Fig. 6. Size, gestation time and diversity of self-reproducing programs.

time and the diversity of the population. We define gestation time as the number of steps executed over the total number of offspring. Simpson's index is used to calculate diversity. We group individuals into species depending on their size. The graph shows Simpson's index scaled with the factor 256/255 because individuals may reach a maximum size of 256 bytes.

5 Conclusion

Experiments with a three-dimensional environment of self-replicating computer programs are reported. The environment consists of a cube of CPUs. Each CPU may be running at most one individual. Individuals have read as well as write access to the memory of neighboring CPUs. They replicate by copying their instruction to a neighboring CPU and setting the instruction pointer appropriately. Individuals need to move in order to find new CPUs which may be used for replication. Speciation sets in after all CPUs have been filled and individuals need to compete in order to use CPUs for replication. Thus, selection comes about naturally in our environment.

Instead of using a highly simplified instruction set, we have chosen to experiment with a high level machine language. Redundant mappings have been used in order to increase evolvability. We have shown that artificial evolution can also be carried out in a complex assembly language. In addition we have used a mutation operator which takes the structure of the machine language into account to ensure strong causality. This type of environment aids speciation and leads to a highly diverse population of individuals. In future experiments we plan to investigate the impact of different redundant mappings on the evolvability of the population. A more detailed investigation of the evolutionary dynamics would also be interesting, i.e. do we get a similar diverse ecosystem with parasites and the like as in Tierra?

References

1. C. Adami. *Introduction to Artificial Life.* Springer-Verlag, New York, 1998.
2. W. Banzhaf, P. Nordin, R. E. Keller, and F. D. Francone. *Genetic Programming - An Introduction: On The Automatic Evolution of Computer Programs and Its Applications.* Morgan Kaufmann Publishers, San Francisco, CA, 1998.
3. R. Davidge. Looking at life. In F. J. Varela and P. Bourgine, eds., *Toward a practice of autonomous systems: Proc. of the 1st Europ. Conf. on Artificial Life*, pp. 448–454, Cambridge, MA, 1992. The MIT Press.
4. A. K. Dewdney. Computer recreations: In the game called core war hostile programs engage in a battle of bits. *Scientific American*, 250(5):15–19, 1984.
5. A. K. Dewdney. Computer recreations: A core war bestiary of viruses, worms and other threats to computer memories. *Scientific American*, 252(3):14–19, 1985.
6. A. K. Dewdney. Computer recreations: A program called mice nibbles its way to victory at the first core war tournament. *Scientific American*, 256(1):8–11, 1987.
7. A. K. Dewdney. Computer recreations: Of worms, viruses and core war. *Scientific American*, 260(3):90–93, 1989.
8. P. Dittrich, M. Wulf, and W. Banzhaf. A vital two-dimensional assembler automaton. In C. C. Maley and E. Boudreau, eds., *Artificial Life VII Workshop Proceedings*, 2000.
9. M. Ebner. On the search space of genetic programming and its relation to nature's search space. In *Proc. of the 1999 Congress on Evolutionary Computation, Washington, D.C.*, pp. 1357–1361. IEEE Press, 1999.
10. M. Ebner, P. Langguth, J. Albert, M. Shackleton, and R. Shipman. On neutral networks and evolvability. In *Proc. of the 2001 Congress on Evolutionary Computation, Seoul, Korea.* IEEE Press, 2001.
11. D. E. Goldberg. *Genetic Algorithms in Search, Optimization, and Machine Learning.* Addison-Wesley Publishing Company, Reading, MA, 1989.
12. J. L. Hennessy and D. A. Patterson. *Computer Architecture: A Quantitative Approach.* Morgan Kaufmann Publishers, San Mateo, CA, 1990.
13. J. H. Holland. *Adaptation in natural and artifical systems: an introductory analysis with applications to biology, control, and artificial intelligence.* The MIT Press, Cambridge, MA, 1992.
14. M. A. Huynen. Exploring phenotype space through neutral evolution. *Journal of Molecular Evolution*, 43:165–169, 1996.
15. S. A. Kauffman. *The Origins of Order.* Oxford University Press, Oxford, 1993.
16. M. Kimura. *Population Genetics, Molecular Evolution, and the Neutral Theory: Selected Papers.* The University of Chicago Press, Chicago, 1994.
17. J. R. Koza. *Genetic Programming.* The MIT Press, Cambridge, MA, 1992.
18. J. R. Koza. *Genetic Programming II.* The MIT Press, Cambridge, MA, 1994.
19. M. M. Mano. *Computer System Architecture.* Prentice-Hall, Englewood Cliffs, NJ, 3rd ed., 1993.
20. M. Mitchell. *An Introduction to Genetic Algorithms.* The MIT Press, Cambridge, MA, 1996.
21. A. N. Pargellis. The evolution of self-replicating computer organisms. *Physica D*, 98:111–127, 1996.

22. A. N. Pargellis. The spontaneous generation of digital "life". *Physica D*, 91:86–96, 1996.
23. S. Rasmussen, C. Knudsen, and R. Feldberg. Dynamics of programmable matter. In C. G. Langton, C. Taylor, J. D. Farmer, and S. Rasmussen, eds., *Artificial Life II: SFI Studies in the Sciences of Complexity, Vol. X* , pp. 211–254, Redwood City, CA, 1991. Addison-Wesley.
24. S. Rasmussen, C. Knudsen, R. Feldberg, and M. Hindsholm. The coreworld: emergence and evolution of cooperative structures in a computational chemistry. *Physica D*, 42:111–134, 1990.
25. T. S. Ray. An approach to the synthesis of life. In C. G. Langton, C. Taylor, J. D. Farmer, and S. Rasmussen, eds., *Artificial Life II: SFI Studies in the Sciences of Complexity, Vol. X* , pp. 371–408, Redwood City, CA, 1991. Addison-Wesley.
26. T. S. Ray. Is it alive or is it ga? In R. K. Belew and L. B. Booker, eds., *Proc. of the 4th Int. Conf. on Genetic Algorithms, University of California, SD*, pp. 527–534, San Mateo, CA, 1991. Morgan Kaufmann Publishers.
27. T. S. Ray. Synthetic life: Evolution and optimization of digital organisms. In K. R. Billingsley, H. U. Brown III, and E. Derohanes, eds., *Scientific Excellence in Supercomputing*, pp. 489–531, Athens, GA, 1992. The Baldwin Press.
28. T. S. Ray. Evolution and complexity. In G. Cowan, D. Pines, and D. Meltzer, eds., *Complexity: Metaphors, Models, and Reality. SFI Studies in the Sciences of Complexity, Proc. Vol. XIX* , pp. 161–176. Addison-Wesley, 1994.
29. T. S. Ray. Selecting naturally for differentiation. In J. R. Koza, K. Deb, M. Dorigo, D. B. Fogel, M. Garzon, H. Iba, and R. L. Riolo, eds., *Genetic Programming 1997, Proc. of the 2nd Annual Conf., July 13-16, 1997, Stanford University*, pp. 414–419, San Francisco, CA, 1997. Morgan Kaufmann Publishers.
30. T. S. Ray and J. F. Hart. Evolution of differentiation in multithreaded digital organisms. In M. A. Bedau, J. S. McCaskill, N. H. Packard, and S. Rasmussen, eds., *Artificial Life VII, Proc. of the 7th Int. Conf. on Artificial Life*, pp. 132–140, Cambridge, MA, 2000. The MIT Press.
31. I. Rechenberg. *Evolutionsstrategie '94*. frommann-holzboog, Stuttgart, 1994.
32. M. Shackleton, R. Shipman, and M. Ebner. An investigation of redundant genotype-phenotype mappings and their role in evolutionary search. In *Proc. of the 2000 Congress on Evolutionary Computation, La Jolla, CA*, pp. 493–500. IEEE Press, 2000.
33. R. Shipman. Genetic redundancy: Desirable or problematic for evolutionary adaptation? In A. Dobnikar, N. C. Steele, D. W. Pearson, and R. F. Albrecht, eds., *4th Int. Conf. on Artificial Neural Networks and Genetic Algorithms*, pp. 337–344, New York, 1999. Springer-Verlag.
34. R. Shipman, M. Shackleton, M. Ebner, and R. Watson. Neutral search spaces for artificial evolution: A lesson from life. In M. A. Bedau, S. Rasmussen, J. S. McCaskill, and N. H. Packard, eds., *Artificial Life: Proc. of the 7th Int. Conf. on Artificial Life*. MIT Press, 2000.
35. M. Sipper. Fifty years of research on self-replication: An overview. *Artificial Life*, 4:237–257, 1998.
36. J. Skipper. The computer zoo - evolution in a box. In F. J. Varela and P. Bourgine, eds., *Toward a practice of autonomous systems: Proc. of the 1st Europ. Conf. on Artificial Life*, pp. 355–364, Cambridge, MA, 1992. The MIT Press.
37. E. van Nimwegen, J. P. Crutchfield, and M. Huynen. Neutral evolution of mutational robustness. *Proc. Natl. Acad. Sci. USA*, 96:9716–9720, 1999.

Pareto Optimality in Coevolutionary Learning

Sevan G. Ficici and Jordan B. Pollack

DEMO Lab—Department of Computer Science
Brandeis University, Waltham Massachusetts 02454 USA
www.demo.cs.brandeis.edu

Abstract. We develop a novel coevolutionary algorithm based upon the concept of Pareto optimality. The Pareto criterion is core to conventional multi-objective optimization (MOO) algorithms. We can think of agents in a coevolutionary system as performing MOO, as well: An agent interacts with many other agents, each of which can be regarded as an objective for optimization. We adapt the Pareto concept to allow agents to *follow gradient* and *create gradient* for others to follow, such that coevolutionary learning succeeds. We demonstrate our Pareto coevolution methodology with the *majority function*, a density classification task for cellular automata.

1 Introduction

The challenge facing an agent situated in a coevolutionary domain can be described as a form of multi-objective optimization (MOO) where every other agent encountered constitutes an objective to be optimized. That the techniques of conventional MOO, most notably those incorporating the notion of *Pareto optimality*, may be usefully applied to coevolution has recently been suggested [4, 20], but has yet to be deeply explored. This paper begins our investigation into the connection between coevolution and multi-objective optimization, paying particular attention to how the Pareto optimality concept may lead to methods that address issues unique to coevolution. (See Noble and Watson [13] for another example of a "first-generation" Pareto coevolutionary algorithm.)

All coevolutionary systems involve concurrent processes of gradient creation and gradient following. The central themes of coevolution research concern the ways in which these processes interact to dynamically modify the scope of interactions that occur between coevolving entities. Hillis' [6] seminal work on coevolving a population of sorting networks against a population of input vectors gives a compelling illustration of how these processes may interact. Hillis recognizes that the feedback loops between the processes of gradient creation and gradient following can behave like an optimizer with a dynamically adjustable evaluation function; by judiciously selecting the range of test cases applied to the sorting networks, coevolution can make more effective use of finite computational resources to achieve better results.

Nevertheless, the ability to dynamically alter a learning landscape does not by itself guarantee that coevolution will lead to effective learning. Indeed, conventional coevolutionary methods are known frequently to exhibit a number of

J. Kelemen and P. Sosík (Eds.): ECAL 2001, LNAI 2159, pp. 316–325, 2001.

irksome modes of behavior that hinder learning. Intransitive superiority relationships [3] and mediocre-stable states [17] are two examples.

Our experiments involve the coevolution between cellular automata (CA) rules for a density classification task (the *majority function*) and initial condition densities. We choose this problem because an extensive literature on the majority function exists (e.g., [14,21,1,11,12]) and the best rules currently known derive from coevolution (by Juillé and Pollack, with success rates of 85.1%, 86.0% [10], and 86.3% [9]), allowing us to more meaningfully ascertain the significance of our results. We are able to discover rules with a competitive success rate of 84.0%. While not superior to the results of Juillé and Pollack, our rule is significantly better than all results published elsewhere, and we anticipate improved results, as we discuss below. More importantly, we argue that our method offers a potentially more general and comprehensive approach towards coevolution.

This paper is organized as follows: Section 2 details how we employ the Pareto optimality criterion in our coevolutionary algorithm; Section 3 describes the majority function and how we deploy our Pareto coevolutionary algorithm to work on it; Section 4 gives additional details of our experiment setup; Section 5 reviews results; Section 6 discusses plans for new experiments and concludes.

2 Pareto Coevolution Methodology

2.1 Learning from Teaching

We equate the processes of gradient following and gradient creation, discussed above, with the roles of *learning* and *teaching*, respectively. Every agent with which a learner interacts is a teacher, every agent with which a teacher interacts is a learner. An agent in a coevolutionary framework usually, though not always, plays both roles simultaneously; and, when both roles are played by an agent, they are usually not performed with equal success. Our approach to coevolution considers the roles of learning and teaching to be orthogonal.

The only evidence that a learner has to indicate success at following gradient comes from achieving good outcomes (high payoffs) through interaction with agents (teachers). But, the learning gradient is typically of very high dimension in coevolution, since each agent with which a learner can interact represents a dimension to be optimized. Therefore, some principled method of integrating this space is required, and this is the reason we look towards the Pareto optimality concept for help.

We define the role of the teacher to be one that is awarded fitness for providing gradient for learners. But, what evidence can we gather to infer success at this job? We define the *learnability* of a teacher, with respect to a particular learner, to be the likelihood that the learner can be transformed, over some number of variation steps, to become competent (or more competent) at the task posed by the teacher. (Note that learnability becomes sensitive to the state of an entire population if recombinative methods are used in variation.) Thus, the task posed by a teacher is unlikely to be learned if the learner is too remote from regions in variation space where learners are competent at the task—the teacher is "too difficult." Teachers that are completely mastered by a learner also have a low learnability in the sense that the learner has no chance of improving its performance on the task, since it is already perfect.

Rather than try to measure teacher learnability directly, our approach is to discover teachers that can demonstrate *gaps* in learner competence—i.e., show one learner to be more proficient than another at a particular task. We operate on the intuition that teachers that fill competence gaps are likely to be learnable because they expose and explore pre-existing gradients of learner ability in different dimensions of behavior. We rely on the variation process to open new dimensions. This way, we can hope always to have relevant challenges that are of appropriate difficulty. Note that, if an evolving population contains a learner (call it L^*) that is superior to all other learners in the population with respect to every teacher, then no competence gaps exist "above" L^* to fill with teachers that are certain to challenge it. We must wait until some variation occurs to generate a new learner that outperforms L^* in some dimension(s) and creates new gaps in competence. This approach to creating and maintaining gradient for coevolutionary learning is substantially different from those of Rosin [18], Juillé [7], Olsson [15], and Paredis [16].

2.2 Learning: Following Gradient

This section describes how we measure success at following a high-dimensional gradient using the Pareto optimality concept. We name the set of learners $\mathcal{R}$ and the set of teachers $\mathcal{S}$. The payoff matrix $\mathbf{G}$ describes the performance of all learners against all teachers, where $\mathbf{G}_{i,j}$ is the payoff earned by learner i when interacting with teacher j.

- Learner x *Pareto dominates* learner y with respect to the set of teachers $\mathcal{S}$, denoted as $x \overset{\mathcal{S}}{\succ} y$, iff: $\forall w \in \mathcal{S} : \mathbf{G}_{x,w} \geq \mathbf{G}_{y,w} \quad \wedge \quad \exists v \in \mathcal{S} : \mathbf{G}_{x,v} > \mathbf{G}_{y,v}$.
- Learners x and y are *mutually non-dominating*, denoted as $x \overset{\mathcal{S}}{\simeq} y$, iff: $\exists w, v \in \mathcal{S} : \mathbf{G}_{x,w} > \mathbf{G}_{y,w} \quad \wedge \quad \mathbf{G}_{x,v} < \mathbf{G}_{y,v}$.
- The *Pareto front* of a set of learners $\mathcal{R}$, denoted as $F^0(\mathcal{R})$, is the subset of all non-dominated learners in $\mathcal{R}$: $F^0(\mathcal{R}) = \{x \in \mathcal{R} : \nexists w \in \mathcal{R},\ w \overset{\mathcal{S}}{\succ} x\}$.
- The *dominated subset* of $\mathcal{R}$, denoted as $D^0(\mathcal{R})$, is the subset of learners that are dominated by some learner in $\mathcal{R}$: $D^0(\mathcal{R}) = \{x \in \mathcal{R} : \exists w \in \mathcal{R},\ w \overset{\mathcal{S}}{\succ} x\}$.
- Note that a learner belongs exclusively either to the front or the dominated set: $F^0(\mathcal{R}) \cap D^0(\mathcal{R}) = \emptyset$ and $F^0(\mathcal{R}) \cup D^0(\mathcal{R}) = \mathcal{R}$.

Once we identify the Pareto front F^0 and the set of dominated learners D^0, we may compute the next *Pareto layer*, $F^1 = F^0(D^0(\mathcal{R}))$—the set of learners that are non-dominated once we exclude the Pareto front. We may continue this process until every learner is understood to belong to a particular Pareto layer. Pareto layers indicate both generality and uniqueness in learner competence: Every learner in F^n is less broad in competence than some learner in F^{n-1}, and every learner in F^n can do something better than some other learner in F^n.

Intra-Layer Ranking. High-dimensional spaces of competing objectives are known to cause problems for Pareto ranking in ordinary (non-coevolutionary) EAs [5]. As we note above, a large number of teachers gives a high-dimensional gradient. This creates potentially many ways in which a learner may excel and earn a place in a particular Pareto layer. Indeed, we find in our experiments that the number of learners in F^0 tends to increase over evolutionary time and may ultimately include as much as 75% of the entire population.

We therefore require some form of *intra-layer ranking* to differentiate learners within a particular (possibly crowded) layer. We consider two approaches, both stemming from diversity-maintenance techniques, and find them to behave similarly (the experiments reported in this paper use the second approach). Our first approach is similar to Juillé and Pollack's [8] *competitive fitness paradigm.* In comparing two learners from the same layer, we give each learner a point for each dimension (teacher) in which it out-scores the other learner. (Alternatively, if learners x and y out-score each other in n_x and n_y dimensions, respectively, then the better of the two agents gets $|n_x - n_y|$ points and the other gets 0 points.) We accumulate points over all pair-wise comparisons of learners *that belong to the same layer*, and then rank learners according to the sums. Our second approach applies Rosin's [18] *competitive fitness sharing* method within each layer. The learners within a layer are then ranked with respect to each other according to the results of the fitness sharing. Regardless of which intra-layer ranking approach is used, *inter*-layer ranking is achieved by giving the highest-ranked learner(s) of Pareto layer F^n a global rank just below the lowest-ranked learner(s) of Pareto layer F^{n-1}.

2.3 Teaching: Creating Gradient

This section describes how we measure success at creating gradient for learning. We begin with the m by n payoff matrix $\mathbf{G}$, where m is the number of learners and n is the number of teachers. Matrix entry $\mathbf{G}_{i,j}$ is the payoff received by learner i when it interacts with teacher j. If learner x performs better than learner y with respect to teacher j (i.e., $\mathbf{G}_{x,j} > \mathbf{G}_{y,j}$), denoted $x \overset{j}{>} y$, then we say that teacher j *distinguishes* the learner pair (x, y) in favor of x. By causing a pair of learners to receive different payoffs, a teacher exposes a gap in the proficiencies of the two learners. We are interested to identify all such proficiency gaps in the population of learners, as made apparent by the population of teachers.

To identify all learner pairs that are distinguished by each teacher, we construct a new n by $m^2 - m$ matrix $\mathbf{M}$. Each column of this matrix corresponds to a particular pair-wise comparison of learners across all n teachers. We exclude self-comparisons, since there cannot exist any proficiency gaps between a learner and itself, and we treat the learner pairs (A, B) and (B, A) as distinct—(A, B) is reserved for teachers that distinguish A and B in favor of A, while (B, A) is for teachers that distinguish in favor of B. The matrix entry $\mathbf{M}_{j,k}$ equals one if teacher j distinguishes the learners in pair $k = (x, y)$ in favor of x. Clearly, if entry $\mathbf{M}_{j,k}$ is non-zero, then entry $\mathbf{M}_{j,k'}$ must be zero, where $k = (x, y)$ and $k' = (y, x)$.

$$\mathbf{G} = \begin{array}{c|ccc} & \alpha & \beta & \gamma \\ \hline \mathrm{A} & 1 & 1 & 3 \\ \mathrm{B} & 2 & 3 & 2 \\ \mathrm{C} & 1 & 2 & 1 \end{array} \qquad \mathbf{M} = \begin{array}{c|cccccc} & (\mathrm{A},\mathrm{B}) & (\mathrm{A},\mathrm{C}) & (\mathrm{B},\mathrm{A}) & (\mathrm{B},\mathrm{C}) & (\mathrm{C},\mathrm{A}) & (\mathrm{C},\mathrm{B}) \\ \hline \alpha & 0 & 0 & 1 & 1 & 0 & 0 \\ \beta & 0 & 0 & 1 & 1 & 1 & 0 \\ \gamma & 1 & 1 & 0 & 1 & 0 & 0 \end{array} \tag{1}$$

Once we obtain matrix **M**, we have identified all the ways in which each teacher demonstrates utility as a touchstone of learner competence. But, how shall we use this information to create a selective pressure for teachers? To begin with, a teacher that fails to reveal any variation in learner ability is dubious.

In matrix **M** (Equation 1), we see that teacher β not only distinguishes all the learner pairs that teacher α does, namely pairs (B, A) and (B, C), but also distinguishes another pair, (C, A). If we were to apply Pareto ranking to matrix **M**, then we would conclude that teacher β is superior to α because its ability to reveal variations in learner competence is in some sense more general. Let us take this example to an extreme and imagine teacher α to distinguish only a single pair (a, b) while β distinguishes (a, b) and very many more. What these two dramatically different teacher profiles tell us is that the kind of challenge offered by teacher α must be very unlike that offered by β. Therefore, even though β supposedly dominates α, teacher α reveals a dimension of variation in the learner pair (a, b) that β does not. For this reason, Pareto ranking of matrix **M** is inappropriate.

We may instead give all teachers that distinguish at least one learner pair a score of one, and then divide each teacher's score by the number teachers with an identical profile (to better maintain diversity). But, we feel this method to be too coarse-grained. Our compromise approach is to perform fitness sharing much like Rosin [18].

Equation 2 shows that the score received by teacher j is the sum, across all learner pairs distinguished by it, of the *value* of a learner pair divided by the pair's *discount factor*. The discount factor of a pair is the total number of teachers that distinguish it. For the value of a pair, we experiment with two possibilities. Our first approach (Equation 3, left) simply gives every pair an equal value ($v_k = 1$). Our second approach (Equation 3, right) recognizes certain pairs as more significant than others. For example, we may argue that a teacher should get more reward for distinguishing between two good learners than for distinguishing between two poor ones. Further, a teacher should be rewarded more for showing a generally good learner to do something worse than a generally poor learner than the other way around. Therefore, our second approach assigns the value of a learner pair $k = (x, y)$ to be the fitness of the loser y.

$$s_j = \sum_k \mathbf{M}_{j,k} \frac{v_k}{d_k} \qquad d_k = \sum_i \mathbf{M}_{i,k} \tag{2}$$

$$v_k = 1 \quad \textbf{OR} \quad v_k = \mathrm{fitness}(y), \text{where } k = (x, y) \tag{3}$$

3 Majority Function

3.1 Description

Density classification tasks are a popular area of study in cellular automata research [14,21,9,10,1,11,12]. The objective is to construct a rule that will cause a one-dimensional, binary CA to converge, within some pre-determined number of time-steps, to a state of all ones if the percentage (or *density*) of ones in the initial condition (IC) is greater than or equal to some pre-established value, $\rho \in [0, 1]$. Otherwise, the rule should cause the CA to converge to a state of all zeros. The majority function uses a value of $\rho = 0.5$.

Though Land and Belew [11] prove that no rule correctly classifies all initial conditions, the highest possible success rate remains unknown. Currently, the best rules (of radius three, operating on a CA lattice of 149 bits) achieve success rates of 85.1%, 86.0%, and 86.3% over a uniform sampling of initial conditions, and were discovered by Juillé and Pollack [10,9] through coevolution. These rules represent a significant improvement over all earlier rules, for example ABK (82.4%), DAS (82.3%), and GKL (81.5%), as discussed in [10,14].

3.2 Coevolution of Rules and Densities

Following Juillé and Pollack [10,9], we use two-population coevolution to find CA rules of radius three that operate on a one-dimensional lattice of 149 bits. A radius of n means that lattice positions $i - n$ through $i + n$ (with wrap-around) are used by the CA rule to determine the next state of lattice position i. A radius of three gives a "window" of seven bits, meaning the rule must contain $2^7 = 128$ bits, one for each possible window state. This gives us a search space of 2^{128} possible rules. A lattice of 149 bits has 2^{149} possible states, and therefore initial conditions. While one population evolves rules (bit-strings of length 128), the other population evolves initial condition *densities* (floating-point numbers)—not actual initial conditions.

Our experiments depart from those of Juillé and Pollack by using our Pareto coevolution methodology, outlined above. As with many two-population coevolutionary domains, ours has an intrinsic asymmetry in the difficulties faced by the two populations: the discovery of good rules is much harder than the discovery of challenging IC densities. Though the space of possible initial conditions is much larger than the space of possible rules, there exist only 150 distinct IC densities (from all-zeros to all-ones). Further, densities generally become more difficult as they approach 0.5 [12,11], so the space in some sense approximates a uni-modal landscape. For these reasons, rules are assigned only the role of learners and IC densities are assigned only the role of teachers. We can imagine other coevolutionary domains where coevolving entities would be called upon to fulfill both roles, such as Tic-Tac-Toe.

3.3 Derivation of Payoff Matrix: Test Case Sampling

All ranking methods potentially impose severe non-linearities by expanding small differences in performance and compressing large ones. Pareto ranking is no exception and may even be considered more extreme in some ways. Because of this non-linear behavior, the payoff matrix **G**, which forms the basis of our approach, should be as accurate as possible. Thus, our approach is at least as sensitive to non-deterministic domains as conventional coevolutionary methods.

Our CA domain, however, is deterministic. Because we evolve rules against densities rather than actual initial conditions, rule performance against a particular density is expressed as the percentage of correctly classified ICs of that density class. But, we generally cannot afford to sample a particular density class exhaustively. This impedes our ability to compare rules of similar ability. Further, as rules improve in performance, we generally rely on densities closer to 0.5 to distinguish them. Yet, the distribution of initial conditions over densities is binomial, which makes our sampling of ICs significantly more sparse as densities approach 0.5 (and our estimation of rule performance less accurate).

We desire a method to extract meaningful information about rule performance with relatively few samples. Fortunately, for our Pareto coevolution methodology to work, we do not need to know precisely how well a particular rule (learner) performs against a particular IC density (teacher); we only need relative ranking information. Our solution is to once again turn to the Pareto optimality criterion. Note that our use of Pareto ranking to derive the payoff matrix (described here) is not to be confused with our use of Pareto ranking to rate learners once payoffs are known (described in Section 2.2).

We compute each column j of the payoff matrix **G** in the following manner. We generate some number q of initial conditions ($q = 40$ in our experiments) that are representative of density (teacher) j; all m rules (learners) are tested on this set of ICs. The test results are placed in an m by q matrix **H**, where $\mathbf{H}_{u,v} = 1$ if rule u correctly classifies IC v and $\mathbf{H}_{u,v} = 0$ otherwise. We then perform Pareto ranking of rule performance based on **H**. Rules on the Pareto front F^0 are given the highest payoff, those in layer F^1 the next highest, and so on. These payoffs form column j of the payoff matrix **G**.

All rules cover some subset of ICs that belong to a particular density class. These subsets may overlap in any number of ways. We require that a rule dominate another in terms of *measured coverage* in order to receive a higher payoff; this is more stringent than mere comparison of *measured success rates*. For example, two rules that each cover 50% of some density class may overlap entirely on the one extreme, or not at all on the other extreme. Regardless of the amount of actual overlap, the chance that our measured coverage will indicate one rule to dominate the other is extremely small, even though the measured success rates (number of ICs solved by the two rules) will very likely be unequal. But, as the actual performance rates of two rules grow apart, the more likely it becomes for the better of the two to dominate the other in measured coverage (especially if actual coverage of the stronger rule overlaps heavily with that of the other).

Our method of test-case sampling removes the need for an explicit similarity metric—similarity is guaranteed by the process of generating multiple ICs from a density value. Juillé and Pollack [10,9] require a similarity metric to cluster IC densities so that rule performance can be gauged with respect to density.

4 Experimental Setup

A rule is given 320 time-steps to converge the CA to the correct state. The sizes of our populations of rules and IC densities are $N_R = 150$ and $N_{IC} = 100$, respectively. All rules are tested against all densities in every generation. Each density is sampled 40 times, giving a total of 6×10^5 evaluations per generation.

The initial population of rules is composed of random bit-strings, distributed uniformly over the range of string densities (0 to 128 ones). The initial population of IC densities is distributed uniformly over the interval [0, 1]. Rules are varied by one-point crossover and a 2% per-bit mutation rate. IC densities are varied by the addition of Gaussian noise of zero mean and standard deviation of 0.05.

Rule ranks are squared before they are normalized for the roulette wheel. The next generation of rules is created by using Baker's [2] SUS method to select 75 rules that remain unaltered and another 75 rules to which the variation operators are applied. The next generation of IC densities is created by using SUS to select 50 densities that remain unaltered and another 25 that are varied. The remaining 25 densities are picked at random from a uniform distribution.

An IC density is converted to an actual initial condition by first adding Gaussian noise of zero mean and standard deviation of 0.05. We multiply the result by 149 and take the floor to arrive at an integer between 0 and 149. The initial condition will be a random bit-string of exactly that number of ones. All rules see the same set of initial conditions.

5 Results and Discussion

We have conducted six runs of our experiment, three for each of our two methods of valuating learner pairs (see Section 2.3 and Equation 3). Our best result to-date is a rule that correctly classifies 84.0% of initial conditions, shown in Table 1. We determine this success rate by testing the rule against 4×10^7 randomly generated ICs (that is, a binomial distribution of IC densities). The rule took approximately 1300 generations to evolve. Two of the other runs each exceed 81% success; our worst result is 78.8% success. While our best result comes from our first method of valuating learner pairs (see Equation 3), the data do not distinguish the performance of the two methods.

In experiments where $N_R = N_{IC} = 400$ [10], Juillé and Pollack discover a rule that achieves 85.1% success (some runs give $\leq 76\%$). Their best results (86.0% and 86.3% success) are discovered in experiments where $N_R = N_{IC} = 1000$ [10, 9] (experiments last 5000 generations; all exceed 82.0% success). They test all rules against all densities, but each density is sampled only once. In contrast, our experiments use much smaller population sizes ($N_R = 150, N_{IC} = 100$),

Table 1. Currently best evolved rule using Pareto coevolution.

Rule 1	00010000 01010011 00000000 11010010 00000000 01010001 00001111 01011011
84.0%	00011111 01010011 11111111 11011111 00001111 01010101 11001111 01011111

but we sample each density 40 times. Thus, the total amount of computation in our experiments falls in between those of Juillé and Pollack. But, in exchange for a more expensive IC density sampling procedure (see Section 3.3), we avoid the explicit similarity metric required by Juillé and Pollack to classify ICs, and thereby arrive at an approach that should more easily generalize to other problem domains (e.g., sorting networks). Indeed, we intend ultimately to apply our Pareto coevolution methodology to variable-sum games, in addition to zero-sum games such as those studied by Rosin [18] and Juillé [7].

While we do not improve upon the results of Juillé and Pollack [10,9], we improve significantly upon all results published elsewhere. The next most effective rule is by Andre, et al [1], which performs at 82.4%. This rule was discovered with genetic programming in experiments using a rule population size of 51,200, each tested on 1000 different ICs (ICs were not coevolved) per generation. We are confident that we can improve our results with larger populations.

6 Conclusion and Future Work

We propose a novel coevolutionary algorithm based upon the concept of Pareto optimality. Our algorithm distinguishes the role of the gradient follower from that of the gradient creator, even though both may coexist within the same agent, and utilizes Pareto-inspired metrics of success for both roles. We use our algorithm to coevolve cellular automata rules for the majority function and discover a rule that correctly classifies 84.0% of initial conditions. Though our result is encouraging, we clearly have many more experiments to perform. We must try larger populations, and perhaps different inter-generational replacement schemes. We require control experiments to identify the contributions of each component of our overall methodology. In these controls, we will substitute one or two of our methods (for rewarding learners and teachers, and for sampling densities) with more conventional mechanisms, for example using a problem-specific similarity metric instead of our more expensive sampling method. We are conducting a deeper analysis of our algorithm's dynamics. Particularly, a more game-theoretic review of our algorithm is necessary to fully expose its behavior in variable-sum games and zero-sum games with intransitive superiority relationships. Finally, we are investigating various ways to integrate our metrics of learner and teacher success for single-population coevolution.

Acknowledgments. The authors thank Anthony Bucci, Hugues Juillé, Norayr Vardanyan, Richard Watson, and members of the DEMO Lab.

References

1. D. Andre et al. Evolution of intricate long-distance communication signals in cellular automata using genetic programming. In C. G. Langton and K. Shimohara, editors, *Artificial Life V*, pages 16–18. MIT Press, 1996.
2. J. E. Baker. Reducing bias and inefficiency in the selection algorithm. In *Proc. Second Int. Conf. on Genetic Algorithms*, pages 14–21. Lawrence Earlbaum, 1987.
3. D. Cliff and G. F. Miller. Tracking the red queen: Measurements of adaptive progress in co-evolutionary simulations. In F. Moran et al., editors, *Third Euro. Conf. on Artificial Life*, pages 200–218. Springer, 1995.
4. S. G. Ficici and J. B. Pollack. A game-theoretic approach to the simple coevolutionary algorithm. In Schoenauer et al. [19], pages 467–476.
5. C. M. Fonseca and P. J. Fleming. An overview of evolutionary algorithms in multiobjective optimization. *Evolutionary Computation*, 3(1):1–16, 1995.
6. D. Hillis. Co-evolving parasites improves simulated evolution as an optimization procedure. In C. Langton et al., editors, *Artificial Life II*, pages 313–324. Addison-Wesley, 1991.
7. H. Juillé. *Methods for Statistical Inference: Extending the Evolutionary Computation Paradigm.* PhD thesis, Brandeis University, 1999.
8. H. Juillé and J. B. Pollack. Co-evolving intertwined spirals. In L. J. Fogel et al., editors, *Proc. Fifth Annual Conf. on Evolutionary Programming*, pages 461–468. MIT Press, 1996.
9. H. Juillé and J. B. Pollack. Coevolutionary learning: a case study. In J. Shavlik, editor, *Proc. Fifteenth Int. Conf. on Machine Learning*, pages 251–259. Morgan Kaufmann, 1998.
10. H. Juillé and J. B. Pollack. Coevolving the "ideal" trainer: Application to the discovery of cellular automata rules. In J. R. Koza et al., editors, *Proc. Third Annual Conf. on Genetic Programming*, pages 519–527. Morgan Kaufmann, 1998.
11. M. Land and R. K. Belew. No perfect two-state cellular automata for density classification exists. *Physical Review Letters*, 74(25):1548–1550, 1995.
12. M. Mitchell et al. Evolving cellular automata to perform computations: Mechanisms and impediments. *Physica D*, 75:361–391, 1994.
13. J. Noble and R. A. Watson. Pareto coevolution: Using performance against coevolved opponents in a game as dimensions for pareto selection. In L. Spector et al., editors, *Proc. 2001 Genetic and Evo. Comp. Conf.* Morgan Kaufmann, 2001.
14. G. M. B. Oliveira et al. Evolving solutions of the density classification task in 1d cellular automata, guided by parameters that estimate their dynamic behaviour. In M. A. Bedau et al., editors, *Artificial Life VII*, pages 428–436. MIT Press, 2000.
15. B. Olsson. NK-landscapes as test functions for evaluation of host-parasite algorithms. In Schoenauer et al. [19], pages 487–496.
16. J. Paredis. Towards balanced coevolution. In Schoenauer et al. [19], pages 497–506.
17. J. B. Pollack and A. D. Blair. Co-evolution in the successful learning of backgammon strategy. *Machine Learning*, 32(3):225–240, 1998.
18. C. D. Rosin. *Coevolutionary Search Among Adversaries.* PhD thesis, University of California, San Diego, 1997.
19. M. Schoenauer et al., editors. *Parallel Prob. Solv. from Nature 6.* Springer, 2000.
20. R. A. Watson and J. B. Pollack. Symbiotic combination as an alternative to sexual recombination in genetic algorithms. In Schoenauer et al. [19], pages 425–434.
21. J. Werfel et al. Resource sharing and coevolution in evolving cellular automata. *IEEE Transactions on Evolutionary Computation*, 4(4):388–393, 2000.

On Emergence in Evolutionary Multiagent Systems

Aleš Kubík

Institute of Computer Science, Silesian University,
Bezručovo nám. 13, 746 01 Opava, Czech Republic
ales.kubik@fpf.slu.cz

Abstract. The emergence in multiagent systems is studied from various perspectives. In this paper we first briefly present the formal framework of the basic type of emergence (also called combinatoric). It will be shown that in some cases time dimension in the functionality of multiagent systems can be reduced to the basic emergence. We focus on how evolutionary processes in multiagent systems influence their functionality primarily concerning the emergent phenomena.

1 Introduction

Informal treatment of emergence can be found in number of literature on multiagent systems (MAS), see e.g. [15], [3], [14], [6], [12], [1], [2], [13]. These works concern the emergence from the various contexts of multiagent systems but lack formal treatment of emergence and touch upon the topic only from the intuitive or psychological point of view.

Recently the authors [19], [18] try to catch the essence of the emergence by means of testing it in the framework of definition and comparison of behavior of the multiagent system as a whole with the behavior of the individual agents. The observation of their behaviors should lead to the conclusion about the uprise (or denial) of emergence in the MAS. Even if the authors get to the point in the way of putting the MAS and its individual parts (their behavior) face to face, the testing framework nevertheless lacks more precise definitions of its building blocks or interpretations. We feel that we need formal framework of the emergent multiagent systems in order to:

1. understand basic principles of emergence,
2. describe and categorize the different kinds of emergent phenomena, and
3. use the designed concepts to implement, manage and control artificial multiagent systems or understand real ones.

2 Towards Formalization of Emergence

The work of [5] mentions the *combinatorics emergence* and *creative emergence*. The former occurs as a recombination of existing symbols (agent properties) in

J. Kelemen and P. Sosík (Eds.): ECAL 2001, LNAI 2159, pp. 326–337, 2001.

the multiagent system. The latter focuses on the situation when new symbols (properties) are added into the system giving rise to new recombinations. In this part the formal definition of so called basic type of emergence is given in the framework of grammar systems theory [20]. By the basic type we mean the *combinatorics emergence* in the sense that neither the agents nor the environment evolves. Under the term evolution in our context we understand the change in the agents behavioral rules or active change in the environment without the influence of the agents or their combination.

2.1 Preliminaries

Our model is inspired by the growing theory of formal languages [20]. Now we give some explanations concerning formal languages and grammar systems. For elements of formal languages theory the reader is referred to [16], [17].

For an alphabet V, by V^* we denote the set of all strings over V. The empty string is denoted by λ, and $V^* - \{\lambda\}$ is denoted by V^+. With the term *pure grammar* we will denote a grammar with alphabet V of symbols not divided to distinct sets of terminal and nonterminal alphabets T and N respectively.

A cooperating grammar system [8] is a construct of grammars mutually rewriting strings of symbols on a common tape. Its principles are inspired by a distributed solution models with blackboard architecture. The grammars work with different cooperation strategies (how the grammars are allowed to rewrite the symbols on the tape) and modes of derivation (when the process of derivation stops).

A *cooperating grammar system* is a construct $\Gamma = (N, T, S, P_1, P_2, ..., P_n)$, where N is a nonterminal alphabet, T is a terminal alphabet, $S \in N$ is a starting symbol and $P_1, P_2, ..., P_n$ are finite sets of (usually regular or context-free) rewriting rules over $N \cup T$.

A *basic derivation step* is a relation $\Longrightarrow_{P_i}$:

$$x \Longrightarrow_{P_i} y \text{ iff } x = x_1 u x_2,\ y = x_1 v x_2,\ x_1, x_2 \in (N \cup T)^*,\ u \to v \in P_i.$$

Then derivations of arbitrary length ($\Longrightarrow^*_{P_i}$) can be defined, and for $k \geq 1$ derivations of exactly k steps , ($\Longrightarrow^{=k}_{P_i}$), of at least or at most k steps ($\Longrightarrow^{\geq k}_{P_i}$, $\Longrightarrow^{\leq k}_{P_i}$, resp.), as well as the maximal derivation: $x \Longrightarrow^t_{P_i} y$ iff $x \Longrightarrow^*_{P_i} y$ and there is no $z \in (N \cup T)^*$ such that $y \Longrightarrow_{P_i} z$.
Let $F = \{*, t\} \cup \{\leq k, = k, \geq k \mid k \geq 1\}$. For any cooperating grammar system Γ with at most n components and $f \in F$, the language generated by Γ in the cooperative mode f is

$$L_f(\Gamma) = \{x \in T^* \mid S \Longrightarrow^f_{P_{i_1}} x_1 \Longrightarrow^f_{P_{i_2}} x_2 \ldots \Longrightarrow^f_{P_{i_m}} x_m = x,\ m \geq 1,\ 1 \leq i_j \leq n,\ 1 \leq j \leq m\}.$$

In the next part we introduce the necessary notions for array grammars. For further readings see e.g. [7], [10], [9], or [11]. Array grammars are characterized by two-dimensional tape of symbols. Informally an array is a set of pairs of point coordinates and the corresponding symbol placed at that point on the tape. Let us proceed with a more formal definition.

Let Z denote the set of integers, and let V be an alphabet. Then an array A over V is a function $A : Z^2 \to V \cup \{\#\}$, where

$$support(A) = \{v \in Z^2 \mid A(v) \neq \#\}$$

is a set of points with coordinates that don't contain blank (background) symbol $\# \notin V$. We can write $A = \{(v, A(v) \mid v \in support(A)\}$. The set of all two-dimensional non-empty arrays over V will be denoted by V^{+^2}. Any subset of V^{+^2} is called an array language. The arrays are words of array languages.

For an array $A \in V^{+^2}$ and a finite pattern α of symbols over $V \cup \{\#\}$, we say that α is a subpattern of A, if we can superimpose α over A in such a way that all vectors of α marked with symbols from V coincide with the corresponding symbols in A and each symbol $\#$ in α correspond to a symbol $\#$ in A.

A construct $G = (N, T, \#, P, \{(v_0, s)\})$ is an *array grammar*, where N, T are alphabets comprised of nonterminals and terminals respectively, $\#$ is a blank symbol and $\{(v_0, s)\}$ is the start array with the start vector $v_0 \in Z^2$ and the start symbol $S \in N$ and P is a finite set of rewriting rules of the form $\alpha \to \beta$, where α, β are array patterns over $N \cup T \cup \{\#\}$. The condition of identical shapes of α and β is satisfied (we say they are *isometric*).

For an array grammar $G = (N, T, \#, P, \{(v_0, s)\})$ and two words $A, B \in (N \cup T)^{+^2}$ we define the relation $A \Longrightarrow B$, if there is a rule $\alpha \to \beta \in P$ such that α is a subpattern of A and B is obtained by replacing α in A by β. By $\Longrightarrow^*$ we denote reflexive and transitive closures of $\Longrightarrow$.

By the construct $L(G) = \{A \in T^{+^2} \mid \{(v_0, S)\} \Longrightarrow^* A\}$ we denote the language generated by G.

A *cooperating array grammar system* is a cooperating grammar system with array grammars (grammars with array rewriting rules). Formally it is an (n + 4)-tuple $G = (N, T, \#, \{(v_0, S)\}, P_1, P_2, ..., P_n)$, with the components defined as above.

2.2 The Basic Type of an Emergence – A Formal Framework

After a short introduction of basic notions and definitions from the theory of formal languages we give a formal definition of a basic emergence in this subsection. In our definition we do not consider the terminal and non-terminal alphabet but distinguish agent and environment symbols instead (but these categories do not correspond with each other).

Let $MAS = (V_A, V_E, \#, \{(v_0, S_0), (v_1, S_1), ..., (v_{r\times s-1}, S_{r\times s-1})\}, A_1, ..., A_n)$ be a *multiagent system* modeled by a modified cooperating grammar system in a following way: V_A is an alphabet of agent symbols, V_E is an alphabet of the environment, $V = V_A \cup V_E$ is an alphabet of MAS description, $\#$ is a blank symbol, $\{(v_j, S_j) \mid 0 \leq j \leq (r \times s - 1),\ r, s \in Z,\ S_j \in V \cup \{\#\}\}$ is a start description of the *environment* in the form of an array over V^{+^2} and $A_1, A_2, ..., A_n$ is a finite set of agents. An *agent* is a construct (pure array grammar)

$$A = (V_i, \#, P_i, \{(v_0, S_0), (v_1, S_1), ..., (v_{r\times s-1}, S_{r\times s-1})\}),$$

where $V_i \subseteq V$ is an alphabet of an agent, # is a blank symbol,

$$\{(v_0, S_0), (v_1, S_1), ..., (v_{r\times s-1}, S_{r\times s-1})\}, r, s \in Z, S_j \in V_i \cup \{\#\}, 0 \leq j \leq r \times s - 1$$

is a start description of an environment and P_i is a set of rewriting rules defined in the form as follows: The agent is capable to derive $u \Longrightarrow v$, where $u, v \in V^{+^2}$ and there is a rule $\alpha \rightarrow \beta \in P_i$ such that α is a subpattern of A and B is obtained by replacing α in A by β.

The behavior of an agent A_i is the following language generated by an agent:

$$L(A_i) = \{w \in V_i^{+^2} \mid \{(v_0, S_0), (v_1, S_1), ..., (v_{r\times s-1}, S_{r\times s-1})\} \Longrightarrow^* w, \\ r, s \in Z,\ S_j \in V, 0 \leq j \leq r \times s - 1\}.$$

The events in the environment (derivation of words on the tape) proceeds in the *MAS* in parallel manner such that the set of agents can rewrite the part of the tape (environment) without disturbing each other.

For the $GROUP = \{A_i \mid 1 \leq i \leq n\}$ and two arrays D_1 and $D_2 \in V^{*^2} \cup \{\#\}$ a direct derivation step is defined by $D_1 \Longrightarrow_{GROUP} D_2$ if and only if there exist array productions $p_j \in P_i, 1 \leq i \leq n,\ 1 \leq j \leq l_i$ (j-th rule of i-th agent), $p_j = \alpha_j \rightarrow \beta_j$, such that if for any array area $\omega_{j1_{k_1}}, 1 \leq k \leq l_i$ that is a subpattern of β_{j_1} and is not a subpattern of α_{j_1} and for any array area $\omega_{j2_{k_2}}, 1 \leq k \leq l_i$ that is a subpattern of β_{j_2} and is not a subpattern of α_{j_2} it holds that these areas are disjoint: $\tau_{v_{j_1}}(\omega_{j1_{k_1}}) \cap \tau_{v_{j_2}}(\omega_{j2_{k_2}}) = \emptyset, j_1 \neq j_2, k_1 \neq k_2, 1 \leq j_1, j_2, k_1, k_2 \leq l_i$; for some $v_j \in Z^2, 1 \leq j \leq l_i$.

Informally the above condition means that if there are rewriting rules of agents that have to be applied in parallel manner on the environment they cannot simultaneously rewrite symbols on the same position (with the same index). Observe that it is not necessary that the left and right sides of rewriting rules applied in parallel manner must be disjoint.

What follows is a definition of summation of behaviors of individual agents in the environment.

Let W_1, W_2 and W_{supimp} are array words over $(V_A \cup V_E)^{+^2}$. The superimposition of the word W_1 over the word W_2 is a function *superimpose* : $\{W_1, W_2\} \rightarrow W_{supimp}$, and it holds:

(1) if $W_1(v_j) \in V_A$ then $W_{supimp}(v_j) = W_1(v_j)$

(2) if $W_1(v_j) \in V_E$ and $W_2(v_j) \in V_E$ then $W_{supimp}(v_j) = W_1(v_j)$

(3) if $W_1(v_j) \in V_E$ and $W_2(v_j) \in V_A$ then $W_{supimp}(v_j) = W_2(v_j)$

for some $v_j \in Z_2, 1 \leq j \leq l_i$. We shall write W_1 *superimpose* W_2. For n array languages $L_1 = \{W_1 \in (V_A \cup V_E)^{+^2}\}$, $L_2 = \{W_2 \in (V_A \cup V_E)^{+^2}\}$, ... , $L_n =$

$\{W_n \in (V_A \cup V_E)^{+^2}\}$ their summation is an array language that is a result of their superimposition:

$$L_{sum} = \{W_{sum} \in (V_A \cup V_E)^{+^2} \mid W_{sum} =$$
$$W_1 \textit{ superimpose } (W_2 \textit{ superimpose } (W_3 \ldots (W_{n-1} \textit{ superimpose } W_n)))$$
$$\cup \ldots \cup$$
$$W_n \textit{ superimpose } (W_{n-1} \textit{ superimpose } (W_{n-2} \ldots (W_2 \textit{ superimpose } W_1)))\}.$$

Given the MAS comprised of n agents $A_1, A_2, ..., A_n$ generating languages L_1, $L_2, ..., L_n$ according to definition given above the *basic emergence* is a property of MAS such that the language generated by the MAS $L(MAS)$ contains words that can not be generated by the summation of languages generated by individual agents. In other words the behavior of the whole multiagent system is more than the summation of the behaviors of individual agents. Any multiagent system is emergent if and only if it has the property of emergence.

The modeled MAS is nonevolutionary in that we compare the behavior at the specified level of agents' rationality. The rationality which can be understood as a capability to fulfill specified tasks is fixed and doesn't evolve. It means that only the *possibility* of occurrence of some kind of behavior is explored. In the framework given above we consider only the the consequences of the interactions of agents in their common environment. We try to filter out the influence of the independent environmental change or the change in agents' rationality.

3 The Role of Time in MAS

In this contribution we hope to explore different kind of emergence than in the above mentioned case. We will focus on the time dimension of the functionality of multiagent systems including evolutionary processes. The inspiration of emergence comes from study of living systems, especially animal and human societies. Nevertheless the researchers strive to understand its nature in order to be able to build *artificial* systems. The existence of every artificial system is timely constrained and its functionality can deteriorate with time. From this standpoint we must consider not only the question whether some behavior can occur in the society of agents but also for how long time the behavior is pertinent or to the contrary what can be the time savings when performing distinct tasks in the MAS.

3.1 Modifications of the Basic Emergence Model

In this section we explore scenarios of multiagent systems from the point of view of time and show that the time factor in the MAS can be modeled by means of our basic emergence principles.

For this purpose we modify the formal model of the MAS in that we introduce the notion of the memory for the agents where they can send each other messages. The messages will be words over the alphabet of agent and environment symbols. The application of the productions then will be conditioned by the contents of the agents memories. So we have to change the basic rewriting step of the agents and the configuration of the MAS. The rest of the definitions is left untouched.

The production rules of the agents will have the form:

$$p = (b, (r, q) : A \longrightarrow w, d, m_1, \ \dots \ , m_n),$$

where $d, m_1, \ \dots \ , m_n$ are words over $(V_A \cup V_E)^*$, b is the word (possibly empty string) over $(V_A \cup V_E)^*$ and $(r, q) : A \longrightarrow w$ is a semi-conditional production, where w is an array word.

In the production p the c is the word contained in the memory of the agent (grammar), r and q are permitting and forbidding conditions of p; r, q are strings over $(V_A \cup V_E)^+$ or in case $r = 0$ or $q = 0$ the corresponding condition is ignored; $A \longrightarrow w$ is a production, d is the word that has to be deleted from the memory of A_i and $m_1, \ \dots \ , m_n$ are messages to be sent to the memory of $A_1, \ \dots \ , A_n$. The agents rewrite the environment array words, (possibly) delete the words from their own memory and send each other messages that will stack up in their memory.

Let $MAS = (V_A, V_E, \#, \{(v_0, S_0), (v_1, S_1), ..., (v_{r\times s-1}, S_{r\times s-1})\}, A_1, ..., A_n)$ be a *multiagent system* modeled by a modified cooperating grammar system. A tuple $\gamma = (w, c_1, \ \dots \ , c_n)$, where $w \in (V_A \cup V_E)^+$, $c_i \in (V_A \cup V_E)^*, 1 \le i \le n$ is a configuration of MAS; w is the current sentential form and $c_i, 1 \le i \le n$ is the current contents of the $i - th$ memory.

Let $MAS = (V_A, V_E, \#, \{(v_0, S_0), (v_1, S_1), ..., (v_{r\times s-1}, S_{r\times s-1})\}, A_1, ..., A_n)$ be a *multiagent system* and let $\gamma = (w, c_1, \ \dots \ , c_n)$ and $\gamma' = (w', c'_1, \ \dots \ , c'_n)$ be two configurations of MAS. Then a direct derivation step is a relation $\gamma \Longrightarrow_{MAS} \gamma'$, if there is a component $A_i, 1 \le i \le n$ with the production

$$p = (b, (r, q) : A \longrightarrow w, d, m_1, \ \dots \ , m_n),$$

such that the following conditions hold:

1. $w = x_1 u x_2$, $w' = x_1 v x_2$, $x_1, x_2 \in (V_A \cup V_E)^*$;
2. if $r \neq 0$, then $w = x_1 r x_2$ for some $x_1, x_2 \in (V_A \cup V_E)^*$;
3. if $q \neq 0$, then $w \neq x'_1 r x'_2$ for any $x'_1, x'_2 \in (V_A \cup V_E)^*$;
4. $c_i = b z_i$ for some $z_i \in (V_A \cup V_E)^*$;
5. $c'_k = m_k c_k$ $(c_k m_k)$ for all $k, 1 \le i, k \le n$ with $i \neq k$, if the $k - th$ memory is organized as a stack;
6. $c'_i = m_i c''_i$ and $d c''_i = c_i$ or $(c'_i = c''_i m_i)$, if the $i - th$ memory is organized as a stack.

3.2 Scenario 1

Take there is an environment with robots that gather blocks and pile them in the selected place - let us call it a home zone. Suppose each robot has exactly the same holding capacity as is the weight of every block. We are interested not only whether the emergent behavior occurs in the multiagent system but also what will be the time to fulfil the task. Let us suppose we have two robots and three blocks in the environment with home zone where the agents are supposed to gather the blocks. Formally the multiagent system will be the construct

$$MAS = (V_A, V_E, \#, \{(v_0, S_0), (v_1, S_1), ..., (v_{r\times s-1}, S_{r\times s-1})\}, A_1, A_2), \text{ where}$$

- $V_A = \{F_1, F_2, H_1, H_2\}$ is the set of agent symbols,
- $V_E = \{B, Z, Z_1, Z_2, f, X\}$ is the set of environment symbols,
- $\{(v_0, S_0), (v_1, S_1), ..., (v_{r\times s-1}, S_{r\times s-1})\}$ is a start description of an environment, and
- A_1, A_2 are the agents represented by the modified array grammars.

The semantics of the agent and environment symbols is as follows:

- F_1, F_2 are symbols of free (nothing holding) agents,
- H_1, H_2 are symbols of agents holding the block,
- B is a symbol of block,
- Z is the symbol of the home zone where the agents are to pile up the blocks,
- Z_1, Z_2 are representing the agents bringing the block into the home zone,
- f is the symbol of free space, where no agent or block is placed, and
- X is the symbol of steps the agents take in the course of fulfiling the task (the number of symbols X, possibly 0 represents the number of steps).

The initial configuration of the MAS is $\gamma = (w, \epsilon, \epsilon)$, where

$$\text{w} = \begin{array}{cccc} Z & f & f & f \\ f & f & f & F_2 \\ f & f & f & B \\ f & F_1 & B & B \end{array}$$

is the initial state of the environment (tape). The agents A_1, A_2 are modeled by array grammars and contain set of productions P_1, P_2. The production rules can be devided into the rules representing searching behavior (free agents look for the block), the rules representing the situations when the agent finds the block and takes it, the rules representing the agents moving with the blocks and the entering and leaving of the home zone, respectively. We now give some of the rewriting rules contained in the P_1[1]: $P_1 =$

[1] Due to the limited space it is not possible to enumerate the wholeset of productions.

$$\{(X^d,(0,0):F_1f \longrightarrow fF_1,X^d,X^{d+1},\epsilon),(X^d,(0,0):F_1B \longrightarrow fH_1,X^d,X^{d+1},\epsilon),$$
$$(X^d,(0,0):H_1f \longrightarrow fH_1,X^d,X^{d+1},\epsilon),(X^d,(0,0):ZH_1 \longrightarrow Z_1f,X^d,X^{d+1},\epsilon),$$
$$(X^d,(0,0):Z_1f \longrightarrow ZF_1,X^d,X^{d+1},\epsilon),\ \ldots\}$$

Every time the agent rewrites the tape (it makes a step) it updates its memory where it stores the information about the number of steps it took. This is done by deleting the old information and storing the new one consisting in adding one symbol X to the memory string. The productions of the set P_2 are analogous in that we use the subscript 2 to denote the actions of the second agent.

Results. The agents simultaneously change the conditions in the environment and make "time stamps" in their own memory. The sequence of derivations could be as follows:

$$\begin{matrix} Z & f & f & f \\ f & f & f & F_2 \\ f & f & f & B \\ f & F_1 & B & B \end{matrix} \Longrightarrow_{MAS} \begin{matrix} Z & f & f & f \\ f & f & f & f \\ f & f & f & H_2 \\ f & f & H_1 & B \end{matrix} \Longrightarrow_{MAS} \cdots \Longrightarrow_{MAS}$$

$$\begin{matrix} Z & H_2 & f & f \\ H_1 & f & f & f \\ f & f & f & f \\ f & f & f & B \end{matrix} \Longrightarrow_{MAS} \begin{matrix} Z_2 & f & f & f \\ f & f & f & f \\ H_1 & f & f & f \\ f & f & f & B \end{matrix} \Longrightarrow_{MAS} \cdots \Longrightarrow_{MAS}$$

$$\begin{matrix} Z_1 & f & f & f \\ f & f & f & f \\ f & f & F_2 & f \\ f & f & f & B \end{matrix} \Longrightarrow_{MAS} \cdots \Longrightarrow_{MAS} \begin{matrix} Z & f & f & f \\ f & H_2 & f & f \\ f & f & f & f \\ f & F_1 & f & f \end{matrix} \Longrightarrow_{MAS}$$

$$\begin{matrix} Z_2 & f & f & f \\ f & f & f & f \\ f & F_1 & f & f \\ f & f & f & f \end{matrix} \Longrightarrow_{MAS} \begin{matrix} Z & F_2 & f & f \\ f & f & F_1 & f \\ f & f & f & f \\ f & f & f & f \end{matrix}$$

The configuration after the eleven steps will be the construct (number of symbols X represents the number of steps the agents took:

$$\gamma_{12} = (\begin{matrix} Z & F_2 & f & f \\ f & f & F_1 & f \\ f & f & f & f \\ f & f & f & f \end{matrix} ,XXXXXXXXXXX,XXXXXXXXXXX)$$

The language generated by the Agent A_1, i. e. in the situation it is the only agent that interacts with the environment is the set $L(A_1)$ containing the array words with symbols $\{F_1, H_1, Z, Z_1, B, f\}$. The language generated by the second agent is the construct $L(A_2)$ containing the symbols $\{F_2, H_2, Z, Z_2, B, f\}$. If we superimpose the two languages we get the language L_{sum} containing the symbols $\{F_1, H_1, F_2, H_2, Z, Z_1, Z_2, B, f\}$. We can see that the language generated by the whole MAS is the set $L(MAS)$ that contains no other words in comparison with the language L_{sum}. As for the time one agent takes to fulfil the task it is 17 steps for each agent (take the agent must step out of the home zone when it delivers the block).

3.3 Scenario 2

Now we modify slightly the previous scenario in the way that the agents have the holding capacity 1, 5 times bigger than is the block wheight. The second case will be the comeup of the simple cooperation. The agents can bear the blocks simultaneously and even can hold three blocks at one time. We add the symbols $\{C_1, C_2, C_3, Z_{C_1}, Z_{C_2}, Z_{C_3},\}$ of the cooperation to the alphabet of the MAS. Their interpretation is such that

- symbols C_1, C_2, C_3 represent one, two, and three blocks respectively the agents hold (and bear) simultaneously,
- and $Z_{C_1}, Z_{C_2}, Z_{C_3}$ represent agents bringing one, two, and three blocks respectively into the home zone.

We must as well add new productions to the agent action rule sets. The set P_1 will contain the following rewriting rules:

$$\{(X^d, (0,0) : F_1 f \longrightarrow f F_1, X^d, X^{d+1}, \epsilon), (X^d, (0,0) : F_1 B \longrightarrow f H_1, X^d, X^{d+1}, \epsilon),$$

$$(X^d, (0,0) : H_1 f \longrightarrow f H_1, X^d, X^{d+1}, \epsilon), (X^d, (0,0) : Z H_1 \longrightarrow Z_1 f, X^d, X^{d+1}, \epsilon),$$

$$(X^d, (0,0) : Z_1 f \longrightarrow Z F_1, X^d, X^{d+1}, \epsilon),$$

$$(X^d, (0,0) : F_1 B F_2 \longrightarrow f C_1 f, X^d, X^{d+1}, X),$$

$$(X^d, (0,0) : F_1 B H_2 \longrightarrow f C_2 f, X^d, X^{d+1}, X),$$

$$(X^d, (0,0) : H_1 B H_2 \longrightarrow f C_3 f, X^d, X^{d+1}, X),$$

$$(X^d, (0,0) : C_1 f \longrightarrow f C_1, X^d, X^{d+1}, X),$$

$$(X^d, (0,0) : C_2 f \longrightarrow f C_2, X^d, X^{d+1}, X),$$

$$(X^d, (0,0) : C_3 f \longrightarrow f C_3, X^d, X^{d+1}, X),$$

$$(X^d, (0,0) : Z C_1 \longrightarrow Z_{C_1} f, X^d, X^{d+1}, X),$$

$$(X^d, (0,0) : \begin{matrix} Z_{C_1} & f \\ f & \end{matrix} \rightarrow \begin{matrix} Z & F_1 \\ F_2 & \end{matrix}, X^d, X^{d+1}, X),$$

$$(X^d, (0,0) : \begin{matrix} Z_{C_1} & \\ f & f \end{matrix} \rightarrow \begin{matrix} Z & \\ F_2 & F_1 \end{matrix}, X^d, X^{d+1}, X),$$

$$\ldots \}$$

We have added the rewriting rules that symbolize the cooperative bearing of the blocks, the movement of the agents when holding blocks simultaneously, entering and leaving the home zone in common. Similarly the set of rewriting rules P_2 of the second agent will be modified. Notice that in case we deal with cooperation derivations the agent that rewrites must send the other agent the message containing the symbol X because it makes the step as well even if the rule is applied only from the point of one of the agents.

Let's take a look at the simultaneous actions of the agents in the environment when the cooperation is involved and the holding capacity of the agents is $1,5$ times bigger than is the wheight of one block. The following derivation sequence represents the cooperative gathering of the blocks by the two agents. It is the best solution of the task measured in steps the agents take:

$$\begin{matrix} Z & f & f & f \\ f & f & f & F_2 \\ f & f & f & B \\ f & F_1 & B & B \end{matrix} \Longrightarrow_{MAS} \begin{matrix} Z & f & f & f \\ f & f & f & f \\ f & f & f & H_2 \\ f & f & H_1 & B \end{matrix} \Longrightarrow_{MAS} \begin{matrix} Z & f & f & f \\ f & f & f & f \\ f & f & f & f \\ f & f & f & C_3 \end{matrix} \Longrightarrow_{MAS}$$

$$\begin{matrix} Z & f & f & f \\ f & f & f & f \\ f & f & C_3 & f \\ f & f & f & f \end{matrix} \Longrightarrow_{MAS} \begin{matrix} Z & f & f & f \\ f & C_3 & f & f \\ f & f & f & f \\ f & f & f & f \end{matrix} \Longrightarrow_{MAS} \begin{matrix} Z_{C_3} & f & f & f \\ f & f & f & f \\ f & f & f & f \\ f & f & f & f \end{matrix} \Longrightarrow_{MAS}$$

$$\begin{matrix} Z & F_1 & f & f \\ F_2 & f & f & f \\ f & f & f & f \\ f & f & f & f \end{matrix}$$

The agents when acting simultaneously generated the language $L(MAS)$ containing the words consisting of symbols as in the scenario 1 plus the set of words containing the symbols $\{C_1, C_2, C_3, Z_{C_1}, Z_{C_2}, Z_{C_3}, \}$. This language contains the same word set than the language L_{sum} plus the new words containing new symbols added to the model. We face the basic type of emergent property in the MAS.

The same is true when we take time into consideration. After the task is completed by the agents the configuration of the MAS will be

$$\gamma_7 = (\begin{matrix} Z & F_2 & f & f \\ f & f & F_1 & f \\ f & f & f & f \\ f & f & f & f \end{matrix}, XXXXXX, XXXXXX)$$

Both agents in cooperation can complete the task in 6 steps that is less than 1/2 of the time one agent with the same capabilities spends in achieving the same task instance.

The presented scenario is the example of creative emergence according to [5] when new symbols give rise to new recombinations. We can see that the evolutionary processes in agents' behaviors can be filtered out and brought to the basic type of emergence (without the evolutionary changes) by splitting the situation into more multiagent systems with the agents of given rationality (symbolic description). Also if we observe "time emergence" there is to appear basic behavior emergence as well.

4 Conclusions

In this paper we focused on the evolutionary changes that affect the rationality of individual agent behaviors with the emphasis these processes have on the origin of the emergent properties of multiagent systems. We showed an example that conforms to the idea of basic emergence appearing in multiagent systems and how we tackle the question of "time emergence" in this case. There are nevertheless cases (e.g with the evolution of stigmergic cooperation) that don't meet the criteria of basic emergence test but still reveal interesting behavior from the engineering point of view.

In the future we hope to tackle this problem by means of formal description of multiagent systems coupled with software engineering techniques.

References

1. Bonabeau, E., Dessalles, J. L., Grumbach, A.: Characterizing emergent phenomena (1): A critical review. Revue Internationale de Systémique **9** (1995) 327 - 346
2. Bonabeau, E., Dessalles, J. L., Grumbach, A.: Characterizing emergent phenomena (2): A conceptual framework. Revue Internationale de Systémique **9** (1995) 347 - 371
3. Brooks, R. A.: Intelligence Without Reason. In: Proceedings of the 12th International Joint Conference on Artificial Intelligence (IJCAI '91). Morgan Kaufmann, San Francisco, 1991 569 - 595
4. Cariani, P: Towards an evolutionary semiotics: the emergence of new sign-functions in organisms and devices. In: Van de Vijver, G., Salthe S., Delpos, M. (eds.): Evolutionary Systems. Kluwer Academic Press, Dordrecht, 1998 359 - 377

5. Cariani, P.: Emergence of new signal-primitives in neural networks. Intellectica **2** (1997) 95 - 143
6. Cariani P.: Emergence and artificial life. In: Langton, C. G., Taylor, C., Farmer, J.D., Rasmussen S. (eds.): Artificial life II. Sante Fe Institute studies in the sciences of complexity, vol X. Addison-Wesley, Reading, MA., 1991 775 - 798
7. Cook, C. R., Wang, P. S.-P.: A Chomsky hierarchy of isotonic array grammars and languages. Computer Graphics and Image Processing **8** (1978) 144 - 152
8. Csuhaj-Varju, E., Dassow, J. , Kelemen J., Păun G.: Grammar Systems. Gordon and Breach, London, 1994
9. Dassow, J., Păun, G.: Regulated Rewriting in Formal Language Theory. Springer Verlag, Berlin, 1989
10. Fernau, H., Freund, R., Holzer, M.: Regulated array grammars of finite index, Part I: Theoretical investigations. In: Păun, Gh., Salomaa, A. (eds): Grammatical models of multi-agent systems. Gordon and Breach, Reading, 1998
11. Freund, R.: Array Grammar Systems. In: Kelemenová, A. (ed.): Grammar Systems '98. Silesian University, Opava, 1998 91 - 116
12. Holland, J. H.: Emergence: From Chaos to Order. Addison-Wesley, Reading, MA, 1998
13. Kubík, A.: The Lowest-Level Economy as a Grammar Colony. Journal of Economics **5/48** (2000) 664 - 675
14. Matarić, M.: Interaction and Intelligent Behavior. MIT, AI Lab., Cambridge, Mass., 1994
15. Minsky, M.: The Society of Mind. Simon and Schuster, New York, 1986
16. Salomaa, A.: Formal Languages. Academic Press, 1973
17. Simovici, D. A., Tenney, R. L.: Theory of Formal Languages with Applications. World Scientific, Singapore, 1999
18. Ronald, E. A., Sipper, M.: Engineering, Emergent Engineering, and Artificial Life: Unsurprise, Unsurprising Surprise, and Surprising Surprise. In: Bedau, M. A., Caskill, J. S., Packard, N. H., Rasmussen, S. (eds.): Artificial Life VII, 2000 523 - 528
19. Ronald, E. A., Sipper, M., Capcarrère, M. S.: Design, Observation, Surprise! A Test of Emergence. Artificial Life **5** (1999) 225 - 239
20. Rozenberg, G., Salomaa, A. (eds.): The Handbook of Formal Languages. Springer-Verlag, Berlin, 1996

Division of Labour in Simulated Ant Colonies Under Spatial Constraints

Jesper Bach Larsen

EvAlife Group, Dept. of Computer Science, University of Aarhus, Denmark
jesper@larsen.cc, www.larsen.cc/jesper

Abstract. Division of labour in ant societies is a kind of role model for distributed problem solving. They are so because ants are simple, non-cognitive, distributed and autonomous and yet they solve an optimisation problem that is very complex and dynamic. This is very desirable in computer science, but as of yet not much research has gone into explaining the underlying mechanisms of division of labour. The venue in this paper is to, by means of evolutionary algorithms, find the implications spatial constraints play in the division of labour in a foraging task of virtual ants. The ants differ only in size, and size implies constraints regarding to ease of movement in an environment where obstacles of other ants and clay exists. The results show that spatial constraints do play a major role in the job-task division evolved, in that test setup with increasing constraints exhibited division of labour in that ants of different sizes occupy different spatial areas of the task domain. In the process, we have evolved the behaviour of the ants that underlie the division of labour. This was done via mapping functions and motivation networks.

1 Introduction

The effectiveness of an ant society can be highly contributed to the fact that division of labour occurs in the colony. Seen from the outside the foraging task seems to be centrally coordinated, or at least controlled by some stringent and predetermined rules or patterns [3]. Yet the colony is not directed by some hive mind, and the underlying mechanisms of division of labour is of yet mostly unexplained. As inspiration for division of labour in computer systems, ant societies are brilliant. They are so because ants are simple entities, non-cognitive, distributed and autonomous and yet the overall colony performs fault tolerant, flexible and seemingly intelligent. Biology research explains division of labour as evolutionary optimisation, and their models are top-down aggregated approaches based on observations. This is the case in age-polyethism in which the age is the major determinant of subtask assignment [3,4]. Many studies have shown strong evidence against this postulated correlation, by showing great flexibility in task assignment under radical changes in the population, whereby older ants would revert to tasks previously performed, and younger ants would accelerate the transition to new tasks as well [1,2]. Other mechanisms must thus underlie the mechanisms of division of labour. In this paper we shall pursue the theory of physical polymorphism in which the morphology is used as the determining parameter of task assignment. The importance of morphology towards division of

J. Kelemen and P. Sosík (Eds.): ECAL 2001, LNAI 2159, pp. 338-348, 2001.

labour is investigated. The importance seems plausible since fit to tasks depend on morphology. In our experiments we have tailored a virtual ant environment with 2 physical different worker castes and a foraging task that requires 2 opposing physical characteristics of the ants. A larger body size implies certain problems when moving in narrow places, as is the case in the constructed nest-structure. The big ants have an advantage where the spatial constraints are low, as in the food collecting areas, whereas the smaller ants move slower, and have a disadvantage here. In the nest the smaller ants have the advantage of being able to move more freely around without colliding into nest-walls, whereas the big ants just merely can move around. We thus seek a spatial pattern in the foraging task of the big versus the small ants. As such it is the emergent effects in the foraging task under spatial constraints that we seek. An Evolutionary Algorithm evolves the behaviour used to achieve this. From a computer science perspective, our aim is to obtain a better understanding of how autonomous distributed agents can cooperate in a foraging task in an environment where only local knowledge is available, and where a division of the task is needed, and where no centralized mechanisms can be used to coordinate this division.

This paper is organized in the following way. Chapter 2 describes the AntDOL model, which underlies our simulations. In chapter 3 the experiments performed and results obtained are presented. A discussion of the implications is found in chapter 4, as is also shortcomings and ideas to pursue in further research.

2 The AntDOL Model

The following model is based on a simulation-framework developed in cooperation with Ilja Booij, Leiden University, Holland.

The model for division of labour in virtual ants, hereafter called AntDOL, is an evolutionary algorithm where the population of individuals is a population of ant colonies. As usual fitness is calculated for each of these individuals, and the fittest individuals are allowed to recombine into new individuals for the following generation. To calculate the fitness of an individual (an ant colony) the colony is simulated for a fixed number of time steps, and the number of food items foraged during the simulation is used for the fitness calculation. Each individual has a virtual gene that encodes the behaviour of all ants in the colony. Statistical data is collected from the simulations, which are stored for later analysis. The task is to find the optimal colony in regard to the fitness-function, and then retrospective to observe statistical and spatial emergent patterns of division of labour, i.e. do the different castes have a specific spatial placement.

2.1 The Ant Colony

The colony is represented as a discrete grid of locations that can hold specific objects. There can be 3 different types of objects on a location: ant, food and obstacle. There can be only one ant at a location, and only if no obstacle is placed there. There can be any number of food objects on a single location. The grid uses a 9-cell Moore Neighbourhood, so that each ant, from any interior location, can move in 8 directions.

Movement is restricted to be within borders, i.e. no wrapping, and ants can only pass locations without obstacles and ants.

2.1.1 Ants

Ants come in 2 sizes. The small cover 4x4 locations, and the large cover 6x6 locations. The speed with which the ant can move is defined as $\lfloor AntSize/2 \rfloor$, i.e. a 4x4 ant can move 2 locations at a time, and a 6x6 can move 3 locations at a time. Ants have sensors and actions that enable them to respond to the local environment. Ants can only sense and influence the closest vicinity of the environment. All sensors are real valued in the interval [0,1]. The ants can retrieve data from the environment by these sensors only. The following sensors are implemented in the system. Sensor senseFood returns a direction to a possible food-source in the immediate vicinity of the ant. Food in the direction of movement is preferred. Sensor senseAnt returns a weighted score of the presence of other ants in the direction of movement. Closer ants are weighted heavier. Sensor senseClay returns a weighted score of the presence of clay in the direction of movement. Closer clay objects are weighted heavier. Sensor sensePointingHome returns a direction to the nest (1/360 degrees). Sensor senseCarryFood returns 1 if the ant is carrying food, and returns 0 otherwise. Sensor senseFrustrated returns a frustration level, which is incremented when intended actions are blocked (i.e. by obstacles or other ants), and decreased with time. The ant can only make changes to the environment and to itself by actions. The following actions were implemented in the system. Action MoveForward moves the ant $\lfloor AntSize/2 \rfloor$ steps in the current direction. If the path is blocked by obstacles, ants or borders, the ant is moved only the possible number of locations. Action PickupFood makes the ant pickup a random food object within the nearest vicinity. Food is not consumed, just carried. There is no difference in carrying capacity of the ants. Action DropFood makes the ant drop a carried food item at the current location. Action TurnLeft makes the ant turn anti-clockwise to the next Moore Neighbourhood location. Action TurnRight makes the ant turn clockwise to the next Moore Neighbourhood location. The sensors and actions are coupled together via behaviours, which are described next.

2.1.2 Behaviours and Decision Making

The virtual ants are controlled by a motivation-network, which is a decision-making model of animals that is based on response-curves. As such it is a model that builds upon the idea of response-thresholds by Bonabeau, merely with several thresholds for each behaviour, and the response-curves used are different [0]. In a motivation network a number of behaviours compete for execution based on a maximum score. For sake of planning in the foraging task, we have constructed a number of high level behaviours, that are solely composed of the sensors, and the basic actions presented earlier. The following behaviours were implemented in the system: In behaviour findFood, if food can be sensed, the ant will go to that location. If no food can be sensed, the ant will wander almost randomly around, with a preference for moving away from the nest. If food is found, it is not picked up automatically. In behaviour findHome, if the ant is within the nest and it carries food, it will drop this food item. If the ant is away from the nest it will follow the nest sensor towards the nest, and avoid obstacles underway by use of the ant and clay sensors. In behaviour takeFood, the action PickupFood is called. In behaviour dropFood, the action DropFood is called.

The behaviours each used the following sensors: In behaviour findFood and findHome, the senseCarryFood sensor was used. In behaviour takeFood, the senseFood, senseCarryFood, senseAnt and senseFrustration was used. In behaviour dropFood, the sensors senseFood, senseCarryFood, senseAnt, sensefrustration and senseClay was used. The motivation-network is implemented such that each sensor S_i of a behaviour B_k is mapped to a real value, $f_{k,j}(S_i)$, in the interval [0,1] by a mapping function $f_{k,j}$. The mapped sensor value, or the average of all the mapped sensor values (of the behaviour uses more than one sensor) is called the score of the behaviour. The behaviour with the highest average score is selected for execution at each time step. Our mapping functions are 5 degree polynomials discretized by 6 points in the interval [0,1]. Evaluation of a point in the interval [0,1] is done by interpolation. This coupling of sensors to behaviours via mapping-functions enables the ant to respond in a very differentiated way based on the shape of the mapping functions. The mapping functions (the 6 real numbers) are encoded in virtual genes as real numbers. Please notice that even though all ants in a give colony share the same genes, their behaviour is different. This is due to differences in sizes (which relates to obstacle avoidance), and local sensor stimuli, that e.g. eventually will influence the frustration sensor differently. The reason that we have hand-coded high level behaviours and not evolved them is due to the emphasis on the spatial patterns of division of labour, and only secondary the behaviour itself. See figure 1 for a model of the motivation network.

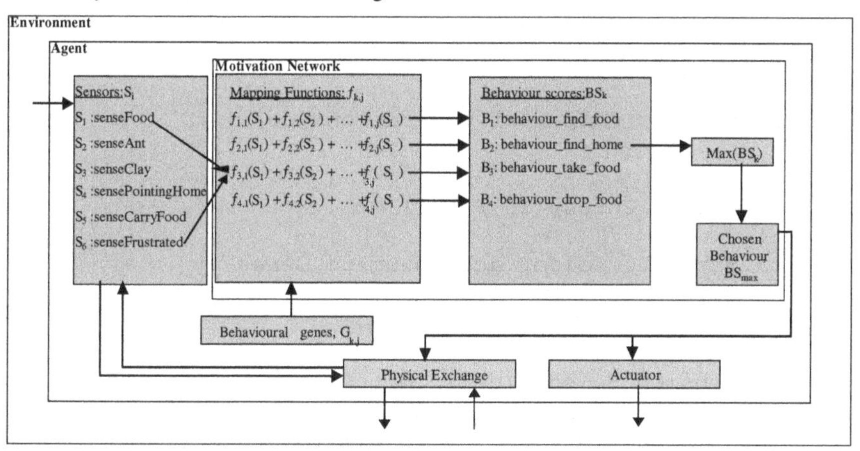

Fig. 1. Motivation Network. The coupling between sensors and behaviour through mapping functions which are encoded as artificial genes.

2.2 The AntColony EA

As noted, the 6 sample-points of every mapping function, are encoded in a gene-string that is used by an individual (a colony) in the EA. Because of this all ants in the colony share the same gene. This model uses 11 mapping functions, and thus maximizes the fitness over 66 real-variables. Mutation and crossover operators are applied to a hand coded gene-seed to initialise the population. The fitness of a colony is calculated on the number of food items the ants drop in the center of the nest (zone 0) relative to the number of ants and the number of simulation steps. Because the simulation is highly stochastic, a sliding average of 5 generations was used for fitness calculation (see Equation 1 below). Please note that fitness is calculated on the aggregate result of a whole colony simulation, not on the individual ant. The individual ant does not exist in the EA model used.

$$\text{Fitness(Colony)} = c * (\text{Foods delivered last 5 generations}) / (\text{Number of Ants} * \text{Simulation steps in colony} * 5) \quad \mathbf{(1)}$$

To further prevent stagnation due to stochastic influence we also copy the best colony to the next generation, unaltered (elitism). An outline for the EA algorithm can be seen below. As such the EA works as usual, but with the little twist that the fitness evaluation of an individual is a simulation of the colony. The colony is simulated according to the genes, which encode the behavior. An outline for the EA algorithm can be seen below.

```
AntColonyEA()
{
  t=0
  Initialise P(t)

  for a fixed number of generations do
  {
    t=t+1
    for each Colony in P(t-1) do
    {
      Simulate Colony according to Genes
      Calculate fitness of Colony
    }

    TournamentSelect (n-1) colonies from P(t-1) to P(t)
    Mutate all colonies in P(t)
    Recombine all colonies in P(t)
    Copy Elitist Colony from P(t-1) to P(t)
  }
}
```

Fig. 2. The AntColony evolutionary algorithm. The TournamentSelect uses size 2. The variable ‘t’ is used as generation counter, and ‘P(t)’ designates the Population set at time ‘t’.

3 Experiments and Results

Different experiments were performed with varying simulation-parameters. For the AntColonyEA a population of 50 colonies was chosen. This might not be a very big population compared to the complexity of the solution-space, but the restriction had to be made due to the very cpu intensive nature of this setup. All AntColonyEA simulations were run for 50 generations. The mutation rate used was $p_m = 0.7$, with the time dependent variance σ^2(generation) = 1/(1+generation)., which is also called simulated annealing For the crossover operator we used $p_c = 0.7$, and it was implemented as a swap operator, i.e. genes were swapped between individuals and not arithmetically combined. On the colony/individual level we used 60 ants in all, 30 of size 4x4, and 30 of size 6x6. Simulations with different ratios between sizes were also done; these are reported briefly in the results section. Each colony was simulated for 2000 time steps. The grid world is of size 300x300 locations. Food was placed randomly in circles in the perimeter of the world, and replenished all the time, so lack of food is not an issue in the optimization. One set of experiments was performed without spatial constraints, and another one with spatial constraints. These sets are then subsequently compared. The spatial constraints in the experiments were small clay objects spaced evenly around the center of the nest with the perimeter of the nest being a little more open than the center. The nest structure is open enough for the big ants to move around, albeit with a relative high probability of colliding with clay objects. The center of the nest, called zone 0 is the destination for the foraging task, and food items dropped there are recorded as delivered. The rest of the nest is called zone 1 and extends 1/3 out of the distance to the world-border. The rest of the world is zone 2. For later comparison around 0,7% of the total area is in zone 0, 9,7% is in zone 1, and the rest 89,6% in zone 2. Please see figure 5 for a picture of the simulation environment.

3.1 Statistical Results

What we are looking for in this section is some pattern of how the two sizes of ants organize themselves spatially. As can be noted in figure 3, there is a not surprising difference in the fitness obtained in the two test set-ups. The setup without spatial constraints outperforms the one with spatial constraints by a factor of 2,5 (a). From (b) we can tell that without spatial constraints the two sizes of ants will drop food items at the goal, whereas in the right graph the big ants settle on an average dropping distance that is somewhat near the nest border, and the small ants drop close to the centre. In this setup the optimal thus seems that the big ants drop the food further away from the centre. It does not pay off to travel the narrow path to the centre. The relative premature flattening of the fitness is probably due to our non-recombining crossover operator. A diversity measure in the start revealed that the population spanned some 60% of the search space, but only 1% in the end.

In table 4 and 5 different statistical numbers for the time spend, and the number of food items dropped is listed according to the zone. Table 4 confirms the plots in figure 3 in that the different ants spend an equal amount of time in the zones, as well as having an equal dropping ratio pattern in the setup without spatial constraints. With introduction of the spatial constraints we see a new pattern in table 5. In zone 0, the small ants dominate in time steps spend, and we can see in the food drops part of the

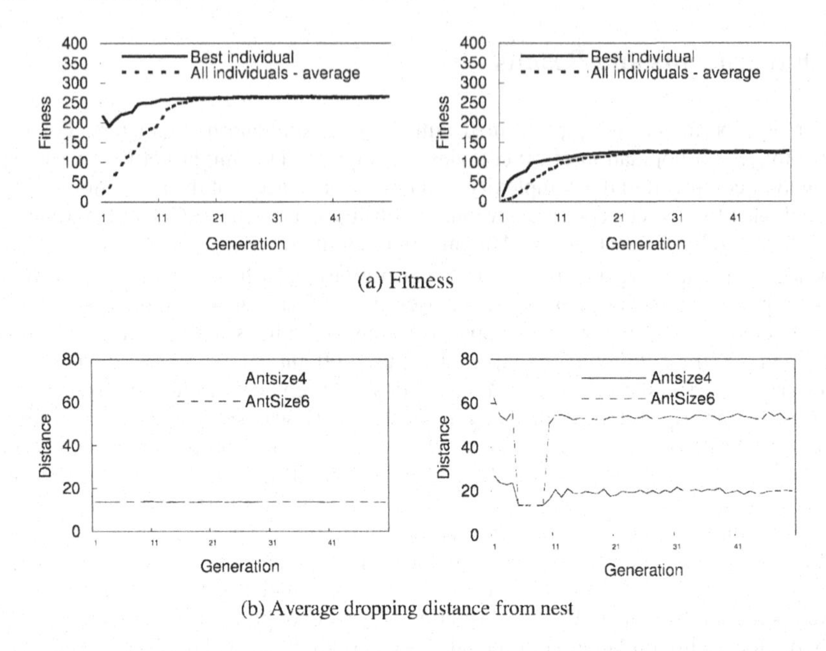

Fig. 3. Statistical results. The left column is results from the test setup without spatial constraints; whereas the right column is with spatial constraints.

table, that they have no less than 96% of the drops in this zone. This is actually quite odd, because if the big ants had carried food in this area they would have dropped it. The only explanation is that they have had a hard time getting out of the nest probably after dropping food further away. In zone 1 we see that the small ants still spend more time here than the big ants but they have only 25% of the food-droppings. This could only indicate that they forage much of their food here, and bring it to zone 0. Seemingly the supply of food around the nest is not abundant enough for the small ants to stay there, so they spend quite some time in zone 2 also. In zone 2 the big ants have most of their droppings and most of their time. In all we see a clear spatial pattern of where the different ants spend their time, and also where they deliver food, namely that the small ants spend most of the time inside the nest, bring food from the nest entrance (zone 1) to the nest centre. The ratio between big ants and small ants does not seem to be optimal though, or the distance from the perimeter to the nest entrance, and from the nest-entrance to the centre is not right. One would expect that the big ants would have the advantage in the obstacle-less zone 2 and moving faster also, whereas the small ants have an advantage with higher dexterity in zone 0 and 1, but they move slower. As the small ants go forage in zone 2, there must be a lack of big ants. In test set-ups with pure big ants, or pure small ants we didn't get as high fitness as the mix population, so this division of the labour in spatial areas must actually represent a better and more optimal solution.

Table 1. Time steps spend and food items dropped in the different concentric zones. Simulation without spatial constraints. Table should be read column wise.

	Time spend			Food drops		
	Zone 0	Zone 1	Zone 2	Zone 0	Zone 1	Zone 2
Antsize 4	49%	48%	50%	41%	46%	0%
Antsize 6	51%	52%	50%	59%	54%	0%

Table 2. Timesteps spend and fooditems dropped in the different concentric zones. Simulation with spatial constraints. Table should be read column wise.

	Time spend			Food drops		
	Zone 0	Zone 1	Zone 2	Zone 0	Zone 1	Zone 2
Antsize 4	63%	60%	40%	96%	25%	9%
Antsize 6	37%	40%	60%	4%	75%	91%

3.2 Evolved Mapping Functions

The mapping-functions that are encoded in the genes of the fittest colonies are listed in this section. Please note that the shape of one mapping function is co-evolved together with other mapping functions. So the mapping functions of all behaviours should be seen as a whole. In figure 4, we see that in both test set-ups, it is an important behaviour for the ants to find food (a) if it is not carrying any. The peak in the left column must be a stochastic flux of evolution as the sensor is binary-valued. Find-Home behaviour is important when the ant is carrying food (b). The takeFood behaviour has highest score at 0 in both set-ups. In the , which is when carry sensor is 0, and the foodsensor is 0. This means that the ant is not carrying food, and that food is 0 degrees just in front of ant and thus able to pickup. It seem from the ant sensor, that this behaviour in the non-spatial constrained set-up is most important when no ants are around (sensor value=0). The dropFood behaviour has optimum at 1 in both set-ups. In the spatially constrained setup the behaviour required values of 1 for all censors. So the ant should carry food, it should be very frustrated, sense a lot of food, and movement should be blocked by clay. This is a behaviour that could be coined kamikaze, in that only when all odds are against it, it will drop the food. (note the foodsensor is 1=360 degrees). Later we shall see a typical simulation snapshot where this is the case, namely at the edge of the nest. In the non-spatial constrained set-up the most notable difference being that the ant sensor need to have value 0, which indicate that crowding around the drop zone is not an issue in this set-up.

The emerged sensors seem quite reasonable. The dropping of food items has evolved to depend on high values of the ant and clay sensors, as well as high value for the frustration sensor. This means that the ants will not drop food prematurely in the outer zones. Dropping in zone 0 happens automatically, so this need not be handled by the behaviour. In terms of division of labour we see that the foraging task is split up when the spatial constraints are too high.

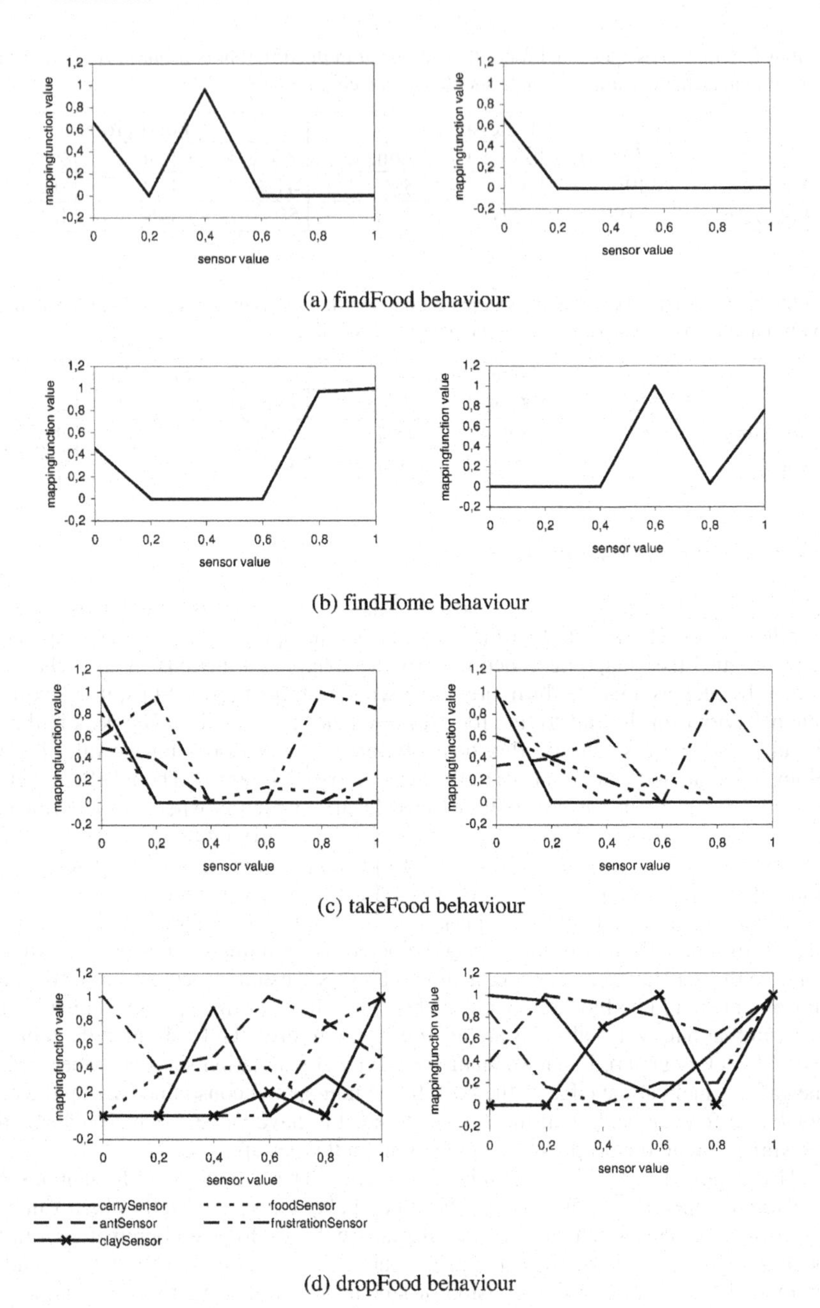

Fig. 4. The resulting mapping functions encoded in the fittest colonies. The functions in the left column is from simulations without spatial constraints, whereas the functions in the right column has spatial constraints. The legend in the bottom-left explains the lines. To calculate a score for a behaviour one would take the average of all mapping functions applied to their sensor input.

3.3 Emergent Patterns

As a typical view of a client simulation we see figure 5. We see roughly 3 times the number of small ants in the nest compared to big ants. Big ants do get into zone 0, as at least one food carrying big ant is heading for the nest centre, and another big one is leaving out bounds for more food. Note the food scattered around the nest boundary, which is ready for small ants to pickup. In the left part of the nest area not much food is placed around the nest, so here quite a few smaller ants have gone for food further out. This supports the observations in earlier sections on the 40% time used in zone 2 for the small ants. With this, relative low, number of ants in the world, we will primarily see emergent effects caused by the clay objects, as the ants are too few to really constrain each other spatially.

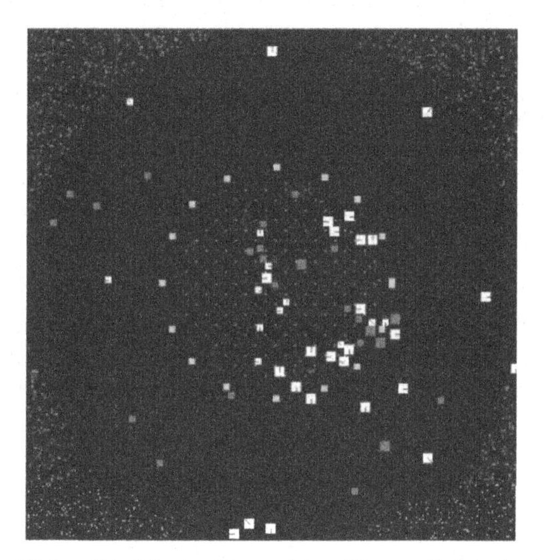

Fig. 5. The AntColonyEA Client with graphics. The small dots in the perimeter are food, whereas the square and small dots in the centre are clay blocks. The squares with a direction-vector drawn on them are ants. The darker ones carry food, the light ones carry no food.

4 Discussion

In the results sections we have seen good evidence for a spatially division of the ants performing a foraging task both in a non-constrained, and in a constrained environment. The difference between the two tests set-ups show that when spatial constraints are introduced the division of the labour changes, such that the individuals that are better fit for a specific subtask performs this. The model introduced, assumes very little, and yet evolves priority mapping-functions that encode a dynamic behaviour that divides the labour. So what was important in the model? For one, the ability of the ants to sense their immediate fit for the foregoing task at a specific location with a specific setup of objects in the vicinity seemed very important. Test simulations without the frustration sensor confirmed that the right behaviours were difficult to evolve without this introspective sensor. Many observations in nature justify this introspective-sensor, as e.g. bees estimate their own value at a specific task by measuring waiting-times. If they wait too long they will give up, and revert to other tasks. As the only difference between ants were their size, which in turn implied differences in their sensory inputs in certain situations, this sensor could trigger dynamic behaviour that depended on the specific ant. This idea seems to underlie many theories of division of labour. Namely that it is an emergent spatial effect that comes about by the interactions of the ants, the structure of the task and the ant's ability to transfer to subtasks if needed [6,7]. That the small ants eventually ended up in the nest area primarily is not something we coded, but this was an emergent effect of the big ants being quicker at fetching food, and the availability of food in the nest area for the small ants, otherwise they would transfer to foraging in the outer zones. It is also clear from our testing that the setup of the spatial constraints means a lot to the emergent patterns.

Implications for computer science could be in distributed autonomous software-agents. This study suggest that if these agents are equipped with measuring/sensory capabilities that enable them to estimate their own fit/success of a particular job, and then change behaviour or task accordingly to obtain better fit, then dynamic behaviour could be the result. The good thing about the ants in this study is that they are not smart by themselves, but they change behaviour based on own measurements, e.g. via frustration sensors, and yet the emergent patterns of cooperation at the overall level seems intelligent.

Acknowledgments. The author would like to thank Ilja Booij for cooperation on the simulation framework. Thiemo Krink for advice and guidance (and for generally being optimistic), Rasmus K. Ursem, Rene Thomsen and Peter Rickers for constructive advice.

References

1. Bonabeau, E. 1999. Swarm Intelligence , From Natural to Artificial Systems. Oxford University Press.
2. Bourke, A.F.G. & Franks, N.R. 1995. Chapter 12. Social evolution in Ants. Monographs In Behavior And Ecology. Princeton.
3. McDonald, P. & H. Topoff. 1985. Social regulation of behavioral development in the ant. Journal of Comparative Psychology 99: 3-14.
4. Wilson, E.O. 1985a. Between-caste aversion as a basis for division of labour in the ant *Pheidole pubiventris*.
5. Wilson, E.O. 1985b. The principles of caste evolution. Behavioral Ecology and Sociobiology 16: 89-98.
6. Krink T. 1999. Cooperation and Selfishness in Strategies for Resource Management. Marine Environmental Modeling Seminar,Lillehammer, Norway.
7. Tofts, C. & N.R. Franks. 1992. Doing the right thing: ants, honeybees and naked mole-rats. Trends in Ecology and Evolution 7:346-349.
8. Tofts, C. 1993. Algorithm for task allocation in ants (A study of temporal polyetism). Bulletin of Mathematical Biology 55:891-918.
9. ferber, J. 1999. Multi-agent Systems – an introduction to distributed artificial intelligence. Addison Wesley.

Emergent Organisation in Colonies of Simple Automata

I.W. Marshall and C.M. Roadknight

B54/124, Adastral Park, Martlesham Heath, Ipswich, Suffolk, UK. IP5 7RE
ian.w.marshall@bt.com, christopher.roadknight@bt.com

Abstract. We have simulated a colony of approx 3000 simple automata, with interaction behaviours that attempt to emulate those observed in stromatolite building microbes. The colony is able to maintain a diverse genome that can metabolise large quantities of up to 100 different nutrient molecules simultaneously. Group organisations that process short lived molecules more rapidly than others readily emerge under the correct conditions. We have applied the lessons to the control and management of a distributed communications system.

Introduction

Stromatolites [1] are rock mounds built by heterogeneous colonies of bacteria [2]. Colonies of this type can be found today in Shark Bay, on the coast of Western Australia. Fossil colonies can also be found throughout earth history – currently the oldest known was found in rocks dated at 3.5Ga, also in Western Australia [1]. The colonies exhibit sophisticated internal organization that emerges from interactions between individual colony members. In comparison with higher organisms such as ants and termites, the interactions are relatively simple. It is thus possible that they could be completely understood and modeled. In the work we report here, we have made no attempt to model a realistic colony. Instead we have taken the known interaction types from microbial colonies, and modeled a colony of autonomous automata that exhibit similar interaction styles. The aim was to assess whether the known interactions are sufficient to produce stable colonies with useful emergent properties. Our ultimate objective is to discover simple and effective ways of enabling self-configuration of large-scale complex structures, such as next generation communication networks. Previously [3] we have described how the colony can evolve and how this could be applied. As an illustration of the benefits of emergent colony behaviours, we have also demonstrated how the colony can be used to generate differentiated quality of service in an autonomous and emergent manner [4]. In this paper we focus on the relationship between our simulated colony and the interactions in real colonies, and show for the first time that our colony exhibits emergent structure similar to that found in real stromatolite building communities. The structure enables the colony to metabolise available food sources extremely efficiently. We go on to demonstrate how this can be used to balance system load in a computational environment

J. Kelemen and P. Sosík (Eds.): ECAL 2001, LNAI 2159, pp. 349-356, 2001.

Interaction Model

Bacteria have no differentiated sensory organs, so interaction is necessarily limited. Our simulation allows for limited motility (as exhibited by *Spirochaetes*) and the resultant physical interaction between individuals attempting to occupy the same space, plasmid interchange (used in previous bacteria inspired genetic algorithms [5,6]), and chemical exchanges via the environment. The chemical exchanges contain no explicit messages from one individual to another. The interaction is simply that if an individual metabolises a chemical it reduces the local concentration of that chemical, and increases the local concentration of the metabolite that it excretes. These changes in turn affect the behaviour of near neighbours, and the colony wide diffusion gradients can lead to rich behavioural patterning [7]. Chemical diffusion is modeled by treating molecules as packets, and routing the packets across a dense grid of 4 port switches, always away from their source (where concentration is highest). This allows the diffusion model to be discrete, and easily combined with the individual automata we use to model the microbes. It also makes application of the results into a communication network very straightforward. We believe that our approach to modeling chemical interaction is unique. Motility is constrained to slow movement along chemical gradients, where movement is not blocked by other microbes, and is thus much less significant than in models based on tools such as SWARM [8]. The microbes evolve using a modified version of the distributed GA we described in earlier work [3] that uses plasmid interchange to play the role of crossover. The modifications make the GA completely decentralized, and are presented here for the first time.

Experimental Details

The bacterial evolution algorithm has already been discussed in several papers [3,4]. To summarise, our model randomly places automata (with random initial genomes) at some of the vertices of a 2-dimensional grid that forms their environment. The genome of each automaton (crudely simulated bacterium) contains plasmids that enable the host to metabolise different kinds of available nutrients, and earn a reward (energy) for each molecule metabolised. For the purposes of this research a plasmid is the specific software that requires a certain nutrient plus a set of subjective rules as to whether the nutrient should be accepted (eg. Software A will be run if the unit is less than X% busy and has a queue smaller than Y). There are up to 100 different nutrient molecules. A random subset of these molecules is injected at each vertex along one side of the grid, at the beginning of every epoch (an epoch is 100 timesteps or 1 hop). The size of the injected subset is a variable of the model, and can frequently be close to zero. Once injected, nutrient molecules are forwarded (away from or parallel to the side they entered) automatically from vertex to vertex until a vertex is reached with a resident 'bacterium'. If a suitable gene is present, and the bacterium has not already got enough nutrients, the molecule is metabolized. Otherwise it is forwarded after a small delay. Each movement from one vertex to another is referred to as a hop. The bacteria can maintain a'queue' of up to 100 nutrient molecules before they are forced to forward nutrients. If a lot of molecules are decaying in the queue the bacterium can move one place up the grid if there is a vacant vertex. If no molecules are decaying the bacterium can move one place down the grid if there is a vacant vertex. They communicate with neighbours immediately above and below, so swaps are also possible. The time it takes to process 4 molecules is referred to as an epoch (about 1 hop). The colony evolves through the exchange and mutation of plasmids. Bacteria that accumulate sufficient energy replicate. Bacteria that do not accumulate energy die. It is possible for some individuals to survive without ever replicating, but they must do some metabolic processing, since energy is deducted each cycle (to simulate respiration).

The network simulation here differs to the previous versions in that the colony is partially subdivided. Nutrient molecules are forwarded with some randomised lateral movement as before, but when forwarding would take the nutrient beyond the bottom of a sub-colony of microbes the nutrient is forwarded to the top of another sub-colony. Routing down the colony has also been changed slightly to speed up the diffusion process. Molecules can now be forwarded to the next OR the next but one layer. These changes enable us to spread the model across several physical computers and increase the size of colony that can be modeled real time, since links between the concentrations are less rich and slightly slower than within concentrations. Our test topology had 6 semi-independent sub-colonies. The 6 sub-colonies are connected in a star formation (1 forwards to 2 and 4, 2 forwards to 1 and 5, 3 forwards to 4 and 6, 4 forwards to 3 and 1, 5 forwards to 6 and 2 and 6 forwards to 5 and 3). Each sub-colony runs on identical hardware and has the same available resources at it's disposal.

To allow the automata to be truly autonomous, as they would be in a real colony, plasmids are no longer obtained from a globally accessible gene pool by stressed individuals. Instead randomly chosen plasmids attach to molecules that are being forwarded, in line with the following rules.

1. If an individual FAILS to process a molecule (and so must forward it) AND a random number between 0-100 is less than the no. of nutrient molecules it has already stored for later processing (max 100) THEN a 'plasmid' from it's genome is attached to the request. In simpler terms; if do not process this item but I AM busy then attach a plasmid from my genome to the item.
2. If an individual succeeds OR fails to process a food item AND its busyness (plus a constant, currently 5%) is less than a random number between 0-100 THEN take a 'plasmid' from the 'plasmid attachment' space, if there is one. In simpler terms; regardless of if I process the item or not, if I am NOT BUSY then take a 'plasmid' from the request if there is one.

For the experiments, groups of 25% of the different nutrient molecules were given lifetimes of 2, 5, 10 and 20 hops respectively, and the injection rate was around 25% of the maximum allowed by the model. This is enough to allow around 50% of the vertices to be inhabited. The processing efficiency is compared to a simulation in which evolution was blocked by stopping plasmid interchange, so the initial random gene distribution persists

Results

The evolution of one of the six sub-colonies is illustrated in Fig 1 below. The unlabelled microbes are generalists, the marked microbes are specialists adapted to handle foods with particular lifetimes (in this sub-colony) as follows; 1 =2 hops or less, 2 = 2-5 hops , 3 = 5-10 hops, 4 = 10-20 hops. The snapshots are taken after 15, 50, 200, 400, 800 and 1600 epochs. Food is being injected at the top of the colony. A substantial degree of layering is clear after 200 epochs, with individuals adapted to metabolise short lived nutrients concentrating closest to the source. This slowly improves as the colony continues to evolve.

To illustrate the benefit of the limited motility Figure 2 shows the sub-colony after 1600 epochs of an equivalent simulation when no movement was allowed. The structure is similar but somewhat less organized, and it took longer for the structure to emerge. This is because without motility it is hard for genomes adapted to process the short lived nutrients to supplant those adapted to more persistent nutrients residing at the top of the colony.

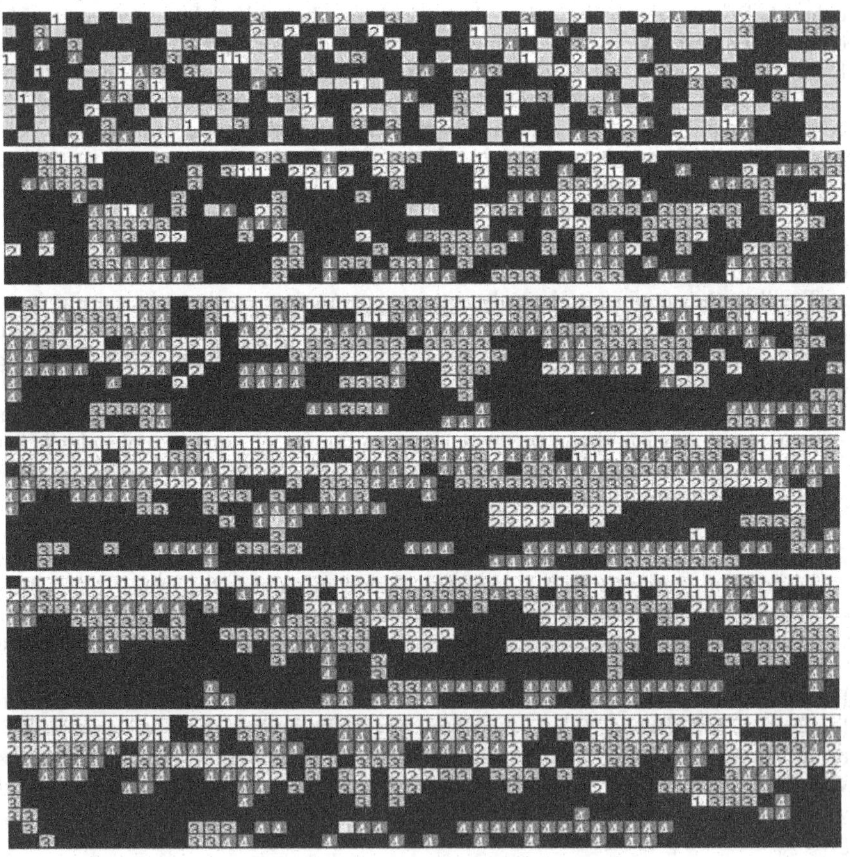

Fig. 1. Emergence of layering in a 500 member sub-colony. Initial state is at the top, images down the page are at successively later stages in the evolution of the colony (see text for details).

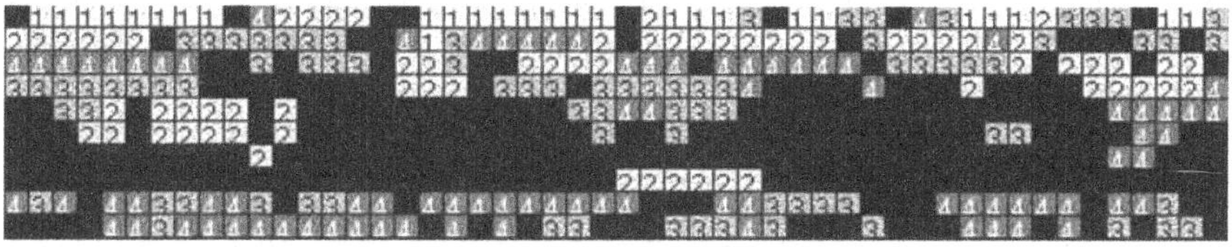

Fig. 2. Evolved colony with no motility. Annotation as for fig 1

In figure 3 we show the structure of the whole colony after 1500 epochs with a greater supply of nutrients. Each sub-colony is being supplied with 750 molecules per epoch, with an average lifetime of 8 hops. The molecules that are not processed, and do not decay are passed on along the arrows in the figure once they reach the base of the colony. Around 50 molecules per epoch are being passed on along each arrow and their average residual lifetime is 7 hops (annotated as ttl i.e. time to live). One of the sub-colonies is expanded, however in this case the bacteria are annotated differently from the first two figures. The numbers represent the average lifetime of the nutrients being processed by that individual. Layering of the genomic capability is still evident, despite the fact that in this case the colony is being supplied with more food than it can consume and as a result there are very few vacant vertices in the grid. In fact 20% of the molecules are decaying before being metabolized, largely as a result of the bacteria being less free to move to optimal positions.

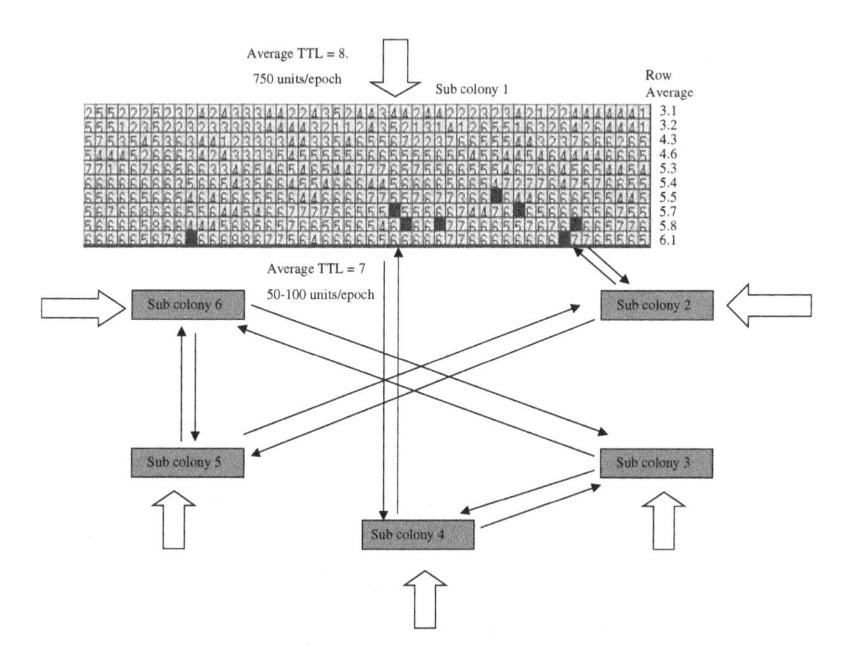

Fig. 3. View of the 3000 member colony when food is plentiful

To illustrate the temporal evolution of the colony characteristics we show the consumption efficiency of the colony against simulated time in figure 4 (for the same conditions as in figure 3). The evolved colony is compared with a colony that cannot evolve. This comparison is valid as current networks are non-adaptive, a comparison with other adaptive techniques would be useful but is beyond the scope of this paper. As might be expected the properties of the random colony do not change with time. In the evolved case however, 2 timescales of evolution are apparent. A fast initial transient due to the sharing of initial plasmids (plasmid learning) and a slow improvement due to the continuing evolution and sharing of novel plasmids via random mutations. Reassuringly the evolved colony performs rather better (80%) than the random colony (70%) with far less living individuals (1000 evolved, 2000 random). Perhaps more significantly when a new nutrient is introduced to the evolved colony it adapts, whereas the random colony does not. The graph does not show this since the new nutrient was only 1% of the total input and the small performance changes cannot be resolved at this scale.

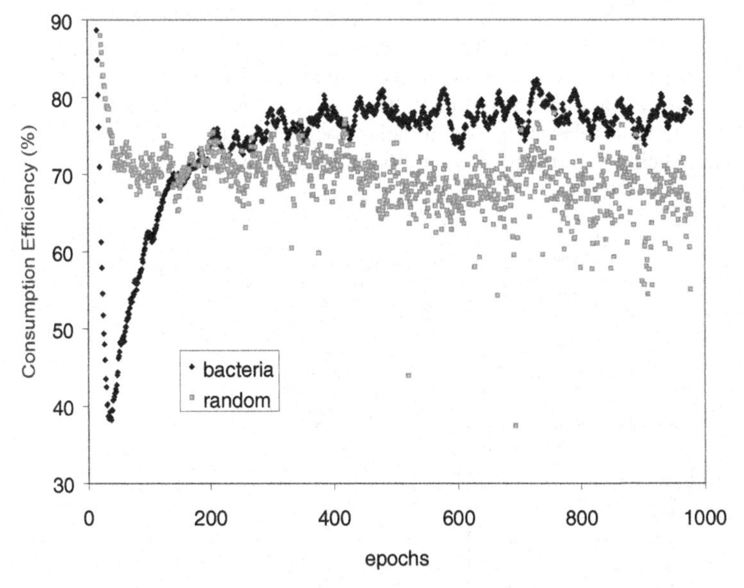

Fig. 4. Temporal evolution of the colony, comparing evolved efficiency with efficiency of a random distribution (averaged over 6 colonies)

The evolved colony also copes well with uneven distributions of the injected nutrients. In figure 5 we show the effect of progressively concentrating nutrient input into fewer sub-colonies. The x-axis has an even distribution of input (at 80% max as in figures 3 and 4) into the top edge of all six sub-colonies at the left. The figures shown are the average of all 6 members of the network. The input is concentrated until the supply to one colony is the max possible, and the other colonies are fed so that the total nutrient supply remains constant. It is clear that at all times the evolved colony is more effective than the random colony. This final graph is particularly interesting since the colony is behaving in exactly the way we would like a good distributed load balancing algorithm to. It is spreading load across available resources, ensuring a high probability of success (comparable to real computing systems at high load), whilst minimizing the resources required. We are confident that by appropriate adjustment of the simulation parameters the success rate could be increased to any desired value, and required resources would always be less than for a static distribution of resources.

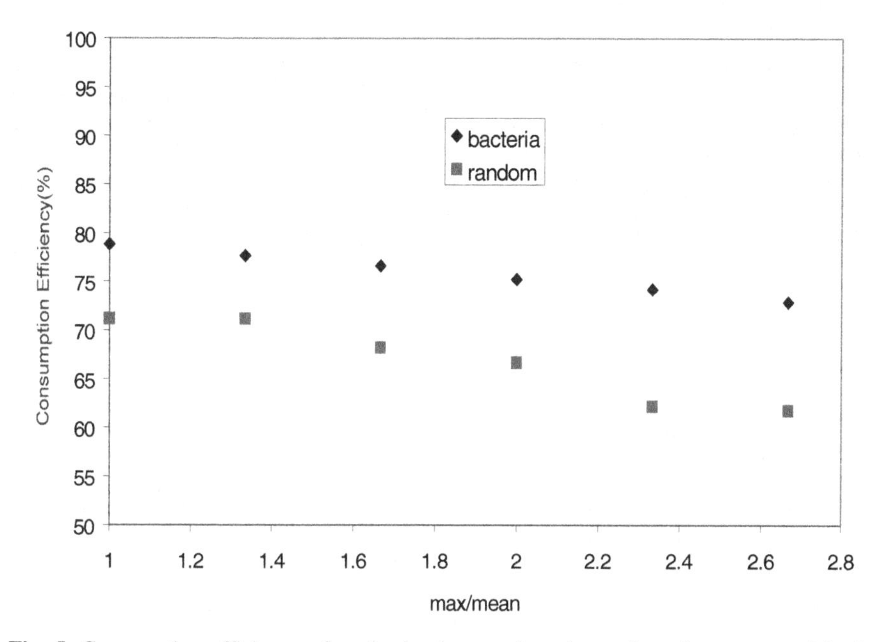

Fig. 5. Consumption efficiency of evolved colony and random colony for a range of load distributions

Discussion

The evolution of colony structure, and colonial learning that we have observed is consistent with the thinking and observations of Shapiro [9,10]. Colonies of symbol processing devices have also been investigated in the grammer system area [11]. It is our hope that this work will inspire radical new approaches to decentralized control in distributed systems, and also shed light on the fundamental issues in microbial genetics raised by Shapiro and others. Our objectives are thus similar to those of the vast majority of a-life researchers. Our methods however are a little different. In particular, the choices in our model are driven by the needs of distributed systems applications rather than a desire to emulate biology. The packet network model for distribution is one example of this, that builds on our background in networking. We believe that this approach is a productive way of avoiding the complexities of true biological versimilitude, whilst at the same time enabling us to build simulations that do more than simply replay sequences of events based on current (often highly simplified) biological knowledge. In other words our model has emergent properties because we have built into it some detailed knowledge of a real system (the Internet) with emergent properties, that is nevertheless probably considerably simpler than a real biological system.

Conclusions

We have demonstrated that enabling individuals in a distributed individual based evolutionary colony simulation to interact via a simple discrete distribution model is sufficient to produce emergent colony properties. Adding limited motility speeds up the self-organisation and stabilizes it somewhat, but does not alter the qualitative properties of the organized colony. The organized colony is a highly efficient consumer of a wide range of resources with different lifetimes. Control algorithms based on our observations can be applied to load balancing and QoS in distributed computing systems.

References

[1] S.Golubic "Stromatolites of Shark Bay" pp103-149 in Environmental evolution ed. Margulis and Olendzenski, MIT press 1999
[2] S.Sonea and M.Panisset, "A new bacteriology" Jones and Bartlett, 1983.
[3] I.W.Marshall and C.Roadknight, "Adaptive management of an active services network", *British Telecom. Technol. J.*, 18, 4, pp78-84 Oct 2000
[4] I.W.Marshall and C.Roadknight "Provision of quality of service for active services" Computer Networks, April 2001
[5] I.Harvey, "The Microbial Genetic Algorithm", unbublished work, 1996, available at ftp://ftp.cogs.susx.ac.uk/pub/users/inmanh/Microbe.ps.gz
[6] N.E.Nawa, T.Furuhashi, T.Hashiyama and Y.Uchikawa, "A Study on the Relevant Fuzzy Rules Using Pseudo-Bacterial Genetic Algorithm" Proc IEEE Int Conf. on evolutionary computation 1997
[7] E.Ben-Jacob, I.Cohen and H.Levine "Cooperative self-organisation of microorganisms" Advances in Physics, 49, 4, 395-554, 2000
[8] G.Booth, "Gecko: A Continuous 2D World for Ecological Modeling", Artificial Life, 3, pp. 147-163, Summer 1997.
[9] J.A.Shapiro, "Thinking about bacterial populations as multicellular organisms" Ann. Rev. Microbiol. 52, 81-104. 1998
[10] J.A.Shapiro, "Genome system architecture and natural genetic engineering in evolution", In Molecular Strategies in Biological Evolution, L. Caporale, ed., Annal. NY Acad. Sci. 870, 23-35 1999
[11] E. Csuhaj-Varju, J. Dassow, J. Kelemen, G. Paun, Grammar Systems, Gordon and Breach, London, 1994

Evolving Communication without Dedicated Communication Channels

Matt Quinn

Centre for Computational Neuroscience and Robotics,
University of Sussex, Brighton, U.K.
matthewq@cogs.susx.ac.uk

Abstract. Artificial Life models have consistently implemented communication as an exchange of signals over dedicated and functionally isolated channels. I argue that such a feature prevents models from providing a satisfactory account of the origins of communication and present a model in which there are no dedicated channels. Agents controlled by neural networks and equipped with proximity sensors and wheels are presented with a co-ordinated movement task. It is observed that functional, but non-communicative, behaviours which evolve in the early stages of the simulation both make possible, and form the basis of, the communicative behaviour which subsequently evolves.

1 Introduction

The question of how communicative behaviour might have originated is an interesting one, and the transition from non-communicative to communicative behaviour has long been of interest to ethologists [2,4]. Artificial Life techniques, such as agent-based simulation models, are potentially useful tools for exploring questions and hypotheses related to this transition. In particular, they enable the simulation of co-evolving, interacting organisms at the level of changes in behaviour and perception. There are a number of models in the ALife literature which simulate the evolution of an organised communication system in an initially non-communicating population of agents (e.g., [11,6,1,5,3]). In all these models, communication is restricted to an exchange of signals over dedicated and functionally isolated communication channels. This feature, I wish to argue, severely reduces the explanatory value of a model of the evolutionary origins of communication in natural systems.

Dedicated channels are a reasonable feature of a model which assumes that individuals are already able to communicate. However, explaining the *origins* of communicative behaviour typically involves explaining how it could have evolved from originally non-communicative behaviours [2,4,7]. This kind of explanation is not possible with a model which restricts all potential communication to dedicated and functionally isolated channels. However, this problem is avoided if a model allows potentially communicative behaviour to be functional (and hence acquire selective value) in contexts other than communication. In order to illustrate this point, I present a model in which there are no dedicated communication channels. Agents are evolved to perform a non-trivial co-ordinated

J. Kelemen and P. Sosík (Eds.): ECAL 2001, LNAI 2159, pp. 357–366, 2001.

movement task and are equipped only with proximity sensors and wheels. They are controlled by neural networks which map low-level sensory input directly onto motor output. It is observed that the functional but non-communicative behaviour which evolves in the early stages of the evolutionary simulation forms the basis of (and makes possible) the communicative behaviour which subsequently evolves to play an integral role in the successful completion of the task.

2 Originary Explanations

At first sight, the mere existence of communication appears somewhat paradoxical when viewed from an evolutionary perspective. "Communication", as Wilson writes, "is neither the signal by itself, nor the response, it is instead the relationship between the two" [12, p.176]. But how might such an obviously co-dependent relationship originate? What good would a signal be before there is a response for it to trigger? But equally, what possible use is a response to a signal unless that signal already exists? The apparent paradox is summed up succinctly by Maynard Smith: "It's no good making a signal unless it is understood, and a signal will not be understood the first time it is made" [7, p.208]. Of course, this paradox is only apparent. There are at least two possible parsimonious explanations for the evolution of a communicative interaction. Firstly, organisms can often benefit from attending to many aspects of other organisms' behaviour, even when that behaviour is not a signal. For example, there are obvious benefits to responding to movements indicative of, for example, imminent attack, rather than actually having to respond to the attack itself. Thus organisms may evolve to react to informative, but non-signalling behaviours; these behaviours may in turn evolve to capitalise on the response they now evoke, thereby becoming signals in their own right [2,4]. Secondly, most organisms are constantly responding to all manner of environmental stimuli. Hence a signal might evolve because it is able to trigger one such pre-existing response. For example, it has been argued that the courtship signal of the male water-mite *Neumania papillator*—namely, rapid leg movements causing water-borne vibrations—evolved to exploit an existing vibration response in the female which served to locate prey [8]. Note that both types of explanation resolve the apparent paradox by postulating an originally non-communicative role for either the signal or the response.

Bearing this in mind, consider the simulation models of the evolution of communication mentioned in the introduction [11,6,1,5,3]. Despite their many differences, these models share important common features. Firstly, in all of these models, agents are equipped with a set of behaviours that are designated as potential signals. These potential signals have no function or consequence in any context except signalling. Secondly, agents have one or more dedicated sensory channels with which to perceive these signals. These channels are sensitive to no other stimuli except signals, and signals register on no other sensory channels. What makes these features so problematic? It is not just the rather anomalous presence of specialised communication apparatus in a non-communicating population. More problematically, it is that these models clearly prevent all potentially communicative behaviour from having any non-communicative function. Since neither the production nor the reception of a designated signal can confer any selective advantage outside of a signal-response relationship, these models can tell us nothing about why either the signal or the response exist before the existence of communication. Clearly then, if we wish to produce models that can provide such an explanation, we need to allow for the possibility of potential signals and responses initially evolving in non-communicative contexts. In the next section, I introduce a model which meets this requirement.

3 The Model

For this experiment, a co-operative co-ordination task previously implemented in [9] was adopted. Agents are simulated Khepera robots, equipped only with short-range active infra-red (IR) sensors and two motor-driven wheels (the simulator is described in [9]). The body plan of a Khepera robot is shown in figure 1. Each animat is controlled by an evolved neural network (described below). This network takes input directly from the sensors and uses this input to specify motor-speed outputs. Clearly there are no dedicated communication channels incorporated in this model. Furthermore, neither signals nor responses are initially incorporated into the model (indeed nor is any other sort of meaningful behaviour).

Agents are evolved in a single population, and are evaluated in pairs. Their task is as follows: The pair are placed in an obstacle-free environment in one of a number of possible initial configurations such that each is within sensor range of the other. They are then given 10 seconds in which to move at least 25 cm (approximately 10 agent radii) from their initial position whilst staying within sensor range and not colliding with one another. Whilst easy to describe, this task presents a non-trivial distributed control problem. Covering the required distance is unproblematic (agents have a top speed of 8cm/s). However, a number of other difficulties must be overcome. The IR sensors provide an animat with a noisy and non-linear indication of its distance from an object and have a range of just under 5cm. Should agents move beyond a distance of 5cm apart, they will have no indication of each other's location. Simply moving away and 'hoping' to be followed would thus be a bad strategy. Furthermore, since agents are cylindrical, their IR sensors do not reveal the direction in which another agent is facing. This means that orientation matching strategies, which form an integral component of flocking algorithms [10], cannot be utilised in this task. From [9] it appears that successful completion of this task appears to require some form of role differentiation between agents (e.g., a 'leader' and a 'follower'), and thus some interactive mechanism by which roles can be allocated. Communication is clearly one way in which this could be achieved.

The nature of the task facing the agents means that an individual's ability to perform well is significantly affected by its partner's behaviour. In addition, and particulary in the initial stages of evolution, variation in starting positions also has a large impact on the success of a pair of agents. For this reason, agents

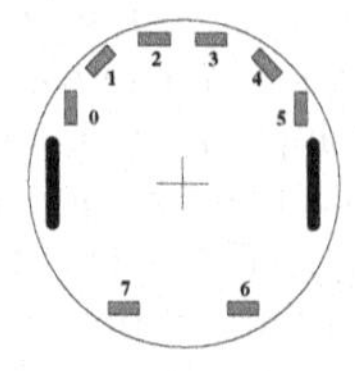

Fig. 1. Left: Khepera mini-robot, showing the 8 IR sensors, and two motor driven wheels. **Right:** Parameters defining the agents' starting positions.

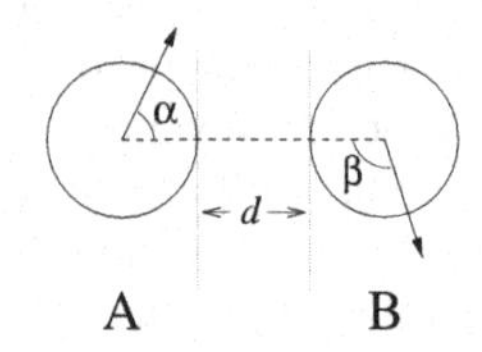

were given a large number of evaluations with many different partners and a fixed selection of starting positions. Starting positions were defined by three variables, d, the agents' distance apart, and α and β, the orientation of each agent relative to the other, as shown in figure 1. A sample set of starting configurations was chosen such that, $d \in \{1.5, 2.5, 3.5\text{cm}\}$ and $\alpha, \beta \in \{0, \frac{2\pi}{5}, \frac{4\pi}{5}, \frac{6\pi}{5}, \frac{8\pi}{5}\}$. Of these, 30 rotationally symmetrical duplicates were removed, leaving a set of 45. Each agent in the population was evaluated twice from each starting configuration, with each of these 90 trials undertaken with a different partner. The fitness score attributed to an agent was simply the mean of all the scores it achieved in these trials. In an individual trial, each agent of the evaluated pair received an equal score. There was therefore no conflict of interest between partners in an evaluation; each individual pairing was a co-operative venture. The score given to both agents at the end of a single trial is given by the following function:

$$\left| P.\left(\sum_{t=1}^{T} \left[g_t.(1 + \tanh(s_t/10)) \right] \right) \right|$$

Here s_t is the amount by which the distance between the agents exceed the 5cm sensor range at time step t (if this is not exceeded $s_t = 0$). P is a collision penalty scalar, such that $P = \max(1 - c/(1 + c_{max}), 0)$, where c is the number of collisions that have occurred, and c_{max} is the maximum number of collisions allowed, here $c_{max} = 10$. Finally the function g_t is a measure of any improvement the pair have made on their previous best distance from the starting point. This is a function of d_t, the euclidean distance between the pair's location (i.e., its centre-point) at time t and its location at time $t = 0$, and D_{t-1} is the largest value that d_t had attained prior to time t. The value of g_t is zero if the pair are not improving on their previous best distance or if they have already exceeded the required 25cm, otherwise $g_t = d_t - D_{t-1}$. Note that scores are normalised, so the maximum possible score is 1.

Agents are controlled by artificial neural networks. These networks comprised 8 sensor nodes, 4 motor nodes and some number of artificial neurons, connected together by directional excitory and inhibitory weighted links. The thresholds, weights, decay parameters and the size and connectivity of the network were genetically determined (as detailed in [9]). At any time-step, a neuron's output, O_t, is determined by the value of its 'membrane potential', m_t. If m_t exceeds the neuron's threshold then $O_t = 1$ (the neuron fires) otherwise $O_t = 0$. Here m_t is a function of a neuron's weighted, summed input(s), and the value of m_{t-1}

scaled by a temporal decay constant, such that:

$$m_t = \begin{cases} (1-\gamma_A)m_{t-1} + \sum_{n=0}^{N} w_n i_n & \text{if} \quad O_{t-1} = 0 \\ (1-\gamma_B)m_{t-1} + \sum_{n=0}^{N} w_n i_n & \text{if} \quad O_{t-1} = 1 \end{cases}$$

where the decay constants γ_A and γ_B are real numbers in the range [0:1] and w_n designates the weight of the connection from the nth input, i_n. Each sensor node outputs a real value in the range [0.0:1.0], which is simple linear scaling of its associated IR sensor. Motor outputs consist of a 'forward' and a 'reverse' node for each wheel. If the summed weighted inputs to an output node are positive its output will be 1, otherwise 0. The output for each wheel is attained by subtracting its reverse node output from its forward node output.

4 Analysis

A simple generational evolutionary algorithm was employed to evolve populations of 180 (initially random) genotypes. A total of 30 runs were carried out, with each population being allowed to evolve for 2000 generations. Of these runs 27 produced successful solutions to the task (where success was defined as an evaluation score consistently in excess of 0.975). The solutions found by the successful runs had a number of similarities. In each case a pair were successful primarily because one agent adopted a 'leader' role whilst its partner adopted the role of 'follower'. However, these roles were not pre-assigned. All successful runs ultimately produced homogeneous solutions; thus, neither agent was intrinsically biased toward adopting either role. Investigation revealed that it was the interactions between agents which served to establish which role each would adopt. This section focusses in some detail on a single run. It starts with a description of the evolved behaviour of the agents, with particular reference to those aspects which appear to function as signal and response and to co-ordinate the allocation of 'follower' and 'leader' roles. Next, analysis of the evolved neural network controller is presented to confirm that the signal and response do indeed perform the causal roles ascribed to them. Having established that communicative behaviour has indeed evolved its origins are then addressed. Analysis of the early stages of the evolutionary run is presented in order to show that the evolution of communication in this simulation affords a satisfactory explanation of the kind set out in section 2.

4.1 Evolved Behaviour

The successful achievement of the task can be very briefly described as follows. Initially, each agent rotates counter-clockwise until it is almost facing its partner. I shall refer to this process as 'alignment'. Once agents have become aligned, and after a (generally) brief interaction, one agent reverses away from the other, and is almost immediately followed by its partner which moves forwards (see figure 2(iv)). For the remainder of the task the pair maintain this configuration, with the leader moving backwards and the follower moving forwards, together achieving the required distance. How are the roles of leader and follower allocated? Observation of the team from each of the starting positions shows that whenever there is a difference in alignment times, this difference plays an important part in role allocation. Figure 2 shows an example of two agents which become aligned at different times. The first agent to become aligned moves very close to its still-rotating partner and then waits, oscillating back and forth. Subsequently the second agent becomes aligned and reverses away, closely followed by its partner.

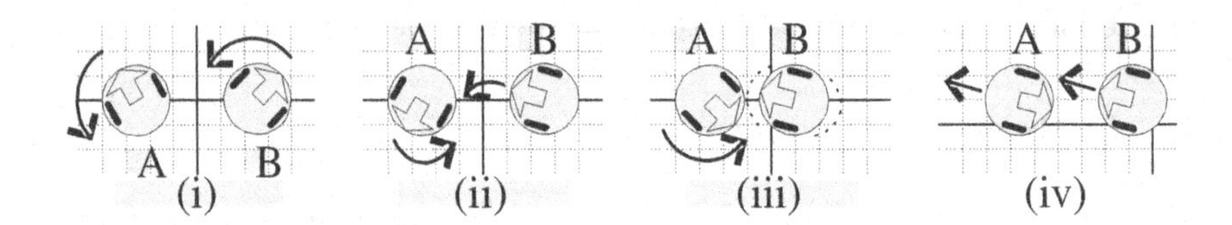

Fig. 2. An example interaction: (i) Both agents rotate anti-clockwise; (ii) Agent B becomes aligned first, and moves toward A; (iii) Agent B then remains close to A, moving backward and forward staying between 0.25–2.0cm (aprox.) from A. (iv) Once agent A becomes aligned it reverses and is followed by B.

It seems then, that the actions of the first aligned agent serve as a signal to the second. If an agent perceives the signal whilst still rotating, it will adopt the leader role. However, if it becomes aligned without having perceived the signal, it will itself perform the signalling action and subsequently take the follower role. Such a strategy would clearly serve to coordinate role allocation and aid in the successful completion of the task[1].

Analysis of the evolved neural network was undertaken to ensure that the behaviours identified as signal and response did indeed perform the causal roles suggested above. Figure 3 shows the evolved neural network after all non-functional neurons and connections have been removed. Note that the network only utilises two sensors. Sensor 0 is the sensor immediately in front of the left wheel, and sensor 3 is the right-hand sensor of the front-most pair of sensors (see figure 1). Since agents rotate counterclockwise, sensor 0 will normally be the first to register the presence of another agent. What the analysis presented below will demonstrate is this: In cases where there is a difference in alignment times, if sensor 0 is saturated (i.e. fully activated) *prior* to the activation of sensor 3, the result is that an agent will reverse, this constitutes perception of, and response to, the signal. However if sensor 3 is activated without the prior saturation of

[1] From a minority of starting positions there is insufficient difference in alignment times for this strategy to be effective; the procedure by which roles are then allocated is more complicated. However, analysis of the general case is sufficient for the purposes of this paper.

sensor 0 then an agent will move forwards, thereby producing the signal. This clearly does not constitute a full explanation of the network's functionality, but is sufficient for the purposes of the argument being presented.

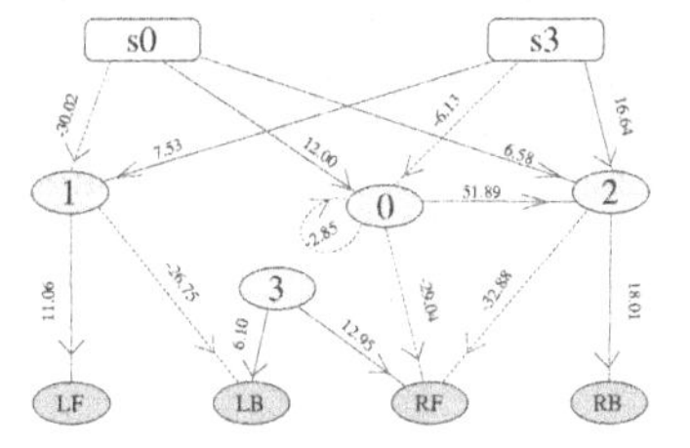

Neuron	T	γ_A	γ_B
0	24.36	0.00	0.00
1	0.14	1.00	0.00
2	10.22	1.00	0.84
3	-34.07	1.00	1.00

Fig. 3. The evolved neural network after pruning. Solid lines are excitory links and dashed are inhibitory. The table above gives the threshold, T, and decay constants, γ_A and γ_B, of each neuron. All values shown to 2 decimal places.

The first thing to note about this network is that in the absence of any sensory input, it will cause an agent will rotate counterclockwise. This constitutes an agents 'base behaviour', occurring because neuron 3 fires constantly (it has a negative threshold and no inputs) and excites the LB and RF motor nodes. Let us consider how the base behaviour is modulated by sensory input. As an agent rotates toward its partner, sensor 0 will be the first sensor to be activated. This sensor, it was suggested, receives the signal. Note that it strongly inhibits neuron 1, thus preventing any forward movement. It may also activate neuron 0. Given its non-firing (γ_A) decay rate and its recurrent inhibitory connection, neuron 0 will fire if sensor 0 is saturated (or near saturation) for 3 consecutive timesteps. If this occurs, neuron 0 will in turn activate neuron 2 and these, combining with always active neuron 3, will cause the agent to reverse. Thus, the reversing response of the 'leader' occurs because the extreme proximity of its partner saturates sensor 0. If however, an agent's partner is not close enough to trigger the response, the agent will continue rotating and sensor 3 will subsequently become active. Only minimal activation of sensor 3 is required for neuron 1 to fire, and when this occurs the agent will move forwards. Since the agent is not completely facing its partner, forward movement causes sensor 3's activation to decrease until rotation recommences, turning the agent back toward its partner, reactivating sensor 3, thus causing further forward movement. Repetition of this cycle moves the agent closer to its partner, producing the initial arcing motion shown in figure 2(ii). Collision is averted because, at extreme proximity, sensor 0 will become sufficiently active to cause the agent to reverse briefly, producing the 'oscillating' movement shown in figure 2(iii).

4.2 Origins

From the previous section it is evident that agents have evolved behavioural sequences which function as signal and response. This section aims to show the non-communicative origins of this behaviour through a description of the first 400 generations of the evolutionary simulation.

Successful agents in initial generations of the run were, unsurprisingly, those which were able to produce any form of consistent movement. Such behaviour received some score, since it displaced the pair's centre-point, but it also resulted in agents rapidly losing sensor contact with one another. By around generation 20, and through until around generation 50, agent behaviour consisted essentially in switching on both motors and proceeding, ballistically, in a straight line. The initial spread of this behaviour is reflected in figure 4b in the rapid increase, over the first 20 generations, of both the mean distance travelled by a pair during a trial, and the mean distance between them at the end of a trial.

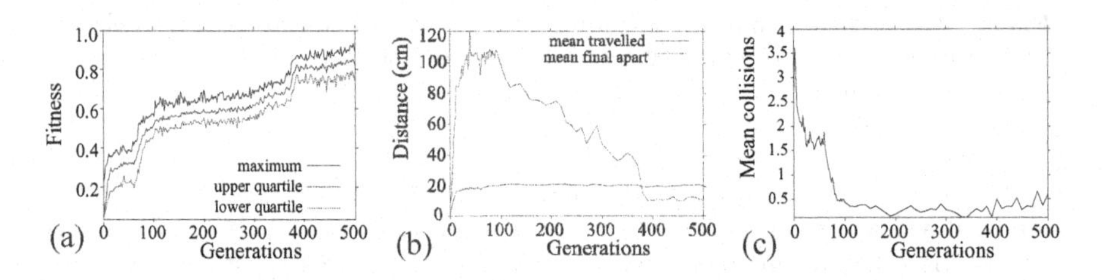

Fig. 4. Fitness and performance statistics over the first 500 generations

In the period between generations 50 and 100 there is an increase in population fitness which is strongly correlated with a drop in the collision rate (figures 4a and 4c). This reflects the progressive development of basic collision avoidance strategies. Collision avoidance generally took the form of rotating, halting both motors, or halting one and reversing the other. It increased in effectiveness over this period as more sensors came to be used to alter the underlying ballistic behaviour. One consequence of these developments is particularly relevant here. This occurred at positions where both agents' underlying ballistic behaviour led them directly toward each other. In such cases the various halting and turning responses to close proximity typically led to a form of 'deadlock' situation, in which each agents remained close to the other, moving towards and then backing away from its partner, whilst turning to the left and right.

At generation 110, the situation from most starting positions is unchanged. However, a new behaviour has appeared which allows the deadlock situation to be broken. This takes place as follows. The deadlock situation commences and initially proceeds as described above. However, after some time, one agent backs away from its partner. The partner, its collision avoidance behaviour effectively deactivated, recommences its forward progress. The pair move away jerkily, with one agent moving forward and the other moving in reverse. Such pairings rarely maintain formation for the full extent of the required distance. In addition, the reversing agent and the following agent are phenotypically different. Only one type of agent was capable of backing away from a deadlock situation, and then only when in combination with the other type of agent which itself was not capable of this reversing behaviour. Combinations of two agents of the same type remained deadlocked. Nonetheless, this behaviour—reversal in response to sustained proximity—is the origin of the response behaviour described in the previous section. Although modified in form over the remainder of the run, it is to be observed in all subsequent generations.

From this point, up until around generation 370, the polymorphism of two interdependent types persists. The population comprises those which do reverse in response to proximity ('leader' types) and those that do not ('follower' types). This period is marked by a progressive specialisation of each type to its role. The ability of mixed-type pairs to maintain the follower-leader formation improves markedly. Another significant development occurs. Agents begin to circle in the absence of sensory input, and in later generations, to rotate. This behaviour serves to increase the probability of encountering what was previously a 'deadlock' situation, but which is now potentially an opportunity for an improved score. The gradual increase in population fitness over this period is somewhat deceptive. Individuals engage in both same-type and mixed-type interactions; the increase in mixed-type performance is offset by a decrease in same-type scores over this period. The increasing tendency for pairs to either remain deadlock (same-type) or move away in formation (mixed-type) is reflected in the steady drop in the final distance between agents at the end of each trial.

In summary, agents initially evolved ballistic behaviour and then basic obstacle avoidance. In combination these led to 'deadlock' situations. Agents which subsequently came to reverse in response to sustained proximity capitalised on both of these behaviours. Situations involving sustained proximity only arose after the evolution of obstacle avoidance, and reversal was only adaptive because the original ballistic behaviour caused the non-reversing agent to 'follow'. Once the response was in existence, it cames to play an important role in shaping future behaviour. Agents began to evolve strategies which increased the probability that they would be able to trigger the response, and strategies which increased the probability that they would be in a position to respond. It was another 1000 generations before the behaviours described in the previous section were fully formed. However, it should be clear from the above that origins of those behaviours can be satisfactorily explained within the context of this model.

5 Conclusion

Open any textbook which gives an evolutionary account of animal behaviour and you will find hypothetical reconstructions of the processes surrounding the origins and early evolution of communicative behaviour which revolve around non-communicative behaviours acquiring selective value in a communicative context (for example, [2,4,7]). Artificial Life models are potentially very well suited to exploring these and related hypotheses and to critically evaluating the assumptions on which they are based. However, it is difficult see how such models can even begin to do this if they are implemented in a way that prevents non-communicative behaviour from acquiring a communicative function.

The model described in this paper has not set out to test any particular hypothesis. It is intended simply as a proof of concept. It demonstrates firstly that it is possible to evolve communication in a model without dedicated channels, and secondly, that an explanation of how communication evolves in such a model is far more relevant to the evolution of communication in natural systems than those afforded by previous models.

Acknowledgements. Many thanks to Jason Noble and Seth Bullock for their comments.

References

1. D. Ackley and M. Littman. Altruism in the evolution of cooperation. In R. Brooks and P. Maes, editors, *Proc. Artificial Life IV*. MIT Press, 1994.
2. R. Hinde. *Animal Behaviour*. McGraw-Hill, London, 1966.
3. H. Kawamura, M. Yamamoto, K. Suzuki, and A. Ochuchi. Ants war with evolutive pheromone style communication. In D. Floreano, J-D. Nicoud, and F. Mondada, editors, *Proc. 5th European Conf. on Artificial Life*. Springer Verlag, 1999.
4. J. Krebs and N. Davies. *An Introduction to Behavioural Ecology*. Blackwell, 1981.
5. M. Levin. The evolution of understanding: A genetic algorithm model of the evolution of communication. *BioSystems*, 36:167–178, 1995.
6. B. MacLennan. Synthetic Ecology: An approach to the study of communication. In C. Langton et. al., editor, *Proc. Artificial Life II*. Addison-Wesley, 1991.
7. J. Maynard Smith. *The Theory of Evolution*. C.U.P., 3rd edition, 1997.
8. H. Proctor. Courtship in the water-mites *neumania papillator*: males capitalize on female adaptations for predation. *Animal Behaviour*, 42:589–598, 1991.
9. M. Quinn. A comparison of approaches to the evolution of homogeneous multi-robot teams. In *Proc. Int. Congress on Evolutionary Computing*, 2001. (In press).
10. C Reynolds. Flocks, herds, and schools: A distributed behavioral model. *Computer Graphics*, 21(4), 1987.
11. G. Werner and D. Dyer. The evolution of communication in artificial organisms. In C. Langton et. al., editor, *Proc. Artificial Life II*. Addison-Wesley, 1991.
12. E.O. Wilson. *Sociobiology: The New Synthesis*. Belknap Press, Harvard, 1975.

Modelling Animal Behaviour in Contests: Conventions for Resource Allocation

Matt Quinn[1] and Jason Noble[2]

[1] Centre for Computational Neuroscience and Robotics
University of Sussex, Brighton BN1 9QG, UK
matthewq@cogs.susx.ac.uk
[2] Informatics Research Institute, School of Computing
University of Leeds, Leeds LS2 9JT, UK
jasonn@comp.leeds.ac.uk

Abstract. The selective pressures affecting animal contest behaviour are investigated with an evolutionary simulation model. Two agents of differing fighting ability compete for control of an indivisible resource. Results indicate the evolution of coordinated behaviour that avoids unnecessary fighting. Detailed examination of one run shows the use of an arbitrary convention which makes ability irrelevant to contest outcome. Implications for theories of animal conflict are explored.

1 Introduction

In their quest to survive and reproduce, animals of the same species inevitably compete for resources such as food, territory, and mating opportunities. Contests may occur when two animals simultaneously attempt to gain control of a single indivisible resource. If both individuals want the resource, but only one can have it, then their genetic interests conflict, and we might expect the question of possession to be settled aggressively, through fighting. However, one of the curious facts about animal contests is how often all-out violence is avoided. From spiders [1] to elephants [2], we find that most of the time contests stop short of serious injury and are settled by what appear to be threats, signals of strength or determination, or even arbitrary conventions.

How and why do animals manage to settle contests peacefully so much of the time? Ethologists used to argue that animal contests were often non-violent because too much fighting would be bad for the species. For instance, Huxley [3] believed that animals produced and attended to ritualized threat displays in order to "reduce intra-specific damage." Unfortunately this idea fails to recognize that animals are not interested in the survival of their species, but rather in the propagation of their own genes. The difficulty is to explain why selfish individuals should be expected to show restraint in a contest. Applying game theory to evolutionary biology, Maynard Smith and Price [4] used the Hawk-Dove game to show that a tendency to stop short of all-out fighting could be explained in terms of evolved self-interest. Given the reasonable assumption that the contested resource is valuable but not worth risking serious injury for, the

J. Kelemen and P. Sosík (Eds.): ECAL 2001, LNAI 2159, pp. 367–376, 2001.

Hawk-Dove game demonstrates that aggressive and peaceful strategies will both do well when they are rare in a population, which means that in the long run we should expect a stable state in which both are present.

Communication is one way of minimizing violence in contests: in theory, if both competitors exchange information about their fighting ability, and then the weaker defers to the stronger, most fights are avoided. (Equivalent mechanisms might be based on other relevant variables, such as the subjective value of the resource for each animal.) Game-theoretically inclined biologists have debated at length whether such a signalling system would be evolutionarily stable. There is widespread agreement that if information about the opponent's fighting ability cannot be faked, as when one animal is simply much bigger than the other, the information will be attended to. Much more problematic is the case where a signal indicating strength could in theory be given by a weak animal, i.e., it is possible for weak competitors to bluff or cheat. One school of thought [5] says that there would be nothing to prevent the evolution of bluffing in such cases, and that the signalling system would therefore collapse—once bluffing becomes prevalent, there is no longer any point in attending to your opponent's signal. An opposing view [6] suggests that under certain circumstances it is not worthwhile for a weak animal to signal dishonestly, as bluffs might be challenged by stronger competitors to disastrous effect. All parties agree that honest signalling, if it is to occur, needs to be stabilized by costs of some sort—it must not be worthwhile to bluff. A number of cost regimes that could conceivably underlie a signalling system advertising fighting ability have been identified [7], e.g., a vulnerability handicap, whereby stronger animals place themselves in a position where they are vulnerable to attack, a risk they could not afford to take if they were weak. However, due to the difficulties involved in measuring the costs, risks, and benefits of contests empirically, it is still an open question as to whether animals really use signalling systems to minimize violent conflicts.

We wanted to find out more about the selective pressures impinging upon animal contest behaviour, and have therefore constructed an artificial life model of contests between two agents over an indivisible resource. In our model the agents are controlled by recurrent neural networks which are shaped over generational time by a genetic algorithm. The world is two-dimensional and complies with basic physics such as inertia and friction. Contests take place in a circular arena; the agents can fight by ramming each other, and must physically occupy the centrally located resource in order to win possession of it. Agents differ in their fighting ability, and although they can sense their own ability, they cannot directly detect the ability of their opponent. By allowing behavioural strategies to evolve over time in this model, we hoped to answer various questions about contest behaviour. For example, should agents behave differently depending on their fighting ability? If the opponent makes an aggressive approach, should an agent take the threat seriously? When a contest is settled peacefully, how do the two agents reach a decision as to who gets the resource and who backs down? Will it be evolutionarily stable for the agents to exchange truthful signals as to their fighting ability?

One of our aims in this paper is to interpret the evolved strategies in the light of current biological theory. If a particular theory is useful in understanding the evolved behaviour of an agent controlled by a recurrent neural net, that should increase our confidence that it will be useful in understanding the behaviour of real animals. At the same time, we believe that our semi-realistic simulation may suggest novel ways of thinking about contest behaviour. Most of the biological ideas discussed so far are based on insights from game-theoretic models, and, for reasons of mathematical tractability, such models reflect simple situations with a very small range of strategies, and, at best, a minimal treatment of space and time. For instance, in a model suggesting honest signalling of strength by weaker competitors [6], all individuals are either weak or strong, and they make only two decisions: which of two arbitrary signals to make, and then whether to flee, attack, or attack after a pause. In our model, contests are played out over space and time, introducing possibilities such as chases and stand-offs. The need to ram the other agent in order to cause damage means that aggressive intentions can be perceived; simulating momentum automatically implements a notion of commitment to an attack, as an agent may be going so fast that it could not possibly slow down before impact. Factors like these may well lead to the evolution of strategies that cannot be expressed in a simple game-theoretic model, and the same factors may be critical to an understanding of behaviour in real animal contests. One of the authors [8] has previously looked at the evolution of contest behaviour in a richer context than that afforded by paper-and-pencil game theory, and found that honest signals of fighting ability were unlikely to evolve. However, the simulation environment in this earlier work was only of intermediate complexity.

Another goal for the work is to illustrate the potential for artificial life models to examine the evolution of communication from genuinely non-communicative origins. Movement is the only way that one agent can impinge upon the sensory world of the other, and if the agents come to exchange signals of fighting ability, these signals will have to develop from movements that do not originally have signalling as their function. We want to move away from the provision of arbitrary signalling channels, as found in most artificial life models of the evolution of communication. If you start with agents that already have the ability to send and receive signals on a dedicated channel, then you cannot be investigating the *origin* of communication, only the parameters under which it might be evolutionarily stable once established (see Quinn, this volume, for further discussion of this point).

2 The Simulation

Contests commence with two agents placed at random positions[1] and orientations in a simulated arena 80cm across (Fig. 1). The circular body of each agent is 10cm in diameter. The contested resource is also 10cm across and is located in the centre of the arena. Agents have a fighting ability drawn from the range {0.1, 0.3, 0.5, 0.7, 0.9}, and an energy level that starts at one and is reduced through damage; an energy level of zero indicates serious injury. Agents can

[1] Agents must start at least 5cm away from the arena edge, 5cm away from the resource, and 10cm away from each other.

accelerate to a maximum speed of approximately 15cm/s. Time is simulated in slices of 0.1s duration. Inertia, friction and elastic collisions are implemented in the simulation.

- 7 distance sensors (d0-d6) for detecting the opponent.
- 2 sensors (Rl, Rr) for detecting the resource.
- An omnidirectional distance sensor (Rd) and a contact sensor (Ro) for detecting the resource.
- A bumper sensor (bp) for detecting collisions.
- Sensors for fighting ability (A) and energy level (E).

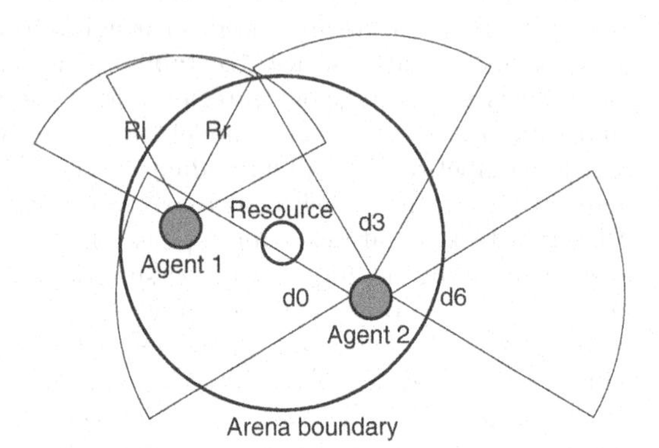

Fig. 1. A list of the agents' sensory inputs, plus a diagram showing the arena, the resource and the competing agents, to scale. Both agents are facing north. Agent 1 illustrates the arrangement of the two resource sensors, Rl and Rr. Note that agent 1 cannot see the resource from the position shown. Agent 2 illustrates 3 of the 7 proximity sensors for detecting the opponent. The sensors are equally spaced around the front half of the agent's body; in the interests of clarity, sensors d1, d2, d4 and d5 are not shown. Note that agent 2 can detect agent 1 with d0 (and also with d1, not shown).

Agents can inflict damage by ramming each other. The agents' bodies are equally vulnerable at all points, but in a collision A will only damage B if A is moving towards B at the time. Thus, in a head-on collision, damage is mutual, but if A rams B while B is moving away or stationary, only B will be damaged. Damage is subtracted from the target's energy level, and is equal to $K \times F_{\text{attacker}} \times (1 - F_{\text{target}}) \times \frac{\text{speed}}{\text{max. speed}}$; K is a constant set equal to 4, and F is fighting ability. The essence of this is that fighting ability is valuable for both attacking and defending, and high speed means more damage. Fighting is a risky proposition: for example, if two agents of only average ability (0.5) ram each other at full speed, they will inflict 1 unit of damage, meaning serious injuries to both. Once an agent has suffered a serious injury, it is out of action for that contest and disappears immediately from the arena.

Agents can run away instead of fighting. The boundary of the arena is not a wall, and if an agent crosses the boundary it is assumed to have fled the contest, and will disappear from its opponent's sensors. (Note that this does not mean that the agent remaining in the arena automatically gains the resource.)

A contest will normally end when one contestant gains possession of the resource, which an agent achieves by keeping the centre of its body continuously within the resource zone for 10s. Contests can also end if both agents flee the arena, or if a 30s time limit expires. (If an agent is within the resource zone at the end of 30s, the contest is extended for up to 10s to give the agent a chance to gain possession.)

Agents have 14 sensory inputs (detailed in Fig. 1) allowing them to perceive each other, the resource, and their internal state. Agents are controlled by arbitrarily recurrent artificial neural networks. The thresholds, weights and decay parameters, and the size and connectivity of the network are all genetically determined. Agents are equipped with left and right motors, much like a Khepera robot, and four binary outputs from the neural network control forward and backward drives for each motor (if the forward and backward drives for a motor are both activated, output is zero).

Each generation, an agent participates in 25 contests, one for each of the possible combinations of their own and their opponent's fighting ability. Agents also participate in a further five trials in which there is no opponent present in the arena; these trials were included in order to ensure selection pressure for the ability to take an uncontested resource.

In terms of fitness, gaining control of the resource is worth 100 units. Injury is a fitness cost that scales with the damage an agent has suffered, to a maximum of 200 units for a serious injury. There is also a modest benefit for fleeing a contest promptly, scaled according to the time remaining when the agent flees, up to a maximum of 10 units for instantaneous flight. These costs and benefits have been chosen such that the resource is worth having, but not worth risking serious injury for. Previous models [5,6,8] have shown that this balance between resource benefit and injury cost poses the most interesting strategic problem for the contestants. (If the resource is made too valuable, competitors will fight to the death. If the resource is not valuable enough, competitors will not be tempted to fight at all.) The inclusion of a benefit for leaving the contest early captures the idea that participating in a contest always entails an opportunity cost. The costs of movement, including attacks, are assumed to be negligible: this reflects empirical work on spider contests [1] showing that the long-term fitness costs of serious injury—and, of course, death—are orders of magnitude greater than other costs such as energetic expenditure associated with threat displays.

The evolutionary engine for the simulation was a genetic algorithm operating on a population of 200 agents. Selection was fitness-proportional, based on mean fitness per contest, with the elite individual retained unchanged in the next generation. Crossover was used 75% of the time. The agents' neural networks were represented in the genotype using a topological encoding scheme, which necessitated several different types of mutation. The rate of non-structural mutations (e.g., weight changes) was 1.0 per genotype, the probabilities for adding or deleting a neuron were 0.0025 and 0.02 respectively, and the probabilities for adding, deleting and reconnecting a synapse were 0.01, 0.05, and 0.03.

3 Results and Interpretation

Ten runs of at least 20000 generations were conducted, each with a different random seed value. Initial controllers consisted of three neurons with randomly determined connections and parameters. We will first summarize general findings across the ten runs, and then examine a single run (run zero) in greater detail.

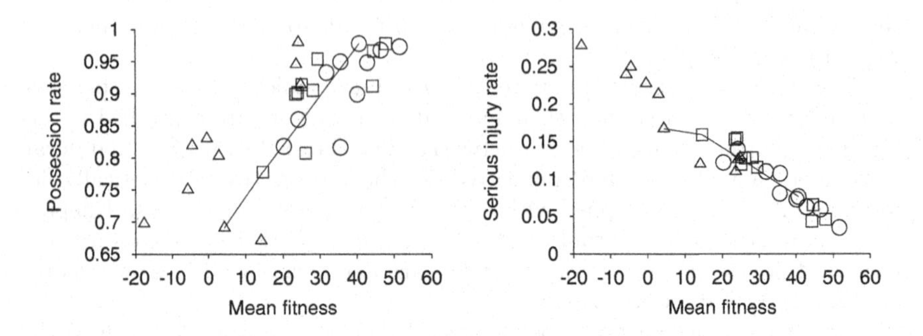

Fig. 2. Left: proportion of contests in which an agent takes possession of the resource, plotted against mean fitness at three points in time across ten runs. **Right:** rate of serious injury (per agent per contest) by mean fitness score, at three points in time across ten runs. **Both:** in these and subsequent figures, the mean state of a population between generations 200–400, 5000–5200, and 18000–18200 is represented with a triangle, a square and a circle respectively. The relevant points for run zero are joined with a line to show progress over time.

One question we can ask is whether or not the agents managed to settle contests with a minimum of violence, as animals do. Figure 2 (left) shows the proportion of contests in which one agent or the other took possession of the resource, plotted against mean fitness scores. Focusing on the situation after 18000 generations (points marked with circles) we can see that the possession rate is at least 80% in all ten runs. This indicates that the agents are at least managing to allocate the resource most of the time, rather than, for example, both fleeing, or becoming locked in a stalemate. Note also that possession rates and fitness both tend to increase over generational time. Figure 2 (right) shows that the prevalence of serious injury is negatively linked to fitness, that it decreases over time, and that by 18000 generations it is as low as around 5% in some runs. If the agents' method for resolving contests was to fight until one or the other was seriously injured, the rate would be 50% (or even higher given that sometimes both would be injured simultaneously). Thus, rates of 5–10% indicate that a more peaceful system has evolved.

This last point is reinforced if we consider the simple fact that fitness scores after 5000 generations are all positive. In a perfect world, agents might toss a coin and allocate the resource accordingly, without a fight. Each agent could therefore expect the best possible mean fitness[2] of $100/2 = 50$. If fighting is occurring, injuries will mean that fitness scores are lower—for example, if agents were to inflict an average of 50% damage on each other, then mean fitness would fall to -50. So the observation that fitness scores are greater than zero suggests that fighting has been minimized.

[2] In fact the true theoretical optimum in our simulation is 62.5 due to the bonus for early fleeing and the 5 trials out of 30 conducted with no opponent.

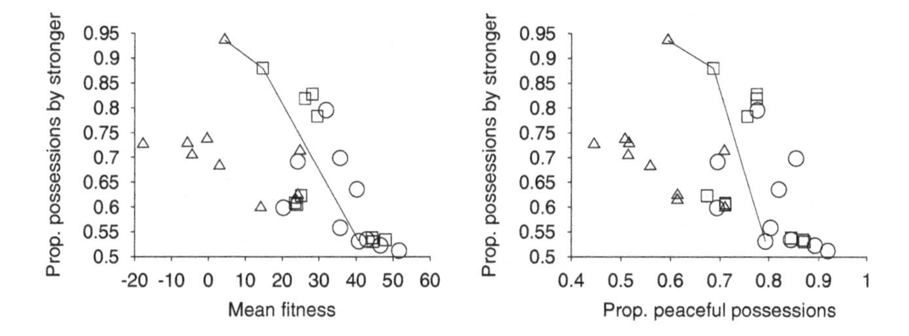

Fig. 3. Left: proportion of possession events in which the stronger agent takes the resource, plotted against mean fitness at three points in time across ten runs. **Right:** proportion of possession events in which the stronger agent takes the resource, plotted against the proportion of peaceful possessions (i.e., one agent takes the resource without either agent being damaged), at three points in time across ten runs. See Figure 2 for an explanation of symbols used.

Could the agents be signalling their fighting ability to each other, with the weaker deferring to the stronger? Figure 3 suggests that this is not the case for most runs. If such a signalling system had evolved, the stronger of the two agents should usually take the resource without a fight. Figure 3 (left) shows that, early on in a run (points marked with triangles) the stronger agent wins about 70% of the time. But as evolution progresses, this figure falls, and in about half of the runs it falls close to the chance level of 50%. Furthermore, Figure 3 (right) shows that high rates of possession by the stronger agent are associated with fewer peaceful possessions, i.e., that when victory is to the strong it is not without a fight. The general profile that emerges from late in the runs (points marked with circles) is one of peaceful allocation of the resource, high mean fitness scores, and fighting ability not being an important predictor of success. The latter point does not match what we would expect if a signalling system was in place.

If the agents have evolved a way of allocating the resource without violence, and if fighting ability has become unimportant, then what might be happening? We extended run zero to 68000 generations and looked at the results in greater detail in order to find out—Figures 2 and 3 show that, at 18000 generations, run zero is reasonably representative of the ten runs.

Figure 4 shows various statistics for run zero. Unsurprisingly, given their randomly connected networks, agents in the initial generation were not likely to gain possession of the resource; they tended to move in a random direction and thus flee the arena without fighting. However, we see that the ability to move inwards and gain possession of the resource develops very quickly, and by about generation 8 we see the highest levels of serious injury and the lowest levels of peaceful possession as both agents try to move onto the resource. Between generation 10 and generation 75, contests become more peaceful and the stronger

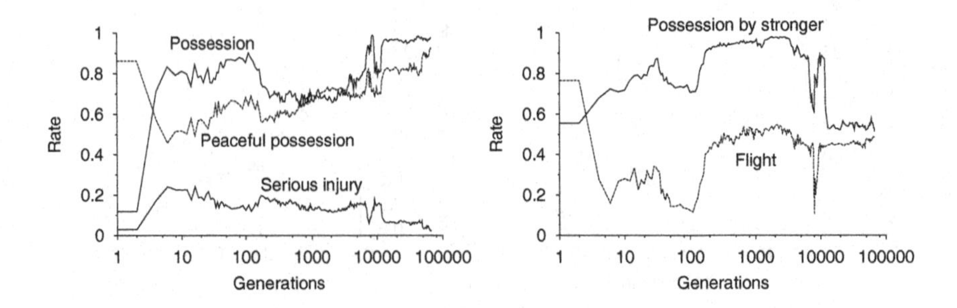

Fig. 4. Left: proportion of contests ending in possession of the resource, proportion of peaceful possessions, and rate of serious injury, plotted over generations for run zero. **Right:** proportion of possessions by stronger agent, and rate at which agents fled the arena, plotted over generations for run zero. Note that the time axes are log-scaled.

agent becomes more likely to win. This is because the agents have started to use their fighting ability sensor, and will move in more confidently if they are strong, and may flee immediately when very weak. By generation 100 we see a slow, clockwise spiralling motion in towards the resource; the bias towards right turns is maintained throughout the run. As the generations pass the behaviour of the agents becomes more and more strongly modulated by their fighting ability. Around generation 8000 there is a spike in the possession rate, and a drop in the rate at which the stronger agent wins. This occurs because the agents adopt a slow, waggling motion, and become very reluctant to collide with each other. Around generation 9000 fast, smooth motion is recovered. Typically, both agents orbit the resource zone, and each avoids the other, but the stronger is prepared to turn in more tightly, and thus gains the resource. Then, around generation 12200, fighting ability rapidly becomes less useful as a predictor of victory. The agents circle each other, and one seems to get a positional advantage on the other, irrespective of being weak or strong. The first agent chases the second out of the arena, and then returns for the resource. Given this behaviour, the rate of serious injury drops, and the rate of peaceful possessions jumps to 80% or more. The pattern is maintained for the rest of the run.

This behaviour appears to be a convention for settling most contests without violence. We analyzed one of the evolved neural nets in order to see how the agents managed to do this (Fig. 5). The evolved controller represents a balance between two tendencies: a spiralling taxis towards the resource, and movement relative to the opponent depending on orientation. The "straight-ahead" neuron is alway firing, due to a negative threshold and connections to various inputs; it tries to drive both forward motors. However, in the absence of significant sensory input, the "right turn" neuron dominates, switching on the right-backwards motor and switching off the right-forwards motor, resulting in rotation to the right. When the resource is detected with the Rr sensor, the right turn neuron is inhibited, and the agent moves forward. When the resource is no longer visible,

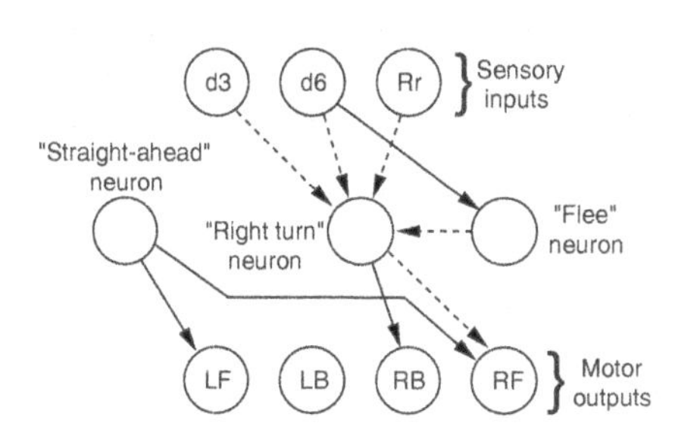

Fig. 5. Schematic diagram of an evolved neural network from generation 17,900 in run zero. Excitatory and inhibitory connections are indicated by solid and broken lines respectively. Neurons and connections that are non-functional or relatively unimportant are not shown.

the agent turns right again—this is enough to implement a spiralling movement in towards the resource.

Detecting the opponent with either the d3 (straight ahead) or the d6 (90° right) sensor inhibits right turns and causes forward movement. This will have two quite different results: if the opponent is detected dead-ahead, the agent will advance aggressively. If the opponent is detected to the right side, the agent will advance but this will ultimately lead to its running away, chased by the opponent. This happens because the d6 sensor also activates the "flee" neuron, which has a very slow decay rate, and functions as a switch. Once the flee neuron is firing, it permanently inhibits the right turn neuron, and so the agent keeps moving in a straight line, which will always take it out of the arena. The opponent, having a very similar neural net, will advance aggressively, "chasing" the fleeing agent from the arena, and then return to the resource when its right-turning taxis regains control.

Interestingly, the evolved networks use the random placement of the agents in the arena as an arbitrary way of settling the contest without a fight. Both agents will turn right from their initial positions, the lucky one will catch sight of the other dead ahead, and the unlucky one will first notice the opponent to its right. The unlucky agent will be chased away; the lucky one will take the resource uncontested. The convention is not quite arbitrary, however: it appears to have evolved because having another agent approach you from the side does indeed put you at a disadvantage, given that you would need to turn and build up speed again before you could attack that agent.

4 Conclusion

Our simulation has shown that a population of selfish agents can evolve relatively peaceful strategies for resource allocation. The agents exploit a convention that is, on the one hand, rooted in facts about their sensory and motor capacities, and, on the other, rendered arbitrary because of random starting positions. We had not anticipated the evolution of a convention for avoiding conflict: the evolved agents found a third way between the rival game-theoretic predictions of honest signalling [6] and poker-faced inscrutability [5]. Arbitrary conventions have been discussed in the animal-contest literature before, but were generally not thought to be relevant to situations where the competitors differed in fighting ability. Our model suggests that it may be worth looking for such conventions in a wider range of animal contest situations. More generally, we believe that our findings illustrate the value of an open-ended approach to modelling animal communication and interaction.

Acknowledgements. We acknowledge computing support from the Sussex High Performance Computing Initiative.

References

1. S. E. Riechert. Spider interaction strategies: Communication vs. coercion. In P. N. Witt and J. Rovner, editors, *Spider Communication: Mechanisms and Ecological Significance*, pages 281–315. Princeton University Press, 1982.
2. J. H. Poole. Announcing intent: The aggressive state of musth in African elephants. *Animal Behaviour*, 37:140–152, 1989.
3. J. S. Huxley. Ritualization of behaviour in animals and men. *Philosophical Transactions of the Royal Society of London: Biological Sciences*, 251:249–271, 1966.
4. J. Maynard Smith and G. R. Price. The logic of animal conflict. *Nature*, 246:15–18, 1973.
5. P. G. Caryl. Communication by agonistic displays: What can games theory contribute to ethology? *Behaviour*, 67:136–169, 1979.
6. P. L. Hurd. Is signalling of fighting ability costlier for weaker individuals? *Journal of Theoretical Biology*, 184:83–88, 1997.
7. S. L. Vehrencamp. Handicap, index, and conventional signal elements of bird song. In Y. Espmark et al., editors, *Animal Signals: Signalling and Signal Design in Animal Communication*, pages 277–300. Tapir Academic Press, 2000.
8. J. Noble. Talk is cheap: Evolved strategies for communication and action in asymmetrical animal contests. In J.-A. Meyer et al., editors, *Proc. Sixth International Conference on Simulation of Adaptive Behavior*, pages 481–490. MIT Press, 2000.

A Model of Human Mate Choice with Courtship That Predicts Population Patterns

Jorge Simão[1]* and Peter M. Todd[2]

[1] CENTRIA — Computer Science Department, FCT — New University of Lisbon, 2825 - 114 Monte da Caparica, Portugal
jsimao@di.fct.unl.pt

[2] Center for Adaptive Behavior and Cognition, Max Planck Institute for Human Development, Lentzeallee 94 — 14195 Berlin, Germany
ptodd@mpib-berlin.mpg.de

Abstract. We present a new model of human mate choice incorporating non-negligible courtship time. The courtship period is used by individuals to strategically swap to better partners when they become available. Our model relies on realistic assumptions about human psychological constraints and the specifics of the human social environment to make predictions about population level patterns that are supported by empirical data from the social sciences.

1 Introduction

Human mate choice, when you're immersed in it, seems like a long and complicated affair. Vast amounts of time and energy are spent thinking about who to attract, how to attract them, and how to keep them attracted (or how to terminate the attraction and escape) once a relationship is underway. Yet attempts to understand the process of mate choice through formal models typically throw away most of this complexity, reducing everything to a simple sequence of decisions about whom to mate with. How realistic can such models be if the heart of the romantic process—courtship—has been removed?

Typical assumptions of models of mutual mate choice are that individuals search and encounter mates sequentially (usually without the ability to go back to, or "recall," earlier mates), and that individuals make their mating choices as a single, *irreversible* decision whether to mate an individual or not [4,8,6,5, 3]. This conflicts with the fact that humans use extensive courtship periods to establish long-term sexual/romantic relationships, and that this allows individuals to engage in relationships in tentative ways—possibly switching to better alternatives if they become available in the future [2]. Many of these models also make the unrealistic assumption that complete and accurate information is available to all individuals [1,6,5,3]. This includes information about the distribution of qualities of potential partners, about the searching individual's own

* This work was supported by a PRAXIS XXI scholarship. Thanks also to Chantilly Lodewijks for emotional and financial support.

J. Kelemen and P. Sosík (Eds.): ECAL 2001, LNAI 2159, pp. 377–380, 2001.

quality, and sometimes even about the preferences of other individuals [1]. Given that such information is often not available in the real world, it is not surprising that most of these models are of limited empirical validity.

In this paper, we present an individual-based model of human mate choice based on an evolutionary functional analysis. Our model relies on more realistic assumptions about human psychological constraints and the specifics of the human social environment. In particular, we assume that individuals build networks of acquaintances gradually instead of having complete information about all potential alternative partners instantaneously, and we consider the strategic impact of long courtship and investment on mate switching decisions. We argue that individuals can make simple, efficient, and robust mate decisions by exploiting the specifics of the adaptive problem domain rather than attempting to perform complex optimizations, thus constituting an example of *ecological rationality* [7].

2 A Model Built Around Courtship

To model the mate search process, we begin with a population of constant size $2 * P$ with a fixed sex ratio of 50% (so P is the number of males or females). Individuals of both sexes have a one dimensional quality q, randomly generated from a normal distribution with mean μ and variance σ^2 ($0 < Q_{min} \leq q \leq Q_{max}$). When two individuals mate, they are removed from the population and replaced by two new individuals of random quality (one of each sex). Time is modeled as a sequence of discrete steps. Pairs of males and females meet at a certain stochastic rate: in each time step each individual has a probability Y of meeting a new individual of the opposite sex. Each individual maintains a list of the potential mates already met (the *alternatives list*).

Within the alternatives list, one member can have the "special status" of being the individual's current *date*. This happens when both individuals previously agreed to date and have not changed partners in the mean time (see below). It is also possible for an individual not to be dating anybody (i.e., in the beginning of its "life", or when it gets "dumped"). The length of time that two individuals are dating is regarded as the courtship or dating time c_t. If a pair of individuals remain dating for a period of K time steps (i.e., when $c_t > K$), they are deemed to mate and are removed from the population (and replaced). Every individual has a maximum lifetime of $L(> K)$ time steps. If individuals are unable to mate during that period they are removed from the population (they "die"). To replace the dead individual, a new one is created with the same sex.

In each time step, individuals decide whether to continue to date the same partner or to try to date another individual that they know, that is, someone from their list of alternatives. To make this decision, several pieces of information are used: the quality of each alternative q_a (all alternatives are considered independently), the quality of the current date q_d, the current courtship time c_t, the age of the individual t, the age of each alternative t_a, and the age of the date t_d. If an individual is not dating someone else, q_d and c_t are set to 0. A binary output function S with these input variables is used to make courtship decisions, returning 1 iff a dating proposal is to be made.

We assign mated individuals a fitness value $f = q_d^{Q_S} * (L - t)$, where q_d is the quality of the individual's partner, Q_S is a scaling factor that relates quality values to effective mating potential, and t is the individual's age at mating. Thus we reward a preference for high quality, and introduce time pressure to motivate individuals to mate early (in addition to the limited life-time). Given this fitness function, agents can use the following reasonable (although not optimal) strategic decision rule for deciding whether to try to switch from the current date of quality q_d to another alternative of quality q_a (with this switch only happening if the alternative individual decides to switch from their current date as well):

$$\mathbf{S}(q_a, q_d, c_t, t, t_a, t_d) = \begin{cases} q_d = 0 & \longrightarrow 1 \\ t + K > L & \longrightarrow 0 \\ t_a + K > L & \longrightarrow 0 \\ q_a^{Q_S} * [L - (t + K)] > & \\ q_d^{Q_S} * [L - (t + K - c_t)] & \longrightarrow 1 \\ \text{otherwise} & \longrightarrow 0 \end{cases} \quad (1)$$

2.1 Simulation Results

We evaluated the performance of this model by seeing if it could produce realistic mating patterns matching those observed in human populations. Specifically, we looked for high rates of assortative mating (i.e., roughly equally attractive individuals mating with each other, as reflected by within-pair attractiveness correlations of around 0.6 in many human studies [4]) achieved with relatively little search. We set population size $2 * P = 100$, lifetime $L = 200$, and scaling factor $Q_S = 0.75$. The (quasi) normal quality distribution was set with $\mu = 10$, $\sigma^2 = 4$, $Q_{min} = 0.001$, and $Q_{max} = 20$. Each simulation run consisted of $10 * L = 2000$ steps. We varied two parameters likely to have a wide range of values across different real settings: the rate-of-meeting Y and courtship time K. Figure 1a depicts the linear correlation between the qualities of individuals in mated pairs as a function of Y and K, while Fig. 1b shows the mean number of alternatives met before settling with the last date (both averaged across 5 runs). The reasonably high correlation coefficients observed (mostly between .5 and .6) suggest that individuals are making good use of their mating potential even though they have no direct knowledge of their own mate value or the distribution of qualities in the population. This is also achieved by most individuals with rather little search: for example, with $K = 15$ and $Y = .2$, we obtained a correlation of qualities of .53, and observed that 87% of the individuals in the population were able to mate, even though they only met a small number of alternatives (6.9 before settling with the last date and 10.0 before mating), and only entered into 2.0 dates on average (including the very last one). This realistic combination of statistics was never obtained in previous models of human mate choice [4,8].

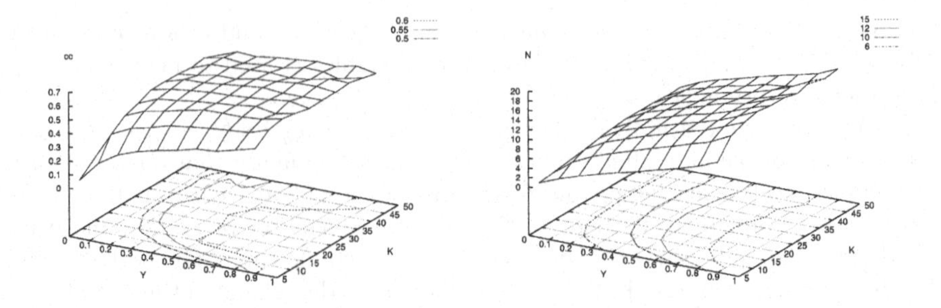

Fig. 1. a) Mean correlation of qualities in mated pairs. **b)** Mean number of alternatives seen before settling with the last date.

3 Conclusions

In this paper, we have briefly demonstrated the use of an individual-based artificial life modeling approach to gain insights into the processes underlying human sexual/romantic relationships. In particular, by building extended courtship processes into our model of the mating game, we have accounted for existing data in ways unattained by earlier models. This work is part of a larger project to build realistic models of human mating systems that explain individual-level behavior and predict population-level patterns. We hope that our work will motivate researchers in the fields of artificial life and evolutionary psychology to begin courting each other to build explicit computational models of the mechanisms guiding human social behavior.

References

[1] C. Bergstrom and L. Real. Towards a theory of mutual mate choice: Lessons from two-sided matching. *Evolutionary Ecology Research*, 2(4):493–508, 2000.
[2] D. Buss. *The Evolution of Desire.* Basic Books, 1994.
[3] R. Johnstone. The tactics of mutual mate choice and competitive search. *Behavioral Ecology and Sociobiology*, 40(1):51–59, 1997.
[4] S. M. Kalick and T. E. Hamilton. The matching hypothesis reexamined. *Journal of Personality and Social Psychology*, 51(4):673–682, 1986.
[5] J. McNamara and E. Collins. The job search problem as an employer candidate game. *Journal of Applied Probability*, 27(4):815–827, 1990.
[6] G. A. Parker. Mate quality and mating decisions. In P. Bateson, editor, *Mate Choice*, pages 141–166. Cambridge University Press, 1983.
[7] P. M. Todd, L. Fiddick, and S. Krauss. Ecological rationality and its contents. *Thinking and Reasoning*, 6(4):375–384, 2000.
[8] P. M.Todd and G. F. Miller. From pride and prejudice to persuasion: Satis cing in mate search. In *Simple Heuristics that Make Us Smart*, pages 287–308. Oxford University Press, 1999.

Establishing Communication Systems without Explicit Meaning Transmission

Andrew D.M. Smith

Language Evolution and Computation Research Unit,
Department of Theoretical and Applied Lingustics, University of Edinburgh, UK
andrew@ling.ed.ac.uk

Abstract. This paper investigates the development of experience-based meaning creation and explores the problem of establishing successful communication systems in a population of agents. The aim of the work is to investigate how such systems can develop, without reliance on phenomena not found in actual human language learning, such as the explicit transmission of meaning or the provision of reliable error feedback to guide learning. Agents develop individual, distinct meaning structures, and although they can communicate despite this, communicative success is closely related to the proportion of shared lexicalised meaning, and the communicative systems have a large degree of redundant synonymy.

1 Introduction

There is a growing body of literature in which investigations into the evolution of language[1] are carried out by computer simulation [8,1,4]. For most of these researchers, the evolution of language is regarded as essentially being equivalent to the evolution of syntax, because the use of syntactic structure is seen as the main difference between animal and human communication systems. For example, vervet monkeys have a well-known communication system which allows them to distinguish different predators [5], but they do not combine their signals to convey complex meanings. Kirby [9] has shown that the simple ability to create general rules, by taking advantage of coincidental correspondences between parts of utterances and parts of meanings, can result in the emergence of syntax, as general rules generate more utterances than idiosyncratic rules, and are therefore replicated in greater numbers in following generations. Similar accounts [2,10] also show syntax emerging as a consequence of the recognition and coding of regularities between signals and meanings.

Nehaniv [14] has pointed out, however, that syntax only develops successfully from unstructured signals because the signals are coupled with meanings which are already structured, and it is no coincidence that the emergent syntactic structure parallels the pre-existing semantic structure. In these simulations,

[1] This field is concerned not with the evolution of particular languages, such as English, from their ancestor languages, but rather with the general capacity, apparently unique to humans, for using infinitely expressive communication systems [13].

J. Kelemen and P. Sosík (Eds.): ECAL 2001, LNAI 2159, pp. 381–390, 2001.

the meanings are also explicitly part of the linguistic transfer from speaker to hearer, therefore obviating the critical problem, exemplified by Quine [17](p. 29–30), of how a learner determines the meaning which a signal intends to convey. Furthermore, attempts to develop learnt communication systems frequently involve some sort of reinforcement learning process [20,6], which has the primary role in guiding the learning mechanism. Oliphant [15] points out, however, that such error signals, which work well on an evolutionary timescale, are less useful over an individual's lifetime where failure might mean immediate death, and indeed even the very existence of reliable error signals is questioned by many authors on child language acquisition [3].

If we try to define the meaning of a word, we find ourselves caught in a kind of lexical web, where words can only be defined by their relationship to other words, and in terms of other words. There is no obvious way of entering this web, unless at least some words are grounded in reality [7], such that they can be used to point out actions and objects in the real world. It is reasonably uncontroversial to say that meanings must capture patterns of categorisation (whether categories are defined in classical terms of shared features or prototypes [21]) which enable us to state, for instance, which things are *rabbits* and which are not. Furthermore, meanings are not innate, but are created anew in each language learner, who creates an individual system of meaning based on their experiences [3].

Our aim is to model, in a population of agents, the creation of meanings by explicit categorisation, and then to investigate the spread of meanings through the population, without the meanings themselves being transferred between agents, and without any error signals to reinforce the learning process.

2 Meaning Creation by Object Discrimination

In order to develop a model of independent, grounded meaning creation, we establish a simple world of agents and objects, similar to that described by Steels [19], in which the objects can be described in terms of their features[2], which are intrinsically meaningless, but which can be thought of in terms of more imaginable language-like features such as *colour, height* or *smell.* The agents in the model world interact with the objects by using *sensory channels*, which are sensitive to the corresponding features of objects, and can detect whether a particular value of a feature falls between two bounds. Initially, the channels can only detect that a value falls between 0.0 and 1.0, but the agents have the power to split the sensitivity range of a channel into two discrete segments, resulting in a *discrimination tree* [20]. The nodes of a discrimination tree can be considered categories or *meanings*,[3] as seen in the sensory channel in figure 1, which has been refined twice, and has four new meanings.

[2] Feature values are represented as pseudo-randomly generated real numbers which are normalised to lie between 0.0 and 1.0

[3] Meanings are given in the notation *sc-path*, where *sc* identifies the sensory channel, and *path* traces the path from the tree root to the node in question, where 0 signifies a lower branch and 1 an upper branch.

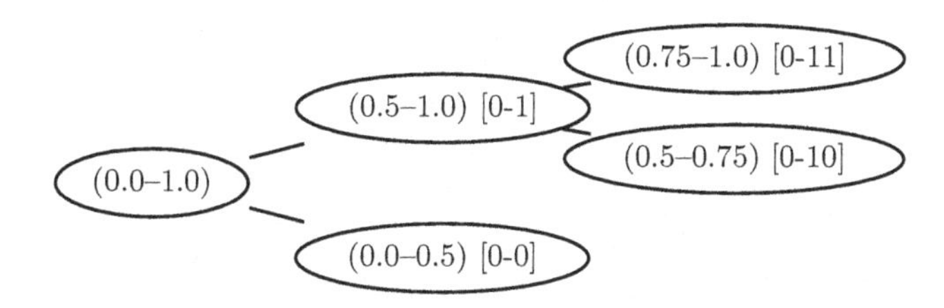

Fig. 1. A discrimination tree (channel 0) which has been refined twice. Each node shows the bounds between which it is sensitive, and the meaning to which it corresponds (following Steels).

In order to provide a framework for the unguided refinement of the sensory channels based on observation, we follow Steels [19] in using discrimination games, in which an agent attempts to distinguish one object from a larger set of objects. Each game proceeds as follows:

1. An agent considers a random set of objects (the *context*), one of which is chosen at random to be distinguished from the others and is called the *topic*.
2. The agent investigates all its sensory channels to categorise the objects.
3. If the topic is uniquely identified by any category, the game succeeds.
4. If the game fails, the agent refines a randomly-chosen sensory channel.

Object	Categories/Meanings		
A	0-0	1-00	2-111
B	0-11	1-1	2-110
C	0-0	1-1	2-111
D	0-10	1-01	2-10

The above table shows an agent categorising objects as part of a discrimination game. The agent has four objects A-D, and has categorised them with three sensory channels. If the aim of this game is to discriminate B from the context ACD, then the game can succeed, as both $0-11$ and $2-110$ are possible distinguishing categories. On the other hand, if the aim is to distinguish C from the context ABD, then the game will fail, as none of the categories which C falls into distinguish it from all the other objects. Failure triggers the refinement of a random channel, creating more detailed categories, which *may* be useful in future games. Over time, the agents develop their sensory channels such that the discrimination games nearly always succeed, though the extent to which an individual channel is refined depends on the number of channels which the agent has: the more channels, the fewer refinements on each are necessary.

Figure 2 shows the idiosyncratic meaning representations of two agents in the same world. The first agent has developed the first three channels to a greater extent than the second agent, who in turn has developed the fourth and fifth channels more extensively. It is helpful to quantify the amount of difference between two trees t_1 and t_2, which we can do by averaging the proportion of

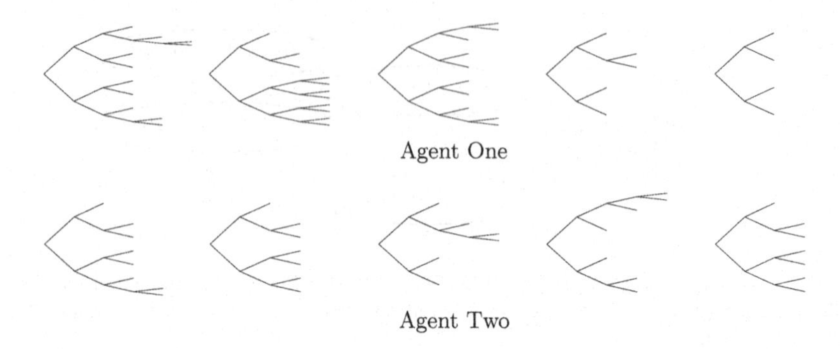

Fig. 2. Two agents each have five sensory channels, with which they construct different representations of the same world.

nodes in tree t_1 which are also in t_2, and the proportion in t_2 which are also in t_1. Averaging over all the trees in figure 2, the two meaning representations have a *meaning similarity* measure of 75%. It is important to note that both agents are successful in the discrimination games, and so their representations are equally good descriptions of their world. This model, then, satisfies one of our goals, namely that the agents are not given innate meanings, but can create inventories of basic concepts individually, based on their own experiences.

3 Communication

The next step is to investigate whether the agents can communicate with each other, using the meanings they have constructed. Clearly the agents must be able to use some sort of signals, and so they are endowed with the ability to create signals from random strings of letters, and to express and understand these signals without error. In addition, they maintain a dynamic lexicon of associations between signals and meanings, which develops as they participate in the experiments, and which they use in order to make decisions about their communicative behaviour. *Communicative success* occurs if the speaker and hearer are both referring to the same object, but it is not necessary for them to use the same meaning to do so.

A *communicative episode* is played between two agents chosen at random, the *speaker* and the *hearer*. Figure 3 shows a model of the speaker's role, which begins with a discrimination game, in which meanings which can distinguish the topic (filled circle) from the rest of the context (dashed area) are collated. One of these meanings is chosen at random and then looked up in the speaker's lexicon. If the speaker cannot find a word for the meaning it is trying to convey, then it creates a random string of letters and stores this in its lexicon with the required meaning. Having obtained a word to convey the meaning, the speaker utters the word, and the focus passes to the hearer, who receives the word, and can observe the context in which it was uttered, shown in figure 4.

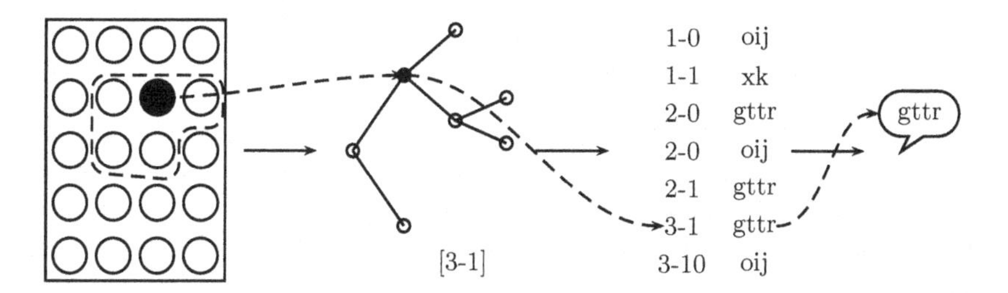

Fig. 3. A communicative episode begins with an agent chosen at random to be the speaker, who finds a meaning to distinguish the topic from the context, and utters a word to convey this meaning.

The word is decoded via the hearer's lexicon into a meaning, and the hearer then establishes which object in the context (if any) is uniquely identified by the meaning it has chosen. If the referent (object) identified by the hearer corresponds to the speaker's original topic, then the communication episode succeeds. The success or failure of a communication game has no effect on the internalised representations of either agent. This model of communication conforms to our initial assumptions, as the internal meanings are explicitly *not* transmitted with the signals, and the agents do not receive feedback from each other about the success of their communicative or learning processes.

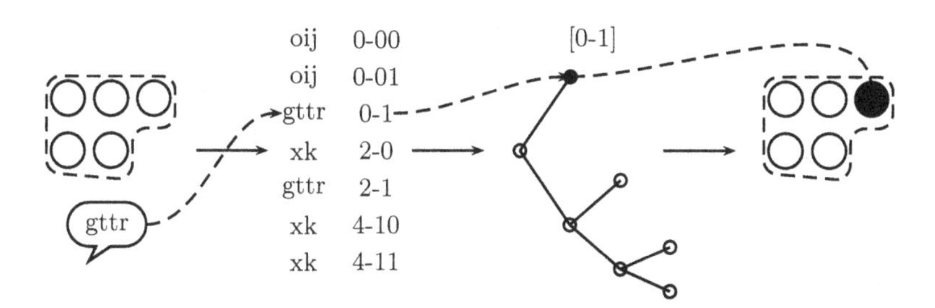

Fig. 4. The communicative episode continues with the hearer, who, given the context, decodes the word into a meaning which identifies an object.

4 The Lexicon

The mappings from meaning to signal and vice-versa are at the heart of the communication process, and are handled via a lexicon, which stores associations between meanings and signals, as well as a count of how often the signal-meaning pair has been used (either uttered as a speaker or understood as a hearer), and a confidence probability, which represents the agent's confidence in the association between the signal and the meaning.

The confidence probability of the signal-meaning pair, consisting of signal s and meaning m, represents the history of all associations between words and meanings an agent has ever made, and is defined as the proportion of the times s has been used in which it has been associated with m, or $\frac{Usage(s,m)}{\sum_1^l Usage(s,l)}$ where l is the number of entries in the lexicon. A short extract from an example lexicon is given below, only showing the entries for two of the signals (*gttr* and *oij*), and the meanings associated with them.

Signal	Meaning	Usage	Conf. Prob.
gttr	0-0	1	0.083
gttr	0-1	2	0.167
gttr	0-11	1	0.083
oij	1-0	9	0.600
gttr	2-0	4	0.333
oij	2-0	6	0.400
gttr	2-1	1	0.083
gttr	3-1	2	0.167
gttr	4-00	1	0.083

How does the speaker decide which signal to choose, when it is trying to express a particular meaning (say $2-0$)? Given the lexicon above, the signal *oij* would seem a reasonable choice for two reasons: it has been associated with $2-0$ on six occasions, compared to *gttr*'s four, and the agent is more confident in the association with *oij* (0.4) than *gttr* (0.33). However, Oliphant and Batali [16] have demonstrated an ideal strategy for achieving an accurate communication system, known as *obverter*, where the speaker chooses words which he knows the hearer will understand. Unfortunately, true obverter learning assumes that the speaker can read the lexicons of the other members of the population, to calculate the optimal signal to use for any meaning. Such mind-reading is not only unrealistic, but even avoids the need for communication at all, and so an alternative is needed. It seems reasonable to assume that the only lexicon the speaker has access to is its own, and so we assume that the speaker uses this as an approximation to that of the hearer. Instead of explicitly choosing the word that the hearer will understand, the speaker chooses the word that *it* would be most likely to understand if it was the hearer. Returning to the lexicon above, we can see that although *oij* has been associated with the meaning $2-0$ on more occasions than *gttr*, if heard, it would actually be interpreted as $1-0$ (because $1-0$ is the meaning which maximises the confidence probability for *oij*), whereas *gttr* would be interpreted with the correct $2-0$ meaning.

Interestingly, the agent would not find a word from its lexicon to express many meanings which do have some associations (e.g. $0-0$, $3-1$ etc.). One of the outcomes of obverter learning is the avoidance of ambiguity, so we find that, at any one time, each word in the lexicon is only used with one meaning, although the particular meaning can of course change as the associations in the lexicon are updated. This means that, although there are eight meanings in the lexicon extract, only two of them are actually used by the speaker, and so only these can be regarded as being truly *lexicalised.*

We have seen how the speaker tries to second-guess the hearer and chooses words which are likely to be understood before uttering them, but a greater problem is faced by the hearer in understanding the meaning which is being conveyed. On hearing a signal, the hearer's only guide in determining the intended meaning is the observation of the context (which of course includes the target topic object). From this, the hearer constructs a list of all the possible meanings, that is, *all* meanings which categorise only one of the objects in the context. All these possible meanings are equally plausible, so the hearer associates each of them with the signal in its lexicon, adjusting its confidence probability for each accordingly. Over time, the interpretation of each word will tend to the speaker's intended meaning, if the two agents have identical meaning structures [18].

5 Results

The meaning structures constructed by the agents in our model world, however, are of course not only not identical, but also change over time. Under these circumstances, is it possible for the agents to communicate? Figure 5 (left) shows

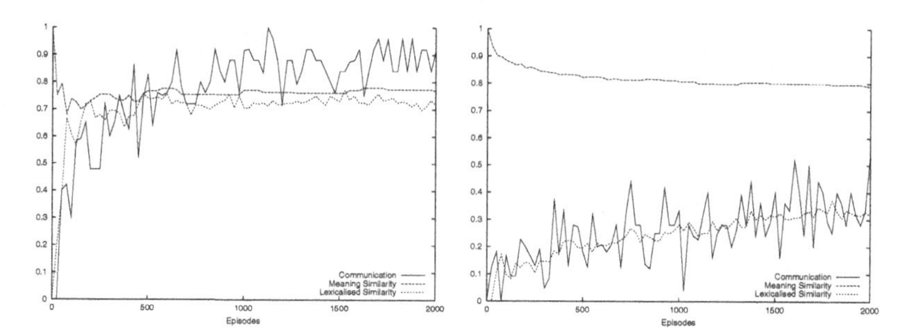

Fig. 5. Communicative success, meaning similarity, and lexicalised similarity for a population of two agents and 100 objects. Each discrimination game is played with a context size of five objects. The number of sensory channels available to each agent is five (left) and 100 (right).

that communication is successful a large percentage of the time, although it is not optimal, and does not appear to increase significantly after the initial rise to around 90%. The similarity of the agents' meaning structure drops initially, as the agents refine their sensory channels individually and separately, and then does not change significantly. This occurs because the pressure to develop meaning structure comes only from failure in discrimination games, and after an initial flurry, the agents all have sufficiently detailed meanings to succeed in nearly all discrimination games. Once this state is achieved, the communication rate stops improving and remains fairly constant. If the number of sensory channels available is increased substantially (figure 5: right), a similar result is found, except that the rate at which communication stops improving is much lower. It can also be seen that the communication success rate is closely parallelled in both cases by the *lexicalised similarity* of the agents, which is defined in the same way as meaning similarity (see section 3), but only taking into account tree nodes which are lexicalised.

An interesting phenomenon which occurs in these kind of simulations is the large amount of synonymy which pertains in the lexicons, where more than one word is interpreted with the same meaning. As an example, after 1000 communicative episodes, two agents have the meaning structures shown in figure 6. Attached to each node on the discrimination trees is the number of words which

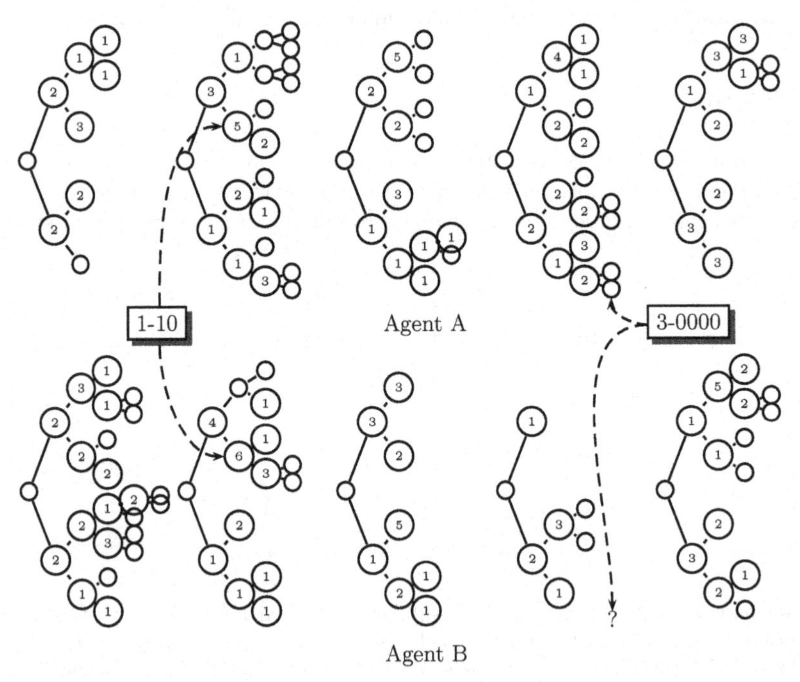

Fig. 6. Two agents each have five discrimination trees numbered 0-4. Each lexicalised node is marked with the number of words which would be interpreted as that meaning.

this agent would interpret as the meaning denoted by the node, or the number of *synonyms* attached to the meaning. For instance, we can see that there are five words which would each be interpreted by agent A as $1-10$, and six which would be interpreted as this by agent B. Further inspection (not shown) indicates that four of these synonyms have been lexicalised by both agents, suggesting a high level of redundancy, which is caused by *meaning drift*:

The interpretation of a word, of course, changes over time as the agents develop their experience of the word's use. Words are only created when an agent wants to express a meaning which isn't lexicalised. For example, in figure 6, agent A might wish to express the meaning 3 – 0000, but it does not have a word which it would interpret correctly, so it creates a new word *ujszo*. Agent B hears the new word, and creates a list of possible meanings. This list, however, cannot include A's meaning 3 – 0000, because B's meaning structure does not contain this meaning, and so B will lexicalise *ujszo* with a different meaning. Over time, B's preferred meaning is likely to be a more general meaning, which is shared by A.[4] There is now a difference of opinion over the meaning of *ujszo*, but crucially, agent A can continue to associate it with B's meaning, while B cannot associate it with A's original meaning. A's association between *ujszo* and the shared meaning gradually increases, until it eventually exceeds that of the original meaning. Both agents will now use *ujszo* for the more general meaning: the word's meaning has drifted. As a direct consequence, A no longer has a word with which it can express the meaning 3 – 0000. If it does need to convey this meaning, it must create another new word, and the cycle begins again.

Meaning drift is an inevitable characteristic of systems in which the agents' conceptual systems are not the same, if there are an unlimited number of signals, and there is little pressure to modify meaning structure. Inducing the meanings of words from context inevitably biases the meanings towards those meanings which are more general, and shared by the agents. Words which refer to specific meanings which are not shared will see their meanings drift to those which are shared, resulting in a large number of synonyms for the shared meanings, and few, if any, words at all for the agent-specific meanings.

6 Discussion

We have developed a world in which agents can communicate about their environment, without explicitly transferring meanings, without knowing exactly what the speaker is referring to, and without providing the learner with any feedback about communicative success, all criteria motivated by research into how human children acquire language [3]. Although communication can succeed in cases where agents refer to the same object with different meanings, the overall success of communication seems to be directly related to the amount of shared meaning structure in the agents. The communication system has a great deal of synonymy, caused by the differences in meaning structure and the unlimited number of possible signals. Work is under way to extend the model, focusing on ways to reduce synonymy, for instance by implementing the *principle of contrast* [12], and to investigate the effects of specific biases in meaning induction,

[4] Because general meanings are created before more specific meanings on the discrimination trees, they are more likely to occur in both agents' meaning structures.

such as the *shape bias* [11]. It is claimed that such biases explain the learning of meanings [3], and this work will go some way to showing where these claims are feasible.

References

[1] J. Batali. Innate biases and critical periods: Combining evolution and learning in the acquisition of syntax. In R. Brooks and P. Maes, editors, *Artificial Life 4: Proceedings of the Fourth International Workshop on the Synthesis and Simulation of Living Systems*, pages 160–171. Addison-Wesley, Redwood City, CA, 1994.
[2] J. Batali. The negotiation and acquisition of recursive grammars as a result of competition among exemplars. In E. Briscoe, editor, *Linguistic Evolution through Language Acquisition: Formal and Computational Models*. Cambridge University Press, Cambridge, 2001.
[3] Paul Bloom. *How Children Learn the Meanings of Words*. MIT Press, 2000.
[4] E. Briscoe. Grammatical acquisition: Inductive bias and coevolution of language and the language acquisition device. *Language*, 76(2):245–296, 2000.
[5] D. Cheney and R. Seyfarth. *How Monkeys See the World: Inside the Mind of Another Species*. University of Chicago Press, Chicago, IL, 1990.
[6] Edwin D. de Jong. *Autonomous Formation of Concepts and Communication*. PhD thesis, Vrije Universiteit Brussel, 2000.
[7] S. Harnad. The symbol grounding problem. *Physica*, D 42:335–346, 1990.
[8] James R. Hurford. The evolution of critical period for language acquisition. *Cognition*, 40:159–201, 1991.
[9] S. Kirby. Learning, bottlenecks and the evolution of recursive syntax. In Ted Briscoe, editor, *Linguistic Evolution through Language Acquisition: Formal and Computational Models*. Cambridge University Press, Cambridge, 2001.
[10] V. Kvasnička and J. Pospíchal. An emergence of coordinated communication in populations of agents. *Artificial Life*, 5(4):319–342, 1999.
[11] Barbara Landau, Linda B. Smith, and Susan S. Jones. The importance of shape in early lexical learning. *Cognitive Development*, 3:299–321, 1988.
[12] E. M. Markman. *Categorization and naming in children*. MIT Press, 1989.
[13] John Maynard Smith and Eörs Szathmáry. *The major transitions in evolution*. Oxford University Press, 1995.
[14] Chrystopher L. Nehaniv. The making of meaning in societies: Semiotic and information-theoretic background to the evolution of communication. In *Proceedings of the AISB Symposium: Starting from Society — the application of social analogies to computational systems, 19–20 April 200*, pages 73–84, 2000.
[15] M. Oliphant. The learning barrier: Moving from innate to learned systems of communication. *Adaptive Behavior*, 7(3/4):371–384, 1999.
[16] M. Oliphant and J. Batali. Learning and the emergence of coordinated communication. *Center for Research on Language Newsletter*, 11(1), 1997.
[17] W. V. Quine. *Word and Object*. MIT Press, 1960.
[18] A. D. M. Smith. The role of shared meaning structure in communication. Paper given at the 7th Human Behaviour & Evolution Society Conference, London, 2001.
[19] L. Steels. Perceptually grounded meaning creation. In M. Tokoro, editor, *Proceedings of the international conference on multi-agent systems*. MIT Press, 1996.
[20] L. Steels. *The Talking Heads Experiment*, volume I. Words and Meanings. Laboratorium, Antwerpen, 1999. Special pre-edition.
[21] J. R. Taylor. *Linguistic Categorization*. Oxford University Press, 1995.

The Difficulty of the Baldwinian Account of Linguistic Innateness

Hajime Yamauchi

Language Evolution and Computation Research Unit,
Department of Theoretical and Applied Linguistics,
The University of Edinburgh, Edinburgh, UK
hoplite@usa.net

Abstract. Turkel [16] studies a computational model in which agents try to establish communication. It is observed that over the course of evolution, initial plasticity is significantly nativised. This result supports the idea that innate language knowledge is explained by the Baldwin effect [2][14]. A more biologically plausible computational model, however, reveals the result is unsatisfactory. Implications of this new representation system in language evolution are discussed with a consideration of the Baldwin effect.

1 Introduction

For decades, the innate capacity of language acquisition has been one of the central issues of the study of language. How heavily does language acquisition rely on innate linguistic properties? This question, often called the '*nature & nurture problem*', brings endless debates in linguistics and its adjacent fields. Indeed, a number of phenomena that occur during language acquisition are quite puzzling when one tries to determine what parts of language acquisition are innate or attributed to postnatal learning. An intensive array of studies has gradually revealed that this twofold structure of language acquisition never appears as a clear dichotomy. Rather, the intriguing interaction between innate and learning properties of language acquisition seems to require a new avenue of linguistic studies.

1.1 The Baldwin Effect and Language Evolution

James Mark Baldwin [2] assumed that if an individual is capable of acquiring an adaptive behavior postnatally, addition of such a learning process in the context of evolutionary search potentially changes the profile of populational evolution; the learning paves the path of the evolutionary search so that evolution can ease its burden of search. In addition, this special synergy of learning and evolutionary searches has a further effect, known as 'genetic assimilation' [18]. This is a phenomenon in which "a behavior that was once learned may eventually become instinctive" [17].

J. Kelemen and P. Sosík (Eds.): ECAL 2001, LNAI 2159, pp. 391–400, 2001.

Then this learning-guided evolution scenario, known as *the Baldwin effect*, possibly provides a strikingly attractive perspective to the *nature-nurture* problem in linguistics. It has been attested by a number of computer simulations in the field of computer science that if an environment surrounding the population is prone to shift to a new environment, some part of the behavior is better preserved for learning. If those environments do not share any commonality, an individual who relies in every aspect of behavior on learning will be the most adaptive. However, if those environments hold some universality, an individual who has partially nativised and partially learned behavior will be the most adaptive; for example, the nativised part of the behavior covers the universality and the learned part of the behavior covers the differences. Consider this in the case of language evolution. The whole human population is well divided into a number of sub-populations in many aspects; races, cultures, and so forth. Boundaries of language diversities often coincide with those of the sub-populations. Then, for children, it is a great advantage to keep some part of the linguistic knowledge for learning while the other is innately specified. This helps the child even if he is reared in a different linguistic society from his parents; he still may acquire the society's language. Therefore, the *nature-nurture* problem in linguistics can now be considered in the context of the evolution of language. Universality of the world's languages may correlate to the evolution of nativised linguistic knowledge while linguistic diversities are correlated to learning. Since this universality-*nature*, diversity-*nurture* correlations are perfectly compatible with Chomsky's *L*anguage *A*cquisition *D*evice theory [4], and as the Baldwin effect and the LAD theory both involve genetics, the study of the Baldwin effect in the domain of LAD becomes particularly appealing.

The Baldwin effect in linguistics may also provide an attractive solution for a long-standing problem. Preliminary studies suggest that language evolution is out of the scope of natural selection mainly because of its dysfunctional nature. For those researchers, language evolution is a consequence of exaptation or a big leap in evolution [13]. This no-intermediate scenario would be, however, explicable by natural selection when it is guided by learning since learning can smooth the no-intermediate landscape. Subsequently, it has been a popular idea that the Baldwin effect is a crucial factor in the evolution of language (e.g., [14][16]).

1.2 The Principles and Parameters Approach

Given its logical complexity, researchers agree that linguistic input is the most important ingredient of language acquisition. Counter-intuitively, however, such vital linguistic input employed to construct knowledge of a language is importantly often insufficient [3]. In other words, children have to acquire their target languages under qualitatively and quantitatively insufficient circumstances. Absence of "Negative Evidence" in language acquisition is one of the clearest examples of this. As a part of the insufficiency, usually children are not provided negative feedback for their grammatical mistakes while such information is vital for any second language learners.

To reduce this complication, Chomsky has claimed some special synergy of innate linguistic knowledge and the acquisition mechanism is required. The basic concept of his original formulation of the nature of language acquisition, called Principles & Parameters theory, [6] is as follows. In the P&P approach, two types of limited innate linguistic knowledge are accessible, called 'principles' and 'parameters'. Principles are universal among all natural languages and considered as genetically endowed. Parameters are partially specified knowledge which are encoded in binary parametric values. Setting of each parameter is triggered by post-natal linguistic experiences. We can conceive the possible mechanism of the LAD as an incomplete learning device in which certain binary information is missing

2 Implementation of the LAD in Dynamic Systems

The combination of genetically hardwired features and postnatal learning processes in the Baldwin effect is perfectly compatible with Chomsky's P&P theory of the LAD. Together with its "genetic assimilation" process [18], the Baldwin effect may shed light on the nature of the current relationship between innateness and postnatal learning in language acquisition.

Precisely because of this compatibility it is crucial to pay careful attention to the implementation of the P&P approach in a genetic search. Given an assumption that the LAD is one of the most elaborated cognitive abilities, it is highly unlikely that such ability is DIRECTLY coded in the genes. Rather it is more plausible to assume that linguistic innateness relies on some degree of polygenic inheritance [1].

More specifically, principles and parameters are not coded by a simple concatenation of genes. Rather a *combination* of those genes expresses one principle/parameter. This genetic mechanism is called "epistasis". Epistasis is a situation in which the phenotypic expression of a gene at one locus is affected by the alleles of a gene at other loci. Pleiotropy, in a very crude form, means that one gene contributes to express more than one phenotypic character. Thus, one gene in the model will affect an expression of one phenotypic trait, but also will determine other traits.

In the next section, we examine the effect of the two phenomena in the study of the evolution of the LAD.

3 The Experiments

To test the effect of epistasis and pleiotropy on the Baldwin effect, we conducted two different types of simulations. The basic part of our model is adapted from the study of Turkel [16] to appear). First, an exact replication of Turkel's simulation was tested. Then modified versions were tested. In those modified simulations, Stuart Kauffman's [11] NK-Landscape model was introduced to implement epistasis and pleiotropy. The specific explanation of NK-Landscape in these simulations is given later. Here the basic structure of the model is explained. In Kauffman's NK-Landscape model, unlike ordinary GA models where one gene expresses one trait of a phenotype, a SET of genes determines one trait of a phenotype. In other words, one specific part of the phenotype (a phenotype consists of 12 traits in this simulation) may be decided by two or more distinctive genes. How many genes are required to express one trait is specified in the value of K. The values of K are always between 0 and N-1 where N designates the number of the genes. Dependency of genes is either contiguous or non-contiguous. In the case of contiguous dependency, a gene forms a concatenation with other adjacent genes. Note that in the contiguous dependency case, which we employ in this paper, both ends of a chromosome are considered as neighbors of each other so that K-dependency of phenotypes is available in all loci.

In terms of evolutionary search, the increase of the value of K toward N means that the fitness landscape becomes increasingly rugged. In a rugged landscape, evolutionary search tends to be trapped in local optima. The correlation between the fitness and similarity of genotypes (typically measured by Hamming distance) is also kept low in the landscape. Therefore, an identical phenotype of two agents does not guarantee for them to have an identical genotype. In a simulation using this model, a look-up table is created at the beginning of the simulation. The size of tables corresponds to N times 2^K since each allele is affected by 2^K possible combinations of other genes.

In the next section, we look at the result of Turkel's original study, then make a comparison to our obscured phenotype model. All results of these simulations are averages of 100 runs.

3.1 Simulation1: Replication of Turkel

Based on Hinton & Nowlan's simulation [10], Turkel conducts an experiment that holds a populationally dynamic communication system. While Turkel mostly adopts Hinton & Nowlan's genetic encoding method (fixed, and plastic genes), he provides an external motivation for it according to P&P approach. Turkel considers those fixed genes —0s and 1s— as '*principles*', and the plastic genes —?s— as '*parameters*'.
The algorithm of Turkel's simulation is quite straightforward and mostly intuitive. Most parts of the algorithm are quantitatively the same as Hinton & Nowlan; initially 200 agents are prepared. The ratio of 0:1:? In Turkel is different in his four different configurations of simulations —2:2:8 (High-plasticity), 4:4:4 (Equal ratio), 3:3:6 (Original), and 6:6:0 (No-plasticity)— respectively. Distribution of these genes in an individual agent is randomly decided initially. In the initial population, generally there is no case that two agents hold the same genotype. The reproduction process includes one-point crossover with 20% probability. Considering the spirit of GA, it is somewhat odd but mutation is not included [10] mutation was not included also). Two agents (one is selected from 1st agent to 200th in order, and its partner is randomly selected) compare their genotype. If those two agents' genotypes are exactly the same pattern including loci of ?s, the first-chosen one is assigned 2 fitness points. If the agents do not exactly match but those no-matching alleles have 0-? Or 1-? Combinations, they are considered as potentially communicable. Then they are sent to learning trials. By changing all ?s into either 1s or 0s randomly, the two agents attempt to es-

tablish communication within 10 trials. If the agents succeed to possess exactly the same phenotype within ten trials, communication is considered to be established. In each trial, the agents reset their phenotype and express new phenotypes from their genotypes. During the learning process, learning cost is introduced implicitly. The size of decrement per trial is 1 from the highest fitness value of 12. The range of the fitness values is, thus, from 12 (immediate success) to 1 (complete failure). If two agents have any 1s and 0s combination in the same allele, they are assigned the fitness value of 1 since it would be impossible to establish communication.

In our replication experiment, we choose Turkel's "Original" configuration where the number of ?s is 6 and the number of both 1s and 0s are 3 each.
The result obtained from our simulation was, as expected, almost identical to Turkel's original simulation. Fig. 1 shows the average number of 0s, 1s, and ?s in the evolved population.

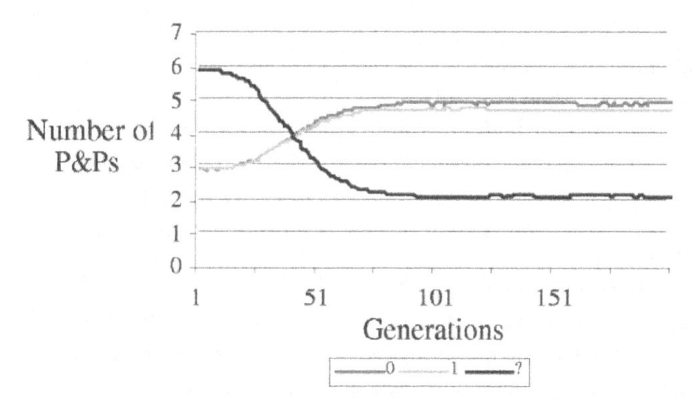

Fig. 1.

In the figure, a steep descent of ?s is observable in an early period. Once the population reaches the "plateau" condition, no further change takes place. On the plateau, virtually all agents have one unified genotype. The reason for this is the lack of mutation; the one-point-crossover reproduction process does not produce any turbulence under the unified situation.

It is often the case that before the Baldwin effect eliminates all plastic genes, a population reaches this plateau. This was especially salient in his preliminary studies where populations were more plastic. In those situations, the Baldwin effect did not have enough space to enjoy its power; before doing so, the populations typically converged to one genotype. Thus, at the end of each run, a comparatively large number of plastic genes remained, although the number of plastic genes was fewer than in the initial populations in almost all cases. To make it clearer, consider the following points. First, when the entropy of genotypes in a population is high (as in an initial period), high plasticity is advantageous; the more plastic, the more chance an agent has of proceeding to the learning trials. On the contrary, a fixed agent suffers great difficulty in this kind of situation; the fitness value is most likely 1 since the chance of exact match is extremely slim.

Too much plasticity cannot increase the actual fitness value either. Although highly plastic agents can often potentially communicate with other agents, the actual probability of establishing communication is quite low as the number of possible 0 and 1 combinations increases exponentially.

If the agent fails to establish a communication, the fitness value is 1. Thus, although it is somewhat contradictory, the best strategy to maximise fitness value is to keep the number of parameters as small as possible. It effectively means increasing the chance of establishing communication within 10 learning trials. To do this, it is necessary to reduce the number of plastic genes —genetic assimilation. Genetic assimilation, however, increases the number of fixed genes. Since the penalty for discrepancy of fixed genes on the same locus is most fatal (one Hamming distance is enough), this elimination process has to be done by increasing the identical genotype except in the loci of plastic genes. In other words, low plastic agents have to make sure that they meet either agents who have exactly the same genotype or all-the-same-but-partially-plastic agents. This turns out to be a selective pressure toward a uniform genotype. Therefore, genetic assimilation must intrinsically go hand in hand with convergence to identical genotypes. Importantly, however, these two processes are quasi-independent processes; although the force of both pressures comes from natural selection through the reproduction process, genetic assimilation is required from the learning trial *per se* while the convergence pressure comes into the place by more general requirement, "parity". As noted above, when two agents are compared their pre-learning phenotype (= genotypes), discrepancy of principles is strongly malign —even with one discrepancy in their principles, the two agents have no possibility of establishing communication— while parameters always match with any principles or other parameters. As long as any loci that have principle-principle pairs match, an agent can have any number of parameters on any locus; although a lot of parameters indeed decrease the chance of communication but never reduce the chance completely while discrepancy between principles extinguishes it. In this regard, parameters are more benign than principles. Thus the pressure of convergence is generally greater than that of genetic assimilation. Since the pressure of convergence drives the agents to align their genotypes, consequently the population typically converges into a single genotype before complete genetic assimilation takes place. This is the reason why when the population is highly plastic, the absolute number of plastic gene remains higher than in a population.

3.2 Simulation2: Implementation of NK-Landscape Model

Our next simulations incorporate the NK-Landscape models while most of Turkel's algorithms are untouched. A brief description of the simulation is given.

First, we determine the number of gene dependency regarding the expression of the phenotype. K designates the number. The value of K is fixed within a simulation; the same value is always applicable to any locus (this means that at any locus, the degree of gene dependency is not affected), any agents, and any generation. Since the range of K is from 2 to N-1, the maximum value is 11 (N=12) in these simulations.

Then, we prepare 200 agents. All agents consist of 12 genes. This time, instead of the three types of genes —0, 1, ?— only two types of genes exist, namely 0 and 1. Thus, at this level, there is no plasticity. These genes are equally shuffled into the 12 loci. The number of the two types of genes are the same in one agent, 6 each. These 12 genes are randomly distributed into 12 loci.

Thirdly, a look-up table is generated. This table correlates a genotype and phenotype. Below, an example is provided (Table 1).

Table 1.

	000	001	010	100	011	101	110	111
Locus1	0	**?**	1	1	?	?	0	?
Locus2	?	?	?	0	1	1	0	?
Locus12	?	1	1	0	?	?	0	?

The number of rows corresponds to the number of loci —12. The number of column corresponds to the number of possible combinations of genes. If K=3, the number of column is 2^3. To project a principle/parameter in the first position of a phenotype, we have to check the first row –"Locus1". If three genes from the first locus are 0, 0, 1, respectively in the genotype, we put ? in the first position of the phenotype (the cell in the table is emphasized). To project a principle/parameter in the second position of the phenotype, the second row is referred to. At the end of this projection process, the phenotype contains 12 principles/parameters in total. This is compatible with Turkel's genes. To make the simulations comparable to the former simulation, the ratio of 0, 1, and ? is set as 1:1:2. This is done by controlling the ratio of 0, 1, ? in look-up tables. Once this process is done, the rest of the simulation is exactly the same as Turkel's.

Although all possible values of K are tested, here we pick up three of the results; K=2, K=7, and K=11. All are in Fig. 2. First, we look at the result of K=2. The graph shows that genetic assimilation is still saliently observed.
6 parameters at the initial population are eliminated up to 2.9 (recall all results are an average of 100 runs) around 90th generation. This is one parameter more than the original simulation. Correspondingly, the position of the "plateau" shifts slightly to the right hand side. This means that slightly more generations are required to reach a single genotype. Secondly, K=7 is tested. The decrescent curve of the parameters is much shallower than that of K=2. As a consequence, the left edge of the plateau shifts more to the right. At this point, no decrement is observed. Rather, a small increase of plasticity is observed. This is because the increase of plasticity may improve the chance to obtain the fitness value of 2 or more. On the other hand, decrease of parameters is a tougher demand since it has to come with genetic convergence; a parameter cannot be replaced with 1-principle or 0-principle randomly; it must be par with other agents.

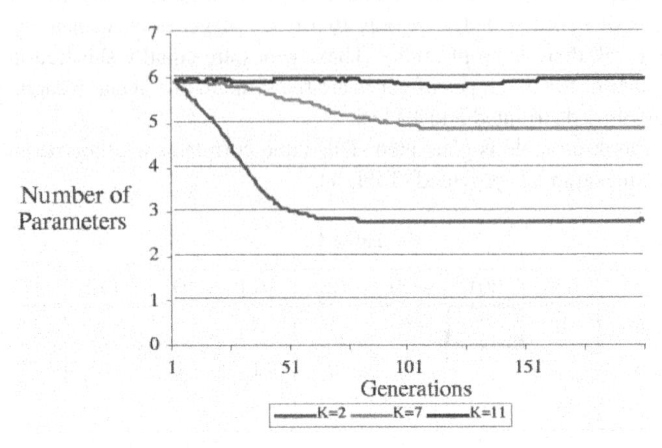

Fig. 2.

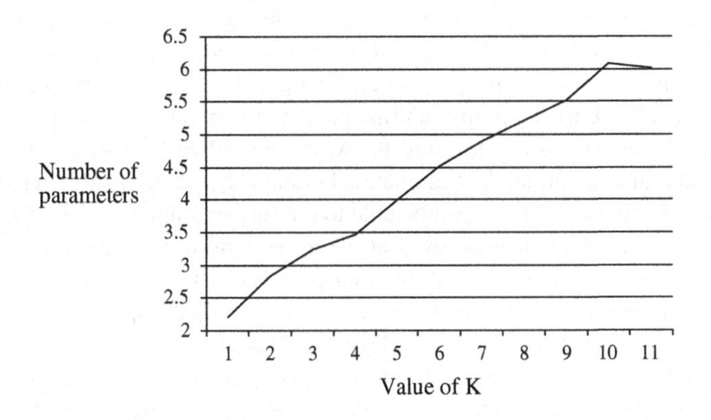

Fig. 3.

Fig. 3 shows the relationship between the value of K and the number of parameters at the end of each simulation. The consequence is crystal clear —as the value of K increases, more parameters remain in the population.

From these, it is apparent that the Baldwin effect is progressively weakened as the genetic dependency increases. In other words, the Baldwin effect is highly sensitive to epistasis and pleiotropy.

The results shown above beautifully reveal how epistasis and pleiotropy affect the Baldwin effect in populational dynamic communication. These results strongly suggest that parameters are hardly eliminated, even if keeping high plasticity may be a costly option. From these, it is now clear that under these circumstances, the scenario of the evolution of the LAD may severely undermine its elimination of parameters.

4 Conclusion

The experiments show that pleiotropy and epistasis effectively dampen the emergence of the Baldwin effect in the dynamic communication system. Although the modification is simple and quite straightforward regarding its technical complexity, the actual outputs are radically different. This has to be taken as a serious caution for our future studies. In sum, epistasis and pleiotropy in genes for the LAD, thus, may require a radical re-interpretation of the scenario of the evolution of the LAD.

However, there are some points we should improve the models to make a firmer claim. For example, in the simulations presented here, during the communication period, agents convert their ? characters to either 0 or 1 characters. We interpret this attempt to establish communication as learning. Strictly speaking, it is difficult to consider it as learning in a linguistic sense. In the simulations, learning takes place without any input from previous generations or even from the same generation. Usually, language acquisition takes place with linguistic inputs in a linguistic community. Adults' utterances are learners' primary linguistic inputs. When the learners become adults, their speeches become the next generation's inputs. Thus, linguistic inputs generally come down from previous generations to next generations. Such inputs are independent from genetic inheritance. Furthermore, the process does not include any update process of an agent's internal state.

Recently, more and more scholars have begun to reconsider the exact mechanism of the Baldwin effect. Most of the studies of the Baldwin effect itself share their roots in either Waddington's studies *in vivo* or Hinton & Nowlan's computer simulation *in silico*. Although the Baldwin effect is alleged to be observed in both studies, it is also true that the actual mechanisms for the Baldwin effect working in these studies are quite different. As Simpson [15] and Depew [9] argue, the Baldwin effect is easily dissected into its parts, and possibly the effect is simply just the sum of these parts. If we strictly follow this point of view, there is no need to invoke the sum as "a new factor in evolution [2]". In his exploration of language evolution, a biologist T. Deacon [8], however, has recently proposed a new type of mechanism of the Baldwin effect. This new mechanism, called "niche construction" has a self-organizing, emergent aspect in its core. This self-organizing, emergent type of mechanism seems to be particularly attractive for the case of language evolution, as it might provide a solution by which language evolution can circumvent the problem of pleiotropy and epistasis raised here.

References

1. Atmar, W. (1994). Notes on the simulation of evolution. *IEEE Transactions on Neural Networks* 5(1).
2. Baldwin, J. M. (1896). A New Factor in Evolution. *The American Naturalist*, *30*, 441-451, 536-553.
3. Chomsky, N. (1965). *Aspects of the theory of syntax.* Cambridge, MA: MIT Press.

4. Chomsky, N. (1972). *Language and mind*, enlarged edition. New York: Harcourt Brace Jovanovich.
5. Chomsky, N. (1981a). *Lectures on government and binding*. Dordrecht: Foris.
6. Chomsky, N. (1981b). Principles and parameters in syntactic theory. In N. Hornstein & D. Lightfoot (Eds.), *Explanations in Linguistics*. London: Longman.
7. Chomsky, N. (1986). *Knowledge of language*. New York: Praeger.
8. Deacon, T. (1997) *The Symbolic Species*. New York: W.W. Norton.
9. Depew, D. (2000) The Baldwin Effect: An Archaeology. *Cybernetics And Human Knowing*, *7*, (1), 7-20.
10. Hinton, G. E., & Nowlan, S. J. (1987). How learning can guide evolution. *Computer Systems*, *1*, 495-502.
11. Kauffman, S. A. (1989). Adaptation on rugged fitness landscapes. In D. L. Stein (Ed.), *Lectures in the science of complexity, 1*. Redwood City, CA: Addison-Wesley.
12. Niyogi, P., & Berwick, R. C. (1996). A language learning model for finite parameter spaces. *Cognition*, *61*, 161-193.
13. Piatelli-Palmarini, M. (1989). Evolution, Selection and Cognition: From "learning" to parameter setting in biology and the study of language. *Cognition*, *31*, 1-44.
14. Pinker, S., & Bloom, P. (1990). Natural language and natural selection. *Behavioral and brain sciences*, *13*, 707-784.
15. Simpson, G. G. (1953) The Baldwin Effect. *Evolution, 7*, 110-117.
16. Turkel, J. W. (to appear) The Learning Guided Evolution of Natural Language. In T Briscoe (Ed.), *Linguistic Evolution through Language Acquisition: Formal and Computational Models*. New York: Cambridge University Press.
17. Turney, P., Whitley, D., & Anderson, R.W. (1996). Evolution, learning, and instinct: 100 years of the Baldwin effect, *Evolutionary Computation*, *4*, (3), iv-viii.
18. Waddington, C. H.: *The evolution of an evolutionist*. Edinburgh: Edinburgh University Press.

Making Evolution an Offer It Can't Refuse: Morphology and the Extradimensional Bypass

Josh C. Bongard and Chandana Paul

Artificial Intelligence Laboratory
University of Zürich
Zürich, Switzerland
{bongard, chandana}@ifi.unizh.ch

Abstract. In this paper, locomotion of a biped robot operating in a physics-based virtual environment is evolved using a genetic algorithm, in which some of the morphological and control parameters of the system are under evolutionary control. It is shown that stable walking is achieved through coupled optimization of both the controller and the mass ratios and mass distributions of the biped. It was found that although the size of the search space is larger in the case of coupled evolution of morphology and control, these evolutionary runs outperform other runs in which only the biped controller is evolved. We argue that this performance increase is attributable to extradimensional bypasses, which can be visualized as adaptive ridges in the fitness landscape that connect otherwise separated, sub-optimal adaptive peaks. In a similar study, a different set of morphological parameters are included in the evolutionary process. In this case, no significant improvement is gained by coupled evolution. These results demonstrate that the inclusion of the correct set of morphological parameters improves the evolution of adaptive behaviour in simulated agents.

1 Introduction

In the field of robotics, much work has been done on optimizing controllers for biped robots [1,12,22]. Similarly, genetic programming [8] and genetic algorithms [7] have been used to evolve controllers for hexapod robots. Genetic algorithms have also been used to evolve recurrent neural networks for bipedal locomotion: Fukuda *et al* [6] employed a dynamic simulator; Reil and Husbands [19] employed a three-dimensional physics-based simulator. However, in all of these approaches, little or no consideration was paid to the mechanical construction of the agent or robot.

Alternatively, Brooks and Stein [3] and Pfeifer and Scheier [18] have pointed to the strong interdependence between the morphology and control of an embodied agent: design decisions regarding either aspect of an agent strongly bias the resulting behaviour. One implication of this interdependence is that often, a good choice of morphology can lead to a reduction in the size or complexity of the controller. For example, Lichtensteiger and Eggenberger [13] demonstrated

J. Kelemen and P. Sosík (Eds.): ECAL 2001, LNAI 2159, pp. 401–412, 2001.

that an evolutionary algorithm can optimize the sensor distribution of a mobile robot for certain tasks, while the controller remains fixed. As an extreme case, the study of passive dynamics has made clear that a careful choice of morphology can lead to locomotion without any actuation or controller at all [17].

Examples now abound that demonstrate the evolution of both the morphology and control of simulated agents [20,21,11,4,16], as well as real-world robots [15,10,14] is possible. However, we argue in [2] that the coupled evolution of both morphology and control of adaptive agents is not as interesting in and of itself: rather, the implications of such studies open up a host of research questions regarding the evolution of adaptive behaviour that are not amenable to study solely through the optimization of control. Virtual Embodied Evolution (VEE) was introduced in [2] as a systematic methodology for investigating the implications of evolving both the morphology and control of embodied agents. In this paper we show not only that coupled evolution of both morphological and control parameters of a bipedal agent can facilitate the discovery of stable locomotion—despite the increased size of the search space necessitated by the inclusion of the additional morphological parameters—but also that only certain sets of morphological parameters facilitate evolutionary search.

The following section introduces the mechanical construction and neural controller of the biped agent, as well as the genetic algorithm used to evolve locomotion. Section 3 presents the results obtained from evolving only the neural networks for a bipedal agent, as well as evolutionary runs in which morphological parameters were included in the genome. Section 4 provides some discussion and analysis as to why coupled evolution of morphology and control can outperform the evolution of control. In the final section we conclude by stressing the importance of incorporating morphological considerations into the evolutionary investigation of adaptive behaviour.

2 The Model

For all of the evolutionary runs reported in this paper, the agents act within a physically-realistic, three-dimensional virtual environment[1]. The agent is a simulation of a five-link biped robot with six degrees of freedom. The agent has a waist, and two upper and lower leg links as shown in Fig. 1 a). Each knee joint, connecting the upper and lower leg links, has one degree of freedom in the sagittal plane. Each hip joint, connecting the upper leg to the waist, has two degrees of freedom: one in the sagittal plane and one in the frontal plane. These correspond to the roll and pitch motions. In the second set of experiments reported in section 3, a second type of biped is used, in which five mass blocks are attached to the lower legs, upper legs and waist as shown in Fig. 1 b).

The joints are limited in their motion using joint stops, with ranges of motion closely resembling those of human walking. The hip roll joint on each side has a

[1] The environment and biped agents were constructed and evaluated using the real-time physics-based simulation package produced by MathEngine PLC, Oxford, UK, `www.mathengine.com`.

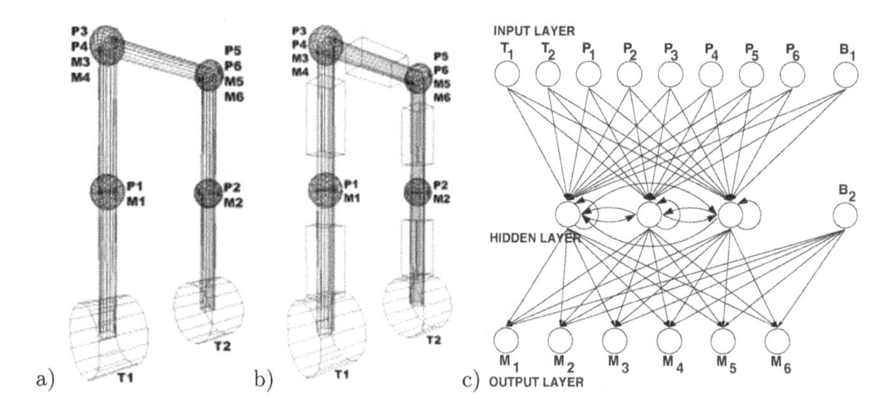

Fig. 1. Agent construction and neural network topology. a) shows the biped agent without the attached masses. b) shows the agent with the attached masses. c) gives a pictorial representation the neural network used to control both types of agents. `T1` and `T2` correspond to the two touch sensors, `P1` through `P6` indicate the six proprioceptive sensors, and `M1` through `M6` indicate the six torsional motors of the biped. `B1` and `B2` indicate the two bias neurons.

range of motion between $-\frac{\pi}{7}$ and $\frac{\pi}{7}$ radians with respect to the vertical. The hip pitch joint has a range of motion between $-\frac{\pi}{10}$ and $\frac{\pi}{10}$, also with respect to the vertical. The knee joint has a range of motion between $-\frac{\pi}{2}$ and 0 with respect to the axis of the upper leg link to which it is attached. Table 1 summarizes the morphological parameters for both types of bipeds.

The agent contains two haptic sensors in the feet, and six proprioceptive sensors and torsional actuators attached to the six joints, as outlined in Figs. 1 a) and b). At each time step of the simulation, agent action is generated by the propagation of sensory input through a recurrent neural network; the values of the output layer are fed into the actuators as desired positions. The input layer contains nine neurons, with eight corresponding to the sensors, and an additional bias neuron. All neurons in the network emit a signal between -1 and 1: the haptic sensors output 1 if the foot is in contact with the ground, and -1 otherwise; the proprioceptive sensor values are scaled to the range $[-1, 1]$ depending on their corresponding joint's range of motion; and bias neurons emit a constant signal of 1. The input layer is fully connected to a hidden layer composed of three neurons. The hidden layer is fully and recurrently connected, plus an additional bias neuron. The hidden and bias neurons are fully connected to the eight neurons in the output layer. Neuron activations are scaled by the threshold function $\frac{2}{1+e^{-a}} - 1$. The values at the output layer are scaled to fit the range of their corresponding joint's range of motion. Torsion is then applied at each joint to attain the desired joint angle.

Evolution of bipedal locomotion is achieved using a floating-point, fixed-length genetic algorithm. Each genome encodes weights for the 60 synapses in the neural network, plus any additional morphological parameters. All values

Table 1. The default size dimensions, masses and joint limits of the biped. Parameters set in boldface indicate those parameters that are modified by evolution in the experiments reported in section 3. The valid ranges for these parameters are also given.

Index	Object	Dimensions	Mass
1	Knees	r = 1ul	1um each
2	Hip sockets	r = 1ul	1um each
3	Feet	r = 2ul, w = 3ul	1um each
4	Lower Legs	**r = [0.2,0.8] ul**, h = 8ul	0.25um each
5	Upper Legs	**r = [0.2,0.8] ul**, h = 8ul	0.25um each
6	Waist	**r = [0.2,0.8] ul**, w = 8ul	0.25um
7	Waist Block	**l = [0.4,3.6] ul, w = h = [0.2,3.0] ul**	0.103um
8	Lower Blocks	**l = [0.4,3.6] ul, w = h = [0.2,3.0] ul**	0.103um each
9	Upper Blocks	**l = w = [0.2,3.0] ul, h = [0.4,3.6] ul**	0.103um each
Index	**Joint**	**Plane of Rotation**	**Range (rads)**
10	Knee	sagittal	$-\frac{\pi}{2} \rightarrow 0$
11	Hip	sagittal	$-\frac{\pi}{7} \rightarrow \frac{\pi}{7}$
12	Hip	frontal	$-\frac{\pi}{10} \rightarrow \frac{\pi}{10}$

in the genome range between −1.00 and 1.00. Each evolutionary run reported in this section is performed using a population size of 300, and is run for 300 generations. Strong elitism is employed in which 150 of the most fit genotypes are preserved into the next generation. Tournament selection, with a tournament size of three, is employed to select genotypes from among this group for mutation and crossover. 38 pairwise one-point crossings produce 76 new genotypes. The remaining 74 new genotypes are mutated copies of genotypes from the previous generation: an average of five point mutations are introduced into each of these new genotypes, using random replacement.

In the set of experiments using the agent shown in Fig. 1 a), three additional morphological parameters are included in the genome. These parameters dictate the radii of the lower legs, upper legs and waist, respectively. The range of possible radii for these segments is $[0.2, 0.8]$ unit length[2] In the second set of experiments, eight morphological parameters are included in the genome: the first three values dictate the widths of the lower mass block pair, upper mass block pair and waist mass block, respectively, each of which can range between 0.2 and 3.0 ul. The next three values indicate the lengths of the lower mass block pair, upper mass block pair and waist mass block, respectively, which range between 0.4 and 3.6 ul. The final two values indicate the vertical placement of the lower and upper block mass pairs, which can range between 0.8 to 7.2 ul

[2] All lengths and masses reported in this paper are relational: the unit length (`ul`), and the default mass (`um`), are set equal to the radii and masses of the knee and hip sockets, respectively.

Table 2. Experimental regime summary.

Run Set	Morphology	Blocks	Total block mass	Genome length	Number of independent runs
1	Fixed	Absent	N/A	60	30
2	Variable	Absent	N/A	63	30
3	Fixed	Present	0.512um	60	20
4	Variable	Present	0.512um	68	20

above the centre of the foot: in this way, all four blocks can be attached to the upper or lower pairs of legs. The horizontal position of the waist block mass remains centred, and is not changed. In the case of agents without block masses, the morphological parameter settings can affect the total mass, mass distribution and moment of inertia of the agent. In the case of agents with block masses, the morphological parameter values can affect only the mass distribution and the moment of inertia, although more degrees of freedom of the rotational moment of inertia are subject to selection pressure in this case. For the variable morphology evolutionary runs, the three or eight morphological parameters are distributed evenly across the length of the genome in order to maximize recombination of these values during crossover.

The fitness of a genome is determined as follows. The weights encoded in the genotype are assigned to the synapses in the neural network, and in the case of the variable morphology bipeds without mass blocks, the radii of the waist, lower and upper legs are set based on the additional three values in the genome. In the case of the variable morphology bipeds with the mass blocks, the dimensions and positions of the blocks are set based on the additional eight parameters. The agent is then evaluated for up to 2000 time steps in the physical simulator. Evaluation halts prematurely if both of the feet leave the ground at the same time (this discourages the evolution of running gaits); the height of the waist passes below the height of the knees; or the waist twists more than 90 degrees away from the desired direction of travel. The northern distance of the agent's hip at the termination of the evaluation period is then treated as the fitness of the genome.

3 Results

Four sets of evolutionary runs were conducted using the parameters given in Table 2. Fig. 2 summarizes the evolutionary performance of the two sets of runs using agents without mass blocks, and Fig. 3 reports the evolutionary performance of the two sets of runs using agent populations with mass blocks. It can be seen in Fig. 2 that in both fixed and variable morphology agent populations, there is a roughly uniform distribution of fitness performance achieved by the most fit agents at the end of the runs. However Figs. 2 b) and d) indicate that

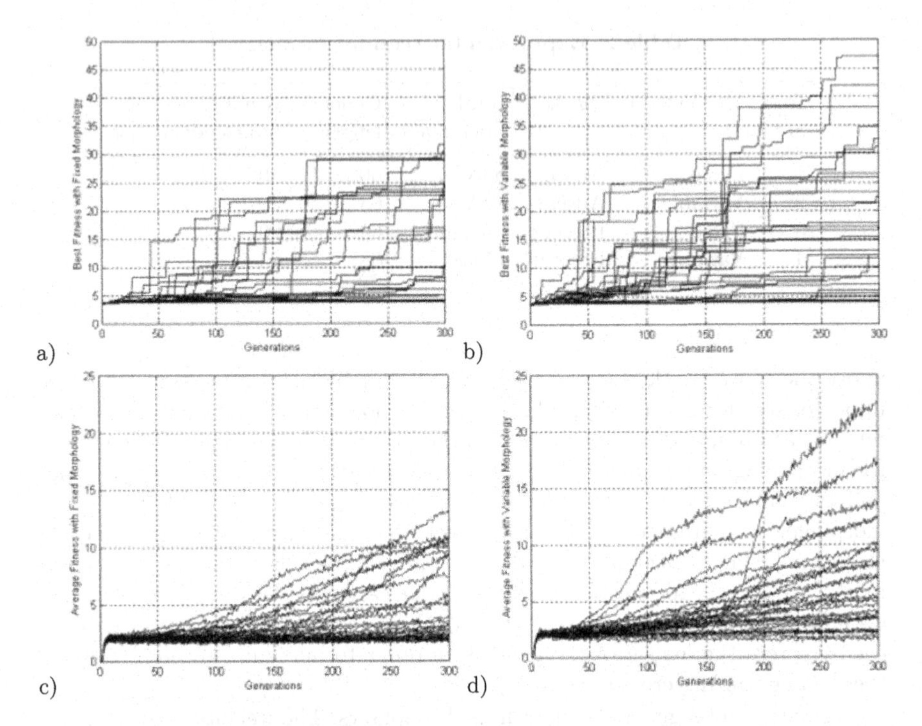

Fig. 2. Evolutionary performance of fixed and variable morphology agent populations without mass blocks. a) and b) report the highest fitness values attained by agents with fixed and variable morphologies, respectively, from 30 independently evolving populations of each agent type. c) and d) report the average fitness of these populations.

variable morphology populations repeatedly achieved higher fitness values than the fixed morphology populations.

In contrast, Fig. 3 indicates that stable locomotion is more difficult for evolution to discover for agent populations with mass blocks, compared to agent populations without mass blocks, irrespective of whether or not the size and position of the blocks is under evolutionary control. Only two of the 20 populations achieve stable locomotion in both cases; the remaining runs do not realize any significant fitness improvements over evolutionary time.

4 Discussion

It is clear from Fig. 2 that agent populations with varying leg widths tend to outperform agent populations with fixed leg widths. This stands in contrast to the intuitive notion that in the variable morphology case, the increased dimensionality of the search space—corresponding to the additional three morphological parameters—will degrade search.

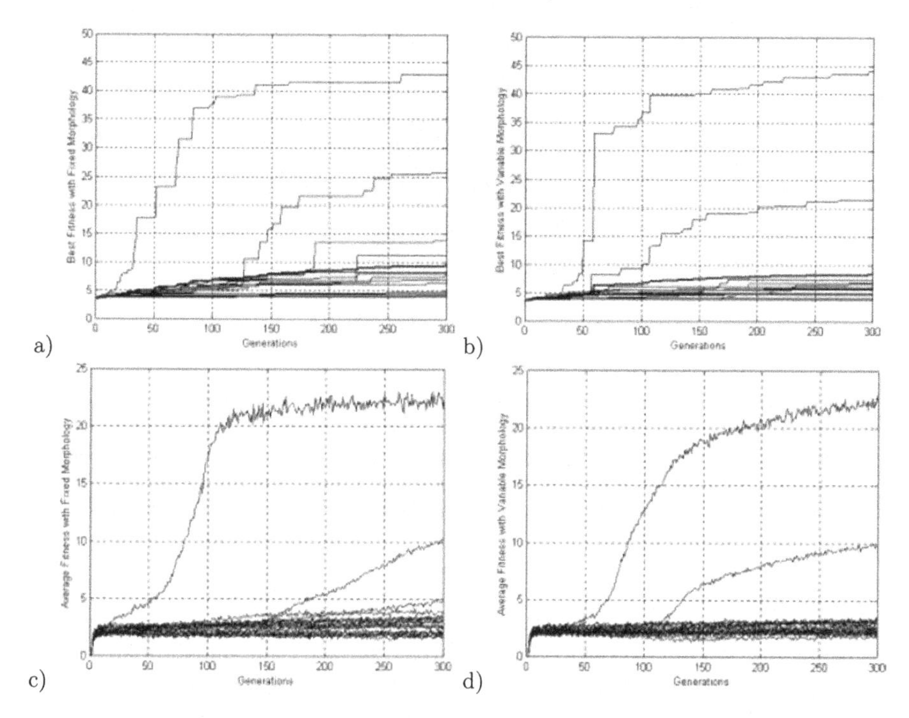

Fig. 3. Evolutionary performance of fixed and variable morphology agent populations with mass blocks. a) and b) report the highest fitness values attained by agents with fixed and variable mass blocks, taken from 30 independent evolutionary runs. c) and d) report the average fitness values of these populations.

We did not find evidence that the variable morphology populations tended to converge on any particular mass distribution. On the contrary, the morphological parameters of the most fit agents at the end of each run fall within their possible ranges with a roughly uniform distribution. This suggests that for our particular instantiation of bipedal locomotion and choice of controller, no one mass distribution is better than another. In other words, evolution of variable morphology agents does not perform better because evolution is able to discover a "good" morphology: rather, the addition of morphological parameters transforms the topology of the search space through which the evolving population moves, creating connections in the higher dimensional space between separated adaptive peaks in the lower dimensional space. These connections are known as extradimensional bypasses, and were introduced by Conrad in [5].

Using a Euclidean topology to represent a fitness landscape, the cross-section within the vertical plane in Fig. 4 indicates a one-dimensional landscape in which the value of a single phenotypic trait `P1` dictates fitness `F`. This landscape contains two separated adaptive peaks, A and B: a population centred around peak A cannot easily make the transition to the higher fitness peak at B. However,

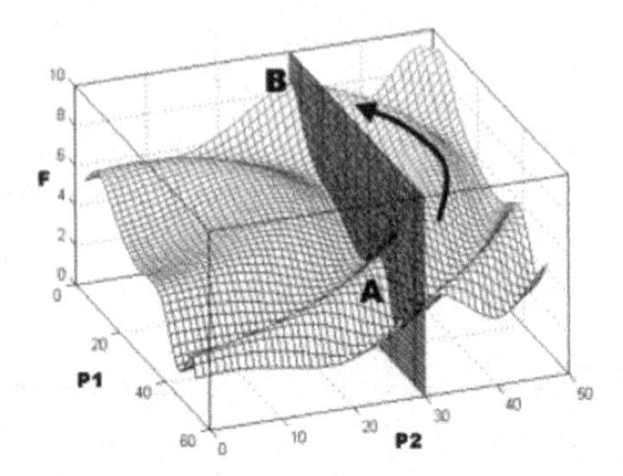

Fig. 4. Schematic representation of an extradimensional bypass. In the one-dimensional Euclidean fitness landscape indicated by the cross-section within the vertical plane, the adaptive peak A is separated by a wide gulf of low fitness phenotypes from the higher peak B. In the higher dimensional fitness landscape indicated by the surface, an extradimensional bypass, represented by the curved surface, connects peaks A and B.

through the addition of a second phenotypic parameter P2, the landscape is expanded to two dimensions (indicated by the surface), and an adaptive ridge—indicated by the upward sloping arrow—provides an opportunity for an evolving population to move from peak A to B via this extradimensional bypass. Using the Euclidean space metaphor here has made it easy to visualize the way in which morphology is exploited to improve evolution. However, the concept of an extra dimensional bypass can be generalized to non-Euclidean representations of fitness topologies, such as the fitness graphs described in [9].

We hypothesize that although the additional morphological parameters increase the dimensionality of the search space, in this case they introduce more adaptive ridges between local adaptive peaks, thereby smoothing the fitness landscape and facilitating evolutionary search. In other words, given a particular morphology, any combination of control changes does not confer increased fitness, but a change in morphology, coupled or followed by control changes does confer increased fitness. This is supported by the variable morphology populations, some of which do not converge at the end of the evolutionary run on morphologies far removed from the default case.

More direct evidence for the presence of extradimensional bypasses has been found in additional evolutionary runs, in which the genetic algorithm again optimizes the control parameters and the leg widths of the simulated biped without mass blocks. In these runs the population is seeded with random control parameters, but the three morphological parameters in the genome are all set to the default leg width of 0.5 unit length. Figs. 5 a) and b) report the morphological history of the best individuals from two successful evolutionary runs. Figs. 5 c) and d) report the corresponding best and average fitness of the two populations. The patterns of morphological change in these populations is typical of those found in the additional runs.

In both populations, the most fit individual at the end of the evolutionary run has a morphology identical or very similar to the starting, default case. These findings disprove the alternative hypothesis that search is improved when both the controller and morphology are evolved because there is one or several morphologies that are better suited to locomotion than the arbitrarily chosen default configuration. In other words, the morphological parameters do not introduce new, high peaks in the fitness landscape that did not exist in the lower dimensional space.

On the other hand, there are long periods in which the most fit individual has a morphology far removed from the default case. This indicates that the morphological parameters are useful for evolutionary search for this task, and are being incorporated into the fitter individuals in the population. This indicates that the evolving population is moving along the extra dimensions introduced by the morphological parameters.

Both of these findings support the hypothesis that the additional morphological parameters smooth the fitness landscape to some degree, and allow for more rapid discovery of stable bipedal locomotion. Moreover, Figs. 5 c) and d) show that several of the periods dominated by agents with the non-default morphological parameter values (indicated by the dark bands on the best fitness curves) are succeeded by rapid fitness increases. Future phylogenetic studies are planned to investigate how the genetic material from these periods are incorporated into the subsequent genotypes that confer increased fitness on the agent.

The evolving agent populations with affixed mass blocks, indicated in Fig. 3, presents a much different picture. In these populations, the addition of eight morphological parameters does not improve evolutionary search. In the 20 fixed morphology populations and 20 variable morphology populations, only two instances of stable locomotion were discovered in each. It is clear that bipedal locomotion using agents with mass blocks, using our experimental set-up, is a more difficult task for the genetic algorithm, but the appearance of stable walking indicates it is not impossible for either the fixed or variable morphology regime to discover stable locomotion.

From our current experiments it is not clear why evolutionary search is not improved in this case, but it seems likely that there are two factors hindering improvement in the variable morphology populations. First, it seems plausible that the ruggedness of the lower dimensional fitness landscape, in the case of agents with fixed blocks, is greater than in the landscape for agents without mass blocks and fixed leg widths, because of the decreased evolutionary performance shown in Figs. 3 a) and c), compared with the performance shown in Figs. 2 a) and c). Second, the dimensionality of the search space for agent populations with mass blocks increases from 60 to 68, as compared with an increase of only 60 to 63 for agent populations without mass blocks.

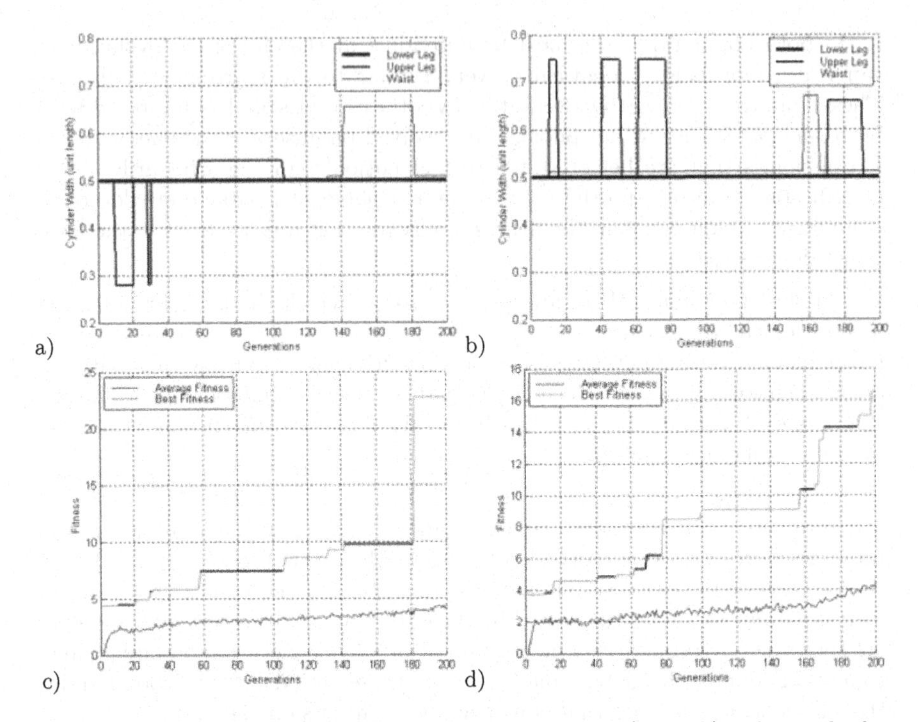

Fig. 5. Morphological change in two populations. a) and b) indicate the leg widths for the most fit agent from two evolving populations. c) and d) indicate the best fitness and average fitness of these populations. The dark bands on the best fitness lines indicate periods in which those agents have morphologies far removed from the default case.

5 Conclusions and Future Research Directions

In this paper, stable locomotion was evolved in embodied, bipedal agents acting within a three-dimensional, physically-realistic virtual environment. It has been demonstrated that, for the case of locomotion in these agents, the subjugation of certain morphological parameters to evolutionary search increases the efficacy of the search process itself, despite the increased size of the search space.

Preliminary evidence was provided which suggests that artificial evolution does not do better in the case of the variable morphology populations because it is able to discover better morphologies than those imposed in the fixed morphology populations, but rather because the type of parameters included in the search create adaptive ridges linking previously separate adaptive peaks.

However, a control set of experiments was provided in which a different set of morphological parameters were included in the genomes of the evolving populations. In these experiments, there was no performance increase in the search ability of the genetic algorithm. This suggests that for the artificial evolution of adaptive behaviour, the arbitrary inclusion of morphological parameters does not always yield better results.

In future experiments we plan to conduct phylogenetic studies and adaptive walks to investigate in more detail how the inclusion of morphological parameters transforms the fitness landscape of the evolving populations. Moreover, we hope to formulate a systematic method for predicting which morphological parameters of embodied agents can augment the evolutionary discovery of adaptive behaviour.

References

1. Benbrahim, H., Franklin, J. A.: Biped Dynamic Walking Using Reinforcement Learning. In: Robotics and Autonomous Systems **22**: (1997) 283–302.
2. Bongard, J. C., Paul, C.: Investigating Morphological Symmetry and Locomotive Efficiency using Virtual Embodied Evolution. In: , J.-A. Meyer et al (eds.), From Animals to Animats: The Sixth International Conference on the Simulation of Adaptive Behaviour: (2000) 420–429.
3. Brooks, R. A., Stein, L. A.: Building Brains for Bodies. In: Autonomous Robots **1**:1 (1994) 7–25.
4. Chocron, O., Bidaud, P. Evolving Walking Robots for Global Task Based Design. In: Proceedings of the IEEE Congress on Evolutionary Computation. (1999) 405–412.
5. Conrad, M.: The Geometry of Evolution. In: Biosystems **24**: (1990) 61–81.
6. Fukuda, T., Komota, Y., Arakawa, T.: Stabilization Control Of Biped Locomotion Robot Based Learning With GAs Having Self-Adaptive Mutation And Recurrent Neural Networks. In: Proceedings of 1997 IEEE International Conference on Robotics and Automation. (1997) 217–222.
7. Gallagher, J.C., Beer, R.D, Espenschied, K.S., Quinn, R.D.: Application of Evolved Locomotion Controllers to a Hexapod Robot. In: Robotics and Autonomous Systems **19**: (1996) 95–103.
8. Gruau, F., Quatramaran, K.: Cellular Encoding for Interactive Evolutionary Robotics. Technical Report, University of Sussex, School of Cognitive Sciences, Brighton, UK. (1996)
9. Jakobi, N.: Encoding Scheme Issues for Open-Ended Artificial Evolution. In: Parallel Problem Solving from Nature IV, Springer-Verlag, Berlin, (1996) 52–61.
10. Juárez-Guerrero, J., Muñoz-Gutiérrez, S., Mayol Cuevas, W. W.: Design of a Walking Machine Structure using Evolution Strategies. In: IEEE Intl. Conf. on Systems, Man and Cybernetics. (1998) 1427–1432.
11. Komosinski, M., Ulatowski, S.: Framsticks: Towards a Simulation of a Nature-Like World, Creatures and Evolution. In: Proceedings of Fifth European Conference on Artificial Life. Springer-Verlag, Berlin, (1998) 261–65.
12. Kun, A., Miller, W. T. III: Adaptive Dynamic Balance of a Biped Robot using Neural Networks. In: Proceedings of the IEEE International Conference on Robotics and Automation. (1996) 240–245.
13. Lichtensteiger, L., Eggenberger, P.: Evolving the Morphology of a Compound Eye on a Robot. In: Proceedings of the Third European Workshop on Advanced Mobile Robots. Piscataway, NJ. (1999) 127–34.
14. Lipson, H., Pollack, J. B.: Automatic Design and Manufacture of Artificial Lifeforms. In: Nature **406** (2000) 974–78.

15. Lund, H. H., Hallam, J., Lee. W.-P.: Evolving Robot Morphology. In: Proceedings of of the IEEE Fourth International Conference on Evolutionary Computation. IEEE Press (1997).
16. Mautner, C., Belew, R. K.: Evolving Robot Morphology and Control. In: Sugisaka, M. (ed.), Proceedings of Artificial Life and Robotics 1999 (AROB99), Oita (1999).
17. McGeer, T.: Passive dynamic walking. In: Int. J. Robotics Research **9**:2 (1990) 62–82.
18. Pfeifer, R., Scheier, C.: Understanding Intelligence. MIT Press, Cambridge, MA (1999).
19. Reil, T., Husbands, P.: Evolution of Central Pattern Generators for Bipedal Walking in a Real-time Physics Environment. IEEE Transactions on Evolutionary Computation, submitted (2001).
20. Sims, K.: Evolving 3D Morphology and Behaviour by Competition. In: Artificial Life IV. MIT Press, Cambridge, MA (1994) pp. 28–39.
21. Ventrella, J.: Explorations of Morphology and Locomotion Behaviour in Animated Characters. In: Artificial Life IV. MIT Press, Cambridge, MA (1996) 436–441.
22. Vukobratovic et al.: Biped Locomotion: Dynamics, Stability, Control and Applications. Springer Verlag, Berlin (1990).

Model of Evolutionary Emergence of Purposeful Adaptive Behavior. The Role of Motivation

Mikhail S. Burtsev[1], Vladimir G. Red'ko[2], and Roman V. Gusarev[3]

Keldysh Institute of Applied Mathematics, 4 Miusskaya sq., Moscow RU-125047, Russia
[1]mr.bur@beep.ru, [2]redko@keldysh.ru, [3]gusarer@mail.ru

Abstract. The process of evolutionary emergence of purposeful adaptive behavior is investigated by means of computer simulations. The model proposed implies that there is an evolving population of simple agents, which have two natural needs: energy and reproduction. Any need is characterized quantitatively by a corresponding motivation. Motivations determine goal-directed behavior of agents. The model demonstrates that purposeful behavior does emerge in the simulated evolutionary processes. Emergence of purposefulness is accompanied by origin of a simple hierarchy in the control system of agents.

1 Introduction

The purposefulness is very non-trivial feature of intelligent animal behavior. What are the particularities of purposeful adaptive behavior? How could goal-directed behavior emerge? We try to investigate the most general aspects of these questions by means of a simple Alife model.

The main notion that we use to characterize purposefulness is that of *motivation.* The importance of concept of motivation was underlined by the number of authors [1-5]. In particular, Tsitolovsky [1] studied very simple model of motivation in connection with neuron adaptation, Balkenius [2] analyzed different models of motivation, Spier and McFarland [3] investigated the model of two motivations. We try to analyze the question "How could purposeful behavior (which is due to motivations) origin in an evolutionary process?" We imply that a motivation is a quantitative characteristic that is directly related to a corresponding need.

The design of agents in our model was inspired by PolyWord model [6]. The idea of special evolving neural network, that is intended to give rise to agent's goal [7], is to some extent analogous to special neural network inputs, controlled by motivations, in our model.

2 The Model

The key assumptions of our model are as follows.

There are agents, which have two natural needs: the need of energy and the need of reproduction.

J. Kelemen and P. Sosík (Eds.): ECAL 2001, LNAI 2159, pp. 413–416, 2001.

The population of agents evolves in the simple environment, where patches of grass (agent's food) grow. The environment is a linear one-dimensional set of cells. The agents receive some information from their environment and perform some actions. Possible actions are moving (to a neighboring cell), jumping (over several cells), eating grass, resting and mating. Mating results in birth of new agents. An agent has internal energy resource; the resource is increased during eating. Performing an action, the agent spends its resource. When the resource of the agent goes to zero, the agent dies.

Any need of an agent is characterized quantitatively by a motivation. If there is a certain motivation, the agent can search for a solution to satisfy the need according to the motivation. E.g., if the energy resource of an agent is small, there is the motivation to find food and to replenish the energy resource by eating. This type of behavior can be considered as purposeful (there is the purpose to satisfy the need).

Motivations are defined as follows:

$$M_R = \min\left\{\frac{R}{R_1},\ 1\right\},\ M_E = \max\left\{\frac{R_0 - R}{R_0},\ 0\right\},$$

where M_R and M_E are motivations, corresponding to needs of reproduction and energy, respectively; R is internal energy resource of the agent; R_0 is some "optimal" value of energy resource; R_1 is the value of energy resource, which is the most appropriate for reproduction.

An agent behavior is controlled by a one-layer neural network. At the given iteration of time any agent executes only one action.

The population of agents evolves. A genome of an agent codes the synaptic weights of the agent's neural network. The main mechanism of the evolution is the formation of genomes of newborn agents by means of crossovers and mutations.

More detailed description of the model can be found in [8].

3 Computer Simulations

To analyze the influence of motivations on behavior of agents, we performed two series of simulations. In the first series, the agents had motivations (as described above) and in the second series, the agents had no motivations (neural network inputs from motivations are specially suppressed). The simulations in both series were performed for different values of food amount in the external environment.

All agents of the initial population had the same synaptic weights of neural networks. These weights determined some reasonable initial instincts of agents.

The first instinct was the instinct of food replenishment. This instinct ensured two types of actions: 1) if an agent sees a grass in its own cell, it eats this grass, 2) if an agent sees a grass in a neighboring cell, it moves into this cell. The second instinct was the instinct of reproduction. This instinct implies that if an agent sees another agent in one of the neighboring cells, it tries to mate with this neighbor. The synaptic weights from motivational inputs in the neural network were equal to zero for agents of initial population.

The main quantitative characteristic that we used in order to describe the quality of an evolutionary process was the total number of agents in population N. We obtained the dependencies $N(t)$ on time t for both series of experiments: for population of agents with motivations and for population of agents without motivations (Fig. 1). One can see that motivations provide an opportunity for the population to find better survival strategy for agents in the course of evolutionary search.

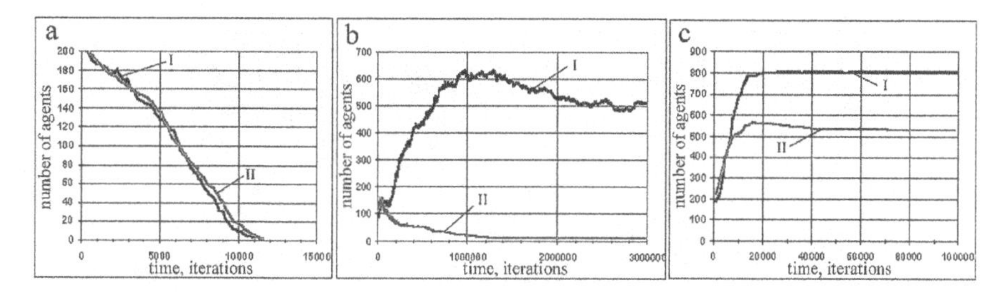

Fig. 1. Dependencies of number of agents in population with motivations (I) and without motivations (II) on time at small (a), intermediate (b) and large amount (c) of food in external environment (size of world in all simulations was equal to 900 cells).

We performed detailed analysis of agents' actions evolution for population with and without motivations. Based on this analysis we interpreted the behavioral control of agents as follows.

The scheme of behavioral control of an agent *without motivations* that was discovered by evolution is shown at Fig. 2a. This scheme consists of three rules:

1. If the agent sees a grass patch, it seeks to eat this food.
2. If the agent sees a neighbor, it tries to give birth to an offspring.
3. If there is nothing in the field of vision, the agent decides to rest.

The first two rules are just instincts, which we forced upon the agents of the initial population. The evolution confirmed that they are useful and adaptive. The third rule was discovered by evolution, and, of course, this rule has certain adaptive value. These three rules can be considered as simple reflexes.

Let's consider the control system of the agent *with motivations*. In this case hierarchy in the control scheme emerges. Three rules described above constitute the lower level of the control system. The second level is due to motivations (Fig. 2b). If the energy resource of an agent is low, the motivation to search food is large, and the motivation to mating is small, so the agent uses only two of mentioned rules, the first and the third – the mating is suppressed. If the energy resource of the agent is high, the motivation to reproduction is turned on.

So, the transition from the scheme of control without motivations to the scheme with motivations can be considered as the emergence of a new level of hierarchy in the control system of the agents. This transition is analogous to the metasystem transition from simple reflexes to complex reflex in the Metasystem Transition Theory by V. Turchin [9].

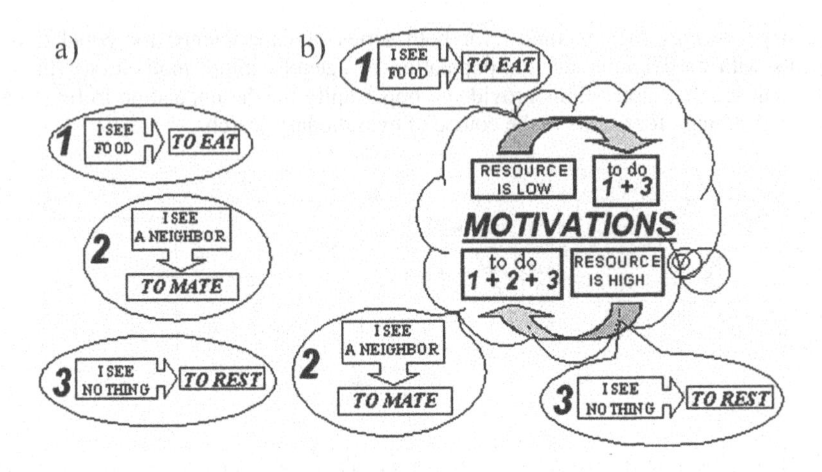

Fig. 2. Scheme of behavioral control of agents without (a) and with (b) motivations.

Thus, the model demonstrates that 1) simple hierarchical control system, where simple reflexes are controlled by motivations, can emerge in evolutionary processes, and 2) this hierarchical system is more effective as compared to behavioral control governed by means of simple reflexes only.

References

1. Tsitolovsky, L.E.: A Model of Motivation with Chaotic Neuronal Dynamics. Journ. of Biological Systems, V. 5. N.2, (1997), pp. 301-323.
2. Balkenius, C.: The Roots of Motivations. In J.-A. Mayer, H. L. Roitblat and S. W. Wilson (eds.): From Animals to Animats II, Cambridge, MA: MIT Press, (1993), http://www.lucs.lu.se/People/Christian.Balkenius/Abstracts/ROM.html
3. Spier, E., McFarland, D.: A Finer-Grained Motivational Model of Behaviour Sequencing. In P. Maes, M. Mataric, J. Meyer, J. Pollack and S. Wilson (Eds.), From animals to animats 4, MIT Press, (1996), pp. 421-429.
4. Donnart, J.Y. and Meyer, J.A.: Learning Reactive and Planning Rules in a Motivationally Autonomous Animat. IEEE Transactions on Systems, Man, and Cybernetics - Part B: Cybernetics, V. 26, N.3, (1996), pp. 381-395.
5. Nolfi S., Parisi D.: Learning to Adapt to Changing Environments in Evolving Neural Networks. Adaptive Behavior V.5, N.1, (1997), pp. 75-98.
6. Yaeger, L.: Computational Genetics, Physiology, Learning, Vision, and Behavior or PolyWord: Life in a New Context. In Langton, C. G. (ed): Artificial Life III. Addison-Wesley, (1994), pp. 263-298.
7. Ackley, D., Littman, M.: Interactions Between Learning and Evolution. In Langton, C. G., Taylor, C., Farmer, J. D., and Rasmussen, S. (eds.): Artificial Life II. Addison-Wesley, (1992), pp. 487-509.
8. Burtsev, M., Gusarev, R., Red'ko, V.: Alife Model of Purposeful Adaptive Behavior. Technical report, (2000), http://www.keldysh.ru/mrbur-web/pab
9. Turchin, V. F.: The Phenomenon of Science. A Cybernetic Approach to Human Evolution. Columbia University Press, New York, (1977), http://pespmc1.vub.ac.be/POSBOOK.html

Passing the ALife Test: Activity Statistics Classify Evolution in Geb as Unbounded

Alastair Channon

Department of Electrical and Electronic Engineering
Anglesea Road, University of Portsmouth, PO1 3DJ. UK
alastair@channon.net
http://www.channon.net/alastair

Abstract. Bedau and Packard's evolutionary activity statistics [1,2] are used to classify the evolutionary dynamics in Geb [3,4], a system designed to verify and extend theories behind the generation of evolutionary emergent systems. The result is that, according to these statistics, Geb exhibits unbounded evolutionary activity, making it the first autonomous artificial system to pass this test. However, having passed it, the most prudent course of action is to look for weaknesses in the test. Two weaknesses are identified and approaches for overcoming them are proposed.

1 Introduction

Perhaps the most important goal in Artificial Life is the generation of an autonomous system (that is excluding systems such as the Internet, which owes its evolution to human endeavour) that exhibits unbounded evolution. "Life-as-it-could-be" [5], as opposed to models of life-as-we-know-it, can only be achieved within such a system, for life is a product of evolution. Thanks to Bedau and Packard's evolutionary activity statistics [1,2], we have a test that a candidate system should pass before claims of unbounded evolution are taken seriously. The test is based on the following statistics: component activity increment Δ_i, component activity a_i, diversity D, total cumulative activity A_{cum}, mean cumulative activity $\bar{A}_{\text{cum}}$ and new evolutionary activity per component A_{new}.

$$\Delta_i(t) = \begin{cases} 1 & \text{if component } i \text{ exists at } t \\ 0 & \text{otherwise} \end{cases} \qquad D(t) = \#\{i : a_i(t) > 0\}$$

$$a_i(t) = \begin{cases} \sum_{\tau=0}^{t} \Delta_i(\tau) & \text{if component } i \text{ exists at } t \\ 0 & \text{otherwise} \end{cases} \qquad A_{\text{cum}}(t) = \sum_i a_i(t)$$

$$A_{\text{new}}(t) = \frac{1}{D(t)} \sum_{i: a_i(t) \in [a_0, a_1]} a_i(t) \qquad \bar{A}_{\text{cum}}(t) = \frac{A_{\text{cum}}(t)}{D(t)}$$

For A_{new} to be a good measure of new activity, the range $[a_0, a_1]$ should be chosen such that component activities within it can be considered both adaptively significant (so a_0 should be high enough to screen out most non-adaptive

J. Kelemen and P. Sosík (Eds.): ECAL 2001, LNAI 2159, pp. 417–426, 2001.

Table 1. Classes of evolutionary dynamics and their statistical signatures, based on table 1 from [2]. Rows 3b and 3c have been added to class 3 (see text).

Class	Evolutionary Dynamics	Statistical Signature D	A_{new}	$\bar{A}_{cum}$
1	none	bounded	zero	zero
2	bounded	bounded	positive	bounded
3a	unbounded (D)	unbounded	positive	bounded
3b	unbounded ($\bar{A}_{cum}$)	bounded	positive	unbounded
3c	unbounded (D & $\bar{A}_{cum}$)	unbounded	positive	unbounded

activity) and not amongst the highest activities (so a_1 should be low enough that a good proportion of activities lie above it).

For artificial systems, a "shadow" should be run, mirroring the real run in every detail except that whenever selection operates in the real system, random selection should be employed in the shadow. Shadow statistics can then be used to determine a_0 and levels of activity that can be considered significant.

After determining long-term trends in these statistics, the system being examined can be classified according to table 1. The hallmark of class 3 (unbounded evolutionary dynamics) is unbounded A_{cum} in combination with positive A_{new}.

Other possibilities exist with zero A_{new}, but these belong in class 1 (no evolutionary activity) because such cases have no significant new components. Table 1 in [2] only shows the first row (3a) for class 3, but footnote 1 in [2] mentions the other rows (3b and 3c). Note that table 1 includes all possibilities for positive A_{new}, because zero $\bar{A}_{cum}$ implies zero A_{new}. So any system with unbounded evolutionary dynamics will belong to class 3 (one of 3a, 3b and 3c).

Previously, only the biosphere has passed this test,[1] although a number of Artificial Life systems have been evaluated. The following quote from their discussion section summarises Bedau, Snyder and Packard's conclusion.

> "We also suspect that no existing artificial evolving system has class 3 dynamics. In our opinion, creating such a system is among the very highest priorities of the field of artificial life. From one perspective, this is a negative result: Echo, and perhaps all other existing artificial evolutionary systems, apparently lack some important characteristics of the biosphere – whatever is responsible for its unbounded growth of adaptive activity. But at the same time this conclusion calls attention to the important constructive and creative challenge of devising an artificial model that succeeds where all others have failed." [2, p. 236]

[1] Maley [6] makes the claim that two of his models exhibit unbounded evolutionary activity. However, Urmodel 3 shows less new activity than its shadow (with no reason to think that it would become greater), Urmodel 4 shows a lower mean activity than its shadow and both are only examined during their initial growth stages.

Nehaniv [7] defined open-ended evolution as unbounded growth in a measure "cpx". However in [8] he proceeded to show that a trivial system exhibits open-ended evolution according to this definition.

2 Implementing the Statistics in Geb

Full details of Geb are available in [3,4]. Here I include just enough for an explanation of the evolutionary components used in the statistics. Geb is a virtual world containing autonomous organisms, each controlled by a neural network. Each neuron has a bit-string label, or 'character', which is used during development and for matching the neural outputs of one organism with basic behaviours (turning, killing, etc.) and with inputs of other organisms. An organism is born with a simple axiom network that results in reproduction. This develops through the application of a genetically determined Lindenmayer system (L-system) [9]. Each L-system production rule has the following form:

$$\mathcal{P} \rightarrow \mathcal{S}_r, \mathcal{S}_n \; ; b_1, b_2, b_3, b_4, b_5, b_6$$

$\mathcal{P}$ Predecessor (initial bits of node's character)
$\mathcal{S}_r$ Successor 1: *replacement* node's character
$\mathcal{S}_n$ Successor 2: *new* node's character
bits: $b_1, b_2, b_3, b_4, b_5, b_6$ specify linkage details

The *successors* (1 and 2) are characters for the node(s) that replace the old node. If a successor has no character (0 length) then that node is not created. Thus, the predecessor node may be replaced by 0, 1 or 2 nodes. Necessary limits on the number of nodes and links are imposed.

An evolved genotype contains a large number of production rules (once decoded), but only the rules found to match neuron's characters most closely are used during development. In this way, increasingly specific production rules can evolve, with regressive rules existing as fall-back options should a rule be damaged by crossover or mutation, and as material for further evolutionary search.

When a new organism is 'born', all possible production rules are decoded from its genotype. Then the developmental process is part-simulated in advance of it truly taking place, as a means of filtering out all the production rules that would never be used, either because they would never match any possible node's character, or because more specific rules exist for each node that could develop. All rules that remain will be used if the organism lives long enough.

2.1 Choosing the Component Class

As Geb's genotypes both change length and contain a high degree of neutrality[2] the genotype is not a good choice of component class. Production rules, the alleles

[2] In order to avoid confusion, I only use the term *neutral* to refer to genetic variations that are phenotypically equivalent, and not in relation to shadow runs.

from a genotype, are a much better choice. It can be expected that if a production rule has an adaptive advantage, then it will persist. Better still though is the choice "production rules that survive the filtering process at birth", for these are actually used in the developmental process; the idea behind activity statistics is to measure the degree to which components both *persist* and are *used*.

When mutation causes a component to not be expressed (currently present), the activity count of the original component is no longer included in the total activity of the system, even if the mutation is functionally neutral. At first I implemented the activity statistics on production rules directly. But there is often a high degree of neutrality in a production rule, especially when its 'successors' relate to neurons that are over-specified (have excess bits at the end of their characters) or development-terminal (not matched by any production rule). The predecessor and link-bits sections of production rules are more plastic. If a predecessor bit is mutated, then the rule will most likely either fail to match or be less specific to its target neuron than another rule. If a link detail bit is mutated, then the result will more often than not be a damaged network, and organisms with that production rule active (not filtered out) will be driven to extinction. So the choice of component used here is 'predecessor plus link details' $(\mathcal{P}, b_1, b_2, b_3, b_4, b_5, b_6)$. This can be thought of as a disjoint grouping of alleles, with each group being a component. Which individual a component is from is irrelevant: two identical production rules in two different organisms result in two instances of the same component.

This grouping does not completely remove the neutrality problem. As successor lengths increase, neural character lengths increase, and so the number of predecessors that can potentially match a typical neural character increases. If two rules have the same successors and link-details (or neutral variants), then it makes no difference to development which one is used. So, as component lengths increase, we can expect the level of neutrality to increase.

Having chosen this component class, there is a clear consequence for the possible classifications of evolutionary dynamics. Because the number of neurons that an organism can have is limited (for practical reasons), the number of production rules that can survive filtering is limited. And because the population size is small (a maximum of four hundred organisms), there is little room for more than a couple of species at a time. So diversity of these components will certainly be bounded, and we can rule out class 3a and 3c dynamics.

2.2 Implementation Details

Geb's shadow mirrors the real run in every detail except that selection is random. Whenever a real organism is killed, a randomly chosen shadow organism is also killed. Whenever a real organism is born (as the product of two real organisms), a new shadow organism is born as the product of two randomly chosen shadow organisms, using the same reproduction procedure with the same rates of crossover and mutation.

It is not feasible to gather the statistics at every timestep. So snapshot existence records are taken at regular intervals and the evolutionary statistics are calculated from these. In the results reported here, snapshots were taken every one thousand timesteps. To put this in context, the run reported lasted six million timesteps, during which time there were over five hundred and eighty million organism reproductions. Because activity is intended as a measure of how much a component both is used (already covered above) and *persists*, I screen out (in each of the real and shadow populations) isolated occurrences: when a component occurs in the current snapshot but not the previous one.

In previously published work on Geb, total extinction (population size dropping to one individual) was not mentioned because it had not been encountered. However, during the long trial runs undertaken when experimenting with evolutionary statistics, I encountered occasional runs in which total extinction occurred. So for the set of runs from which the example reported here is taken, I set a minimum number of organisms to twenty. The fact that total extinction is so rare despite the population size being so small (a maximum of four hundred organisms) indicates that there is no serious problem here. Once population sizes can feasibly be increased, the problem should in practice disappear rapidly.

3 Results and Discussion

This section contains the results from a typical run, drawn from the full set of twenty runs. Atypical variations are discussed at the end of this section.

3.1 Activity Waves

In order to gain an understanding of the dynamics behind the higher level evolutionary statistics, it is a good idea to look first at the activity wave diagrams, which simply show all components' activities plotted against time. Figure 1 shows the activity wave diagrams for the real and shadow populations.

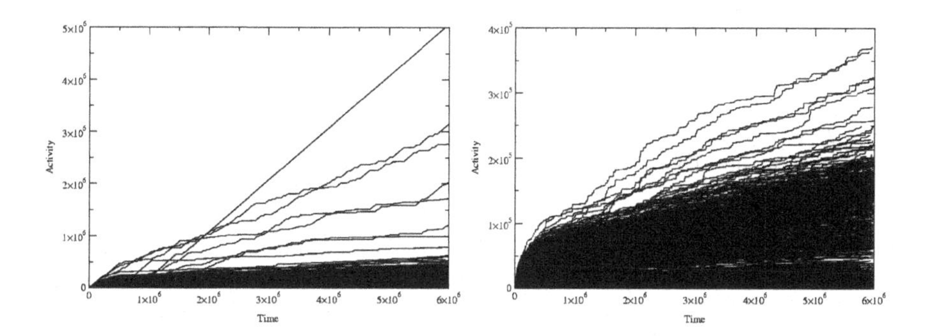

Fig. 1. Activity wave diagrams for the real (left) and shadow (right) runs. Note the different scales for real and shadow.

The most obvious feature of the real run's activity waves in figure 1 is that many of them keep increasing. This would also be true in a similar analysis of genes from the biosphere's evolution. Genes that are beneficial to life tend to remain in the population of genes and be used by many species: humans have a significant proportion of genes in common with mice, flies and even plants. So because the components here are (groups of) genes, not whole genotypes, this feature does not imply a quasi-stable ecosystem.

In systems without neutrality new components initiate an activity (by presence) wave that increases with a constant slope and then stops when the component goes extinct. Here however, that increase is often shared between two or (perhaps many) more phenotypically equivalent components, with interchanging presence. If the population size was larger, then there would be greater scope for more than one of these neutral variations to be expressed in the population at any time. But with a population of less than four hundred, and short lifetimes, genetic variation spreads quickly through the population so wave transitions between neutral variants show up almost as either-or events.

We can also observe that as time goes by, the level of neutrality increases: the average rate of increase for a new component decreases and the number of components in each neutrality-group increases. This is consistent with the expectation from section 2.1 that as component lengths increase, we would see the level of neutrality increase.

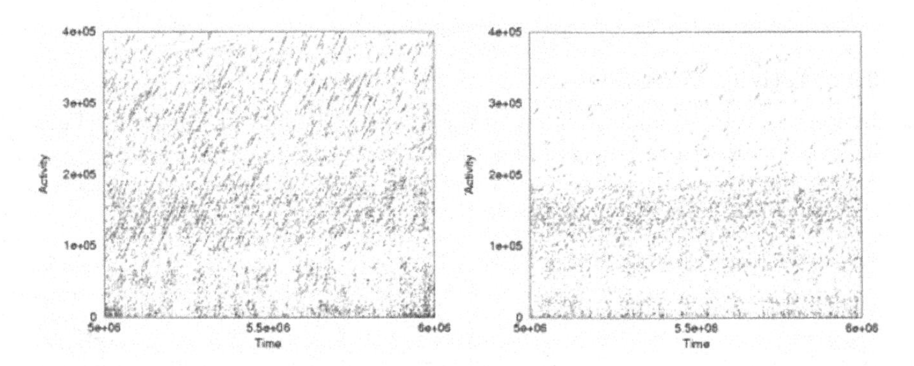

Fig. 2. Activity point-plots for the real (left) and shadow (right) runs in the last million timesteps, within the shadow's activity range.

Because of this increase in neutral-group size and decrease in average component activity rate of growth, most activity falls within the solid black regions at the bottom of each graph. So it is instructive to look in more detail at the bottom-right corner of the activity wave diagrams. Figure 2 shows the activity waves in the last one million timesteps, with just a point for each recorded activity value. Its scale covers the shadow run's full range only, so that the real and shadow data can be easily compared. Notice the long runs of consecutively increasing activity in the real run, and the lack of them in the shadow.

3.2 Determining the New-Activity Range

In order to measure new activity (A_{new}), we must first determine the range $[a_0, a_1]$ of component activity values that should be considered both adaptively significant and not amongst the highest activities. The method given in [2] involves finding the activity value which is equally likely to have occurred in the real run as in the shadow run, and setting $[a_0, a_1]$ to be a narrow band that surrounds it. A log-log plot of the real and shadow component activity distributions for the run reported here shows that they cross at an activity of approximately 1.42×10^5. However, it is clear from figure 1 that the component activity distributions are far from constant over the run, and that this value increases during the course of the run. Looking again at figure 2 shows that by the end of the run, activities of approximately 1.42×10^5 are common in the shadow.

Here I have used a fixed range that screens out the majority of the shadow activity in the final million timesteps of the run. This results in artificially low values for A_{new} early on in the run, but the results are still positive despite this. In the last million timesteps, less than 3.5% of the shadow activity is above 2.8×10^5, and approximately 27% of the real activities are above 3.2×10^5. So the results that follow were calculated using a new-activity range of $[2.8 \times 10^5, 3.2 \times 10^5]$.

3.3 Evolutionary Statistics and Classification

Figure 3 shows both total and mean activity increasing rapidly in the real run, and much slower in the shadow run. New activity is positive in the real run, and much higher than in the shadow, which exhibits only occasional blips of new activity. Diversity is bounded in both the real and shadow, as expected (see section 2.1).

Figure 4 shows the activity difference and "excess activity", for both total and mean activity, between the real and shadow statistics. Excess activity is defined in [10] as activity difference divided by shadow activity.

According to this classification system, these results clearly fall into class 3b: unbounded evolutionary activity.

3.4 Atypical Runs

These results are typical of the twenty runs that were carried out for this set of experiments. However, three of the runs encountered problems, causing their results to be atypical. Two of these effectively met total extinction. In section 2.2 I mentioned that I imposed a minimum limit on the number of organisms, in an attempt to avoid total extinction. However, if population size hits this limit and does not increase rapidly, then many reproductions may occur with selection effectively random. This causes evolutionary activity to plummet as adaptive traits are lost. Once lost, this activity cannot be regained, except by the evolution of new adaptive components. These results should not be a cause for concern, for the same reasons mentioned in section 2.2: once population sizes can feasibly be increased, the problem should in practice disappear rapidly.

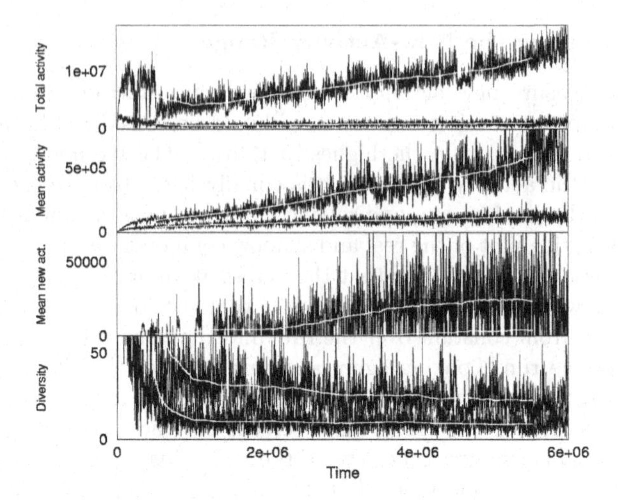

Fig. 3. Total activity, mean activity, new activity and diversity from a typical Geb run and its shadow. Running averages are shown in white.

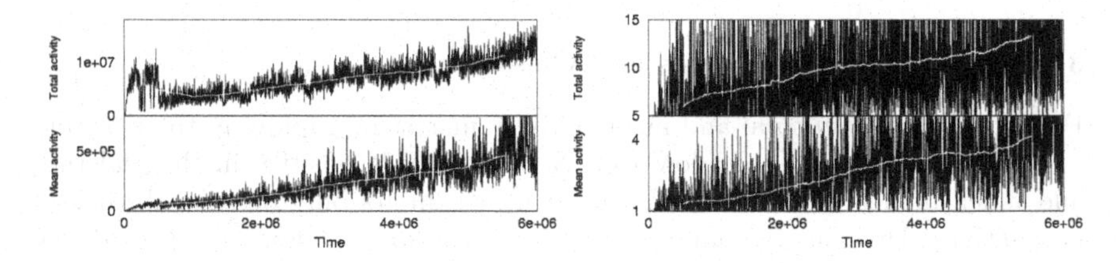

Fig. 4. Normalised total and mean activity: difference (left) and "excess" (right). Running averages are shown in white.

In the third atypical run, it appears that a freak mutation has caused the only existing species to take on a behaviour of never reproducing or moving forward and always turning and trying to kill. Of course this would ordinarily be a very poor strategy. It is easy to imagine how the bad gene (production rule) could have spread through a population of just one species as fit individuals reproduced with the new unfit ones, causing their children to pick up the dominant bad gene. However, one would not expect this to pose a threat to a different species. This is easily verified: introducing just a few organisms from any of the other evolved populations (from the other runs), causes the old organisms to be rapidly displaced by the newcomers. So this result is also not a cause for concern, for the same reason: it is due to the small population size, which cannot support more than one or two species at a time.

4 Criticisms and Conclusions

Geb has demonstrated class 3 behaviour, and so passed the test. Does this mean that Geb truly exhibits unbounded evolution? Possibly, for it was designed to verify and extend theories of evolutionary emergent systems generation and so a number of potential pitfalls have been avoided. However, having passed the test the most prudent course of action is to look for weaknesses in the test.

The main concern that I have at this time is that the test relies on normalisation (or validation) from a shadow that can drift away from core aspects of the real run that it is intended to shadow. For example, the components that exist in the real population at any one time (well into evolution) are almost certainly more densely clustered than those in the shadow. So the mutation of a real component is more likely to produce another high-activity component than the mutation of a shadow component. Once the real and shadow populations have been allowed to evolve, we are no longer comparing the real run with a true shadow. One way around this problem would be to develop a method of comparing the real run with a shadow that is regularly reset (both components and activity history) to be identical to the real run but which evolves using random selection between resets. The normalised activity increment between resets would be determined by comparing the real and shadow increments.

My other criticism of the test as it stands is in its use of mean activity when looking for unbounded activity growth, especially when classifying a system as belonging to class 3b. When diversity is bounded, the retention (forever) of a single component results in unbounded mean activity. The test should not be so influenced by such components, and should rather look for trends in typical components. So it is median activity, not mean activity, that should be measured, and required to be unbounded for a system to be classified as within class 3b. The activity waves from Geb's runs indicate that median activity is also unbounded. However, when median activity is measured in both real runs and their shadows, it shows up as unbounded in both, and normalised median activity appears to be bounded (figure 5). However, in light of my main concern above, just as the positive results cannot be trusted, this negative result cannot be trusted either. The correct course of action is to proceed as outlined above, by developing a shadowing method that regularly resets the shadow state to the real state, and then look at the results again, including median activity.

The main conclusion of this paper is that these results are encouraging but, until the above concerns and any others that the Artificial Life community may raise are resolved, the test results cannot be considered conclusive. Whether or

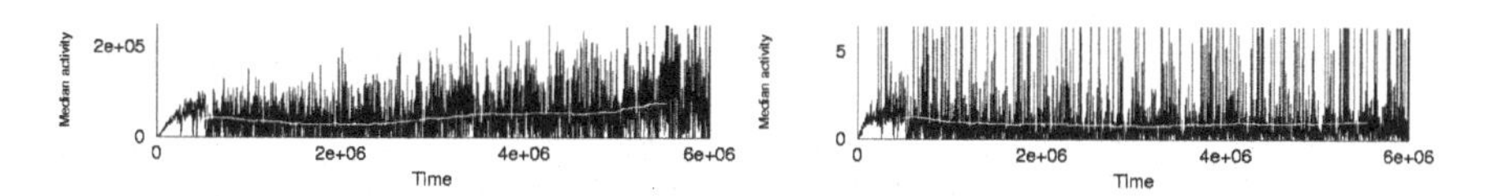

Fig. 5. Normalised median activity: difference (left) and "excess" (right). Running averages are shown in white.

not Geb exhibits anything that deserves to be thought of as unbounded evolutionary activity remains to be determined. We do however now have an Artificial Life system with which to test the Artificial Life test.

Acknowledgements. Thanks to the anonymous reviewers for helpful comments.

References

[1] Bedau, M. A., and Packard, N. H. Measurement of evolutionary activity, teleology, and life. In *Proceedings of Artificial Life II* (Redwood City, CA, 1991), C. G. Langton, C. Taylor, J. D. Farmer, and S. Rasmussen, Eds., Addison-Wesley, pp. 431–461.

[2] Bedau, M. A., Snyder, E., and Packard, N. H. A classification of long-term evolutionary dynamics. In *Proceedings of Artificial Life VI, Los Angeles* (Cambridge, MA, 1998), C. Adami, R. Belew, H. Kitano, and C. Taylor, Eds., MIT Press, pp. 228–237.

[3] Channon, A. D., and Damper, R. I. Perpetuating evolutionary emergence. In *From Animals to Animats 5: Proceedings of the Fifth International Conference on Simulation of Adaptive Behavior (SAB98), Zurich* (Cambridge, MA, 1998), R. Pfeifer, B. Blumberg, J.-A. Meyer, and S. Wilson, Eds., MIT Press, pp. 534–539.

[4] Channon, A. D., and Damper, R. I. Towards the evolutionary emergence of increasingly complex advantageous behaviours. *International Journal of Systems Science 31*, 7 (2000), 843–860.

[5] Langton, C. G. Artificial life. In *Santa Fe Institute Studies in the Sciences of Complexity, Vol VI: Proceedings of the Interdisciplinary Workshop on the Synthesis and Simulation of Living Systems (Artificial Life I)* (Redwood City, CA, 1989), C. G. Langton, Ed., Addison-Wesley, pp. 1–47.

[6] Maley, C. C. Four steps toward open-ended evolution. In *Proceedings of the Genetic and Evolutionary Computation Conference (GECCO99)* (San Francisco, CA, 1999), W. Banzhaf, J. Daida, A. E. Eiben, M. H. Garzon, V. Honavar, M. Jakiela, and R. E. Smith, Eds., Morgan Kaufmann, pp. 1336–1343. Volume 2.

[7] Nehaniv, C. L. Measuring evolvability as the rate of complexity increase. In *Artificial Life VII Workshop Proceedings* (2000), C. C. Maley and E. Boudreau, Eds., pp. 55–57.

[8] Nehaniv, C. L. Evolvability in biologically-inspired robotics: Solutions for achieving open-ended evolution. In *Sensor Fusion and Decentralized Control in Robotic Systems III, SPIE Proceedings 4196* (2000), P. McKee, G. T. Schenker, Ed., pp. 13–26.

[9] Lindenmayer, A. Mathematical models for cellular interaction in development. *Journal of Theoretical Biology 18* (1968), 280–315. Parts I and II.

[10] Rechtsteiner, A., and Bedau, M. A. A generic neutral model for measuring excess evolutionary activity of genotypes. In *Proceedings of the Genetic and Evolutionary Computation Conference (GECCO99)* (San Francisco, CA, 1999), W. Banzhaf, J. Daida, A. E. Eiben, M. H. Garzon, V. Honavar, M. Jakiela, and R. E. Smith, Eds., Morgan Kaufmann, pp. 1366–1373. Volume 2.

On the Evolution of Artificial Consciousness

Stephen Jones

387 Riley St
Surry Hills, NSW,
Australia

Abstract. In this paper I suggest how we might conceptualise the development of some kind of artificial consciousness. While evolutionary algorithms and neural nets will be at the heart of the production of artificial intelligences there is a particular architectural organisation that may be *necessary* in the production of conscious artefacts. This is the operations of multiple layers of feedback loops in the anatomy of the brain, in the social construction of the contents of consciousness and in particular in the self-regulation necessary for the continued operation of metabolically organised systems.

1 Characteristics of Consciousness

A conscious system is an "organism" that must be capable of adaptive interaction with its environment. It needs to sense its environment and to produce appropriate effects in it. This amounts to its capacity to gather information from its context through its sensors and to process that information into a useful form. These sensors transduce whatever it is in the environment that they are set up to handle. The results are a transform of the input through those modes for which the sensor is enabled. This transform may then be presented to some sort of interpretive processor, and subsequently to an effector, or it may be directly wired to an effector or output device of some sort. Effectors allow an organism to do things in the world, from getting out of the way, to catching prey, to producing books or experiments. Effectors also allow an output to be placed in the environment that may be recognised as a signal by some other entity also within that environment. This recognition will be species and experience dependent, as determined by the kind of sensors available to that particular species and the recognition filters that have developed through the individual organisms' experiential history. These recognition filters become embedded in the overall neural connectedness of the system as it evolves through its experience, ie. the dynamics of its relations with the world (its situatedness [1]). Meaning can be partly a function of existential significance (say the recognition of food) or emotional significance (chemical constraint satisfaction) or, say, intellectual/cultural significance. Recognition is also necessary for the establishment of communication.

A conscious entity must act independently and initiate procedures by doing things like gathering information for itself and using that information correctly [2]. Through attentional and memory mechanisms it should receive, store and reflect on perceptual input showing understanding and adaptive response. It needs to be able to report upon the contents of its consciousness [3], interact with others in general, and project into

J. Kelemen and P. Sosík (Eds.): ECAL 2001, LNAI 2159, pp. 427-431, 2001.

the world for purposes of generating feedback. Its attentional and memory processes, the day-to-day maintenance of its (dynamical) state and its recognition of its own position in the environment are all achieved through self-regulatory processes established through feedback loops re-entering the data of its behaviour back into its system so that it knows what it has done. This is a generalised version of our process of conversation where we correct ourselves on the fly by listening to what we are saying and interrupting it as it comes out.

The main thing that differentiates consciousness from the merely mechanical is subjective or phenomenal experience; the *experience* of what it *feels* like to *be* something [4]. This issue of subjectivity is highly contentious, see eg [3, 5]. Given the privacy of subjective experience, how could we possibly know whether a machine, no matter how good a reproduction of human behaviour it produces, could have anything like subjectivity? Since the Turing Test only requires that an AI be able, by its free interaction with us, to convince us that it is in fact conscious, and as I use an operational statement of consciousness and subjectivity, perhaps we can only trust as we do with each other. For me and for others, eg. [6], subjectivity is the internal state of the system experienced by the system itself in the transformed data flow through its anatomy/hardware.

Consciousness *accrues* as the infant matures. We don't come fully equipped at birth, but we have the capacity to become fully equipped over the early years of our lives. It is through *experience* that we gain knowledge of the world; though there are anatomical characteristics of brains that determine which aspects of the world we are able to experience. These structure our consciousness and what we are able to know via our senses. But the brain needs experience to flesh out consciousness and provide meaning or interpretability to those input transforms as sensations that it enables, thus making the world.

The genome cannot specifically establish all the synaptic connections that the brain needs to function in the world and thus cannot explain the development of any particular consciousness. There are too many variables in the world. For example the fact that we have the capacity for language and the architecture to receive, understand and produce it is a function of the genome but the fact that we are able to acquire *any one* of the numerous languages human societies use, shows a massive flexibility of potential which could not be pre-specified. Adaptability and biological plasticity are essential.

2 Architecture of Consciousness

Conscious brains have an organisational structure that has evolved over a very long period, building on the structures of lower level organisms going back to the beginning. This architecture has been singularly successful in supporting the survival of particular groups of species leading to us as well as, quite possibly, to different kinds of "consciousness" in other large brained mammals such as chimpanzees and dolphins. We can talk about this organisational structure from two angles: (a) the anatomy and physiology of the brain and (b) the brain's cognitive and control functions. The physical body/brain is the biological substrate of consciousness, but its activities are sensing and experience, interpreting and reporting, calculating and emoting. These functional and emotional levels run on the biological substrate and have similar architectural relations to that which can be seen in the anatomy.

This can be seen as a mapping between the cognitive architecture and the anatomical architecture and appears in the work of Bernard Baars [7] and the late James Newman [8]. They have proposed a direct and consistent mapping between the cognitive function and the anatomy of the brain. The cognitive aspects are represented in Baars' "global workspace" or working memory, essentially a theatre metaphor in which consciousness is the stage and the audience is all the unconscious processing which prepares things for the stage; and the anatomical aspects are represented in Newman's thalamo-cortical loops system involving the neural connections of the thalamus with its ramifications into the cortex, coupled with the cortex's ramifications back into the thalamus and its various associated neuro-anatomical subsystems. This mapping provides a model of the neural systems thought most likely to embody the major processes of day-to-day consciousness and of the integration and control of the informational structure through which we have our place in the world.

2.1 Thalamo-Cortical System

The thalamus is a sensory relay station in the centre of the brain and acts somewhat as the hub in a wheel, the spokes of which are nerve bundles carrying sensory data from the body periphery, which are then relayed up into the cortex and cortical association areas. For example, the optic tract runs from the retina, through the optic chiasm and thence into the lateral geniculate of the thalamus from where it is distributed into the visual cortex at the back of the brain. These are ascending pathways. At the same time there is a vast array of nerve bundles descending from all areas of the cortex, particularly feature processing and pre-frontal areas onto the intralaminar nuclei and the nucleus reticularis in the thalamus. These descending pathways act to *gate* the sensory data being presented to the cortex via inhibitory neurons in the nucleus reticularis. The descending pathways convey feedback from the cortex onto the thalamus as a sensory relay module, providing a self-regulatory capacity in which the cortex controls what sensory data it is being sent at any moment. It is here that we can find an anatomical structure which permits a mechanism for selective attention, Baars' "spotlight", on the stage of the global workspace.

This thalamo-cortical loop system, or some operationally similar version of it, offers a hardware basis for an internal "representation" or "cognitive map" of the world and will be needed in the development of an artificial consciousness. The thalamo-cortical loops system provides a means for integrating the flow of information in the brain, for providing self-regulated attention, for providing working memory through the presence of the past that these feedback loops establish and for concentrating that information flow into the whirlpool of our conscious being.

3 Implementation

The temptation with much AI is to suppose that hard wiring will bring about the necessary relations between structures. This is a top-down approach and suffers the basic problem of top-down: that you can never predict all the possible situations in which the entity might find itself and have to deal with as new variations in the environment. The alternative then requires that our neural network system, as a cascaded structure of multiple hidden layers, can be enabled to evolve its connections, both as actual connections and their weightings so that things which it encounters are things which it has means for understanding and knows how to deal with or at least can make a guess at dealing with.

With feedback structures previous states re-enter the neural circuit at various levels bringing previous experience in on the current process. A memory of what happened in the neural weighting process during previous experience will also carry with it the details of the states entered in dealing with the event. Each hidden layer is an input layer for the next hidden layer as well as possibly being an input to some other prior layer in the same sensory pathway, say for focussing and attention, or some other sensory pathway, say for naming and identification or grip control in handling. For example the visual pathway bifurcates and visual information is sent to visual processing and subsequent recognition layers as well as being sent to hand position control layers for reaching and gripping. Proprioceptive, visual and tactile feedback circuits throughout the reaching and gripping process all provide real-time data necessary for the appropriate completion of the task.

Getting these systems to work in real environments I, and others, eg. [9], suggest is a matter of creating a basic body with recurrent neural networks handling its experiential sensory data flow. It is then caused to evolve in the actual environment within which it has to operate. That way the behaviour of the system converges on appropriate solutions to handling the vagaries of the environment with inadequate solutions not being reinforced. It is this reinforcement that is the feedback in this system. In the multiple internal tasks of maintaining state in a dynamic environment, ie, self-regulation, the weights on connections in the networks grow and decay according to use and the functional value of the behaviour. Obviously there is no point in "bashing your head against a brick wall". Internal feedback is as necessary to self-regulation as external "social" feedback. If the behaviour of the system cannot emphasise a certain sensory data stream by which it can focus on a threat then that system will fail (and probably be eaten). A hardware architectural equivalent of the thalamo-cortical recurrent network is logically the way to do this.

4 Conclusion

I have outlined an operational definition of consciousness in so far as it might be applied to artificial entities and have focussed on a way in which the physical architecture of biological consciousness might be taken as a model for the development of artificially conscious entities such as robots and the like. Feedback is a primary factor in the self-regulation of biological entities and, I suggest, will be found to be necessary in the homeostatic self-maintenance of intelligent robots and similar systems.

References

1. Brooks, R.A. (1999) Cambrian Intelligence: The Early History of the new AI. A Bradford Book, MIT Press, Cambridge, Mass.
2. Kirk, R. (1996) "On the Basic Package" Robert Kirk talks to Stephen Jones - The Brain Project <http://www.culture.com.au/brain_proj/kirk.htm>.
3. Dennett, D. (1991) Consciousness Explained. Little, Brown and Co.
4. Nagel, T. (1974) "What is it Like to Be a Bat" Philosophical Review, 1974, pp.435-450.
5. Chalmers, D. (1996) The Conscious Mind, Oxford University Press.
6. Aleksander, I. (1996) Impossible Minds. Imperial College Press, London.
7. Baars, B.J. (1997a), In the Theater of Consciousness: The Workspace of the Mind. New York: Oxford University Press.
8. Newman, J. (1997) "Putting the Puzzle Together: Part I: Toward a General Theory of the Neural Correlates of Consciousness." in Journal of Consciousness Studies, 4, No.1, 1997, pp47-66.
9. Husbands, P., Harvey, I., Cliff, D. and Miller, G., (1997) "Artificial Evolution: A New Path for Artificial Intelligence?" Brain and Cognition 34, 130-159.

Some Effects of Individual Learning on the Evolution of Sensors

Tobias Jung, Peter Dauscher, and Thomas Uthmann

Institut für Informatik
Johannes Gutenberg-Universität
55099 Mainz, Germany
{tjung, dauscher, uthmann}@informatik.uni-mainz.de

Abstract. In this paper, we present an abstract model of sensor evolution, where sensor development is only determined by artificial evolution and the adaptation of agent reactions is accomplished by individual learning. With the environment cast into a MDP framework, sensors can be conceived as a map from environmental states to agent observations and Reinforcement Learning algorithms can be utilised. On the basis of a simple gridworld scenario, we present some results of the interaction between individual learning and evolution of sensors.

1 Introduction

Nature has produced a variety of sense organs strikingly adapted to the environment of the respective animal, an observation, which leads to the question which driving forces might have been involved in the evolutionary process. The Artificial Life approach might be able to shed light on some aspects of this question (cf. [1], [2]). The focus of this article is an investigation of the influence of individual learning processes on the evolution of sensors in a simple scenario such that quantitative results may be obtained.

2 The Model

The scenario considers an evolving population of agents; for simplicity, each agent is situated in a simple environment of its own such that direct and indirect agent-agent interactions are avoided by construction. An agent consists of two parts: (1) A genetically determined sensor provides information from the environment and (2) a control unit chooses one of a set of possible actions based on the sensory input. The mapping from sensory input signals to actions is adapted during lifetime in an individual Q-Learning process (cf. [3]). The scenario used here is a simple, time discrete predator-prey scenario (1 predator individual, 1 prey individual) on a toroidal gridworld (11×11 sites).

The considered evolving population only consists of predator individuals, the prey individuals follow a fixed simple evasion strategy and are considered as part of the environment. The fitness of a predator, used for selection, depends on how

J. Kelemen and P. Sosík (Eds.): ECAL 2001, LNAI 2159, pp. 432–435, 2001.

many times it catches the prey within a given time period. When the predator catches the prey, a new episode starts with the predator and a new prey situated randomly in the gridworld.

The state of the environment is the position of the prey relative to that of the predator, restricted by a limited field of vision which is further partitioned into three subsections, as indicated in Fig. 1a). Thus, at most 64 environmental states can be distinguished: 63 corresponding to the possible locations of the prey if it is in the field of vision, and the one additional environmental state in which the prey is not visible.

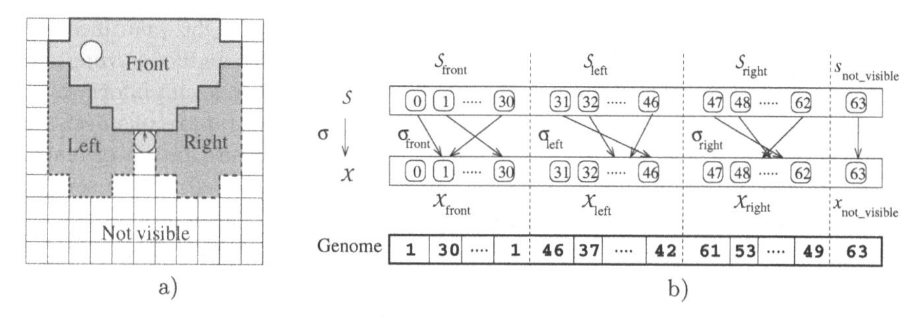

Fig. 1. a) An agent (grey circle in the center with $\uparrow$ indicating its orientation) and its field of vision in the gridworld environment. b) Mapping $\sigma : \mathcal{S} \to \mathcal{X}$ of the agent. The corresponding genome is shown below.

Within the model, a sensor is conceived as a mapping σ from environmental states $\mathcal{S} = \{s_0, \ldots, s_{63}\}$ to internal agent states $\mathcal{X} = \{x_0, \ldots x_{63}\}$. It is this sensor map σ which is subjected to evolution, with its genetic representation shown in Fig. 1b). On the one hand, it is well possible that this map is not injective such that in general information is lost. On the other hand, actions of the predator are based solely on observations $x_i \in \mathcal{X}$, it is therefore clear that a proper mapping σ is crucial for the predator to take the right actions.

A *policy* is learned with the Q-Learning algorithm (TD(0)) during lifetime. At each time step t the control unit chooses an action a_t only dependent on the current internal agent state $x_t = \sigma(s_t)$, yield by the sensor and therefore implements a Markov Decision Process[1] (MDP). This action leads to a new external state s_{t+1} being perceived as an internal state x_{t+1} and a reward r_{t+1}.

[1] In general, the sensor does not return the complete state of the environment, such that the learning agent faces a POMDP (see e.g. [4]). Whereas the theory and convergence results of Q-Learning only apply to the Markovian case, it seems to perform rather satisfying in this special scenario.

3 Simulation Experiments and Results

The primary aim of the simulation experiments was to investigate the influence of the number of learning steps on the efficiency of the evolved sensors. Therefore two interesting measures were considered: (1) The fitness of the individuals and (2) the relevant information conveyed by the sensors.

In a first simulation series, individuals were evolved with the lengths of learning periods varied. The best individual of each generation then was tested by coupling it to learning processes of a constant length, chosen independently of the length in the evolutionary scenario. These results are shown in Fig. 2.

In a second series, the amount of relevant information transmitted by the sensors was computed. The concept of relevant information [5] quantifies the amount of information between two dependent signals as the *mutual entropy* and can be readily applied to decision processes [6]. In this scope, relevant information describes the amount of information useful in finding an advantageous decision and thus provides a simulation and fitness independent measure for sensor maps. Fig. 3a) shows the amount of relevant information transmitted by the sensor; Fig. 3b) shows the ratio of relevant to total (i.e. Shannon) information.

Varying learning periods within the evolutionary process showed that longer learning duration facilitates evolution and leads earlier to better suited sensors, if the learning duration does not factor into the evolutionary task. Furthermore it could be shown that a different learning duration obviously results in the development of different kinds of sensors optimal in the respective environment but performing poorly under differing conditions. This observation could be attributed to the fact that the learning efficiency is determined by the number of internal states in two opposing ways: There is a tradeoff between compressing the representation, which facilitates the learning process, and preserving meaningful information, which grows more important in the long run. These qualitative observations are reflected by relevant information in a quantitative manner: a direct dependence of relevant information and learning duration could be observed (for further details, see [7]). Although at first sight, these results seem to

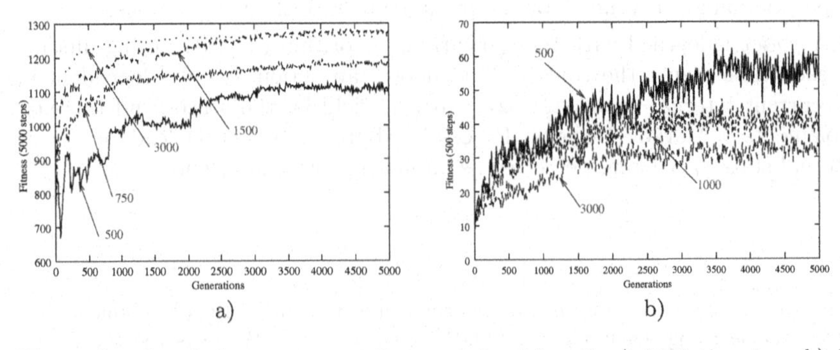

Fig. 2. Results of the constant testing periods of length a) 5000 timesteps b) 500 timesteps. Numbers above curves denote respective learning durations.

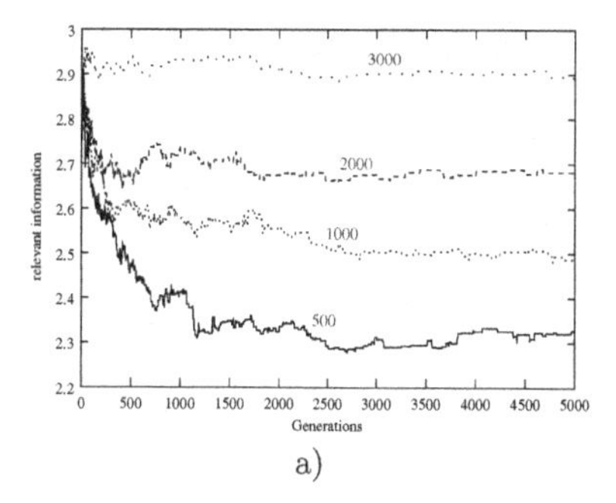

a)

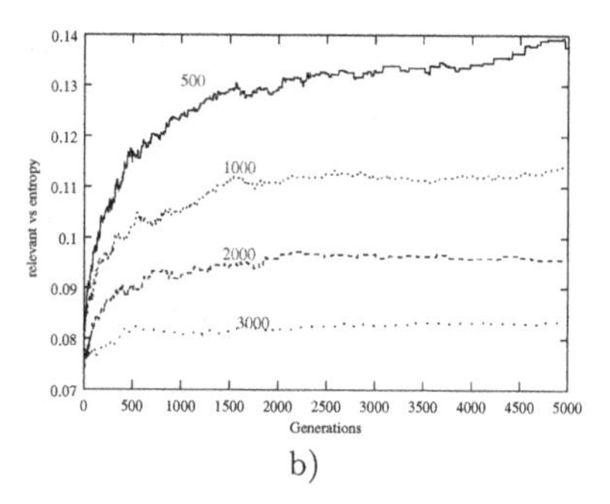

b)

Fig. 3. Entropy measures for varying lengths of the learning period. Numbers above the curves denote the respective length. a) relevant information b) the ratio of relevant to total information.

bear some resemblance to the *Baldwin-Effect* [8], it is yet unclear, if individual learning being independently modeled, can affect evolution in a similiar way.

Acknowledgements. We wish to thank Daniel Polani from the Institut für Neuro- und Bioinformatik, Medizinische Universität Lübeck for useful hints.

References

1. A. Mark, D. Polani, T. Uthmann. *A framework for sensor evolution in a population of braitenberg vehicle-like agents.* In: C. Adami et al. (eds.) *Artificial Life VI*, pp. 428–432, MIT Press. 1998.
2. A. Liese, D. Polani, T. Uthmann. *On the development of spectral properties of visual agent receptors through evolution.* In: D. Whitley et al. (eds.) *Proc. GECCO 2000, Las Vegas, Nevada* pp. 857-864. Morgan Kaufmann, 2000.
3. R. Sutton, A. Barto. *Reinforcement Learning: An Introduction.* MIT Press, Cambridge, MA, 1998.
4. L. P. Kaelbling, M. L. Littman, A. R. Cassandra. *Planning and acting in partially observable stochastic domains.* Articial Intelligence, **101**, 1998.
5. N. Tishby, F. Pereira, W. Bialek. *The Information Bottleneck method.* In *Proc. of the 37th Allerton Conf. on Communication, Control, and Computing 1999.*
6. D. Polani, T. Martinetz, J. Kim. *On the Quantification of Relevant Information* Presented at the 7th Scandinavian Conf. on AI, Odense, Denmark, Feb. 2001
7. T. Jung, P. Dauscher, T. Uthmann. *On Individual Learning, Evolution of Sensors and Relevant Information* Accepted for Workshop "Evolution of sensors in nature, hardware, and simulation", GECCO 2001.
8. J. M. Baldwin. *A new factor in evolution.* American Naturalist, **30** pp. 441-451, 1896.

Transitions in a Simple Evolutionary Model

T. Lenaerts, A. Defaweux, P. Beyens, and B. Manderick

Computational Modeling Lab
Department of Computer Science (DINF)
Vrije Universiteit Brussel
{tlenaert, adefaweu, pbeyens}@vub.ac.be
bernard@arti.vub.ac.be
http://como.vub.ac.be/

Abstract. We report on the construction of a simple alife model to invesitgate the theories concerning evolutionary transitions. These theories can be considered as biological metaphors for cooperative problem solving. The benefit of this model for computer science results from the model's property to assimilate, in evolutionary fashion, complex structures from simple ones. It is this assimilation process we want to capture in an algorithm in order to apply it to learning and optimization problems.

1 Introduction

A natural approach to solve complex problems is to construct a solution incrementally through the combination of components based on some heuristic. The underlying basis of this approach is the fact that a solution is composed of cooperating parts through some form of interaction. The performance and efficiency of the problem solver will depend on the chosen heuristic in terms of its usefulness for that particular problem instance and effectivness on the solution represenation.

The heuristics in which we are interested are called an Evolutionary Algorithms (EAs)[6]. In EAs, solutions to a particular problem are discovered through a set of algorithms which mimic natural evolution. In order to solve a particular problem, all the building blocks which form the entire solution have to be present in one individual at some point in time. Based on this fact it might be useful to allow these individuals to interact and maybe cooperate in order to find the solution since a group of cooperating individuals has a higher functionality than an individual in isolation. In most configurations there is no interaction between these individuals except during the recombination process when building blocks, in the form of genetic information, are exchanged to produce better offspring. This is opposed to research in Biology, since in this context evolution is always influenced by interaction.

So, one would like to incorporate this interactive approach into artificial evolutionary systems. During the evolutionary process, individuals are evolved that represent solutions to the different subproblems and these individuals can then

J. Kelemen and P. Sosík (Eds.): ECAL 2001, LNAI 2159, pp. 436–439, 2001.

cooperate to solve the overall problem together. Unfortunately, these individuals face a serious dilemma. On the one hand, they have to compete among themselves in order to reproduce. On the other hand, they need to cooperate in order to solve together the overall problem. We are investigating a solution to the above dilemma based on a theory of evolutionary transitions as described in Michod[97] and Maynard Smith and Szathmáry[97].

Here, we report on our first attempts to capture the underlying mechanisms of transitions in a model of cooperative problem solving. In Section 2, we describe a general approach for the construction of our problem solver. Next, we present the different construction steps of our Alife model in Section 3 and finally we conclude in Section 4.

2 Evolutionary Transitions

Due to the limited size of this report, we refer to [5] and [9] for further information on the existing theories concerning Evolutionary Transitions. Briefly, these theories describe the formation of groups of cooperating entities with extended functionality in a system of selfish replicators and how these groups become new units of selection on a higher level. Some well-known examples are the transition from genes to gene networks and the transition from single-cell to multi-cellular organisms.

Evolutionary Transitions can be considered as biological metaphors for cooperative problem solving. The benefit of the biological model of evolutionary transitions for computer science results from the model's property to assimilate, in evolutionary fashion, complex structures from simple ones. The creation of complex structures is guided by the principles behind cooperation and conflict. There are different relevant area's in which the model can be applied, for instance, one domain could be Multi-agent systems[3]. In order the examine the usefulness of the model in computer science we perform two steps:

1. Build a simple artificial life model through which we can evaluate the theory concerning transitions and fitness.
2. Build a problem solver, based on the theory of evolutionary transitions and the expertise we obtained from the previous model, for learning and optimisation problems.

Current work focuses on the first step of our study and is described in the next section.

3 Alife Model

In order to evaluate Michod's ideas we rebuilt an Alife model which is based on the Alife model by Packard[7,1]. This model contains the essential ingredients of a biological evolutionary system and it was originally introduced to illustrate the difference between intrinsic and extrinsic adaptation. It consists of organisms which are allowed to traverse a two-dimensional world. The elementary actions of the organisms are to consume food at the current location and to reproduce when a certain threshold is reached. These organisms vary genetically in the number of offspring they can produce (g_{off}) and the food threshold required for reproduction (g_{th}). The size and distribution of the population of organisms depends entirely on the configuration of the food field and the way it is refreshed. Bedau called these organisms tropic bugs since they invariably climb the food gradient[1]. We extended this model in subsequent steps to model Evolutionary transitions.

Step 1 : Group Formation; explicit competition An important requirement for a transition to occur is the emergence of non-selfish behaviour in an evolutionary system. Therefore, in this first step, we examine whether this behaviour can originate in the Packard model.
The behaviour was implemented through a common pool resource game which can be modelled through a prisoners-dilemma game[4]. Two forms of interaction (game playing) were evaluated; random pairwise interaction and a n-player configuration. The experiments performed in this step are similar as those done by [8] in their study of feeding restraint. In these experiments it is hard to identify which bugs belong to which group since their relation is not explicit. Nevertheless we can quantify the different levels of selection through Price's covariance equation[5,4,8].

Step 2 : Group Formation; explicit interactions The problem in the previous step was the identification of the groups. This identification is necessary since we also want to know the structure of these groups. So instead of performing an n-player game or random pairwise matching we allowed a bug to choose its partner explicitely[2].
The game played here is similar to an evolutionary game with two types of agents, a selfish and altruistic agent. By cooperating an agent produces a fitness increment $b > 0$ for its partner, at a personal cost $c < b$. By defecting an agent produces zero at zero cost. Thus, this is also a prisoners dilemma where selfishness is the dominant strategy. Groups are in this case identified as the networks of bugs which are linked together and again we can investigate whether there is selection at different levels through Price's equation.

Step 3 : Group Formation; explicit functionality An important part in the discussion whether groups of cooperating replicators can increase in numbers, concerns the idea that a group has more functionality than a single replicator. This argument is used to motivate why higher level networks keep collaborating and eventually become new evolutionary units.
To introduce this concept we need variation in the behaviour of bugs. To do this, we removed the gradient search from the model and allowed the bugs to use individual strategies for finding food. This extension is a simplied version of Bedau's model of strategic bugs[1].

Step 4 : Individuality; aggregation of groups Once a group of cooperating individuals emerged that group has to be protected against defecting members or external parasites which reap off the benefits of the group. This step is not completed yet, but once coordinated group movement is established we can examine the emergence of new evolutionary entities. In this context we are examining aggregation techniques which merge the group into a single unit. This aggregation should occur when the group remains active over a period of time, i.e. all components remain connected, eat enough to allow survival of the network and move in union.

4 Conclusion

We presented an Alife model under construction which can be used to examine evolutionary transitions. The architecture is based an simple model which simulates intrinsic adaptation[7,1]. We extended it with explicit interactions and examine extended group functionality in this context.

There are still many questions to be answered and decisions to be made before the entire model is finished. For instance: How should the aggregate unit reproduce and what will be the threshold? How will the new unit interact with other simple or complex units in the population? How should these aggregated networks mutate? Furthermore a more extensive analysis of the experimental data has to be performed.

References

1. Mark A. Bedau and Norman H. Packard. Measurement of evolutionary activity, teleology, and life. Artificial Life II, SFI Studies in the Sciences of Complexity, **X** (1992) 431–461
2. Ben Cooper and Chris Wallace. Evolution, partnerships and cooperation. In Journal of Theoretical Biology, **195** (1998) 315 – 320.
3. A. Defaweux, T. Lenaerts, S. Maes, B. Manderick, A. Nowé, K.Tuyls, and P. van Remortel nad K. Verbeeck. Niching and evolutionary transitions in multi-agent systems. *Workshop on Evolutionary Computation and Multi-agent Systems.* In Proceedings of the Genetic and Evolutionary Computation Conference (GECCO). San Fransisco, California. USA, (2001)
4. H. Gintis. Game Theory Evolving; A problem-centered introduction to modeling strategic interaction. Princeton University Press, Princeton, New Jersey. (2000)
5. Richard E. Michod. Darwinian Dynamics; evolutionary transitions in fitness and individuality. Princeton Paperbacks, New Jersey. (1997)
6. M. Mitchell. An Introduction to Genetic Algorithms. A Bradford Book, MIT Press, 3th edition, 1997.
7. Norman H. Packard. Intrinsic adaptation in a simple model for evolution. Artificial Life, SFI Studies in the Sciences of Complexity, (1988) 141–155
8. John W. Pepper and Barbara B. Smuts. The evolution of cooperation in ecological context: An agent-based model. Dynamics in Human and Primate Societies: Agent-Based Modeling of Social ands Spatial Processes. (2000)
9. John Maynard Smith and Eörs Szathmáry. The major transitions in evolution. Oxford University Press, Great Britan. (1997)

Towards the Implementation of Evolving Autopoietic Artificial Agents

Barry McMullin and Dominique Groß

Research Institute for Networks and Communications Engineering (RINCE),
Dublin City University, Dublin 9, Ireland
McMullin@rince.ie
http://www.eeng.dcu.ie/~alife/

Abstract. We report modifications to the SCL model[2], an artificial chemistry in Swarm designed to support autopoietic agents. The aim of these modifications is to improve on the longevity of the agents and to implement growth. We demonstrate by means of two simulation runs that the improvements indeed have the desired effects.

1 Background: SCL

A computational model of autopoiesis was originally described by Maturana and Varela[3]. This was re-implemented by McMullin in Swarm[1] as the SCL system [2,1]. SCL contains three different types of particles: substrate (S), catalyst (C) and links (L). In this model an autopoietic agent or organisation consists of a closed (self-repairing) chain of doubly bonded links which encloses a C particle. In order to qualify as an interesting instance of autopoiesis, its longevity has to exceed that of its components (the links) which are designed to be unstable. In the experiments presented in [2] the minimal autopoietic organisation could be sustained for up to approximately 1800 time-steps, but was eventually irreparably corrupted by some decaying links which could not be replaced quickly enough.

2 New Features: SCL-GRO

We report a modified version of SCL, called SCL-GRO[2]. This aims at an improvement of the self-repair mechanism of the autopoietic agents (without trivializing the problem by, for example, turning off the decay of links), and the introduction of growth. The motivation for these improvements derives from an overall long-term goal of realising evolution of artificial autopoietic agents. The following additional features have been implemented.

- **Affinity:** Links have been given an affinity for chains of bonded links (closed or open); if a link is in the neighborhood of a chain it may drift along the chain but will not move away from it.

[1] http://www.swarm.org/
[2] http://www.eeng.dcu.ie/~alife/src/scl-gro/

J. Kelemen and P. Sosík (Eds.): ECAL 2001, LNAI 2159, pp. 440–443, 2001.

- **Smart repair:** Free links actively scan for decaying links in a chain, i.e., links which will decay as soon as there are enough free lattice sites around them to do so. If they find them, they replace these links.
- **Improved mobility of links:** Whereas the L particles in the original implementation were immobile once they were bonded, we allow bonded links to move as well (subject to the bond constraints).
- **Displace motion:** If there is a free link in the neighborhood of a doubly bonded link, then with some user-specified probability it triggers the breaking of one of the bonds of the bonded link, and "pushes" this (now singly-bonded) link in the direction away from the free link; the latter then moves into the newly vacated lattice position. In this configuration the free link can be integrated into the chain through a chain splicing event, which also restores the closure of the membrane.

In earlier versions of SCL a special feature, so-called bond inhibition had to be introduced in order to ensure that the free links inside the membrane did not bond and thus become unavailable for repair of the membrane. With growth now enabled this measure became ineffective in SCL-GRO; bond inhibition only works in the immediate neighborhood of the membrane (thus preserving the principle of locality!). With a larger membrane most of the free links inside will not be in such a neighborhood. This problem has been resolved by differentiating the bonding interaction into three cases, depending on the prior bonding states of the link particles. The distinct cases are for bonding between two free links (chain initiation), between a free link and a single bonded link (chain extension), and between two singly bonded links, possibly via an additional free link (chain splicing). The probability for the first of these (chain initiation) is then set to zero. This ensures that free links will not spontaneously bond to each other, but can still bond to pre-existing chains (in order to maintain—and grow—the membrane).

3 Results

We will now shortly summarise results from selected, but typical runs in SCL-GRO. The results given below hold approximately for all experiments we have performed.

Longevity: We turned off growth (by turning off the displace motion) in order to compare the longevity of the autopoietic agents in SCL-GRO with those in the original SCL. We seeded the SCL-GRO world with a medium sized autopoietic agent. In the course of the simulation the membrane undergoes some change and also inserts a few additional links. Despite continuous ruptures of the membranes, in the specific experiment described here the self-repair mechanism of the autopoietic agent remains effective until about time-step 5400, when a terminal rupture happens.

Taking into account that the probability of decay per time-step of a link is set to 0.002 and that the initial configuration consisted of 42 links, then the expected life-time of the membrane—in the absence of repair—is about 12 time

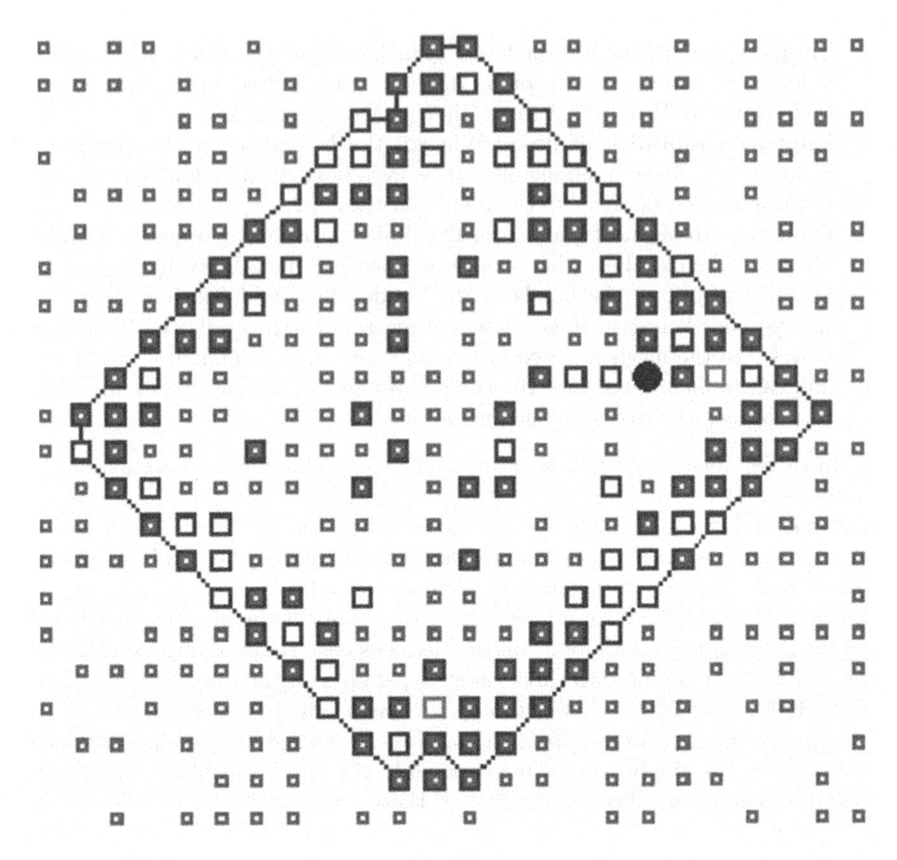

Fig. 1. This is a screenshot from the experiment with growth turned off: after 3130 time-steps the autopoietic agent is still intact. The small squares are S particles; the large and large dark squares are L and L^+ particles (without and with absorbed substrates respectively); the filled circle is the catalyst. Three distinct layers of the agent are clearly visible.

steps. This is to be contrasted with the actual life-time of the membrane of over 5000 which is more than 400 times longer. The improvement over the original SCL model becomes clear if one compares this result to the analogous number of the earlier reported experiments [2] where the autopoietic agent lived for about 22 times its expected (unrepaired) life-time (decay probability of 0.001, 12 links in the membrane and an absolute life time of no more than 1800 time-steps). It is clear that in experimental settings like this, quantitative results are only of limited value in the absence of more extensive coverage of the parameter space. However, there is a factor of 20 between the two results, which has some suggestive value.

Growth: Let us now turn on the displace motion and start the experiment again with an initial configuration, consisting of a small membrane enclosing a catalyst and four free links such as in [2]. Starting the simulation, the membrane immediately starts to deform into irregular shapes. Occasionally new links are inserted into the chain, which leads to a growth process. Inside the membrane the catalyst continuously produces new free links, which tend to drift outwards, ready to repair a potential fracture of the membrane. Once the agent has attained a certain size three functionally distinct parts or layers can be discerned. First there is the most inner layer, the "reactor", characterized through the presence of the catalyst and substrate which has drifted through the chain of links into the interior of the membrane. The middle-layer consists of free links, which were produced in the "reactor" and are "waiting" to repair a potential rupture. The third layer is the "membrane", which has already been described above. Once the membrane attains a certain size, it usually takes a rectangular shape; this is apparently partly due to the affinity between free links and the membrane and the consequent inhibition of its inward motion, and partly a reflection of the underlying lattice geometry.

Dependent on the chosen disintegration probability of the links, the chain sooner or later suffers a terminal rupture. Especially if the chain is already quite large, then locally there might not be enough repair material, i.e., free links, available to effectively perform a repair operation. This effect seems to be a cause for the terminal rupture in this run, where one clearly sees the lack of free links in the neighborhood of where the membrane is broken (not shown). Under this setting the terminal rupture of the membrane occurs at time step 1168. Various experiments (not reported here) suggest that the coarse qualitative features of the behaviour of the model remain invariant under a significant variation of the link-decay probability.

4 Discussion and Conclusion

The SCL-GRO model, as described in this article, clearly adds to the qualitative phenomenology of the original version of SCL in that it considerably extends the longevity of the autopoietic agents and supports their systematic growth. These might be useful steps in the direction of realising reproduction of autopoietic agents by growth and fission.

References

1. B. McMullin. Modelling autopoesis: Harder than it might seem!, 1997. Dublin City University, School of Electronic Engineering, Technical Report: bmcm9704, URL: http://www.eeng.dcu.ie/ alife/bmcm9704/.
2. Barry McMullin and Francisco J. Varela. Rediscovering computational autopoiesis. In Phil Husbands and Inman Harvey, editors, *Proceeedings of the Fourth European Conference on Artificial Life*, Series: Complex Adaptive Systems, Cambridge, 1997. MIT Press. URL: http://www.eeng.dcu.ie/~alife/bmcm-ecal97/.
3. F. Varela, H. Maturana, and R. Uribe. Autopoiesis: The Organisation of Living Systems, its Characterization and a Model. *Biosystems*, 5:187–196, 1974.

Verification of Text Transcription History by Using Evolutionary Algorithms

Yuji Sato[1], Masafumi Kawamoto[2], and Motoi Chizaki[3]

[1] Faculty of Computer and Information Science, Hosei University
3-7-2 Kajino-cho, Koganei-shi, Tokyo 184-8584, Japan
yuji@k.hosei.ac.jp
[2] Graduate School of IT Professional Course, Hosei University
2-17-1, Fujimi, Chiyoda-ku, Tokyo 102-8160, Japan
it003314@itpc.i.hosei.ac.jp
[3] Faculty of Engineering, Hosei University
3-7-2 Kajino-cho, Koganei-shi, Tokyo 184-8584, Japan
chizaki@ogw.ei.hosei.ac.jp

Abstract. The application of ALife methods to the field of historical document genealogy is reported. At first, the conventional analytical approaches to Greek and Latin text transcription history are described and the problems associated with them are discussed. Next, focusing on the similarities between text transcription history and the Tierra system, we propose the use of evolutionary algorithms for the verification of text transcription genealogies with the purposes of breaking through the limitations of conventional analysis of Greek and Latin text transcription history and increasing the reliability of those methods. In this report, as the first step, we deal only with the mutations involved in copying, and attempt to make use of them in the verification of genealogy. We conducte computer simulation experiments based on existing manuscript data and demonstrate the feasibility of effectively using evolutionary algorithms as one means of verifying the currently proposed Greek and Latin text genealogies.

1 Background

Tierra project of Thomas Ray [1] demonstrates that computer programs can evolve in a computer in the same way as living organisms do in nature. Tierra defines digital organisms, which have self-replicating programs written as series of machine language instructions as their genetic material, as an analogy to the process of biological evolution in nature. The self-replication process, which involves mutation and natural selection, is executed repeatedly, beginning with a single digital organism (ancestor species). A mutation is defined as bit-flipping, in which the 1s and 0s of machine language are randomly reversed with a given probability. Here, we consider two types of mutations. One is a copying error that occurs with a given probability at the time of self-replication; the other is the operation of bit-flipping, in which a certain proportion of all of the bits in the memory area are reversed. Natural selection corresponds to the death of an individual for which the genetic material has changed by mutation in such a way that the individual fails to find an empty region in the

J. Kelemen and P. Sosík (Eds.): ECAL 2001, LNAI 2159, pp. 444-453, 2001.

memory space or self-replication is not possible. This repeated process of reproduction through mutation and natural selection results in the creation of a variety of programs in the computer. As an end result, it has been reported that the evolution of a self-replicating program that reduced the number of program steps to 1/4 that of the ancestor species and reduced the execution time to 1/6 was observed as a structure most suitable for survival in an environment of limited energy and common resources. The demonstration by this research that a roughly designed program can autonomously evolve into an optimum program in a computer in a manner similar to the evolution of biological organisms in nature has opened up new possibilities in the field of information processing. On the other hand, because this is an artificially constructed system in a virtual world that is referred to as a computer space, some researchers have wondered whether it would be necessary to intentionally set the problem in advance in order to obtain the desired result.

Consider, on the other hand, the history of the Greek and Latin text transcription process. In the present time, text can be preserved for long periods of time by using storage media such as floppy disks and CD-ROMs. Furthermore, printing technology has made the accurate reproduction of text possible. In ancient times, however, documents were recorded on media such as papyrus, goat skin and stone. Furthermore, before the widespread use of printing technology, documents were reproduced by scribes who copied the text by hand. That is to say, documents written on such media have been passed down through the Middle ages to modern time by means of hand-copying by scribes. Because documents written on papyrus, goat skin, stone and other such media are easily affected by fire, weathering, etc., few original manuscripts written by classical authors remain in existence. Accordingly, the content of the original document is inferred on the basis of a number of existing manuscripts. All of the existing manuscripts have been affected to some degree by damage to the manuscript itself, copying errors made by scribes, mistakes introduced by intentional changes, and other such effects of the process of being passed down through the ages. Accordingly, the restoration of the original text from existing manuscripts is a difficult and complex task and the accuracy of that process continues to be a problem.

In this report, focusing on the similarities of Tierra and text transcription history, we propose the use of evolutionary algorithms [2] in the verification of text transcription genealogies with the purpose of breaking through the limitations of conventional analysis of Greek and Latin text transcription history.

In the following chapter 2, the conventional analytical approaches to Greek and Latin text transcription history are described and the problems associated with them are discussed. In chapter 3, a method of using evolutionary algorithms to verify genealogy is proposed. In chapter 4, the evaluation method, evaluation results, and discussion are presented, followed by the conclusion.

2 Greek and Latin Text Transcription History

2.1 Text Transcription History

In ancient Greece, the standard medium for written records was papyrus. An example of text written on papyrus [3] is shown in Fig. 1. Papyrus is a kind of paper that is made from the fibrous pith of a reed that grows wild in the Nile river delta. When papyrus was in short supply, the Ionians sometimes used goatskin (or horsehide) instead. Literary texts written on papyrus used a form of writing that was more difficult for readers to understand than the forms used in modern books. As we can see from Fig. 1, that difficulty comes from the fact that punctuation marks were a later development and the fact that the text was written without spaces between words. In the Greek and Latin languages, which do not separate written words with spaces, serious efforts to reform the text format did not begin until the Middle Ages. To make up for the difficulty in reading text written without word spacing, the method of attaching accent symbols is used in the Greek language. The general use of accent symbols, however, did not begin until the early part of the Middle Ages. Before the advent of printing technology, text written in such hard to read format had been passed down to modern times through successive hand copying by scribes.

Fig. 1. An example of text written on papyrus. The text was written without spaces between words and punctuation marks were a later development.

2.2 Text Criticism

Works that were written on papyrus, goatskin, stone and other such media are easily affected by fire, weather and other such deteriorating forces, so few original manuscripts written classical authors remain. Also, the documents are often partially damaged by weathering [4], as shown in Fig. 2. Therefore, in order to know the text as originally written by a classical author when the original handwritten manuscript of the classical author no longer exists, attempts are made to restore the original text on the basis of several existing manuscripts, which are the result of a process of repeated copying. The task of restoring original text from existing manuscripts consists of two main processes. The first step is "collation (recensio)". In collation, multiple existing manuscripts are used as evidential materials for the reconstruction of the oldest restorable text. For that purpose, the correlations among the existing manuscripts are clarified in order to create a "manuscript genealogy" that can be used in the reconstruction of the lost manuscripts from which the existing manuscripts are derived. The "manuscript genealogy" is represented in the form of a family tree. An example of a manuscript genealogy [3] is shown in Fig. 3, where existing manuscripts are represented by letters of the English alphabet (A, B, C, D, X, Y and Z). With the existing manuscripts as archetypes, the lost intermediate manuscripts are represented by letters of the Greek alphabet (α, β and γ). The letter ω represents the prototype manuscript. In the second step of the process, the genealogy obtained by the collation process is examined for unnecessary branches and logical contradictions. In Fig. 3, for example, if manuscript B is derived only from manuscript A, then manuscript the text. If it is judged to be not authentic, that text is subjected to "emenda B is judged to have greater text corruption than manuscript A and so manuscript B is removed from the genealogy. Also, the original text obtained from the genealogy after reconstruction is studied and a judgement is made concerning the authenticity of tion (emendatio)" to correct the errors in the text. The attempt to restore the text to a form that approximates the original as closely as possible by using the two processes of collation and emendation in this way to retrace the path of text transcription history is referred to as "text criticism".

Fig. 2. An example of existing manuscripts, which are the result of a process of repeated copying. The documents are often partially damaged by weathering.

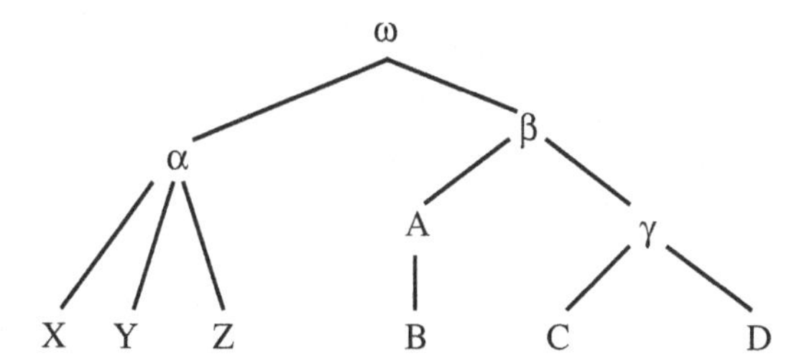

Fig. 3. An example of a manuscript genealogy. The English alphabet (A - D) represent existing manuscripts. The Greek alphabet (α, β and γ) represent the lost intermediate manuscripts. The letter ω represents the prototype manuscript.

2.3 Limits of Genealogical Methods

Here, we provide an overview of the genealogical methods that are used in the process of collation and briefly explain the limits of those methods. The theory of genealogical methods in classical studies has been described by Paul Maas. In genealogical methods, attention is first turned to the errors made by scribes when copying manuscripts and a genealogy is constructed. For example, the close-distant relationship among manuscripts is judged according to whether or not the errors in a given manuscript appear in other manuscripts, the correlations between manuscripts are measured, and a genealogy that covers the overall manuscript transcription history is constructed. Then, the constructed genealogy is examined for unnecessary branches and logical contradictions. Here, a tacit premise in genealogical methods is that the transcription history is "closed". That is to say, it is assumed that the various manuscripts each were each the result of a direct transcription of the reading and errors of one archetype document; accordingly, the assumption is that the work can be traced back to a single original document. Actually, however, the scribes of ancient times and the Middle Ages did not necessarily copy text from single archetypes. It also happened that multiple transcripts were compared and different works that were easy to read or interesting were incorporated into the scribe's own manuscript. That is to say, problems of reliability often remain when genealogical methods are applied to the text transcription history. Even if we were to accept, for the moment, that transcription history is "closed", the construction of a highly reliable genealogy is still a rather difficult task. That is because the scribes made many different kinds of mistakes when copying documents and multiple errors intertwined in complex ways. What is more, in addition to the errors made when transcribing, factors such as loss of part of the text from fire or weathering may also be involved. Examples of the types of errors include those described below [3].

(a) Confusion of two terms that have similar form or spelling: These are mistakes that involve words of similar form or spelling due to lack of spacing between words, multiple types of abbreviated forms, etc.
(b) Rewriting of text due to changes in spelling and pronunciation: For example, in later-period Latin, b was not pronounced as a fricative and was confused with v. In the Greek language, several vowels and double-vowels changed to the single sound, iota (ι).
(c) Incorrect abbreviation or omission: These are mistakes in the number of times that characters are repeated consecutively, omission of an entire line of text, etc.
(d) Repeated copy errors: These are errors that involve several characters or syllables written repeatedly.
(e) Transposition errors: These include the transposing of characters, copying of lines in the wrong order, etc.
(f) Context-dependent errors: These are errors in which the change in a word form is mistaken for the change in form of an adjacent word and assimilated, etc.
(g) Unsuccessful intentional changes: Although an attempt was made to correct a part of the text that is judged to be difficult to understand or to contain an error, the correction was not successful, either because the attempt was inappropriate or because the scribe had insufficient knowledge to make the correction.

3 Proposal for the Use of Evolutionary Algorithms in Verification of Genealogy

3.1 The Similarities and Differences of Text Transcription History and Tierra

The correspondence between text transcription history and Tierra is illustrated in Fig. 4. Regarding a literary text as one program in the Tierra system, many similarities between text transcription history and Tierra can be seen. It is possible to make correspondences between the original text and the ancestor species, text copying and program self-replication, transcription errors in copying and the copying errors that occur with a given probability in program self-replication, text damage caused by fire, etc. and bit-flipping of a certain proportion of the entire data memory space. That is to say, although Tierra is an artificially constructed system in the virtual world of computer space, that same system can be taken to exist in the process of Greek and Latin text transcription through the ages.

On the other hand, there are also points of difference between text transcription history and Tierra. Whereas the copying errors that occur in Tierra at the time of self-replication are only one type of mutation, there are multiple types of mutation in the text transcription process, including simple copying errors, transpositions due to changes in spelling or pronunciation, and intentional changes. Furthermore, as opposed to the Tierra mutations, which are accomplished by bit-flipping of a certain proportion of the entire memory space, the mutation rate in text transcription history varies according to the material, region, time period, and so on. In Tierra, there is no crossover operation [5], but in text transcription history, the act of copying text on the basis of multiple transcripts can be viewed as crossover. In Tierra, the size of the memory space is fixed, so the number of self-replicating programs is limited; text manuscripts, however, can increase in number without bound. That is to say, the process of Greek and Latin text transcription history lives in a more complexly evolved form.

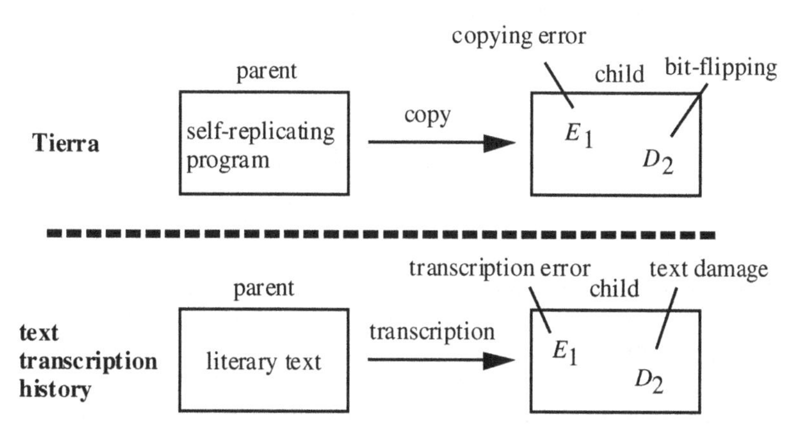

Fig. 4. The correspondence between text transcription history and Tierra.

3.2 Modeling Text Transcription History with Evolutionary Algorithms

Here, focusing on the similarities of text transcription history and Tierra that were described above, we propose the use of evolutionary algorithms for the verification of text transcription histories (genealogies) with the purposes of overcoming the limits of genealogical methods and increasing the reliability genealogical methods. In this report, as the first step, we deal only with the mutations involved in copying, and attempt to make use of them in the verification of genealogy. The use of evolutionary algorithms to model text transcription history is illustrated in Fig. 5. One page of a manuscript is considered to be one individual. The text written on that page corresponds to a chromosome. The act of copying performed by a scribe is considered to be self-replication. Multiple manuscripts are copied from one transcript, and both the original manuscript and the reproduced manuscripts are passed down to the next generation. The mutation types are defined here as simple errors in transcribing a character, mistaking a word, omission of the verb 'be' and vowel replacement due to changes in pronunciation. Concerning the loss of text due to fire, weathering, etc., a lifetime is set for each individual to substitute for such effects.

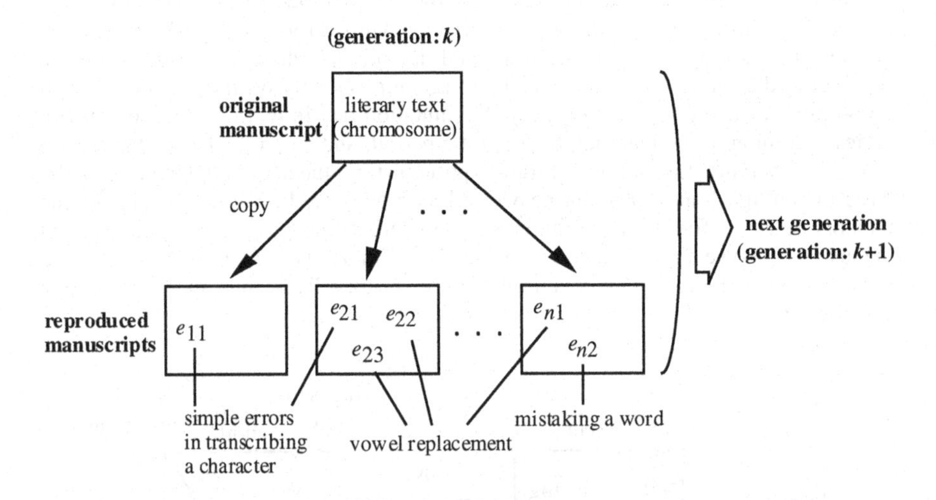

Fig. 5. The use of evolutionary algorithms to model text transcription history.

4 Evaluation Experiments

4.1 Evaluation Method

The genealogy proposed by Cyril Baily [6] is shown in Fig. 6. This genealogy refers to letters on atomism "Epicurus to Herodotus [6]". This is one genealogy that is generally referenced in the field of Greek and Latin text transcription history. However, no technique for quantitative evaluation of the reliability of this kind of genealogy has been proposed. Furthermore, because of the limits of genealogical methods, the situation in which text is copied from multiple transcripts is not considered in the text transcription history. In this report, we attempt to verify the genealogy proposed by Baily by means of computer simulation using a evolutionary algorithm. For that purpose, we focus on the similarities among the manuscripts in the genealogy shown in Fig. 6, and use part of the genealogy as training data for the purpose of adjusting the parameters for performing the genetic manipulations. Concerning similarity, manuscripts in the genealogy that have a parent-child relationship are judged to be highly similar if there is a high proportion of cases in which the same expressions are used for terms for which multiple types of substitution are possible. As a result, the text from page 40 to page 50 of manuscripts P and H was selected as the training data. Using the mutation program after parameter adjustment, the reliability of the parent-child relationships among the manuscripts in the genealogy is investigated. The child individuals obtained by applying the mutation program to a given parent manuscript in the genealogy are examined for similarity to the child manuscripts in the genealogy. Or, conversely, the individuals obtained by applying the mutation program to multiple child manuscripts in the genealogy are examined for similarity to the parent manuscripts in the genealogy. If inconsistencies are found between the results of the computer simulations in which were employed and the genealogy shown in Fig. 6, the inconsistent parts are judged to require further study with respect to the transcription history relationship. Alternatively, the inconsistent parts are judged to have a high probability of having been copied from multiple original documents.

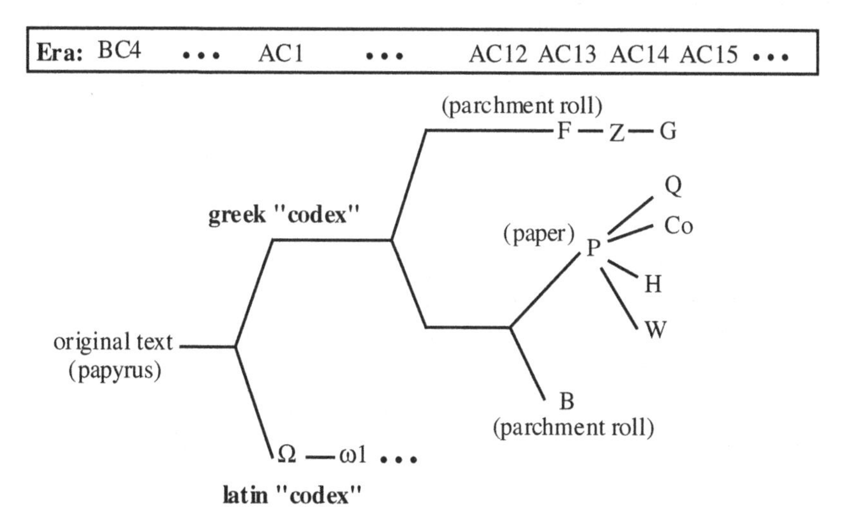

Fig. 6. The genealogy proposed by Cyril Baily. This is one genealogy that is generally referenced in the field of Greek codex.

4.2 Experimental Results and Discussion

An example in which manuscript P was used as input for the mutation program after parameter adjustment is shown in Fig. 7 and an example of an output manuscript is shown in Fig. 8. In Fig. 8, the underlined words are the parts that have been transposed, in relation to the input manuscript. The child manuscript shown in Fig. 8 exhibits many points of similarity to manuscript Q in the genealogy shown in Fig. 6. That is to say, the reliability of the parent-child relationship between manuscripts P and Q in the genealogy proposed by Baily can be considered to be high. On the other hand, no manuscript that was similar to manuscript G was found among the output manuscripts that result when manuscript Z in the genealogy is input to the mutation program. That is to say, a stricter examination of the parent-child relationship between manuscript Z and manuscript G in the Baily genealogy is considered to be necessary. Alternatively, it is necessary to investigate the possibility that manuscript G was transcribed from multiple copies in addition to manuscript Z. Judging from these experimental results, we believe that there is a possibility that evolutionary algorithms can be applied effectively as one means of verifying the currently proposed Greek and Latin text genealogies.

However, in parallel, it is necessary to further study the appropriateness of the training data that was used in these experiments. Additional future experiments based on more data are also needed. We could then give some of the copies to experts in document genealogy, and ask them to draw up their best guess as to the family tree. We could then compare the expert judgments to the results of evolutionary computation analysis, and see which was better in a situation where the real family tree is known. Furthermore, it is necessary to investigate the extent to which the problem of the inability of genealogical methods to cope with text transcription histories that may involve transcription from multiple copies can be dealt with by the addition of a crossover mechanism to the evolutionary algorithm.

,80 και ου δει νομιζειν την υπερ τουτων χρειαν ακριβειαν μη απειληφεναι, οση προς το αταραχον και μακαριον ημων συντεινει. ωστε παραθεωρουντας ποσαχως παρ' ημιν το ομοιον γινεται, αιτιολογητεον υπερ τε των μετεωρων και παντος του αδηλου, καταφρονουντας των ουδε μοναχως εχον η γινομενον γνωριζοντων ουτε το πλεοναχως συμβαινον, την εκ των αποστηματων φαντασιαν παραδιδοντων, ετι τε αγνοουντων και εν ποιοις ουκ αταρακτησαι. αν ουν οιωμεθα και ώδι πως ενδεχομενον αυτο γινεσθαι [και ε]ν ποιοις ομοιως [] αταρακτησαι], αυτο το οτι πλεοναχως γινεται γνωριζοντες, ωσπερ καν οτι ώδι πως γινεται ειδωμεν, αταρακτησομεν.

Fig. 7. An example in which manuscript P was used as input for the mutation program after parameter adjustment.

,80 και ου δει νομιζειν την υπερ τουτων χρειαν ακριβειαν μη απειληφεναι, οση προς το αταραχον και μακαριον ημων συντεινει. ωστε παραθεωρουντας ποσαχως παρ' ημιν το ομοιον γινεται, αιτιολογητεον υπερ τε των μετεωρων και παντος του αδηλου, καταφρονουντας των ουδε μοναχως εχον η γινομενον γνωριζοντων ουτε το πλεοναχως συμβαινον, την εκ των αποστηματων φαντασιαν παραδιδοντων, ετι τε αγνοουντων και εν ποιοις ουκ [εστιν] αταρακτησαι. αν ουν οιωμεθα και ώδι πως ενδεχομενον αυτο γινεσθαι [και ε]ν ποιοις ομοιως αταρακτησαι], αυτο το οτι πλεοναχως γινεται γνωριζοντες, ωσπερ καν οτι ώδι πως γινεται ειδωμεν, αταρακτησομεν.

Fig. 8. An example of an output manuscript. The underlined words are the parts that have been transposed, in relation to the input manuscript.

5 Conclusion

We have proposed the application of ALife methods to the field of historical document genealogy. Specifically, focusing on the similarities between text transcription history and the Tierra system, we attempted to make use of evolutionary algorithms to overcome the limitations of genealogical methods and to increase the reliability of those methods. We conducted computer simulation experiments based on existing manuscript data and demonstrated the feasibility of effectively using evolutionary algorithms as one means of verifying the currently proposed Greek and Latin text genealogies. In this report, as the first step, we deal only with the mutations involved in copying. In future work, further investigation of the extent to which the problem of the inability of genealogical methods to cope with text transcription histories based on transcription from multiple copies can be dealt with by the addition of a crossover mechanism is needed.

References

1. Ray T.: An Approach to the Synthesis of Life. In Langton C.G., Tayler C., Farmer J.D., and Rasmussen S., (eds.): *Artificial Life II*, Addison Wesley (1992)
2. Baek T., Hammel U., and Schwefel H-P.: Evolutionary Computation: Comments on the History and Current State. *IEEE Trans. on Evolutionary Computation*, Vol. 1, No. 1, April (1997)
3. 3. Reynolds L.D. and Wilson N.G.: *Scribes and scholars: A Guide to the transmission of Greek and Latin Literature*. Oxford University Press (1991)
4. 4. http://www.humnet.ucla.edu/humnet/classics/philodemus/RhetI.html
5. 5. Goldberg D.E.: *Genetic algorithms in search, optimization, and machine learning*. Addison Wesley (1989)
6. 6. Bailey C.: *Epicurus, the extant remains: with short critical apparatus, translation and notes*. Clarendon Press, Oxford (1926)

Genetic Algorithm as a Result of Phenomenological Reduction of Natural Evolution

Eva Schwardy

Department for Cybernetics and Artificial Intelligence
Faculty of Electrical Engineering Technology
042 01 Košice, Slovakia
schwardy@poprad.fei.tuke.sk

Abstract. What is the result of the phenomenological reduction of natural evolution (if any)? The problem of the nature of evolution is traditionally neglected by philosophers, being considered for replied well by biologists in the intentions of reductive mechanistic definition of life as a form of proteins motion, or higher level of otherwise non-living matter organization, and not interesting for the computer scientists at all. Philosophy is simply afraid of anything as unpredictable and deliberate as living beings are, biology does not regard (or understand?) philosophical or cybernetic deductions and as for the computer science, it is important, THAT an optimizing algorithm works, not WHY it does, if so.

1 Evolutionary Studies

The computer simulation of evolving populations is important in the study of ecological, adaptive, and evolutionary systems (Taylor et al., 1989 in Collins et al., 1991). The other main approaches are the study of the fossil records, molecular studies, observational and experimental studies, and mathematical analysis. These approaches are inherently limited in ways that computer simulations are not. First, the fossil record is incomplete and difficult to interpret. Second, while molecular studies can determine the underlying genetic similarities and differences of various species, these studies are time-consuming, and the results are often difficult to interpret due to the complexity of the biochemistry of natural life. Third, evolutionary experiments in the laboratory or field are usually limited to at most a few dozen generations because natural organisms (other than microbes) grow and reproduce slowly. In addition, such experiments are difficult to control and repeat, because of the complexity of the interactions between an organism and its environment. Fourth, only the simplest genetic systems can be completely described and understood analytically.

In contrast, simulated evolution makes it possible to study simplified models of nontrivial evolutionary systems over thousands of generations. Evolutionary simulations require many thousands of organisms to avoid small population effects (such as random genetic drift), and in most cases, hundreds or even thousands of generations must be simulated. Although these models are simplified,

J. Kelemen and P. Sosík (Eds.): ECAL 2001, LNAI 2159, pp. 454–457, 2001.

they are much more complex and realistic than those that can be approached with mathematical analysis. By their very nature, computer simulations are easily repeated and varied, with all relevant parameters under the full control of the experimenter. Of course, an inherent weakness of computer simulations is the inability to attain the fill complexity of natural life [1].

1.1 Simulation Approaches

Possible approaches to modeling – simulations of natural systems can be viewed as (1) analytical, (2) searching, (3) creationistic and (4) synthetic.

Analysis. Analytical models provide an effective tool for mathematical treatment of such systems. Their fundamental limitations are as follows:

(a) either they are too "strong", limited by a number of input conditions, which determine their validity for a narrow area of modeled phenomena, or "weak", broad and liberate to such limits, that their results are trivial and telling nothing interesting;
(b) analytical description of system complex, non-linear and synthetic to such a measure natural systems being modeled are seems to be too reductionistic to me;
(c) my personal experience confirmed that however expressive they may be, to the majority of biologists, they remain undisclosed, not understood and thus neglected and of no use.

Searching. Under searching ones, I understand models testing efficiency, fitness or whatever of every possible input parameter combination, in GAS every bit string, in natura every possible DNA sequence. Obviously, we would find the fittest solution, but model like this would lead to a hardware combinatorical explosion not managable in a real time, and again, it is clear that neither natural selection works like this, not every possible combination is tested. As an optimizing technique, such an approach is useless.

"Creationist". In creationist model, such a strong antropic principle is used that modeling is changed to a puzzle solving according to a blueprint, without any feedback from the environment, with intention as conscious as a creator's one may be. If living and evolving systems are modeled, absence of any surprise and invention points that this approach cannot be accepted neither as a model of natural evolution, nor as an optimizing algorithm inspired by nature.

Synthetic. Synthetic models are hopeful either as (a) "nature-like" optimizing techniques exploiting known characteristics of evolution to "evolve" – find optimal solutions or (b) plausible models of evolutionary processes they are based on. I'll discuss both the aspects on GAS [7].

2 Genetic Algorithms

Genetic algorithms are optimizing techniques for non-linear multi-variable problems, where a minute change in any input variable can be far-reaching when shifting the result. They have adapted basic features of evolution, as we know it, such as mutation, evaluating fitness function, selection proportional to the quality of the solving proposed, and recombination as their operators to treat the file of input data arranged in character or bit strings – the "chromosomes" – in a way which resembles the process of natural evolution, not fully understood even by their authors (similar to the evolutionary theories authors). They perform phenomena which can be described as emergent, similar to those that take place in the "real world" evolution but cannot be handled for reasons listed above [3].

If I try to solve problems like these I can be happy with them as optimizers and view them as "black-boxes" where input data are transformed into an optimal solution regardless the way how I reached them.

However, if one suggests that there is an internal similarity between GAS and evolution, a homomorphy (contrary to an external, isomorphic one), it is inevitable to be aware of what one is going to interpret [7].

Any successful models based on evolution as we know it must have more features in common with it than those explicitly defined by a programmer. It is useful to keep in mind that as models differ from their specifications on one side, they differ from our suggestions and expectations on the other side. However doubtful is the statement that they do model natural evolution (based on the selection type – natural vs. artificial – in either of the systems), there is no doubt that their behaviour is unexpected, undisclosed, emergent, i.e. their evolution, even if artificial, is not specified – takes place if not as an inventive and spontaneous one, at least as non-trivial (so to say, as if biological operators did their job even when taken out of nature – evolved).

3 Phenomenology

Phenomenology is a philosophical approach to things as to phenomena, i.e. as they seem to be, without anything that is not inevitable. Let natural evolution try to undergo such an approach. At the first glance, it will resist it, full of unclear developmental and ecological relations between organisms to each others, organisms and their environment... How to neglect them. . 13 Traditionally, this question is avoided by philosophy, eventually considered to be replied well in a scale of various kinds of reductionism and anthropocentrism. Don't ltt us get discouraged. What can be neglected from the evolution without loosing its principles. '7 For sure all the animals and plants, not to mention their abiotic environment.

On the other side, if we start to neglect all that living and non-reliable from the natural evolution, it is not the blank sheet of paper we finish at but just the principles of evolutionary algorithms.

GA thus is a result of eidetic reduction of natural evolution. Reductionistic objection may be legal: on the one side this is mine against many other approaches, on the other side hereby I commit such an extensive one – how comes? – phenomenological reductionism does not damage the most fundamental feature of evolution, it is conservative and "good" – it simplifies the phenomenon but does not reduce it to a hierarchically lower level of description – does not cancel the emergence.

4 Perspectives

I have presented the result of my paper: the "eidos" of evolution being searched for is genetic algorithm. There are plenty of biological data to confirm it.

An algorithm processing is to be evaluated statistically to corroborate (or falsify) the starting point proposed and if confirmed, it offers a great opportunity to:

- build a calculus for evolutionary biology other than mathematical analysis – which is:
  - restricted strongly by multitude of input constraints and
  - too broad to describe particular problems or too narrow to be of broader use – a needed synthetic one;
- show that:
  - there is no need for philosophy to "be afraid" of computers and/or living matter and
  - the question of the phenomenon of evolution can be plausibly solved or at least understood.

References

[1] Collins, Robert J., David R. Jefferson: The Evolution of Sexual Selection and Female Choice, pp. 327–336. Towards the Practice of Autonomous Systems, MIT Press/Bradford Books, 1991.
[2] Hoffineister, Frank, Thomas Btick: Genetic Self-Learning, pp. 227–235. Towards the Practice of Autonomous Systems, MIT Press/Bradford Books, 1991.
[3] Holland, John H.: Genetic Algorithms. Scientific American July 1992, pp. 42–50.
[4] Hughes, Mark: Why Nature Knows Best About Design. The Guardian, 14 September 1989.
[5] Kauffman, Stuart A.: The Origins of Order. Self-Organisation and Selection in Evolution. Oxford University Press, New York, Oxford 1993.
[6] Langton, Christopher G.: Artificial Life, pp. 39–94. The Philosophy of Artificial Life, Oxford University Press, 1996.
[7] Schwardy, Eva: Genetic Algorithms. Survey of Evolutionary-Biological Implementations of Artificial Life Models. Master Thesis. Prague, 1998.
[8] Schwardy, Eva: Phenomenology of Natural Evolution by Genetic Algorithms, pp.300–305. Cognition and Artificial Life. Bratislava, 2001.

String Rewriting Grammar Optimized Using an Evolvability Measure

Hideaki Suzuki

ATR International, Information Sciences Division
2-2-2 Hikaridai Seika-cho Soraku-gun Kyoto 619-0288 Japan
hsuzuki@isd.atr.co.jp

Abstract. As an example of the automatic optimization of an Artificial Life system design, a string rewriting system is studied. The system design is represented by a set of rewriting rules that defines the growth of strings, and a rule set is experimentarily optimized in terms of maximizing the evolvability measure, that is, the occurrence ratio of self-replicating strings. It is shown that the most optimized rule set allows many strings to self-replicate by using a special character able to copy an original string sequentially. In the paper, a man-made rule set is also presented and is compared to the optimized rule set.

1 Introduction

In the design of an Artificial Life (ALife) system, the system's 'evolvability', defined as the possibility (potential ability) of evolving a variety of genotypes or phenotypes, is one of the most important properties [1,2,21,12,10]. The evolvability determines the system's final performance; hence, to design a 'good' ALife system, we have to answer the following three primary questions about evolvability.

- What is an objective measure for evolvability?
- What are the necessary conditions for an ALife system to be highly evolvable?
- How can we optimize an ALife system design in terms of evolvability?

Recently, focusing on the evolvability of natural/artificial proteins, the author began a study addressing the first two questions [16,17,18]: a new evolvability measure ρ, defined as the density of functional genotypes in the protein genotype space, was proposed, and a necessary condition for high evolvability $\rho > \rho_c$ was established, where ρ_c is a critical density above which the functional protein genotypes are strongly connected in the genotype space with unit mutational steps.

The third question, on the other hand, has been previously studied on 2D-CA (two-dimensional cellular automata) by Reggia et al. [9,3,13]. They optimized a transition table in CA and succeeded in bringing about the spontaneous emergence of self-replicating structures in a CA space.

J. Kelemen and P. Sosík (Eds.): ECAL 2001, LNAI 2159, pp. 458–468, 2001.

Following these previous studies, the present paper introduces another example in which a basic ALife design is optimized using the measure of evolvability. The system consists of strings (words) and a set of rewriting rules (production grammar) that defines the growth of the strings. With this rule set, a string is rewritten recursively and replicated if the original string has an appropiate sequence of characters. The evolvability of the system is measured by the ratio of self-replicating strings out of all possible strings, and in terms of the maximization of this measure, a rule set is experimentarily optimized. It is shown that the optimal rule set obtained from the experiments is equipped with a novel strategy that allows a large number of original strings to self-replicate.

The organization of the paper is as follows. After describing the basic model (Section 2) together with the conditions for self-replication (Section 3), Section 4 gives a man-made rule set and a self-replication process under this rule set. In Section 5, the experimental results are shown and several different rule sets are compared. Section 6 offers a discussion.

2 The Model

2.1 Elementary Character Set

We prepare K elementary characters denoted by '`.`', '`0`', '`1`', $\cdots$. Among them, '`.`' is a *quiescent* character, and the others are *nonquiescent* characters. An original string (sequence of characters) on which a production rule set is operated consists only of nonquiescent characters and is written, for example, as `101342`, `51144`, etc. The nonquiescent characters can be compared to biological molecules that react in solvent, and the quiescent characters can be compared to a solvent or water in which chemical reactions take place.

2.2 Production Grammar

The production grammar is a set of R rewriting rules, each of which includes a left side string (with length less than or equal to L) and a right side string that is created by the modification of the left side string. Both strings can include a quiescent character, nonquiescent characters, or *wild-card* characters denoted by '`?`' and '`*`'. `?` is a specific character able to match any single character, and `*` is a specific character able to match an arbitrary sequence of characters. Using these characters, a production rule set with $R = 6$ is written, for example, as

$$r = 1 : \quad \texttt{5} \rightarrow \texttt{05} \tag{1a}$$

$$r = 2 : \quad \texttt{4?} \rightarrow \texttt{4??} \tag{1b}$$

$$r = 3 : \quad \texttt{321} \rightarrow \texttt{32} \tag{1c}$$

$$r = 4 : \quad \texttt{2?} \rightarrow \texttt{20} \tag{1d}$$

$$r = 5 : \quad \texttt{?3} \rightarrow \texttt{3?} \tag{1e}$$

$$r = 6 : \quad \texttt{0?*1} \rightarrow \texttt{0*?1} \tag{1f}$$

Here, Rule (1a) stands for "insert the character 0 before 5", Rule (1b) stands for "double the character after 4", Rule (1c) stands for "delete 1 of 321", Rule (1d) stands for "substitute 0 for the character after 2", Rule (1e) stands for "transpose the character before 3 to after 3", and Rule (1f) stands for "transpose the character after 0 to before the next nearest 1". These rewriting rules define *insertion*, *deletion*, *substitution*, and *transposition* of any single character, and we require that the right side string of every rule should be of a form made by at most three modifications from the left side string. The regulations of a rewriting rule in this paper are summarized as follows.

- A left side string is an arbitrary string with length less than or equal to L.
- A left side string includes at least one nonquiescent character, at most one ?, and at most one *. * cannot be the first or last character of a left side string.
- A right side string is a string of the form created by at most three modifications of the left side string.
- A modification of the left side string is the insertion, deletion, substitution, or transposition of a character, provided that the insertion of * is not allowed and the insertion of ? is allowed only when the left side string includes a ?.

The rewriting of a substring can be compared to a chemical reaction process that changes biological molecules into other molecules. The use of the wild-card characters enables interaction between a pair of distant characters, which can be compared to a reaction between a pair of molecules that meet via chemical diffusion processes. This type of matching is also similar to the matching by 'nop' instructions used in Tierra [11].

As shown in Rules (1a)$\sim$(1f), all rewriting rules of a rule set are lined up in the order of the priority value r. When we apply the rule set, we first check to determine whether the 1st rule (the $r = 1$ rule) is applicable to the current position of a string, and if it is not applicable, we next check to determine whether the 2nd rule is applicable, and so on.

2.3 Production of a String

A rewriting rule set defines the production of a string. See Fig. 1. In this example, the original string is changed into the produced string by Rules (1a)$\sim$(1f). First, focusing on the leftmost substring '21···', a rule that matches the substring and has the largest priority (the smallest r value) is searched. Here, the $r = 4$ rule is selected and the first two character '21' is changed into '20'. Next, focusing on the substring '30···', a matched rule is searched. Because there is no matched rule for this substring, the character '3' is directly copied to the produced string. Then, focusing on the substring '03···', the $r = 6$ rule is selected and the substring '03551' is changed into '05531'. This rewriting procedure is repeated until it reaches the rightmost character.

Note that in this procedure, the rewriting operations themselves are independent and parallel, and yet the division of the string is a sequential one and

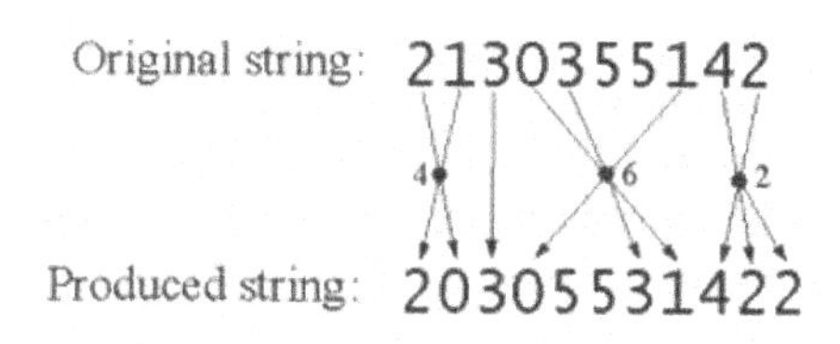

Fig. 1. An example of string production using Rules (1a)~(1f). The numbers next to the black dots represent the r values of the applied rules. An arrow with no black dot represents a direct copy of a character.

the rewriting of the right parts of the string might be affected by that of the left parts. We could avoid this dependency by introducing some specific characters operated preferentially, but here we use the above implementation to assure an even opportunity for nonquiescent characters.

To count the number of rewriting processes, we introduce a time clock t. With each time clock, the entire string is updated using the above procedure, and during the recursive production of several clocks, produced strings are made that are used for the judgment of self-replicability.

3 The Conditions for Self-Replication

Strictly speaking, judging the self-replicability of a string under a production rule set is a difficult task. An appropriate condition for self-replication must preclude 'trivial' self-replicating strings (Fig. 2), and to meet this requirement, here we extend the conditions for self-replicating structures in 2D-CA (two-dimensional cellular automata) given by Reggia et al. [9], and present six conditions for self-replication. A string S is self-replicating if S satisfies:

Condition 1: [compactness] S is a consecutive sequence of nonquiescent characters.

Condition 2: [size] The number of characters in S (denoted by I_o) is sufficiently large. ($I_o = 5$ is taken in this paper.)

Condition 3: [changeability] S changes its character sequence during the self-replication process.

Condition 4: [self-replication] Two or more replicants (copies) of S appear in a produced string after several time clocks.

Condition 5: [isolation] The replicants of S are isolated (delimited by the quiescent characters).

Condition 6: [self-reference] The replicants of S are created by referring to the original characters of S. Or, in other words, the 'reference index' α is large enough (close to 1) for the replicants of S.

The reference index in Cond. 6 is defined as follows. As shown by the arrows in Fig. 1, one rewriting makes a mapping from an original string to a produced string. By inverting this mapping, we can determine the reference relation from a character in the produced string to the character(s) in the original string. For example, in Fig. 1, the first character '2' in the produced string refers to the leftmost '21' in the original string, and the fourth character '0' in the produced string refers to the three characters '0', '3', and '1' in the original string. When we find a replicant of S in a produced string, using this inverse mapping relation, we can identify a subset of characters in S which are referred to by the characters in the replicant. The reference index α of the replicant is then defined as

$$\alpha = \frac{[\text{number of characters in the original } S \text{ referred to by the replicant}]}{[\text{number of characters in the original } S]}. \quad (2)$$

Among the above six conditions, those other than Cond. 4 are for precluding trivial self-replicating strings (patterns). Figure 2 shows a couple of examples of such trivial self-replicators obtained from preliminary experiments. Most trivial self-replicators can replicate themselves so swiftly that if they emerge in an evolutionary experiment of a string rewriting system, more complex and life-like self-replicators will become extinct instantly. To avoid this situation, a production grammar has to be designed so that it suppresses the occurrence of trivial self-replicating strings.

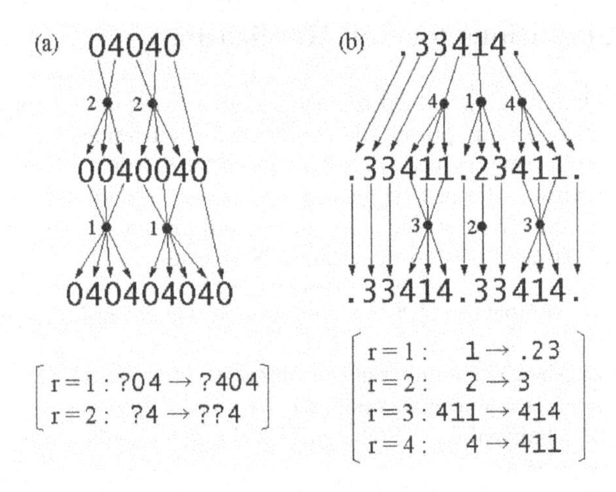

Fig. 2. Examples of trivial self-replicating strings and their rewriting rule sets obtained from preliminary experiments. (a) '0404' satisfies Conds. 1, 3, and 4 but does not satisfy Conds. 2, 5, and 6. The reference indices for the two replicants are $\alpha = 2/4 = 0.5$. (b) '33414' satisfies Conds. 1, 2, 3, 4, and 5 but does not satisfy Cond. 6. The reference indices for the two replicants are $\alpha = 3/5 = 0.6$ and $\alpha = 2/5 = 0.4$. Both of the replicator's replication rates are very high (they are doubled every two clocks).

4 Man-Made Rewriting Rule Set

Using the string production model described above, we can design a rewriting rule set that allows particular strings to self-replicate in a *life-like* way. Figure 3 shows such a man-made design. A key rule of this rule set is the $r = 3$ rule that enables the character 5 to read the next neighboring character and make a copy before the next nearest 4. This operation is repeated until 5 copies a stopping character 3, and after that, the $r = 1$ and $r = 2$ rules copy 5 and 4. With this process, a string of the form '51#34' (where # is an arbitrary substring not including 5, 1, 3, or 4) can self-replicate, so that the occurrence rate of self-replicating strings is expected to be fairly high for this rule.

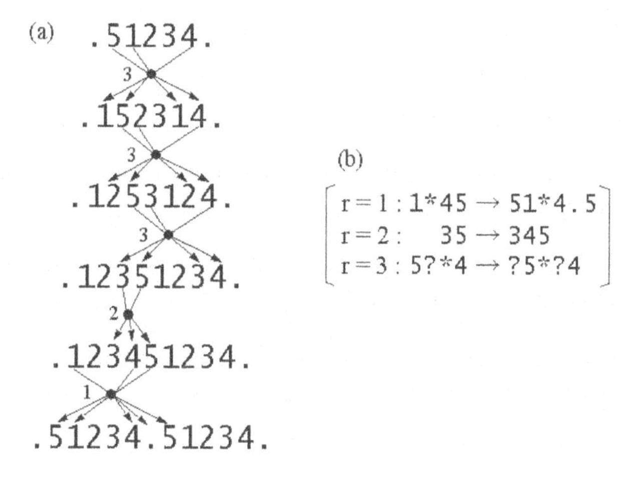

Fig. 3. A string that can self-replicate under a man-made rule set. (a) The self-replication process, and (b) the rewriting rule set. The string satisfies all of Conds. 1 to 6. The reference indices for the left and right replicants are $\alpha = 0.8$ and 1, respectively.

5 Experimental Optimization of the Rule Set

5.1 Optimization of the Production Grammar

Using the conditions for self-replication described in Subsection 3, we can assess the production grammar in terms of evolvability. For a fixed number of the string length I_o, we define the evolvability measure ρ as the number of strings that satisfy Conds. 1 to 6 divided by the number of strings that satisfy Conds. 1 and 2. ρ is evaluated using an exhaustive-search method: we generate all possible strings with length I_o satisfying Conds. 1 and 2 (whose number is $(K-1)^{I_o}$), attach a nonquiescent character '.' on both ends of each string, rewrite it with the production grammar within the limits of the maximum string length I_m and the maximum time clock number T, and judge whether the produced strings satisfy Conds. 3 to 6.

In this paper, the production grammar is optimized using a simple hill-climbing method. We first generate an initial random rule set that satisfies the regulations described in Subsection 2.2. Several modified rule sets are created from this rule set, and if a modification increases ρ, the new rule set is adopted. We take the substitution of a pair of strings or the rearrangement of the priority of a subset of rules as a modification. The procedure is repeated until a particular stopping criterion is satisfied. In a typical numerical experiment, an optimization run is performed several dozens of times using different random number sequences, and the best rule set (the rule set with the largest ρ) is selected. The simulation is conducted on a personal computer with a Pentium III processor (900 MHz).

5.2 Results of Experiments

Table 1 shows the champion rule set obtained from a four-day simulation run. This rule set allows 137 strings to self-replicate for string length $I_o = 5$, which is by far the best result among about 130 trials using different random number sequences. Figure 4 shows a typical example of the self-replication process of these strings. Here, the character '2' plays a key role. By the $r = 0$ rule, 2 makes a pair of copies of the leftmost character in a substring on both sides, which enables a sequential copy of the original string. Interestingly, the original string is completely lost and a pair of copies is created before two 2s during this process.

Table 1. The champion rewriting rule set obtained from experiments

Pri. (r)	Rewriting rule	Pri. (r)	Rewriting rule	Pri. (r)	Rewriting rule
0	?*2 → *?2?	7	3?*4 → ?3*?	14	?5 → ??
1	2 → 05	8	0? → ?0?	15	?0 → 0?
2	3 → 3	9	?*2 → 5*2	16	3 → 4
3	3? → ?3	10	4? → 4	17	1 → 12
4	3? → 3??	11	?3 → 53	18	?1 → 1
5	0 → 0	12	?*1 → .??*1	19	5 → 52
6	4 → 4	13	?4 → 4		

To obtain this set, the following parameter values are used: the number of elementary characters (including the quiescent character) is $K = 7$, the original string length is fixed at $I_o = 5$, the maximum length of the produced string is $I_m = 6 \times I_o = 30$, the number of rewriting rules is $R = 20$, the maximum length of the left side string in a rule is $L = 4$, the maximum time clock number for judgment of self-replication is $T = 0.8 \times {I_o}^2 = 20$, and the maximum number of rules modified at once is $0.3 \times R = 6$.

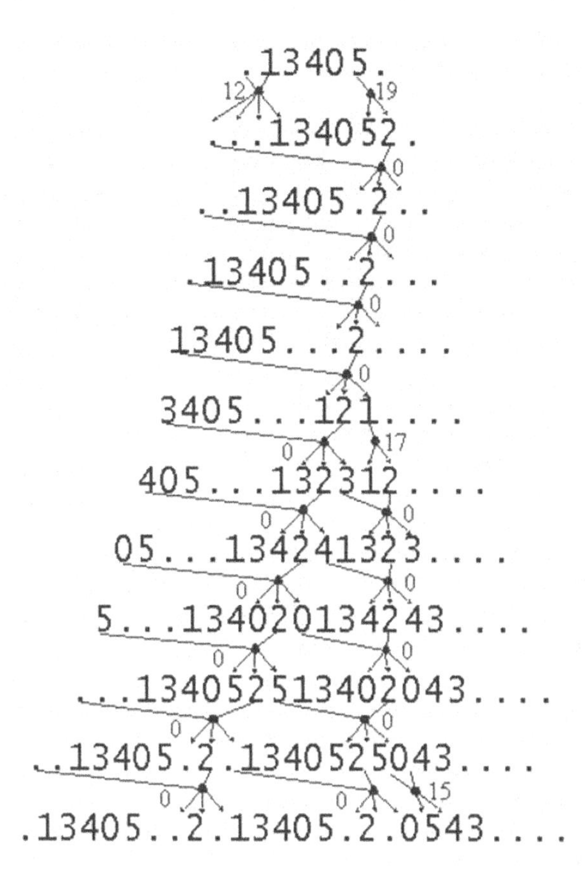

Fig. 4. Self-replication process under the rule set of Table 1. The string satisfies all of Conds. 1 to 6, and the reference index is $\alpha = 1$ for both replicants.

Because this process is completely a self-referential one ($\alpha = 1$ for both replicants), the rule set allows a great number of other strings with various lengths to self-replicate. Table 2 shows this result. According to this table, the rule set for trivial self-replicating strings (Fig. 2(b)) does not have self-replicating strings with other string lengths, whereas both the man-made and the champion rule sets have many self-replicating strings with different lengths. This result suggests that a good rewriting rule set has a large ρ value for strings with various lengths.

6 Discussion

A set of string rewriting rules was optimized so that it might enable as many strings to self-replicate as possible. To preclude trivial self-replicating strings, several conditions for self-reference and self-replication were established, and the

Table 2. NSRSs (number of self-replicating strings) and ρ (evolvability measure) for different values of the original string length l_o

l_o	Trivial rule set (Fig. 2(b))		Man-made rule set (Fig. 3(b))		Champion rule set (Table 1)	
	NSRSs†	ρ	NSRSs	ρ	NSRSs	ρ
4	0	0	1	0.000772	22	0.0170
5	3	0.000386	4	0.000514	137	0.0176
6	0	0	13	0.000279	587	0.0126
7	0	0	40	0.000143	2683	0.00958
8	0	0	120	0.0000714	11341	0.00675
9	0	0	363	0.0000360	45276	0.00449

NSRSs is calculated using all strings with length l_o († includes strings not satisfying Cond. 6), and ρ is calculated from $\rho = \text{NSRSs}/(K-1)^{l_o}$.

obtained rule sets were assessed under this criteria. The champion rule set and a man-made rule set use a particular character that reads and copies an original string sequentially, allowing many strings with different lengths to self-replicate.

We now discuss some implications of the result.

Meaning for the artificial chemistry: String rewriting is one of the most common experimental systems that are used in artificial chemistry [5,6,7,4,20]. In the previous approaches, however, the reaction rules between informational objects (characters, strings, or λ-terms) are defined by the humans or specified by the axioms of mathematical calculation. The system evolvability is determined by these predefined rules and cannot be optimized. The present approach is the first one that optimized the basic reaction rules between the objects.

Application to the core-based programming system: The proposed method for the optimization of the rewriting rules can be used for the design optimization of a core-based programming system [11,14,15]. Among several possibilities, here we focus on the dynamic core used in 'SeMar' [14,15]. Since the string treated in this paper and the SeMar core have the same data structure (both allow insertion/deletion of a data unit), if we obtain a highly functional rewriting rule set from optimization, we can design functions of SeMar proteins with reference to the rewriting rule set. This approach might enable us to design a programming system that can exhibit open-ended evolution.

Evolvability of language: The motivation of the present study is to enhance the evolvability of an Artificial Life system. The evolvability is principally determined by the *language* that defines the correspondence between genotypes and phenotypes [19], and the author believes that the production grammar is one of the most appropriate tools to represent an ALife system design in a general way. The same method of optimization might be applied to the design of terminal functions used in Lisp trees of genetic programming [8] or other computational genetic systems.

Acknowledgements. Mr. Nawa of ATR-ISD labs provided helpful discussions for this study. Dr. K. Shimohara of ATR labs also actively encouraged this research.

References

1. Altenberg, L.: Evolvability checkpoints against evolutionary pathologies. In: Maley, C.C., Boudreau, E. (eds.): Artificial Life 7 Workshop Proceedings (2000) 3–7
2. Bedau, M.A., Packard, N.H.: Measurement of evolutionary activity, teleology, and life. In: Langton, C.G. et al. (eds.): Artificial Life II: Proceedings of an Interdisciplinary Workshop on the Synthesis and Simulation of Living Systems (Santa Fe Institute Studies in the Sciences of Complexity, Vol. 10). Addison-Wesley (1992) 431–461
3. Chou, H.H., Reggia, J.A.: Emergence of self-replicating structures in a cellular automata space. Physica D **110** (1997) 252–276
4. Dittrich, P., Banzhaf, W.: Self-evolution in a constructive binary string system. Artificial Life **4** (1998) 203–220
5. Fontana, W.: Algorithmic chemistry. In: Langton, C.G. et al. (eds.): Artificial Life II: Proceedings of an Interdisciplinary Workshop on the Synthesis and Simulation of Living Systems (Santa Fe Institute Studies in the Sciences of Complexity, Vol. 10), Addison-Wesley (1992) 159–209
6. Fontana, W., Buss, L.W.: 'The Arrival of the Fittest': Toward a Theory of Biological Organization. Bull. Math. Biol. **56** (1994) 1-64
7. Ikegami, T., Hashimoto, T.: Coevolution of machines and tapes. In: Morán, F. et al. (eds.): Advances in Artificial Life (Third European Conference on Artificial Life Proceedings), Springer, Berlin (1995) 234–245
8. Koza, J.R.: Genetic Programming: On the Programming of Computers by Means of Natural Selection. MIT Press, Boston (1992)
9. Lohn, J.D., Reggia, J.A.: Automatic discovery of self-replicating structures in cellular automata. IEEE Transactions on Evolutionary Computation **1** (1997) 165–178
10. Nehaniv, C.L., Rhodes, J.L.: The Evolution and Understanding of Biological Complexity from an Algebraic Perspective. Artificial Life **6** (2000) 45–67
11. Ray, T.S.: An approach to the synthesis of life. In: Langton, C.G. et al. (eds.): Artificial Life II: Proceedings of an Interdisciplinary Workshop on the Synthesis and Simulation of Living Systems (Santa Fe Institute Studies in the Sciences of Complexity, Vol. 10). Addison-Wesley (1992) 371–408
12. Ray, T.S., Xu, C.: Measures of evolvability in tierra. In: Sugisaka, M., Tanaka, H. (eds.): Proceedings of The Fifth International Symposium on Artificial Life and Robotics (AROB 5th '00) Vol. 1 (2000) I-12-I-15
13. Reggia, J.A., Lohn, J.D., Chou, H.H.: Self-replicating structures: evolution, emergence, and computation. Artificial Life **4** (1998) 283–302
14. Suzuki, H.: An Approach to Biological Computation: Unicellular Core-Memory Creatures Evolved Using Genetic Algorithms. Artificial Life **5** N.4 (2000) 367–386
15. Suzuki, H.: Evolution of Self-reproducing Programs in a Core Propelled by Parallel Protein Execution. Artificial Life **6** N.2 (2000) 103–108

16. Suzuki, H.: Minimum Density of Functional Proteins to Make a System Evolvable. In: Sugisaka, M., Tanaka, H. (eds.): Proceedings of The Fifth International Symposium on Artificial Life and Robotics (AROB 5th '00) Vol. 1 (2000) 30-33
17. Suzuki, H.: Evolvability Analysis Using Random Graph Theory. Proceedings of AFSS 2000 (The Fourth Asian Fuzzy Systems Symposium) Vol. 1 (2000) 549-554
18. Suzuki, H.: Evolvability Analysis: Distribution of Hyperblobs in a Variable-Length Protein Genotype Space. In: Bedau, M.A. et al. (eds.): Artificial Life VII: Proceedings of the Seventh International Conference on Artificial Life. MIT Press, Cambridge (2000) 206–215
19. Suzuki, H.: Optimal Design for the Evolution of Composite Mappings. In: Sugisaka, M., Tanaka, H. (eds.): Proceedings of The Sixth International Symposium on Artificial Life and Robotics (AROB 6th '01), Vol. 2 (2001) 373-376
20. Suzuki, Y., Tanaka, H.: Chemical evolution among artificial proto-cells. In: Bedau, M.A. et al. (eds.): Artificial Life VII: Proceedings of the Seventh International Conference on Artificial Life, MIT Press, Cambridge (2000) 54–63
21. Taylor, T.: On Self-reproduction and Evolvability. In: Floreano, D. et al. (eds.): Advances in Artificial Life (5th European Conference on Artificial Life Proceedings) Springer-Verlag, Berlin (1999) 94-103

A Visually-Based Evolvable Control Architecture for Agents in Interactive Entertainment Applications

Andrew Vardy

Computer Science Deptartment
Carleton University, Ottawa, Canada
avardy@scs.carleton.ca

Abstract. A visually-based evolvable control architecture for agents in interactive entertainment applications is presented. Agents process images of their local surroundings according to evolved image processing operators. A behaviour-based framework of *action rules* encapsulates image processing parameters and action parameters into discrete behavioural modules. These modules are interconnected and retain internal state through the dynamics of these internal connections. This novel control architecture has a wide behavioural range and is specified in an evolvable framework which allows agents for entertainment applications to be evolved as opposed to explicitly designed. The results of several demonstrations and experiments are presented to showcase the possibilities of this control architecture.

1 Introduction

We present an evolvable control architecture for computer-controlled agents in interactive entertainment applications. The proposed control architecture is intended to support a wide range of behaviour of the type often sought after by computer game designers and animators. The focus is on behaviour that is more reactive than deliberative. 'Pursuit', 'evasion', 'obstacle-avoidance', and 'wandering', are examples. The control architecture presented here has the capacity to implement these kinds of behaviours individually, in combination, and in sequence and it allows these multitudinous permutations of steering behaviours to evolve using a genetic algorithm (GA). The task to which we set an individual agent involves playing a game of survival against a hostile opponent agent which represents the human player in a video game. Artificial evolution is envisaged to take the place of explicit programming work via the *off-line* (i.e. pre-release) production of sophisticated behavioural controllers.

Current controllers apply a mix of rigid sequential behaviours along with randomized movement patterns to produce their character's behaviour [15,16]. These mechanisms are difficult to program, inflexible, and non-adaptive. They rely heavily on the human tendency to anthropomorphise [16]. In order to provide a viable alternative to current game AI techniques our system must support

J. Kelemen and P. Sosík (Eds.): ECAL 2001, LNAI 2159, pp. 469–479, 2001.

controllers that are challenging and interesting to human players. The only information available to an agent in our system is a ray-traced image of its immediate surroundings. It cannot see behind walls or consult global data structures and is therefore more intuitively understandable. Other intended attributes of this system are comprehensibility of evolved controllers and low computational burden. The control architecture described here is tailored towards games where the principal relationships existing between agents are spatial—as opposed to games where various internal and external resources need to be managed. Commercial games such as Quake, Tomb Raider, Baldur's Gate and numerous others are prime candidates for the kind of agent controllers described here. Note that most ostensibly three-dimensional games are actually fundamentally two-dimensional in nature.

The next section of this report provides an overview of the core concepts of the control architecture followed by a discussion of related work. We then go on to describe a number of refinements on the core architecture before presenting experimental results. A concluding section provides commentary on the results as well as future directions. Note that space limitations necessitate a highly selective and condensed style of coverage. Please see [14] for a more complete discussion.

1.1 System Overview

Agents occupying a cell in a two-dimensional grid apply evolved image processing operations on one-dimensional images of their local surroundings. Parameters for image processing are specific to each action rule. These parameters include the *color translation table* and the *convolution mask*. The action rule also includes parameters for action rule interconnection and a parameter that indicates what action the agent should take when it is active. The direction of action is determined by the image processed by the winning action rule. The pixel in this image with the highest response gives the direction. An agents entire genetic code is comprised by concatenating the codes for its constituent action rules. The code is comprised of integers, each in the range [-3, 3]. For the experiments below all agents have four action rules. See figure 1 for illustration of each of the following processes:

Image Generation. 1-D *retinal images* are generated by tracing rays out from the agent to all points on the circumference of a rasterized octagon. The agent's field of vision is omni-directional and range-limited and exhibits such phenomena as occlusion and perspective.

Color Translation. The values of various grid objects in the retinal image are used as an index into the color translation table. Application of this lookup table yields the pre-processed or translated image.

Convolution. The convolution mask for the active action rule is applied to the pre-processed image. It is an array of values swept along the image and multiplied with image elements at corresponding positions, with the sums taken as the processed image [1]. Circular convolution is applied so that the original and processed images maintain the same length. The mask is also applied in reverse, generating an additional processed image (the image with the lower maximum is discarded). This allows masks to be orientation independent.

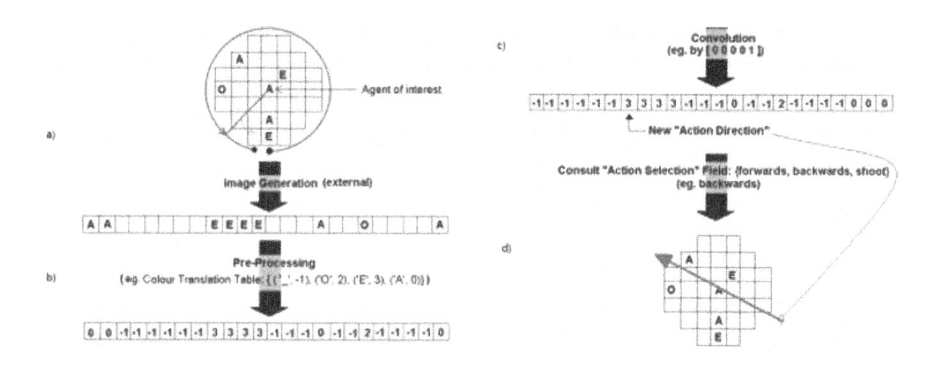

Fig. 1. Agent processing: (a) Image generation (obstacles=O, enemies=E, agents=A). (b) Colour translation. (c) Convolution. (d) Action direction selection and action selection.

Action Rule Selection. The current retinal image is processed by all action rules to determine which rule should become active. The maximum value from each of these processed images must exceed a threshold value for the corresponding action rule to be further considered. Each action rule has an activation level which decays to zero at an evolvable rate. The currently active action rule has a weight field which is added to another action rule (possibly itself) selected according to a displacement field. An importance field is added to the activation level to yield the overall strength. The rule with the highest strength is then selected as active. An action role also contains an action selection field indicating one of {accelerate, decelerate, null, launch projectile}. A refractory period exists for firing projectiles such that a 'shooting' action rule is prevented from becoming active for a certain fixed period after the agent has launched a projectile.

Action Direction Selection. Retinal images are generated such that their left-hand edge corresponds to the direction immediately beneath the agent. Therefore the highest valued pixel in the processed image corresponds uniquely to a particular action direction. The highest valued pixel location closest to the last action direction is chosen as the new action direction.

1.2 Related Work

Applications such as *Creatures* [9] allow the user to educate and assist virtual animals in their growth and exploration of a virtual world. See [9] for a review of commercial artificial life related entertainment applications. Funes *et al.* describe a system whereby Internet users play an on-line game of 'Tron' against a community of automatic players adapting via genetic programming [8]. Reynolds [13] designs autonomous agents explicitly for behavioural animation and games. He presents techniques designed to navigate agents through their worlds in a "life-like and improvisational manner" using *steering behaviours* such as seeking, evasion, obstacle avoidance, and various others. This contrasts with our approach which is to find an evolvable control architecture within which all of these types of behaviour can find expression.

A study on visual processing in the fly [6] describes a hypothesized early layer in fly visual system where the entire visual field is processed by local operators known as Elementary Motion Detectors (EMD's), similar in spirit to the convolution masks used here. Also, a full-length convolution mask is essentially an image matcher, similar in concept to the idea of retinotopic images hypothesized for landmark identification and homing in insects [5]. This system adheres to a behavioral decomposition of behaviour—specifically, one whose behavioural modules are evolved [7] [11], rather than explicitly engineered [2].

The use of artificial evolution for image processing has been attempted in various forms by a number of researchers. Poli employed genetic programming to develop operators for image segmentation and feature detection using medical imagery as an example [12]. Of particular interest is work by Harris and Buxton who use genetic programming to evolve linear filters for edge detection [10]. The end product of their system is a symbolic function which is sampled to produce a convolution mask. Here we evolve the masks directly.

1.3 Refinements

Range Thresholds. Early experiments revealed that the lack of range information was a severe impairment. A table of *range thresholds* was added as an image preprocessing stage. The distance of objects from the agent is thresholded according to an entry in this table corresponding to the object type. If the object's distance is greater than threshold then the pixel is translated as 'empty'. Figure 2 shows a trail of an agent run by only a single active 'wandering' action rule. The wander behaviour is simply achieved with a colour translation table that gives a high weight to 'empty', has a somewhat long and flat convolution mask (i.e. $\{11111111111111111111\}$), and a forwards action selection field. Wandering behaviour with and without a range threshold on 'obstacles' is shown. Clearly, the intended behaviour is better achieved with a range threshold.

Differential Processing. The ability to predict the future direction of a moving target is provided by *differential processing.* If an agent's genotype specifies that this feature is active, then the current retinal image is pre-processed via

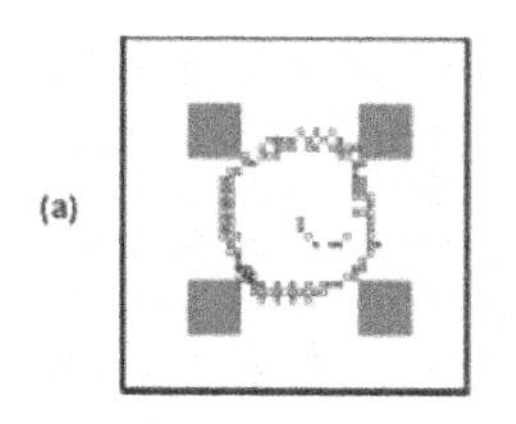

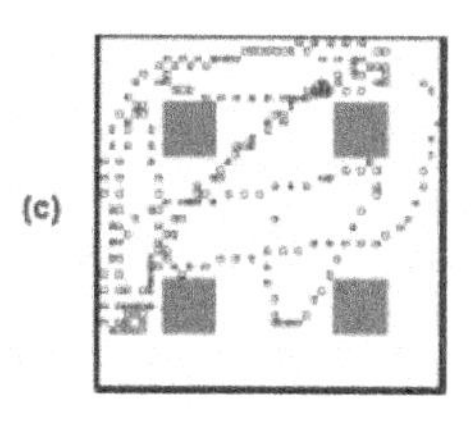

Fig. 2. Two 'wander' behaviours: (a) Without range thresholds; (b) Range thresholding obstacles at distance of 5. Grey cells and borders are 'obstacles'. Unfilled squares indicate past positions of agent. Final position indicated. Duration = 250 iterations.

range thresholding and colour translation as usual. The previous retinal image is then subjected to the same pre-processing operations and is subtracted from the current pre-processed retinal image. Figure 3 illustrates simple pursuit and predictive pursuit (with and without differential processing, respectively). A single forwards action rule implements pursuit. For simple pursuit the rule's convolution mask is simply '1'. For predictive pursuit the mask used is the 48-long string {-1, -1, -1, -1, -1, -1, 1, 1, 1, 1, 1, 1, 1, 0, 0, 0, 0, 0, 0, 0, 0, 0, 0, 0, 0, 0, 0, 0, 0, 0, 0, 0, 0, 0, 0, 0, 0, 0, 0, 0, 0, 0, 0, 0, 0, 0, 0, 0}. Retinal image length is 224. The -1's match the previous position and the 1's match the current position while the 0's pad out the mask so that the best match occurs in the center—accelerating the agent towards the target's approximate future position. Differential processing supports other features such as predictive firing and movement detection.

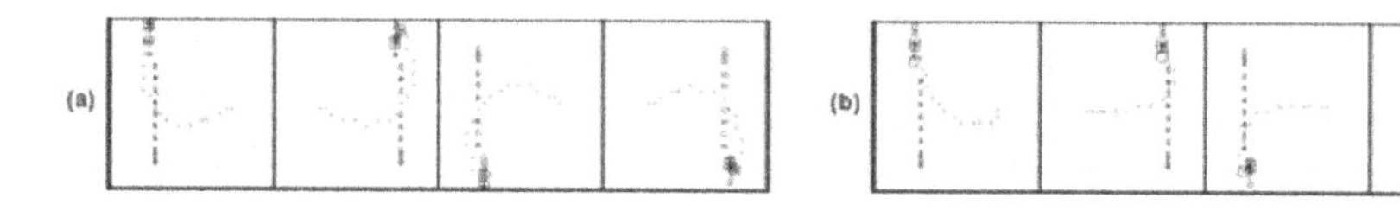

Fig. 3. Two 'pursuit' behaviours: (a) Simple pursuit without differential processing; (b) Predictive pursuit with differential processing. Unfilled diamonds indicate the path of the pursuer. Unfilled squares indicate the path of the evader. The evader's behaviour here is blind and unchanging.

2 Experiments

A collection of agents, known as the *animals* is placed in a bounded arena together with a competing agent known as the *avatar* (representing a human player). Launched projectiles are damaging only to the opposing species. There are five types of objects in the world: obstacles, animals, animal missles, avatars, avatar missles. All objects occupy a full grid square. Properties of the physics model are: Velocity proportional to force; Viscosity; Random perfectly elastic collisions. The phase of agent 'thinking cycles' is randomized so that only half of the population will be thinking on any given iteration.

For evolutionary experiments the individual being tested is duplicated (six times for animals, once for the avatar) into the arena and evaluated via competition with the adversary species. For co-evolutionary experiments the avatar co-evolves with the animals. Each is evaluated against the best individual from the previous generation, as in [4]. For single-species evolution the evolving animals are pitted against a hand-coded avatar. The fitness function rewards damage done to the opposing species, as well as high survival time for the losing species and low elimination time for the winning species (damage factor outweighs time factor) [14]. Fitness is averaged from three evaluations with half of the standard deviation subtracted to obtain the final score for an individual. A steady-state tournament selection style GA is applied with a tournament size of three for both reproduction and removal. Two-point crossover is used to create a single new population member which replaces an existing member. All population members are subjected to creep and jump mutation. Figure 4(a) illustrates the evaluation environment. The avatar begins its life in a randomized location near the center while the animals begin in randomized locations around the perimeter.

Visualization. A particular environmental configuration known as *the course* has been designed to present an agent with a rich variety of stimuli, but in a standardized and regulated fashion. A *movie* was recorded of the retinal images presented to a dummy agent as it was moved throughout the course. Figure 4(b) illustrates the course and the path taken by the dummy recording agent as it passed through it. The recorded movie is presented in figure 4(c) as an array of vertical one-dimensional images flowing from left to right through time.

2.1 Results

Currently, the only meaningful way of investigating the results of an evolutionary run with this system is qualitative description with the aid of the visualisation tool just presented. Some trials with human players were attempted but no conclusive results could be drawn—likely because of small sample size [14]. Therefore, there is only sufficient space to describe a single run. Figure 5(a) shows the fitness profile for evolved animals and avatars for the coevolutionary run, KA24. These figures reveal the ambiguity of measuring progress in coevolutionary simulations. It appears that some form of competitive coevolution is occurring, but this is not certain. Figures 5(b) and 5(c) shows the decoded action rules for the best members from the final generation (621) of animals and avatars, as well as the activity of those rules when presented with the movie. Figure 6 shows the sequence of processed images generated by the same animal and avatar when presented with the movie. Overlaid on the processed images are symbols to indicate the type of action rule being applied.

The animal depicted in Figures 5(b) and 6(c) exhibits no interesting dynamic between its two action rules. It effectively has only one rule because the first is

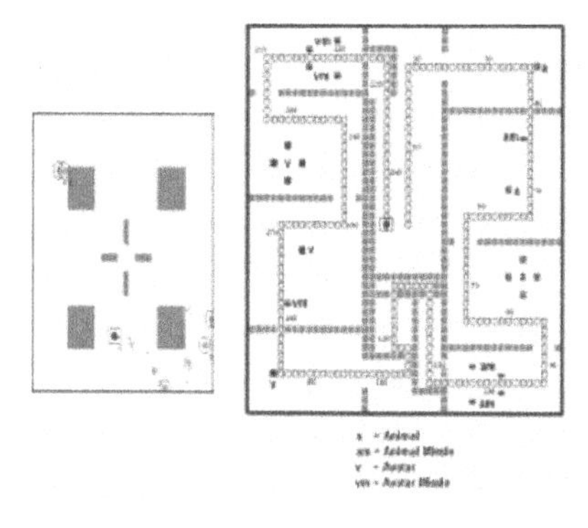

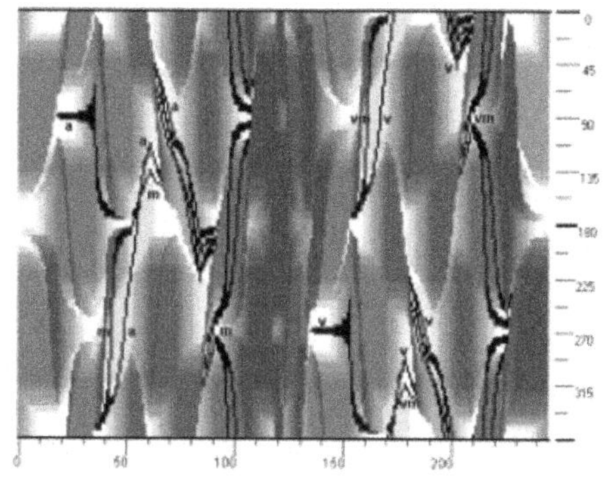

(a) Evaluation of animals (circles) vs. avatar (square).

(b) *The course* overlaid with the dummy recording agent's path through it. The trail is annotated with the frame number.

(c) Movie recorded by dummy agent. Frame number on horizontal axis. Vertical axis gives angular position in degrees where 0° represents the direction directly beneath the agent. Black shows indicated type. Grey shows obstacles with grayscale proportional to distance from agent.

Fig. 4.

a 'null' rule which does not even influence the other. Figure 6(a) shows that the animal's shooting rule aims to one side of the avatar. Strangely, its settings for colour translation are weighted more towards avatar missiles. By observing the animal in the evaluation environment it was noticed that the animal would actually fire in a predictive fashion. Differential processing enables the animal to fire at incoming avatars and avatar missiles at a point just ahead of and in the path of the target's current position. This did not hold for all conceivable orientations and velocities but predictive firing was generally evident.

The avatar depicted in figures 5(c) and 6(b) exhibits interesting and complex behaviour. The first action rule has a zero length convolution mask. What this rule does is continue to move the avatar in the direction selected by the last rule. Presumably the advantage of this rule is, in the absence of other sensory input, to keep going in the current direction. In a closed arena this is a simple but effective strategy for covering a lot of space. The second and third action rules are entrained in a 'see-saw' dynamic. Both rules have self-inhibitory connections such that whenever either becomes active it allows the other to take over. Both respond to similar stimuli except that the third rule has a higher response to animals and avatars (animals and avatars present a strong stimulus around frames 50 and 150—see 4(c)). It is believed that the third rule's use of differential processing allows movement detection. The fourth rule, a shooter, has a fairly narrow convolution mask with a high center weight. It translates animals to the high weight of 2, as expected, but also translates avatar missles to the highest weight of 3. This is believed to be a way of externally representing

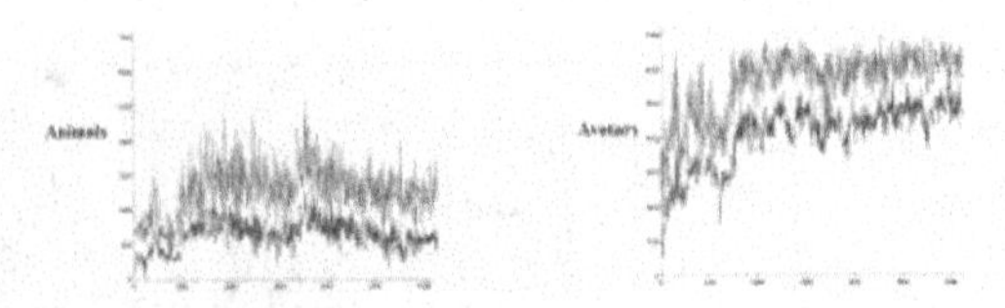

(a) Fitness profile for run KA24. Top trace is peak, bottom is mean.

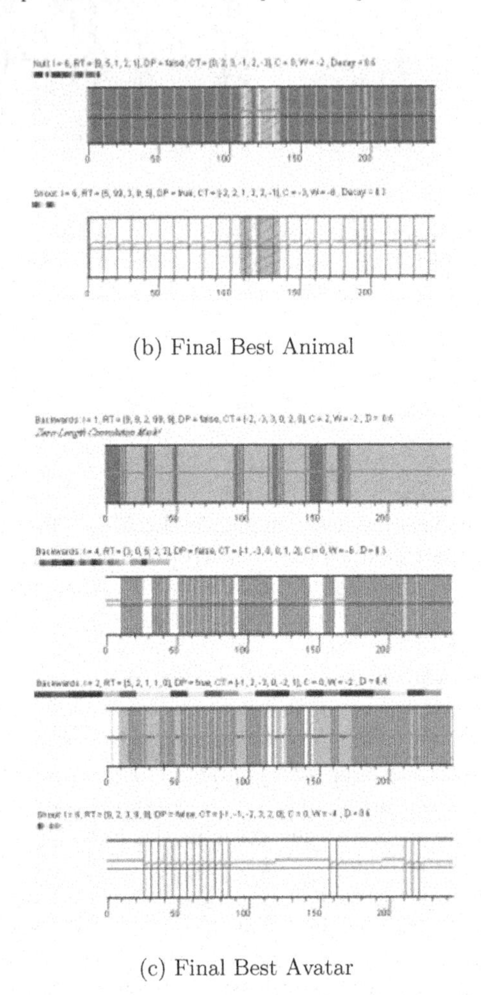

(b) Final Best Animal

(c) Final Best Avatar

Retinal Image Length (Maximum Convolution Mask Length)

Fig. 5. Each row in (b) and (c) shows: The interpreted action rule; An image of the action rule's convolution mask, with lighter pixels corresponding to higher-valued entrys; A graph of the action rule's activity over the course of the movie. I = importance; RT = range thresholds; DP = differential processing; CT = colour translation; C = connection; W = weight; D = decay. RT array: {obstacle, avatar, avatar missile, animal, animal missile}. CT array: {empty, obstacle, avatar, avatar missle, animal, animal missle}. Rule activity (activation level + importance) is plotted as a light line. Dark grey bands indicate where the rule was active; Light grey, where it could be active; Hatched, where no action rule is active. Horizontal axis corresponds to movie frame number.

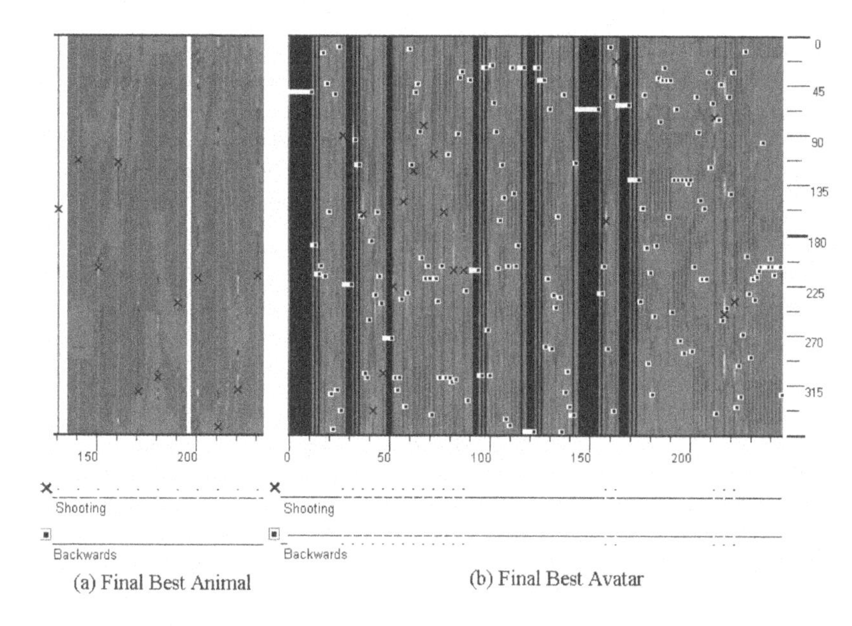

Fig. 6. Processed images of evolved agents presented with the movie. Higher grayscale values correspond to higher image value. Symbols are overlaid upon the processed images to indicate applied action: Black squares outline in white indicates backwards movement; 'X' indicates shooting. See Figure 4(c) for labelling of axes. Note that only a segment of the response for the animal is shown.

state—if the avatar sees its own missles it will fire at them 'thinking' that it must have had something good to fire at in the first place. Matching the processed images in figure 6(b) with figure 4(c), as well as examining corresponding positions in figure5(c) shows that this avatar indeed tends to shoot directly at animals and avatar missles.

Note that all of the various runs conducted exhibited similar interesting examples of behaviour as described above. Also, the single-species runs could be said to be successful in that they showed steadily increasing fitness profiles.

2.2 Future Work

It will be necessary to devise and implement new means to analyze evolved agents. Some form of 'obstacle course' would allow isolated competences of agents to be evaluated objectively. However, even this may not reveal all of the tools and tricks that allow an evolved agent to succeed against its adversaries. An ethological approach of observing the agent in its evaluation environment can reveal more but is more difficult to carry out objectively. Ignoring the issue of mechanisms we could instead focus on raw performance by evaluating the agents in competition with humans and with controllers coded using other methods, such as neural networks and state machines. Again in terms of analysis, more investigation into the specific properties of evolved convolution masks is required (fourier domain analysis). Also, we require additional means to understand the coevolutionary process. Many such methods are described in [3]. Finally, the action rule arbitration scheme can be seen conceptually as being decoupled from the evolved image processing aspects of the system. More investigation into alternate action selection schemes is warranted.

3 Conclusions

This report has detailed the design, motivations, and experimental results for a system that can be used to evolve agents for use in interactive entertainment applications. We have presented a novel visually-based control architecture that was specifically designed for its target application. As well as extending and analysing the current system we hope to explore using some of the same concepts for use in mobile robot navigation. Also, although this has been an engineering project it would be interesting to see how the system presented here could be adapted to serve as a model of visual processing in simple animals. While further investigation is certainly warranted, this work has uncovered an attractive new alternative control mechanism for agents in interactive entertainment.

Acknowledgments. This work has been partially supported by NSERC Postgraduate Scholarship B - 232621 - 2000.

References

1. Bassmann, H., Besslich, P., "Ad Oculos", Thompson, 1995.
2. Brooks, R.A. "Intelligence Without Reason". In L. Steels and R. Brooks (Eds.) *The Artificial Life Route to Artificial Intelligence*, Erlbaum, pp. 25-81, 1995.
3. Cliff, D., Miller, G.F., "Tracking the Red Queen", In F. Foran *et. al.* (Eds.) *Third European Conference on Artificial Life*, Springer, pp. 200-218, 1995.
4. Cliff, D., Miller, G.F., "Co-evolution of Pursuit and Evasion II", In P. Maes *et. al.* (Eds.) *SAB4*, MIT Press, Cambridge MA, 1996.
5. Dill, M., Wolf, R., Heisenberg, M., "Visual Pattern Recognition in Drosphilia involves retinotopic matching", Nature Vol. 365, p. 751-753, 1993.
6. Egelhaaf, M., Borst, A. "Motion Computation and Visual Orientation in Flies", Comp. Biochem. Physiol, Vol. 104A, No. 4, p. 659-673, 1993.
7. Floreano, D. "Evolutionary Robotics in Artificial Life and Behavior Engineering", In T. Gomi (Ed.), *Evolutionary Robotics*, Ontario (Canada): AAI Books, 1998.
8. Funes, P., Sklar, E., Juilli, H., Pollack, J. "Animal-Animat Coevolution". In R. Pfeifer *et. al.* (Eds.) *SAB5*. MIT Press. p. 525-533, 1998.
9. Grand, S., Cliff, D. Malhotra, A. "Creatures: Artificial Life Autonomous Software Agents for Home Entertainment", CSRP 434, University of Sussex, 1996.
10. Harris, C., Buxton, B., "Evolving Edge Detectors", Research Note RN/96/3, UCL, Gower Street, London, WC1E 6BT, UK, Jan. 1996.

11. Harvey, I., Husbands, P., Cliff, D., Thompson, A., Jakobi, N. "Evolutionary Robotics", In *Robotics and Autonomous Systems*, v.20, p. 205-224, 1997.
12. Poli, R., "Genetic Programming for Feature Detection and Image Segmentation", In T. Fogarty (Ed.) *AISB'96 Workshop on EC*, Brighton UK, pp. 110-125, 1996.
13. Reynolds, C.W., "Steering Behaviors For Autonomous Characters", In *Proceedings of Game Developers Conference 1999*, Miller Freeman, San Francisco CA, 1999.
14. Vardy, A., "A Visually-Based Evolvable Control Architecture for Agents in Interactive Entertainment Applications", M.Sc. Dissertation, University of Sussex, 2000. (http://www.scs.carleton.ca/~avardy/vardy_msc_diss.pdf)
15. Lamothe, A., "Windows Game Programming for Dummies", IDG Books, 1998.
16. Woodcock, S., "Game AI: The State of the Industry", `http://www.gamasutra.com/features/19990820/`.

Symbiotic Composition and Evolvability

Richard A. Watson and Jordan B. Pollack

Dynamical and Evolutionary Machine Organization
Volen Center for Complex Systems – Brandeis University – Waltham, MA – USA
richardw@cs.brandeis.edu

Abstract. Several of the Major Transitions in natural evolution, such as the symbiogenic origin of eukaryotes from prokaryotes, share the feature that existing entities became the components of composite entities at a higher level of organisation. This composition of pre-adapted extant entities into a new whole is a fundamentally different source of variation from the gradual accumulation of small random variations, and it has some interesting consequences for issues of evolvability. In this paper we present a very abstract model of 'symbiotic composition' to explore its possible impact on evolvability. A particular adaptive landscape is used to exemplify a class where symbiotic composition has an adaptive advantage with respect to evolution under mutation and sexual recombination. Whilst innovation using conventional evolutionary algorithms becomes increasingly more difficult as evolution continues in this problem, innovation via symbiotic composition continues through successive hierarchical levels unimpeded.

1 Introduction

The Major Transitions in evolution [1,2,3] involve the creation of new higher-level complexes of simpler entities. Summarised by Michod for example [3], they include the transitions "from individual genes to networks of genes, from gene networks to bacteria-like cells, from bacteria-like cells to eukaryotic cells with organelles, from cells to multicellular organisms, and from solitary organisms to societies". There are many good reasons to be interested in the evolutionary transitions: they challenge the Modern Synthesis preoccupation with the individual as the unit of selection, they involve the adoption of new modes of transmitting information, they address fundamental questions about individuality, cooperation, fitness, and not least, the origins of life [1,2,3].

In several of the transitions "entities that were capable of independent replication before the transition can replicate only as part of a larger whole after it" [2]. Although Maynard Smith and Szathmary identify some transitions which do not fit what they describe as "symbiosis followed by compartmentation and synchronised replication", several of the transitions including the origin of eukaryotes from prokaryotes [4], and the origin of chromosomes from independent genes, do involve the quite literal union of previously free-living entities into a new whole. This paper focuses on the evolutionary impact of this mechanism, which we shall refer to as 'symbiotic composition', or simply 'composition'. We are concerned with an *algorithmic* understanding of this mechanism the Major Transitions: What kind of adaptation does

J. Kelemen and P. Sosík (Eds.): ECAL 2001, LNAI 2159, pp. 480-490, 2001.

the formation of higher-level complexes from simpler entities afford? And we seek to understand the class of adaptive problem, the kind of fitness landscape, for which this mechanism is well suited.

Composition presents some obvious contrasts with how we normally understand the mechanisms of neo-Darwinist evolution. The ordinary (non-transitional) view of evolutionary change involves the accumulation of random variations in genetic material within an entity. But innovation by composition involves the union of different entities, each containing relatively large amounts of genetic material, that are independently pre-adapted as entities in their own right, if not in their symbiotic role. This immediately suggests some concepts impacting evolvability.

First, a composition mechanism may potentially allow 'jumps' in feature space that may cross 'fitness saddles' [5] in the original adaptive landscape (defined by the mutation neighbourhood). Moreover, since these higher-level aggregations of features are not arbitrary but rather are shaped by prior adaptation, these jumps are not equivalent to large random mutations, but rather are 'informed' by prior or parallel adaptation.

Crossing fitness saddles has been a central issue in evolvability. There are many possible scenarios for how adaptation may overcome a fitness saddle: for example, genetic drift and 'Shifting Balance Theory' [5,6], exaptation [7], neutral networks [8], extra-dimensional bypass [9], ontogenic processes [10], or landscape dynamics [11]. Each of these affords an increase in the width of fitness saddle that might be crossed (with respect to that which may be crossed by simple mutation). And conceivably, some of them may produce saddle-crossing ability that is not arbitrary, but informed by prior or parallel adaptation. But, what is the size of fitness saddle that we should expect to encounter? It seems likely that as one scale of ruggedness is overcome, a larger scale of ruggedness becomes the limiting characteristic of a landscape. Under composition, the entities resulting from one level of organisation provide a new 'unit of variation' for compositions at the next level, and thus the size of jumps is proportional to extant complexity. In this sense, composition suggests a scale-invariant adaptive mechanism.

Second, in composition, the sets of features that are composed are pre-adapted in independent entities. The components of the union arise from entities at a lower level 'unit of selection'. This independence provides a 'divide and conquer' treatment of the feature set. Intuitively, the hope is that a generalist entity, utilising two different niches, resources, or habitats, for example, can be created by the composition of two specialist entities each independently adapted to one of these niches, resources or habitats. Thereby, the problem of being well adapted to the general habitat is divided into the independent problems of being well adapted to component habitats. This decomposition of a problem into smaller problems is know algorithmically as 'divide and conquer' (e.g. [12]); so named because of the significant algorithmic advantage it offers when applicable. Such divide and conquer advantage is not available to natural selection when features are adapted within a single reproductive entity.

The model we describe below develops these two concepts – a scalable mechanism of saddle-crossing, and divide and conquer advantage – both applied in a scale-invariant hierarchical fashion. We do not attempt to model biological mechanisms in any detailed way; our model is deliberately very abstract. For example, we assume a mechanism of symbiotic composition that simply produces the union of features from two organisms. And, the fitness landscape that we use for our experiments is deliberately chosen to exemplify the adaptive utility of composition as contrasted with conventional evolutionary algorithms. However, by using an abstract model we can focus on the combinatorial aspects of the processes, and an algorithmic model such as this provides an important facet to our understanding of the Major Evolutionary Transitions and the adaptational significance of symbiotic composition.

The remainder of the paper is structured as follows: Section 2 describes a scale-invariant fitness landscape; Section 3 describes our composition model, the Symbiogenic Evolutionary Adaptation Model (SEAM); Section 4 describes the results of some experiments with SEAM and this scale-invariant fitness landscape; Section 5 concludes.

2 A Scale-Invariant Fitness Landscape

In this section we examine a fitness landscape that we will use to exemplify the characteristics of the composition model we describe later. Of interest to us here is that this landscape has saddles at all scales, resulting from its hierarchical construction [13].

Sewell Wright [5] stated that "the central problem of evolution... is that of a trial and error mechanism by which the locus of a population may be carried across a saddle from one peak to another and perhaps higher one". This conception of evolutionary difficulty, and the concept of evolution as a combinatoric optimisation process on a rugged landscape [14], provides the now familiar model at the heart of the issues addressing evolvability. Ruggedness in a fitness landscape is introduced by the *frustration* of adaptive features, or *epistasis* when referring to the interdependency of genes – that is, it occurs when the 'selective value' of one feature is dependent on the configuration of other features. Fitness saddles are created between local optima. The simplest illustration is provided by a model of two features, F_1 and F_2, each with two possible discrete values, a and b, creating four possible configurations: F_1a/F_2a, F_1a/F_2b, F_1b/F_2a, F_1b/F_2b. Table 1, below, gives four exemplary cases for selective values, or fitnesses, for these four combinations. The overlayed arrows in each case show possible paths of adaptation that improve in fitness by changing one feature at a time.

Table 1. Example selective values for combinations of two features.

Case 1
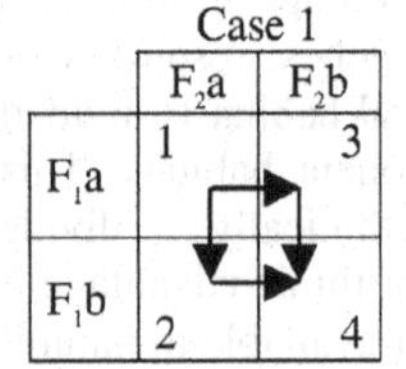

Case 2
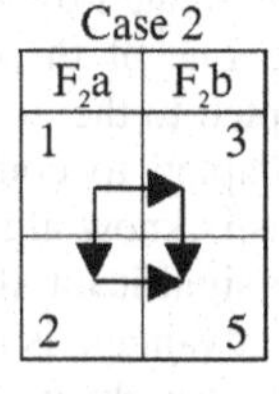

Case 3
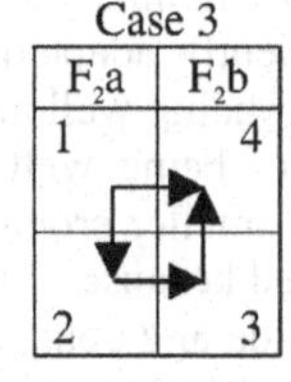

Case 4
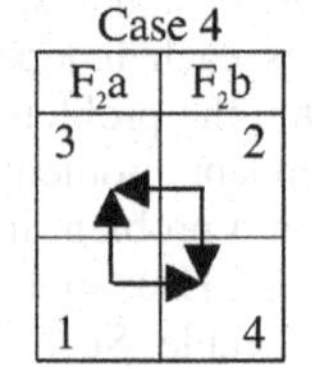

Case 4b
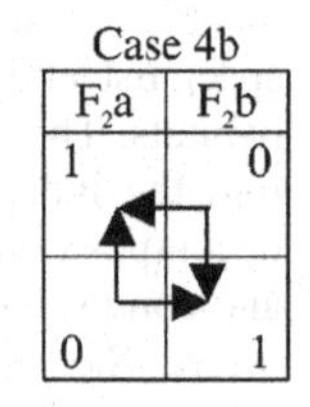

Case 1 shows no epistasis: the difference in selective value between F_1a and F_1b is the same regardless of the value of F_2; and the difference in selective value between F_2a and F_2b is the same regardless of F_1. Cases 2, 3 and 4 each show some epistasis but with different effects. In Cases 2 and 3, the landscape is not planar, and the possible routes of single-feature variation are different in Case 3, but the landscapes still only have one optimum. Only in Case 4, where preference in selective value between F_1a and F_1b is reversed depending on the value of F_2, *and* the preference in selective value between F_2a and F_2b is reversed depending on F_1, does epistasis create two optima and a resultant fitness saddle. Changing from F_1aF_2a to F_1bF_2b without going through a lower fitness configuration requires changing two features at once. Lewontin ([11], p84) identifies this same problematic case (for two diploid loci) in a concrete biological example. Accordingly, this form of epistasis provides the base case for the landscape we will use, but for further simplification, we make the fitness values symmetric (Case 4b).

Having established an appropriate two-feature epistasis model, we need an appropriate way to extend it to describe epistasis between a larger number of features. In particular, we want to re-use the same structure at a higher level so as to create the same kind of epistasis between *sets* of features as we have here between *single* features; in this way, we can create a principled method for producing fitness saddles of larger scales. Our approach is to describe the interaction of four features F_1, F_2, F_3, F_4, using the interaction of F_1 with F_2 in one pair, as above, the interaction of F_3 and F_4 as a second pair similarly, and then, at a higher level of abstraction, describe the interaction of these two pairs in the same fashion. To do this abstraction we treat the two possible end states of the F_1/F_2 subsystem, i.e. its two local optima (labelled c and d, in Table 2), as two discrete states of an 'emergent variable', or 'meta-feature', MF_1. Similarly, the two possible end states of the F_3/F_4 subsystem (e and f) form two states for MF_2. If the original, 'atomic' features are interpreted as low-level features of an entity, then a meta-feature may be interpreted as a higher-level phenotypic feature of an entity, or some higher-level property of an entity that determines its interaction with other entities and/or its environment.

In this manner we may describe the interaction of the two subsystems as the additional fitness contributions resulting from the epistasis of MF_1 and MF_2. Since each meta-feature includes two 'atomic' features, we double the fitness contributions in the inter-group interaction. Table 2 illustrates.

Table 2. Abstracting the interaction of two pairs of features, F_1/ F_2 and F_3/F_4, into the interaction of two 'meta-features' MF_1/MF_2.

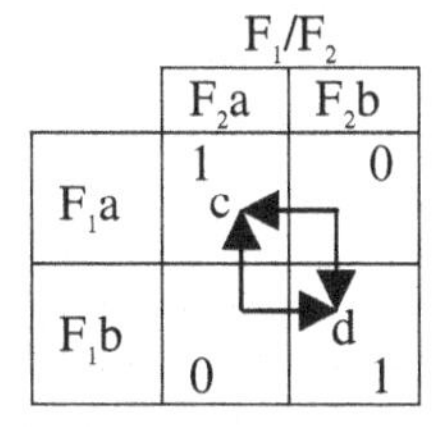

F_1/F_2

	F_2a	F_2b
F_1a	1 c	0
F_1b	0	d 1

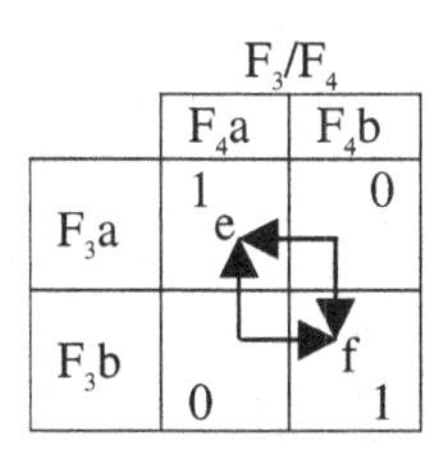

F_3/F_4

	F_4a	F_4b
F_3a	1 e	0
F_3b	0	f 1

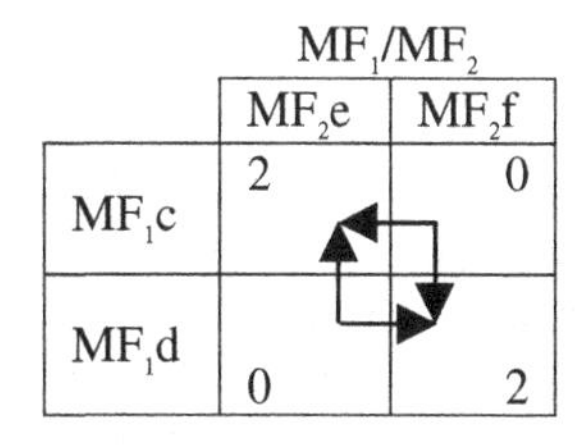

MF_1/MF_2

	MF_2e	MF_2f
MF_1c	2	0
MF_1d	0	2

The fitness landscape resulting from this interaction at the bottom level, together with the interaction of pairs at the abstracted level, produces four local optima altogether. Using a=0 and b=1, we can write these four optima as the strings 0000 and 1111, which are equally preferred, and, 0011 and 1100, which are equally preferred but less so. Naturally, we can take the two best-preferred configurations from the $F_{1...4}$ system and describe a similar interaction with an $F_{5...8}$ system, and so on. Equation 1 below, describes the fitness of a string of bits (corresponding to binary feature values, as above) using this construction. This function, which we call Hierarchical If-and-

Only-If (HIFF), was first introduced in previous work as an alternative to functions such as 'The Royal Roads' and 'N-K landscapes', (see [15]).

$$F(B)=\begin{cases} 1, & \text{if } |B|=1, \\ |B| + F(B_L) + F(B_R), & \text{if } (|B|>1) \text{ and } (\forall i\{b_i=0\} \text{ or } \forall i\{b_i=1\}) \\ F(B_L) + F(B_R), & \text{otherwise.} \end{cases} \qquad \textbf{Eq.1}$$

where B is a set of features, $\{b_1,b_2,...b_k\}$, $|B|$ is the size of the set=k, b_i is the ith element of B, B_L and B_R are the left and right subsets of B (i.e. $B_L=\{b_1,...b_{k/2}\}$, $B_R=\{b_{k/2+1},...b_k\}$). The length of the string evaluated must equal 2^p where p is an integer (the number of hierarchical levels).

A 64-feature landscape using HIFF, as used in our experiments, has 2^{32} local optima (for adaptation changing one feature at a time) [16], only two of these give maximal fitness. To jump from either of the second-best optima (at half-ones and half-zeros) to either global optimum requires changing 32 features at once. Thus, an algorithm using only mutation cannot be guaranteed to succeed in less than time exponential in the number of features [16]. A particular section through the fitness landscape is shown in Figure 1 – the section runs from one global optimum to the other at the opposite corner of the hyperspace (see [13]). As is clearly seen in the fractal nature of the curve in Figure 1, the local optima create fitness saddles that are scale-invariant in structure: that is, the nature of the ruggedness is the same at all resolutions.

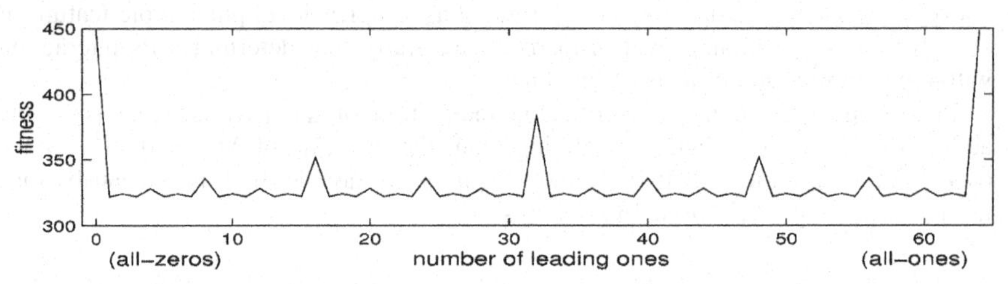

Fig. 1. A section through a 'scale-invariant' HIFF fitness landscape.

The hierarchical structure of the HIFF landscape makes the problem recursively decomposable. For example, a 64-feature problem is composed of two 32-feature problems, each of which has two optima. If both of these optima can be found for both of these subproblems then a global optima will be found in 2 of their 4 possible combinations. Thus, if this decomposition is known, then the search space that must be covered is at most $2^{32} + 2^{32} + 4$. Compared with the original 2^{64} configuration space, even this two-level decomposition is a considerable saving. In addition, each size 32-problem may be recursively decomposed giving a further reduction of the search space. In [16] we describe how an algorithm having some bias to exploit the decomposition structure (using the adjacency of features on the string) can solve HIFF in polynomial time. Here however, we are interested in the case where the decomposition structure is not known. We call this the 'Shuffled-HIFF' landscape [15] because this preferential bias is prevented by randomly re-ordering the position of features on the string such that their genetic linkage does not correspond to their epistatic structure (see [17]).

In summary, this landscape exhibits local optima at all scales, which makes it very challenging to adaptation, and fundamental to the issues of saddle-crossing and scalable evolvability. Moreover, it is amenable to a 'divide and conquer' approach *if* the decomposition of the problem can be discovered and sub-solutions of successive hierarchical levels can be manipulated and recombined appropriately.

3 The Composition Model

In this section we examine a simple abstraction of composition, the Symbiogenic Evolutionary Adaptation Model, or SEAM, (to invoke the idea of symbiotic union). This model was first introduced in [17] as a variant of abstract genetic algorithms.

SEAM manipulates strings of binary features, corresponding to feature combinations as discussed in the previous section. These strings, each representing a type of entity, have variable 'size' in that the number of specified features on the string is variable: for example, in an 8-feature system, F_3bF_5a is represented by the string "`--1-0---`", where "-" represents a "null" feature for which the entity has no value, or is neutral. The primary mechanism of the model is a mechanism of composing two partial strings together to form a new composite entity. This is illustrated in Figure 2. Such a mechanism assumes the availability of a means of symbiotic union such as the formation of a cell membrane, horizontal gene-transfer [19,20], endosymbiosis [4], allopolyploidy (having chomosome sets from different species) [21], or any other means of encapsulating genetic material from previously independent entities such that they will be co-dispersed in future. However, assuming the availability of such a mechanism does not preclude the interesting biological question of when such a mechanism would provide an adaptive advantage, and when such a union will be selected for.

```
  A: ----1-----0–1--          A: ----1----00-1--
  B: --1--0---0------         B: --1-0---0-1----
------------------------    ------------------------
A+B: --1-10---00--1--       A+B: --1-1---000-1--
```

Fig. 2. 'Symbiotic composition'. Left) Union of two variable size entities, A and B, produces a composite, A+B, that is twice the size of the donor entities with the sum of their characteristics. The composite is created by taking specified features from either donor where available. Right) Where conflicts in specified features occur we resolve all conflicts in favour of one donor, e.g. A.

Algebraically, we define the composition of two entities A and B, as the superposition of A on B, below. $A=\{A_1,A_2,...A_n\}$, is the entity where feature F_i takes value A_i. $S(A,B)$ is the superposition of entity A on entity B, and $s(a,b)$ is the superposition of two individual feature values, as follows:

$$S(A,B)= S(\{A_1,A_2,...A_n\},\{B_1,B_2,...,B_n\}) = \{s(A_1,B_1),s(A_2,B_2),...,s(A_n,B_n)\}, \quad \textbf{Eq.2.}$$

where, $s(a,b)=a$, if $a \neq$ null, $s(a,b)=b$, otherwise. Having defined a variation operator that defines a union of two entities we need to determine whether such a union would be selected for. Our basic assumption is that the symbiotic relationship must be in the 'selfish' interest of both the component entities involved. That is, if the fitness of either component entity is greater without the proposed partner than it is with the proposed partner then the composite will not be selected for. If, on the other hand, the fitness of both component entities is greater when they co-occur then the relationship is deemed stable and will persist. But, the fitness of any entity is dependent on its environmental context; possibly, in one environment an entity may be fitter when co-occurring with the proposed symbiont, and in another context the symbiont may depreciate their fitness. Thus whether a symbiotic relationship is preferred or not depends on what environmental contexts are available. For our purposes, the set of possible environmental contexts is well defined: an environmental context is simply a complete set of features (Figure 3).

`---0-11---110---` x, an entity specifies a partial set of feature values.
`0110101100010011` θ, an 'environmental context' is a complete set of features.
`0110111100110011` $S(x, \theta)$, the entity x superimposed on the context θ.

Fig. 3. A partially specified entity must be assessed in a context.

We assume that the overall fitness of the entity will be a sum of its fitness over different environmental contexts weighted by the frequency with which each environment is encountered. But, we would not generally suppose that the frequencies with which different environments are encountered by one type of entity would be the same as the frequencies relevant to a different type of entity. For example, a biased distribution over environmental contexts may be 'inherited' by virtue of the collocation of parents and offspring, or affected by the behavioural migration of organisms during their lifetime, or the selective displacement of one species by another in short term population dynamics. We did not wish to introduce such factors and accompanying assumptions into our model. But fortunately, the concept of *Pareto dominance* is specifically designed for application in cases where the relative importance of a number of factors is unknown [21]. Put simply, this concept states intuitively that, even when the relative weighting of dimensions is not known, the overall superiority of one candidate with respect to another can be confirmed in the case that it is non-worse in all dimensions and better in at least one. More exactly, 'x Pareto dominates y' is written '$x >> y$', and:

$$x >> y \Leftrightarrow \{\forall\theta: csf(x,\theta) \geq csf(y,\theta)\} \wedge \{\exists\theta: csf(x,\theta) > csf(y,\theta)\}$$

where, for our ecological application, $csf(p,q)$ is the 'context sensitive fitness' of entity p in context q. So crudely, if x is fitter (or at least as fit as) y in all possible environments then, regardless of the weightings of the environments for each entity we know that the overall fitness of x is greater than that of y. This pair-wise comparison of two entities over a number of contexts will be used to determine whether a symbiotic union produces a stable composite. If we write the union of entities a and b as $a+b$, then using the notion of Pareto dominance, $a+b$ is stable if and only if $a+b >> a$, and $a+b >> b$. In other words, $a+b$ is *unstable* if there is any context in which either a or b is fitter than $a+b$.

i.e. $stable(a+b, a, b) \equiv (a+b >> a) \wedge (a+b >> b)$,
i.e. $unstable(a+b, a, b) \Leftrightarrow \{\exists\theta\in C: (csf(a,\theta) > csf(a+b,\theta)) \vee (csf(b,\theta)) > csf(a+b,\theta))\}$,

where C is a set of complete feature specifications. For our purposes, $csf(x,\theta) > csf(y,\theta) \Leftrightarrow f(S(x,\theta)) > f(S(x,\theta))$, where $f(w)$ is the objective fitness of the complete feature set $\boldsymbol{w}$ as given by the fitness function. Thus our condition of instability becomes:

$$\boldsymbol{unstable}(\boldsymbol{a}+\boldsymbol{b}, \boldsymbol{a}, \boldsymbol{b}) \Leftrightarrow \{\exists \theta \in \boldsymbol{C}: (f(\boldsymbol{S}(\boldsymbol{a}, \theta)) > f(\boldsymbol{S}(\boldsymbol{a}+\boldsymbol{b}, \theta))) \vee (f(\boldsymbol{S}(\boldsymbol{b}, \theta)) > f(\boldsymbol{S}(\boldsymbol{a}+\boldsymbol{b}, \theta)))\} \quad \textbf{Eq.3.}$$

We will build each context in the set of contexts by the temporary superposition of other members of the ecosystem. Algebraically, we define a context using the recursive function $\boldsymbol{S}^*$, from a set of n≥2 entities $X_1, X_2, \ldots X_n$, as follows:

$$\boldsymbol{S}^*(X_1, X_2, \ldots X_n) = \begin{cases} \boldsymbol{S}(X_1, \boldsymbol{S}^*(X_2, \ldots X_n)), & \text{if } n>2, \\ \boldsymbol{S}(X_1, X_2), & \text{otherwise.} \end{cases} \quad \textbf{Eq.4.}$$

where $\boldsymbol{S}(X_1, X_2)$ is the superposition of two entities as per Eq.1 above. Some contexts may require more or less entities to provide a fully-specified feature set. In principle, we may use all entities of the ecosystem, in random order, to build a context - but, after the context is fully-specified, additional entities will have no effect. This allows us to write a context as S*(E), where E is all members of the ecosystem in random order. Implementationally, we may simply add entities until a fully-specified set is obtained.

An interesting analogy for this group evaluation is made with the Baldwin effect [22], and 'Symbiotic Scaffolding' [23,24]. That is, these scenarios have in common the feature that rapid non-heritable variation (lifetime learning or the temporary groups formed for contexts) guides a mechanism of relatively slow heritable variation (genetic mutation or composition). In other words, evaluation of entities in contextual groups 'primes' them for subsequent joins, or equivalently, solutions found first by groups are later *canalised* [9] by permanent composite entities.

Figure 4 uses Equations 2, 3 and 4 to define a simple version of the Symbiogenic Evolutionary Adaptation Model.

- Initialise ecosystem, E, with random, single-feature, entities.
- Repeat until *stopping condition*:
  - Remove two entities at random from the ecosystem → ***a*** & ***b***.
  - Produce their union, ***a+b=S(a,b)***, using composition (see Eq.2).
  - If ***unstable***(***a+b***, ***a***, ***b***) return ***a*** and ***b*** to ecosystem, else add ***a+b*** to ecosystem.

where (as in Eq. 3)
$\boldsymbol{unstable}(\boldsymbol{a}+\boldsymbol{b}, \boldsymbol{a}, \boldsymbol{b}) \Leftrightarrow \{\exists \theta \in C: (f(\boldsymbol{S}(\boldsymbol{a},\theta)) > f(\boldsymbol{S}(\boldsymbol{a}+\boldsymbol{b},\theta))) \vee (f(\boldsymbol{S}(\boldsymbol{b},\theta)) > f(\boldsymbol{S}(\boldsymbol{a}+\boldsymbol{b},\theta)))\}$
and ***C*** is a random set of contexts each built using $\boldsymbol{S}^*$(E) (see Eq.4).

Fig. 4. Pseudocode for a simple version of SEAM.

In the implementations used in the following experiments the maximum number of contexts used in the stability test is 50 (although unstable joins are usually detected in about 6 trials on average). Since the Pareto test of stability abstracts away all population dynamics, only one instance of each type of entity need be maintained in the ecosystem. Initialisation needs to completely cover the set of single-feature 'atoms' so that all values for all features are available in the ecosystem. This can be done systematically, or as in our experiments, by over-generating randomly and then removing duplicates.

4 Experimental Results

We show empirical results of SEAM applied to a 64-bit Shuffled HIFF. Our intent is to illustrate the qualitative difference in the way that composition operates in this scale-invariant problem as compared to the operation of accumulation of mutation and conventional abstractions of population-based evolutionary search. Accordingly, we contrast the operation of SEAM with the results of a mutation only algorithm, Random Mutation Hill-Climbing, RMHC, and a genetic algorithm, GA, using sexual recombination. RMHC repeatedly applies mutation to the features of a single string and accepts a variant if it is fitter [25]. We show results for various mutation rates (probability of assigning a new random value {0,1} to each feature). The GA is steady state using Deterministic Crowding to maintain diversity in the population (see [17], for details). The GA is tested using uniform and one-point crossover. A population size of 1000 is used; crossover is applied with probability 0.7; and mutation is applied with 0.03 probability of assigning a new random allele to each feature (0 or 1 with equal probability). SEAM uses the algorithm described in Figure 4. An ecosystem is initialized to the 128 unique entities (for a 64-bit problem). Symbiotic composition (Figure 2) is applied in all cases, no mutation is required. Performance is measured by the fitness of the best string evaluated (in the preceding 500 evaluations) averaged over 30 runs for each algorithm. The problem size of 64 bits gives a maximum fitness of 448. The data for SEAM are terminated when all runs have found both global optima.

As Figure 5 shows, the results for SEAM are clearly qualitatively different from the other algorithms: Whereas innovation by mutation and by conventional evolutionary algorithms becomes increasingly more difficult as evolution continues in this problem, innovation by composition progresses unimpeded through successive hierarchical levels, and actually shows an inverted performance curve compared to the other methods.

5 Discussion and Conclusions

Scale-invariance is a property observed in many natural systems, but whether the natural adaptive environment has characteristics like those of the particular landscape that we use is an empirical matter. Also, SEAM abstracts away all population dynamics and uses a simple multi-context test to determine whether a composite will be stable. Resulting joins are compatible with a selfish model of the entities, but whether this is an appropriate model for multi-species competition in an ecosystem needs to be justified. The experiments reported here involve only one adaptive landscape, selected to exemplify the utility of composition, and composition is modeled using one particular set of assumptions. The behaviour of the models must be examined in more general cases. In the meantime, the model we have outlined provides a concrete illustration of some important concepts raised in the introduction. Specifically: informed jumps in feature space – the fitness saddles in the landscape require large jumps in feature space and no amount of random mutation can provide this appropriately; a variation mechanism (and contextual evaluation) that scales-up with the size of extant entities; and divide and conquer problem decomposition by combining together solutions to small sets of features to find solutions to larger sets of features.

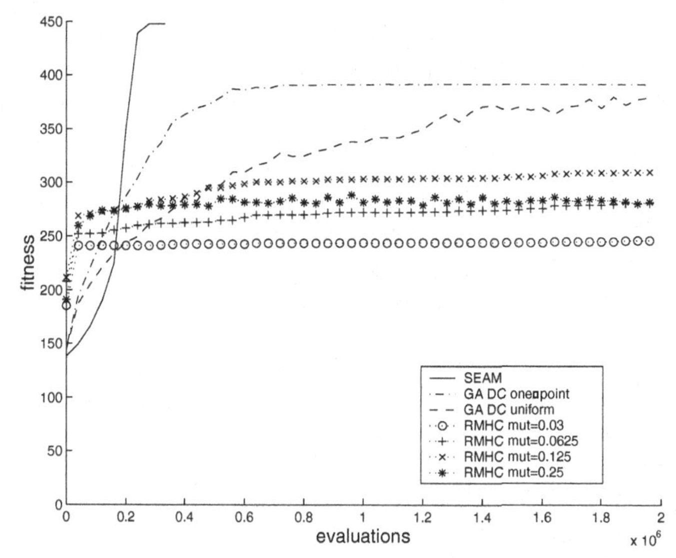

Fig. 5. Performance of SEAM, regular GA with Deterministic Crowding (using one-point and uniform crossover), and Random Mutation Hill-Climbing, on Shuffled HIFF. Error bars are omitted for clarity, but the differences in means are, as they appear, significant.

Some or all of these characteristics could conceivably be provided by other mechanisms. Investigation of these concepts provides valuable insight for issues of evolvability and our model shows that there is a way in which these interesting characteristics could be provided by composition. We suggest that this algorithmic perspective on the formation of higher-level entities, from the composition of simpler entities, provides a useful facet in our understanding of the evolutionary impact of the Major Evolutionary Transitions.

References

1. Buss, LW,1987, *The Evolution of Individuality*, Princeton University Press, New Jersey.
2. Maynard Smith, JM & Szathmary, E, 1995 *The Major Transition in Evolution*, WH Freeman.
3. Michod, RE 1999. *Darwinian Dynamics, Evolutionary Transitions in Fitness and Individuality*. Princeton Univ. Press.
4. Margulis, L, 1993 *Symbiosis in cell evolution*, 2nd ed, WH Freeman and Co.
5. Wright, S. 1931, Evolution in Mendelian populations. *Genetics* **16**: 97-159.
6. Wright, S, 1977, *Evolution and the Genetics of Populations, Volume 3*, U. of Chicago Press.
7. Gould, SJ & VRBA, E, 1982, “Exaptation - a missing term in the science of form”, *Paleobiology*, **8**: 4-15. Ithaca
8. Kimura, M. 1983, *The Neutral Theory of Molecular Evolution*, Cambridge University Press.
9. Conrad, M, 1990, "The Geometry of Evolution." *BioSystems* **24** (1990): 61-81.
10. Waddington, CH, 1942, "Canalization of development and the inheritance of acquired characters", *Nature* **150**: 563-565.
11. Lewontin, RC, 2000 *The Triple Helix: Gene, organism and Environment,* Harvard U. Press.

12. Cormen, TH, Leiserson, CE, Rivest, RL, 1991 *Introduction to Algorithms,* MIT Press.
13. Watson, RA, & Pollack, JB, 1999, "Hierarchically-Consistent Test Problems for Genetic Algorithms", *procs. of 1999 CEC,* Angeline, et al. eds. IEEE Press, pp.1406-1413.
14. Wright, S, 1967, "Surfaces of selective value", *Proc. Nat. Acad. Sci.* **58**, pp165-179 (1967).
15. Watson, RA, Hornby, GS & Pollack, JB, 1998, "Modeling Building-Block Interdependency", *procs. of PPSN V,* Springer 1998, pp.97-106 .
16. Watson RA, 2001, "Analysis of Recombinative Algorithms on a Non-Separable Building-Block Problem", *Foundations of Genetic Algorithms VI,* 2000, Martin WN, & Spears WM, Morgan Kaufmann.
17. Watson, RA, & Pollack, JB, 2000, "Symbiotic Combination as an Alternative to Sexual Recombination in Genetic Algorithms", *procs. PPSN VI,* Schoenauer et al. Springer, 425-434.
18. Fonseca, CM, Fleming PJ, 1995, "An Overview of Evolutionary Algorithms in Multiobjective Optimization", Evolutionary Computation, Vol. 3, No.1, pp.1-16.
19. Mazodier, P, and Davies, J, 1991, "Gene transfer between distantly related bacteria", Annual Review of Genetics 25:147-171.
20. Smith, MW, Feng, DF, & Doolittle, RF, 1992, "Evolution by acquisition: the case for horizontal gene transfers", Trends in Biochemical Sciences 17(12):489-493.
21. Werth, C.R., Guttman, S.I. & Eshbaugh, W.H. (1985) "Recurring origins of allopolyploid species in Asplenium" Science 228, 731-733.
22. Baldwin, JM, 1896, "A New Factor in Evolution," *American Naturalist,* **30**, 441-451.
23. Watson, RA, & Pollack, JB, 1999, "How Symbiosis Can Guide Evolution", *procs. of ECAL V,* Floreano, D, Nicoud, J-D, Mondada, F, eds. Springer, 1999.
24. Watson, RA, Reil, T, & Pollack JB 2000, "Mutualism, Parasitism, and Evolutionary Adaptation", *procs. of Artificial Life VII,* Bedau, M, et al. eds.
25. Forrest, S & Mitchell, M, 1993, "What makes a problem hard for a Genetic Algorithm? Some anomalous results and their explanation" *Machine Learning 13,* pp.285-319.

nBrains
A New Type of Robot Brain

Andy Balaam

COGS, University of Sussex, Brighton, UK.
andrewba@cogs.susx.ac.uk

Abstract. The design and implementation of a possible alternative to Artificial Neural Networks (ANNs) for agent control is described. This alternative, known as the nBrain, uses the phase-space representation as the inspiration for a controller. The agent progresses to different states due to the presence of attractors in phase space, whose locations are set by the robot's genome. Several experiments with simulated agents are described. The experiments were successful in that tasks which have been performed by ANNs in the past were successfully accomplished by nBrains under evolution. The possible advantages and disadvantages of nBrains over ANNs are discussed, and directions for future work are presented.

1 Introduction

This document describes the design, implementation and testing of a new type of agent controller, known as the 'nBrain'. The nBrain uses the phase-space representation to make the dynamical-systems approach to cognition explicit. The aim of the project is to show that nBrains are useful controllers for autonomous agents, and that they are capable of producing some of the behaviours which are produced by ANNs.

1.1 Motivation and Background

Recently it has been widely suggested that the best way to view the brain is as a dynamical system which interacts with other dynamical systems. This view has several advantages over the alternative computational view, and many researchers are now producing work which reflects it, constructing controllers for real and simulated agents which perform many different tasks. These researchers include Beer [1,2,3], Cliff [4,5], Harvey [5,6], Husbands [5], Jakobi [5], Miller [4], Thompson [5] and van Gelder [7]. It is hoped that nBrains may be a useful tool in investigating cognitive phenomena and producing useful autonomous robots.

2 nBrains: An Explanation

This section explains the general features that constitute an nBrain. The state of an agent at any time may be represented by a point in phase space. The

J. Kelemen and P. Sosík (Eds.): ECAL 2001, LNAI 2159, pp. 491–494, 2001.

genotype of an agent with an nBrain controller specifies a vector 'force' field in the space of some of its variables which causes the agent to trace a path through that space.

Imagine that we have an extremely simple agent, with one sensor and one motor. Then its phase space representation is a plane (see figure 1 (left)).

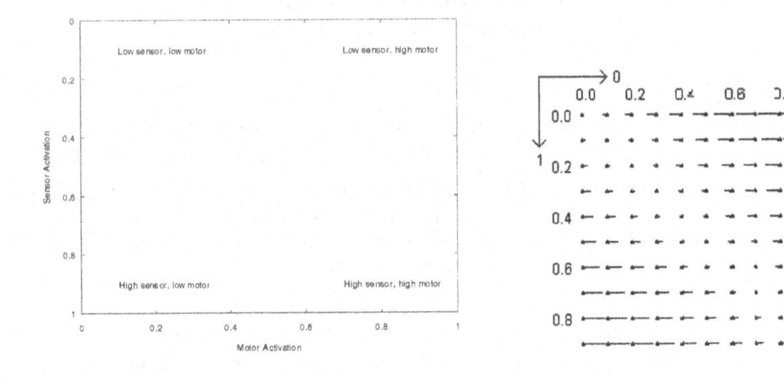

Fig. 1. Left: Phase space of a simple agent with one motor and one sensor. Note that for consistency the vertical axis is zero at the top and one at the bottom. Right: A vector field in phase space which causes the agent to trace a path through phase space such that it moves when light is present and stops otherwise. Dimension 0 (horizontal) represents motor activation, and dimension 1 (vertical) represents sensor activation. When the sensor is active (near the bottom of the graph), the motor tends to become more active (movement towards the right), and when the sensor is inactive (near the top) the motor tends to become less active (movement towards the left). Note that the lines should be taken as arrows pointing from the dot, towards the end of the line.

If we were to design a simple controller for this agent, so that it moved when it saw light, and stayed still otherwise, we could plot this behaviour as a vector field (see figure 1 (right)).

The lines on figure 1 (right) indicate the direction (in the motor dimension) that the robot will move through phase space if it finds itself in any particular state. The lines do not point in the sensor direction as movement in this dimension is not determined by the robot controller but is fed in from the environment.

An nBrain is a vector field like the one shown in figure 1 (right), so that now rather than simply using the phase space diagram to describe the action of an agent controller, we use it to specify that action. In the experiments described the vector field in the space was determined by specifying a number of points which behaved like attractors, causing attractive 'force' towards them.

These concepts may be extended to robots with more sensors and effectors, but they become more difficult to picture, as the diagrams become many-dimensional. They may also be extended to build non-reactive agents using hidden dimensions with no direct physical meaning.

The aim of this project was to show that an agent controller built on these principles can perform many of the same tasks as an ANN, and possibly at a smaller computational cost.

3 Experiments and Results

Experiments were performed using simulated agents in a two-dimensional arena. Agents had sensors which could see lights and other agents, and 'jet-like' motors. In the evolutionary experiments the agents' morphologies were kept fixed.

3.1 Experiment 1

The first evolutionary experiment was designed to try to evolve an agent which can approach a fixed light. The morphologies of the agents were fixed with one forward-facing motor on either side, and two sideways-facing sensors. The agents' genotypes specified the positions of any number of point attractors in the nBrain, and any number of hidden dimensions. Each run began with a random population of size 100. Fitness awarded to an agent every time step was proportional to the reciprocal of the exponential of its distance from the light.

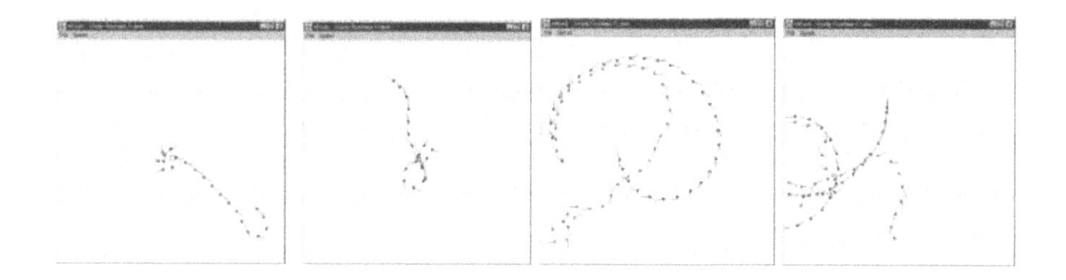

Fig. 2. Left: Experiment 1. By generation 299 agents perform light-approaching behaviour. Right: Experiment 2. The generation 499 agents clearly perform pursuit-evasion behaviour.

Evolution appears to have converged on a simpler and more effective solution than that designed in an earlier hand-designed test. Certainly the experiment was successful in showing that useful behaviour can be evolved in a population of nBrains.

3.2 Experiment 2

The second experiment was inspired by the work of Cliff and Miller [4] on the co-evolution of pursuit and evasion in autonomous agents. Two populations were used, each of size 100. In each fitness evaluation, one agent from each population was evaluated against one from the other population. The fitness of the pursuer was as in experiment 1 with the agent viewed as a light. The fitness of the evader was the negative of this score.

Some evolved individuals may be seen in figure 2. This experiment successfully produced, by co-evolution, agents which are seen qualitatively to perform pursuit and evasion strategies.

4 Conclusions

4.1 Success

The project was successful in that evolution succeeded in finding good solutions to the problems that were set for it. However, the tasks themselves were relatively simple and the solutions produced were mainly reactive controllers. These experiments show that the hypothesis that nBrains might be useful for robot control is worth investigating.

It is possible to speculate that nBrains may have certain advantages over ANNs as robot controllers. These may include computational efficiency, the tendency to lead the researcher down the dynamical systems line of thought, their very straightforward genetic encoding (and the possibility that this might lead to smoother fitness landscapes). None of these possibilities has been tested in this study.

Studying a new type of agent controller is a large subject, and there is a great deal of ground which is unexplored. Some possible areas for future study include testing nBrains in much more complex environments, examination of the computational costs of nBrains in comparison with the computational costs of equivalent ANNs, and testing nBrains against problems with which ANNs have had difficulty in the past. This project has confirmed that nBrains are worthwhile topics for investigation.

References

1. Beer, R. 'Framing the Debate Between Computational and Dynamical Approaches to Cognitive Science.' in *Behavioural and Brain Sciences* 21, 630 1998
2. Beer, R. 'A dynamical systems perspective on autonomous agents'. Technical Report CES-92-11, Case Western Reserve University, Cleveland, Ohio, 1992.
3. Beer, R., Gallagher, J. 'Evolving dynamic neural networks for adaptive behaviour'. *Adaptive Behaviour*, 1(1):91-122, 1992
4. Cliff, D., Miller, G. 'Co-evolution of pursuit and evasion II: Simulation methods and results'. Tech Rep. CSRP377, COGS, University of Sussex, 1995.
5. Harvey, I., Husbands, P., Cliff, D., Thompson, A., Jakobi, N. 'Evolutionary Robotics: the Sussex Approach'. *Journal of Robotics and Autonomous Systems* 1997.
6. Harvey, I. 'Cognition is not computation, evolution is not optimisation'. In *Artificial Neural Networks - ICANN97*, 1997.
7. van Gelder, T. 'The Dynamical Hypothesis in Cognitive Science' in *Behavioural and Brain Sciences* 1998

Can Wheeled Robots Illuminate Adaptive Behaviour?

Jon Bird

Centre for Computational Neuroscience and Robotics
School of Biological Sciences
University of Sussex
Brighton BN1 9QG, UK
jonba@cogs.susx.ac.uk

Abstract. This paper evaluates Bedau's proposal that the fundamental properties of living systems are best investigated using unrealistic models. Two issues raised by this position are considered. Firstly, how can unrealistic models be validated? Secondly, can unrealistic models produce, rather than just simulate, the phenomena they aim to explain? The discussion is focussed by considering wheeled robot models of adaptive behaviour.

1 Introduction

Many biological phenomena are the result of complex non-linear interactions between processes operating at different levels and time scales; consequently it is difficult to formulate, let alone answer, questions about them. Bedau highlights three particular examples: multi-level emergent activity in systems such as cells and organisms; open-ended adaptive evolution; unbounded complexity growth in evolving systems [1]. Artificial Life (AL) distinguishes itself from biology by adopting a synthetic approach to understanding complex biological systems such as these. Following this methodology, Bedau [1] suggests that when trying to investigate the fundamental properties of living systems, a useful way to proceed is to abstract away from as much of the complex detail as possible and use *unrealistic* models. These aim to capture the minimal set of properties that can *produce* the biological phenomenon being investigated. Bedau emphasizes that the only way that an unrealistic model can explain a particular natural phenomenon is by generating „actual examples of the phenomenon in question; it is not sufficient to produce something that represents the phenomenon but lacks its essential properties“ [1; p.21]. This paper briefly describes the conventional approach to modelling, in particular the explanatory role that models can play and how they are validated. It then considers how unrealistic models can be validated and whether unrealistic models can produce, rather than just simulate the phenomena they aim to explain. In order to focus the discussion the paper considers wheeled robot models of adaptive behaviour. There are good reasons for this focus. Firstly, it is the type of problem and approach that Bedau bases his argument on: how organisms generate adaptive behaviour is a deep question in biology. Secondly, wheeled robot models have sensory-motor systems that are considerably simpler than most adaptive organisms; Keijzer [2] has recently argued that wheeled robot models are unsuitable tools for investigating adaptive behaviour, precisely because they are

J. Kelemen and P. Sosík (Eds.): ECAL 2001, LNAI 2159, pp. 495-498, 2001.

unrealistic. Thirdly, the majority of autonomous agents research uses either wheeled robots or simulations of wheeled robots; given their ubiquity, it is important to consider what explanatory role wheeled robot models can play in any theory of adaptive behaviour

2 The Conventional Modelling Approach

Di Paolo et al [3; p.505] make clear that in a conventional modelling approach „[e]ntities in the model represent theoretical entities metaphorically or analogically". The model can then be used to investigate predicted hypotheses about natural phenomena and relate the findings back to the predictive theory for explanatory purposes. For example, Webb [4] used a wheeled robot as a physical model to test a hypothesis about cricket phonotaxis, namely that there was no need for separate song recognition and orienting processes in the cricket. Her robot model showed that both processes could be a consequence of the close coupling of the motor system to the sensory system: effectively, only a mating song of a specific frequency would elicit an orienting response. The robot model was validated by further experiments that showed that ear directionality could account for frequency selectivity in phonotaxis to *real* cricket song in a biologically plausible environment [5].

3 Unrealistic Models

In contrast to cricket phonotaxis, Bedau argues that „we may well have no significant theories of the behavior of systems exhibiting deep unexplained phenomena" and therefore require unrealistic models to act as „emergent thought experiments" [1; p.22]. By distinguishing unrealistic models from conventional thought experiments Bedau is emphasizing that their role is to *generate* complex phenomena and not just explore the predefined boundaries of a theoretical framework. Using unrealistic models to investigate these phenomena still requires some „objective and operational measure" [1; p22] to validate that the model behaves analogously to the biological system being investigated. Bedau suggests using statistical measures of the macro behaviour of the model; however, the selection of the appropriate statistical tool would have to be determined on some arbitrary, non-theoretical basis and, as Bedau recognizes, this introduces ambiguity into the validation process. The epistemological status of unrealistic models is therefore currently unclear. This is also the case for many robot models used in embodied cognitive science, whose purpose is not to test a particular theory but is rather to explore mechanisms that generate certain types of adaptive behaviour [6].

Unrealistic models are an example of strong AL, which aims to build simulations that are *instantiations* [7] or *realizations* [8] of the natural phenomenon under study. Essentially, the aim is to produce an artificial system that could be a functional replacement of the phenomenon being investigated. For example, an artificial heart made of plastic is a physical model of a biological heart. Furthermore, it could be said to realize or instantiate a biological heart because it has the same functionality: it can pump blood around a vascular system. Simulations are symbolic forms, whereas the processes that they simulate consist of material substances and in many cases it is clearly a category error to confuse the two. Sober [7] uses the example of a simulated bridge to make this point: no matter how detailed the bridge simulation is, it is never more than a symbolic representation and quite different from a real bridge, which it could never functionally replace. However, in other cases, it is a controversial issue as to whether there is such a clear cut distinction. There is a particular difficulty when the simulation is modelling an emergent process. For example, Tierra is considered by some to not merely simulate but actually produce the evolution of diversity [10]. Ray applies Cariani's [11] hierarchy of different levels of emergent behaviour to Tierra and argues that the model demonstrates the strongest form [10]. In direct contrast, Pattee, who is sceptical about strong AL, uses similar criteria to argue that symbolic environments can in principle *never* support the emergence of the sort of novelty that would be necessary for Tierra to produce diversity [8]. This highlights the difficulty in determining whether an unrealistic model has the capacity to generate the phenomenon it is trying to explain. Bedau himself suspects that, „artificial life models are qualitatively different from the biosphere in that the models lack the capacity for new kinds of niches to be continually created as an intrinsic consequence of the course of adaptive evolution"; however, he does not know „how to test for this sort of property" [1; p.22]. This is only a weakness in the approach if it is necessary for the unrealistic model to realize the phenomenon it is trying to illuminate. Wheeled robot models illustrate how unrealistic simulations can also play valid roles in investigating complex biological phenomena.

4 Conclusion

Di Paolo et al [3] describe Bedau's unrealistic model methodology as, „a courageously different understanding of thought experiments, and models in general". They emphasize that the onus of proof is on Bedau to, „show how a simulation, which always starts from a previously agreed upon theoretical stance, could ever work like a source of new empirical data about natural phenomena" [3; p.501]. This paper has highlighted the difficulties in both validating unrealistic models and unambiguously determining that they generate examples of the phenomena being investigated. Unrealistic models cannot be evaluated as though they were conventional simulations because they are used to investigate phenomena where there are no established theories and testable hypotheses. Bedau's unrealistic models have a role analogous to wheeled robot models in embodied cognitive science: they enable an experimenter to carry out bottom-up reverse engineering [13]. The aim is to move toward descriptions of complex phenomena by starting with as simple a mechanism as possible and observing what behaviours it can generate. Unrealistic models are a step towards theories of complex biological phenomena and one of their roles is to help formulate the right sorts of questions for investigating complex biological phenomena. For example, the unrealistic model approach might hopefully lead to the development of clearer criteria for distinguishing simulations and realizations of emergent phenomena. In this way wheeled robot models can illuminate adaptive behaviour and unrealistic models can illuminate other complex biological phenomena.

Acknowledgements. This research was sponsored by a BBSRC CASE studentship with British Telecom. Many thanks to two anonymous referees, Phil Husbands and members of the CCNR and BT Future Technologies group for valuable discussions that have shaped some of these ideas.

References

1. M.A. Bedau. Can Unrealistic Computer Models Illuminate Theoretical Biology? In A.S. Wu, Editor, *Proceedings of the 1999 Genetic and Evolutionary Computation Conference Workshop Program, Orlando, Florida, July 13*, 20 – 23, 1999.
2. F. A. Keijzer. Some Armchair Worries about Wheeled Behavior. In R. Pfeifer, B. Blumberg, J.-A. Meyer and S.W. Wilson, Editors. *From Animals to Animats 4: Proceedings of the Fourth International Conference on Simulation of Adaptive Behavior.* MIT Press/ Bradford Books, 13 – 21, 1998.
3. E.A. Di Paolo, J. Noble and S. Bullock. Simulation Models as Opaque Thought Experiments. In M.A. Bedau, J.S. McCaskill, N.H. Packard and S. Rasmussen, Editors, *Artificial Life VII: The Seventh International Conference on the Simulation and Synthesis of Living Systems, Reed College, Portland, Oregon, USA, 1 - 6 August*, 497 – 506, 2000.
4. B. Webb. Using Robots to Model Animals: A Cricket Test. *Robotics and Autonomous Systems, 16,*117 –134, 1995.
5. H.H. Lund, B. Webb and J. Hallam. A Robot Attracted to the Cricket Species *Gryllus buimaculatus*. In P. Husbands, and I. Harvey, Editors, *Proceedings of the Fourth European Conference on Artificial Life*, MIT Press, 246 – 255, 1997.
6. R. Pfeifer and C. Scheier. *Understanding Intelligence.* MIT Press, 1999.
7. E. Sober. Learning from Functionalism – The Prospects for Strong Artificial Life. In [9], 361 – 378, 1996.
8. H.H. Pattee. Simulations, Realizations, and Theories of Life. In [9], 379 – 393, 1996.
9. M.A. Boden, Editor. *Philosophy of Artificial Life.* Oxford University Press, 1996.
10. T.S. Ray. An Approach to the Synthesis of Life. In [12], 371 – 408, 1991.
11. P. Cariani. Emergence and Artificial Life. In [12], 775 –797, 1991.
12. C.G. Langton, C. Taylor, J.D. Farmer and S. Rasmussen, Editors, *Artificial Life II*, Santa Fe Studies in the Sciences of Complexity, Proceedings, 10, Addison-Wesley, 1991.
13. D. Dennett. *Brain Children: Essays in Designing Minds.* Penguin, 1998.

Evolution, Adaption, and Behavioural Holism in Artificial Intelligence

T.S. Dahl[1] and C. Giraud-Carrier[2]

[1] Machine Learning Research Group, Department of Computer Science, University of Bristol, United Kingdom,
tdahl@cs.bris.ac.uk
[2] ELCA Informatique SA, Lausanne 13, Switzerland
cgc@cs.elca.ch

Abstract. This paper presents work on reproducing complex forms of animal learning in simulated Khepera robots using a behaviour-based approach. The work differs from existing behaviour-based approaches by implementing a path of hypothetical evolutionary steps rather than using automated evolutionary development techniques or directly implementing sophisticated learning. Following a step-wise approach has made us realise the importance of maximising the number of behaviours and activities included on one level of complexity before progressing to more sophisticated solutions. We call this inclusion behavioural holism and argue that successful approaches to complex behaviour based robotics must be both step-wise and holistic.

1 Introduction

After abandoning traditional top-down machine learning (ML) methods for learning novel robot behaviours because of their inherent difficulties with expressing learning biases and background knowledge, we have taken a behaviour-based (BB) [2] approach where complex learning behaviours are implemented through step-wise increases and modifications to already adaptive behavioural foundations.

In order to conduct a number of experiments implementing basic adaptive animal behaviours, we developed a framework of programmable, learning, artificial neural circuits (PLANCS) [8]. It provides a neural circuit class which emulates an independent computational node. The neural circuit abstraction is also a neuron-inspired extension of an object oriented BB architecture called Edmund [5], that supports circuit level cognitive modelling.

During this work we have developed two guidelines that extend traditional BB robotics with respect to developing complex behaviours, in particular learning behaviours. We call the development approach described by The guidelines a *step-wise* and *behaviourally holistic* approach. The terms *step-wise* and *behaviourally holistic* are further described in Section 2. Two of the experiments we conducted on reproducing animal behaviours are presented in Section 3. The experiments are done using the Webots Khepera robot simulator [7]. Section 4

J. Kelemen and P. Sosík (Eds.): ECAL 2001, LNAI 2159, pp. 499–508, 2001.

suggests how an analysis of human evolution can be used to provide a road-map for step-wise, holistic, BB robotics. Finally, Section 5 discusses the relationship between the holistic and the ALife approaches to robotics, and answers some common criticisms of our work.

2 A Step-Wise, Holistic Approach to Robotics

2.1 Defining the Approach

Our step-wise, holistic approach to implementing complex behaviours in robots contains two additions to the recommendations of behaviour based robotics:

1. Implement a hypothesised path of evolutionary steps to a desired animal behaviour.
2. Include a maximum number of different behaviours and activities on each evolutionary level.

2.2 A Step-Wise Approach

Brooks' original rules for BB robotics [3], suggested taking inspiration from evolution. We suggest a more extreme approach where rather than just trying to implement a desired set of behaviours inspired by evolution, each step of a hypothesised historical evolutionary path to a desired set of behaviours is implemented.

The motivation for retracing evolution in implementation is that the complex behaviours found in animals and humans are so poorly understood that robust and efficient direct implementations are impossible. In these cases, retracing evolution forces an investigation of the evolutionary history of the behaviour in question.

Undertaking a complete behavioural investigation can be preferable to implementing a brittle or inefficient approximation, something which has often been the result of trying to implement complex behaviours directly.

What we call complex behaviours are behaviours that are not directly implementable. Our failure to implement automated learning of novel robot behaviours directly was originally the inspiration for researching a behaviour-based approach to such learning.

An Evolutionary Theory of Learning. One of the main sources of inspiration for our holistic development is the growing amount of knowledge of the physiology and evolutionary history of biological systems that is found in areas such as ethology, neuro-science, and the cognitive sciences. There is currently a wide scope for using this knowledge in AI implementations.

In particular Moore [12] presents a clear theory of how increasingly complex forms of learning might have developed. Below we list the main types of learning as presented by Moore, with contributions from other theories added in italics.

- Imprinting
- Alpha Conditioning
- Pavlovian Conditioning
- Operant Conditioning
- Skill Learning
- Imitation
- *Language Learning*

This list of learning types is not complete, and interesting questions are raised concerning types of learning not included in this hierarchy such as classification and insight.

2.3 Behavioural Holism

The second recommendation reflects the realisation that complex behaviours that involve learning always relate to more basic underlying behaviours supporting a number of activities. Complex behaviours cannot be thoroughly explored without being emerged in a rich behavioural context. The evolutionary histories of behaviours are highly interrelated, and looking at a limited number can not reveal all the details necessary for a comprehensive understanding.

Learning is also a problem that needs to be strictly biased if it is to be successful. The way biases are introduced in biological systems is through a hierarchical structuring of data and control [6]. This kind of structuring is done by pre-existing neural circuitry, and the more effective biases we need, the more underlying circuitry we must provide.

Below we present the three main arguments for the need of behavioural holism.

Conclusions from Our Own Work. As we analysed increasingly complex forms of conditioning, it became difficult to design natural learning problems to test the different learning types due to the poverty of the underlying controllers. In designing an experiment for demonstrating alpha conditioning, we needed the robot to recognise that a certain stimulus would regularly occur together with food. The underlying controller could only recognise other robots and food, so we had to invent an artificial pink box sense.

If a holistic approach had been taken, we would have had a number of senses related to other basic behaviours to choose from so that our alpha conditioning experiment would have been more natural and perhaps brought up issues of behaviour integration that were missed because of the artificial nature of our pink box sense.

Our BB analysis of conditioning points out that increasingly complex learning behaviours learning need an increasing number of underlying behaviours to support it.

Arguments from Cognitive Robotics. In 1997, Brooks criticised work in BBAI for not having a wide enough *behavioural repertoire* [4]. He recognises the vastly richer set of abilities needed by robots in order to act like a human, and suggests work be done on activity interaction and control.

Brooks explicitly lists *coherence* as an issue to be considered in cognitive robotics. Coherence is a complex and poorly understood problem that involves many different sub-systems. A step-wise, holistic approach provides a study of increasingly complex manifestation of problems spanning many sub-systems, such as coherence, learning and communication. For complex behaviours, this kind of study is necessary to provide solutions of acceptable quality.

In Section 4 we suggest analysing human evolution in order to create a roadmap of human behavioural evolution as a means to support a holistic approach.

Arguments from Evolution. Zoologists have provided one of the strongest arguments for a holistic approach to AI:

> No single characteristic could evolve very far toward the mammalian condition unless it was accompanied by appropriate progression of all the other characteristics. However, the likelihood of simultaneous change in all the systems is infinitesimally small. Therefore only a small advance in any one system could occur, after which that system would have to await the accumulation of small changes in all the other systems, before evolving a further step toward the mammalian condition.
> *T.S. Kemp [11]*

This quote was also used in [1], which in addition presents the following example. In order to maintain a constant body temperature and extend their periods of activity, warm blooded animals need to consume an order of magnitude more food than cold-blooded animals. As a result, they have changed the way they chew food, their breathing, their locomotion, their parenting behaviour, their senses, their memory capacity, and their brain size.

In cognitive modelling, we can make simultaneous changes, but we cannot make large changes to some types of behaviour without appropriately advancing others.

3 Experiments on Adaptive Behaviours

3.1 Reproducing Animal Learning

What our experiments show is that it is possible to implement certain types of learning using a step-wise BB approach rather than a direct implementation. The goal of our work is to provide new efficient solutions to learning problems where current solutions are inefficient or brittle solutions, in particular the learning of novel behaviour patterns, but also traditional problems such as natural language acquisition.

As necessary in a step-wise approach, we first looked at low level learning mechanisms. We conducted four experiments on habituation learning, spatial learning, behaviour recognition, and basic association. The experiments were chosen in order to reflect both different types of learning and different types of activities. The habituation and spatial learning experiments are concerned with navigation and feeding, the basic association experiment concerns with feeding and avoiding danger by recognising poison. The behaviour recognition is concerned with fighting and courtship displays. From the four F's of animal behaviour, feeding, fleeing, fighting and procreation, we have touched on all but procreation.

Below we present only the experiment on behaviour recognition, as it best displays the working of the step-wise approach.

3.2 Demonstrating Behaviour Recognition

The Experiment. Our first attempt at modelling a more complex form of learning with two interacting adaptive layers, was a courtship display experiment. In this experiment, two robots used a display behaviour to avoid the injuries of physical fighting. These kinds of displays are common in animals and are one of the simplest forms of animal communication [10].

The Environment. In order to simulate conflict behaviours, it was necessary to provide a simulated environment with a number of features. To support physical fighting, we gave each robot a certain strength and we simulated physical damage by making energy levels drop noticeably and proportionally to the opponents strength whenever the Kheperas were in physical contact.

3.3 A Step-Wise Solution

The solution to this restricted form of behaviour recognition consisted of three evolutionary steps or layers: the reactive interaction layer, the learning from fighting layer and the display layer.

Reactive Interaction. As a basis for more complex interactions, we implemented a reactive behaviour where a robot always tries to get in physical contact with, i.e. attack, its opponent when is sees it close by. In Figure 1 we present the circuits that implement the reactive interaction. These circuits illustrate how we build more complex learning behaviours on top of simpler solutions. The *ApproachFeederPositionController* that the reactive interaction behaviour is put on top of is the solution to the mapping experiment. Up to this level, the robots take no notice of other robots.

A Khepera sense was added which recognises when the opponent is in a threatening position, i.e. near by and facing our robot. In such cases, a touch Khepera drive which inhibits all other behaviour, approaches the other robot in a simulated fighting behaviour. This simple reactive behaviour would in the long term lead to the simulated death of the weakest robot.

Learning from Fighting. The first adaptive layer implemented to support behaviour recognition was a layer where the robots learn which one is the strongest by the amount of damage they take. A memory circuit is then used in this

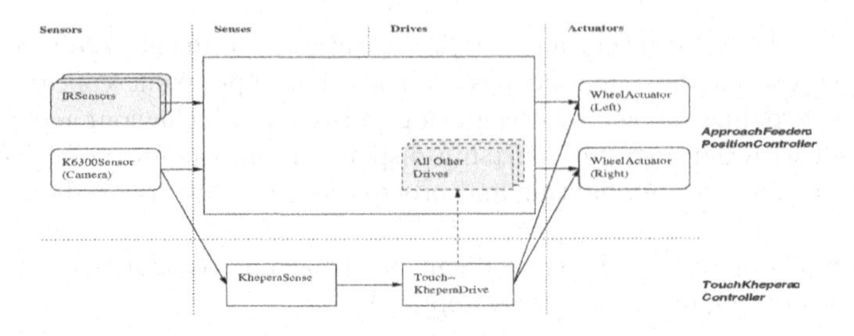

Fig. 1. Reactive Robot Interaction

behaviour to remember the pain of being the weakest robot. This memory is supported by a pain sense which picks up losses of energy and a fear emotion which is activated by the pain sense. After a fear based memory is established, an avoidance drive ensures that the weakest robot avoids its opponent in the future. The circuits involved in the fighting behaviour are presented in Figure 2.

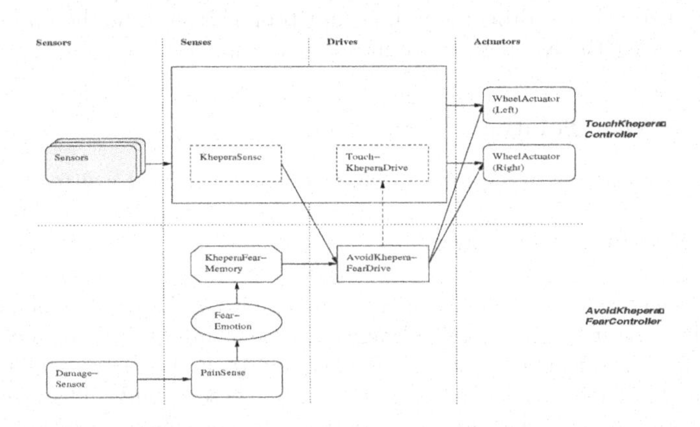

Fig. 2. Learning from Fighting

Learning from Courtship Displays. On top of the fighting layer, we implemented a courtship display layer which took the form of a stand-off initiated by the Khepera sense. In a stand off, the robots remain motionless for an amount of time corresponding to their strength. This behaviour needed a strength sense and a memory circuit to keep track of how long the robot had been displaying. These two circuits were used to support a Khepera stronger sense which was activated when it became clear that the other Khepera was stronger.

This use of memory can be described in the habituation type learning framework as increased sensitisation of a yielding behaviour, where the strength sense acts as a threshold.

The stand-off was over when one robot recognised the opponent as stronger. This recognition fired the fear emotion and a basic memory was created using the same circuit that was used in the physical fighting layer.

The final addition was to let the avoid Khepera fear drive inhibit the display drive in order to yield and as a result break up the stand-off by no longer taking a threatening stand. The circuits used to implement the display behaviour on top of the fighting behaviour are displayed in Figure 3.

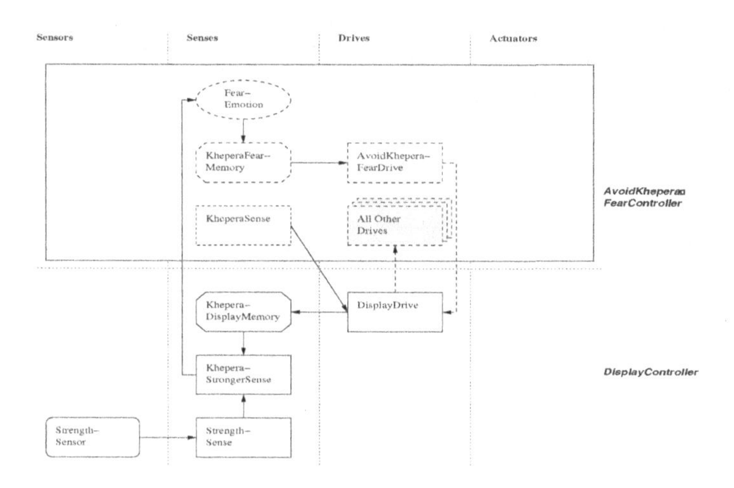

Fig. 3. Learning from Courtship Displays

3.4 Conclusions

Our experiments show that a number of different forms of animal learning can be reproduced in simulated Khepera robots using a step-wise approach. The experiments together with a PLANCS based analysis of the class of learning problems called *conditioning* [15] indicate that a step-wise and behaviourally holistic approach is sufficient for implementing these forms of learning. More complex forms of learning however would need to be studied explicitly before we can evaluate the feasibility of finding a solution using this approach.

4 A Road-Map from Evolution

When hypothesising a path of evolutionary steps to a desired behaviour, it is helpful to have a clear picture of the evolutionary background of that behaviour.

To include appropriate behaviours with sensible levels of sophistication at each step in a holistic manner, it is helpful to have an idea of what types of behaviours and levels of sophistication are likely to have coexisted during evolution.

We suggest that an analysis of human evolution in terms of co-existing behaviours of different evolutionary sophistication as well as bodily complexity can be a helpful road-map for a step-wise, holistic approach to BB robotics.

Figure 4 gives an idea of what such a road map might look like. It presents six numbered evolutionary stages. Between the stages are behaviours and physical attributes that are likely to have coexisted during evolution. Figure 4 is roughly put together from an introductory level text on evolution [16] and is only meant to convey the idea of how a road-map would look. It is not meant to exclude any dimension of behaviour that has an evolutionary history if a better knowledge of that dimension would facilitate development.

	Feeding	Fighting	Fleeing	Procreation	Sensors	Actuators	Adaptability	Habitat	Social Env.
6									
						Speech organ	Symbolic learning		
5									
	Agriculture	Armed fighting	Nesting		Colour vision	Hands	Insight	Settlements	
	Armed hunting								
4									
							Temporal association		Hierarchical Group
				Raise young			Imitation		
3									
		Group fighting	Group protection			Vocal tract			Uniform Group
2									
	Hunters	Displaying	Hiding	Mating	Stereo vision	Legs		Land	Family
			Fleeing pursuer	Release	Stereo hearing				
					Motion compensation	Head			
					Directed vision				
1									
	Grazers	Physical fighting	Moving	Division	Exterio–receptors	Body	Association	Water	Solitary
					Interio–receptors		Habituation		

Fig. 4. Human Evolution as a Road-Map to Robotics

Communication. Communication is a dimension of behaviour which was not included in Figure 4. This aspect of behaviour spans all of evolution, has a dramatically increasing complexity and on a human level constitutes the important areas of production and recognition of speech and text.

The holistic approach was developed to study learning, another type of secondary behaviour. It should also provide results in the area of communication since the evolutionary histories of the two behaviours are closely interrelated.

Presumably, there are other dimensions of behaviour or influences on behaviours that could be added to Figure 4. We continue to seek discussion about our analysis and also aim to update it with new results from the relevant sciences.

Currently Unreachable Problems. In robotics, research is currently taking place on issues on all the different levels of complexity presented in Figure 4 and very little work has been done on integrating behaviours in holistic frameworks. The behaviours on the lower levels might be studied in isolation with some credibility, while in the levels further up, one-dimensional research gets less and less useful in an AI context.

Existing research in areas like planning, reasoning, and language recognition and production, has produced important scientific results and impressive software engineering tools, but it is not obvious that any of these tools have a place in complex robotic systems.

5 Discussion

Step-Wise, Holistic Approaches and ALife. Our reason for not taking an automated search approach to evolutionary robotics the way the ALife community does [14] is primarily because we believe that it is more efficient to implement current theories from the relevant sciences directly, than it is to express those theories in the form of fitness functions and environments in order use automated search to find solutions. It can be argued that the solutions found using automated search are more robust and can take into account parameters that are not known to the developers. Automated search might also provide new knowledge about specific problem domains. We believe that these possible results do not warrant the abandonment of our approach. Our approach has an engineering emphasis rather than an automated search emphasis, but we see the two methods as complimentary and believe that the automated search approach to evolutionary robotics would also benefit from adopting our holistic principles.

Common Criticism. It has been suggested to us, and from our results it is sometimes tempting, to try to create a neural circuit based model for high level learning in order to 'solve' the problem of high level learning. We think the utility of such an effort would be limited. Our experiments are an exploration of basic forms of learning and do not test a pre-formulated hypothesis about learning. We do not want to commit to a general theory of learning such as e.g. artificial neural networks or reinforcement learning. Our effort seeks to restrict the search space for common animal learning problems by providing supporting neural structures. The learning problems that remains should be solvable using any learning technology.

The background for this reasoning is that because of the evolutionary cost of adaptability, it is likely that any evolved learning mechanism will be a simple form of learning placed in a complex behavioural context, rather than a more unconstrained complex learning mechanism with a larger probability of learning the wrong things. This belief is reinforced by the way many form of learning, previously thought to demand complex learning frameworks, such as spatial learning and imitation, turn out to be implementable using simple adaptive circuits strategically placed within complex behavioural circuitry. It is further supported by the presence of the necessary contextual neural circuitry in animals [13,9].

References

1. J. M. Allman. *Evolving Brains.* Scientific American Library, 1999.
2. R. C. Arkin. *Behaviour Based Robotics.* MIT Press, 1998.
3. R. A. Brooks. Intelligence without reason. In *Proceedings of IJCAI 91*, pages 569–595. Morgan Kaufmann, 1991.
4. R. A. Brooks. From Earwigs to Humans. *Robotics and Autonomous Systems*, 20(2–4):291–304, 1997.
5. J. J. Bryson and B. McGonigle. Agent Architecture as Object Oriented Design. In *Intelligent Agents IV, Proceedings of the Fourth International Workshop on Agent Theories, Architectures, and Languages (ATAL'97), LNAI 1365*, pages 15–30. Springer Verlag, 1997.
6. J. J. Bryson and L. A. Stein. Modularity and Specialized Learning: Mapping Between Agent Architectures and Brain Organization. In *Proceedings of the EmerNet International Workshop on Concurrent Computational Architectures Intergrating Neural Networks and Neuroscience.* Springer Verlag, 2000.
7. H. I. Christensen. The WEBOTS Competition. *Robots and Autonomous Systems Journal*, 31:351–353, 2000.
8. T. S. Dahl and C. Giraud-Carrier. PLANCS: Classes for Programming Adaptive Behaviour Based Robots. In *Proceedings of the 2001 Convention on Artificial Intelligence and the Study of Simulated Behaviour (AISB'01), Symposium on Nonconscious Intelligence: From Natural to Artificial*, 2001.
9. V. Gallese, L. Fadiga, L. Fogassi, and G. Rizolatti. Action Recognition in Pmotor Cortex. *Brain*, 119:593–609, 1996.
10. M. D. Hauser. *The Evolution of Communication*, chapter 6 Adaptive Design and Communication, pages 450–470. MIT Press, 1996.
11. T. S. Kemp. *Mammal-like Reptiles and the Origin of Mammals.* Academic Press, 1982.
12. B. R. Moore. The Evolution of Imitative Learning. In C. M. Heyes and B. G. Galef, editors, *Social Learning in Animals: The Roots of Culture*, pages 245–265. Academic Press, 1996.
13. R. U. Muller, J. L. Kubie, E. M. Bostock, J. S. Taube, and G. J. Quirk. Spatial firing correlates of neurons in the hippocampal formation of moving rats. In J. Paillard, editor, *Brain and Space*, chapter 17, pages 296–333. Ocford University Press, 1991.
14. S. Nolfi and D. Floreano. *Evolutionary Robotics: The Biology, Intelligence, and Technology of Self-Organizing Machines.* MIT Press, 2000.
15. J. M. Pearce. *Animal Learning and Cognition.* Psychology Press, 2nd edition, 1997.
16. M. W. Strickberger. *Evolution.* The Jones and Bartlett Series in Biology. Jones and Bartlett Publishers, Second edition, 1995.

Evolving Bipedal Locomotion with a Dynamically-Rearranging Neural Network

Akinobu Fujii[1], Akio Ishiguro[1], Takeshi Aoki[2], and Peter Eggenberger[3]

[1] Dept. of Computational Science and Engineering, Nagoya University
Nagoya 464–8603, Japan
{akinobu, ishiguro}@cmplx.cse.nagoya-u.ac.jp
[2] Nagoya Municipal Industrial Research Institute
Nagoya 456–0058, Japan,
aoki@nmiri.city.nagoya.jp
[3] ATR Information Science Division
Kyoto 619–0288, Japan
eggen@isd.atr.co.jp

Abstract. Since highly complicated interaction dynamics exist, it is in general extremely difficult to design controllers for legged robots. So far various methods have been proposed with the concept of neural circuits, so–called Central Pattern Generators(CPG). In contrast to these approaches in this article we use a polymorphic neural circuit instead, allowing the dynamic change of its properties according to the current situation in real time. To do so, we introduce the concept of neuromodulation with a diffusion–reaction mechanism of neuromodulators. Since there is currently no theory about how such dynamic neural networks can be created, the evolutionary approach is the method of choice to explore the interaction among the neuromodulators, receptors, synapses and neurons. We apply this neural network to the control of a 3–D biped robot which is intrinsically unstable. In this article, we will show our simulation results and provide some interesting points derived from the obtained results.

1 Introduction

Legged robots show significant advantages over wheeled robots since they can traverse in uneven and unstructured environments. This high mobility stems from the fact that in contrast to wheeled or tracked robots legged robots *discretely* make contact with their environments via their legs. This, however, inevitably causes highly complicated dynamics between the robots and their environments. Due to this complicated interaction dynamics, it is in general extremely difficult to design controllers for legged robots.

On the other hand, natural agents(i.e. animals and insects) can show adaptive and agile locomotion even under unstructured environments. Investigations in neurophysiology suggest that such remarkable locomotion(e.g. walking, swimming, flying) is generated by specific neural circuits, so–called *Central Pattern*

J. Kelemen and P. Sosík (Eds.): ECAL 2001, LNAI 2159, pp. 509–518, 2001.

Generators(CPG). Based on these findings, so far various methods have been proposed for legged-robots with artificial CPG controllers consisting of a set of *neural oscillators* [1][2][3].

In contrast to these approaches in which *monolithic* CPG neural circuits are used to control locomotion, in this article we use a *polymorphic* neural circuit instead, allowing dynamic change of its properties according to the current situation in real time. This is because what we are intrigued with is not to simply generate stable locomotion patterns but to investigate how sensory information modulate locomotion patterns according to the current situation. To this aim, we introduce the concept of *neuromodulation* with a *diffusion–reaction mechanism* of chemical substances called *neuromodulators*. As there exists no theory about how such dynamic neural networks can be created, the evolutionary approach is the method of choice to explore the interaction among the neuromodulators, receptors, synapses and neurons.

Here as the initial step of the investigation, we attempt to create neural controllers with a neuromodulation mechanism for a 3–D biped robot on flat terrain. Simulations were carried out in order to verify the feasibility of our proposed method. We also provide some interesting points derived from the obtained results. We envision these points will be very crucial particularly in the case of evolving controllers for biped robots which are intrinsically unstable.

2 Related Works

Concerning the evolutionary creation of controllers for legged–robots, so far various methods have been proposed. Beer et al. evolved dynamically recurrent neural networks for a hexapod robot[4]. Jacobi introduced the concept of *minimal simulation* to not only speed up the evolutionary process but also to bridge the gap between simulated and real environments. He applied this concept to evolve a controller for an octopod robot[5]. Gruau et al. proposed the *Cellular Encoding scheme* based on Genetic Programming, and implemented it to evolve a hexapod controller[6]. Kodjabachian proposed a *geometry–oriented encoding scheme* called SGOCE and evolved a neurocontroller for a legged robot[7]. Ijspeert has evolved a controller to mimic a salamanders' locomotion and the resultant controller can smoothly switch between swimming and walking[2].

There are mainly two points to be noted from the above–mentioned works. First, most of these methods are based on *monolithic* neural networks, that is the properties of the controllers such as synaptic weights are fixed once acceptable controllers for the given task are evolved. Second, as mentioned above so far various methods have been proposed for quadruped, hexapod and octopod robots, whilst very few have investigated the evolutionary creation of controllers for biped robots in spite of their remarkable mobility. This is presumably due to their high instability.

3 Dynamically-Rearranging Neural Networks

3.1 Lessons from the Biological Findings

Interestingly, neuroscientific results suggest that biological networks not only adjust the synaptic weights but also the neural structure by blocking or activating synapses or neurons by the use of signaling molecules, so–called neuromodulators. These findings stem from investigations made with lobsters' stomatogastric nervous systems in which certain active neurons diffuse neuromodulators which then rearrange the networks. Note that the effect of a neuromodulator depends not only on these substances, but also on the specific receptors, which are expressed differently in different cells. The effect on a cell depends therefore on the interaction between the neuromodulator and the receptor and not just the neuromodulator alone.

The release of the neuromodulators depends on the activity of the neurons and therefore different sensory inputs may cause different patterns of released neuromodulators. As such dynamic mechanisms yield remarkable adaptation in living organisms, we expect the proposed approach to not only carry promise for a better understanding of adaptive networks, also apply to real–world problems as we already showed in previous work[8].

3.2 Basic Concept of the Dynamically-Rearranging Neural Networks

The basic concept of our proposed dynamically–rearranging neural networks (hereafter: DRNN) is schematically depicted in Fig.1. As in the figure, unlike conventional neural networks, we assume that each neuron can potentially diffuse its specific(i.e. genetically–determined) types of NMs according to its activity, and each synapse has receptors for the diffused NMs. We also assume that each synapse independently interprets the received NMs, and changes its properties(e.g. synaptic weight). The way these changes are exerted on the synapses is also genetically–determined. By selecting regulatory feedback loops(i.e cyclical

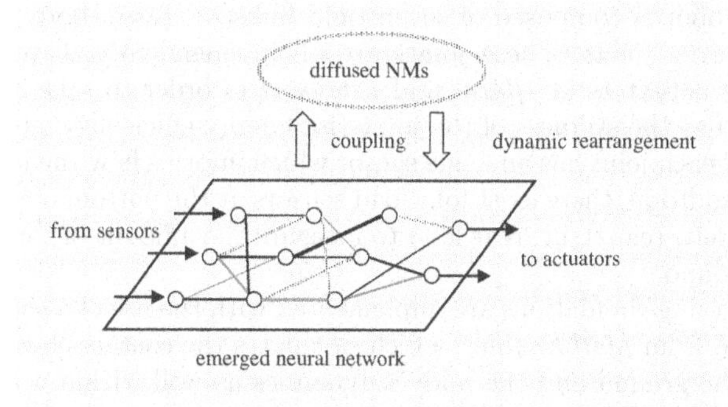

Fig. 1. Basic concept of the DRNN.

interaction between the diffusion and reaction of NMs), we expect to be able to evolve adaptive neural networks, which show not only a seamless transfer from simulations to the real world but also robustness against environmental perturbations(in the figure, the thick and thin lines denote the connections being strengthening and weaking by NMs, respectively). It should be stressed that mechanisms for *situation evaluation*, *behavior generation* and *behavior regulation* can all be embedded within a monolithic neural network. Due to this remarkable feature, we expect that whole genetic information can be reduced rather than straightforward approaches.

In summary, in contrast to conventional Evolutionary Robotics approaches that mainly evolve synaptic weights and neuron's bias of neuro–controllers, in this approach we evolve the following mechanisms instead:

- Diffusion of NMs(i.e. when, and which types of NMs are diffused from each neuron?)
- Reaction to NMs(i.e. how do the receptors on each synapse interpret the received NMs, and modify the corresponding synaptic property?)

To determine the above parameters, we use a Genetic Algorithm(GA). For detailed explanation of the DRNN, see [8][9].

4 Proposed Method

4.1 Task

Our final goal is to create adaptive controllers for biped robots that can appropriately cope with various situations, and to show the feasibility of the neuromodulation mechanism as well. Here as the initial step of the investigation, we take the following task as a practical example: walking on flat terrain in a 3–D simulated environment.

4.2 Biped Robot Model

The 3–D biped robot model used in this study is shown in Fig.2. As in the figure this robot is composed of seven rigid links(i.e. torso, body, thigh, shank, foot) and seven joints. These joints are all independently driven by pairs of antagonistic actuators(i.e. *flexor* and *extensor*) in order to take not only static torque but also the stiffness of the joints(for energy efficiency) into account. We assume that each joint has an angle sensor which informs how the joint concerned rotates. In addition, there exist four load sensors at the bottom of each foot(front right, front left, rear right, rear left) to measure the amount of the vertical force from the ground.

The following simulations are implemented with the use of a simulation package provided from *MathEngine*[1], which calculates the contacts between the rigid body and the ground and the body's dynamics as well within acceptable time. The body parameters used in the following simulations are listed on Table 1.

[1] http://www.mathengine.com

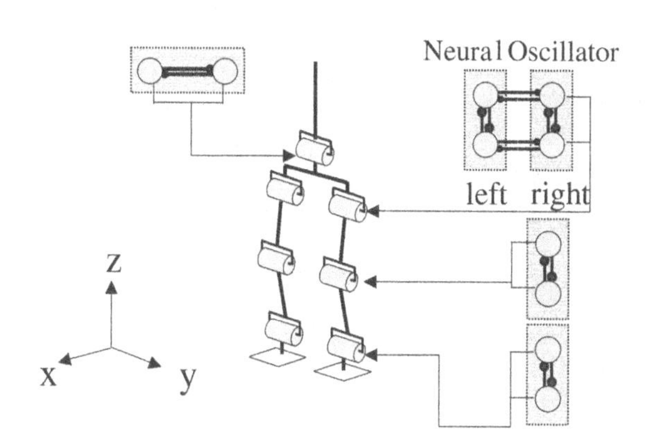

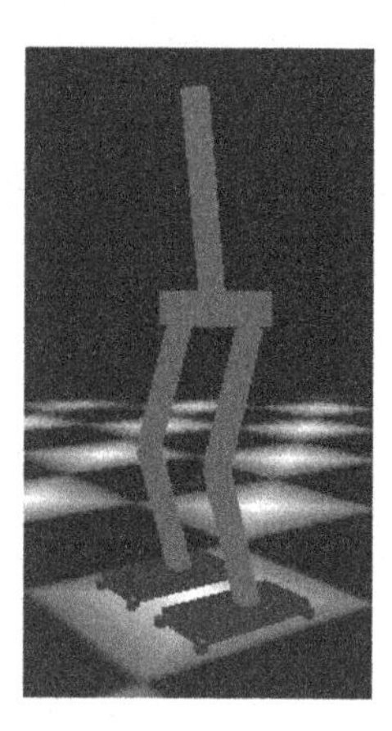

Fig. 2. Model of the 3–D biped robot used in the simulation.

Table 1. Body parameters of the biped robot.

part	size[m]	mass[kg]
torso	0.28	2.0
body	$0.05 \times 0.18(d \times w)$	1.0
thigh	0.2	2.0
shank	0.215	2.0
foot	$0.2 \times 0.1(d \times w)$	1.0

4.3 DRNN Controller

Fig.3 schematically represents the structure of the DRNN controller for the biped robot. We use an oscillator–based neural controller for this purpose inspired by Taga's model[3]. This controller consists of a set of neural oscillators each of which controls its specific joint.

Neural oscillator. In this study, we use a neural oscillator consisting of two neurons proposed by Matsuoka[10]. The dynamics of each neuron in this model is governed by the following differential equations:

$$\tau_r \frac{du_i}{dt} = -u_i - \sum_{j=1}^{n} w_{ij} y_j + s_i - b f_i \quad (1)$$

$$\tau_a \frac{df_i}{dt} = -f_i + y_i \quad (2)$$

$$y_i(u) = max\{0, u\} \quad (3)$$

where, u_i is the membrane potential of neuron i. f is a variable representing the degree of the self–inhibition effect of the neuron concerned. w_{ij} represents the

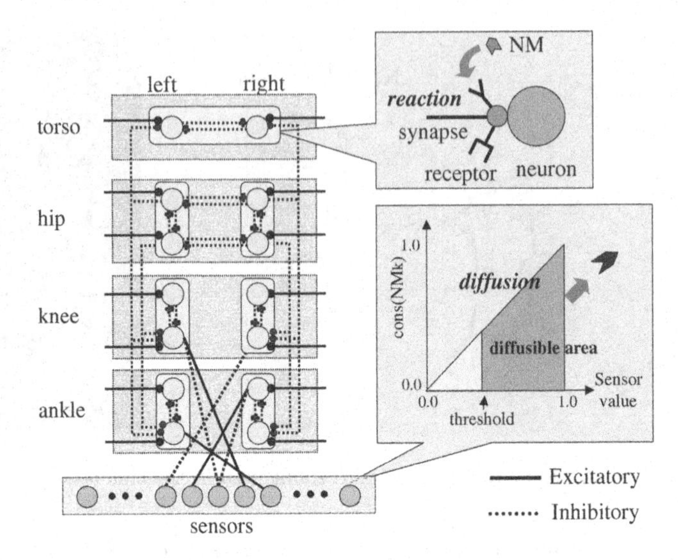

Fig. 3. Controller for the biped robot.

synaptic weight of a connection from neuron j to neuron i. y is the output of the neuron. τ_r and τ_a denote the time constant. s is an external input(e.g. sensory feedback signal).

Diffusion of NMs. In this study, we assume that each sensory neuron is able to diffuse its specific(i.e. genetically–determined) type of NM under a condition which is also genetically–determined(see Fig.3). As the figure indicates, we assume that the concentration of the diffused NM, denoted as $Con(NM)$, is proportional to the activity of the neuron concerned when the activity of the sensor neuron is within the diffusible area. Here, for simplicity, we assume that each neuron can diffuse at most one type of NM. We also assume the number of NM types is four.

Reaction to the diffused NMs. Here we assume that the synapses on the interconnections in each neural oscillator and excitatory connections among the oscillators have receptors for the diffused NMs. Once the NMs are diffused, each synaptic weight will be modified according to its own genetically–determined interpretation as:

$$w_{ij}^{t+1} = w_{ij}^{t} + \eta \frac{dX}{dt} \tag{4}$$

$$X = \sum_{k=1}^{n} R_{ij}(NM_k) \cdot Con(NM_k) \tag{5}$$

where, η is the learning rate, $Con(NM_k)$ denotes the total concentration of the diffused NM of type of k in the network at a given time, and n is the number of NM types. $R_{ij}(NM_k)$ is the parameter which determines how the synapse concerned interprets its received NM of type k and modifies its weight. $R_{ij}(NM_k)$ can take one of four values; $-1, 0, +1$. We also evolve the way the sensors are connected to the neurons.

4.4 Fitness

The evaluation criterion used for the evolutionary process is:

$$fitness = body_x \times footR_x \times footL_x, \tag{6}$$

where, $body_x$ is the resultant distance at the position of the robot body during the evaluation period(30 seconds). $footR_x$ and $footL_x$ represent the resultant distance of the right and left foot, respectively. Thus this equation encourages agents that can successively move forward by alternatively stepping forward. If the robot falls down during the evaluation period, the trial will be forcibly terminated.

5 Results and Discussion

5.1 Simulation Results

Fig.4 shows snapshots of the obtained locomotion. As in the figure, the robot can successively continue the stepping motion on the flat terrain(see from left to right in the figure). Fig.5 shows the torques(hip, knee, ankle) applied to the left and right leg during the stepping. Fig.6 depicts a typical example of the synaptic weights during this locomotion. As the figure indicates, during the stepping the torques applied to the joints are dynamically regulated by changing the synaptic weights. Detailed analysis is currently under investigation.

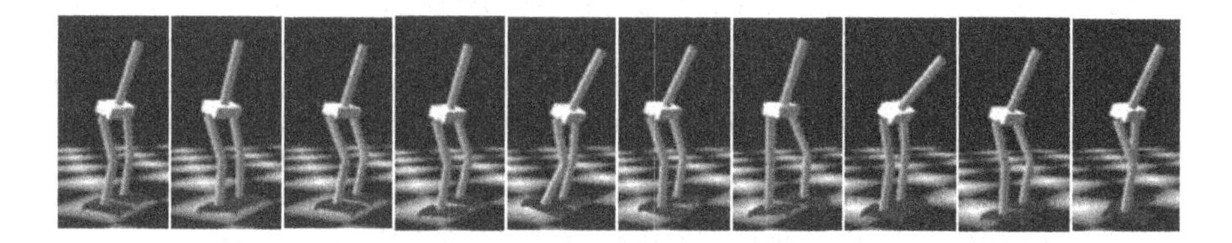

Fig. 4. Obtained locomotion.

5.2 Discussion

In this section, we provide two interesting points to be noted that are derived from the above mentioned results.

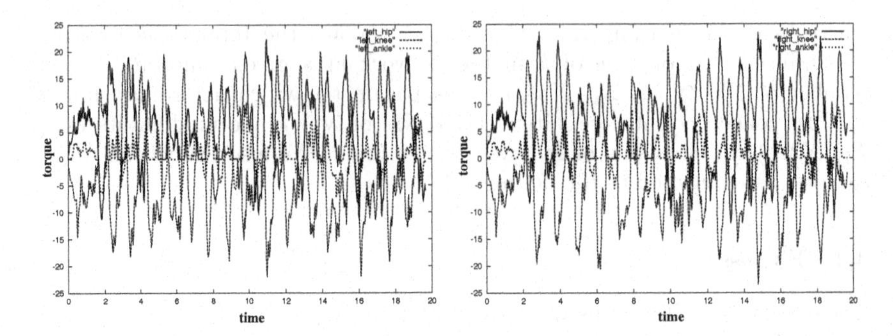

Fig. 5. Transition of the torques(hip, knee, ankle) applied to the left and right legs.

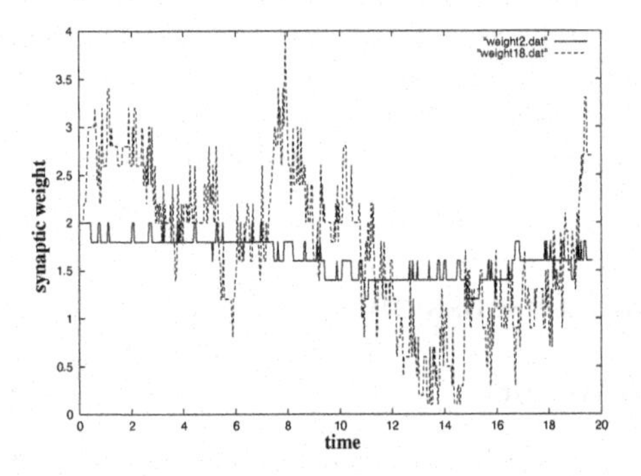

Fig. 6. An example of transition of the synaptic weight.

First, although the obtained neural controller enables successive walking by changing some of the synaptic weights during locomotion as shown in Fig.6, we frequently observed that the evolved controllers end up to possess many synapses with fixed weights. This implies that we could not make the best use of the DRNN under the condition of the simple locomotion on a flat terrain used in this simulation. In other words, under the current environmental setup, even a monolithic controller might be able to be successfully applied. Therefore, it is highly indispensable to evolve the DRNN controller under the existence of environmental perturbation(e.g. external force, slope, obstacles) and compare the resultant adaptability with standard monolithic CPG networks.

Second, we could hardly evolve natural locomotion pattern. Most of the evolved controllers move their bodies slowly by mainly relying on stepping motion as in Fig.4. One way for creating natural gaits is to ameliorate the fitness function by taking not only the resultant distance the robot traveled but also the consumed energy during the locomotion into account. Another way, which is crucially important, is closely related to the embodiment issue, so–called *passivity*. Torsten nicely showed that bipedal robots with passive joints(in his simulation, there exists no actuation at the knee joints) provide high *evolvability* compared to the embodiments without passivity[11]. Although his main stress is to increase evolvability, passivity is expected to contribute to improve energy consumption as well. Therefore, evolving controllers with the use of the embodiment which can potentially realize *passive dynamic walking*[12] would be a nice idea to create natural gaits with low energy consumption as well as high evolvability. Recently, Steven et al. developed a wonderful 3–D passive dynamic walker[13]. Based on this consideration, we are currently constructing a simulated 3–D passive dynamic walker and trying to evolve DRNN controllers with this embodiment(see Fig.7).

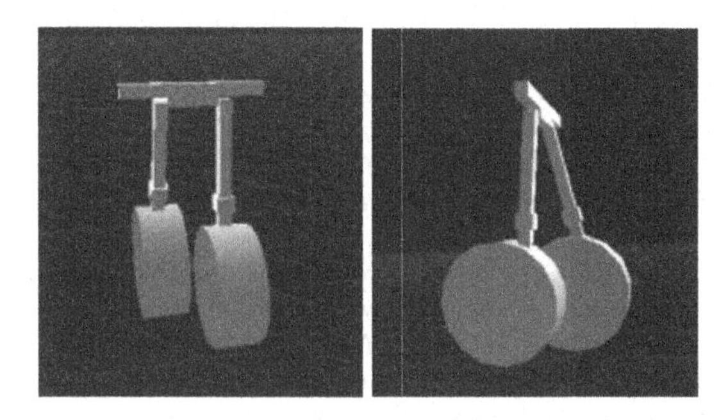

Fig. 7. Passive dynamic walking model.

6 Conclusion and Future Works

In this paper, an evolutionary creation of an adaptive neurocontroller for a biped robot was investigated. To this end we introduced the concept of the neuromodulation widely observed in biological nervous systems.

Currently we are ameliorating the encoding scheme which efficiency creates adaptive DRNN, and investigating adaptability by evolving under environmental perturbations. We are also trying to create more natural gaits based on the embodiment that we potentially achieve passive dynamic walking. Furthermore, in order to persuasively shows adaptability, we are going to investigate whether the same evolved DRNN controller can be seamlessly applied to different embodiment(e.g. body parameters such as mass, length of the leg etc.).

Acknowledgements. The authors would like to thank Auke Ijspeert at University of Southern California, Josh Bongard and Chandana Paul at AI Lab, University of Zurich for their many helpful suggestions for the simulator MathEngine. This research was supported in part by a grant from the Japanese Ministry of Education, Culture, Sports, Science and Technology(No.C:12650442).

References

1. H. Kimura and Y. Fukuoka. Biologically inspired dynamic walking of a quadruped robot on irregular terrain – adaptation at spinal cord and brain stem –. *Proc. of International Symposium on Adaptive Motion of Animals and Machines, AMAM2000, CD-ROM*, 2000.
2. A. J. Ijspeert. A neuromechanical investigaiton of salamander locomotion. *Proc. of International Symposium on Adaptive Motion of Animals and Machines,AMAM2000, CD-ROM*, 2000.
3. G. Taga, Y. Yamaguchi, and H. Shimizu. Self-organized control of bipedal locomotion by neural oscillators in unpredictable environment. *Biological Cybernetics,* **65**, *147-159*, 1991.
4. R. D. Beer and J. C. Gallagher. Evolving dynamical neural networks for adaptive behavior. *Adaptive Behavior, vol.1, no.1, 91–122*, 1992.
5. N. Jakobi. Running across the reality gap:octppod locomotion evolved in a minimal simulation. *Proc. of Evorob98, Springer Verlag, 39–58*, 1998.
6. F. Gruau and K. Quatramaran. Cellular encoding for interactive evolutionary robotics. *Proc.of the Fourth European Conference on Artificial Life, ECAL97, MIT Press, 368–377*, 1997.
7. J. Kodjabachian and J.-A. Meyer. Evolution and development of neural networks controlling locomotion, gradient-following, and obstacle-avoidance in artificial insects. *IEEE Transactions on Neural Networks, 796–812*, 1998.
8. P. Eggenberger, A. Ishiguro, S. Tokura, T. Kondo, T. Kawashima, and T. Aoki. Toward seamless transfer from simulated to real worlds: A dynamically–rearranging neural network approach. *Proc. of the Eighth European Workshop on Learning Robot (EWLR-8), 4–13*, 1999.
9. A. Ishiguro, K. Otsu, A. Fujii, Y. Uchikawa, T. Aoki, and P.Eggenberger. Evolving an adaptive controller for a legged–robot with dynamically–rearranging neural networks. *Proc. of the 6th International Conference on The Simulation of Adaptive Behavior (SAB2000)*, 2000.
10. K. Matsuoka. Mechanisms of frequency and pattern control in the neural rhythm generators. *Biological Cybernetics,* **56**, *345–353*, 1987.
11. T. Reil and C. Massey. Facilitating controller evolution in morpho-functional-machines-a bipedal case study. *Proc. of International Workshop on Morpho-functional Machines,165–181*, 2001.
12. T. McGeer. Passive dynamic walking. *The International Journal of Robotics Research,9(2),62–82*, 1990.
13. S. H. Collins, M. Wisse, and A. Ruina. A 3-d passive–dynamic walking robot with two legs and knees. *The International Journal of Robotics Research*, 2001.

SlugBot: A Robot Predator

Ian Kelly[1] and Chris Melhuish[2]

[1]CORO, Microsystems Laboratory, California Institute of Technology, Pasadena, USA
iankelly@micro.caltech.edu
[2]IASeL, Faculty of Engineering, University of the West of England, Bristol, UK

Abstract. One of the key aspects of most living organisms is their ability to detect and exploit natural sources of energy within their environment. We are currently developing a robotic predator system that will attempt to sustain itself by hunting and catching slugs on agricultural land. A microbial fuel cell will be used to convert the slug bio-mass to electricity thus providing the robot's energy supply. This paper outlines the requirements for such a predator and describes recent progress on the robot. We also present data from trials of the robot hunting and catching slugs in a situation similar to that found in an agricultural field.

1 Introduction

Living creatures are for most, if not all, of their lives truly autonomous in a way that no human produced system is. Most living creatures are able to survive and operate successfully in an unstructured environment without requiring any assistance whatsoever. Autonomy, whether biological or artificial, consists of two major aspects: computational autonomy, and energetic autonomy. Computational autonomy refers to the ability to determine and carry out actions independently, whilst some of these actions may be related to the acquisition of energy, most will be concerned with other aspects of system operation. In the context of a biological system, the other aspects are associated with survival and reproduction, such as avoiding predators, finding shelter, and finding mates. In a robotic system, the other aspects would be associated with performing supplementary tasks as well as avoiding hazards. Energetic autonomy refers to the independent ability to maintain the internal availability of energy above the lethal minimum for sufficiently long periods to enable the system to achieve its mission, which in the biological context corresponds to securing the effective propagation of its genes. Autonomy does not merely involve making correct decisions in order to secure the raw energy; it also includes the conversion of the raw energy source into a usable form.

A typical investigation in artificial life of so-called 'autonomous agents' generally involves simulated agents attempting to survive and reproduce in a world containing spatially localised elements corresponding to 'food' and 'predators', and so on. Although such abstract studies have been genuinely useful in exploring the dynamics of such situations, all fall very far short of the complexities faced by an actual animal in the real world. Recently the design and control of autonomous robots has formed a major area of academic and industrial research. There are now many examples of robots or automated mobile systems (such as missiles, smart torpedoes, and some

J. Kelemen and P. Sosík (Eds.): ECAL 2001, LNAI 2159, pp. 519-528, 2001.

spacecraft) which achieve an apparently high degree of energetic and computational autonomy. Such systems carry enough fuel for their mission or can use radiant energy from their environment, and can control themselves 'intelligently' without human intervention. Some automated cleaning and materials handling AGVs use opportunity battery charging to achieve a degree of autonomy. Several academic research groups have constructed robot environments which feature a 'powered floor', giving the possibility of indefinitely extended operation. However, on reflection, it is clear that most of these so-called 'autonomous' robots still require some explicit or implicit intervention from humans in order to carry out their tasks. Forms of human intervention include supplying information and energy, physically assisting the robot and modifying the environment to suit the robot. The issue of autonomy has been finessed, rather than having been confronted and overcome.

This is by no means the first identification of the lack of autonomy in artificial agents – similar observations in a slightly different context were perceptively articulated by Steels [1] several years ago. However, the project described here represents the first serious attempt to design and construct a robot system with energetic and computational autonomy comparable to an animal system. We have sought to guarantee the comparability by constraining a robotic system to obtain its energy in the same way as most animals - by finding and 'digesting' organic material in an unstructured environment. This decision immediately brings a host of problems in its train. For instance, natural resources of this type are found only in certain types of places, and may only appear transiently, depending on time of day, season, and weather conditions; most importantly, the resources are destroyed by being used. The control of the foraging process must therefore take into account issues such as where and when to look for food, when to revisit an area where food was previously found, when to abandon a site that is not producing food, and so on. Once found, the organic resources, or food, need to be converted to an appropriate form of energy for storage and use. We initially proposed to convert the organic material to electricity by using a mixture of biological and advanced engineering technologies: an initial and quite standard anaerobic fermentation process would have been used to obtain biogas. The biogas would then have been passed through a specially developed tubular solid oxide fuel cell that generates electricity directly. Modern fuel cells of this type can produce electricity from biogas containing methane at concentrations of 20% or less [2]. Our current focus is on microbial fuel cells. In these the electrodes are placed directly into the organic food source, thus eliminating the need for a separate fermentation vessel.

The nature of the organic material to be used imposes further constraints. One option would be to use vegetable matter, which means that the robot would correspond to a grazing animal. Whilst grazing is not a trivial problem, especially when resources are scarce, we opted for the challenge of hunting a mobile animal species. The choice of a prey species was guided by several considerations. Most obviously, the prey species should be reasonably plentiful and not require rapid pursuit, which would be difficult to achieve over soft ground at a reasonable energy cost. It is also likely that there would be some minimum energy cost associated with finding, catching, and consuming any individual creature, however large or small, and so the prey species should be large enough to give a significant margin over the minimum energy expended. Finally, in order to conform to ethical considerations, it should be an invertebrate pest species already subject to lethal control measures, so that the system would be doing something of actual use that would have to be done anyway. All of the above criteria are met by the slugs found on agricultural land in the

UK, especially *Deroceras reticulatum* [3]. They are slow moving, abundant, relatively large, and extremely destructive - UK farmers spend over £20m per annum on buying and spreading molluscicides [4]. Slugs are also potentially more suitable for fermentation than some other possible target species: they do not have a hard external shell or exoskeleton and have a high moisture content.

Slugs are quite general pests, but perhaps do most damage to crops requiring a well cultivated seedbed, such as winter wheat or potatoes [5]. Following cultivation and planting, such ground is very soft; this means that a heavy fermentation vessel could not be moved over it without consuming large amounts of energy. The fermentation vessel/fuel cell system will not therefore be part of the robot, but will be in a fixed location, and the robot will bring slugs to it, and collect power from it. Inevitably, the fermenter will require a certain amount of energy to cover operating overheads, and our initial calculations indicated that a single robot would be unlikely to be able to gather sufficient energy to service both its own and the fermenter's energetic requirements; a multiple robot system must therefore be envisaged. Rather than being a handicap, this confers many potential advantages: the multiple approach gives a potential for achieving system reliability through redundancy. Interestingly, a robot that has run out of power away from the recharging station could be rescued by receiving power from another robot. As regards to hunting strategies we cannot copy an existing animal, because other than microscopic nematode worms, which borrow inside slugs to feed on them, there are no predators that feed exclusively on slugs. Although hedgehogs and some species of birds supplement their diets with slugs when they are available.

2 The Prototype Robot

To allow the robots to operate in unmodified agricultural fields and from the very limited energy resource represented by slugs, the prototype robot has been built taking into account the following considerations: energy efficiency, reliability, and operation outside on soft and irregular ground. Rubber gaiters will be used to protect the currently exposed mechanical systems from the weather, slug slime, and mud. Energy efficiency has been achieved in several ways:

- using light but strong materials like carbon fibre and aluminium
- using decentralised low power controllers and electronics - instead of a single high speed central processor - thus allowing currently unused devices to be shut down
- the use of physical designs and control strategies designed to optimise efficiency

For example, the energy requirements for hunting and catching slugs have been minimised as follows: Any system designed for finding and capturing slugs requires a means of searching the surface of the ground for slugs, and a way to pick them up. In our design, the sensor used for detecting slugs, and the gripper used for catching them, are both located at the end of a long articulated arm. The energy required to move the end of this arm any distance across the surface is much less than the energy required to move the whole robot through the same distance over a soft and muddy field. Although of course the robot will still need to move in-between the hunting patches. The arm is mounted on a turn table which requires less power than that required to move the arm. Hunting for slugs is achieved by firstly moving to the centre of the patch to be searched. The ground around the robot is then scanned by rotating the arm around the robots' base. At the beginning of each revolution the arm is extended by 100mm, until the arm is extended to its maximum, at this point the arm is retracted and the robot will move to the centre of the next patch to be scanned. This step-circling strategy guides the image sensor and hence the gripper, in the most energy efficient method possible. Whilst scanning, the distance above the ground of the sensor and gripper is actively maintained within the band of focus of the image sensor. When a slug is found, it is picked up by the gripper (which is already in the correct location) and transferred to the on-board storage container. The arm is then returned to its former location so that scanning can continue. When all the worthwhile slugs within a patch have been collected, the robot will move to a new location and repeat the procedure. When the onboard slug container is full, or when the robot needs power, or when it appears that more energy will be used in hunting than is expected to be gained from it, the robot will return to the fermenter, where it deposits any slugs caught and recharges itself if necessary.

Fig. 1. Prototype three fingered gripper with wiper blades and compliance gimbal (left), and the arm and gripper system mounted on a turn table (right)

The optimal length of the arm is a function of the spatial density and distribution of slugs, and also of the power required operating arms of different lengths and weights in relation to the power required to move the whole robot. Obviously, the longer the arm, the more ground it can cover without having to move the robot, but the more power will be required both for scanning and for picking up slugs. Our calculations showed that with between 1 and 10 detectable slugs per square metre, an arm with a nominal base length of around 1.5m would be the most efficient, but where there are many more slugs a shorter arm would be more efficient. On the basis of these results and some initial slug counts, we have opted for a 1.5m long arm.

The arm is light (and hence energy efficient), stiff, and easily controlled. Using opto-encoders it is possible to manoeuvre the tip within a few millimetres of the desired location. It has a simple construction, thus increasing its reliability, and making it easier to manufacture and reducing its cost. The final version consists of two 0.75m tubular sections, with a hinged joint between them (see Fig. 1, right). In order to allow the arm to rotate around the whole robot, it is mounted on a powered turntable located in the centre of the robot's chassis. The chassis is large enough to maintain stability in all directions when the arm is fully extended and loaded; the wheelbase dimensions were minimised by mounting the robot's batteries on the turntable on the opposite side from the arm, thus acting as a counterbalance. The arm is constructed from carbon fibre tube to meet the requirements of lightness and stiffness. To keep the arm structure light and the inertia low, the motor and gearbox required to provide movement at the elbow joint are mounted on the turntable; the drive to the elbow joint is transmitted by a lightweight toothed belt inside the arm. Since the numbers of slugs available on the surface peak in the early evening and just before dawn, the rate of gathering them during these periods must be as high as possible. To this end, the arm motors and gearboxes were selected so that the arm can move from fully retracted to fully extended, or vice-versa, in under 1.5 seconds. Self locking worm gearboxes provide the required motor speed reduction, and allow the arm to be held in position without consuming any energy. Most of the motors are fitted with optical encoders to provide position and velocity feedback.

The arm's end-effector is a robust lightweight gripper capable of picking up and releasing both wet and dry slugs, regardless of their size, sliminess, and orientation, and in the presence of irregularities in the substrate. Several design iterations were necessary; the current version consists of three fingers at 120 degrees spacing, operated by a single miniature motor (see Fig. 1, left). As the fingers close, they meet underneath the slug so that it can be lifted; when the gripper is opened, wiper blades ensure the slug's release, irrespective of the amount of slime present.

Slugs are detected and targeted visually, and the camera carrying out this function is mounted in the centre of the gripper, away from slugs and mud, and ideally positioned to provide accurate feedback of the position of the gripper with respect to a target slug. To ensure that the view from the camera is always perpendicular to the ground, regardless of the arm's extension, the whole gripper assembly normally hangs freely on a gimbal. The gripper assembly can be locked in position by fully opening the gripper, thus stopping it from swinging. Underneath each of the three wiper blades is a plate which produces passive alignment with the contours of the ground when the gripper descends over a slug, ensuring that all three blades move under the slug when the gripper closes. The gripper mechanism will be protected from the weather, mud, and slime by a flexible cover.

Since the fermentation and recharging station will be in a fixed location, it is vital that the robots are always able to locate and return to it. The nature of the terrain means that wheel slip and the irregularity of the substrate will preclude any extended use of odometry. Initially for pragmatic reasons we will take advantage of non-biological technology by using a combination of the Global Positioning Satellite (GPS) system, and an active infrared localisation system [6, 7]. GPS will also be used for mapping the locations of grazed areas, so that good patches can be found again, and over-grazing can be avoided. Although using GPS can be considered to be artificially modifying the environment it could be replaced by a location system based on visual landmarks. We have opted for GPS for several reasons: our focus is on hunting for and producing energy from slugs, for the foreseeable future the US will maintain the GPS satellites, and GPS can be considered to be part of the environment since it doe not require the setting up any external hardware. Overgrazing may not be a problem: a study undertaken for this project [8] found that the removal of all surface slugs from a field location every few days does not, in the medium term, appear to reduce the number of available surface slugs. It appears that a large reservoir of underground slugs are always present; as slugs are removed from the surface more slugs migrate upwards to replace them. This is an advantage in terms of energy supply, but would be a disadvantage if SlugBots, or biological predators, are used for slug control.

Although the fields in which trials will be undertaken are quite large and empty, obstacles may be present from time to time, and the robot will carry the normal complement of obstacle avoidance sensors required by any mobile robot. Obstacle detection will be achieved using a combination of ultrasonic sonar and, as a last resort, bump sensors. In addition, two sets of miniature ultrasonic sonar transceivers have been placed in the gripper: one set points downwards so that the sensor for slug detection can be kept at the optimal distance from the ground regardless of any irregularities in the ground, and a second set faces outwards to detect any obstacles in the path of the gripper.

3 The Electronic Control Architecture

The robots' control system is quite complex: there are five motors in total (gripper, elbow, shoulder, turntable, left wheel drive, and right wheel drive) which must be precisely controlled, and several sensors (shaft encoders, imaging, obstacle avoidance, scan height sonar, battery charge remaining, limit switches etc.) which must be monitored. Instead of centralising all the processing, we have opted for a distributed system with the emphasis on local processing, allowing us to use a low-speed three-wire serial (I^2C) bus to connect all the major subsystems. However, the high-level control of the robot is still handled by a single processor. By using this decentralised approach, the overall system complexity has been decreased, and the reliability should have been increased. For example, all the motion controllers, except for the gripper, share the same design – a PIC microcontroller handles communication and basic processing, and a Hewlett-Packard HCTL-1100 closed-loop motion controller manages the shaft encoders and motor drives; each motor controller has a unique address on the I^2C bus.

4 Detecting Slugs

Slugs cause most damage to crops, like winter wheat, during the first few months after the crop has been planted. During this period the crop is sparse with a relatively low uniform height allowing an excellent view of the ground where the slugs will be. The crop is also very flexible and springs back up if flattened, allowing the robots to freely roam around the field without causing damage to the crop. However, field observations and experiments showed that even under these relatively favourable conditions, the visual detection of slugs is a difficult problem, mainly because of the presence of non-slug objects such as living and dying vegetation, stones, and lumps of soil. The slug detection system has to be accurate, since any objects that are collected that are not slugs waste energy. Just as importantly the detection system has to be quick, since most slugs feed on the surface during two short periods each day; for a couple of hours after sunset and a couple of hours before sunrise. During these periods energy returns will be higher since more slugs are available. Like the rest of the robot, the slug detection system has to be 'low-power' since the robot will be getting its energy from the very limited energy resource represented by slugs.

The slug detection system is based around one of the new generation 'single-chip' CMOS image sensors, which is small, lightweight, relatively low power (<175mW), of adequate resolution (164 by 124 pixels), and sensitive (down to 0.1 Lux). It is inexpensive, has a digital interface, and the maximum frame rate of 60 Hz enables high scanning speeds. This image sensor also has adjustable automatic exposure control, and can calculate the average image intensity of the last frame, and perform pixel level thresholding using an adjustable threshold.

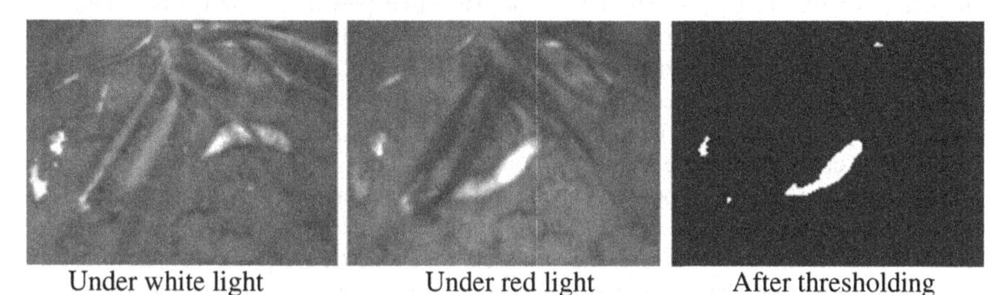

Under white light Under red light After thresholding

Fig. 2. Images of a Deroceras reticulatum slug and grass taken with the chosen image sensor

We have reduced the complexity of the problem of the visual detection of slugs by developing a simple method of filtering out image components deriving from soil, living and dead vegetation, and stones. Since slugs are active mainly from dusk to dawn, some form of illumination of the image area is required. The solution employed exploits the use of a combination of coloured light and optical filtering to increase the relative visibility of slugs whilst decreasing the visibility of vegetation and earth. We have found that this can be achieved by using red light from high brightness LEDs and placing a matching filter in front of the image sensor. Under these lighting conditions both vegetation and soil appear dark, whilst the slug *Deroceras reticulatum* reflects red light and thus appears bright in the received images. Figure 2 (left) shows a 32mm long *Deroceras reticulatum*, together with some grass, under white light - note the relative brightness of the slug and grass. Figure 2 (centre) is the same image (except for some movement of the slug) taken using the red light and filter combination - the grass now appears dark whilst the slug is bright. Figure 2 (right) shows how the image of the slug can be picked out from the background by applying a simple pixel-based threshold function to the red illuminated image of figure 2 (centre). There is an added benefit, in that the threshold scheme does not detect slugs under 15mm in length, and so filters out small slugs that would cost more in energy to retrieve than they can possibly yield. Unfortunately under this lighting condition stones also appear brightly and must therefore be filtered out by the slug detection system.

The final stage is the elimination of stones by the identification of bright patches that are of the correct size and shape for slugs. This is achieved by finding the biggest connected bright patch in each image and by calculating the ratio of bright (slug/stone) pixels to dark (background) pixels within a boxed area around the bright patch. The length and width of this box are equal to the length of the patch. The rationale behind this is that most of the time slugs are stretched out and are long and narrow and hence fill a small proportion of the box, whilst stones tend to be much rounder and fill a much higher proportion of the box. This system has the advantage of being invariant to the orientation of the object under examination and within pre-set bounds of its size. We have deliberately stopped the system correctly identifying slugs under 15mm in length, since they would use more energy to collect than they would return, and slugs over 75mm in length because they are too big for the gripper. The system only looks at the biggest object in each image, because trying to pick up a slug next to a bigger stone would fail with the design of the current gripper. If this ratio is above a predefined threshold then the bright patch is a very likely to be a slug and so is collected by the robot. Any objects that occupy any of the pixels at the boundary of the image are ignored on the basis that part of the image is outside the current field of view and so cannot be accurately identified. On a later scan these objects will be examined when they fall completely within an image. Figure 3 shows some examples of processed images, along with the surrounding box (only the top and bottom lines are shown) and the calculated fill ratio. An object with a ratio greater than 100 is considered to be a slug.

To meet constraints of power, speed, and cost, we have employed an Altera low-power programmable logic (PLD) device to perform the high-speed image collection and compression; the data are then passed to a 50MIPs Scenix low-power 8-bit microcontroller for shape recognition. The average image intensity and pixel-wise thresholding are both performed directly by the image sensor. Each image is buffered into an 8-bit wide 15ns static RAM by the PLD, with four thresholded pixels being stored in each location of the SRAM. Using this detection scheme on the hardware specified above we obtain a detection rate of 8 to 16 frames per second, whist only consuming approximately 1 watt of power. This allows a scan rate of 1m per second.

5 Energetics

There are up to 150 *Deroceras reticulatum* slugs per square metre in fields of crops such as winter wheat [3]. Unfortunately most of these are too small to return an energetic profit. Of the *Deroceras reticulatum* slugs that are detectable by the imaging system there are on average approximately 30 per square metre [9], which weigh more than 150mg, and ten of these weigh 500-700mg [9] and are about 35mm in length. Let us now consider the 'ballpark' amount of energy required harvesting these slugs, from moving to a new harvesting patch then carrying out a step-circling search and then returning to the start point.

The robot takes about 300 seconds to perform each step-circling trajectory with the arm, requiring approximately 5W of power. Once a slug has been detected it takes about 10 seconds to collect it, deposit it and return the gripper to where the slug was found (requiring about 36W). Moving to the next adjacent patch to be scanned takes 25 seconds travelling at 0.15 metres per second and requires 75W. The energy requirement for the complete task, based on a density of 1 slug/m^2 (i.e. 10.8 slugs per patch) is 9,388 Joules. The potential energy available from each slug is about 1500 Joules [3], assuming they weigh 500mg. Again considering the case of a density of 1 slug/m^2 the total calorific energy potentially gained from one complete spiral scan is 1500*10.8 = 16,200 Joules.

If there are 10 slugs per square metre, then each complete spiral scan would require 44,380 Joules; a larger value than above since there are more frequent captures and placements. However, this could return a potential calorific energy content of 162,000 Joules. The above energy returns are based upon the full energy content of the slugs which is quite unrealistic. Looking at the energy required versus the energy returned shows that for a density of 1 slug/m^2 the fermentation process would have to be at least 57% efficient, and 27% efficient for of a density of 10 slugs/m^2.

6 Slug Catching Trials

To test both the hardware and low-level reactive slug catching strategies five trials of the SlugBot attempting to catch ten *Deroceras reticulatum* slugs on real soil with stones of various sizes were conducted. Each trial consisted of four complete 360° scans with arm extensions of 400mm, 475mm, 550mm and 625mm, and scanned a surface area of 1.13m^2. In addition to the stones that were already present in the soil fifteen stones of various sizes were placed on the surface of the soil. To make the tests as realistic as possible the soil was not trodden down thus resulting in a rough surface similar to that found in the real field. At the beginning of each trial the slugs were placed about halfway across the scan area at random places around the robot, they ranged from 20mm to 35mm in length.

In all the trials the robot used the vision system to identify slugs and guide the gripper down to the slug in an attempt to pick it up. As expected, the trials had to be run in the evening out of direct sunlight, otherwise the slugs quickly buried themselves to escape from the heat. Initial tests quickly showed that although the vision system could correctly distinguish between slugs and stones, the current gripper design is ineffective at picking up slugs on uneven ground. Thus it was decided that any slug the robot attempted to pick up, but the gripper failed to actually collect, would be manually removed thus allowing the results to accurately show the effectiveness of the vision system. Without this slug removal, some slugs that are not successfully collected by the gripper move into the field of view of later scans, thus falsely making the vision system look more effective.

Table 1. Slug catching trial results

Slugs			Non slug items			
Re-examined	Identified as slugs	Picked up by gripper	Re-examined	Identified as slugs	Picked up by gripper	Time to complete test (s)
6	5	0	3	1	0	398
7	6	1	4	0	0	532
6	4	1	5	2	0	626
9	8	2	1	0	0	410
6	5	0	2	0	0	296

Table 1 shows the results for the five runs at a scan rate of 200mm per second, with slug removal. The "re-examined" columns show how many times, in each trial, that the turntable was stopped so that a suspected slug could be further examined by the vision system . The "identified as slugs" columns show how many times the robot tried to collect an object thinking that it was a slug, The "picked up by gripper" columns show how many objects that were identified as slugs were successfully collected. The number of slugs correctly identified is lower than it could have been in the first three experiments because three of the slugs moved out of the scan area, in addition another slug buried itself in the third experiment. Therefore of the slugs that remained within the region scanned, 70% were correctly identified. Over 94% of the stones were correctly ignored. Although the gripper only managed to pick up less than 13% of the objects that it attempted to collect.

Conclusions

We have discussed the design and construction of a robot that should be capable of autonomously hunting for slugs within real agricultural fields. We have also given details of the vision system that is employed to detect slugs and shown that it operates with a very high degree of accuracy against a background of the soil from the fields that we will run the robot in. Unfortunately the current gripper design is ineffective under the conditions found within real fields and a modified version is now under consideration.

Acknowledgement. We would like to thank Owen Holland for all his assistance.

References

[1] Steels L (1995), When are robots intelligent autonomous agents? Journal of Robotics and Autonomous Systems 15, pp. 3-9
[2] Staniforth J, Kendall K (1998), Biogas powering a small tubular solid oxide fuel cell. Journal of Power Sources 71, pp 275-277
[3] South A (1992), Terrestrial Slugs: Biology, Ecology, and Control. Chapman and Hall
[4] Glen DM (1994), Ecology of Slugs in Cereals in Relation to Crop Damage and Control. In: Leather SR, Walters KFA, Mills NJ et al (eds), Populations, and Patterns in Ecology, Intercept, pp. 163-171
[5] Glen DM, Milsom NF, Wiltshire CW (1998), Effects of Seed-bed Conditions on Slug Numbers and Damage to Winter Wheat in a Clay Soil. Annals of Applied Biology, pp. 115, 177-190
[6] Kelly ID, Keating DA (1996), Flocking by the Fusion of Sonar and Active Infrared Sensors on Physical Autonomous Mobile Robots. Conf. on Mechatronics, pp. 1/1-1/4
[7] Kelly ID (1997), The Development of Shared Experience Learning in a Group of Mobile Robots. Ph.D. Thesis, Department of Cybernetics, University of Reading, UK, pp 2/4-2/12
[8] Glover C (1998), Reinvasion of Cleared Areas by the Slug Deroceras reticulatum. Internal Report. Long Ashton Research Station, Bristol
[9] Wiltshire C (2000), Personal Communication. Long Ashton Research Station, Institute of Arable Crops Research, University of Bristol, UK

Mobile Robot Control Based on Boolean Logic with Internal Memory

DaeEun Kim and John C.T. Hallam

Institute of Perception, Action, and Behavior
Division of Informatics, 5 Forrest Hill
The University of Edinburgh
United Kingdom
{daeeun,john}@dai.ed.ac.uk

Abstract. The purpose of this paper is to explore the effect of adding known amounts of memory to pure reactive systems in a variety of tasks. Using a finite state machine approach, we construct controllers for a simulated robot for five tasks—obstacle avoidance, wall following, exploration, and box pushing—with two sensor configurations using evolutionary computation techniques, and compare the performance of stateless and memory-based controllers. For obstacle avoidance and exploration no significant difference is observed; for wall-following and box pushing, stateless controllers are significantly worse than memory-based but increasing amounts of memory give no significant increase in performance. The need for memory in these cases reflects a need to discriminate sensorimotor contexts to effectively perform the task.

1 Introduction

The purpose of this paper is to explore the effect of adding known amounts of memory to pure reactive systems in a variety of tasks. The memory elements will be represented in the form of a set of rules with states in our approach, equivalent to a Boolean Logic Network with a fixed number of 1-bit registers. First, we discuss the way memory is generally implemented in behavior-based robotics systems. Then we describe a set of experiments that reveal behavioral and performance differences between a pure reactive controller and a memory-based reactive controller evolved for various particular tasks, and where appropriate we explain the effect of memory on the resulting robot behaviors.

Reactive systems in mobile robots are often seen as a direct coupling between perception and action without any intermediate processing. However, real robot systems following a behavior-based approach actually use various memory elements to handle the control problem correctly and efficiently; for instance, Brooks' subsumption architecture [4] contains memory in the augmented finite state machines. Memory provides a medium for processing of local temporal information in Behavior-based systems. However, there has been little research about the effect of varying amounts of memory in behavior-based robotics.

J. Kelemen and P. Sosík (Eds.): ECAL 2001, LNAI 2159, pp. 529–538, 2001.

In some robotic tasks perceptual information alone is insufficient to determine uniquely the appropriate action. It may be necessary to integrate these partial perceptual cues with previous actions or perceptual information to decide proper actions. Many tasks, such as landmark-based navigation and sequential behaviors, require that an agent also be able to integrate perceptual information over time in order to determine the appropriate course of action [16,17]. In evolutionary robotics, this ability is usually provided by evolving continuous-time recurrent neural networks which, for instance, can be evolved to generate a fixed sequence of ouputs in response to an external trigger occurring at varying intervals of time [17]. Such networks make effective controllers[14,9,6], offering the possibility of creating many possible dynamical systems with a variety of kinds of attractor, but are hard to analyze them theoretically and to quantify the amount of memory a particular network provides to an agent.

Some evolutionary robotics experiments [13] have shown that pure reactive systems without memory can implement various primitive behaviors: for example, object avoidance and exploration behaviors, and box-pushing to a target signalled by an environmental marker need only a pure reactive controller with binary sensory inputs, and more complex combinations of such behaviors can be realized using stateless switchers to arbitrate between the primitives.

Rather than using neural networks, Lee [12] achieves these results by evolving a Boolean Logic Network using genetic programming (GP). Each behavior is a pure reactive controller with no internal state. Its control structure is a combinational logic circuit, evolved using GP, acting on sensor values converted to binary by thresholding or comparison (the thresholds and comparisons are also determined by the GP evolution). Such an approach offers the possibility of adding memory in an easily quantified way.

A Boolean Logic Network with internal memory is equivalent to a finite state machine[10]. Finite state machines have very useful properties for describing aggregations and sequences of behaviors [1]: they can represent the interaction of behaviors, the behaviors active at any given time and the transitions between them. They have been used to accomplish various high-level goals with complex behavioral control systems consisting of a set of primitive behaviors [2,7]. Finite state automata provide a good mechanism to express the relationships between various behavior sets and have been widely used within robotics to express control systems [1]. Brooks [4] also used a variation of finite state automata—augmented finite state machines—to express behaviors within his subsumption architecture.

Bakker proposed a means for counting the number of states required to perform a particular task in an environment[3]. Such state counts extracted from finite state machines can be a measure of the complexity of agents and environments. They applied their methodology to a fixed simple environment such as the Woods7 environment [15] instead of real robotic environments.

Wilson[15] used a zeroth-level classifier system(ZCS) for his animat experiments. The original formulation of ZCS has no memory mechanisms, because the input-output mappings from ZCS are purely reactive. Following Wilson's proposal, adding one-bit and two-bit memory registers to ZCS was tested in Woods environments by Cliff and Ross [5]. They insisted ZCS manipulate and exploit internal states appropriately and efficiently in non-Markovian environments. Their experiments used simplified models and were quite different from real robotic experiments, but they showed the potential and importance of internal states for robot behaviors.

Similar to the ZCS experiments, Lanzi has shown that internal memory is effective with adaptive agents and reinforcement learning, when the agent's sensors provide only partial information about the surrounding environment and particular similar situations require different actions [11].

2 Methods

To investigate the consequences of adding memory to a purely reactive controller, we have run several experiments evolving controllers for tasks such as obstacle avoidance, wall following, exploration, and box pushing using a simulated agent and environment. Our robot model is a Khepera-style robot, with binary sensors in front to detect any wall or obstacle. Those sensors are modeled as infrared sensors without any explicit acceptance cone[1]. The sensor values are binary, derived by thresholding the distance provided by the sensor, and sensor readings just say if any object is near robot within some limit in the direction of the sensor. These binary sensors do not give any quantitative value of how close to any obstacle the robot is. However, we allowed 3 bits, eight possible values, for each wheel action.

As mentioned above, we represent the fixed-memory controller as a collection of rules conditioned on the sensory state and on the controller's internal state. The rules thus implement the transition function of a finite state machine, and are similar to a classifier system [8]; they provide an economical way of representing the controller for evolution using a genetic algorithm. This scheme reduces the length of genes, thus increasing the convergence speed. In the experiments, the chromosome representation follows an integer encoding for output action and state number. When we have n sensors, m motor actions and s states, the length of the chromosome for the direct representation of a finite state machine will be $s(m+1)2^n$, assuming all the sensors are binary-valued. For example, when we need many sensors and several states, the gene encoding size is exponentially increasing.

Using a set of rules, where each rule has state information and a sensor representation of short, fixed, length as mentioned above, k rules in a set, n sensors, m motor actions, and s states results in a genome of length $k(n+s+m)$, which is linearly proportional to the number of sensors.

Genetic operators are used for a set of rules encoded by bit strings. The basic structure can be seen as a rule-based system where preconditions for a set

[1] The sensor detects the intersection of an object with a fixed length ray projected along the sensor's pointing direction and computes the distance to the nearest such intersection along the ray.

of rules are examined to see their applicability, when given the current situation. The preconditions have fixed-length bit encoding with values of 0, 1, #(don't care). The action part of the rule is also a fixed-length encoding with the values of motor action. Unlike a classifier system, the evolutionary computation with a set of rules used here does not use any reinforcement learning: the genetic algorithm creates new rules with genetic operators and a set of rules is tested for a given environment by evaluating the results of the chosen actions.

A set of rules including states is defined as a machine $M = (Q, \Sigma, \Delta, \delta, \lambda, q_0)$ where q_0 is an initial state, Q is a finite set of states, Σ is an input value, Δ is the multi-valued output value, δ is a state transition function from $Q \times \Sigma$ to Q, and λ is a mapping from $Q \times \Sigma$ to Δ, $\lambda(q, A)$ will have one of the output set Δ, where q, A is an element of Q, Σ, respectively. Each rule R_i in machine M is defined as

$$\begin{aligned} R_i(M) &= (A_i, (q_i, q_i^*), D_i) \\ &= ((\sigma_{i1}, \sigma_{i2}, \sigma_{i3}, ..., \sigma_{in}), \\ &\quad (q_i, q_i^*), (d_{i1}, d_{i2}, d_{i3}, ..., d_{im})) \end{aligned}$$

where A_i is an input string or sensor value, $D_i \in \Delta$ is a motor output string, q_i is the current state before rule activation, q_i^* is the next state after rule activation.

In the rule structure, we can divide a set of state rules into several classes depending on the controller state q_i. Each set of rules thus obtained is 'reactive' in that it describes a mapping from sensor signals to motor actions (neglecting state changes). The quantity of memory in an agent is reflected by the number of distinct states used, i.e. $|\{q_i\}|$. We assume each rule must necessarily match the current state to be activated. State transition information can then be generated from the best rule to match the current sensor readings.

The state information can thus be seen as a medium to activate a set of rules, switching the controller between different sensorimotor couplings. When the current state, e.g. q_1, is set, it can activate only a set of rules whose preconditions have state q_1 and the best rule among them will decide the next state q_2, which will activate another set of rules.

3 Experiments

We used a simulation program for robot with Khepera-like kinematics. At each time step, the robot reads a sensor value from the modeled sensors and the control action determined by the chosen rule is applied to the motors. We tested four different tasks—obstacle avoidance, wall following, exploration, and box pushing—with two sensor configurations: two infrared sensors with angle $45°, 135°$, or four infrared sensors with angle $45°, 80°, 100°, 135°$.

Each evolutionary run proceeds thus: N different initial positions for the robot are randomly selected in the arena in Fig. 1, to provide robustness in the evolved controller; evolutionary computation chooses K positions randomly without replacement from the N initial positions at each generation and evaluates the new control strategies at those positions and uses the fitnesses thereby

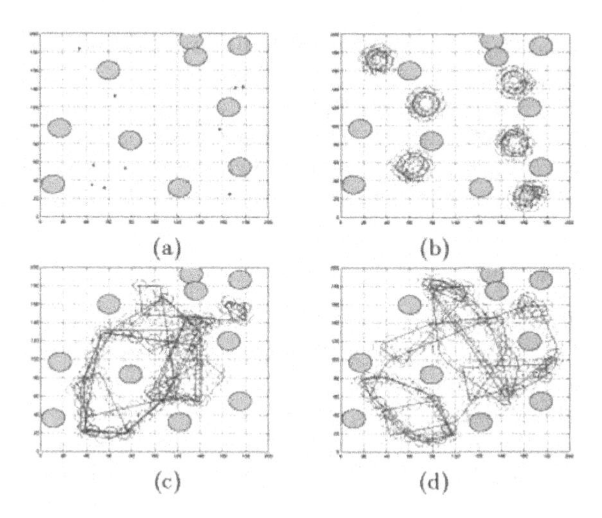

Fig. 1. Obstacle avoidance behavior (a) arena with initial positions (b) reactive controller with 2 sensors (c) reactive controller with 4 sensors (d) 2 state machine controller with 4 sensors

computed for selection of parents in the genetic algorithm; every 10 generations controllers are re-evaluated at all N positions. This N-K sample method reduces the computation time, but also it has the effect of running over all initial positions regularly in the evolutionary run. We used $K = 5$ and $N = 10$ in our experiments. For significance statistics, we run 10 tests for each configuration of controller structures for each of four behaviors, respectively.

3.1 Obstacle Avoidance

For obstacle avoidance behavior, we put circular boxes as obstacles randomly in the arena as shown in Figure 1. The fitness function F_O is defined as a penalty function as follows:

$$F_O = P - \Sigma_{t=1}^{max}(\alpha(M_L(t) + M_R(t)) \\ +\beta|M_L(t) - M_R(t)|/(2 * L) + \gamma C(t))$$

where P is the base of penalty, L is the robot wheelbase, $M_L(t)(M_R(t)$ the left(right) wheel motor output at time t, $C(t)$ is a collision detection function, and $\alpha = 0.1$, $\beta = 0.5$, $\gamma = 5.0$ are scaling coefficients for each term. If a robot collides with any obstacle, $C(t)$ will be set to one from that time onwards; otherwise it is zero. This collision factor encourages the controller to escape obstacles without collision. Usually γ is set to a high value, and when the robot experiences a collision the controller will have a very low fitness. The fitness calculates the difference between forward speed and rotation speed, to encourage motion as straight as possible. This experiment comprised controller evaluations lasting 2000 time steps, a population of 100 controllers and 4000 generations of the genetic algorithm.

For this task, memory-based controllers give a little better performance than reactive, but not significantly so. The number of sensors are more important as shown in Fig. 1. When a robot has only two sensors, it cannot move straight forward due to collision with obstacles. It just staggers around a local area.

3.2 Wall Following

To evolve wall following behavior, we assumed there are many "energy tanks" in the arena the robot is exploring. If the robot finds one of these energy tanks, it acquires energy and can explore more area before its energy is exhausted. We assumed the robot loses its energy when moving. By placing the energy tanks near the walls of the environment and rewarding the agent for visiting them, we encourage the evolution of wall-following behavior: the fitness can be roughly calculated as the number of visited energy tanks and the energy level the robot has.

Thus, a fitness function F_W is a penalty function defined as follows:

$$F_W = P - (1 - 0.7C)\alpha \Sigma_i^N V(i) \\ -(1 - C)(\beta E - \gamma T)$$

$$V(i) = \begin{cases} 1 \text{ if energy tank } i \text{ is visited} \\ 0 \qquad\qquad \text{otherwise} \end{cases}$$

where P is the base of the penalty, C is a collision detection flag, N is the number of energy tanks in the environment, E is the remaining energy after exploration and T is the number of time steps it takes the robot to visit all energy tanks. We assume energy tanks are located at divisions around the wall and obstacles as shown in Figure 2 on a 20cm by 20cm grid (the whole arena was 200cm by 200cm).

If it collides during exploration, robot movement stops and calculates the fitness function F_W with $C = 1$, otherwise $C = 0$. Energy E is changed every time step, by losing a small proportion of energy with initial energy setting.

In this experiment each controller generated by the genetic algorithm was evaluated for 2000 time steps; the population size was 100 and the evolution ran for 4000 generations.

In this task, a 2-state machine is significantly better than the pure reactive controller—see Fig. 2(d), where we run 10 tests for each configuration and the error bar shows 95 % confidence interval. In some initial positions, the reactive controller cannot find the wall and it circles in the free space (see Fig. 2(b)). The memory-based system is able to distinguish whether the robot has found a wall and to switch behavior when this occurs, as in Fig. 2(c). The stateless controller adopts an appropriately curved path to follow a wall, but in free space such a path results in circling motions. With the 2-state controller, if the sensors read clear, the robot moves straight forward until it finds a wall; after detecting a wall the robot switches to a curved path strategy. For this task, the number

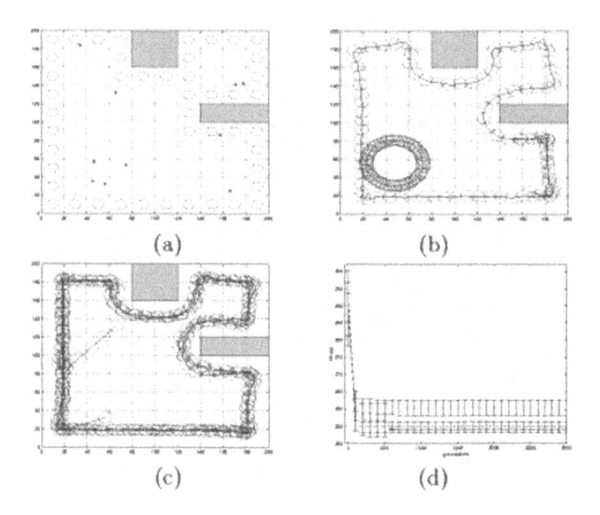

Fig. 2. Wall following behavior (a) arena with initial positions and energy tanks (b) reactive controller (c) 2 state machine controller (d) 10 runs test

of sensors does not improve the behavior, and the robot with 4 sensors shows similar movements to that with 2 sensors.

3.3 Exploration

For exploration behavior, we used the same environment as that in the test of obstacle avoidance behavior as shown in Figure 3. For this behavior, we can apply the above energy distribution idea over the whole environment for the fitness function, but we used a different calculation method. The fitness function is defined as follows:

$$F_E = -\Sigma_i^N p(i) \log p(i)$$

where N is the number of small divisions, $p(i)$ is the probability that the robot has stayed in the i-th section and can be calculated dividing the number of time steps the robot has moved in the i-th area by the total time steps the robot runs.

Using this entropy measurement rewards the robot for visiting all areas of the environment uniformly, while the energy tank mechanism encourages the robot to visit the whole environment. As before, population size 100 with 4000 generations with 2000 time steps for each controller were used in each experiment.

There was no significant difference between reactive and memory-based controller in terms of fitness, but the state machine was a little better than a reactive controller on inspection of its behavior—see Fig. 3. When a robot has two sensors, it shows wall following behavior. 3-state machine and 4-state machine behavior was similar to 2-state machine and in many cases 3-state and 4-state machines evolved into a 2-state machine controller. It appears that the 2-state machine is most desirable in this behavior. The 2-state machine controller shows a very

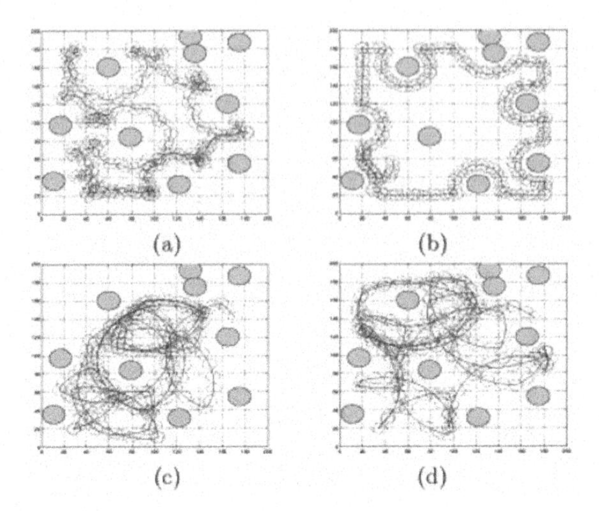

Fig. 3. Exploration behavior (a) reactive controller with 2 sensors (b) 2 state machine controller with 2 sensors (c) reactive controller with 4 sensors (d) 2 state machine controller with 4 sensors

delicate wall following behavior in Fig. 3(b). Increasing the number of sensors allows the robot to move fast around the environment, because it reduces the possibility of collision by seeing the environment well.

3.4 Box Pushing

We assume a circular box is located in the arena and a robot moves the box into a target position indicated by a light. The robot uses its obstacle sensors to detect the circular box and ambient light sensors to detect the light. We simplified the light sensor value to binary values to say whether the light is close to the left or right sensor, or the light is positioned behind the robot. Control structures with a different number of memories were tested to see if the memory causes differences in the robot's performance. The fitness function F_B is defined as a penalty function as follows:

$$F_B = \Sigma_{t=1}^{max}(\alpha W(t) + \beta D(t) + \gamma C(t))$$

where $W(t)$ is the distance between robot and box, $D(t)$ is the distance between box and light, and $C(t)$ is the collision detection function. In the experiment, we set $\alpha = 0.1$, $\beta = 1.0$, and $\gamma = 50.0$.

It was hard to evolve box pushing behavior with random initial positions—in a real application it would be combined with exploration behavior. 1500 time steps for each random gene controller were used with initial robot positions around the circular box. Robot controllers with a different number of sensors have show similar fitness values and behaviors. On varying the sensor pointing

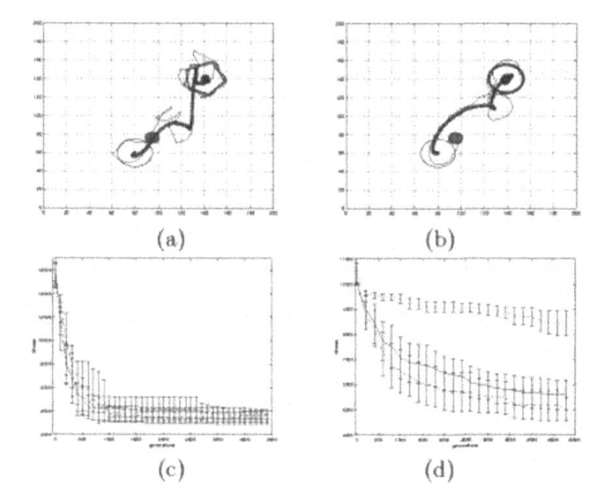

Fig. 4. Box pushing behavior (a) reactive controller with 2 sensors (b) 2 state machine controller with 2 sensors (c) 10 runs test with angle 45 (d) 10 run test with angle 30 (dotted : reactive, dashed-dot : 2 states, solid : 3 states

angles, 60°, 120°, the memory-based controller was better than reactive controller as shown in Fig. 4(d), where we run 10 tests for each configuration and the error bar shows 95 % confidence interval. The fitnesses of 2-state, 3-state, and 4-state machines are not significantly different.

4 Conclusion

In reactive systems, mobile robots work efficiently through a tight state-free coupling between perception and action. We investigated the effect of allowing the controller small amounts of memory in four different behaviors, obstacle avoidance, exploration, wall following, box pushing, and corridor following. In simple behaviors such as obstacle avoidance and exploration, a reactive system without memory elements is sufficient to control a mobile robot. The number of sensors and each sensor position have more influence on performance of such behaviors than the amount of state available.

In some behaviors, such as wall following, box pushing and corridor following, the memory element plays an important role by allowing the controller to specialize its behavior depending on context—for instance, free-space versus close to a wall. The amount of memory available is not significant for these tasks as tested here, because the tasks themselves only require a couple of states to be distinguished.

Future work will look at the effect of memory on behavior and performance when the evolutionary system is able to choose the sensor thresholds as in Lee's [12] work (here the thresholds were determined by the experimenter) and on the effect of incorporating memory into the stateless arbitrators Lee used to combine his primitive behaviors (for instance to combine exploration and box-pushing).

References

1. R.C. Arkin. *Behavior-based Robotics.* MIT Press, Cambridge, MA, 1998.
2. R.C. Arkin and D. MacKenzie. Temporal coordination of perceptual algorithms for mobile robot navigation. *IEEE Transactions on Robotics and Automation*, Vol. 10, No.3:276–286, June 1994.
3. B. Bakker and M. de Jong. The epsilon state count. In *From Animals to Animats 6: Proceedings of the Sixth International Conference on Simulation of Adaptive Behaviour*, pages 51–60. MIT Press, 2000.
4. R. Brooks. A robust layered control system for a mobile robot. *IEEE Journal of Robotics and Autonomation*, Vol. RA-2, No.1:14–23, 1986.
5. D. Cliff and S. Ross. Adding temporary memory to zcs. *Adaptive Behavior*, 3(2):101–150, 1995.
6. D. Floreano and F. Mondada. Evolution of homing navigation in a real robot. *IEEE Transactions on Systems, Man and Cybernetics*, 26(3):396–407, 1996.
7. E. Gat and G. Dorais. Robot navigation by conditional sequencing. In *Proceedings of the IEEE International Conference on Robotics and Automation*, pages 1293–1299, 1994.
8. D. E. Goldberg. *Genetic Algorithms in Search, Optimization, and Machine Learning.* Addison Wesly, Reading, MA, 1989.
9. I. Harvery, P. Husbands, and D. Cliff. Issues in evolutionary robotics. In *From Animals to Animats 2: Proceedings of the Second International Conference on Simulation of Adaptive Behaviour*, pages 364–373. MIT Press, 1992.
10. Zvi Kohavi. *Switching and Finite Automata Theory.* McGraw-Hill, New York, London, 1970.
11. P.L. Lanzi. Adaptive agents with reinforcement learning and internal memory. In *From Animals to Animats 6: Proceedings of the Sixth International Conference on Simulation of Adaptive Behaviour*, pages 333–342. MIT Press, 2000.
12. W-P. Lee, J. Hallam, and H.H. Lund. Applying genetic programming to evolve behavior primitives and arbitrators for mobile robots. In *Proceedings of IEEE International Conference on Evolutionary Computation*, Indianapolis, USA, 1997.
13. Wei-Po Lee. *Applying Genetic Programming to Evolve Behavior Primitives and Arbitrators for Mobile Robots.* Ph. D. dissertation, University of Edinburgh, 1998.
14. J.A. Meyer, P. Husbands, and I. Harvey. Evolutionary robotics: A survey of applications and problems. In *First European Workshop on Evolutionary Robotics, Proceedings of the Second International Conference on Simulation of Adaptive Behaviour*, Paris, France, 1998. Springer.
15. S. W. Wilson. Zcs: A zeroth level classifier system. *Evolutionary Computation*, 2(1):1–18, 1994.
16. B. Yamauchi and R. Beer. Integrating reactive, sequential, and learning behavior using dynamical neural networks. In *From Animals to Animats 3: Proceedings of the Third International Conference on Simulation of Adaptive Behaviour*, pages 382–391. MIT Press, 1994.
17. B. Yamauchi and R. Beer. Sequential behabior and learning in evolved dynamical neural networks. *Adaptive Behavior*, 2(3):219–246, 1994.

Emergence of Cooperative Tactics by Soccer Agents with Ability of Prediction and Learning

Yoichiro Kumada and Kazuhiro Ueda

Department of General System Studies, The University of Tokyo
3-8-1, Komaba, Meguro-ku, Tokyo 153-8902, Japan
kuma@blake.c.u-tokyo.ac.jp
ueda@gregorio.c.u-tokyo.ac.jp

Abstract. This paper describes soccer agents that can learn by themselves to make use of emergent cooperative tactics. Considering the methods that actual coaches of soccer help their players to learn the way of good play, we developed a mechanism of learning to distinguish good tactics from bad ones. The agents adaptively learned the utilities of the situations and the conditional probabilities about the situations in order to predict the opponents' behavior and in order to select good actions. The agents cooperatively created some combinations of passes, such as a wall pass and a one-two pass which we had never taught in advance how to execute.

1 Introduction

In the domain of multiagent or multi-robot systems, one of the most important issues is to clarify the mechanism that an agent could cooperate with the other agents based on the prediction and the estimation of the others' behaviors [2]. For this research target, we are required to introduce the spatially-limited environment in which the learning structure is embedded for the emergence of cooperative tactics. This kind of learning environment is already introduced by actual soccer coaches. For example, Charles Hughes [1], a famous soccer coach, emphasizes that one of the most important instruction is not about "what to learn", but about "how to learn". In other words, various tactics of soccer are embedded in the good learning environment. Based on this idea, we constructed the soccer agents that learn cooperative tactics. We realized the following: (1) accomplishment of training tasks by a small number of agents, (2) acquisition of appropriate cognitive maps by using a grid map, (3) prediction of other agents' behavior and learning of precise prediction itself, and (4) learning of the utilities of the game situations.
In the next section, we will explain the model of the agents that meet the above points.

2 Model of the Agent and Their Learning Environment

If the objective of learning is to acquire an appropriate formation, all the agents are required to participate in the game. Hughes [1] says, however, that the number of players in the training and the size of the training field should be appropriate to the task. In this study, the agents of three attackers and two defenders are repeatedly playing mini games

J. Kelemen and P. Sosík (Eds.): ECAL 2001, LNAI 2159, pp. 539–542, 2001.

in a 3×4 gridized small field. See Figure 3, and Figure 4; A, D, and b mean an attacker, a defender, and a ball, respectively. In this mini-games, [1]

- if some attacker kicks the ball out of the end line, the attacking team wins,
- if some defender kicks the ball out through one of the side lines, the defending team wins.

These are the tasks given to the attackers and the defenders.

2.1 Variables Used in the Model of Each of the Agents

We provides each agent with variables that are needed for realizing the requirements listed in Sect. 1.

Condition Variable: The agent recognizes the world as a grid map in order to get information about which grid the allies, the opponents and the ball are in. We call this world model as "condition", which is denoted by C, in this paper.

Action Variable: The agents are designed to take four types of basic actions, "stay there", "only move", "dribble the ball", and "pass the ball". These actions are denoted with the variable A.

Conditional Probability: The probability that some condition ($C_{t+1} = c_{t+1}$) happens after action(A_t) is taken under any condition ($C_t = c_t$), is denoted with $P(C_{t+1} = c_{t+1}|C_t = c_t, A_t)$ (the suffix t indicates a time step, in consideration of summation taken in calculation of Sect. 2.2). This describes the effect of an action.

Utility: The utility of any condition is denoted with $U(C)$.

2.2 Structure of the Agents

We designed our agents that operate on the decision cycle [4] that repeats the following.

Selection of the Action that Maximizes Estimation of Utility.

$$action \leftarrow argmax_{A_t} \sum_{c_t} [Bel(C_t = c_t) \times \sum_{c_{t+1}} P(C_{t+1} = c_{t+1}|C_t = c_t, A_t) U(c_{t+1})]$$

$Bel(C_t = c_t)$ denotes the probability distribution that the condition at time t is c_t. It is, in other words, the agents' belief. $\sum_{c_{t+1}} P(C_{t+1} = c_{t+1}|C_t = c_t, A_t)U(c_{t+1})$ means the estimation of the utility that an action A_t is taken under the present condition $C_t = c_t$.

By taking summation of the present conditions' probabilities $Bel(C_t = c_t)$, the total estimation of utility is calculated. The output is the action which maximizes this estimation.

[1] See Figure 3. the horizontal, bold-faced line is the end line, and the vertical boundaries are the side lines.

2.3 Learning by the Agents

It is hard for programmers to know beforehand how the probabilities and the utilities should be set for good plays of soccer. Our agents learn these settings during mini games.

Learning of Conditional Probability: Based on the agent's belief that the current condition is c_{t+1}, the conditional probability $P(C_{t+1} = c_{t+1}|C_t)$ is reinforced $(P_{new}(C_{t+1} = c_{t+1}|C_t) = P_{prev}(C_{t+1} = c_{t+1}|C_t) + wBel(C_{t+1} = c_{t+1}))$.

Learning of the Utility: We adopted a method of evolutionary learning. Reward is given for the sequence that constitutes a successful mini-game, while punishment is given for the unsuccessful one $(U_{new}(C_{N-i}) = U_{prev}(C_{N-i}) + \text{reward} \times d^i)$.

3 Result – Emergence of the Basic Cooperative Tactics

What kind of play could the agents learn? For example, they could learn "high quality pass". High quality pass means that the pass-receiver doesn't wait for a pass to come to its own grid but move toward an open space and catch a pass. Figure 1 shows that the number of "high quality pass" increased in the process of learning. This increase in number of "high quality pass" is thought to have a close relation with the increase of the winning percentage of the attackers (Figure 2).

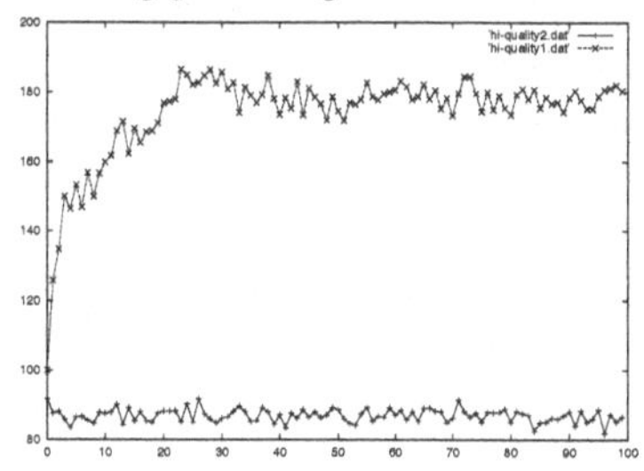

Fig. 1. The numbers of high-quality passes:with learning(high-quality1), without learning (high-quality2), on average of 50 simulations

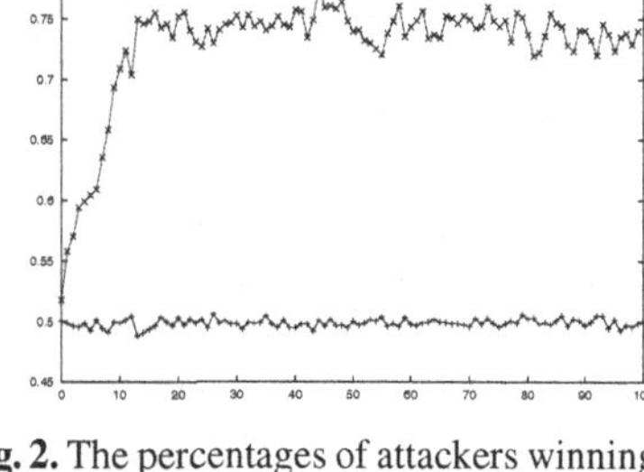

Fig. 2. The percentages of attackers winnings: without learning(mean00), and with learning on reduction ratio 0.85(mean85).

See Figure 3. A2(attacker) ran into a open space to catch a pass from A1(attacker), rather than waited without moving for the ball coming to its grid. In addition, A2 gave a high quality pass for A3(attacker). This combination of the passes was what is called a wall pass. Next, see Figure 4. In this sequence, A1(attacker) gave a pass to A2(attacker) and A2 passed back to A1 again. This combination of the passes was what is called a one-two pass.

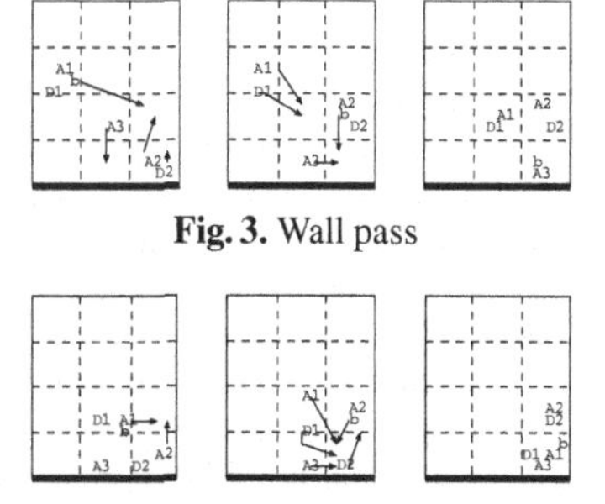

Fig. 3. Wall pass

Fig. 4. One-two pass

The above examples show that, the agents ,which exceeded the opponents in number, could estimate the other agents' tendencies of prediction and utilities, and that they could cooperate with the allies and defeat the opponents.

4 Discussion and Conclusion

In most of the previous studies on soccer agent, cooperations depended on the learning of formation and positioning, on communications between agents(especially in real robot league), and on assigning various roles on agents [3]. But, some researches are trying to realize the emergence of cooperation. For example, Uchibe [5] is trying to acquire cooperative behaviors in the real robot environment using genetic algorithm, but have not yet succeeded in prediction of the other agents.

On the other hand, our method is based on the supposition that, in order for soccer players to find good cooperative tactics, they need not to be taught about good tactics, but to be given good environments for learning. So, we provided the agents with the cognitive model using the grid map and with the limited mini-game environment in which, on execution of the tasks, some reward will be given to the agents.

Then, what did our agents achieve?

In the initial setup of our simulations, each agents had not yet learned any tendencies of behaviors of all agents including itself. These tendencies were represented by the conditional probabilities. As games went repeatedly, the agents learned the utilities and changed their behaviors, which means that the agents could play and learn the tactics in the dynamical environments. We, hence, assume that even if the opponents changed their tactics, the agent could revise the conditional probability to correct the prediction.

And cooperative tactics emerged in our environment, which is important for soccer agents. Generally speaking, soccer environment is too complicated for programmers to describe all possible tactics. Although our agents are not designed to cooperate, they are driven to cooperate by the environment.

Acknowledgment. This research is partially funded by Japan Society for the Promotion of Science (Grant-in-Aid for Science Research (C) No.12680369 and Grant-in-Aid for Encouragement of Young Scientists (A) No.13780270). The research is also supported by Foundation for Artificial Intelligence Research Promotion (No.12AI320-8).

References

1. C. Hughes. Soccer Tactics and Skills. British Broadcasting Corporation, 1980.
2. H. Kitano *et al.* RoboCup: A Challenge Problem for AI and Robotics. H. Kitano(Ed.). RoboCup-97:Robot Soccer World Cup I, Springer, 1998.
3. RoboCup Official Site:RoboCup The Robot World Cup Initiative. http://www.robocup.org/
4. S. Russell, and P. Norvig. Artificial Intelligence A Modern Approach, Prentice-Hall, 1995.
5. Eiji Uchibe, Masateru Nakamura, and Minoru Asada. Co-evolution for Cooperative Behavior Acquisition in a Multiple Mobile Robot Environment. Proc. of IEEE/RSJ International Conference on Intelligent Robots and Systems 1998 (IROS '98), pp. 425-430, 1998.

Patch Sorting: Multi-object Clustering Using Minimalist Robots

Chris Melhuish[1], Matt Wilson[1], and Ana Sendova-Franks[2]

[1]Intelligent Autonomous Systems (Engineering) Laboratory, Faculty of Engineering, University of the West of England, Coldharbour Lane, Frenchay, Bristol, England, BS16 1QY
chris.melhuish@uwe.ac.uk, matthew.Wilson@uwe.ac.uk
[2] Intelligent Computer Systems Centre, Faculty of Computer Studies and Mathematics, University of the West of England, Coldharbour Lane, Frenchay, Bristol BS16 1QY
ana.sendova-franks@uwe.ac.uk

Abstract. This study shows that a task as complicated as patch sorting can be accomplished with a 'minimalist' solution employing four simple rules. The solution is an extension of the object clustering research of Beckers *et al.* [1] and the object sorting research of Melhuish *et al.* [2]. Beckers *et al.* [1] used a very simple mechanism and achieved puck clustering in an arena with simple robots. Melhuish *et al.* [2] extended this technique to sort two objects, again using simple robots and a simple mechanism. The new mechanism reported in this paper, explores the sorting of any number of different objects into separate clusters. The method works by comparing two objects: the object the robot is carrying and, using a special antenna, the object with which the robot has collided. The results in this paper provide a demonstration of the success of this n-colour mechanism.

1 Introduction

Deneubourg *et al.* [3] began research into the idea of sorting objects using minimal rules. They present a simulation which demonstrates a simple mechanism that is sufficient to generate separate clusters of two different objects. The mechanism modulates the probability of dropping objects as a function of the local density of objects near the robot. Lumer and Faieta [4] claim to have extended the algorithm to sort objects, which are all different, into clusters of similarity. The possibility therefore exists to extend the method to sort any different number of discrete object types into separate clusters. However, this algorithm, while successful in simulation, has the major drawback that the agents need to be able to sense the local densities of the different types of objects. While this information is easily made available to simulated robots, it is difficult to transfer to real robots operating with minimum sensing capability.

A simpler mechanism was used by Beckers *et al.* [1] and was successfully implemented on a group of real robots. This mechanism was later modified and tested by Melhuish *et al.* [2]. It involves robots moving in straight lines within an arena. Whenever their scoop is depressed, the robots reverse and turn though a random angle. The robots are able to push a single object (frisbee) in the direction of their motion. If while carrying a frisbee the robots collide with another frisbee, the frisbee

J. Kelemen and P. Sosík (Eds.): ECAL 2001, LNAI 2159, pp. 543-552, 2001.

currently being carried is deposited and the robot reverses and turns through a random angle.

After testing the above mechanism on their robots, Melhuish *et al.* [2] extended the mechanism by the addition of an extra rule allowing two-object segregation. This algorithm used red and yellow frisbees and an infrared sensor able to detect which of these two colours the robots were carrying. The extra rule differentiates actions in situations where the robot is carrying a frisbee and collides with another frisbee: red frisbees to be dropped immediately, while yellow frisbees are first pulled back a distance before being dropped. This pullback mechanism allows the yellow frisbees to be pulled away from red clusters.

This paper describes research into a further extension to the mechanism above to enable the sorting of any number of different types of objects into separate patches. This behaviour was described by Melhish *et al.* [2] as patch sorting and defined as *"grouping two or more classes of objects so that each is both clustered and segregated, and ... each lies outside the boundary of the other"*. To perform the task an antenna has been added to each of the robots, which is used to sense the shade of a frisbee. If a robot is 'carrying' a frisbee and a collision takes place with another frisbee, the antenna moves and senses the shade of the frisbee in front. A comparison is then made with the frisbee being carried so that one of two possible actions can result: if the sensor comparison judges the two frisbees to be of a similar shade, then the frisbee being carried is dropped otherwise the frisbee is taken away.

All of the mechanisms described above could be implemented using a single robot. However, groups of robots are used to increase fault-tolerance and to decrease the time taken to complete the task. Groups of robots also allow better comparisons to be made with social insects, which are a great source of inspiration for the object sorting work described above. For example, behaviour similar to patch sorting (defined below) is clearly visible in the nests of *Otontomachus* ants. Dejean and Lachaud [5] observed that *"the brood was separately stocked: eggs piled together, most larvae clinging to the inner wall of the tube by their dorsal hairs, and pupae in a group distant from the drinking trough."*

The antenna plays an important role both in the robot mechanism described in this paper and in social insect behaviour. Ants have two antennae, which they use to sense differences in pheromone strength [6]. This difference can then be used to follow a pheromone trail: a behaviour, which has also been implemented in real robots [7]. *Apis mellifera* Honey bees can use their antennae to estimate the thickness of nest walls during building [8].West-Eberhard [9] claims that wasps can use their antennae to perceive their local environment, enabling them to make building decisions. Downing and Jeanne [10] provide the first experimental evidence of this in the species *Polistes fuscatus*. When half of a wasp's antenna is amputated, the shape and structure of the nest constructed by the wasp is significantly different from a wasp with an intact antenna. Observations of *Polistes dominulus* wasps by Karsai and Penzes [11] show that these wasps antennate the bottom of cells to distinguish between small cells and large cells. The robots described in this paper behave in a similar way to social insects: they antennate objects directly in front of them and perceive the difference between these objects and the objects being carried.

As with previous research [1,2], this work uses robots with minimal abilities in order to conduct object sorting. There are two ways in which the robots' abilities can be minimized. Firstly, the robot hardware itself can be made with the minimum of components implying minimal sensing and communication ability. This, potentially, allows the production of cheaper homogeneous robot units, which are less prone to malfunction [12]. It also enables the development of rules for future use in micro or nano robots. These robots may need to *"operate in very large groups or swarms to affect the macroworld"* [2]; where for ease of production, only basic hardware can be used and non-communication will be important to scale to the number of robots required [13]. Secondly, the robots can be labelled minimalist because they are only able to run a minimal set of rules. Simple behavioural rules are employed because they are generally more robust than complicated rules. This study employs minimalist robots in the widest sense, where both the physical capabilities and the behavioural rules aim to be the minimum required to complete the task.

This paper is divided into 4 sections. Section 2 describes the method and includes a description of the simulation and performance measures. Section 3 contains the results and Section 4 discusses the implications of, and possible extensions to, this study.

2 Method

2.1 The Real Robot Arena

This study presents the results of experiments conducted both in simulation and in a real multi-robot system. The photographs and full descriptions of the real environment, robots and frisbees can be found in Melhuish *et al.* [2]. In brief, the robots operate in an octagonal arena containing shaded frisbees. The robots are 23 cm in diameter and are easily portable. The robots have been modified, each having an antenna fitted. At the end of this antenna is a cup containing an infrared sensor. During forward movement of the robot, these antennae are kept in an upright position. When a robot is carrying a frisbee and bumps into another frisbee in front of it, the antenna is slowly lowered. Because it contains an infrared sensor, the antenna allows a reading to be taken of the frisbee in front of the robot. Another sensor on the body of the robot takes a reading of the frisbee being carried. If the two sensor readings give similar values, then the frisbee being carried is dropped.

2.2 The Simulation

Physical robot experiments are extremely time-consuming and usually plagued by technical constraints due to the unreliability of robot hardware [14]. It was therefore decided that before validation took place on real robots, the patch sort mechanism should first be tested in simulation to determine the likelihood of success [15]. In designing the simulation, two different methods were considered: a probabilistic approach which considers the system as a whole, and a more direct simulation modeling each of the robots individually and employing deterministic rules, with the addition of noise.

Global probabilistic modelling as described by Martinoli *et al.* [16], involves the representation of robot clustering as a sequence of probabilistic events. Cluster sizes are modified on the basis of simple geometric considerations and robot control parameters. There are two advantages, given by Martinoli *et al.* [16], in using this method. The first advantage is that it enables the researcher to investigate and determine which of the characteristics of the experiment are the most influential to the clustering process. The second advantage is that probabilistic models run faster than direct simulations. Despite these advantages, the probabilistic model was rejected. While it is true that sometimes the main characteristics of the simulation may be determined with this method, it is also possible that characteristics which are not responsible for the behaviour in the real robots but just happen to lead to the same desired outcome could mistakenly be chosen as being relevant. Instead of choosing the probabilistic model, a direct simulation was chosen for implementation. We argue that, in this case, a direct simulation gave us better 'feel' for the behaviour of the robots.

The problems of using simulations to test control mechanisms designed for real robots are well known [e.g. 17, 18, 19]. In order to increase reliability of control mechanisms developed in simulation, Jakobi *et al.* [20] and Jakobi [21] studied the effect of introducing noise. They found that even minimal simulations were sufficient to provide solutions capable of crossing the reality gap provided a large enough amount of random variation is included in this simulation in the right way. In our simulation noise is included in two ways. Firstly, the motions of the robots are subjected to random variation and secondly robots sometimes make mistakes (probability of 0.02) when comparing the colour of the frisbee they are carrying with the frisbee directly in front.

Figures 1a and 1b below provide an illustration of the simulation. Here six robots are participating in the sorting of six different types of objects. The colours of the frisbees are represented by different shapes.

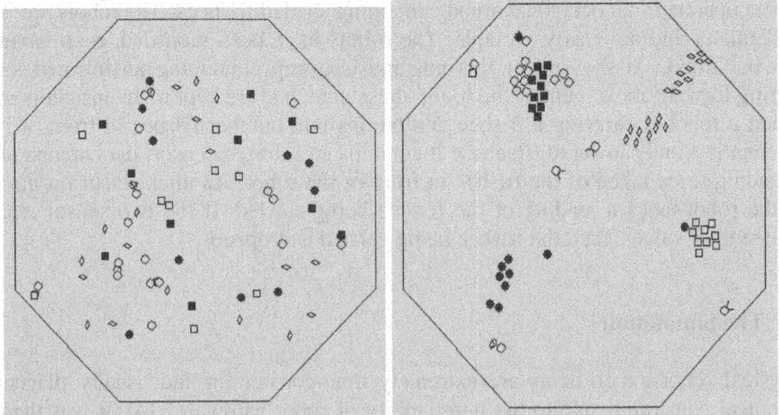

Fig. 1. a) Start of a simulated trial. b) After 100,000 time steps.
Key: ○< Robot. All other shapes are different colours of frisbees.

The simulation was first tested using two algorithms that have already been shown to be successful on real robots: the clustering of Beckers *et al.* [1] and the segregation of Melhuish *et al.* [2]. For both algorithms the number of robots (six) and the number of frisbees (22 for clustering and 44 for segregating) were set to be the same as those used by Melhuish. The same criterion was used as a measure of success (i.e. 90% clustering – see below). For the clustering algorithm, the mean number of iterations taken to achieve the 90% clustering was 97000, with a standard error of 12200 and for the segregation algorithm the mean time taken to reach 90% clustering for the red frisbees was 149000 with a standard error of 15500. The conversion rate works out to be approximately 10000 iterations in simulation to half an hour of robot time, which allows us to make a comparison with results available in [2]. For example, our mean time for the segregation algorithm in simulation converts to 7 hrs and 27 mins, whereas from the mean calculated from four experiments (outlier is ignored) conducted by Melhuish *et al.* [2] is 6 hrs and 38 mins. This shows that the simulation is successful in producing comparable results.

2.3 Definitions and Performance Measures

To examine the progress and final success of a patch sorting experiment a metric was devised using the definition of a cluster as *"a group of frisbees in which any member was within a frisbee radius of at least one other member"*[2]. During each iteration, the largest cluster size was calculated for each colour of frisbee. For our metric, we estimated the overall progress of the clustering process by summing the largest cluster size for each colour of frisbee and giving this as a percentage of the overall total number of fribees in the arena.

The rules below define the patch sort mechanism used in the present paper:

Rule 1: ***If*** *(gripper pressed & Object ahead) then*
Make a random ***turn*** *away from the object.*

Rule 2: ***If*** *(gripper is pressed & no Object ahead & colour in front different from colour carried) then*
Reverse *for pull-back distance & Make a random* ***turn*** *left or right*

Rule 3: ***If*** *(gripper is pressed & no Object ahead & colour in front same as colour carried) then*
Drop *object and reverse small distance & Make random* ***turn*** *left or right.*

Rule 4: *Go* ***forward***

3 Results

The experiments conducted for this study aimed to explore the success of the patch sorting algorithm. The algorithm's performance was investigated as the number of different types of objects to be sorted is increased. Six robots were used throughout the experiments. The value six was chosen for the simulation so that the number of robots used was feasible for physical implementation. Experiments 4.1 and 4.2 were conducted in simulation using 60 frisbees and experiment 4.3 was conducted with real robots using 30 frisbees to decrease the time needed to complete the task.

3.1 Experiment 1: Simulation (Time to Achieve 92% Completion)

The first set of experiments explore the time the robots take to sort one to seven colours. Twenty trials were conducted for each of the seven colours. A value of 92% was chosen, because it allows four misplaced frisbees, the same number used by

Melhuish *et al.* [2] for their segregation experiments. Each trial was deemed to be successful and was therefore terminated at this 92% completion rate. For 60 frisbees, a 92% completion rate corresponds to 56 out of the 60 frisbees being within the largest cluster of their particular colour (see Figure 2).

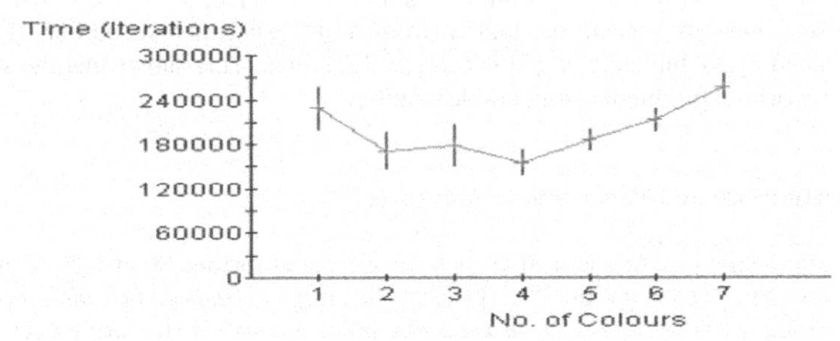

Fig. 2. Mean time (± Standard Error) to complete 92% sorting for 1 to 7 colours.

It can be seen from the graph that the clustering of one colour is more time consuming than sorting two and three colours. This is due to two large clusters forming which are stable for a period. During this time, both clusters have frisbees removed from their perimeter by the robots. The smaller cluster is more vulnerable, because it is more likely to be disrupted by robot collisions. These frisbees can then be added to the other cluster or deposited back in the cluster from which they were removed. The probability of these frisbees being added to the larger cluster is slightly greater than the probability of them being added to the smaller cluster, because the larger cluster has a greater perimeter. This mechanism means that gradually the smaller cluster reduces in size with its frisbees being added to the larger cluster – a time consuming process. This effect explains the shorter completion time when sorting more than one colour, where the aim is to produce two or more separate clusters. However, as the number of colours increases beyond four, the graph indicates that the frisbees become more difficult to sort. Possible reasons for this increase in difficulty as the number of frisbees increases are:

1. The chance that a robot that is carrying a frisbee and collides with a frisbee of the same colour is reduced because there are less frisbees of each colour within the arena.
2. Isolated individual or groups of frisbees can become encapsulated within clusters of other colours.
3. Small clusters tend to be unstable and so clusters of different colours form together. There is a tendency for these clusters to merge due to the disruption caused by collisions of the robots with the clusters.

Experiment 2: Simulation (Average Percentage at Plateau)

As discussed in experiment 1, it was not possible to obtain results beyond seven colours for the 92% completion criterion. To allow comparisons between the sorting performances of many more colours, a new experiment was devised. This experiment involved waiting for the clustering performance to plateau and then taking the mean percentage completion value for a period of 100,000 iterations. This is the mean percentage value around which the performance of the sorting process was able to stabilize. A total of four hundred trials took place; twenty trials for each of one to twenty different colours.

The results from experiment 1 were used to obtain a value at which we could be confident that the trial had reached the plateau level. For each colour a value two standard deviations above the mean was calculated. The calculated value represents a confidence of approximately 98% that a trial had reached its plateau. This was the value from which percentage performance readings started to be taken. Where results were not available from the previous experiment (i.e. 8,9 and 10 colours), readings began at 1,000,000 iterations, to err on the side of caution. Readings were taken for every iteration from this point in the trial onwards and for a further 100,000 iterations. At the end of each trial, the mean of the readings found was calculated, then at the end of the twenty trials for each colour the mean and the standard error were determined. A graph of these results can be seen in Figure 3, which shows how up to a point (about 9 colours) the performance of the patch sorting degrades as the number of colours being sorted increases. The standard error of these values also increases as the number of colours being sorted increases. The reason for the poorer preformances may lie in the increased number of patterns, that can be created from an increased number of colours: each pattern has a different level of complication and therefore different sorting difficulty.

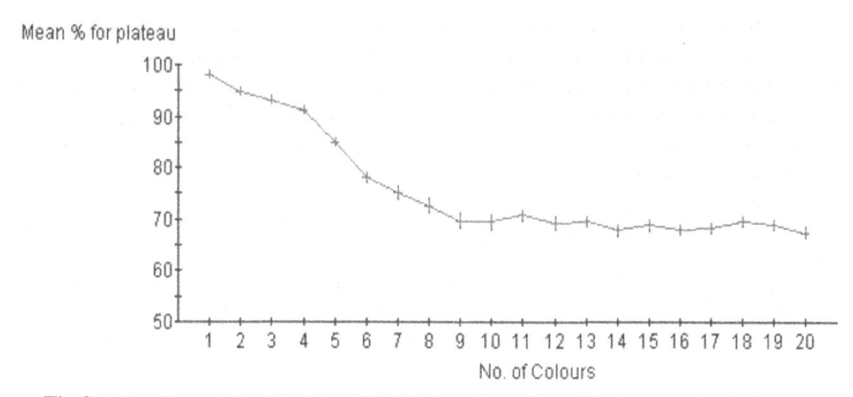

Fig 3. Mean (± s.e.) for 20 trials, 100, 000 iterations after trials have reached plateau.

3.3 Experiment 3: Real Robots

To validate the results produced in simulation, we conducted three experiments with real robots operating in a physical arena. The physical experiments were conducted with six robots and 30 frisbees, which were split evenly between three different shades (whites, greys and blacks). Three trials of the experiment were conducted and the best performance the mechanism was could achieved in all three trials was 70%. This corresponds to 21 out of the 30 frisbees, being in the correct clusters. To reach this level of success, took 1 hour 19 minutes, 2 hours 40 minutes and 3 hours; giving a mean of 2 hours 20 minutes and a standard error of 25 minutes.

The success of the experiments is clearly visible in Figure 4 below. The pictures have been taken from the beginning and the end of the second of the three trials. The picture on the left hand side shows the frisbees evenly distributed at the start of the experiment and the picture on the right shows the frisbees clearly separated into three separate patches, where each patch consists mainly of one colour.

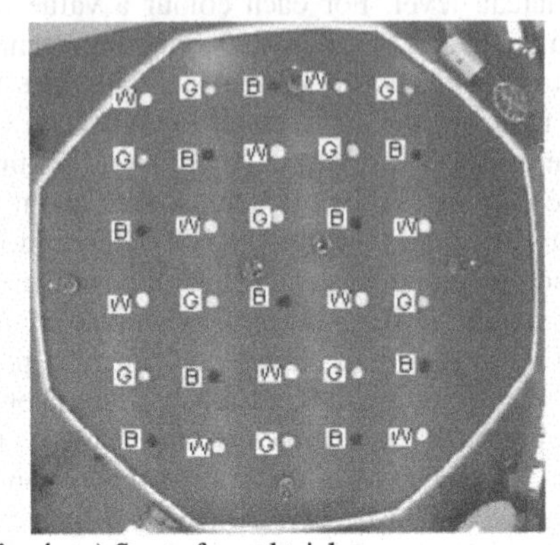

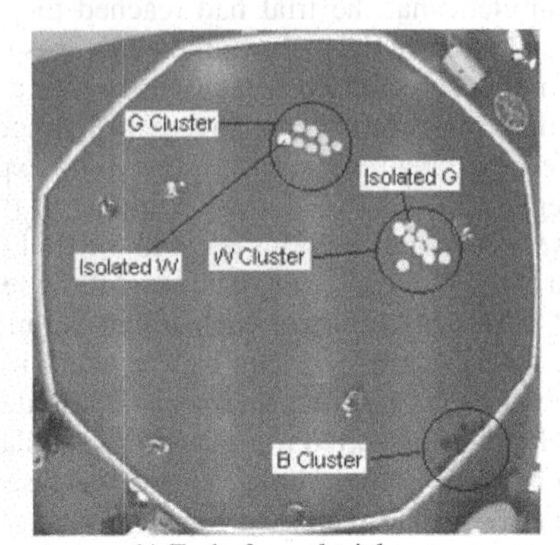

Fig. 4. a) Start of a real trial b) End of a real trial.
Labels added to pictures for clarity: W – White frisbee; G – Grey frisbee; B – Black frisbee

The slightly poorer performance achieved by the sorting mechanism on real robots, is probably attributable to the difficulties involved with using infrared sensors, which were affected by changing light levels within the arena. Another reason for the performance metric showing poorer performance could be due to the proportion of robots able to carry frisbees. Once the frisbees are patch sorted, the robots are still able to pick off frisbees from the clusters. With less frisbees, each frisbee represents a higher percentage of the total number of frisbees. If 50% of the robots are carrying frisbees at any one time, the performance level is already below the 92% used in the simulation.

By demonstrating that a good level of performance on three colour patch sorting is possible using real robots, we can now have more confidence that our patch sorting mechanism would be successful if we attempted to transfer all our simulated results to real robots.

5 Discussion

This study extended the object clustering research of Beckers et al. [1] and the object sorting research of Melhuish et al. [2]. It has shown that patch sorting can be achieved by systems of simulated robots, with limited sensing ability and a limited set of rules. The performance of the patch sorting mechanism was tested using 1 to 20 different colours of frisbee in the virtual arena at one time. It was found that for 1 to 7 different coloured frisbees the robots could achieve a high level of clustering for every test.

Following the success of the mechanism in simulation, it was tested on physical robots. The mechanism was successful at sorting three different shades of frisbees into separate clusters. However, the quality of this sorting was not quite as high as that achieved in simulation. It was noted in observations made during the physical trials that this poorer quality of sorting was mainly due to fluctuations in the infrared sensor reading, which were affected by changes in light levels in the arena. An additional factor which affected the performance metric for the physical experiments was that there were less frisbees being used. This meant that each frisbee was worth a higher percentage. Robots 'steal' frisbees from sorted clusters and each frisbee being carried by a robot reduces the performance rating of the system by a larger amount.

Encapsulated frisbees and merging clusters (see section 3.2) were identified as hindering the sorting performance of both the real and simulated systems of robots. Both these problems could possibly be reduced or eliminated by altering the default settings. One idea is that the arena size affects the probability of clusters of different colours forming together. Increasing the size of the arena increases the space into which the frisbees can spread out. This may place different clusters at a distance far enough away from each other to prevent the clusters being nudged by the robots. A further improvement in performance may be seen by increasing the total number of frisbees as the total number of colours increases. This could eliminate the need for clusters to form together to become stable, because each colour would have sufficient frisbees to form a stable cluster on its own. With further experiments using different default settings it may be possible to show that the patch sorting mechanism scales up to a large number of colours without a fall in performance.

The work produced in this report takes a contrasting approach to the global equation based simulations often used to study biological effects. The type of model we used is comparable to biological simulations, which are microanalytic in their nature [22]. In this type of simulation each individual organism and environmental effect is separately represented. We claim that our patch sorting algorithm allows robots to exhibit behaviours that are similar to the behaviour of real ants and we invite further studies of sorting behaviour in real ants using microanalytic simulations. The hope is that through such work it will be possible to make direct comparisons with our algorithm.

In conclusion, a task as complicated as patch sorting has been shown to be achievable using the simple mechanism. The mechanism has been shown to work for up to twenty different types of object.

References

1. Beckers R., Holland O., Deneubourg J-L.: From Local Actions to Global Tasks: Stigmergy & Collective Robots, Proc. of the 4th International Conference on Artificial Life (1994)
2. Melhuish C., Holland O., Hoddell S: Collective Sorting and Segregation in robots with Minimal Sensing, 5th International Conference on the Simulation of Adaptive Behaviour. Zurich. From Animals to Animats. MIT Press . (1998)
3. Deneubourg J.L., Goss S., Franks N., Sendova-Franks A., Detrain C., Chretien L.: The Dynamics of Collective Sorting: Robot-Like Ants and Ant-Like Robots, Simulation of Adaptive Behavior: from Anmals to Animats, Meyer J.-A., and Wilson S. (eds), MIT Press, (1991) 356-365
4. Lumer E.D., Faieta B.: Diversity and Adaptation in Populations of Clustering Ants, Proceedings of the Third International Conference on Simulation of Adaptive Behavior: From Animals to Animats 3, Cambridge, MA: MIT Press (1994) 499-508
5. Dejean, A., Lachaud, J.-P.: Polyethism in the ponerine ant Odontomachus troglodytes: interaction of age and interindividual variability. Sociobiology 8(2) (1991) 177-196.

6. Holldobler B., Wilson E.O. (1990), "The Ants", Cambridge, Massachusetts: Harvard University Press
7. Webb B.: Robots, crickets and ants: models of neural control of chemotaxis and phonotaxis, Neural Networks II (1998) 1479-1496
8. Lindauer M. & Martin H.: Uber die orientierung der biene im duftfeld, Naturwissenschaften 50(15), (1963) 509-514
9. West-Eberhard M.J.: The social biology of polistine wasps, Misc. Publ. Mus. Zool. Univ. Mich. 140, (1969) 1-101
10. Downing H. A., Jeanne R.L.: The regulation of complex behavior in the paper wasp Polistes fuscatus (Insecta, Hymenoptera Vespidae), Animal Behavior 39, (1990) 105-124
11. Karsai I., Penzes Z.: Comb building in Social Wasps: Self-organization and stigmergic Script., J. Theor. Biol. 161, (1993) 505-525
12. Johnson P.: Cooperative Control of Autonomous Mobile Robot Collectives in payload Transportation, MSc Thesis Virginia Polytechnic USA (1994)
13. Kube C., Zhang H.: Collective Robotics: From Social Insects to Robots, Adaptive Behaviour 2 (1994) 189-218
14. Cao Y.U., Fukunaga A.S., Kahng A.B., Meng F.: Cooperative Mobile Robotics: Antecedents and Directions, IEEE Int. Conf. On Intelligent Robots and Systems (1995)
15. Kube R.C., Zhang H.: Collective Robotic Intelligence, 2nd Int. Conf. On the Simulation of Adaptive Behavior (1992)
16. Martinoli A., Ijspeert A. J., Gambardella L. M.: A Probabilistic Model for Understanding and Comparing Collective Aggregation Mechanisms, European Conference on Artificial Life (1999)
17. Brooks R.: Artificial Life and Real Robots, Proc. 1st Eur. Conf. On Artificial Life (1992)
18. Krieger M., Billeter J.-B.: The call of duty: Self-organised task allocation in a population of up to twelve mobile robots, Robotics and Autonomous Systems 30 (1990) 65-84
19. Lambrinos D., Moller R., Labhart T., Pfeifer R., Weigner R.: A mobile robot employing insect strategies for navigation, Robotics and Autonomous Systems, 30, (2000) 39-64
20. Jakobi N., Husbands P., Harvey I.: Noise and the reality gap: The use of simulation in evolutionary robotics., In F. Moran, A. Moreno, Merelo J.J, Chacon P., editors, Advances in Artificial Life: Proc. 3rd European Conference of Artificial Life. Springer-Verlag (1995)
21. Jakobi N.: Half baked, Ad-hoc and Noisy: Minimal Simulations for Evolutionary Robotics, 4th Conference on Artificial Life (1998)
22. Collins, Robert J., Jefferson D.R.: Representations for Artificial Organisms, In: Simulation of Adaptive Behavior: from Anmals to Animats (1991)

Behavioural Formation Management in Robotic Soccer

Václav Svatoš

Czech Technical University, Faculty of Electrical Engineering
Technická 2, 166 27 Prague
svatosv@lab.felk.cvut.cz

Abstract. Our decision-making system navigates a robot in the team context. Team goals are accomplished by offensive and defensive tendencies of particular agents. Our distribution utilises dynamically conformed formations that maintain relations among agents. During game, a formation manager estimates actual goals at group level. Goal locations are presented as global attractors. This reduces common knowledge storage. Particular decisions calculate them respecting local relations. At individual level, an agent applies social laws to neighbouring team-mates. Their strategies ensue from an on-line role assignment counting actual team trends.

1 Introduction

Behavioural task decomposition allows maintaining a sub-optimal variant of robot control in partially observed environment. A sensation makes a local description of the situation. The main team knowledge is comprehended in the relations between sensor and actuator. We aim at flexible team co-operation among autonomous robots. In this conception, advanced knowledge is responsible for correlated performance among individual agents.

Below, we demonstrate connectivistic capabilities of hybrid architecture in the multi-agent domain. This article is organised as follows. First, we recapitulate features of behavioural and reactive modules that make together a control system of a particular agent. In the chapter 2, we suppose system interactions performing group strategies. Practical experiments are proposed in the chapter 3. The chapter 4 summarises achieved results in comparison with related work and the chapter 5 concludes the work we have done.

1.1 General

Common approximate goal achievement appears in a suitable decomposition of task by means of force capabilities encapsulating typical activities to behavioural units [1]. There is an individual agent that tries to solve a team goal in a dynamic environment. A parallel performance allows controlling a robot team using the bottom-up architecture. Due to its hierarchical conception, the same competition

J. Kelemen and P. Sosík (Eds.): ECAL 2001, LNAI 2159, pp. 553–562, 2001.

mechanism could be extended at a team level using joint intention [2] or shared plan [3].

Likewise our hybrid system [4] takes advantages of the stimulus – response processing in the behavioural inspiration. Respecting needs of an agent life, instant activity selection passes through internal motivations. This conception applies ethological principles [5], [6]. An appetency flow [7] reduces causality rules to an action selection mechanism (ASM) as well. So far the behavioural control has fitted to navigation tasks [8], [9].

1.2 Methodology

The behavioural model is ready to RoboCup [11] application. Particular agent exploits heterogeneous cues to maximize self prospect. There are two principal kinds of objects in the soccer domain. Static objects bound a working area by landmarks. Dynamic ones (like a location of the ball and players) define actual formations. Input channels are established directly from all message classes to calculus-based system inputs. The ASM also monitors such needs of an agent life as a stamina attribute or a game time.

We deal with heterogeneous agents that make a play for a score. Moreover, agent's teamwork follows maximal group benefit in individual action selection. To do this, a field-player takes one of three roles at the time: defender, midfielder, and forward player.

The set of sensors and basic activities pack the state space of the hybrid system. Reactive mechanism manages critical situations by fast sequences at a low level. Advanced trends emerge at a behavioural level. Both levels manipulate with activity modules counting in following attributes:

1. *Activation energy* $a(t)$. This one takes an accumulator feature that rates a temporal applicability for the behaviour. The activation energy rapidly declines on any satisfaction of the behaviour.
2. *Implementation.* Any behaviour is manifested force activity or a silent appetency. The sequential procedure is fired to satisfy a behavioural module in the force case. In this context, activation energy means desirability to the instinctive action.
3. *Activity conditions.* Physical constrains limit force actions. Inner behaviours are valid until activation energy overflows.
4. *Hierarchical excitation of other modules.* The causal links improve a priority from the most required actions and instant cues to applicable ones.

2 Flexible Hierarchy of Team Control

In soccer team design, we distribute a team control to two hierarchical levels (see Fig.1). A group level stimulates sub-goals performed at an individual level. Functional decomposition maintains autonomous performance that makes a good thing in a local situation.

At the *individual level*, we work with the basic behaviour set. Used nodes are classified by self-behaviour manifestation:

1. *Individual activity.* This node fills a force procedure to interact with the environment.
2. *Internal stimulus.* This evaluates energy depletion by any individual activity. The stamina attribute in the RoboCup simulation makes internal stimuli of the system.
3. *Inner node.* This one makes only appetency. The complex behaviour emerges from several individual activities.

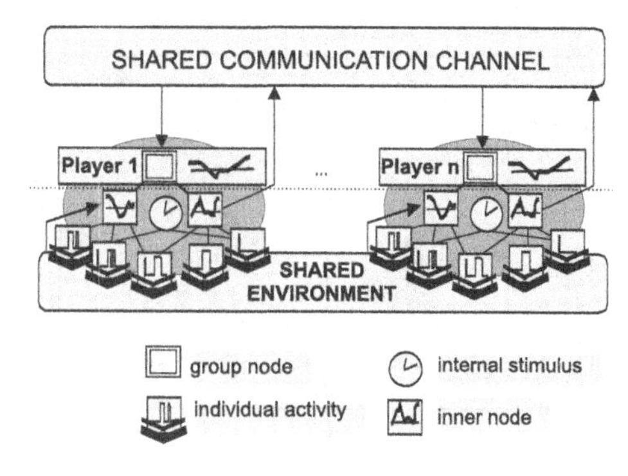

Fig. 1. Hierarchical structure of team control. Single players handle the individual performances autonomously at individual level (bottom). All individual activities are interconnected together through causal relationships. A collective goal is managed at group level (top). The coordination passes both through shared environment and through shared communication channel. The environmental feedback reflects individual activities. The shared channel synchronizes group strategy by exchange of group variables.

Subgoals are shared at the *group level.* This maintains a social distribution of interests by functional domains. An update of individual knowledge is illustrated in the Fig. 1. Broadcast through shared communication channel is used to exchange them. The group node interconnects individual decision systems to the group level and adapts global states to internal local representation from the agent view.

2.1 Global State Representation

We describe functional domain by global vector fields rather then by a list of particular geometrical co-ordinates. This field influences the agent specification to aggregate roles in significant positions. Player layouts are concerned in two principal frames. First, the co-ordination of collective activities counts in locations of actuators. Second, we take the centre of opposite players' area to team tendencies. The team knowledge is stored in three forms:

1. *Formation.* One formation defines role of players. This determines an individual strategy at the lower levels. Such knowledge modifies role assignment in alternative ways.
2. *Global vector field.* The partial fields bring repulsive and attractive trends [13] to specification of group strategy – see Fig.2. Attractive forces cause orientation of players to object in approximate directions. A frontal perception is not necessary. The player speeds up an effect of searching steps in the reactive handling.

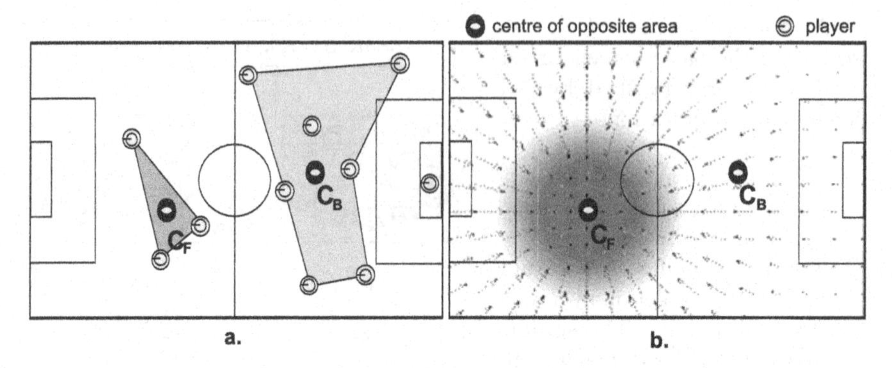

Fig. 2. Attractors based on location of opposite players. The part **a.** illustrates opposite players layout in the game field. The filed map at part **b.** shows partial potential vector field applied to collective decision in "back" formation. Glossary: $\mathbf{C}_\mathrm{F}$ - centre of opposite forward players area (frontal opposite point), $\mathbf{C}_\mathrm{B}$ - centre of opposite defenders area (back opposite point)

3. *Global accumulators.* The group level also accumulates a global activation energy. These attributes influence on formation decomposition.

The sub-goal management is based on low-level performance respecting formation constrains: team strategies are described by spatial attractive trends. The formation assignment distributes these trends to partial directions. The group node interprets knowledge from the formation to internal appetencies.

2.2 Formation

The group level maintains essential knowledge about actual task distribution in formation frames. One formation manages agents coming to given bounds. Such autonomous performance, the agents fluctuate among formations. Principally, formations sort low-level activities using classification of team players to roles. We talk about *formation participants* performing basic motions that are coordinated by means of social distances in the local context. All agents coming to the formation fill the same role in the team hierarchy. The agents take a brief formation assignment including relationships to a predictor team-mate and a successor one.

The dynamic memberships in the formation are the main feature of a proposed role assignment concept. Aggregation and dispersion make the basic coordination tools. Properties of the formation are listed at Fig. 3. The formation outline is itemised as follows:

1. *Role.* The formation uniquely determines a role of a participant that specifies a domain of common knowledge.
2. *Relation parameter.* This property qualifies bases for sort of formation participants. The formation cover of all players with the relation parameter coming to *upper and lower bounds.*

3. *Pivots.* The pivots adjust a formation behaviour. The role is transferred to an individual performances by social law. Agents manifest them in a virtual chain along pivots.
4. *Ordered participants.* Each agent in the chain approximates self-activity from known trends of neighbours.

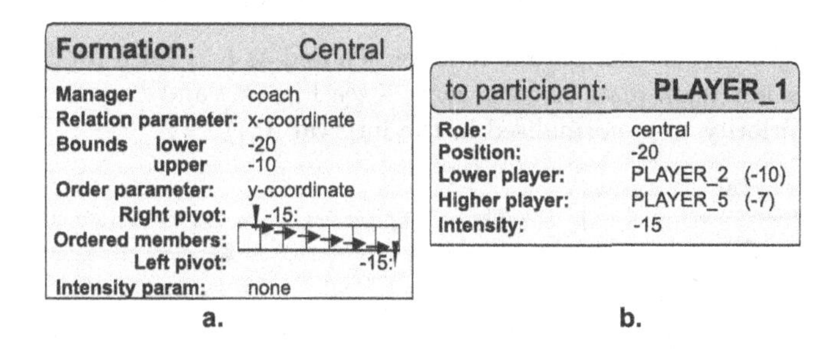

Fig. 3. Formation frame. One formation frame is shown in the part **a.** The formation manager orders observed agents by the order parameter. Formation participants are informed about them by broadcasts. The part **b.** lists a formation broadcast. The partial knowledge is addressed to the participant in the text format. The broadcasts also consist of a recommendation by the relation parameter (position in this case) and intensity of the recommendation.

Formation recommendations are shared by broadcasts. An example of the participation frame is listed in Fig. 3b. The formation concept takes in definition of several bases. We classify any formation by levels of goal availability. A distance to the self goal-line is used to classify field-ordered formations (see Fig. 4a). The "*frontal*" formation covers offensive tendencies by forward players. The "*central*" formation consists in midfielders and the "*back*" formation co-ordinates positional behaviours of defenders. The proposed formation model discards goalie from the "back" formation. This allows to co-ordinate team performance in the goalie view. We introduce angular relations as alternative base for the formation classification. Then, the goalie sorts players (see Fig. 4b) to "*left corner*", "*left*", "*ahead*", "*right*" and "*right corner*" formations. These goalie-centred formations overlap the field-ordered formations.

Two formation bases specify roles of participants inside formations as well as across formations. Both dimensions provide self-contained roles in the team hierarchy. Combination of the bases makes possible a spatial distribution of force elements over game field. Next, we show advantages of the field-ordered formations in the dynamical role assignment.

2.3 Formation Emergence

Our formation concept makes groundwork to balanced performance in the team. Each agent applies a basic social law in a formation arrangement while finds self job at the centre of gravity of reflected co-participants. The implementation of a basic social law iterates through gradient shifts while divergence among co-participants decreases. First, the agent counts a defect δ (1) from the local trend itself. Next, it comes near to an enumerated location respecting an actual defect. Two co-participants are taken to the defect of the agent performance as follows:

$$\delta = (1 - w)\ \ell_{i-1} + w\ \ell_{i+1} \tag{1}$$

An estimated relation parameters are denoted as ℓ_{i-1} for a predecessor participant (a left team-mate in local view – see Fig. 3a) and ℓ_{i+1} for a successor one. The priority $\boldsymbol{w}$ is normalised in the interval $\langle 0;1\rangle$

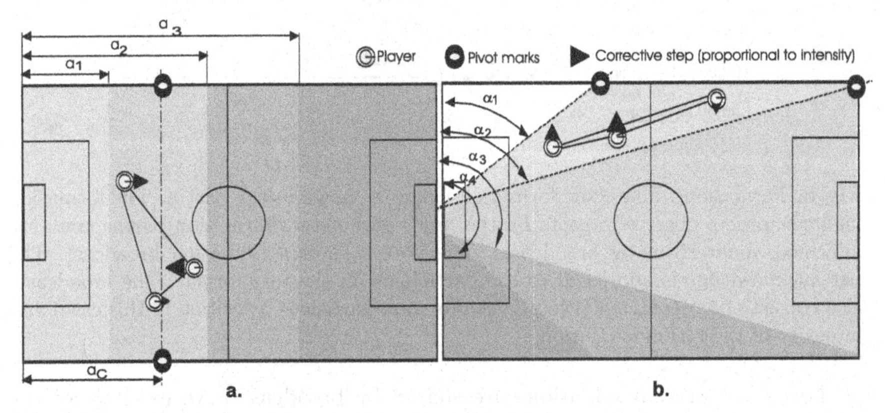

Fig. 4. Formation arrangement. The map in part **a.** illustrates aggregation trend in field-oriented formation. Distances are referenced to goal-line. The filed map at part **b.** illustrates a sectoral dispersion trend in the goalie-oriented formation. Glossary: $\mathbf{a}_1$ – "back" formation, $\mathbf{a}_2$ – "central" formation, $\mathbf{a}_3$ – "frontal" formation, $\mathbf{a}_C$ – distance pivot in "central" formation, α_1 – "left corner" formation, α_2 – "left" formation, α_3 – "ahead" formation, α_4 – "right corner" formation. Players are sorted up by a formation parameter. Goalie classifies location team-mates in angular relationships. During arrangement, the formation participants iterate along the formation parameter base (orthogonal to relation parameter). Actual steps are gained by the defect of the agent.

The formation participants apply the basic social law respecting the actual defect. Over all subscribed formations, the agent superposes defects in recommendations to gain comparative steps in given bases. Used pivots determine one of group trends:

1. *Dispersion of formation participants.* The dispersion over the formation allows often covering a large area by formations. Each participant puts self comparative steps near to a local target. The steps approximate positions of neighbours.
2. *Aggregation of formation participants.* The aggregation is performed as a special case of the dispersion. The aggregate behaviour is determined by identical pivots (the same relation parameter – see Fig. 4a). There is a singular target of the dispersion mechanism. As the target of comparative steps

is shared by the whole formation, their dynamics iterate through the dispersion to the singleton.

Our practical formation arrangement utilises both mechanisms over several formations. The distributions are preferred to cover one functional area more generally. The aggregations concentrate agents along specific roles. We utilise both the field-oriented formations and the goalie-oriented ones in the complex decision. Fig. 4. illustrates both formations in a game situation. One formation base classifies players to exclusive formations by field thresholds $\mathbf{a}_1 \div \mathbf{a}_3$ (respectively $\alpha_1 \div \alpha_4$). The formation management allows switching behaviour parametrically. The group accumulators gain a range of distributions in partial team behaviour.

3 Experiments

The proposed decision-making algorithm has been simulated in the Soccerserver environment [12]. Our client applications brought to practice the hybrid control of individual players. The agents utilised several instrumental behaviours to handle direct manipulations with objects in the robot neighbourhood. We defined final needs that are satisfied by such instrumental behaviour operating with the ball as "kick", "pass" or "tend". Our simulations gave evidence that yet the goal-oriented behaviours satisfy needs of players in combination. The resultant formation arrangement manifested an evolving functional adjustment. Performed agent roles were following a proportional displacement in relation to another team-mates.

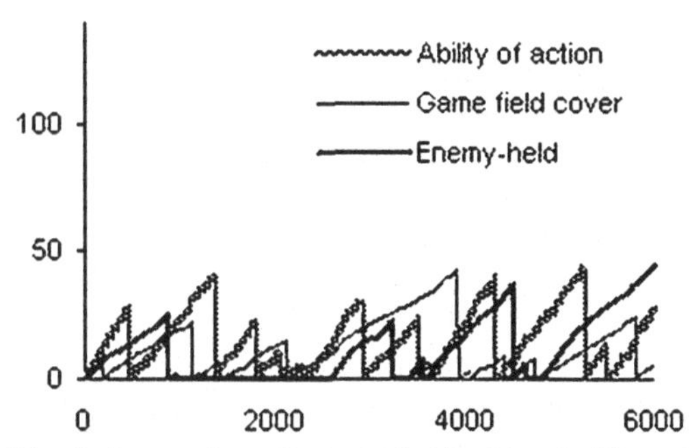

Fig. 5. Integral needs at individual level. There is illustrated accumulation of the criterions counting in geometrical distance that are satisfied quite often.

3.1 Emergence Criterions

The initial reactive base [13] managed instinctive behaviours that are responsible to satisfy robot needs. The reactive base calculated more balanced recommendations from spatial effect of incoming stimuli. Actions were inferred from partial knowledge without keeping of a world map. Used criterions determined environmental impact from an observed situation to the basic behaviour set. Integral needs qualified a goal-orientedness of each player by its manifested temporal effect. Both reactive and behavioural criterions combined following points:

1. *Game field cover.* The formations managed an effective game field cover. An equable displacement of team-mates were denoted as the best cover.
2. *Enemy-held.* The player sought a self-location in context of opposite players. The best enemy-held were found at an area opened by opposite players.
3. *Ability of action.* This criterion qualified a potential manipulation with the ball as distance to it.
4. *Ability of score.* This need enumerated players in an area bounded from the robot to uprights. The optimal criterion counted in higher weight of opposite players that the best criterion was found in an open goal area.

Second, the activation energy calculates temporal effect of basic behaviours till satisfaction. In the hierarchical control structure (see Fig. 1), the inner nodes accumulated the integral criterions. Satisfied integral needs were cleared. Graphs in Fig. 5 illustrate history of the integral needs during experiments.

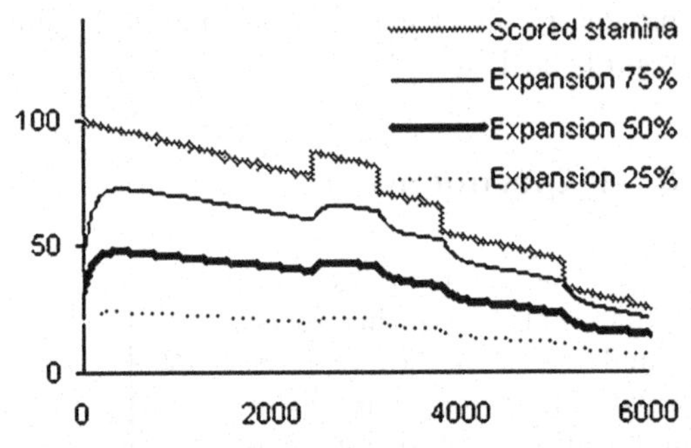

Fig. 6. Group criterions at team level. The impact of the team needs is decomposed to three gains to expansion levels. The expansion factor 25% gains the "back" formation. Transfer from group variables to bounds of "central" formation is given by expansion factor 50%. Idle level of the "frontal" formation follows the expansion factor 75%.

Third, suitable criterions qualified a player performance from the team view. As player aimed to score exclusively, the main goal-oriented criterion qualified their satisfaction in several contexts:

1. *Play time.* This criterion was defined without self-satisfaction in native clock meaning.
2. *Scored stamina.* This one weighted stationary state (balance of actual score) and temporal capacity to force input.

The global variables saved these criterions simultaneously. Moreover, both criterions have combined together to gain actual formation management. Graphs in Fig. 6. illustrate experimental history of the formation bounds. They depend on such team criterion as the scored stamina.

4 Discussion

Balch and Arkin [14] have introduced formations to team co-ordination. They have aimed at uniform space cover rather then force concentration to achieve non-positional goal. Supposed formation have justified geometrical arrangement among each other participants. Agents clustered themselves to given pattern in global view. There was limited number of both patterns and formation participants. Balch and Arkin supposed four robots to formation.

Balch's formation conception was adapted also in our classification role using an order parameter. Parametrical adjustment performed similar patterns in bottom-up way. Our parametrical role assignment also distributed a team strategy more flexibly respecting actual group resources. Our approach took more fault-tolerance communication. A robot solves a team goal more autonomously. A formation manager substituted robots coming out of signal. The manager estimated their temporal contribution to team performance during lost signals.

Homogeneity in role assignment was the main feature of team performance distributed to effective formation domains by Stone and Veloso [15]. The domain specialisation reduced a communication to negotiate among role participants. Similar to our design, they have utilised common knowledge storage at individual level in minimalist principle. On negotiation, we preferred the formation manager that determines roles by means of the formation frames. We also advanced an individual agent perception by group vector fields to enlarge agent's scope. Such team performance results from common team behaviour in defensive trends. Nevertheless offensive trends arise from individual raids, the team performance could also co-ordinate them. The offensive formation cover could be regarded as dual tasks to formation management of defensive strategy.

We also suppose a non-uniform belief distribution along formation participants. Asymmetrical priorities accelerate iterations at one-sided target as well. The major priority speeds up a convergence to target area. Formation elements with identical priority $\boldsymbol{w}$ are formed in an equable displacement to a balanced formation. Asymmetrical priorities concentrate formation participants to one bound side in a geometric progression. The used relation priority among co-participants also forms a centre of formation gravity ℓ_T as follows:

$$\ell_T = \begin{cases} \frac{\ell_{BOUND}}{n} & w = 1 \\ \frac{\ell_{BOUND}}{n} \frac{w-1}{w^n-1} \sum_{0}^{n-1} (n-i)w^i & elsewhere \end{cases} \tag{2}$$

There are ℓ_{BOUND} a formation bandwidth (from the left pivot to right one – see Fig. 3a) and $\boldsymbol{n}$ a total number of co-participants in the formation.

5 Conclusion

Our design fixed one's attention on an interface among group and individual levels. A common knowledge was applied to improve a goal accomplishment. The knowledge was adapted to local views of particular agents. Parametrical interpretation reduced data share content. The teamwork structure was presented as a formation. It framed actual relationships of team-mates and opposite players. The opposite team trends were inferred from centres of clusters. The group level also lets agents know team resources in the sum.

Players come in overlapped formations using two bases – a distance to goal-line and an azimuth to goalie. The formation points out force condensation in significant game field areas. A formation management makes aggregate and distributive behaviours in the team. Group movements result from line shifts of the formation participants.

Besides a vector field generalises opposite trends, the formation model could also map them. A common knowledge is sufficient to classify an opposite team behaviour using opposite formations. Such functional relations will once more reflect the displacement of players. This model will lead up local decisions to a common goal respecting an opposite layout better.

References

1. Arkin, R.: Behavior-Based robotics. MIT Press, Boston (1989)
2. Cohen, P.R., Levesque, H.J., Smith, I.: On Team Formation. In: Hintikka, J., Tuomela, R. (eds.), Contemporary Action Theory, Vol. 2. Dordrecht: Synthese Library, Kluwer Academic Publishers (1997) 87–114
3. Grosz, B.J.: Collaborative Systems, AI Magazine **17**/2 (1996) 67–85
4. Petrus, M., Svatoš, V.: Extended Reactive Behaviour in Mobile Robots. In: Proceedings of Poster 2000, CTU Prague (2000) IC 19
5. Cartwright B. A., Collett, T. S.: Landmark learning in bees. Experiments and models. Journal of Comparative Physiology A **151** (1983) 521-543 Wolfart, E., Fisher, R.B., Walker, A.V.: Position Refinement for a Navigating Robot using Motion Information Based on Honey Bee Stratagies, Proc. Symposium Intelligent Robotic Systems, Pisa, Italy (1995) 257–264
6. Wolfart, E., Fisher, R.B., Walker, A.V.: Position Refinement for a Navigating Robot using Motion Information Based on Honey Bee Stratagies, Proc. Symposium Intelligent Robotic Systems, Pisa, Italy (1995) 257–264
7. Lorenz, K.: Grundlagen der Ethologie. Springer Verlag, New York (1978)
8. Redish, A.D.: Beyond the Cognitive Map. PhD thesis, Dep. Of Computer Science, Carnegie Mellon University, Pittsburgh (1997)
9. Trullier, O., Wiener, S. I., Berthoz, A., Meyer, J.-A.: Biologically based artificial navigation systems: review and prospects, Progress in Neurobiology **51** (1997) 483 – 544
10. Kitano, H., Kuniyoshi, Y., Noda, I., Asada, M., Matsubara, H., Osawa, E.: RoboCup: A chalenge problem for AI. AI Magazine **18**/1 (1997) 73–85
11. Khatib, O.: Real-time Obstacle Avoidance for Manipulators and Mobile Robots. The International Journal of Robotic Research **5**/1(1986) 90–98
12. Noda, I., Matsubara, H., Hiraki, K., Frank, I.: Soccer server: A tool for research on multiagent systems. Applied Artificial Intelligence, **12** (1998) 233–250
13. Svatoš, V., Bureš, M.: Database model of reactive agents. Proceedings of Workshop 2001, CTU Prague. (2001)
14. Balch, T., Arkin, R. C.: Behavior-based formation control for multirobot teams, IEEE Transactions on Robotics and Automation **14**(1998) 926–939
15. Stone, P., Veloso, M.: Task decomposition, dynamic role assignment, and low-bandwidth communication for real-time strategic teamwork. Artificial Intelligence **110**/2(1999) 241–273

Control System of Flexible Structure Multi-cell Robot Using Amoeboid Self-Organization Mode

Nobuyuki Takahashi, Takashi Nagai, Hiroshi Yokoi, and Yukinori Kakazu

Autonomous Systems Engineering, Hokkaido University
North 13 West 8, Kita-ku, Sapporo, 060-8628, Japan
{takanya, nagai, yokoi, kakazu}@complex.eng.hokudai.ac.jp

Abstract. In this paper, we present a distributed multi-cell robot control system that enables autonomous movement. Flexible deformation and actions are features of living things such as amoebae. One amoeba has neither specialized sensitive organs nor motile organs, however, as a whole, a slime of amoebae can present unified movement. That is, organisms such as amoebae own various self-organization abilities. The control system proposed in this study rose from such kind of self-organization, by which cell units move together to give rise to the movement of multi-cell robots, according to only local interaction. Especially, in this study, self-organization phenomena including entrainment of non-linear oscillators were modeled, and effectiveness was shown through computer experiments.

1 Introduction

Flexible structure robots and their control systems are of great importance in recent robotics research field. Robots with soft and flexible structure can be applied to the domestic or welfare usages. However, following features are required for such flexible robots.

1. It should be able to adapt to changing environment.
2. It should own large degrees of freedom (D.O.F.) and be redundant for its structural flexibility.

Various kinds of the flexible structure robots have been developed. The random morphology robot, proposed by Dittrich [1], consists of six servos coupled arbitrarily by thin metal joints. The Genetic Programming was used for the control system of the random morphology robot. The GMD-Snake, developed by Paap [2], consists of numbers of joints and is actuated by motors and strings. In the self-reconfigureable module robot developed by Murata [3], each module may be connected to and detached from another module's surface automatically.

Most of the flexible structure robots were composed of numbers of mechanical parts, and have numbers of actuators. The first difficulty when constructing a flexible structure robot lies in size of hardware, that is, how to construct robot simply and compactly by using or designing good actuators and mechanical components. The second difficulty lies in the construction of the control system

J. Kelemen and P. Sosík (Eds.): ECAL 2001, LNAI 2159, pp. 563–572, 2001.

of the robot, which is responsible for coordinating multi actuators to enable objective behaviors.

To deal with the first difficulty, we proposed a flexible structure robot that is actuated by shape memory alloy actuators (SMA-Net robot). Because this robot uses SMA springs as actuators and skeleton, fewer mechanical components are needed.

To solve the second difficulty, we proposed a control method that uses amoeboid self-organization model. Actually, the features necessary for flexible robots such as adaptation, redundancy, are the features of most living organisms. We took amoebae, well known as primitive and flexible organisms as the model, since amoebae (slime mold) present the ability of basic information processing, although, they does not have any specialized sensitive organ or motile organ. The purpose of this study was to explore a methodology for the control of amoeboid-like flexible structure multi-cell robotics systems.

2 SMA-Net Robot

The SMA-net robot developed consists of several homogeneous cell units. Each unit has six SMA spring actuators and one micro CPU module (Hitachi SH-4). Each SMA actuator is connected to its neighboring cell units (Fig. 1).

SMA actuators have been used in heat engines, flange-bolts for space mission, etc. Recent technology has enabled the formation of a micro SMA actuator on a silicon substrate. In these applications SMA is merely functioned as an actuator. However, in our SMA-net robot, SMA actuators were used not only as actuators but also as the skeleton of the robot. Therefore, SMA actuators worked like muscle fibers. Each single unit of SMA-net robot cannot generate any movement, however, it can contribute to the motion of robot through interaction with its neighboring units. Cooperating or synchronizing with each other, these SMA actuators may generate more powerful force than that of one actuator, and in turn cause movement for nodes of robot.

The SMA-net robot is composed of numbers of cell units and actuators. Therefore the number of control parameters increases. As general research approaches, Genetic Algorithm, Genetic Programming and Neural Networks are

Fig. 1. SMA-net robot

used for such kind of large-scale control problem. However, a great number of trials are necessary to acquire suitable control rules. Moreover, those approaches are not able to adapt to the cases in which environment changes, etc. In this study, a self-organization model referring to actual amoebae was proposed. By self-organization, not only the number of control parameters could be reduced, and learning cost could be decreased considerably, but also the robot would be able to adapt to environment changes.

Another feature of self-organization phenomenon is local interaction. It is self-organization or self-aggregation that produces spatial temporal macroscopic structure through micro local interaction. Accordingly, no central control system is required, so that, the self-organization based control system would be affected neither by increase nor decrease of the number of units.

3 Biochemical Phenomenon of an Amoeba

The objective model described here is related to searching behavior of amoeba (slime mold). Amoebae have both unicellular and multi-cellular periods in their life cycle. Slime mold is a kind of colony of unicellular amoeba. When foodstuffs in the environment are exhausted, amoebae begin to aggregate, after which amoebae form multi-celled slugs.

In spite of not having special sensitive organs and motile organs, slime mold can move towards possible foodstuffs, which is called chemotaxis. The chemotaxis has been investigated by several researchers [5][6][7]. In their studies, it was shown that integral dynamics of physical and chemical processes in amoeba cell protoplasm can be simulated as a self-induced oscillator. Shape change, tension variation and concentration change of intracellular chemical substance of the protoplasm are all relating to the oscillatory phenomena. Though each amoeba cell respectively causes the oscillation, each oscillation synchronizes. As a whole, one unified oscillation is reached.

In addition, each amoeba cell converts stimulation from environment into the oscillation. For example, the frequency of a cell sensing attractive stimulation rises. A mass of slime mold tends to unify oscillation of entire cells through entrainment phenomenon, when the frequency of one cell rises. Similarly, if frequency changes occur at several places within the slime mold, the oscillation would be integrated by entrainment. In the case of entrainment, a phase gradient of the oscillation occurs in the whole slime mold. It is the gradient that causes the integration of the whole slime mold.

Moreover, a foot-like organ called pseudopodium must be formed for moving. Slime mold can specialize its function, even though it is the aggregation of homogeneous amoeba cells. That is, slime mold acts like a distributed system consisting of numbers of self-induced oscillators.

4 Multi-cell Robot Design

In this section, we show the design policy of the amoeboid-like multi-cell robot, and describe the model for computer simulation of the SMA-net robot. This model has following features.

1. Each cell unit is correspondent to one cell of the slime mold.
2. Each cell unit communicates with the others through local interaction among cell units.

In table 1, a comparison was made between amoeboid functions and our robotic model. Firstly, we modeled frequency rises in the cell unit robot. This corresponds to the same effect of amoeba cells receiving an attractive stimulation.

Secondly, we modeled entrainment of oscillations among the cell units. This corresponds to the integration of stimulation to slime mold.

Lastly, we modeled self-organization of function specialization among cell units. With regard to slime mold, this function specialization is similar to the growth of pseudopodium.

We describe model 1 and model 2 in the next section, with the computer simulation results to show the effectiveness of them. Model 3 is briefly introduced in section 7 with experiment results and discussion.

Table 1. Model of multi-cell robot

Processes in amoeba	Equivalent components of robotic model
Sensing the stimulation	Frequency shift of oscillator
Integration of the stimulation	Entrainment of oscillators
Forming pseudopodium	Reconfigration

5 Integration of Stimulation Information by Entrainment

Most fundamental signals taking effort in organism are periodic rhythmic. The pulse of myocardial cell, electroencephalograms, etc., are examples of such signals. In addition, locomotion of a large amount of animals is also dominated by rhythmic signals, which is generated in the low nervous center (e.g., spinal cord and ganglion). Generally, these oscillations are different from the harmonic oscillation such as simple harmonic motion, at the point that these oscillations are not based on an initial condition, and have eigen-amplitude and eigen frequency.

5.1 Oscillator Model of a Cell Unit

The oscillatory phenomenon in amoeba cells is the consequence of concentration-change by reaction of chemical substance. However, a myriad of correlating chemical reactions are involved and it is impossible to describe and model them all. Therefore, in this study, each chemical oscillation was described as a dynamical system model. The nonlinear oscillator of Van der Pol type was utilized for the purpose.

$$\alpha_i \ddot{x}_i - \beta_i \left(x_i^2 - 1\right) \dot{x}_i^2 + x_i = 0 \tag{1}$$

where, x_i: phase of the oscillator.

The reason why we employ this equation is that it has simple and typical behavior of nonlinear oscillator. For engineering purpose such as entrainment of information using oscillator, it is considered that the Van der Pol type oscillator is sufficient and suitable.

When an attractive stimulation is received, the frequency of amoeba cell protoplasm oscillation rises, and it decreases in the case of aversive stimulation. The increase and decrease of the frequency for stimulations can be expressed by changing the parameter α in equation 1.

For example, in Fig. 2, $\alpha = 1.4$ resulted in a normal oscillation state of the amoebic cell. $\alpha = 0.8$ resulted in the oscillation when receiving an attractive stimulation, and $\alpha = 2.0$ showed the state of aversive stimulations.

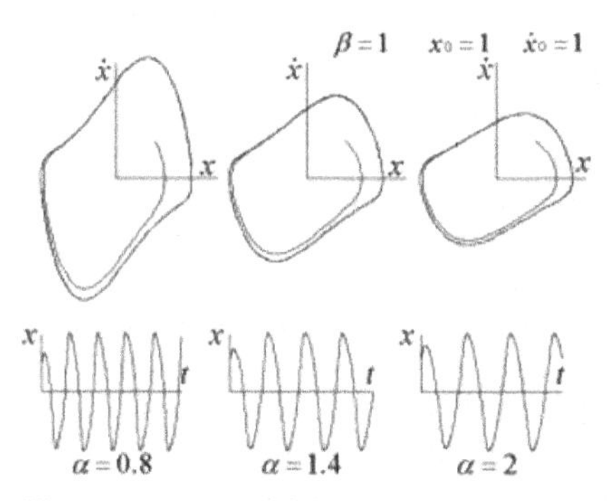

Fig. 2. Van der Pol function as α changes

5.2 Local Diffusion Combination among Units

All oscillations x_i, generated by the Van der Pol oscillator in equation 1, share the same wave form. However, there are phase differences among oscillators. In this section, we describe a coupled oscillator system for achieving synchronization of oscillators.

In slime mold, cells contact directly with each other, or soak completely in liquid solution containing chemical substance. These two intermediations originate the interaction among cells. Under such an interaction condition, when multiple rhythms interact, the phenomenon that cells intend to synchronize to each other occurs, that is, the entrainment. The macroscopic rhythm occurs from micro rhythms through the entrainment, when a large number of oscillators with almost equivalent frequency are spatially distributed. This is called self-organization of spatial temporal patterns.

Additionally, a phase gradient occurs as the consequence of the entrainment, over the whole space. By such a phase gradient, the multi-cell robot would integrate local information sensed by a part of cell units. A large number of related studies have been made on coupled oscillator. Globally Coupled Map (GCM) [8] and Coupled Map Lattice (CML) [9] are most famous. These models were proposed for the analysis of spatio-temporal chaotic phenomena such as rhythmic contraction of a cell group and heat turbulent flow etc. In this study, we presented a simplified form of the CML.

$$x_i(t+1) = (1-\varepsilon)\,x_i(t) + \varepsilon\left(\frac{1}{N_i}\sum_{j=1}^{N_i} x_j(t) - x_i(t)\right) \tag{2}$$

where, ε: coupling coefficient. N_i: a number of cell i's neighbors.

The point different from GCM and CML is that coordinates of cell units are not fixed and change dynamically, so that neighborhood template should be dynamically recalculated.

5.3 Experiment

With the model we described, a computer simulation was realized. Five oscillators were placed in a row. As the frequency of one oscillator increased, the other oscillators changed correspondingly. The transition of attractors and phase differences of the five oscillators was observed. In one case, only the oscillator at the left end received an attractive stimulation, which caused the increase of its frequency. The locus of frequency of the five oscillators was shown in Fig. 3. The initial values of x and $\dot{x}$ were random. Bond length, that is, range inspected by each cell, was set to physical size of one cell. So each cell was combined with its neighboring oscillators. It was clearly shown that the entrainment occurred and all the oscillators were synchronized to one attractor.

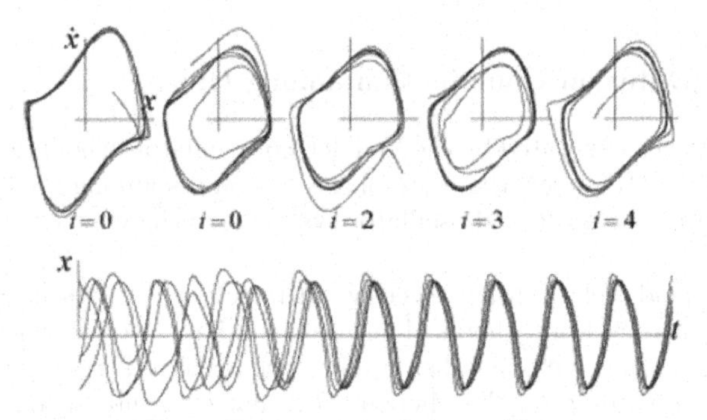

Fig. 3. Entrainment of the oscillators

6 Multi-cell Robot Simulation

To examine the proposed amoeboid self-organization control method, we built a simulator and executed experiments. Two conditions were added in order to simulate the multi-cell robot.

At first, for each cell, an equation on the forces generated according to the phase gradient was added.

$$\overrightarrow{f_{1i}} = -\gamma \sum_{j=1}^{N_i} \left(\sin(\theta_j - \theta_i) \frac{\overrightarrow{r_j} - \overrightarrow{r_i}}{|\overrightarrow{r_j} - \overrightarrow{r_i}|} \right) \tag{3}$$

where, γ: coefficient. N_i: a number of cell i's neighbors. θ_i: phase of cell unit i. r_i: position of cell unit i.

Next, we added a condition for binding cells to avoid dispersion, if forces generated are on different direction. These two conditions correspond to a rule that repulsive force works when cell units get close too much, and attractive force works when cell units get separated away too much. The similar rule was used in other models such as Boids and fish flock, etc. As regards an actual amoeba is concerned, such a force is supplied by surface tension and viscosity of the liquid among cells. In this study, a potential field was used as follows (Fig. 4).

$$\overrightarrow{f_{2i}} = -\phi \sum_{j=1}^{N_i} \left(\left(-e^{-0.02(|\overrightarrow{r_j} - \overrightarrow{r_i}|+d)^2} + k\, e^{-0.01(|\overrightarrow{r_j} - \overrightarrow{r_i}|)^2} \right) \frac{\overrightarrow{r_j} - \overrightarrow{r_i}}{|\overrightarrow{r_j} - \overrightarrow{r_i}|} \right) \tag{4}$$

where, ϕ, k: coefficient.

And the resultant force in each cell becomes as follows.

$$\overrightarrow{F_i} = \overrightarrow{f_{1i}} + \overrightarrow{f_{2i}} \tag{5}$$

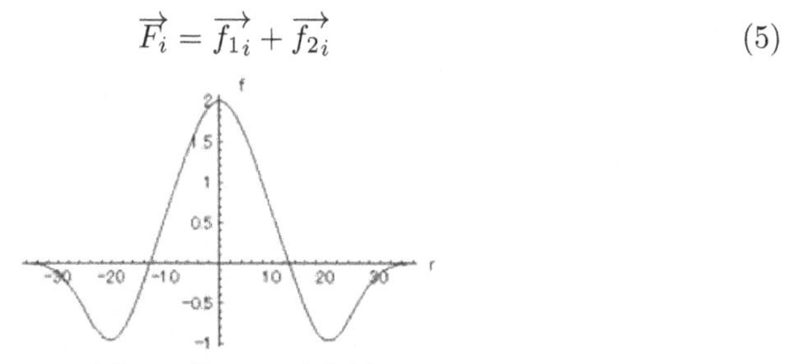

Fig. 4. Potential field

6.1 Experiment of Two-Dimensional Model

The phase gradient caused by the equation 2, and the force by the equation 5 were presented in Fig. 5a. Ten cell units were placed on a plane. Initial values were $\alpha = 1.0$, $\beta = 1.0$ and $\varepsilon = 0.3$. The bond length N was 1.5, when a cell size was 1.0. Similar to the experiment described in section 5, only one cell (arrowed one) received an attractive stimulation (parameter α changed to 0.7). The gray scale of each cell stands for the phase of its oscillator, and the line stretched from the center of each cell shows direction and strength of the force.

6.2 Experiment of the Robots Collide Each Other

Next was an experiment to show the architecture of our multi-cell robot is scalable. In the initial state, there were two multi-cell robots, each of which consists of nine cell units. It was shown that the multi-cell robot works well in spite of changing the number of cell units (Fig. 5b).

The white square on the right side stands for the source of the attractive stimulation. The gray scale of the cell unit expresses the phase, and circles surrounding each cell show the bond length of the cell.

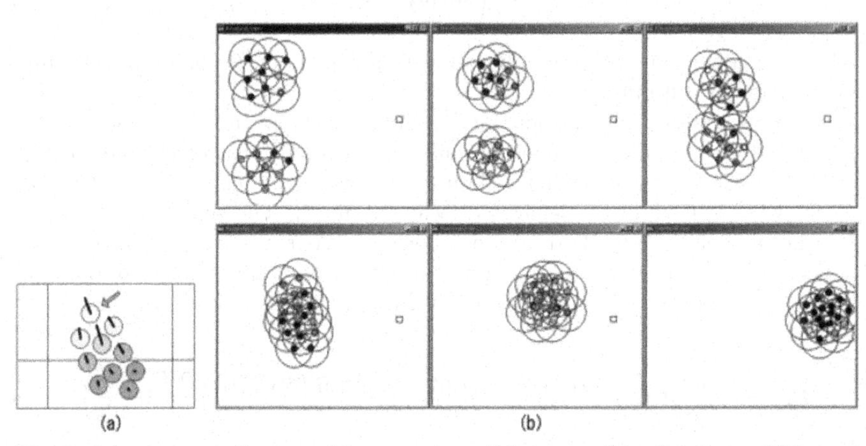

Fig. 5. (a): phase gradients and force vectors. (b): two multi-cell robots collide each other and they become one big robot

7 Movement Function Specialization by Reaction Diffusion System

Like the entrainment, the local interaction that tends to draw states of cells into states of circumstance is a diffusion reaction system. If states of cells merely approach to an average, the state of the whole system would be dissipative, and all cells would tend to be uniform. However, under certain conditions, the local interaction can lead to the structure that transits autonomously from the micro to macroscopic. This is called the self-organization of the spatial pattern or morphogenesis.

7.1 Function Specialization Model among Cell Units

According to A. M. Turing, the morphogenesis occurs in the condition that time variations of inhibitors are slower than activators, and that diffusion of inhibitors is earlier than activators [10]. In the light of this consideration, several researchers have investigated the typical patterns of shellfish and exodermis of some animals, employing computer simulation. We used the morphogenesis in order to model the function specialization among cell units. An amoeba cell moves by interaction with its neighboring cells, instead of moving by a single cell. This mechanism of partial role change among cells was applied to the multi-cell robot.

In this study, we employed a discrete model instead of a continuous diffusion differential equation mode. By employing the discrete model, calculation costs would be decreased. The state of each cell unit was described by w, and $w = 1$ means an activated state. From the state of neighborhood units located r_i, the state w of a center cell unit r was updated as follows.

$$sw = \sum_i w\left(|r - r_i|\right) \tag{6}$$

```
swich(sw) {
 case sw>0: cell changes to a specialization cell
 case sw=0: cell keeps the state
 case sw<0: cell changes to a un-specialization cell }
```

7.2 Experiment

Fig. 6 shows the experimental results. 100 cell units were arranged in random intervals. The initial states were $w = 0$ except for one cell unit. The cell unit of $w = 1$ was displayed black. The white part and the black part show the role of each cell unit. The parameter value was $r1 = 2.9$, $r2 = 6.1$, $w1 = 0.8$, and $w2 = -0.16$. It reached almost the steady state within 15 steps, when the renewal was repeated with the equation 6. Finally, in this experiment, some cell groups appeared in the whole cell robot.

The transition of space patterns appears in the simulation, however, it is our future work to investigate what kind of spatial patterns might be consistent with environment and how these patterns might take effect to the motion of cells. Furthermore, it is also future work to form useful efficient patterns, for example, horizontal stripes toward moving direction (the direction of a phase gradient).

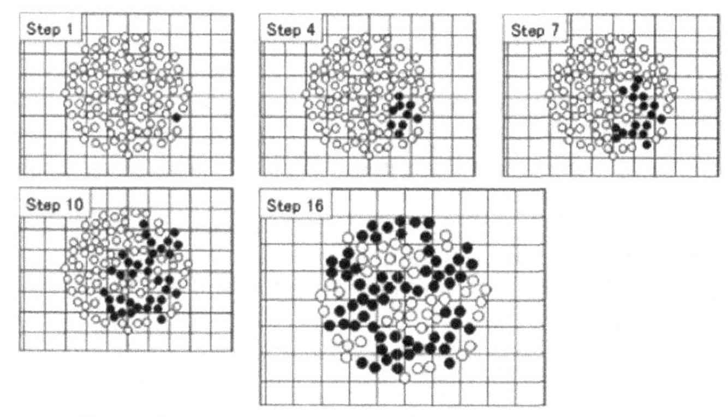

Fig. 6. Function specialization model among cell units

8 Conclusions

In this paper, we proposed a multi-cell robot control system referring to the amoeba chemotaxis, by which our robot interacts with its environment by converting stimulation from its environment into oscillations. The results of the computational simulations showed that our proposed method enables basic behaviors, that is, movement in response to an attractive stimulus, by locally interacted information processing. In the proposed control system, the sensor - action mapping was translated in the change of parameters of the coupled oscillator.

In the experiments, the parameters were manually selected for the objective behavior. However, in future works, the most suitable parameters for objective behaviors should be selected by Genetic Algorithm. We intend to apply the proposed method to a real SMA-Net robot and to increase the number of actuators and extend the SMA-Net robot to a three-dimensional composition. In the future, we also try to expand the model to include separation and coalescence of the slime mold.

References

1. Dittrich, P., Burgel, A., Banzhaf, W.: Learning to move a robot with random morphology. Evolutionary Robotics, Lecture Notes in Conputer Science, Vol.1468, Springer-Verlag, (1998) 165-178
2. Paap, K. L., Dehlwisch, M., Klaasen, B.: GMD-snake. A semi-autonomous snake-like robot, Distributed Autonomous Systems, Vol.2, Springer-Verlag, (1996) 71-77
3. Murata, S., Tomita, K., Yoshida, E., Kurokawa, H., Kokaji, S.: Self-Reconfigurable Robot. Module Design and Simulation, Intelligent Autonomous Systems 6, IOS Press, (2000) 911-917
4. Shu, S. G., Lagoudas, D. C., Hughes, D.: Modeling of a flexible beam actuated by shape memory alloy, Smart Material and Structure, (1997) 265-277
5. Miyake, Y., et al.: Environment-Dependent Self-Organization of Positional Information Field in Chemotaxis of Physarum Plasmodim, J. theor. Biol, Vol 178, (1996) 341-353
6. Yokoi, H., Yu, W., Hakura, J.: Morpho-Functional Machine – Amoeba Like Robot based on Vibrating Potential Method, Intelligent Autonomous Systems 5, IOS Press, (1998) 542-549
7. Dormann D., Vasiev B. and Weijer C.J.: Propagating waves control Dictyostelium discoideum morphogenesis, Bioph, Chem, 72, (1998) 21-35
8. Kaneko, K.: Spatiotemporal chaos in one- and two- dimentional coupled map lattice, Physica D, vol.37, (1989) 60-82
9. Kaneko, K.: Clustering, coding, switching, hierarchical ordering, and control in network of chaotic elements, Physica D, vol.41, (1990) 137-172
10. Turing, A. M.: The chemical basis of morphogenesis, Phil, Trans. Soc, R., London B237, (1952) 37-72

Towards Self-Organising Structure Formations: A Decentralized Approach

Jan Wessnitzer, Andrew Adamatzky, and Chris Melhuish

Intelligent Autonomous Systems Laboratory, University of the West of England,
Coldharbour Lane, Bristol BS16 1QY, United Kingdom
{jan.wessnitzer, andrew.adamatzky, chris.melhuish}@uwe.ac.uk
http://www.ias.uwe.ac.uk

Abstract. We investigate collective decision-making and parallel information processing in regular networks of simple mobile agents. We introduce a novel class of locally connected mobile automata networks where each node changes its internal state vector depending on information received from its neighbours. A configuration of the internal states and local sensor information is translated to a propulsive motion of the network's elements. We show how local rules of automaton internal states transitions can govern the evolution of the network toward task-specific topologies. In this paper, we present simple mechanisms which allow a collective of mobile agents to dynamically organise into simple spatial structures.

1 Introduction

How do coherent global behaviours emerge from the collective activities of relatively simple and locally interacting agents? Nature supplies us with many examples of animal societies which collectively carry out complex tasks, exhibit synchronized motion and make decisions. Social insects such as ants, termites and bees display at the level of the colony a wide range of behaviours. Examples include formations of hunting raids, trails or the building of sophisticated patterns [1]. Such examples provide the inspiration to model a self-organising robotic network, which is to be implemented in hardware at later stages of our research.

Numerous research projects undertaken in the last decade have demonstrated the success of a swarm-intelligence approach in the realization of collective robot systems. Thus, for example, in [2] an idea of object sorting using minimalist robots is investigated. In that research stigmergy is used to achieve self-organizing emergent behaviour in collectives of mobile robots without explicit inter-agent communication. This emergent behaviour is not determined by a global control system but arises from the interactions between robots, each executing the same local rules. No agent in the collective is provided with a blue-print of what the construction should look like. Instead, the macroscopic structures emerge as a result of interactions between simple agents acting by the same simple rules [1].

J. Kelemen and P. Sosík (Eds.): ECAL 2001, LNAI 2159, pp. 573–581, 2001.

What are the salient ingredients required in the design of collective and cooperative behaviour? We assume that some internal states, primitive sensors and local communication could form a necessary base for collective robot behaviour. This paper aims to explore a multi-agent system where each agent shares its internal state vector and sensor readings with neighbouring agents. Essentially, we pursue a quite typical problem of complex systems theory, focussing on the dynamics of interacting agents: *to design a system that while composed of unintelligent units, is capable, as a group to perform tasks requiring intelligence - the so-called 'swarm intelligence'* [3].

Usually, a complex system consists of a large number of entities interacting with each other. Economies, biological cells, road systems, stock markets, cellular automata and collectives of robots are all well-known examples of complex systems (see e.g. [4]). All these examples demonstrate, up to some degree, a phenomenon of self-organisation, when elementary units of a system organize themselves into non trivial structures.

In [5], Melhuish introduced the term of *autostruosis* in an attempt to clarify the vocabulary of construction, which includes terms such as self-organisation, self-assembly, emergence and autopoesis, illustrating that constructions can be made from robot bodies. In this paper, we try to develop a framework and algorithms with which a collective of mobile agents can self-organise into simple structures only using local information.

In the paper we build a model network of locally connected mobile robots. The model is defined in section 2. In section 3 we discuss several algorithms of mobile robot network's behaviour and demonstrate experimental results. A critical view on the problem is offered in section 4.

2 Model

A mobile automata network is a set of elementary processing units, where every unit interacts with other units and can move in a two- or three-dimensional Euclidean space. Each processor of the network communicates with a limited number of other processors on a local basis. The number of neighbours of any processor is several orders less than the total number of processors in the network. Each processor, or an agent, takes a few internal and external states and makes a decision based on the internal states and states of their immediate neighbours. Restricted memory of the processors and diffusion-like information transfer are essential features of mobile excitable networks (see, for example, [7], [8] and [9]).

State and motion updates of every agent are directly coupled to signals and information received from other robots. Updates purely made according to the states of the neighbourhood are insufficient in an attempt to model mobile automata networks. Robots need sensors to enable them to perceive their environment and process external environmental cues.

Let us define a mobile automata network as a tuple $\langle G, Q, W, P, u, f, g, h \rangle$, where G is a regular graph, $G = \langle X, E \rangle$, in which X is a set of nodes, E is a set of edges and $|X| = x$ is the number of agents, or mobile automata [6]. Each node of the network is connected with not more than k nodes, possibly including the node itself, determined by a connection template, or a neighborhood, $u : X \to X^k$. Explicit communication between nodes is only possible according to a specific probability distribution defined in terms of the physical distance between nodes. Let ξ be the communication radius over which agents can establish communication links with other agents. We assume no connection has a significant chance of occurring between nodes that are further than ξ apart. A node may connect uniformly at random to other nodes within this spatial distance ξ.

The nodes update their states simultaneously in discrete time, depending on the states of their neighbors, by node-state transition function $f : Q^k \to Q$. Elements of the set Q are called internal states. They are usual automata states, for example symbols or integers. Each state of the set Q is updated according to various deterministic transition rules. The update functions may vary from model to model and in some versions the rules may only be applied under certain conditions. The functions may be represented in several ways, for example, `if-then` notations, lookup tables or via arithmetical operators if elements of Q are arithmetical numbers.

To move successfully in a complex environment an automata network must detect obstacles, as well as mobile nodes must detect each other to behave in a physical sense. To make our mobile networks behave realistically we supply the automata with sensors and therefore add an additional set I of input states, which represent various sensor readings. Taking into account the sensors we define the node state-to-velocity transition function as follows: $g : Q \times I \to W$, where W is a set of the nodes' velocity vectors. One could also imagine that the sensor readings influence some of the automaton's internal states. Thus, a node-state transition could also be defined as $f : Q^k \times I \to Q$.

The elements of the set P are the positions of network nodes in an Euclidean space. The function h is used for node translations and for collisions as well. The function h and the set P have been included to facilitate the sensing of other units in space in simulation but would possibly become redundant and inapplicable in a real-robot implementation.

3 Simulation Experiments

3.1 Sensing

The problem of structure formation can be regarded as a search for a mapping between the agent's simple stimulus-response rule-sets of local actions based on the local sensor information. Two main processes accomplish formation maintenance. Firstly, the agent's perceptual data is processed and the agent's desired position with reference to its neighbours is determined. For the purpose of these experiments each agent i is capable of sensing the relative distance, r_i, and relative orientation, α_i, of its neighbours at distance not more than ξ.

Secondly, the polar coordinates (r, α) are then translated into the agent's own velocity vector (V, θ). The vector is always oriented in the direction of a desired position relative to neighbouring agents but the vector's length, i.e. the agent's speed, depends on the distance between the agent's current and desired positions. The further an agent is from its desired position, the higher the agent's speed.

A normally distributed error with mean zero and standard deviation σ, $\varepsilon_i \in N(0, \sigma)$, is introduced to the x_i and y_i coordinates of all i neighbours. Varying the standard deviation σ helps to analyze the system's robustness and adds some realism in terms of noisy sensor data to the simulation.

3.2 Experimental Setup

A problem is very simple: *to find rules of local interaction between agents that lead network's evolution toward the formation of lines or squares.*

In computational experiments all agents are randomly placed into an enclosed environment. No agent has a predefined position in any of the formations. An agent's i position and role in the formation process are determined by its few internal states at the time step t: m, n_i^t, c_i^t, k_i^t, where m is a number of agents that are required to build a desired structure, and n_i^t is a current number of recruited agents. Each agent is assigned a unique counter variable c_i^t, which defines the agent's role in the structure; k_i^t represents the number of neighbours an agent actively communicates with.

$r_i^t = (r_1^t, r_2^t, \ldots, r_k^t)$ are the relative distances and $\alpha_i^t = (\alpha_1^t, \alpha_2^t, \ldots, \alpha_k^t)$ are the relative orientations to the k neighbouring agents with which agent i is actively communicating.

3.3 Line Formation

Different approaches to self-organisation have been proposed (e.g. [13], [14],[15]) and the approach discussed in this paper is neighbour-referenced, only using local information.

At the beginning of each experiment all agents are put into an enclosed environment. No communication links exist between agents at the start. Every agent i, except agent z, has the following initial states: $c_i^{t=0} = 0$, $n_i^{t=0} = 0$ and $k_i^{t=0} = 0$. A single, arbitrarily picked, agent z has the initial states $c_z^{t=0} = 1$ and $n_z^{t=0} = 1$. This agent is called a recruiting agent. Let the agents' task be the formation of a line compromising $m = 12$ agents.

The recruiting agent now builds up a communication link with some non-recruited agent which is within its communication radius. Now, two agents are recruited. In order to control the evolution of the network, the agent with the highest counter value always acts as the recruiting agent. The process of recruiting agents is continued until the required number of agents is recruited.

During the line formation, each agent tries to manoeuvre itself between its two topological neighbors. Recruited agents which only have one neighbour try to maintain a certain distance from their neighbour.

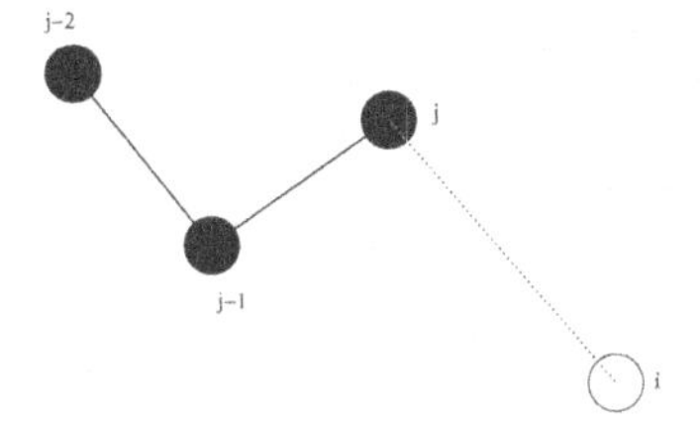

Fig. 1. An example of agents' connectivity.

Some internal state updates only occur under certain conditions. In this case, when two agents are establishing a communication link between each other. A typical situation is shown in figure 1. A non-recruited agent i connects to recruiting agent j if the following condition is true:

$$((k_j = 0) \wedge (c_j \neq 0)) \vee ((k_j = 1) \wedge (c_j > c_{j-1}) \wedge (n_j < m))$$

If the connection takes place the agent i executes the following operations:

$$\begin{aligned} c_i^{t+1} &= c_j^t + 1 \\ n_i^{t+1} &= c_j^t + 1 \\ k_i^{t+1} &= k_i^t + 1 \end{aligned}$$

The agent j simply updates its k_j variable: $k_j^{t+1} = k_j^t + 1$. Now, the agent i acts as the recruiting agent and tries to set up connections to another agent until a line of the required length m is built. The counter variable c is not changed after the initial connection handshake. The internal state n is updated unconditionally at each time step of the simulation as follows:

$$n_i^{t+1} = \max_{j \in u(i)} \{n_j^t\}$$

This rule diffuses the information of how many agents are currently recruited through the whole network. The agent i sets its internal state n_i to the maximum value of n of every agent j in the neighbourhood of i.

The Figure 2 shows a typical simulation run with $m = 12$ agents forming a line at the time steps $t = 100, 200, 300$ and 700.

3.4 Square Formation

In this section, a slight change is made to the connection function. As in the previous section, the agents connect until a line of size m is formed. The same rules as in the previous section apply. However, as soon as $n_j = m$, every agent j additionally tries to connect to the agent z which has $c_z = c_j - 2$. The desired position of agent j only depends on the two agents i and z which have $c_i = c_j - 1$ and $c_z = c_j - 2$. For example, the agent i with $c_i = 5$ will move towards a desired position relative to the agents j and z: $c_j = 4$ and $c_z = 3$.

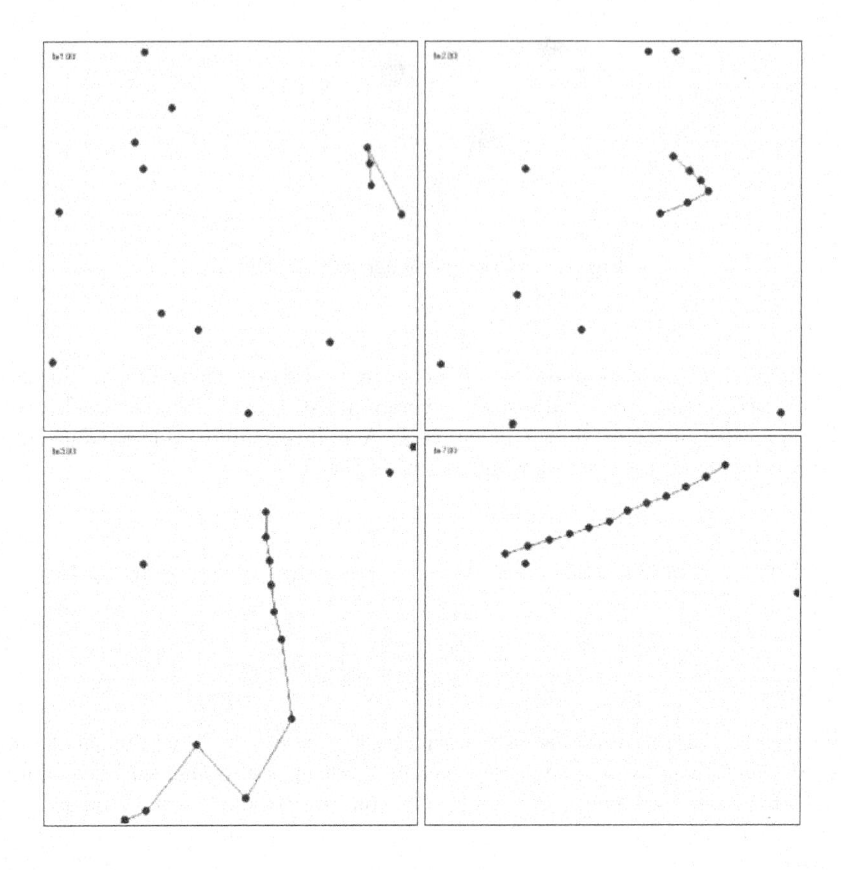

Fig. 2. An example of line formation in a network of 12 mobile agents. The communication links between agents are shown explicitly.

In a square formation, some agents need to form straight line segments whereas others have to form right angles. Therefore it is necessary to distribute these two tasks across the network and to assign them to the agents. For these purposes the internal state c is used. In this case, where $m = 12$, the agents with $c = 5, 8, 11$ will try to form a right angle with their neighbours whereas all others will try to form line segments.

Figure 3 shows 12 agents during the process of forming a square-like structure.

3.5 Analysis

The figures 2 and 3 show that a small number of agents can successfully form simple structures. Can this neighbour-referenced method of structure formation be applied to a large number of agents? How will this method cope in a noisy environment? Let's consider the following situation: the first two agents are

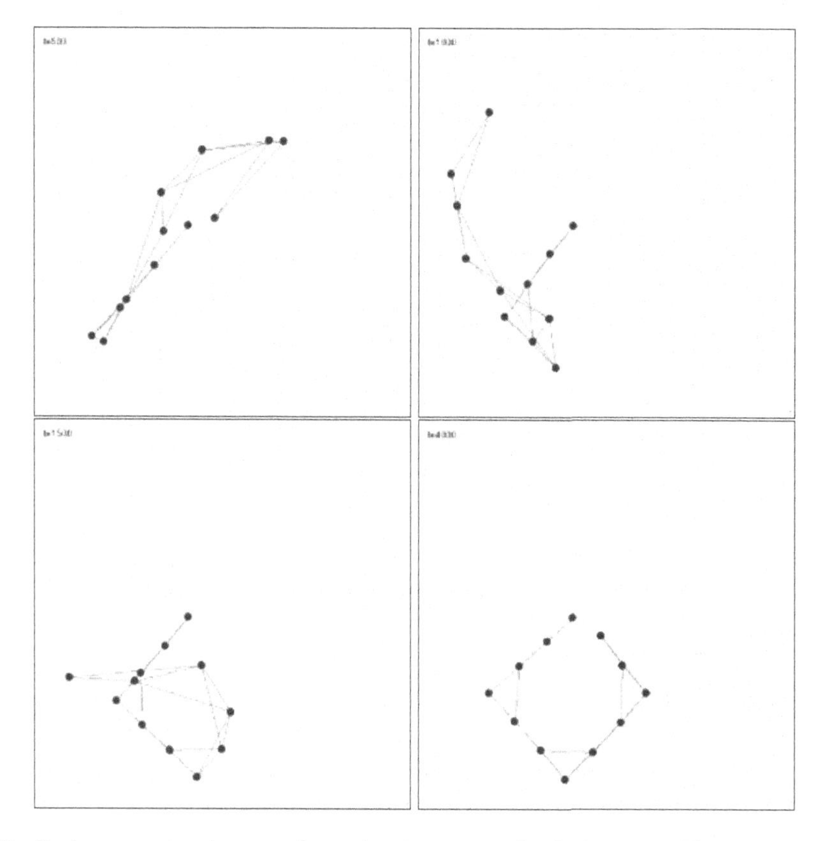

Fig. 3. An example of square formation in a network of 12 agents. The connections between the agents are shown.

immobilised and their positions recorded. Now, the network is allowed to evolve and the agents form a square around these two immobilised agents. No noise is added to the system and the experiment is repeated 100 times. The starting positions of the first two agents are the same for all trials in order to be able to estimate the positions of all other agents. At the end of each trial the final position of every agent is recorded. An expected position for each agent was then extracted from these 100 samples.

The same experiment was then repeated but some arbitrary, normally distributed error with mean zero and a standard deviation of 10 pixels, $\varepsilon \in N(0, 10)$, was added to the true x and y coordinates of every agent. Again, the first two agents are immobilised and the network is allowed to evolve around these agents.

Figure 4 shows the mean distance error of the agents from their expected positions after 100 trials.

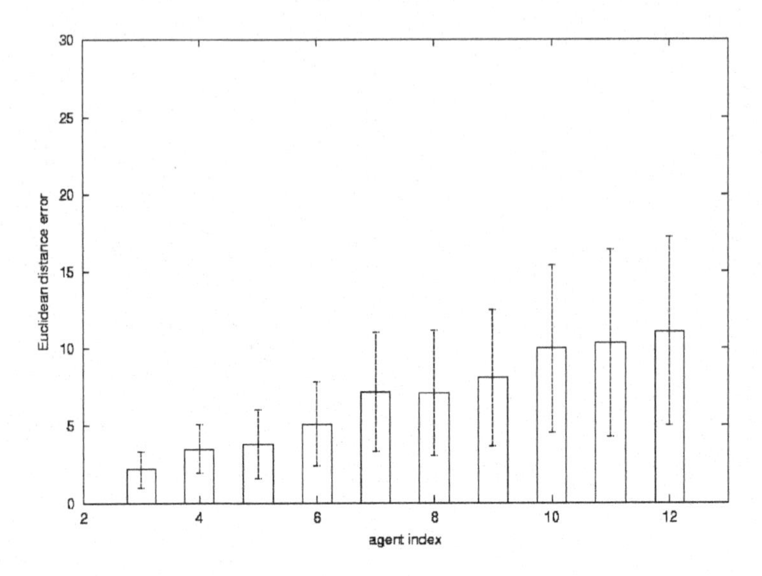

Fig. 4. Euclidean distance error with increasing agent index

4 Discussion

In simulation each agent was updated during every time step. The order in which the agents were updated was randomised at the start of every time step. No failure occurred when the probability of one agent being updated during one time step was set to $\frac{1}{x}$, where x is the total number of agents and therefore every agent has x chances to move during one time step. This experiment shows that synchronous state updates are not necessary with the developed rules in such a massively parallel system.

At present, one arbitrarily chosen agent starts the whole structure formation process but we are developing rules that allow homogeneously initialised agents to achieve the same goal.

In the paper we demonstrated that locally connected agents can form simple structures through diffusion-like information transfer and local parallel computation. We have shown how a task-specific network topology is evolved, using only a small set of internal states and preventing agents from any information about global configuration of the network. So, up to some limit, we offered a simple solution for the problem of coherent activity in a mobile automata network. Our result could benefit a wide range of disciplines and may be particularly important in a design of future distributed robot systems which employ electro-activated polymers [17].

The ultimate goal of further research is to validate present computational experiment results in real-life experiments with mobile robots. Currently, a sensor is being developed which is to be interfaced with our fleet of LinuxBots [16]

to estimate the polar coordinates of neighbouring units. Radio transmitters and receivers will be used to communicate a node's internal states to its neighboring units.

References

1. Bonabeau, E., Dorigo, M., Theraulaz, G.: Swarm Intelligence - From Natural to Artificial Systems. Oxford University Press (1999)
2. Melhuish, C., Holland, O., Hoddell, S.: Collective Sorting and Segregation in Robots with Animal Sensing. In: 5th Internationsl Conference on the Simulation of Adaptive Behaviour, From Animals to Animats, MIT Press (1998)
3. Beni, G., Wang, J.: Theoretical Problems for the Realization of Distributed Robotic Systems. In: Proceedings of the 1991 IEEE International Conference on Robotics and Automation (1991)
4. Chopard, B., Luthi, P. Masselot, A.: Cellular Automata and Lattice-Boltzmann Techniques : an Approach to Model and Simulate Complex Systems. (1998)
5. Melhuish, C.: Autostruosis: Construction without Explicit Planning in Micro-Robots - a Social Insect Approach. In: Evolutionary Robotics Vol III. AAI Books Ed. Takashi Gomi (2000)
6. Adamatzky, A.: Identification of Cellular Automata. Taylor and Francis (1994)
7. Adamatzky, A., Holland, O.: Phenomenology of excitation in 2D Cellular Automata and Swarm Systems. In: Chaos, Solitons and Fractals 9 (1998) 1233-1265
8. Adamatzky, A., Melhuish, C.: Unconventіal Mass-Parallel Controllers for Mobile Robots : Reaction-Diffusion and Excitable Prototypes. In: Proceedings of 1st International Conference on Mechatronics and Robotics (2000) 5-10
9. Adamatzky, A.: Reaction-Diffusion and Excitable Processors : a Sense of the Unconventional. In: Parallel and Distributed Computing Practices (2000)
10. Adamatzky, A., Melhuish, C.: Construction in Swarms : from Simple Rules to Complex Structures. In: International Journal of Systems and Cybernetics: Kybernetes. (2000) V29 No 9/10 1184-1194
11. Beni, G.: The Concept of Cellular Robotic System. In: IEEE International Symposium on Intelligent Control (1988)
12. Holland, O., Melhuish, C., Hoddell, S.: Chorusing and Controlled Clustering for Minimal Mobile Agents. In: 4th European Conference on Artificial Life (1997)
13. Unsal, C.: Self-Organisation in Large Populations of Mobile Robots. Masters' thesis, Virginia Polytechnic Institute and State University (1993)
14. Sugihara, K., Suzuki, I.: Distributed Motion Coordination of Multiple Robots. In: IEEE 5th International Symposium on Intelligent Control (1990)
15. Balch, T., Arkin, C.: Behavior-based Formation Control for Multi-robot Teams. In: IEEE Transactions on Robotics and Automation (1999)
16. Winfield, A., Holland, O.: The Application of Wireless Local Area Network Technology to the Control of Mobile Robots. In: Microprocessors and Microsystems 23 (2000) 597-607
17. Kennedy, B., Melhuish, C., Adamatzky, A.: Biologically Inspired Robots. In: Bar-Cohen, Y. (ed.) Electro-active Polymer (EAP) Actuators - Reality, Potential and Challenges. SPIE Press (2001)

Affective Interaction between Humans and Robots

Cynthia Breazeal

MIT Artificial Intelligence Lab, 200 Technology Square, room 933
Cambridge, MA 02139 USA
cynthia@ai.mit.edu
http://www.ai.mit.edu/projects/kismet

Abstract. This paper explores the role of emotive responses in communicative behavior between robots and humans. Done properly, affective communciation should be natural and intuitive for people to understand. This implies that the robot's emotive behavior should be life-like. The ability to establish and maintain a rich affective dynamic with people has placed important constraints on our robotic implementation. We present our framework, discuss how these constraints have been addressed, and demonstrate the robot's ability to engage naive human subjects in a compelling and expressive manner.

1 Introduction

Motivated by applications such as robotic pets for children or robotic nursemaids for the elderly, rich affective interchanges will become increasingly important as robots begin to enter long-term relationships with people. The majority of social robotics work took inspiration from ants, termites, fish, and other species that exist in anonymous socities. More recently there has been a shift to taking inspiration from species that live in individualized societies, such as primates, dolphins, and humans [1]. In a similar spirit, this work examines human-robot interaction. Whereas past work in robotics and animated life-like characters has explored the role of computational models of emotions in decision making and learning [2,3], this paper focuses on the role of emotions in interacting with people on an affective level. Heavily inspired by the study of emotions and expressive behavior in living systems, our approach is designed to support a rich and tightly coupled dynamic between robot and human, where each responds contigently to the other on an affective level. This property is often overlooked, but is critical for establishing a compelling social interaction with humans. It also places important constraints on the implementation of the `emotion` and `expression` systems. We have implemented and evaluated our work on a highly expressive anthropomorphic face robot called Kismet. Human subjects interact with Kismet in the spirit of a human caregiver, robot infant scenario.

J. Kelemen and P. Sosík (Eds.): ECAL 2001, LNAI 2159, pp. 582–591, 2001.

2 A Functional and Evolutionary View of Emotions

Emotions are an important motivator for complex organisms. They seem to be centrally involved in determining the behavioral reaction to environmental (often social) and internal events of major significance for the needs and goals of a creature [4]. Several theorists argue that a few select emotions are *basic* or *primary* — they are endowed by evolution because of their proven ability to facilitate adaptive responses to the vast array of demands and opportunities a creature faces in its daily life. The emotions of anger, disgust, fear, joy, sorrow, and surprise are often supported as being basic from evolutionary, developmental, and cross-cultural studies [5]. Each basic emotion is posited to serve a particular function (often biological or social), arising in particular contexts (eliciting conditions), to prepare and motivate a creature to respond in adaptive ways. The orchestration of each emotive response represents a generalized solution for coping with the demands of the original eliciting event. Plutchik (1991) calls this stabilizing feedback process *behavioral homeostasis*. Through this process, emotions establish a desired relation between the organism and the environment — pulling toward certain stimuli and events and pushing away from others. Much of the relational activity can be social in nature, motivating proximity seeking, social avoidance, chasing off offenders, etc.

The expressive characteristics of emotion in voice, face, gesture, and posture serve an important function in communicating emotional state to others. This benefits people in two ways: first, by communicating feelings to others, and second, by influencing others' behavior. For instance, the crying of an infant has a powerful mobilizing influence in calling forth nurturing behaviors of adults. Emotive signaling functions were selected for during the course of evolution because of their communicative efficacy. For members of a social species, the outcome of a particular act usually depends partly on the reactions of the significant others in the encounter. The projection of how the others will react to these different possible courses of action largely determines the creature's behavioral choice. The signaling of emotion communicates the creature's evaluative reaction to a stimulus event (or act) and thus narrows the possible range of behavioral intentions that are likely to be inferred by observers.

3 Design of the Emotion System

The organization and operation of Kismet's *emotion system* is strongly inspired by various theories of emotions in humans and animals. Kismet's `emotions`[1] are idealized models of basic emotions, where each serves a particular function (often social), each arises in a particular context, and each motivates Kismet to respond in an adaptive and expressive manner. Taken together, these emotive responses form a flexible system that mediates between both environmental and internal stimulation to elicit an adaptive behavioral response that serves either

[1] As a convention, I will use the boldface to distinguish parts of the architecture of this particular system from the general uses of those words.

social or self-maintenance functions. Summarizing these ideas, an "emotional" reaction for Kismet consists of:

- A precipitating event
- An affective appraisal of that event
- A characteristic expression (face, voice, posture)
- Action tendencies that motivate a behavioral response

Table 1. Summary of the antecedents and behavioral responses that comprise Kismet's emotive responses.

Antecedent conditions	Emotion	Behavior	Function
delay, difficulty in achieving goal of adaptive behavior	anger, frustration	complain	show displeasure to caregiver to modify his/her behavior
presence of an undesired stimulus	disgust	withdraw	signal rejection of presented stimulus to caregiver
presence of a threatening, overwhelming stimulus	fear, distress	escape	move away from a potentially dangerous stimuli
prolonged presence of a desired stimulus	calm	engage	continued interaction with a desired stimulus
success in achieving goal of active behavior, or praise	joy	display pleasure	reallocate resources to the next relevant behavior, (eventually to reinforce behavior)
prolonged absence of a desired stimulus, or prohibition	sorrow	display sorrow	evoke sympathy and attention from caregiver, (eventually to discourage behavior)
a sudden, close stimulus	surprise	startle response	alert
appearance of a desired stimulus	interest	orient	attend to new, salient object
need of an absent and desired stimulus	boredom	seek	explore environment for desired stimulus

Table 1 summarizes under what conditions certain emotive responses arise, and what function they serve the robot. This table is derived from the evolutionary, cross-species, and social functions hypothesized by Plutchik (1991). The table includes the six primary emotions proposed by Ekman (1982) along with three arousal states (boredom, interest, and calm). By adapting these ideas to Kismet, the robot's emotional responses mirror those of biological systems and therefore should seem plausible and readily understandable to people. Figure 1 presents the implementation of the `fear` emotive response to illustrate the relation between the eliciting condition(s), appraisal, action tendency, behavioral response, and observable expression.

For Kismet, some of these responses serve a purely communicative function. The expression on the robot's face is a social signal to the human caregiver, who responds in a way to further promote the robot's "well-being." For instance, the robot exhibits sadness upon the prolonged absence of a desired stimulus. This may occur if Kismet has not been engaged with a toy for a long time. The sorrowful expression is intended to elicit attentive acts from the human caregiver. Another class of affective responses relates to behavioral performance. For instance, a successfully accomplished goal is reflected by a smile on the robot's face, whereas delayed progress is reflected by a frustrated expression. Exploratory responses include visual search for desired stimulus and/or maintaining visual engagement of a desired stimulus. Kismet currently has several protective responses, the strongest of which is to close its eyes and turn away from threatening or overwhelming stimuli. Many of these emotive responses serve a regulatory function. They bias the robot's behavior to bring it into contact with desired stimuli (orientation or exploration), or to avoid poor quality or dangerous stimuli (protection or rejection). Taken as a whole, these affective responses encourage the human to treat Kismet as a socially aware creature and to establish meaningful communication with it.

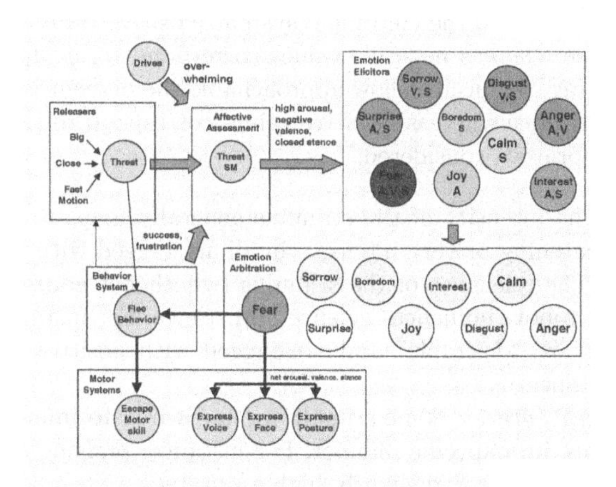

Fig. 1. The implementation of the `fear` emotion. The releaser for `threat` is passed to the affective assessment phase where it is tagged with high arousal, negative valence, and closed stance values. This affective information is then filtered by the corresponding elicitor of each `emotion` process. Darker shading corresponds to a higher activation level. The `fear` process becomes active, causing a fearful expression and evoking an`escape` response.

Emotive Releasers. The input to the emotion system originates from the high-level perceptual system, where each percept is fed into an associated *releaser* process. Each releaser can be thought of as a simple "cognitive" assessment that combines lower-level perceptual features into behaviorally significant perceptual categories. There are many different kinds of releasers defined for Kismet, each hand-crafted, and each combining different contributions from a variety of factors. These factors include the robot's homeostatic state, its current affective state, the active behavior, and the perceptual state (for details, please refer to

[6]). Hence, each releaser is evaluated with respect to the robot's "well-being" and its goals. If the conditions specified by that releaser hold, then its output is passed to the affective appraisal stage where it can influence the emotion system.

Affective Appraisal. Within the appraisal phase, each releaser is appraised in affective terms where the incoming perceptual, behavioral, or motivational information is "tagged" with affective information. There are three classes of tags used to affectively characterize a given releaser. Each tag has an associated intensity that scales its contribution to the overall affective state. The *arousal* tag, A, specifies how arousing this factor is to the emotional system. It very roughly corresponds to the activity of the autonomic nervous system. Positive values correspond to a high arousal stimulus whereas negative values correspond to a low arousal stimulus. The *valence* tag, V, specifies how favorable or unfavorable this percept is to the emotional system. Positive values correspond to a pleasant stimulus whereas negative values correspond to an unpleasant stimulus. The *stance* tag, S, specifies how approachable the percept is. Positive values correspond to advance whereas negative values correspond to retreat. There are four types of appraisals considered:

- *Intensity*: The intensity of the stimulus generally maps to arousal. For instance, threatening or very intense stimuli are tagged with high arousal.
- *Relevance*: The relevance of the stimulus (whether it addresses the current goals of the robot) influences valence and stance. For instance, stimuli that are relevant are "desirable" and are tagged with positive valence and approaching stance.
- *Intrinsic Pleasantness*: Some stimuli are hardwired to influence the robot's affective state in a specific manner. For instance, praising speech is tagged with positive valence and slightly high arousal [6].
- *Goal Directedness*: Each behavior specifies a goal, i.e., a particular relation the robot wants to maintain with the environment. Success in achieving a goal promotes `joy` and is tagged with positive valence. Prolonged delay in achieving a goal results in `frustration` and is tagged with negative valence and withdrawn stance.

Emotion Elicitors. This tagging process converts the myriad of factors into a common currency that can be combined to determine the net affective state. For Kismet, the $[A, V, S]$ trio is the currency the emotion system uses to determine which emotional response should be active. All somatically marked inputs are passed to the *emotion elicitor* stage. Each `emotion` process has as elicitor associated with it that filters each of the incoming $[A, V, S]$ contributions. Only those contributions that satisfy the $[A, V, S]$ criteria for that `emotion` process are allowed to contribute to its activation. Figure 2 summarizes how $[A, V, S]$ values map onto each `emotion` process. This filtering is done independently for each type of affective tag. For instance, a valence contribution with a large negative value will not only contribute to the `sad` process, but to the `fear`, `distress`, `anger`, and `disgust` processes as well. Given all these factors, each elicitor computes its average $[A, V, S]$ from all the individual arousal, valence, and stance values that pass through its filter.

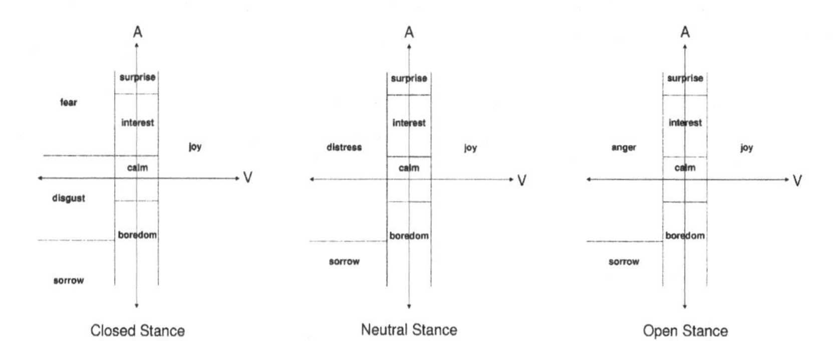

Fig. 2. Mapping of arousal, valence, and stance dimensions, [A, V, S], to `emotions`. This figure shows three 2-D slices through this 3-D space.

Given the net $[A, V, S]$ of an elicitor, the activation level is computed next. Intuitively, the activation level for an elicitor corresponds to how "deeply" the point specified by the net $[A, V, S]$ lies within the arousal, valence, and stance boundaries that define the corresponding `emotion` region shown in figure 2. This value is scaled with respect to the size of the region so as to not favor the activation of some processes over others in the arbitration phase. The contribution of each dimension to each elicitor is computed individually. If any one of the dimensions is not represented, then the activation level is set to zero. Otherwise, the A, V, and S contributions are summed together to arrive at the activation level of the elicitor. This activation level is passed on to the corresponding `emotion` process in the arbitration phase.

Emotion Activation and Arbitration. Numerically, the activation level $A_{emotion}$ of each `emotion` process can range between $[0, A^{max}_{emotion}]$ where $A^{max}_{emotion}$ is an integer value determined empirically. Although these processes are always active, their intensity must exceed a threshold level before they are expressed externally. The activation of each process is computed by the equation:

$$A_{emotion} = \sum(E_{emotion} + B_{emotion} + P_{emotion}) - \delta_t$$

where $E_{emotion}$ is the activation level of its affiliated elicitor process, $B_{emotion}$ is a DC bias that can be used to make some `emotion` processes easier to activate than others. $P_{emotion}$ adds a level of persistence to the active emotion. This introduces a form of inertia so that different `emotion` processes don't rapidly switch back and forth. Finally, δ_t is a decay term that restores an `emotion` to its bias value once the `emotion` becomes active. Hence, the `emotions` have an intense activation period followed by decay to a baseline intensity on the order of a few seconds.

Next, the `emotion` processes compete for control in a winner-take-all arbitration scheme based on their activation level. Each emotive response becomes active under a different environmental (or internal) situation, and each motivates a different observable response in behavior and expression. In a process of behavioral homeostasis as proposed by Plutchik (1991), the emotive response maintains activity through feedback until the correct relation of robot to environment is established.

4 Emotive Expression

Concurrently, the net $[A, V, S]$ of the active `emotion` process is sent to the expressive components of the motor system, causing a distinct facial expression and body posture to be exhibited. The strength of the facial expression reflects the level of activation of the `emotion`.

There are two threshold levels for each `emotion` process: one for expression and one for behavioral response. The expression threshold is lower than the behavior threshold. This allows the facial expression to lead the behavioral response. This enhances the readability and interpretation of the robot's behavior for the human observer. For instance, if the caregiver shakes a toy in a threatening manner near the robot's face, Kismet will first exhibit a fearful expression and then activate the `escape` response. By staging the response in this manner, the caregiver gets immediate expressive feedback that she is frightening the robot. If this was not the intent, then the caregiver has an intuitive understanding of why the robot is frightened and modifies behavior accordingly. The facial expression also sets up the human's expectation of what behavior will soon follow. As a result, the caregiver not only sees what the robot is doing, but has an understanding of why.

Psychologists such as Smith & Scott (1997) posit that facial expressions have a systematic, coherent, and meaningful structure that can be mapped to affective dimensions. It follows that some of the individual features of facial expression have inherent signal value. For instance, raised brows convey attention in both fear as and surprise. This promotes a signaling system that is robust, flexible, and resilient [7]. It allows for the mixing of these components to convey a wide range of affective messages, instead of being restricted to a fixed facial configuration for each emotion. This variation allows fine-tuning of the expression, as features can be emphasized, de-emphasized, added, or omitted as appropriate.

In keeping with this theory, Kismet's facial expressions are generated using an interpolation-based technique over a three-dimensional *affect space* — the same three $[A, V, S]$ attributes used to affectively assess the robot's siutation (see figure 3). The computed net affective state occupies a single point in this space, moving along a trajectory as the robot's affective state changes. The procedure

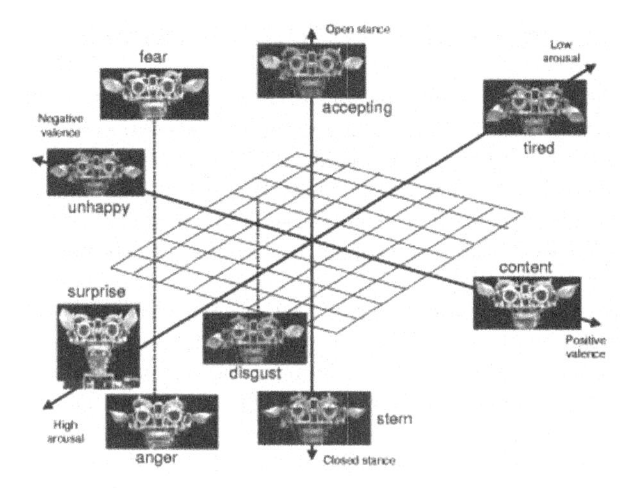

Fig. 3. This diagram illustrates where the basis postures are located in affect space.

runs in real-time, which is critical for social interaction. There are nine *basis* (or *prototype*) postures that collectively span this space of emotive expressions. The basis set of facial postures has been designed so that a specific location in affect space specifies the relative contributions of the prototype postures to produce a net facial expression that faithfully corresponds to the active emotion. With this scheme, Kismet displays expressions that intuitively map to the emotions of anger, disgust, fear, happiness, sorrow, and surprise, and many more. Different levels of arousal can be expressed as well from interest, to calm, to weariness. A similar scheme is used to control affective shifts in body posture.

There are several advantages to generating the robot's facial expression from this affect space. First, this technique allows the robot's facial expression to reflect the nuance of the underlying assessment. Even through there is a discrete number of emotion processes, the expressive behavior spans a continuous space. Second, it lends clarity to the facial expression since the robot can only be in a single affective state at a time (by our choice) and hence can only express a single state at a time. Third, the robot's internal dynamics are designed to promote smooth trajectories through affect space. This gives the observer a lot of information about how the robot's affective state is changing, which makes the robot's facial behavior more interesting. Furthermore, by having the face mirror this trajectory, the observer has immediate feedback as to how their behavior is influencing the robot's internal state. For instance, if the robot has a distressed expression upon its face, it may prompt the observer to speak in a soothing manner to Kismet. The soothing speech is assimilated into the emotion system where it causes a smooth decrease in the arousal dimension and a push toward slightly positive valence. Thus, as the person speaks in a comforting manner, it is possible to witness a smooth transition to a subdued expression.

5 Dynamic Affective Exchanges with Humans

To explore the affective coupling between Kismet and human subjects, we carried out the following experiment. Five female subjects, ranging from 23 to 54 years old, were asked to either praise, scold, alert, or soothe Kismet through tone of voice, and to signal when they felt that Kismet understood them. None had interacted with Kismet previously. All sessions were recorded on video for further evaluations. For each trial, we recorded the number of utterances spoken to the robot, Kismet's expressive feedback cues, subject's responses and comments, as well as changes in tone of voice, if any. Kismet's ability to recognize these affective intents has been reported in [6]. To enduce a change in "emotional" state and to express this state to a human, the output of the affective intent recognzier is fed through the `emotion` and `expression` systems as presented in this paper.

Recorded events show that subjects in the study made ready use of Kismet's expressive feedback to assess when the robot "understood" them. The subjects varied in their sensitivity to the robot's expressive feedback, but all used facial expression and/or body posture to determine when the utterance had been properly communicated to the robot. All subjects would reiterate their vocalizations with variations about a theme until they observed the appropriate change in facial expression. If the wrong facial expression appeared, they often used strongly exaggerated tone of voice to correct the "misunderstanding." The subjects readily discerned intensity differences in Kismet's expression (reflecting different intensities in the underlying `emotional` state) and modulated their tone of voice to influence them. For instance, small smiles versus large grins were often used to discern how "happy" the robot was. Small ear perks versus widened eyes with elevated ears and craning the neck forward were often used to discern growing levels of "interest" and "attention."

During course of the interaction, several interesting dynamic social phenomena arose. For instance, several of the subjects reported experiencing a very strong emotional response immediately after "successfully" scolding Kismet. In these cases, the robot's saddened face and body posture was enough to arouse a strong sense of empathy. The subject would often immediately stop and look to the experimenter with an anguished expression on her face, claiming to feel "terrible" or "guilty." In this emotional feedback cycle, the robot's own affective response to the subject's vocalizations evoked a strong and similar emotional response in the subject as well. Another interesting social dynamic observed involved *affective mirroring* between robot and human. In this situation, the subject might first issue a medium-strength prohibition to the robot, which causes it to dip its head. The subject responds by lowering her own head and reiterating the prohibition, this time a bit more foreboding. This causes the robot to dip its head even further and look more dejected. The cycle continues to increase in intensity until it bottoms out with both subject and robot having dramatic body postures and facial expressions that mirror the other. This technique was employed to modulate the degree to which the strength of the message was "communicated" to the robot.

6 Summary

We have presented a biologically inspired framework for emotive communication and interaction between expressive anthropomorphic robots and humans. This paper primarily pursues an engineering goal to build a robot that can interact with people in familiar social terms, focusing on affective interactions. However a scientific exploration of the emotion models implemented on Kismet is an interesting possibility for future work. By modeling Kismet's emotional responses after those of living systems, people have a natural and intuitive understanding of Kismet's emotional behavior and how to influence it. From our studies, we have found this to be mutually beneficial for both human and robot. It is beneficial for the robot because it can now socially tune the human's behavior to be appropriate for itself – getting the person to bring the desired stimulus into contact at the appropriate time and at an appropriate intensity. It benefits the human because the person do not require any special training to have a comprehensible and rewarding interaction with the robot – knowing when the robot has understood one's affective state and knowing how one's behavior is influencing the robot's affective state. In general, we have found that expressive feedback plays an important role in facilitating natural and intuitive human-robot communication.

References

1. K. Dautenhahn. Getting to know each other – artificial social intelligence for autonomous robots. *Robotics and Autonomous Systems*, 16(2–4):333–356, December 1995.
2. Juan Velasquez. Modeling emotions and other motivations in synthetic agents. In *Proceedings of the 1997 National Conference on Artificial Intelligence (AAAI97)*, pages 10–15, Rrovidence, RI, 1997.
3. S.Y. Yoon, B. Blumberg, and G. Schneider. Motivation driven learning for interactive synthetic characters. In *Proceedings of the Fourth International Conference on Autonomous Agents (Agents00)*, Barcelona, Spain, 2000.
4. R. Plutchik. *The Emotions.* University Press of America, Lanham, MD, 1991.
5. P. Ekman and H. Oster. Review of research, 1970 to 1980. In P. Ekman, editor, *Emotion in the Human Face*, pages 147–174. Cambridge University Press, Cambridge, UK, 1982.
6. C. Breazeal. *Sociable Machines: Expressive Social Exchange Between Humans and Robots.* PhD thesis, Massachusetts Institute of Technology, Department of Electrical Engineering and Computer Science, Cambridge, MA, May 2000.
7. C. Smith and H. Scott. A componential approach to the meaning of facial expressions. In J. Russell and J. Fernandez-Dols, editors, *The Psychology of Facial Expression*, pages 229–254. Cambridge University Press, Cambridge, UK, 1997.

The Survival of the Smallest: Stability Conditions for the Cultural Evolution of Compositional Language

Henry Brighton and Simon Kirby

Language Evolution and Computation Research Unit,
Department of Theoretical and Applied Linguistics,
The University of Edinburgh, Edinburgh, UK
{henryb,simon}@ling.ed.ac.uk

Abstract. Recent work in the field of computational evolutionary linguistics suggests that the dynamics arising from the cultural evolution of language can explain the emergence of syntactic structure. We build on this work by introducing a model of language acquisition based on the Minimum Description Length Principle. Our experiments show that compositional syntax is most likely to occur under two conditions specific to hominids: (i) A complex meaning space structure, and (ii) the poverty of the stimulus.

1 Introduction: Language as a Complex Adaptive System

To what degree can properties of human language be explained by examining the dynamics resulting from the cultural evolution of language? Genetic transmission offers a mechanism for transmitting information down generations [6], and recourse to natural selection for explaining the evolution of language has received much attention [8,7]. The assumption is that the core properties of language are specified by an innate language acquisition device. Recent advances in computational evolutionary linguistics suggest that cultural evolution, too, offers a candidate explanatory mechanism. Here, linguistic information is transmitted down generations through communication. For example, Kirby demonstrates that two linguistic properties unique to human language, compositional and recursive syntax, can be explained in terms of cultural evolution coupled with a general purpose learning mechanism [3,4]. In this article we too treat human language as a complex adaptive system. Central to our analysis is the *iterated learning model* – a framework in which each generation of language user acquires its linguistic competence by observing the behavior of the previous generation. The behavior resulting from the iterated learning model resembles the phenomenon observed in the parlor game *Chinese whispers*, also known as *Broken Telephone*, because the language of each generation can change due to mistakes or misinterpretations in the observation of the language of the previous generation. The chief issue we address is that of *stability*. Language must be, to some degree, stable in order for subsequent generations to communicate.

J. Kelemen and P. Sosík (Eds.): ECAL 2001, LNAI 2159, pp. 592–601, 2001.

Consider all possible languages: some will be more stable than others, and it is precisely the property of stability which offers an advantage to those languages. A stable language will result in a steady state, and will therefore maximize its probability of survival. We aim for a clearer understanding of the conditions for the stability of syntactic language. Before detailing our analysis, we set the scene by discussing the iterated learning model and the role of stability.

Iterated Learning. One generation of the iterated learning model involves a single agent observing a set of meaning/signal pairs produced by the previous generation. First, the agent forms a hypothesis for this observed language. The agent is then called upon to express a random subset of all possible meanings to the next generation. This process is repeated over many generations, with each generation containing a single agent, and each agent operating in two modes: first observation, and then production. The key to the model is the *communication bottleneck* – of all the possible meanings, only a small subset are observed. We can liken this restriction to the language acquisition problem known as the *poverty of the stimulus* – human language learners only ever observe a small subset of all the possible utterances, yet can produce an ostensibly infinite number of utterances. Now, when an agent in the model is called on to express a random subset of the meaning space, some of the meanings may have already have been observed in conjunction with a signal. In this situation expressing the meaning is simple – the agent uses the same signal which accompanied the observed meaning. When the meaning is one which has not been observed, the agent must somehow find an appropriate signal. Here, the hypothesis selected by the agent can help as it may generalize beyond the observed language, and the agent can express the novel meaning by generalizing from the observed language. If the hypothesis does not account for any unseen data, i.e., does not generalize, then some invention scheme must be invoked. Invention must to some degree be unprincipled, and as a result, is likely to deviate from any regularity existing in the observed language. Languages which can be generalized from limited exposure will be termed *learnable*. They must exhibit regularity. Random languages, which by definition do not contain any regularity, are not learnable.

Stability. A stable language is one which is learnable and expressive. Given an appropriate inductive bias, limited exposure to a learnable language can result in generalization to all possible utterances. When all possible meanings can be expressed maximum expressivity results. In this situation the invention scheme is not invoked – unprincipled production does not occur – and the language will persist over many generations. At each generation an agent is called on to express a random subset of meaning space, and as a result, it is possible for a sparse exposure to the language to be observed by the next generation. In extreme situations, the random sample will contain few distinct observations, and a learnable language will not be learnt. Instability will result. However, such a situation is highly improbable and in a sense irrelevant because we view stable languages as *attractors*. In all probability, deviations from stable languages still place the system in the basin of attraction, perturbations can occur but are rare and not destructive.

To summarize, we view language as a complex adaptive system. The process of iterated learning will tend to lead to languages moving to areas of greater stability. Those languages that are learnable and expressive are stable. In this paper we are interested in the conditions under which syntactic (i.e., compositional) languages are attractors. We argue later that the requisite conditions are specific to hominids. Before investigating these conditions we define what we mean by compositionality, and describe the model of language and language acquisition we employ.

2 Language Acquisition Based on the MDL Principle

Compositional syntax is the property of language where the meaning of a signal is some function of the meaning of its parts, and they the way they are put together. We can contrast compositional utterances with holistic utterances, where the meaning of a signal is a function of the signal as a whole – the signal cannot be decomposed to identify fragments of the meaning, only the *whole signal* stands for any kind of meaning. Previous studies which investigate the cultural evolution of compositional syntax (for example, [3] and [1]) have been criticized because the manner in which agents in the simulations select the hypothesis for the observed data is strongly biased – the results are striking yet inevitable [10]. In this section we appeal to a well understood model of induction – the Minimum Description Length Principle – and outline a novel model hypothesis space which can account for compositional and non-compositional languages.

The Minimum Description Length Principle. Ranking potential hypotheses by minimum description length is a highly principled and very elegant approach to hypothesis selection [5]. The MDL principle can be derived from Bayes's Rule, and in short states that the best hypothesis for some observed data is the one that minimizes the sum of (a) the encoding length of the hypothesis, and (b) the encoding length of the data when represented in terms of the hypothesis. A tradeoff then exists between small hypotheses with a large data encoding length and large hypotheses with a small data encoding length. When the observed data contains no regularity, the best hypothesis is one that represents the data verbatim, as this minimizes the data encoding length. However, when regularity does exist in the data, a smaller hypothesis is possible which describes the regularity, making it explicit, and as result the hypothesis describes more than just the observed data. For this reason, the cost of encoding the data increases. MDL tells us the ideal tradeoff between the length of the hypothesis encoding and the length of the data encoding described relative to the hypothesis. We use the MDL principle to find the most likely hypothesis for an observed set of meaning/signal pairs passed to an agent. When regularity exists in the observed language, the hypothesis will capture this regularity, when justified, and allow for generalization beyond what was observed. By employing the MDL principle, we have a theoretically solid justification for generalization.

Finite State Unification Transducers. We extend the scheme of Teal et al to deal with meanings and signals of arbitrary length [9]. Our hypothesis

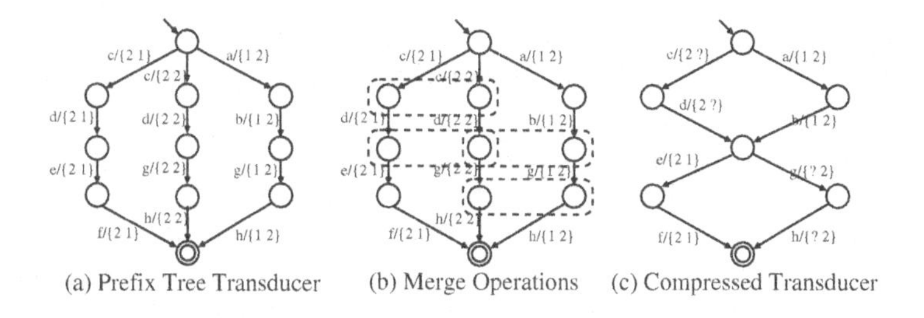

Fig. 1. (a) The prefix tree transducer for L_1. (b) The state merge operations required to induce the compressed transducer shown in (c).

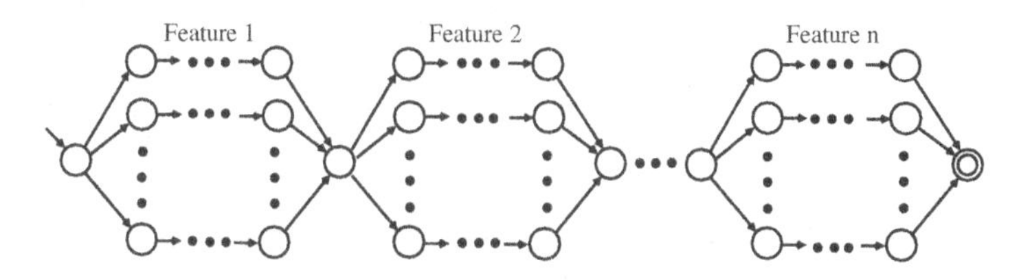

Fig. 2. Transducers of this form are induced by MDL for compositional languages.

space is the set of all Finite State Unification Transducers (FSUTs). A detailed exposition of the FSUT model is beyond the scope of this article. For a more thorough discussion see [2]. In short, a FSUT is a transducer which maps signals (arbitrary length strings of symbols), to meanings (feature vectors). The edges in the transducer are composed of (a) the symbol accepted when traversing this edge, and (b) a meaning, which can contain some wildcard values for feature values. An example set of meaning/pairs is the language L_1:

$$L_1 = \{(\{2,1\}, \text{cdef}), (\{2,2\}, \text{cdgh}), (\{1,2\}, \text{abgh})\}$$

In L_1 meanings are points in a two featured space. Each feature can take one of two values. For clarity, the signals in L_1 are of fixed length and drawn from the alphabet $\{a, b, c, d, e, f, g, h\}$. Now, when an agent in our model observes a language such as L_1, a *prefix tree transducer* is constructed. The prefix transducer is the hypothesis that explains the observed data and only the observed data. Figure 1(a) depicts such a prefix tree transducer. A number of operators can be applied to an FSUT, for example: (a) The *state merge* operator and (b) The *edge merge* operator. When we apply these operators the MDL of the transducer is usually reduced – the transducers are compressed. Compression is performed by repeatedly applying these operators until the MDL of the transducer cannot be reduced further. When the transducer cannot be compressed further, the *compressed transducer* results. Figure 1(b) shows which state merge operations are applied to derive the compressed transducer show in Figure 1(c). The compressed transducer for L_1, due to the simplicity of the example, does not generalize from the observed data. However, with an appropriate language, compression does lead to generalization. Throughout this paper we consider two types of language:

1. *Compositional Languages.* A compositional language is constructed by forming, for each meaning, a signal which is composed of substrings identifying the feature values contained in the meaning. A dictionary detailing which signal fragment to use for each feature value is used. That is, for each feature value we create a unique substring. The signal is then built by concatenating the appropriate substrings for each feature in the meaning.
2. *Non-compositional, or Random Languages.* A random language is holistic. Each signal refers to the whole meaning – no relationship between the parts of the signal to parts of the meaning exists. Each meaning is assigned a random string.

Using a compression algorithm based on the MDL principle, we found a common transducer structure for transducers accepting compositional languages. Our experiments show that these compressed transducers are learnable from compositional input. Figure 2 depicts the general layout for a compressed transducer found by applying the MDL compression algorithm. Each feature is dealt with by separate fragment of the transducer. After a constituent part of the signal has been parsed, the appropriate meaning fragment is logged. After all the features have been parsed, the whole meaning is built up by the transducer by finding the union of the logged meaning fragments. We conducted the same experiments for random languages and found, unsurprisingly, that prefix tree transducers where the most appropriate hypothesis. The simplifying assumption we make is that for non-compositional languages a prefix tree transducer is always the most appropriate hypothesis, and for compositional languages the compressed transducers of the form shown in Figure 2 are always the most appropriate hypothesis. However, feature values are only present in the compressed transducer if they have been observed. Figure 2 illustrates the general layout of compressed transducer, rather than a specific transducer.

We have introduced the MDL principle and stressed that it is a principled model of induction. We then outlined a hypothesis space on which we can apply the MDL principle. The hypothesis space consists of FSUTs. Given an observed language, we can compress the prefix tree FSUT and get a compressed FSUT which can generalize from the observed data, provided the observed language contains regularity. The details of the FSUT compression method have not been discussed. However, the chief point is that compressed transducers are induced from compositional input, and prefix tree transducers are the best hypothesis for non-compositional input.

3 The Relative Stability of Compositionality over Non-compositionality

The issue of stability is only relevant when language users suffer from poverty of the stimulus. If all possible meaning/signal pairs are observed by an agent, then the agent can express all possible meanings. This is not how human language works – we can produce utterances which we have never been exposed to. Stable languages must be learnable by language users. This means that an appropriate hypothesis is recoverable from limited language exposure. The hypothesis induced by the language user must also have high expressivity. High expressivity is the result of generalization – the language user can express meanings for which no appropriate signal has been observed. How stable are compositional languages in comparison to non-compositional languages? The degree to which compositional language is more stable than non-compositional language indicates how likely compositionality is to occur in the iterated learning model. In this section we quantify the stability advantage compositional language provides, and under which circumstances. Our hypothesis is that, within the context of cultural evolution, compositional language has high relative stability under conditions specific to hominids.

Non-compositional languages do not exhibit any regularity in the mapping between meanings and signals. They are not learnable. For this reason, a language user exposed to a non-compositional language can only competently express meanings it has already observed. In this situation, novel meanings can only be expressed through invention. Compositional languages differ markedly in that expressivity is not proportional to number of utterances observed, but instead proportional to the number of *feature values* observed. If all the feature values have been observed hypothesis selection based on MDL tells us that induction to novel meanings containing these previously observed feature values is often justifiable. For example, if meanings are drawn from a 3-dimensional space, with each dimension having 3 values, then there are $3^3 = 27$ possible meanings. But all feature values could be observed with as little 3 meaning observations. It is unlikely that MDL will lead to a hypothesis with maximum expressivity after just 3 exposures, but this example shows how feature values are observed at a much greater rate than whole meanings.

3.1 Monte Carlo Simulations

We use Monte Carlo simulations to establish some foundational results concerning the relative stability of compositional languages over non-compositional languages. In these simulations compositional and non-compositional languages are presented to an agent many times, and under different conditions. We then analyze how these conditions affect the resultant stability of each language type. The two language types are presented in the following manner:

1. A compositional language L, the construction of which is outlined below, is presented to an agent via a communication bottleneck. This means that some number of the meaning/signal pairs in L are picked at random with replacement and given to the agent. This set B of observed meaning/signal pairs is unlikely to contain all the pairs in L. We then use an MDL hill-climbing search to identify the most likely hypothesis. For a compositional language this will result in a FSUT similar to that shown in Figure 2. We

then measure the proportion of the language L the hypothesis can account for. This is the expressivity of the agent, which in turn is a measure of stability. We term this proportion E_{comp}.
2. As in 1, we present a set of observations drawn from a language L at random. However, this time L is not compositional. For non-compositional languages, the hypothesis selected by MDL is the prefix transducer for the observed data. Again, we measure the expressivity of the transducer – the proportion of L the agent can express without recourse to invention. This, again, is measure of stability and we define it as $E_{noncomp}$.

The values E_{comp} and $E_{noncomp}$ measure the degree of stability of the two language types. We can also think of these values as representing the inverse of the mean communicative error between subsequent generations in the iterated learning model. What do the values E_{comp} and $E_{noncomp}$ depend on? The three principle parameters for the Monte Carlo simulations are:

1. *The construction of L.* In Section 2 we described how a compositional language is constructed. There, as in previous work, we refer to a language as a meaning space with a signal attached to every meaning. Here, we change the manner in which a language is constructed to account for a more plausible scenario in which the agents perceive a set of n objects[1]. To the agent, these objects appear as meanings – the meaning corresponds to the perceived properties of the object. We imagine these particular meanings as being provided by the environment and being relatively stable over generations. The correct meaning/signal pair associated with each object is chosen at random from some meaning space for which every meaning has a signal. So, L is set of meaning/signal pairs which correspond to a set of objects.
2. *Meaning Space Structure.* For each object, a random meaning and an appropriate signal is chosen from some meaning space. The dimensions of this meaning space, i.e., the number of features and the number of values per feature is termed the meaning space structure.
3. *Bottleneck Size.* The bottleneck size, b, defines the number of observations of the language the agent is exposed to. The observed set of meaning/signal pairs B is constructed by picking a random meaning/signal pair from the language L, b times with replacement. Note that in order to guarantee seeing all the members of L, b must be infinitely large.

3.2 Requisite Conditions for Stability

Given a compositional language L_{comp} we use the MDL principle to find the most likely hypothesis. This is the transducer T_{comp} which has expressivity E_{comp}, defined above. Similarly, for a non-compositional language $L_{noncomp}$ we find, using MDL, the most likely hypothesis. This is the transducer $T_{noncomp}$ which has expressivity $E_{noncomp}$. Ultimately, we are interested in how much of a stability advantage compositionality confers. We term this quantity the *relative stability*

[1] We could equivalently refer to them as "communicatively relevant situations."

of compositionality and denote it as R: $R = \frac{E_{comp}}{E_{comp}+E_{noncomp}}$ This value depends on the structure of the meaning space and the size of the bottleneck. Instead of thinking of the size of the bottleneck in terms of the number of observations it is more useful to think of it in terms of the *expected object coverage.* That is, how many object observations, when observed at random with replacement, do we have to see before a certain proportion of these objects is observed. For example, a bottleneck size representing a coverage of 0.1 is the average number of random observations required before we expect to see 10% of the objects.

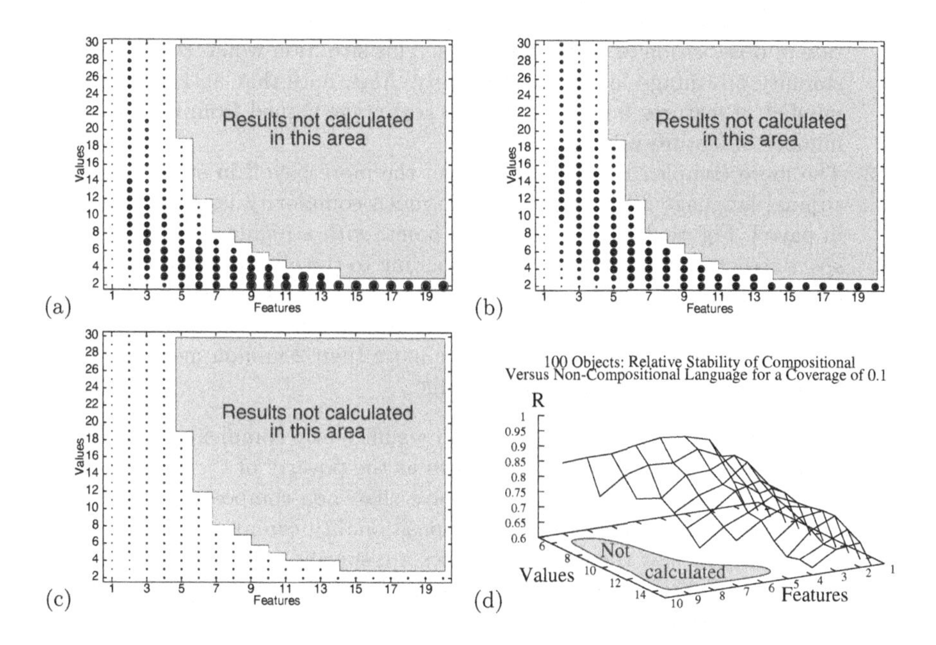

Fig. 3. These figures show the relative stability, R, of compositionality over non-compositionality for different meaning space structures. All values represent averages over 20 independent runs. In (a)-(c) (coverage values 0.1, 0.2, 0.8, respectively) the size of the point for a position in the space reflects the value of R. A small single point corresponds to no preference for compositionality (i.e., $R = 0.5$). The largest points (found in Figure (a)) correspond to $R > 0.9$ Figure (d), which shows a portion of the data depicted in (a), highlights the fact the meaning space structure that maximizes R is a trade-off between high and low complexity.

Figure 3(a)-(c) shows which meaning space structures, for bottlenecks resulting in an expected object coverage 0.1, 0.2, and 0.8, lead to the maximum relative stability of compositional language. The environments contains 100 objects. These results are striking for three reasons:

1. The highest R values occur for small bottleneck sizes. Figure 3(a) and (b) illustrate this point. Figure 3(c) shows little payoff in stability for com-

positional languages. Compositional languages are learnable, and result in high expressivity, even with sparse exposure to the whole language. With a large object coverage they still exhibit these qualities, but so does non-compositional language, so the payoff in compositionality is low.
2. High R values are only possible with a complex meaning space structure. If the perceptual space of the agent is not broken into multiple features or multiple values, then compositionality is not possible. However, even when the conceptual space is cut up into a few features or values, compositionality is still unlikely. The principle reason for this is that a simple meaning space structure results in the rate of observation of feature values to be near the rate of observation of whole meanings. This situation would result in less of stability advantage for compositionality. Also, note that an increase in the number of features far outweighs the advantage gained from increasing the number of feature values.
3. The more complex the meaning space, the more payoff in stability compositional language offers. However, too much complexity leads to a decrease in payoff. Figure 3(d) illustrates this point: with a highly complex meaning space structure the meanings corresponding to the objects are scattered over a vast space, and as a result, regularity in the correspondence between signals and meanings is weakened. For this reason, and reasons of tractability, if the meaning space can discriminate more than 2 million meanings we do not calculate the corresponding R value.

In generations made up of agents with a sufficiently complex conceptual apparatus, coupled with the condition known as the poverty of the stimulus, compositional language is more likely to evolve than non-compositional language. Indeed, under these conditions, non-compositionality cannot result in a stable system. We argue that these circumstances are specific to hominids – compositionality buys us little when (a) during our lifetime we are exposed to a large proportion of the language, or, (b) when our cognitive apparatus restricts us to holistic, or simple, experiences. These results are independent of the number of perceivable objects. For example, when communicating about 1000 or 10,000 objects the same arguments apply.

4 Conclusion

In the model of language evolution presented above two key parameters are present. These parameters, the poverty of the stimulus (the communication bottleneck) and the complexity of the cognitive system present in the individual agents (the meaning space complexity), were varied in an attempt to shed light on the circumstances under which compositionality is most likely to occur. The parameters settings which maximize the likelihood of compositionality, we argue, correspond to conditions specific to hominids:

1. A complex conceptual system. Hominid thought is unlikely to be restricted to holistic experiences.

2. A limited exposure to the range of signals for all pertinent meanings, yet the ability produce signals for a vast number of meanings.

Nowak et al offer a similar argument in their explanation of the evolution of syntactic communication through natural selection [7]. Our analysis strengthens the already compelling argument that syntax can also arise due to the adaptive pressures imposed by communication over many generations. Cultural evolution is a candidate mechanism for explaining the emergence of syntactic language. Central to this analysis is the Minimum Description Length Principle. In our model, only the smallest hypotheses will survive over many generations. The smallest hypotheses arise due to compression, and as a result generalize beyond what was observed.

References

1. J. Batali. The negotiation and acquisition of recursive communication systems as a result of competition among exemplars. In E. Briscoe, editor, *Linguistic Evolution through Language Acquisition: Formal and Computational Models.* Cambridge University Press, 2001.
2. H. Brighton and S. Kirby. Meaning space structure determines the stability of culturally evolved compositional language. Technical report, Language Evolution and Computation Research Unit, The University of Edinburgh, 2001.
3. S. Kirby. Syntax without natural selection: how compositionality emerges from vocabulary in a population of learners. In Chris Knight, Michael Studdert-Kennedy, and James R. Hurford, editors, *The Evolutionary Emergence of Language: Social Function and the Origins of Linguistic Form*, pages 303–323. Cambridge University Press, Cambridge, 2000.
4. S. Kirby. Spontaneous evolution of linguistic structure: An iterated learning model of the emergence of regularity and irregularity. *IEEE Transactions on Evolutionary Computation*, in press.
5. M. Li and Vitányi. *A Introduction to Kolmogorov Complexity and Its Applications.* Springer-Verlag, New York, 1997.
6. J. Maynard Smith and E. Szathmáry. *The Origins of Life: From the Birth of Life to the Origin of Language.* Oxford, 1999.
7. M. A. Nowak, J. B. Plotkin, and V. A. A. Jansen. The evolution of syntactic communication. *Nature*, 404, 2000.
8. S. Pinker and P. Bloom. Natural language and natural selection. *Behavioral and Brain Sciences*, 13:707–784, 1990.
9. T. Teal, D. Albro, E. Stabler, and C.E. Taylor. Compression and adaptation. In D. Floreano, J-D. Nicoud, and F. Mondada, editors, *Advances of Artificial Life*, number 1674 in Lecture Notes in Artificial Intelligence, pages 709–719. Springer, 1999.
10. B. Tonkes and J. Wiles. Methodological issues in simulating the emergence of language. To appear in a volume arising from the Third Conference on the Evolution of Language, in press.

Smooth Operator? Understanding and Visualising Mutation Bias

Seth Bullock

Informatics Research Institute, School of Computing, University of Leeds
seth@comp.leeds.ac.uk

Abstract. The potential for mutation operators to adversely affect the behaviour of evolutionary algorithms is demonstrated for both real-valued and discrete-valued genotypes. Attention is drawn to the utility of effective visualisation techniques and explanatory concepts in identifying and understanding these biases. The *skewness* of a mutation distribution is identified as a crucial determinant of its bias. For redundant discrete genotype-phenotype mappings intended to exploit neutrality in genotype space, it is demonstrated that in addition to the mere *extent* of phenotypic connectivity achieved by these schemes, the *distribution* of phenotypic connectivity may be critical in determining whether neutral networks improve the ability of an evolutionary algorithm overall.

1 Introduction

Mutation operators lie at the heart of evolutionary algorithms. They corrupt the reproduction of genotypes, introducing the variety that fuels natural selection. However, until recently, the process of mutation has taken a back seat to the more dramatic genetic operators. Sexual recombination (the splicing together of genetic material from multiple parents) is often regarded as the major source of an evolutionary algorithm's ability to discover fit phenotypes (the evolutionary programming paradigm being a notable exception). While mutation is required to introduce novel genetic material, it is recombination's role in assembling groups of well-adapted alleles that is often concentrated upon [1]. However, doubts concerning the validity of this "building block hypothesis" have recently undermined the notion that the power of evolutionary algorithms can be identified with the role of sexual recombination [2].

Moreover, recent work on neutrality in genetic encodings [3] suggests that mutation events and the character of an evolutionary algorithm's mutation space may have a greater role to play in the dynamics of evolutionary algorithms than had heretofore been appreciated. Rather than considering evolving populations to be engaging in some kind of hill-climbing via parallel assessment of different combinations of co-adapted alleles, Kimura's [4,5] neutral theory of evolution proposes that a more useful image is of populations percolating through "neutral networks" of adjacent points in genotype space that each code for phenotypes with equivalent fitness. It is contended that such populations, rather than tending to converge at the peaks of local optima, will continue to diffuse through a

J. Kelemen and P. Sosík (Eds.): ECAL 2001, LNAI 2159, pp. 602–612, 2001.

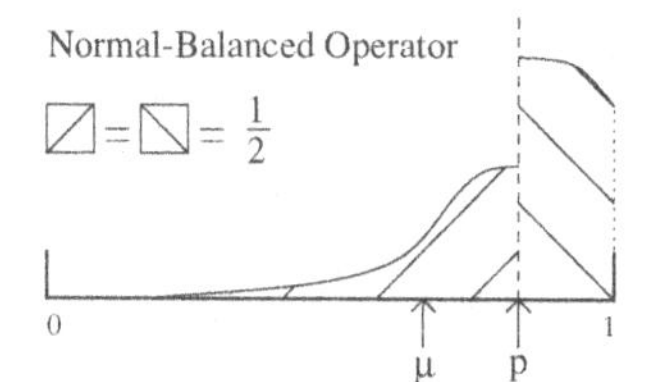

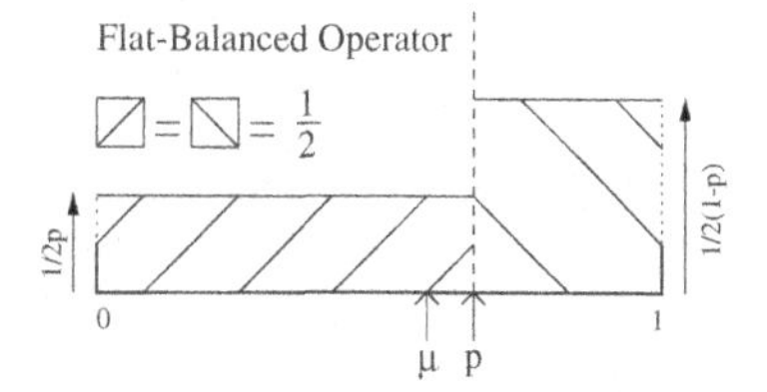

Fig. 1. Probability density functions for two mutation operators that perturb a real-valued parental gene, p, upwards or downwards with equal chance. Mean mutant destination is denoted μ. Mutants must lie within the legal range $[0, 1]$. For the *Normal-Balanced* operator (left), once direction of mutation has been determined, values are repeatedly drawn from the appropriate half of a normal distribution with mean p until a legal mutant value is achieved. For the *Flat-Balanced* operator (right), upwards mutants are drawn from a uniform distribution over all values in the range $[p, 1]$ while downwards mutants are drawn from the range $[0, p]$. Three additional operators are explored in this paper, but are not depicted. *Flat* draws mutants from a flat distribution over the legal range $[0, 1]$. *Absorb* draws mutants from a normal distribution with mean p, replacing illegal mutants by the nearest legal mutant value. *Reflect* behaves as Absorb, but any illegal mutant lying a distance d beyond a legal boundary is replaced by a value lying d within that boundary.

succession of neutral networks of increasing fitness. While it has been suggested that some difficult evolutionary optimisation problems already exhibit neutral networks [6], others have sought to develop genetic encodings that encourage neutral networks in the hope that this will improve the ability of evolutionary algorithms to find optimal solutions in general [7,8].

But how well do we understand the workings of mutation operators? Although they are often very simple to code, they come in many flavours, and previous work [9] has demonstrated that they often exhibit biases that may remain undetected, despite having appreciable effects on the course of artificial evolution — arbitrarily steering populations away from particular areas of the search space and toward others, for example.

This paper will begin by exploring a very simple pair of mutation operators (see Fig. 1). Since these operators, or ones like them, are used widely whenever genotypes feature real values which are constrained to lie within some legal range, analysing them serves a direct purpose in increasing our understanding of the component parts of evolutionary algorithms. But furthermore, in demonstrating their counter-intuitive behaviour and the biases that are inherent in even these simple operators, we may be able to refine our intuitions and develop useful explanatory concepts that allow us to gain insights into the more complex genetic encodings explored in the latter half of the paper.

2 Perturbing Bounded Continuous-Valued Traits

Although the canonical genetic algorithm (GA) employs binary genotypes, many practitioners find it useful to encode members of the evolving population as

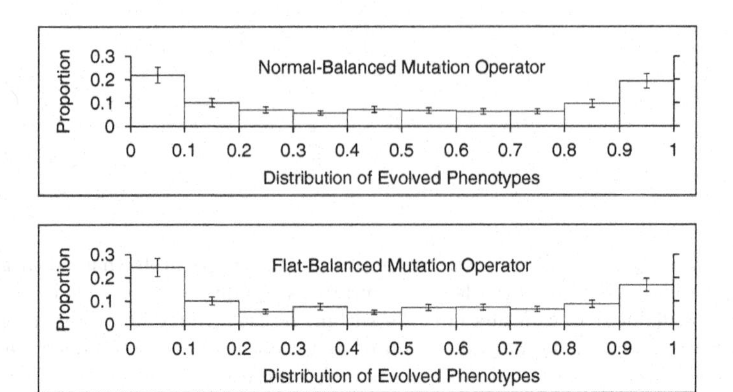

Fig. 2. Each histogram depicts the distribution of trait values for 20 populations of 1000 single-trait organisms after 5000 generations of evolution on a flat fitness landscape. For the Normal-Balanced operator, perturbations were drawn from a normal distribution with zero mean and standard deviation 0.2. An unbiased mutation operator is expected to exhibit a flat distribution, since no value is more likely to evolve than any other. Both of the mutation operators simulated here favour extreme-valued traits.

vectors of real values. Often these values must lie within some legal range. For instance, consider using a GA to evolve a population of images, each encoded as a vector of real-valued RGB triples, e.g., $\{\{0.1, 0.9, 0.3\}, \{0.4, 0.6, 0.9\}, \ldots\}$. Point mutations that *perturb* these RGB components by some small random amount must generate mutant values that remain within the legal range $[0, 1]$. In previous work [9], it was demonstrated that in dealing with illegal values, mutation operators can often introduce a *mutation bias* that either favours trait values close to their legal limits, or values that are far from these limits. In either case, such biases may adversely affect the course of evolution (tending to favour bold or dull images in the example being considered here) perhaps retarding evolutionary optimisation or introducing artefacts into evolutionary simulation models.

One mutation operator, here dubbed *Normal-Balanced* (which was not considered in [9], but was suggested in response to this work) is depicted in Fig. 1 (left), alongside a similar but simpler *Flat-Balanced* mutation operator which proves more amenable to formal analysis. Proponents of the former scheme were of the opinion that it would be able to deal with the possibility of illegal mutants without biasing a GA's evolutionary dynamics.

However, to some extent, intuitions regarding the behaviour of these mutation operators conflict. The fact that mutation rate remains constant across the parental range, and that the likelihood of upwards and downwards mutations are always equal might be taken to indicate that the operators will *not* bias evolving populations. However, the fact that, for any parental value, the expected mutant value (denoted μ in Fig. 1) lies towards the centre of the legal range might suggest that the operators may bias populations *away* from extreme-valued traits.

In order to discover the behaviour of these mutation operators, we can evolve populations on a flat fitness surface, and check whether there exist particular trait values that are more likely to evolve than others (see Fig. 2). In addition, since the Flat-Balanced operator is relatively simple, it is straightforward to *derive* the expected probability density function for the trait values of a population after a period of mutation. Consider the mutant phenotype x of a parental phenotype p (Fig. 1, right). If $p < x$, the probability density function $f(x)$ takes the value $\frac{1}{2(1-p)}$, while if $p > x$, $f(x) = \frac{1}{2p}$. After some period of mutation...

$$f(x) = \int_0^x f(p)\frac{1}{2(1-p)}\mathrm{d}p + \int_x^1 f(p)\frac{1}{2p}\mathrm{d}p$$

That is, the value of the probability density function at x is equal to the contribution from every parental value lower than x added to the contribution from every parental value higher than x. What shape is this function?

$$f'(x) = f(x)\frac{1}{2(1-x)} - f(x)\frac{1}{2x}$$
$$f'(x) = -\frac{1}{2}f(x)\left(\frac{1}{x} + \frac{1}{(x-1)}\right)$$
$$f(x) = C\mathrm{e}^{-\frac{1}{2}\int_0^x \frac{1}{y}+\frac{1}{(y-1)}\mathrm{d}y}$$
$$f(x) = C\mathrm{e}^{-\frac{1}{2}(\ln|x|+\ln|1-x|)}$$
$$f(x) = \frac{C}{\sqrt{(x(1-x))}}$$

Since it must be the case that $f(x) \geq 0 \; \forall \; x \in [0,1]$, it follows that $C > 0$, and thus that $f(x)$ is a convex parabola with a minimum at $x = 0.5$. Hence we would predict from this analysis that the mutation operator would tend to push populations towards the extremes of the legal range.

The same general conclusion could be reached (via a more involved analysis) for the Normal-Balanced mutation operator.

Why do these mutation operators bias populations towards extreme-valued traits? Neither simulating their performance, nor deriving their character formally, gives us much insight into the relationship between the form of the operators and their behaviour. The frequency distributions in Fig. 3 begin to point us in the right direction by clarifying the nature of the over-representation of extreme-valued mutants, and highlighting the relative lack of symmetry in the distribution of mutants generated by the biased operators compared to that of an unbiased operator. Fig. 4 further demonstrates that, unlike considerations of mutation *drift* (calculated as the expected mutant destination) attention to the manner in which *skewness* (calculated as the difference between the mean and median values of mutant distributions) varies across the range of possible parental trait values allows us to distinguish biased from unbiased operators.

When a distribution of mutant values is skewed *away* from the extremities of a range (i.e., the median mutant is further from the centre of the legal range than the mean mutant), mutants falling towards the boundary will also fall closer together than mutants falling further from the boundary. For mutation operators such as Absorb, and the two Balanced operators, the increased *density* of the more extreme mutants creates a "hot-spot", with positive feedback ensuring that more and more mutants fall closer and closer to the boundary *and* to each other. Although, as this is happening, the *mean* mutant may remain far from the legal

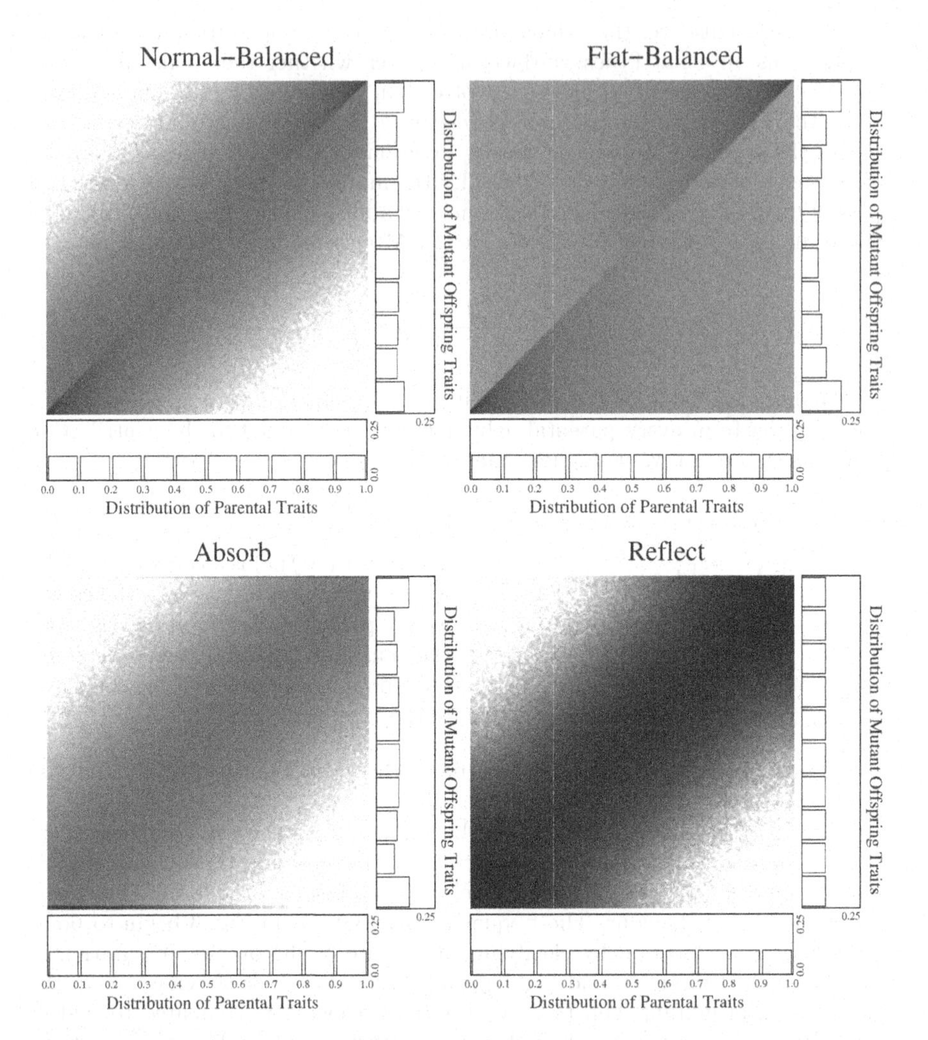

Fig. 3. For each mutation operator, the manner in which the distribution of mutants varies with parental trait value is plotted. The grey-level of each cell (x, y) in a 200×200 array represents the number of mutants in bin y after 10,000 randomly generated parents from bin x were each mutated once. The resulting aggregate distribution of mutant trait values is depicted as a histogram to the right of each array (parental values being depicted by the uniform histogram beneath each array). Deviation from uniformity in this right-hand histogram is indicative of mutation bias. Notice the dense areas in the corners of the Normal-Balanced, Flat-Balanced, and Absorb plots, indicating an accumulation of extreme-valued mutants, and that the Reflect operator has a plot that is symmetrical about $y = x$, and darker overall, indicating a more uniform distribution of mutants. (The values in each array were log-scaled to better reveal the structure in the distribution).

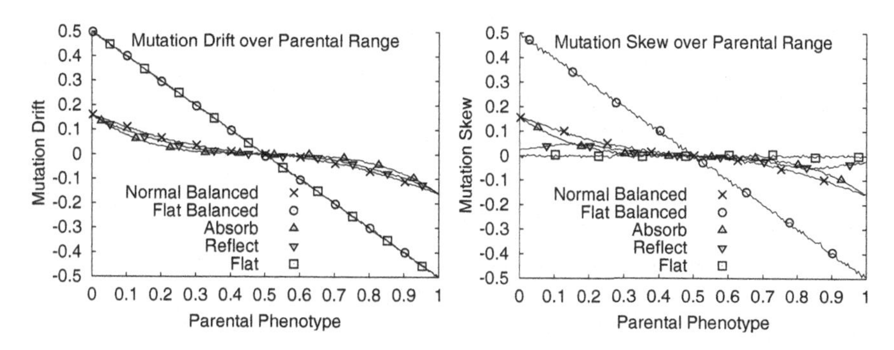

Fig. 4. Depicting the manner in which the strength and direction of mutation *drift* (left) and mutation *skew* (right) vary with parental value. Considerations of mutation drift cannot distinguish those operators biased towards extreme values (Flat-Balanced, Normal-Balanced, and Absorb) from those that are unbiased (Flat and Reflect). However, mutation skew varies inversely with parental value only for those operators favouring extreme values, tending to trap populations near trait boundaries.

boundary, the asymmetrical, skewed distribution of mutants ensures that this has little bearing on the location of the most densely populated mutant destination. In contrast, operators such as Reflect counter the tendency for populations to aggregate at the extreme values of a legal range, by decreasing the skewness of mutant distributions as parents approach trait boundaries.

From this discussion, it is tempting to conclude that certain mutation operators (perhaps Reflect) should be employed, rather than others (Absorb, or either of the Balanced operators). While there may be some truth in this, since certain mutation operators may be less biased than others, previous work [9] strongly suggests that mutation bias in one form or another is part and parcel of the evolutionary process. Rather than attempt to eliminate these biases from evolutionary algorithms, their presence should be anticipated and controlled for.

The analysis and visualisation presented in this section has revealed that our intuitions concerning the behaviour of mutation operators, and importantly the sources of their biases, are often not to be trusted. Measures of behaviour that at first sight appear sensible and instructive (e.g., mean mutant destination as a measure of mutation drift) can be misleading. The skewness of mutation distributions has been identified as a crucial aspect of a mutation operator's behaviour, and has been used to explain the biases of mutation operators that favour extreme-valued traits.

3 Understanding Mutation in Neutral Networks

Discrete-valued genotypes require mutation operators that are perhaps simpler in structure than the less orthodox operators analysed for real-valued genotypes in the previous section. However, like the operators discussed above, discrete (typically binary) genetic encodings also impose a structure upon the mutation space, allowing some transitions but preventing others, and as a result have just as much potential to bias evolutionary dynamics [9]. In this section we will employ some of the techniques introduced in section 2 to explore the biases inherent in a recently proposed massively redundant genetic encoding scheme [7,8].

Recently, encouraged by theoretical [3] and empirical [6] work suggesting that neutrality in fitness landscapes may improve the ability of evolutionary algorithms to discover fit phenotypes, researchers have begun to explore the introduction of *redundancy* into discrete genetic encodings [7,8]. Although redundancy (many genotypes mapping onto the same phenotype) increases the size of the search space, the correct kind of redundancy may populate this search space with neutral networks that increase phenotypic connectivity — enabling evolving populations to more easily explore a wider range of mutant phenotypes than they would be able to under a regular non-redundant encoding scheme. Potentially, neutral networks might increase this connectivity to such an extent that the overall performance of evolutionary algorithms is much improved.

However, is the mere *extent* of phenotypic connectivity the only important consideration here? In this section we will explore how the *distribution* of phenotypic connectivity may impact on the evolutionary dynamics of algorithms employing redundancy in this way.

One manner in which a massively redundant genotype-phenotype mapping has been implemented is through the use of a random Boolean network (RBN) [10], specified using 144 bits of genetic information, to generate an 8-bit phenotype (see Fig. 5). Under this scheme, it has been demonstrated [7,8] that neutral networks connect each of the 256 possible phenotypes to the majority of the 255 possible mutant phenotypes (see Fig. 6 left). However, employing the visualisation technique introduced in the previous section, we can show that the frequency distribution over these neighbours is radically non-uniform (see Fig. 6 right). In fact, under the proposed encoding scheme, a large proportion ($\approx 32\%$) of non-neutral mutations are equivalent to flipping a single bit of the 8-bit phenotype, i.e., identical to those transitions that would be achieved under a regular non-redundant binary encoding (see Fig. 7). The remaining non-neutral mutations are distributed (non-uniformly) across the remaining 247 possible mutants. Whilst it is possible that the RBN scheme may enjoy advantages over a regular binary encoding scheme (perhaps it responds well to crossover, for instance), these results suggest that the cost of moving from a regular 8-bit representation to this redundant 144-bit scheme (an increase of search space size by a factor of 2^{136}) is perhaps not compensated for by the increased phenotypic connectivity that is achieved.

In order that neutral networks improve the performance of evolutionary algorithms, it must be the case that in addition to connecting each phenotype to many others they connect these phenotypes roughly equally, rather than favouring some transitions much more than others. What is important is not just the *number* of phenotypes that neighbour a neutral network, but the *frequency distribution* over this neighbourhood, and whether it is significantly biased.

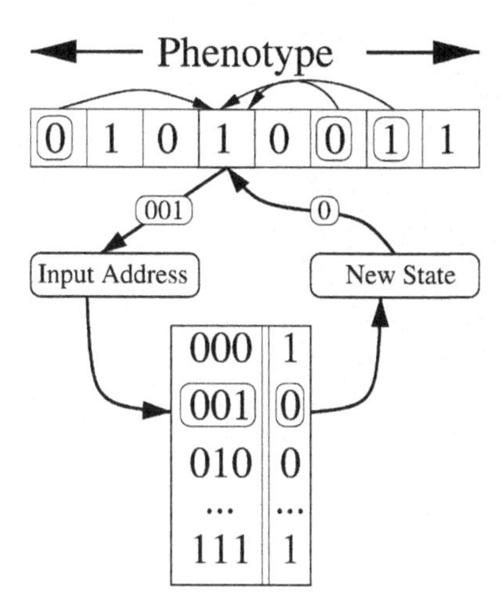

Fig. 5. Schematic representation of a *random Boolean network* as a developmental mapping. Each of the eight bits representing the organism's phenotype is associated with three neighbours, and a look-up table with eight entries. The state of each bit is updated by consulting the entry in it's look-up table (only one look-up table is depicted) determined by the value of its three specified neighbours. All eight bits are updated once per cycle. After 20 cycles, the value of the eight phenotypic bits is taken as the organism's phenotype. A 144-bit genotype represents the initial state of the phenotype (8 bits), the entries in each phenotypic bit's look-up table (8×8), and each phenotypic bit's three neighbours ($8 \times 3 \times 3$). Adapted from [7].

In this section we have discovered that an encoding scheme which promised a massive increase in phenotypic connectivity brings with it a hidden cost in terms of a radically biased distribution over this connectivity. Although many phenotypes are accessible from each neutral network, a population percolating through such a neutral network will tend to repeatedly explore only a few of the possible transitions, with many transitions occurring infrequently if at all.

How can we improve upon this situation? To achieve a more uniform distribution of phenotypic accessibility, we might explore increasing the connectivity of the random Boolean network which underlies the genotype-phenotype mapping, encouraging it to manifest chaotic behaviour which may be more sensitive to single genotypic bit flips. Moreover, if the RBN's behaviour is chaotic, running it for more clock cycles might also allow single genotypic bit flips to exert more influence on the end state of the RBN. However, this would result in a much longer genotype, and much larger neutral networks. Determining whether the trade-off between increased phenotypic connectivity and increased search space favours redundant genotype-phenotype mappings of the kind explored here will require further theoretical and empirical enquiry.

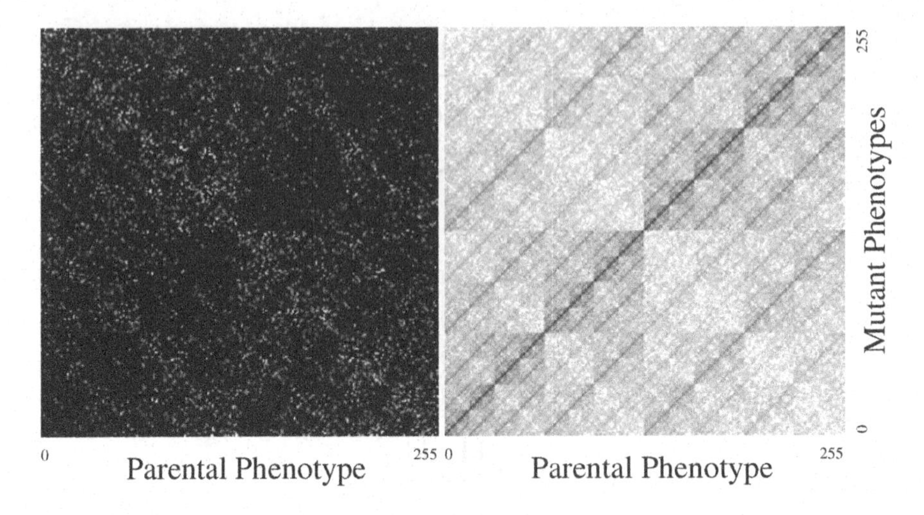

Fig. 6. Depicting the accessibility of mutant phenotypes from parental phenotypes for the RBN encoding scheme. A random 144-bit parental genotype is generated and its corresponding phenotype, x, determined. At random, one of the 144 bits is flipped and the phenotype, y, of the resulting mutant is determined. The cell (x, y) is incremented by one. This process is repeated for 10^6 random parents. *Left*: non-empty cells are depicted in black, empty cells in white. This depicts the *absolute* accessibility of mutant phenotypes from parental phenotypes. *Right*: empty cells are depicted in white, non-empty cells are depicted in grey levels with high-frequency cells shaded heavily. This depicts the *relative* accessibility of mutant phenotypes from parental phenotypes. Note the dark leading diagonal indicating the prevalence of neutral mutations. The uniformity of the left-hand plot indicates that almost all transitions are *possible*, the right-hand plot shows how misleading this conclusion is, since only a limited number of idiosyncratic transitions are at all *probable*. (The values in the right-hand image have been log-scaled to better reveal the structure in the frequency distribution.)

4 Conclusions

Mutation bias is a little-explored and little-understood aspect of evolutionary algorithms. As the role of neutral mutation becomes increasingly significant in our theories of evolution and evolutionary computing, our ability to identify, visualise and understand these biases will itself become increasingly important. This paper has demonstrated that our intuitions regarding mutation bias and genotype-phenotype mappings are currently under-developed and has contributed analytic and visualisation techniques which may help us improve upon this state of affairs.

Acknowledgements. The paper benefited from discussion with Chris Goodyer, Chris Needham, Jason Noble, and Richard Watson, the comments of three anonymous referees, and the 40% advice of Robin Michaels, Paul Rolles, and Jack Rudd.

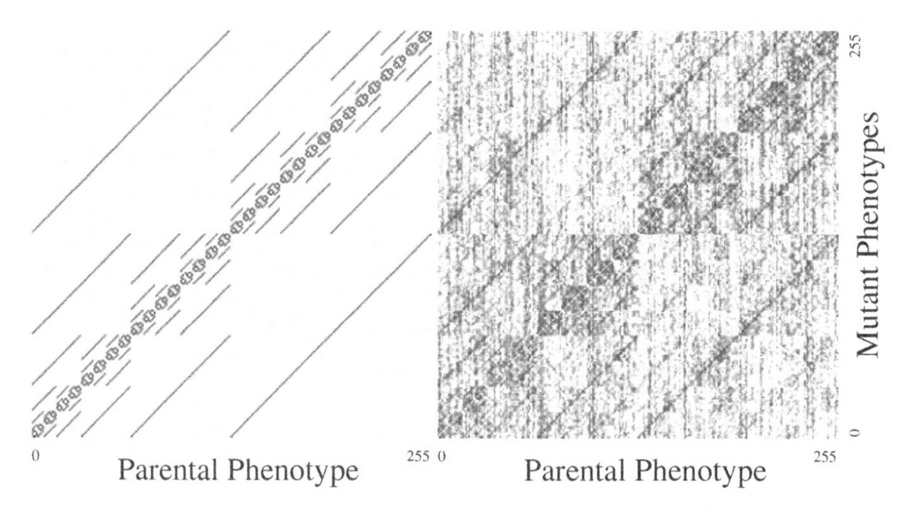

Fig. 7. *Left*: the distribution of the eight one-bit mutant neighbours of each of the 256 possible eight-bit phenotypes under a regular non-redundant *binary encoding* scheme. *Right*: the distribution of non-neutral neighbours collated for 256 *neutral networks* explored under the RBN encoding scheme. For each phenotypic value a random genotype encoding that value was generated. Each of the 144 one-bit mutants of this genotype were generated and their corresponding phenotypes determined. A neutral mutant was chosen at random and its 144 neighbours assessed, and so on. In this way 100 adjacent members of each neutral network were explored. Although each neutral network neighbours many phenotypic values, the similarity between the left- and right-hand plots indicates that a large proportion (32% in this case) of non-neutral mutation events are equivalent to flipping a single bit of the 8-bit phenotype. (The values in the right-hand image have been log-scaled to better reveal the structure in the frequency distribution.)

References

1. Forrest, S., Mitchell, M.: Towards a stronger building-block hypothesis: Effects of relative building-block fitness on GA performance. In Whitley, D., ed.: Foundations of Genetic Algorithms 2, Morgan Kaufmann, San Mateo, CA (1993) 109–126
2. Thornton, C.: The building block fallacy. Complexity International **4** (1997) URL: `http://www.csu.edu.au/ci/vol04/thornton/building.htm`.
3. Barnett, L.: Ruggedness and neutrality — the NKp family of fitness landscapes. In Adami, C., Belew, R., Kitano, H., Taylor, C., eds.: Artificial Life VI, MIT Press, Cambridge, MA (1998) 18–27
4. Kimura, M.: The neutral theory of molecular evolution. Cambridge University Press, Cambridge (1983)
5. Kimura, M.: Population genetics, molecular evolution, and the neutral theory: Selected papers. University of Chicago Press, Chicago (1994)
6. Harvey, I., Thompson, A.: Through the labyrinth evolution finds a way: A silicon ridge. In Higuchi, T., Iwata, M., Weixin, L., eds.: First Int. Conf. Evolvable Systems, Springer-Verlag (1997) 406–422

7. Shipman, R., Shackleton, M., Ebner, M., Watson, R.: Neutral search spaces for artificial evolution: a lesson from life. In Bedau, M.A., McCaskill, J.S., Packard, N.H., Rasmussen, S., eds.: Artificial Life VII, MIT Press, Cambridge, MA (2000) 162–169
8. Shipman, R., Shackleton, M., Harvey, I.: The use of neutral genotype-phenotype mappings for improved evolutionary search. BT Technology Journal **18** (2000) 103–111
9. Bullock, S.: Are artificial mutation biases unnatural? In Floreano, D., Nicoud, J.D., Mondada, F., eds.: Fifth European Conference on Artificial Life, Springer, Berlin (1999) 64–73
10. Kauffman, S.A.: The origins of order: Self-organisation and selection in evolution. Oxford University Press, New York (1993)

The Use of Latent Semantic Indexing to Identify Evolutionary Trajectories in Behaviour Space

Ian R. Edmonds

School of Computing Information Systems and Mathematics
South Bank University, London SE1 0AA, United Kingdom
edmondi@sbu.ac.uk

Abstract. This paper describes the simulation of a foraging agent in an environment with a simple ecological structure, alternatively using one of three different control systems with varying degrees of memory. These controllers are evolved to produce a range of emergent behaviours, which are analysed and compared using Latent Semantic Indexing (LSI): the behaviours are compared between controllers and in their evolutionary trajectories. It is argued that the ability of LSI to reduce large dimensional spaces to a lower dimensional representation which is easier to understand can help in highlighting key relationships in the complexity of interactions between agent and environment.

1 Introduction

This paper describes how Latent Semantic Indexing (LSI) has been applied to help in understanding the emergent behaviour of a foraging agent. This first section introduces the themes which motivated the use of LSI: (i) the complexity of emergent behaviour and its relation to the controller architecture and task, and (ii) related techniques which have been used to help in understanding such complexity. The other sections are as follows: (2) introduces a model of a foraging agent in an environment with a simple ecological structure, alternatively using one of three different sensorimotor controllers with varying degrees of memory, (3) describes the evolution of these controllers, (4) uses a distal perspective to pick out key features of the observed behaviours, (5) applies a fractal view on the behaviours to identify scale factors, and then analyses the behaviours using LSI to provide a lower dimensional representation which is easier to understand, and (6) discusses the evolution of these behaviours and how the trajectories through behaviour space are constrained by the affordances within the environment.

Many systems in Artificial Life and situated robotics have used sensorimotor controllers to evolve a range of emergent behaviours for basic agent tasks, e.g. prey capture [1], obstacle avoidance and foraging [2], corridor navigation [3], garbage collection [4], [6], [7]. Part of the appeal of this approach is that complex behaviour emerges from the interaction of the sensorimotor controller and the agent environment, and an important issue is: what sort of architecture will allow for what sorts of emergent behaviour. This has lead to comparisons of different controllers within the same task environment, e.g. garbage collection [6], rat

J. Kelemen and P. Sosík (Eds.): ECAL 2001, LNAI 2159, pp. 613–622, 2001.

navigation with hippocampal place cells [5], and architectures that explicitly provide for adaptability within the control structure itself, so that an architecture can fit itself to the behaviour needed. In [4], [6], [7] modularity was allowed to emerge in the neural network to correspond with the behavioural modularity required in the garbage collection task, and in [8] a biologically inspired genotype-phenotype mapping is described for neural network architectures that can adapt to a range of behaviours.

A variety of techniques including statistical measures have been used to analyse and understand the emergent behaviour. In [6] two perspectives are used: an observer's perspective (a distal description), is contrasted with proximal descriptions based on detailed local relationships between input (senses), node activations, and actions. Another approach is based on dynamical systems, e.g. used in [9] to uncover attractors in the phase portrait of an agent, used in [10] by developmental psychologists in tracking infant motor skills, and for a review see [11]. A problem with any of the detailed approaches is the high dimensionality of the data: tackled in [6] with multiple charts, in [9] by adopting abstract representations, and in [11] other alternatives are discussed including the use of principle components analysis (PCA) as a data reduction technique. PCA is used in [12], and is similar to the use of LSI below. The work in [12] describes an evolutionary algorithm, SANE, for evolving pools of nodes in neural networks. The evolutionary trajectories of the nodes in their function space has too high a dimension to be easily comprehensible, so PCA is applied to the function space of the nodes and the analysis proceeds using just the first two components.

LSI uses a factor analysis based on singular value decomposition [13], [14] and is used in text based information retrieval. It is particularly suited to domains with large feature sets (high dimensionality), where the occurrence of features in a particular record may be very sparse. For example in [14] it copes with a vocabulary of over 60,000 words in natural language, where individual pieces of text may have only 150 words, and produces useful reduction down to 300 dimensions. LSI has also been used in an ALife context [15] to identify memes within submissions to electronic newsgroups; LSI was applied to individual postings to a newsgroup, and their locations within the reduced dimensional space was used to identify clusters of memes. The co-evolution of these meme clusters was then tracked through time. For a good introduction to the LSI technique (with a worked example) see [13]. Essentially, as in PCA, the factor analysis produces a mapping from the observed features of a data set into a reduced set of orthogonal dimensions (the latent factors); the mapping can then be applied to further data sets.

LSI is used in the work here, to bridge the gap between high level, distal interpretation (where an observer looks for meaning in sequences of actions) and the detailed proximal descriptions based on individual actions and fine granularity. It is related to LSI information retrieval, where the meaning of a text is recovered, solely, from matrix based analysis of the patterns of co-occurrence of individual words.

2 The Model

The agent exists in a simulated world with plants that grow differentially on a water resource gradient; for more detail of the model see [16]. The space is a 100 by 100 toroidal grid with 40 randomly distributed water sources: the water is diffused using a standard lattice diffusion formula leading to a water resource gradient, e.g. fig. 1 (a). Plants can occupy cells in the grid, and are modelled using a lifecycle transition graph. They only grow in the wetter areas, and produce a structured and stable population of approximately 2,000, see fig. 1 (b).

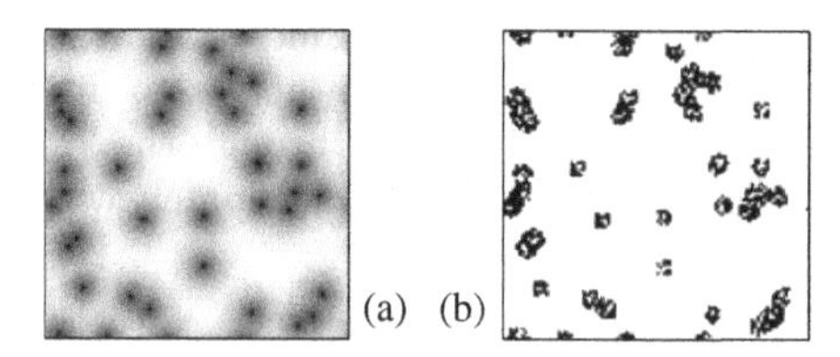

Fig. 1. (a) The water resource forms wet patches separated by arid areas, and (b) the plants grow in the wet patches.

The agent occupies one cell in the grid, and has 4 actions available to it: eat, move forward one step, and rotate by 90 degrees left or right. It has sensory input of the water and the plants: the water sensory input is of the water value at its current location and the water gradient in its forward direction, while the plant sensory input distinguishes the existence (or not) of plants at its current location, the cell directly in front of it, and the cells diagonally to the front left and front right. When a plant is eaten, it is removed from the world, and its biomass is absorbed by the agent. The agent physiology has been modelled so actions have equal energy costs in these experiments and hence evolution will optimise a foraging rate rather than a foraging efficiency [17]. However, initial investigations have indicated that evolution of similar behaviours is robust under a fair range of differential energy costs.

Three controller models have been implemented to investigate how behaviour is effected by different memory structure: (i) a basic neural net, (ii) a neural net with memory, and (iii) a rule based controller. All 3 are essentially sensorimotor controllers, where (i) implements a reactive controller, (ii) provides internal state to the controller, and (iii) provides internal state in the sense of possible sequences of actions.

The basic neural network contains feedforward nodes using a sigmoid activation function with one hidden layer. The input layer receives the 4 bit binary plant sense pattern and the 2 real water values. The 4 output nodes are treated as a stochastic output to select the agent's action in the given sensory state.

The neural net with memory is similar to the basic neural network, but with 5 additional memory input nodes fully connected to the hidden layer. Each memory is represented as a real valued time decaying integration of the stimulus, where the stimulus is each one of the actions performed and the amount of biomass eaten by the agent (see [16]), and is akin to the battery levels used by robot controllers, e.g. [2], or in networks with recurrent loops e.g. [6]. Different values for the memory parameters were explored, and in an alternative recurrent network model the values were allowed to evolve. It was found that the neural network was able to robustly utilise a range of memory responsiveness. The values as set are not claimed to be optimal. For a discussion of the effects of different integrative formulae on foraging search decisions in honeybees see [18] - for example the use of averaging over different numbers of flower visits will alter the effectiveness of a search in highly autocorrelated resource environments.

The rule based controller contains a sequence of 30 rules organised as a loop, plus a memory of its latest position in the sequence. Each rule contains a pattern to match against the sensory input, and an action to fire if the pattern match succeeds. At every step in its life it moves through the loop of 30 rules starting from the last successful rule, trying to find a match with the current sensory input.

3 The Experiments

A simulation consists of placing an agent in a random position in the world and having it make 3000 actions, during which time it may acquire some quantity of biomass. The 3 different controller architectures were evolved in evolutionary runs of 1,000 generations, fig. 2.

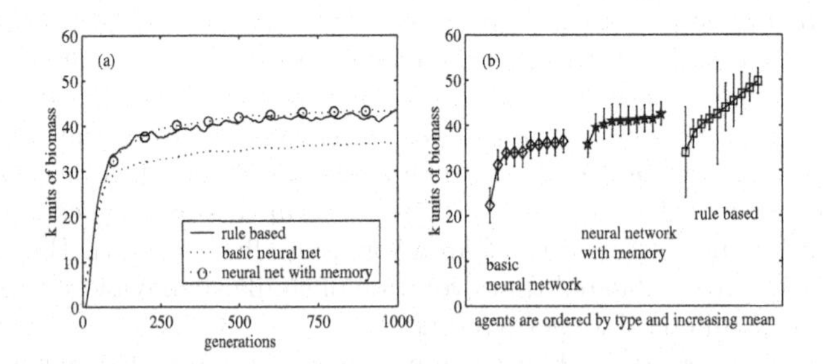

Fig. 2. (a) The means (using a moving average over the previous 20 generations) of the best performers of each of the 10 runs, (b) The means and standard deviations of the best overall performers from each run, ordered according to increasing mean.

In each case the starting population consisted of 5 parent genotypes. Ten mutants of various mutation rates were produced from each parent to give a population of 50. Each mutant was tested in a number of simulations and the mean performance for each mutant was used as the fitness value. The best 5 mutants became the parents for the next generation. Each of the 3 controller architectures were exposed to 10 evolutionary runs and the results (the mean of the best performances of each of the 10 runs) are shown in fig. 2 (a).

The basic neural network performs worse, but the means obscure the differences between the best controllers from individual runs. To show this, the best performing controllers from each of the runs were subject to a more extensive set of 50 simulation tests: the results (means and standard deviations) are shown in fig. 2 (b), with performances ordered according to increasing mean. The rule based controllers provide by far the best performers, although the standard deviations are also highest, perhaps due to their more brittle decision making representation.

4 Results - Observed Behaviours

This section describes some of the behaviours of the best performers and some of their early generation antecedents, from a distal perspective [6], while later sections look at proximal descriptions and LSI. In general the behaviour can be seen as falling within the classic foraging descriptions of exploitation and exploration: the agents move quickly through the arid areas where there are no plants until they find a wetter patch with plants, at which point they start to eat and turn more; fig. 3 shows examples of paths taken by the agents. If they find themselves in an eaten out or barren wet patch, they may turn a little, but move on fairly soon. Perturbation experiments with reduced plant numbers showed a strong correlation between plant numbers and the amount of turning and fits with animal and other agent searching evidence [17].

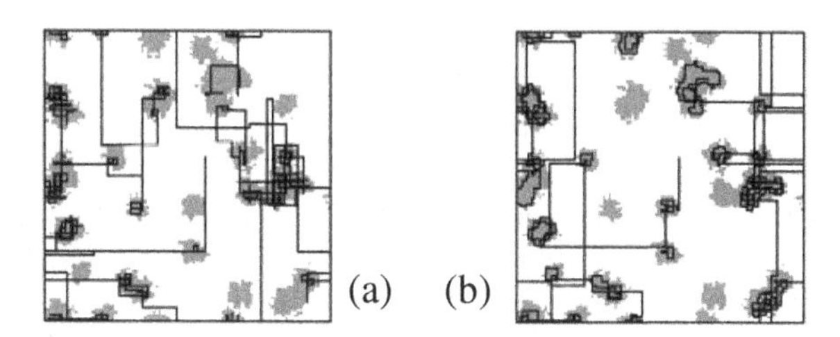

Fig. 3. Movements of the agents during a simulation of 3000 actions, superimposed over grey outlines of the plant areas: (a) basic neural network, (b) rule based controller.

Key differences between the best performers of the 3 controllers were obvious, and most apparent in the exploitation behaviour. The basic neural network had a strategy of tending to head through the middle of a patch of plants and eating as it looped or moved with back and forth actions, see fig. 3 (a) and fig. 4 (a) and (b). This strategy was the least effective of the 3 as it lead to the patch of plants becoming broken up into isolated plants, which became gradually more difficult for the agent to find. The neural network with memory controllers tended to eat by spiraling in to rectangular blocks, see fig. 4 (c) and (d), while the best rule based agents had the most effective strategy which involved eating by following the outer contours of the patch of plants, and gradually spiraling inwards, see fig. 3 (b), fig. 4 (e) and (f). Some of the rule based controllers used zigzag exploration of arid areas, fig. 4 (h), but the strategy of the best agents was to move in long straight lines through the arid areas, fig. 3, even though this meant they sometimes passed by fairly close to a wet patch and didn't turn in towards it.

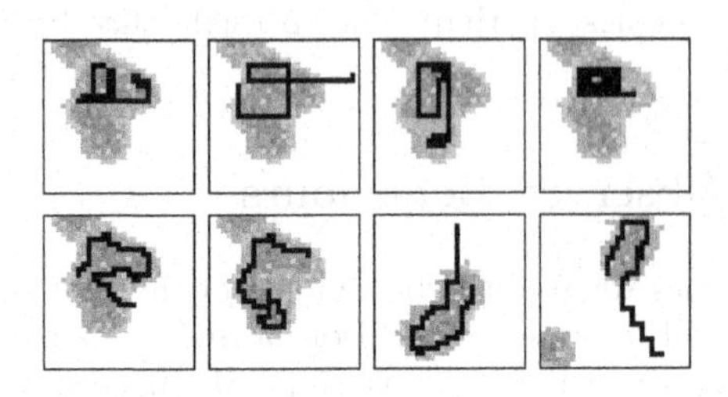

Fig. 4. Paths of the agents over 60 actions. The top row: (a) and (b) are the basic neural network eating by looping or a back and forth movement, (c) and (d) are the neural network with memory eating by spiraling inwards on a rectangular block. The bottom row are all the rule based controller: (e) and (f) eating along the contours of a plant cluster, (g) and (h) are finding a plant cluster using: in (g) a straight line approach, and in (h) a long legged zigzag.

5 Analysis

The next step was to try to find a way to characterise the behaviours of the best performers based on a proximal description. During the simulations, data was collected for each action: the water and plant sense data, the action, and the amount of biomass harvested. A PCA analysis (similar to [12]), showed each action as falling quite clearly into a limited number of clusters, but, it didn't help to bridge the gap to the distal descriptions, as the focus here was still at the grain size of individual actions.

To explore the effect of expanding the focus up to sequences of actions, a technique was adopted from landscape ecology to identify scale factors. In [19] it is shown as being useful to identify fractal dimensions in beetle movement and plant distributions, and in [20] a related but more sophisticated technique is used to identify fractal dimensions in the movement of fish schools foraging on plankton swarms.

A window is moved over the set of actions, and an interesting metric is extracted for all window positions. In the work here the metric was the Euclidean distance travelled from the start to the end of the window, expressed as a fraction of the maximum possible distance (i.e. a straight line walk). The distribution of fractional distances travelled was then separated into 10 bins, and shows, for a particular window size, just how straight or wiggly the walks would look. The process is then repeated for different window sizes: in this case for sizes of 2 to

100 actions. Values in each of the 10 bins are then plotted against window size to show how different fractional walk lengths vary with respect to window size. This helps to identify interesting scale effects, for example, what window size(s) will let us see the most variety of behaviour.

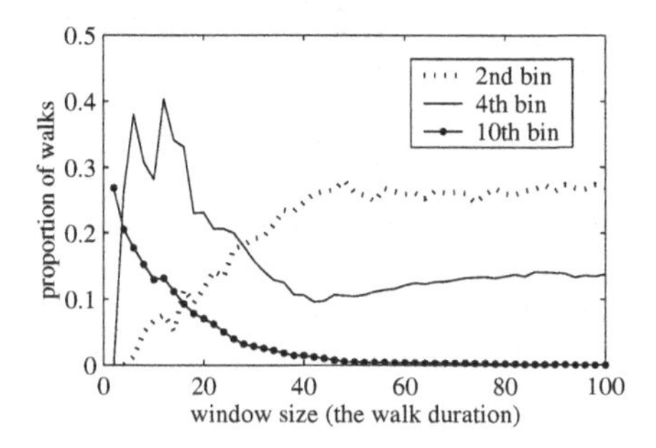

Fig. 5. Analysis of scale factors for the rule based controller. The proportion of walks with fractional walk length falling in the 2nd, 4th and 10th bins (i.e. the 10th bin contains the straight line walks), plotted against window size.

An analysis of the fractional walk length for the rule based controller, fig. 5, shows the 2nd, 4th and 10th bins (i.e. the 10th bin contains the straight line walks): the shapes for the other bins lay between these 3. It can be seen that window sizes of 10 to 20 contain a good mix of fractional walk lengths, and it was decided to use window sizes of 15 in the LSI analysis. Two issues to note, are firstly, that all 3 controllers had similar fractional walk length profiles, because they are related to scale factors in the environment to which all 3 have adapted, and secondly, that other metrics may indicate scale effects for different types of behaviour, although using a metric of *quantity of biomass consumed* did produce a similar scale of interest.

The action data for simulations was compiled into records of sequences of 15 actions, by moving the window over the actions 1 action at a time: for the rest of the LSI analysis, each sequence of 15 actions will be called a behaviour. The data for each behaviour contained 15 chunks of data for the actions plus 16 features containing a count of the number of consecutive actions in the behaviour and 2 features for overall fractional walk length and biomass consumed (i.e. the total number of features was 183). Data for behaviours of the 3 best performing controllers was analysed with PCA and LSI: the PCA wasn't particularly helpful, but the LSI was. Scatter plots, in fig. 6, show the location of each behaviour in the space of the first 2 latent factors, with histograms showing the distribution of points in the vertical and horizontal dimensions.

There are three general points about the interpretation. Firstly, the meaning of the latent factors was determined, as in PCA, by looking at the weights in the vectors that map features onto factors. The axes indicated by the arrows (g) and (h) in fig. 6 (a) have the following general meanings: (g) increasing

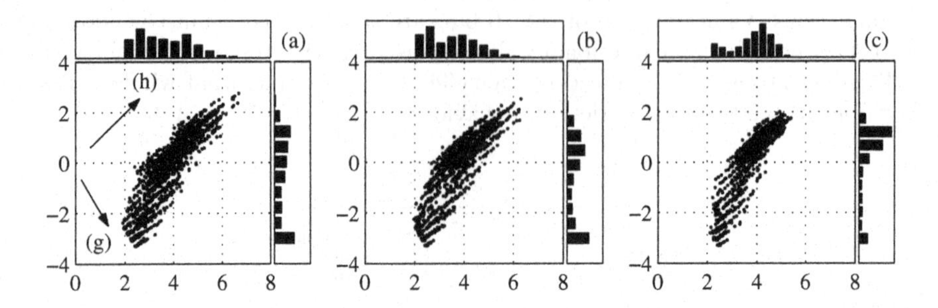

Fig. 6. Scatter plots of the location of each behaviour in the space of the first 2 latent factors from the LSI analysis, with histograms showing the distribution of behaviours in the vertical and horizontal dimensions. (a) shows the basic neural network, (b) shows the neural network with memory, and (c) shows the rule based controller.

with the number of move actions plus a smaller contribution from plants seen and water values, (h) increasing with the numbers of plants seen and eat actions plus smaller contributions from turning actions, water values, and biomass eaten, and a negative contribution from move actions. Secondly, the banded nature of the plots along axis (g) is a consequence of the discrete number of move actions in each behaviour: a behaviour with fewer moves will be in a band closer to the origin. Thirdly, there are 2 loose clusters: at around (4, 1) being associated with seeing plants and eating (i.e. exploitation), and another at (2.5, -3) being associated with maximum moving and seeing no plants (i.e. exploration).

6 Discussion

It can be seen, in fig. 6, that the basic neural network has the most dispersed footprint, indicating more varied behaviour, while the rule based controller has the tighter of the clustering around points (4, 1) and (2.5, -3), more effectively focussing its behaviours into exploitation and exploration. The neural network with memory lies between the other two. Also, there are considerably more behaviours appearing further out from the origin than (5, 2) for the neural network agents, and this is a consequence of them seeing more plants as they plunge through the middle of the plant patches, while the rule based controller follows the contours of the outside of a patch, tending to keep the plants always to one side of its peripheral plant vision.

Further analysis was conducted on the evolutionary trajectories of the behaviours for the neural network with memory, fig. 7. The factor mapping vectors from the original analysis were used to plot the behaviours of the earlier generations into the same latent factor space as the best performers, allowing the interpretation to be consistent across generations and controllers. The generations shown in fig. 7, and their approximate performances relative to the best are: g(1) 5%, g(16) 25%, g(40) 50%, g(920) 100%. The g(1) controller is quasi-random, and although it's footprint shows the same banding as the other generations, the majority of behaviours lie close to the origin (0, 0) indicating little movement and seeing very few plants. The next snapshot, at g(16) shows the behaviours spreading out with a wider flatter distribution, effectively exploring the behaviour space of both moving more and seeing more plants. At the g(40) snapshot, the distribution of behaviours is beginning to focus in on the exploration cluster at (2.5, -3), but has still to settle into the exploitation cluster. After another 880 generations, the final best performer, g(920), has more distinctly settled into the 2 clusters, although it should be noted that the asymptotic approach to this final performance would have given much earlier generations a similar behavioural footprint.

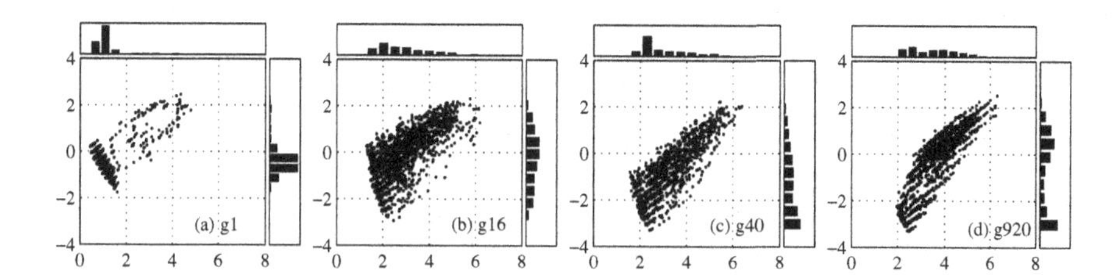

Fig. 7. Scatter plots of the location of each behaviour in the space of the first 2 latent factors from the LSI analysis, with histograms showing the distribution of behaviours in the vertical and horizontal dimensions, for successive generations of the neural network with memory: (a) is g(1), (b) is g(16), (c) is g(40), and (d) is g(920).

The trajectories of the other 2 controllers were fairly similar to the neural network with memory and indicate that the affordances in the environment push a controller into evolving robust exploration behaviour, before exploitation will be honed.

7 Conclusion

This paper has described various strategies of emergent behaviour in a foraging agent. The use of LSI helped in discovering meaning in the detailed data from sequences of actions, thereby providing a bridge between the distal descriptions where the observer decides what is important and the proximal descriptions based on data from individual actions. Furthermore, the ability to use the LSI (feature to factor) mapping vectors to produce a consistent representation of behaviours across controllers and generations also aided in the interpretation of the evolutionary trajectories of the behaviours.

References

1. Gomez, F., and Miikkulainen, R., 1997, Incremental Evolution of Complex General Behavior, Adaptive Behavior, vol. 5, no. 3/4, 317-342

2. Seth, A. K., 1998, Evolving Action Selection and Selective Attention Without Actions, Attention, or Selection, in Pfeifer, R., Blumberg, B., Meyer, J-A., and Wilson, S. W., (eds), Animals to Animats 5, Proc. 5th Int. Conf. on the Simulation of Adaptive Behavior, Bradford Book, MIT Press
3. Shipman, R., 1999, Genetic Redundancy: Desireable or Problematic for Evolutionary Adaption?, The 4th International Conference on Artificial Neural Networks and Genetic Algorithms (ICANNGA '99), April 1999
4. Nolfi, S., 1997, Evolving non-trivial behaviours on real robots: A garbage collecting robot, Robotics and Autonomous Systems, 22, 187-198
5. Foster, D. J., Morris, R. G. M., and Dayan, P., 2000, Models of Hippocampally Dependent Navigation, Using The Temporal Difference Learning Rule, Hippocampus, vol. 10, issue 1
6. Nolfi, S., 1997, Using Emergent Modularity to Develop Control Systems for Mobile Robots, Adaptive Behaviour, vol. 5, no. 3/4, 343-363.
7. Calabretta, R., Nolfi, S., Parisi, D., and Wagner, G. P., 1998, Emergence of Functional Modularity in Robots, in Pfeifer, R., Blumberg, B., Meyer, J-A., and Wilson, S. W., (eds), Animals to Animats 5, Bradford Book, MIT Press
8. Rolls, E. T., and Stringer, S. M., 2000, On the design of neural networks in the brain by genetic evolution, Progress in Neurobiology, 61 (2000), 557 - 579, Pergamon.
9. Husbands, P., Harvey, I., and Cliff, D., 1995, Circle in the round: State space attractors for evolved sighted robots, Robotics and Autonomous Systems, 15, 83-106
10. Thelen, E., 1995, Motor Development, American Psychologist, Feb 95, 79-95
11. Beer, R. D., 2000, Dynamical approaches to cognitive science, Trends in Cognitive Sciences, vole 4, no 3, 91-99
12. Moriarty, D. E., and Miikkulainen, R., 1998, Forming Neural Networks Through Efficient and Adapted Coevolution, Evolutionary Computation, 5(4), 373-399
13. Deerwester, S., Dumais, S. T., Furnas, G. W., Landauer, T. K., and Harshman, R., 1990, Indexing by Latent Semantic Analysis, Journal of the American Society for Information Science, 41 (6), 391-407
14. Landauer, T. K., and Dumais, S. T., 1997, A Solution to Plato's Problem: The Latent Semantic Analysis Theory of Acquisition, Induction and Representation of Knowledge, Psychological Review, 104 (2), 211-240
15. Best, M. L., 1997, Models for Interacting Populations of Memes: Competition and Niche Behaviour, in Husbands, P., and Harvey, I., (eds), Fourth European Conference on Artificial Life, Cambridge, MA, MIT Press
16. Edmonds, I. R., 2001, Tracking the Evolution of a Foraging Agent, Technical Report, SBU-CISM-01-07, South Bank University, London
17. Gelenbe, E., Schmajuk, N., Staddon, J., and Rief, J., 1997, Autonomous search by robots and animals: A survey, Robotics and Autonomous Systems, 22, 23-34
18. Real, L. A., 1994, Information Processing and the Evolutionary Ecology of Cognitive Architecture, in Real, L. A., (ed.) Behavioural Mechanisms in Evolutionary Ecology, University of Chicago Press, 99-132
19. Milne, B. T., 1991, Lessons from Applying Fractal Models to Landscape Patterns, in Turner, M. G., and Gardner, R. H., (eds), Quantitative Methods in Landscape Ecology, Springer-Verlag, 199-235
20. Tikhonov, D. A., Enderlein, J., Malchow, H., and Medvinsky, A. B., 2001, Chaos and fractals in fish school motion, Chaos, Solitons and Fractals 12, 277-288

Data Visualization Method for Growing Self-Organizing Networks with Ant Clustering Algorithm

Tsuyoshi Mikami[1] and Mitsuo Wada[2]

[1] Tomakomai National College of Technology
Nishikioka-443, Tomakomai, 059-1275, Japan
mikami@jo.tomakomai-ct.ac.jp
[2] Graduate School of Engineering, Hokkaido University
Kita-13, Nishi-8, Sapporo, 060-8628, Japan
wada@complex.eng.hokudai.ac.jp

Abstract. The growing self-organizing networks are useful tools suitable for data analysis in which networks learn the topology of the high-dimensional data by inserting/deleting neurons. However, these methods cannot represent the high-dimensional clusters on the lower-dimensional intuitive space. In this paper, we proposed the visualization method by ant clustering to construct the two-dimensional feature map for the growing self-organizing networks.

1 Introduction

Self-Organizing Maps(SOM), proposed by Kohonen, are widely used for many data clstering and its visualization techniques for data mining or pattern recognition. The advantage of SOM different from traditional data clustering (e.g. k-means) is to visualize the complicated cluster distribution onto the two-dimensional plane [3]. However, the SOM has limitations of the cluster generation because the size of output neurons (codebook vectors) is fixed continuously [4, 5].In order to improve such limitations, there have been proposed many growing self-organizing networks, which learn the topology of the data structure and is self-tuning the number of cells (codebook vectors) adaptively and automatically. Dynamic Topology Representing Networks(DTRN), proposed by Si, et al., is one of the most recent research of growing self-organizing networks, which is faster than the other typical growing algorithm [2]. However, these mechanism cannot visualize the topology onto the intuitive space because their output neurons are growing on the high-dimensional space. This is a reason that the SOM is still used generally to many applications regardless of its limitations.

This paper describes new visualization methods so as to improve the drawbacks of growing self-organizing networks. Concretely, we integrated the growing self-organizing algorithms with other data visualization techniques, which is based on the ant clustering algorithm proposed by Lumer and Faieta (called L-F algorithm in this paper) [1]. However, the ant clustering mechanism can

J. Kelemen and P. Sosík (Eds.): ECAL 2001, LNAI 2159, pp. 623–626, 2001.

deal with only few training data (e.g. less than one thousand data) because it spends a long time until task fulfillment due to its random-walk and carriage mechanism by ants. In this paper, we modified the traditional L-F algorithm and applied our modified algorithm to two-dimensional feature map constituted of output neurons (codebook vectors) of the DTRN.

2 Modification of Ant Clustering

According to ecological studies, real ants follow the simple behavioral rules by their local sight to perform the global task fulfillment. Deneubourg, et al., described the collective mechanism in which ants are foraging and gathering foods by simple behavioral rules and their simulation results are very similar to the work real ants performed. Lumer and Faieta proposed the ant clustering algorithm in which the foods are defined as the numerical data having n-dimensional attributes and the ants are gathering them according to dissimilarity (n-dimensional Euclidean distance) in the local sight. By using this method, it is not necessary to define the number of clusters previously [1]. Moreover, since each ant does not know the global state but his local sight and his task is done continuously, the ants can make the clusters adaptively even if the target data are increasing/decreasing [6]. It is also useful for data analysts to investigate the massive data by using this method because the ants and their target data are distributed on the intuitive two-dimensional grid. Such property is suitable for our purpose.

First we considered that it spends a long time for the ants to perform the gathering task fulfillment and modified the ant clustering algorithm to realize faster calculation. In our consideration, the ant is not carrying the datum but the datum is moving and allocated directly on the feasible location. This is not a behavior derived from ecology, but our method can realize faster algorithm. The datum that should be moved is selected randomly from the index number of neurons. The selected datum finds some local regions (e.g. 8*8 grids) randomly on the field and finds the best matching region by calculating the evaluative function as follows.

$$f(i) = \frac{N^2}{N + \eta \sum_{j \in \Pi(i)} d(i,j)}$$

where j is an index of data existing on the local region, $d(i,j)$ is n-dimensional Euclidean distance, η is constant value, N is the number of data on the local sight, and $\Pi(i)$ is a set of data existing on the local sight. Each datum is allocated to the best matching region. The topology of output neurons is n-dimensional space and too complicated (e.g. growing/removing mechanism) to visualize on the intuitive space. It has been noticed that these drawbacks are common properties in the growing self-organizing networks, such as Growing Cell Structures, Dynamic Topology Representing Networks, and so on. In our model, the target datum is regarded as the growing output neurons of the DTRN, that is, codebook vectors are allocated on the two-dimensional grid and their distribution is useful for intuitive feature map.

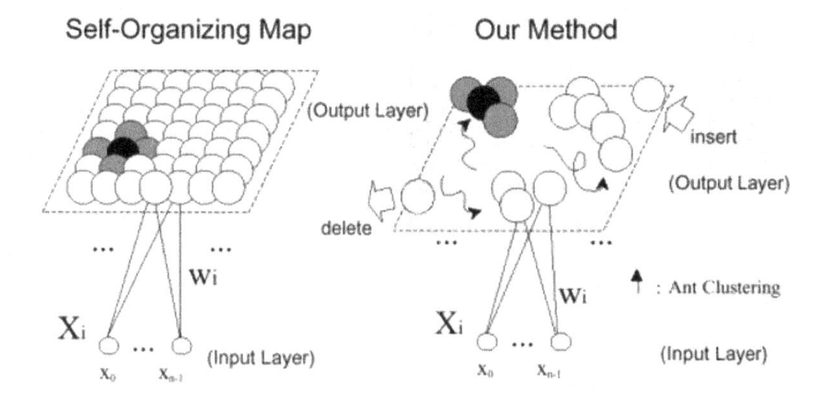

Fig. 1. Kohonen's Self-Organizing Map and Our Method

3 Experiments

We adopted the four-dimensional artificial data as the training set. Three clusters are distributed on the four-dimensional space where the gravity of clusters is (-2,-2,-2,-2), (0,0,0,0), (2,2,2,2) respectively. The dispersion σ is 1.2 as a common property of three clusters by Gaussian distribution. First, only one output neuron is created by the DTRN algorithm and located on the two-dimensional grid space randomly. Then, new data are given from training set and the DTRN process is executed until stagnation. In our experiments, the output neurons are growing up to the limit pre-defined number 400. The output neurons are dispersed randomly on the two-dimensional grid field.

In figure 2, output neurons are located on the two-dimensional grid by using our method (right) and our result is similar to SOM in which the grid size is 40*40 (left) except the boundary representaions. It is able to understand three clusters intuitively according to two-dimensional map. We carried out some experiments by changing random number, but the results are similar to figure 2 (right) mostly.

Moreover, we compared the algorithmic time between the L-F algorithm and our method for 400 data set created by two-dimensional Gaussian distribution according to the reference [1]. In the L-F algorithm, it is necessary to spend 500,000 cycles until convergence, but each cycle is one iteration of the algorithm (namely, 400 cycles mean that 400 output neurons are moving from current location to new location just once). By using our method, it is obvious that output neurons are aggregated themselves into same clusters for only 4700 cycles (each neuron is moving to new location for only about 12 times). Concerning real time, we used the PC/AT compatible computer with Celeron 500MHz CPU and our simulation is executed on the Java Virtual Machine. It spends only one or two seconds until convergence by using our method, meanwhile, about 10 seconds when SOM used for this example. Hereafter, we will investigate the computational cost for many cases.

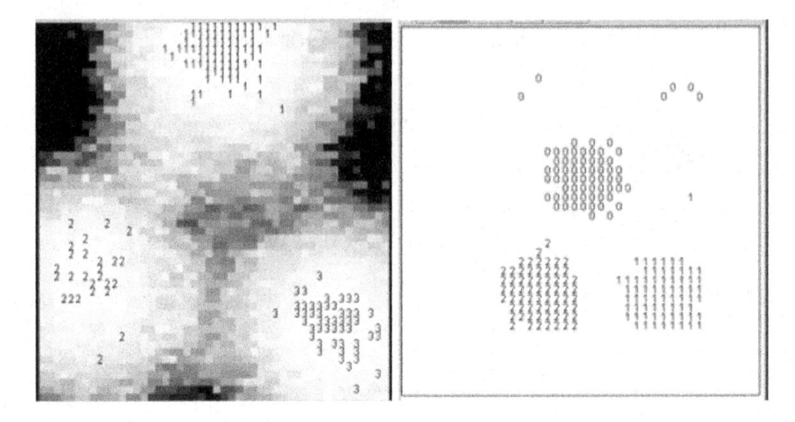

Fig. 2. Comparison of the feature map between the SOM (Left) and our method (right).

4 Discussion

We modified the ant clustering in order to realize data visualization and applied our method to DTRN. In our method, the output neurons are allocated on the two-dimensional grid and aggregate automatically according to the dissimilarity among them. Our algorithm can also apply to the other self-organizing networks. However, our method should define the grid size of two-dimensional map previously, thus, output neurons cannot grow over the grid size. But in our algorithm, it is not necessary to determine the number of output neurons previously though the output neurons are growing within the limit of the grid size. In the future works, we will investigate the property of feature map for more complicated data (e.g. hundred-dimensional) or the massive and non-stationary real data. It is also necessary to compare the SOM with our method in such cases.

References

1. Lumer,E., et al., "Diversity and Adaptation in Populations of Clustering Ants", *From Animal to Animats 3: Proceedings of 3rd International Conference on Simulations of Adaptive Behavior*, The MIT Press (1993)
2. Si, J., et al., "Dynamic Topology Representing Networks", *Neural Networks*, 13 (2000)
3. Kohonen, T., "Self-Organizing Maps", Springer-Verlag (1995)
4. Fritzke, B., "Growing Cell Structure - A Self-Organizing Network for Unsupervised and Supervised Learning", *Neural Networks*, 7 (1995)
5. Fritzke, B., "Growing self-organizing networks - why?", *Proceedings of European Symposium on Artificial Neural Networks : ESANN'96* (1996)
6. Proctor, G., et al., "Information Flocking : Data Visualisation in Virtual Worlds Using Emergent Behaviours", *Virtual Worlds: First International Conference*, Springer-Verlag (1998)

Insect Inspired Visual Control of Translatory Flight

Titus R. Neumann and Heinrich H. Bülthoff

Max Planck Institute for Biological Cybernetics,
Spemannstraße 38, 72076 Tübingen, Germany
titus.neumann@tuebingen.mpg.de
http://www.kyb.tuebingen.mpg.de

Abstract. Flying insects use highly efficient visual strategies to control their self-motion in three-dimensional space. We present a biologically inspired, minimalistic model for visual flight control in an autonomous agent. Large, specialized receptive fields exploit the distribution of local intensities and local motion in an omnidirectional field of view, extracting the information required for attitude control, course stabilization, obstacle avoidance, and altitude control. In open-loop simulations, recordings from each control mechanism robustly indicate the sign of attitude angles, self-rotation, obstacle direction and altitude deviation, respectively. Closed-loop experiments show that these signals are sufficient for three-dimensional flight stabilization with six degrees of freedom.

1 Introduction

Experimental results from insect biology suggest that flying insects use a variety of highly efficient visual strategies for flight control and navigation (e.g. [3],[10]). Considering the extremely small size – less than one cubic millimeter in many insects – as well as the low weight and energy consumption of insect brains, they outperform any existing technical system. It is assumed that the highly specialized, parallel feed-forward information processing in the insect visual system is essential for the speed and robustness of these behaviors. Modeling these strategies on artificial agents can improve performance compared to traditional approaches while reducing the computational effort.

Previous studies of biologically motivated visual control of self-motion and obstacle avoidance in artificial systems were limited to motion in a horizontal or vertical plane with one or two degrees of freedom. Mura and Francheschini (1994) simulated vertical obstacle avoidance and altitude control behavior assuming pure forward motion in the vertical plane with fixed attitude angles [7]. Huber and Bülthoff (1997) demonstrated the simulated evolution of two-dimensional obstacle avoidance and tracking behavior in an artificial agent inspired by the visual system of the fly [5]. Srinivasan et al. (1999) applied several principles of insect vision such as rangefinding by "peering", centering behavior, obstacle avoidance, and visual odometry to robot navigation on the ground plane [9].

However, motion in three-dimensional space has six degrees of freedom which cannot be controlled independently from each other due to the anisotropy of the environment determined by gravity [4]. Body rotations about the vertical axis and altitude changes

J. Kelemen and P. Sosík (Eds.): ECAL 2001, LNAI 2159, pp. 627–636, 2001.

leave the lift force aligned with the direction of gravity (Fig. 1 d), whereas deviations from neutral attitude (i.e. roll and pitch angles not equal to zero) cause lateral motion and erroneous alitude estimates (Fig. 1 e). Thus, flying in terrestrial environments requires attitude control.

We present a biologically inspired, minimalistic model for visual flight control that incorporates attitude control, course stabilization, obstacle avoidance, and altitude control. After a description of the model in the following section, results from open-loop and closed-loop simulations are shown and discussed.

2 Simulation Model

2.1 Agent and Environment

The flight control model is experimentally evaluated using a simulated autonomous agent flying through a three-dimensional virtual environment with a daylight sky model, a textured, uneven surface, and textured obstacles (Fig. 1 b,c). The omnidirectional visual input of the agent (Fig. 1 a) is determined by ray casting.

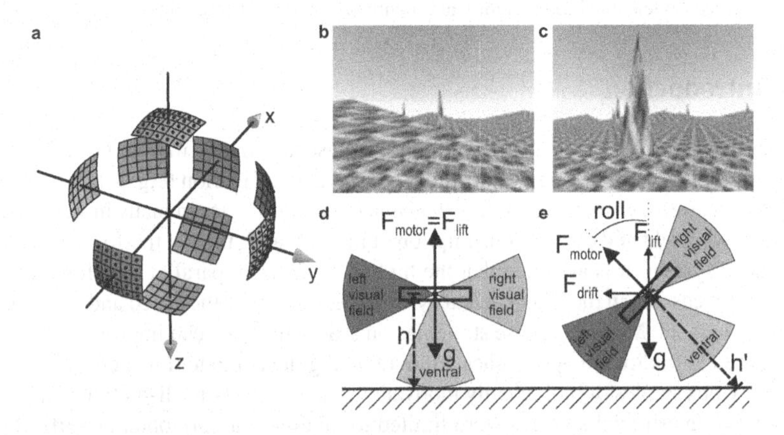

Fig. 1. Omnidirectional receptor configuration (*square centers*) in agent body coordinate system (*a*). Virtual environment: Uneven surface (*b*) with obstacles (*c*). Misalignment of body and world verticals leads to lateral drift and erroneous distance and altitude measurements (*d, e*)

The instantaneous velocity of the simulated agent is set proportional to the force generated by the motor system, ignoring its mass and inertia. This approximation is valid for small flying insects such as *Drosophila*, since they experience strong drag from air viscosity [8], entirely compensating for the propulsion force after a short initial acceleration. Therefore, in this simulation the vertical velocity of the agent is set proportional to the difference of lift force and gravity.

2.2 Visual Receptors and Local Processing

The agent uses 225 receptors to determine the local intensities I_i. The omnidirectional distribution of local viewing directions $\mathbf{d}_i$ is shown in Fig. 1 a. To reduce spatial aliasing, nine samples surrounding the receptor main axis are averaged for each receptor using Gaussian weighting with a half width of $4.0°$ and an inter-receptor angle of $6.0°$. For each pair of adjacent viewing directions $\mathbf{d}_1$ and $\mathbf{d}_2$ with local intensities I_1 and I_2, the local optic flow Φ is estimated using elementary motion detectors (EMD) of the Reichardt correlation type [1]:

$$\Phi(I_1, I_2) = L_{11}(H_1(I_1)) \cdot L_{22}(H_2(I_2)) - L_{12}(H_1(I_1)) \cdot L_{21}(H_2(I_2)). \tag{1}$$

L_{11}, L_{12}, L_{21} and L_{22} are standard discrete time IIR temporal lowpass filters

$$L(I_t) = \left(1 - \frac{1}{\tau}\right) L(I_{t-1}) + \frac{1}{\tau} I_t \tag{2}$$

with time constants τ_{11}=τ_{21}=1.25 and τ_{12}=τ_{22}=5.0. H_1 and H_2 are temporal highpass filters containing a lowpass L_H with time constant τ_H=100.0

$$H(I_t) = I_t - L_H(I_t). \tag{3}$$

Information required to control specific behaviors is extracted from the local intensities I_i and the local motion Φ_j by specialized receptive fields.

2.3 Receptive Fields for Local Intensities

During daylight, most natural open environments exhibit an intensity gradient which is (a) perpendicular to the local average surface, (b) aligned with the direction of gravity, and (c) invariant with the observer's position on the surface. Many animals use this anisotropy of open environments to align their vertical body axis with the direction of gravity, particularly if they do not have direct contact to the surface. In flying insects and fish this mechanism is known as the Dorsal Light Response (DLR), since these animals orient their back towards the region of maximum brightness in their visual environment [4].

Fig. 2 a shows the attitude sensor used in this simulation. The vertical intensity gradient of the environment is averaged separately by opposite hemispherical receptive fields. The response of each receptive field is maximal when the corresponding hemisphere is oriented towards the region of maximum brightness. Thus, the roll angle can be estimated from the difference of these signals by

$$s_{\text{roll}} = \sum_{\mathbf{d}_i \cdot \mathbf{e}_y \neq 0} \frac{\mathbf{d}_i \cdot \mathbf{e}_y}{|\mathbf{d}_i \cdot \mathbf{e}_y|} I_i, \tag{4}$$

where $\mathbf{e}_y = (0, 1, 0)$ is the y axis of the agent's body coordinate system, and I_i is the local intensity in viewing direction $\mathbf{d}_i$. The pitch angle estimate s_{pitch} is computed likewise, replacing $\mathbf{e}_y$ by $\mathbf{e}_x = (1, 0, 0)$.

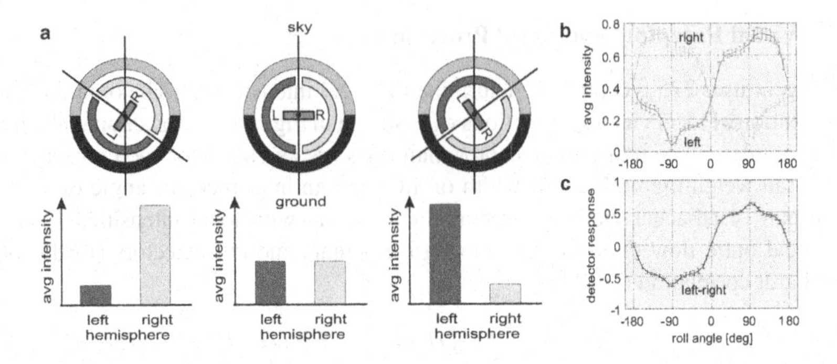

Fig. 2. Dorsal Light Response (DLR). Visual attitude control using a vertical intensity gradient of the environment (*a, outer ring*), comparing the average intensities of two opposite hemispheres (*a, inner ring*). Open-loop responses as functions of roll angle, averaged over 100 randomly selected positions and heading angles: Signals from left and right receptive fields (*b*), difference of left and right signals (*c*)

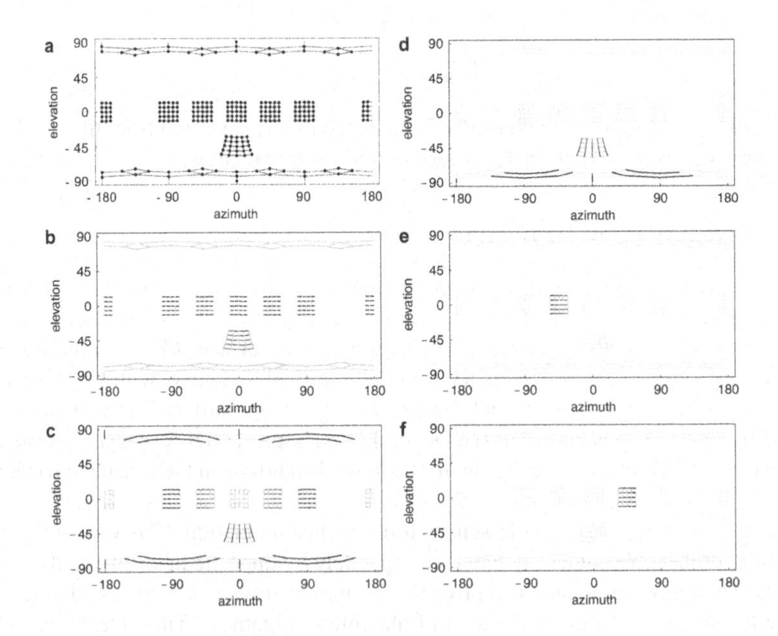

Fig. 3. Distribution of light receptors (*black dots*) and elementary motion detectors (*black lines between dots*) in a spherical field of view (*a*). Receptive fields evaluating local motion: Rotatory self-motion about vertical body axis (*b*), translatory self-motion about longitudinal body axis (*c*), frontoventral translatory flow used for altitude control (*d*), frontolateral left (*e*) and right (*f*) translatory flow used for obstacle avoidance

2.4 Receptive Fields for Local Motion

In a stationary environment, an observer translating with velocity $\mathbf{T}$ while rotating with velocity $\mathbf{R}$ about the origin of the body coordinate system experiences a local optic flow $\mathbf{p}_i$ in viewing direction $\mathbf{d}_i$:

$$\mathbf{p}_i = -\frac{(\mathbf{T} - (\mathbf{T} \cdot \mathbf{d}_i)\,\mathbf{d}_i)}{D_i} - \mathbf{R} \times \mathbf{d}_i. \tag{5}$$

D_i is the distance to the object seen in the local viewing direction [6].

Projecting the local optic flow vectors $\mathbf{p}_i$ expected for a particular self-motion onto the preferred EMD directions $\mathbf{u}_i$ yields a receptive field s (matched filter) responding maximally when this particular flow field occurs as output Φ_i of the local EMDs [2]:

$$s = \sum_i \Phi_i \delta(\mathbf{d}_i)(\mathbf{p}_i \cdot \mathbf{u}_i). \tag{6}$$

The optional window function $\delta(\mathbf{d}_i)$ restricts the receptive field to a specific region of the input image.

Fig. 3 b shows a receptive field for rotations about the vertical axis (s_{vYaw}). It is used for course stabilization against unintended body rotations, simulating the Optomotor Response (OMR) of flies [3]. Assuming pure translatory motion with a constant velocity, the local optic flow is proportional to $1/D_i$ (cf. Eqn.5 and Fig. 6 a-c). Thus, relative distances to objects in the environment can be estimated using receptive fields for translatory flow. The obstacle avoidance (OA) mechanism compares the frontolateral left (s_{vTxLeft}) and right (s_{vTxRight}) optic flow, turning the agent away from close objects. Altitude is controlled by frontoventral translatory flow ($s_{\mathrm{vTxVentral}}$). The corresponding receptive fields are shown in Fig. 3 e,f and d, respectively.

2.5 Flight Control Loop

The motor system of the simulated agent generates roll, pitch and yaw torque as well as a lift force along the vertical body axis. At each simulation step, the motor activation vector

$$\mathbf{m} = (m_{\mathrm{roll}}, m_{\mathrm{pitch}}, m_{\mathrm{yaw}}, m_{\mathrm{lift}}) \tag{7}$$

is updated by the vector of current sensor signals

$$\mathbf{s} = (s_{\mathrm{roll}}, s_{\mathrm{pitch}}, s_{\mathrm{vYaw}}, s_{\mathrm{vTxVentral}}, s_{\mathrm{vTxLeft}}, s_{\mathrm{vTxRight}}) \tag{8}$$

using the connection weight matrix $\mathbf{W}_{\mathrm{sm}}$:

$$\mathbf{m} = \mathbf{s}\mathbf{W}_{\mathrm{sm}} = \mathbf{s} \begin{pmatrix} -w_{\mathrm{roll}} & 0 & 0 & 0 \\ 0 & -w_{\mathrm{pitch}} & 0 & 0 \\ 0 & 0 & -w_{\mathrm{OMR}} & 0 \\ 0 & 0 & 0 & w_{\mathrm{lift}} \\ 0 & 0 & w_{\mathrm{OA}} & 0 \\ 0 & 0 & -w_{\mathrm{OA}} & 0 \end{pmatrix}. \tag{9}$$

3 Results and Discussion

3.1 Attitude Control

Visual attitude control balances the average intensities of opposite hemispheres. Fig. 2 b shows open-loop signals of the left and right receptive fields. The response of each receptive field is maximal when the hemisphere is oriented towards the region of maximum brightness. The difference between the two signals reliably indicates the sign of the current attitude angle (Fig. 2 c).

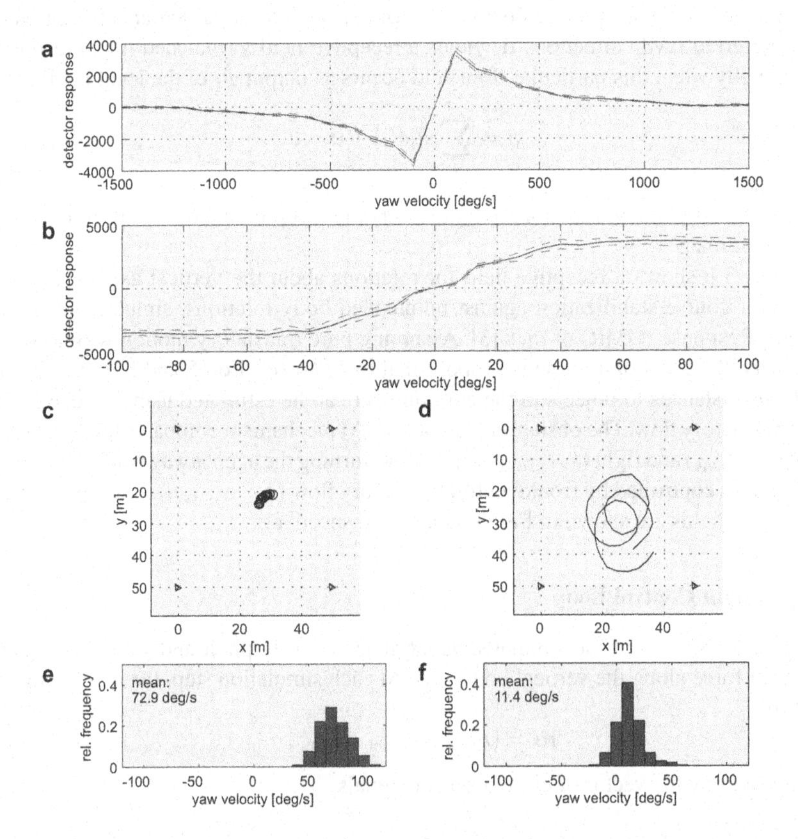

Fig. 4. Optomotor Response (OMR). Detection of rotatory self-motion for course stabilization. Open-loop tuning curve of corresponding receptive field (cf. Fig. 3 b), indicating the correct direction of motion over a large range of velocities (*a*). Small linear range for true velocity measurement (*b*). Closed-loop course stabilization, reducing an asymmetric rotational offset velocity: OMR disabled (*c, e*), OMR enabled (*d, f*)

The described visual attitude sensor uses a global intensity gradient that usually exists in open environments. Thus, it detects the average surface normal which needs to be approximately aligned with the direction of gravity. Additional mechanisms for attitude stabilization are required if the entire visible surface is slanted, if the agent is very close to a large object, or if the intensity gradient does not exist or is reversed, e.g. in closed rooms or over a reflecting snow or water surface.

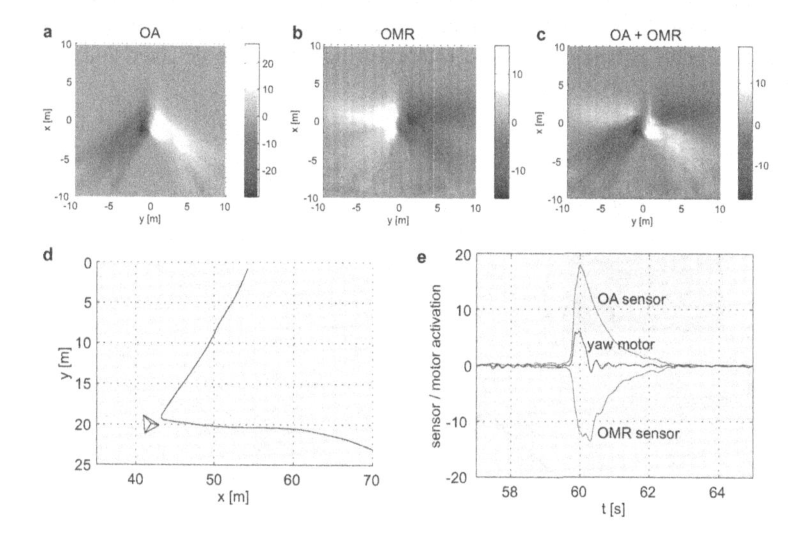

Fig. 5. Obstacle avoidance (OA). Open-loop signals from an agent passively moved over the ground plane with a velocity of 2 m/s on straight trajectories in the x-axis direction, passing an obstacle located at (0,0) with different lateral distances. Difference of left and right frontolateral translatory flow (*a*, cf. Fig. 3 e,f), apparent rotatory flow (*b*, cf. Fig. 3 b), and sum of OA and OMR signals (*c*). In a closed control loop, the sum signal modulates the yaw rotation of the agent, positive values (*bright*) inducing right turns, negative values (*dark*) left turns. Closed-loop saccadic obstacle avoidance maneuver (*d*), activation of corresponding receptive fields and motor signal (*e*)

3.2 Course Stabilization

The open-loop response of the receptive field for yaw rotation (Fig. 4 a) is a typical correlation-type EMD tuning curve. Beyond a relatively small linear range (Fig. 4 b), the signal decreases with increasing velocity, still indicating the correct direction of motion. This signal is sufficient for course stabilization in a closed-loop experiment shown in Fig. 4 c-f. An agent moving over the ground plane is impaired by a unilateral motor malfunction causing an asymmetrical offset of the rotational velocity about the yaw axis. In the absence of any mechanism for course stabilization the rotation cannot be inhibited (Fig. 4 c,e). Enabling the OMR reduces the rotation by a factor of 7 (Fig. 4 d,f). Higher gain factors would improve course stabilization, but deteriorate intended rotations, e.g. for obstacle avoidance.

3.3 Obstacle Avoidance

In a closed control loop, the obstacle avoidance (OA) and course stabilization (OMR) mechanisms (open-loop responses shown in Fig. 5 a-c) generate saccadic obstacle avoidance maneuvers (Fig. 5 d). When approaching an obstacle, the OMR suppresses rotatory motion until the translatory flow, which increases with proximity to the object, exceeds the OMR signal. This initiates a fast rotation which is in turn detected and suppressed by the OMR (Fig. 5 e).

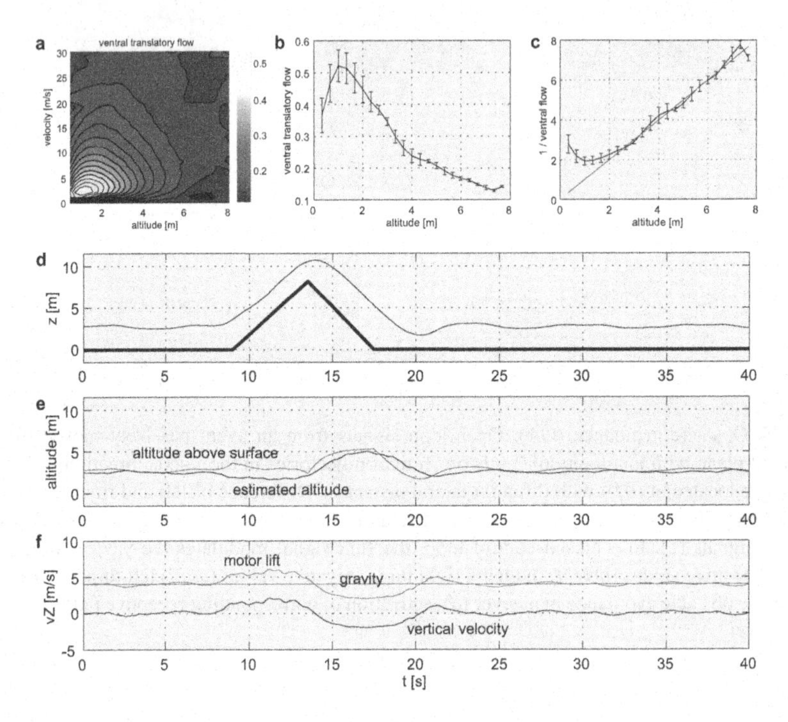

Fig. 6. Altitude control (AC). Open-loop response of receptive field for frontoventral translatory flow (cf. Fig. 3 d) as a function of altitude and velocity (*a*). Receptive field response (*b*) and reciprocal signal (*c*) for a velocity of 2 m/s, showing that for a fixed velocity and within a certain range, true altitude estimation is possible. Time course of closed-loop altitude control: Vertical component of flight trajectory (*d, thin line*) and surface elevation (*d, thick line*), estimated (*e, black line*) and true (*e, gray line*) altitude above surface, motor activation (*f, gray line*) and vertical velocity (*f, black line*) set proportional to the measured ventral translatory flow

3.4 Altitude Control

The average translatory optic flow in the frontoventral visual field is used for altitude control. Fig. 6 a shows the flow signal as a function of translatory velocity and altitude above ground. The signal has a maximum at $v_x = 2$ m/s and decreases slowly with increasing velocity, showing a typical correlation-type EMD tuning curve. Since the signal decreases with increasing altitude h for $h \geq 1$ m (Fig. 6 b), it can be used to estimate altitude (Fig. 6 c) by computing the reciprocal value. For altitude control in a closed control loop, this division is not required since the signal can be used immediately to modulate the lift force (Fig. 6 f).

The described strategy for altitude control assumes translatory motion of the agent. Therefore, it cannot be used for hovering. Simultaneous rotations can corrupt the ventral translatory flow, impairing the ground distance estimation. Additional mechanisms such as the optomotor response can be used to inhibit rotatory motion. If rotations are inevitable, e.g. for attitude correction or obstacle avoidance, the duration can be minimized by fast, saccadic turns. During these saccades the flow signal can be suppressed, or a rotational velocity beyond the sensitivity of the detectors for translatory motion can be chosen.

3.5 Flight Control

Fig. 7 shows closed-loop three-dimensional trajectories of a simulated agent flying autonomously through a virtual open environment containing pyramid-shaped obstacles. The agent maintains its altitude during obstacle avoidance maneuvers (b) and follows the terrain elevation (d).

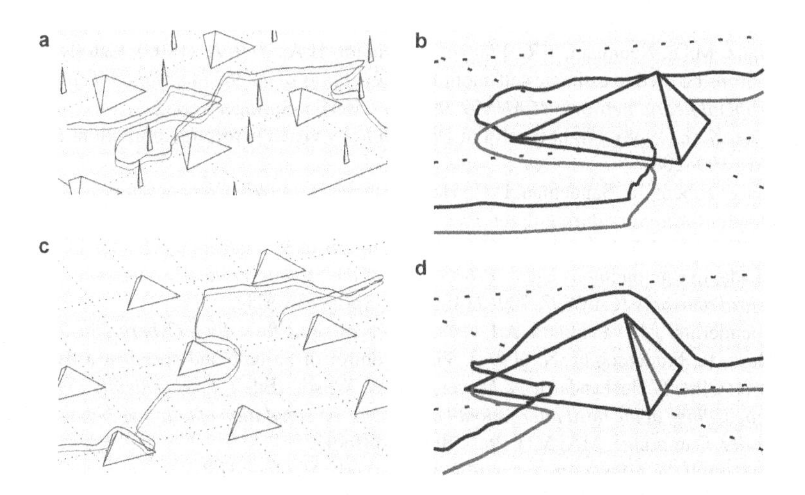

Fig. 7. Closed-loop three-dimensional flight in virtual environments with all six degrees of freedom. Horizontal (*b*) and vertical (*d*) obstacle avoidance. Trajectories (*black*), projection of trajectories on surface (*gray*), surface texture is not shown

4 Conclusion

Simulation results show that simple, biologically inspired visual strategies can establish three-dimensional flight behavior in an artificial autonomous agent. We conclude that in a closed control loop, qualitative signals can be sufficient for flight stabilization. Furthermore, the control of apparently complex behaviors such as 3D flight can be facilitated by functional decomposition into elementarty tasks.

The presented mechanisms are purely reactive and do not require working memory or trajectory planning. They are based on a massive parallel feed-forward connectivity with few sequential processing steps. The connection and weighting schemes are highly adapted to specific tasks and do not change during computation, facilitating hardware implementations such as aVLSI (analog Very Large Scale Integration). Simple, robust control algorithms will be crucial for autonomous vehicle guidance and robotics, especially in applications with strong constraints in size, weight, and energy consumption, such as aerospace and nano-robotics.

Acknowledgments. This work was supported by the Flughafen Frankfurt Main Stiftung and the Max Planck Gesellschaft.

References

1. Egelhaaf, M., Borst, A. (1993). Movement detection in arthropods. In F.A. Miles & J. Wallman (Eds.), *Visual motion and its role in the stabilization of gaze*, 53–77. Amsterdam: Elsevier.
2. Franz, M.O., Neumann, T.R., Plagge, M., Mallot, H.A., Zell, A. (1999). Can fly tangential neurons be used to estimate self-motion? *Proceedings of the 9th international conference on artificial neural networks ICANN99*, 994–999. Berlin: Springer Verlag.
3. Götz, K.G. (1968). Flight control in *Drosophila* by visual perception of motion. *Kybernetik*, **4**(6), 199–208.
4. Hengstenberg, R., Sandeman, D.C., Hengstenberg, B. (1986). Compensatory head roll in the blowfly *Calliphora* during flight. *Proc. R. Soc. Lond.*, B **227**, 455–482.
5. Huber, S.A., Bülthoff, H.H. (1997). Modeling obstacle avoidance behavior of flies using an adaptive autonomous agent. *Proceedings of the 7th international conference on artificial neural networks ICANN97*, 709–714. Berlin: Springer Verlag.
6. Koenderink, J.J., van Doorn, A.J. (1987). Facts on optic flow. *Biol. Cybern.*, **56**, 247–254.
7. Mura, F., Franceschini, N. (1994). Visual control of altitude and speed in a flying agent. In D. Cliff, P. Husbands, J.-A. Meyer, & S.W. Wilson (Eds.), *From Animals to Animats 3: Proceedings of the third international conference on simulation of adaptive behavior SAB94*, 91–99. Cambridge, MA: MIT Press/Bradford Books.
8. Nachtigall, W. (1968). Insects in Flight. New York: McGraw-Hill.
9. Srinivasan, M.V., Chahl, J.S., Weber, K., Venkatesh, S., Nagle, M.G., Zhang, S.W. (1999). Robot navigation inspired by principles of insect vision. *Robotics and Autonomous Systems*, **26**, 203–216.
10. Srinivasan, M.V., Zhang, S.W., Chahl, J.S., Barth, E., Venkatesh, S. (2000). How honeybees make grazing landings on flat surfaces. *Biol. Cybern.*, **83**, 171–183.

The Importance of Rapid Cultural Convergence in the Evolution of Learned Symbolic Communication

Kenny Smith

Language Evolution and Computation Research Unit
Department of Theoretical and Applied Linguistics, The University of Edinburgh
Adam Ferguson Building, 40 George Square, Edinburgh EH8 9LL
kenny@ling.ed.ac.uk

Abstract. Oliphant [5,6] contends that language is the only naturally-occurring, learned symbolic communication system, because only humans can accurately observe meaning during the cultural transmission of communication. This paper outlines several objections to Oliphant's argument. In particular, it is argued that the learning biases necessary to support learned symbolic communication may not be common and that the speed of cultural convergence during cultural evolution of communication may be a key factor in the evolution of such learning biases.

1 Introduction

Language is unique among the communication systems of the natural world - it is culturally transmitted, the relationship between basic lexical tokens and their meanings is arbitrary and those basic lexical tokens are combined to form structured forms which are used to communicate complex structured meanings. How did language come to be as it is and why is it unique?

Much recent work in the field has focused on the evolution of syntactic communication. Explanations of the human capacity for syntax have placed emphasis on two contrasting adaptive processes:

Genetic adaptation of the genetically-encoded human language acquisition device to support syntactic communication due to fitness advantages offered by syntactic communication (e.g. [4,7]). Explanations of this kind appeal to a unique set of selection pressures favouring the evolution of syntax in humans to explain the uniqueness of language.

Cultural adaptation of language in favour of compositionality, due to cultural selection resulting from language learner biases during cultural transmission of communication (e.g. [1,3]). Such models are not primarily concerned with the origin of the language learner's biases but appeal to a uniquely human preexisting mental capacity to explain the uniqueness of language.

Recent work by Oliphant [5,6], building on pioneering work by Hurford [2], focuses on the more basic issue of the emergence of arbitrary and conventionalised word meaning. Oliphant works within the cultural adaptation framework and makes two claims. Firstly, human language is the only learned symbolic communication system. Secondly,

J. Kelemen and P. Sosík (Eds.): ECAL 2001, LNAI 2159, pp. 637–640, 2001.

language is unique in this respect due to the human capacity to read the communicative intentions of other language users. Once this capacity is established optimal, learned symbolic communication reliably follows through cultural evolution of communication systems under a commonly-occurring learning bias.

This paper is primarily concerned with integrating the genetic and cultural adaptation styles of explanation and applying this integrated approach to an investigation of Oliphant's second claim. The integrated model suggests that the type of learning bias identified by Oliphant may not in fact be commonly occurring and the speed of cultural convergence may be a critical factor in the evolution of such a learning bias.

2 The Model

The model is an extension of the model outlined in [6] and is described in full in [9]. Briefly, a population of communicative agents is simulated over time. Agents in the population breed according to communicative accuracy to produce new agents who inherit the genetically-encoded learning rule of their parents.[1] The new agents then make N observations of the communication systems[2] of members of the population and acquire their own communication system based on these observations and their genetically-encoded learning rule. There are therefore two interacting selection pressures at work in the model - natural selection in favour of genes encoding learning rules which result in acquisition of communication systems which allow successful communication, and cultural selection in favour of communication systems which conform to the biases of the learning rules present in the population.

3 Objections to Oliphant's Conclusions

This model suggests three objections to Oliphant's conclusion that the capacity to accurately observe meaning, unique to humans, combined with a commonly-occurring learning rule (a variant of Hebbian learning) results in the evolution of optimal learned symbolic communication.

Firstly, as discussed in [8], the types of learning rule which result in cultural selection for optimal communication systems (which will be termed *constructor* rules) have some very specific biases regarding allowable relationships between meaning-signal pairs and may not be widespread in the natural world. Secondly, as discussed in Section 4, constructor rules may be unlikely to evolve in a population even under apparently ideal circumstances. Finally, work in progress suggests that the combination of constructor rules and a diminished capacity to observe meaning may still result in the emergence of near-optimal communication, and may in fact create selection pressures to improve the

[1] Agents are bidirectional associative networks mapping from unit vector meanings to unit vector signals. The learning rules for these agents specify the conditions under which connection strengths are increased, decreased, or left alone depending on the two inputs to that connection. This leads to 81 possible learning rules.

[2] Agents observe meaning-signal pairs - the ability to observe meaning accurately is assumed, as in [6].

capacity to observe meaning - communication may precede and lead to the capacity to observe meaning accurately.

4 N and Speed of Cultural Convergence

If, as suggested in [8], constructor rules are not commonly occurring, a slightly modified version of Oliphant's argument seems appealing - given a preexisting capacity to observe meaning such learning rules will emerge reliably under natural selection for communicative success due to the fitness payoff they confer. Results generated by the model outlined in Section 2 suggests that this hypothesis is incorrect.

Table 1 indicates the proportion of simulated populations converging on optimal or near-optimal communication systems after a fixed number of generations for various values of N. It is clear that optimal communication systems do not reliably emerge under these conditions, although they are more likely to emerge given more learning by immature agents (larger N).

Table 1. The number of successful runs (out of 100) for various values of N

N	1	3	5	10	15	20	25	30
Successes	0	4	15	39	38	41	56	57

The unreliable emergence of optimal communication is due to the delay between the emergence of genotypes encoding constructor rules and any fitness advantage to agents with such genotypes. Agents using constructor rules need time to converge on an optimal communication system - cultural selection over repeated cultural transmission gradually moves the communication systems of agents using constructor rules into increasingly optimal overlapping areas of communication system space until all such agents have converged on an optimal system.

In the early stages of this construction process individuals using constructor rules have little fitness advantage over other individuals. As a consequence the genetic transmission process will be essentially random - the population will undergo genetic drift. In successful runs genetic drift preserves constructors, by chance, in sufficient numbers for sufficient time to allow the construction process to get well under way. Constructors then show increased communicative accuracy which leads to steady selection for constructor genes, constructor numbers in the population increase and the population converges on an optimal communication system. In unsuccessful runs genetic drift never provides constructor rules in sufficient numbers for the construction process to take off.

Increasing N increases the speed at which cultural convergence occurs between constructor agents, reducing the dependence on benevolent genetic drift and increasing the likelihood of populations arriving at optimal communication systems. While allowing runs to continue for longer or changing the mutation rate used in the model might make optimal communication more likely to emerge for lower N, such tinkering would obscure the essentially contingent nature of the emergence of constructor rules and the importance of rapid cultural convergence in evolving such learning biases.

5 Conclusions

Oliphant [6] argues that optimal, learned symbolic communication will trivially emerge given a capacity to observe meaning during cultural transmission and a commonly-occurring learning bias. This paper raises three objections to Oliphant's proposal. Firstly, the learning bias necessary to construct an optimal, learned communication system may not in fact be commonly-occurring. Secondly, even if other species were capable of observing meaning, the correct learning bias might be unlikely to evolve due to a delay between the emergence of the bias and a fitness payoff to individuals possessing it. The rapid cultural convergence resulting from large N is shown to reduce this delay and increase the likelihood of appropriate learning biases evolving in the population. Thirdly, near-optimal communication systems may emerge and be maintained in populations of agents incapable of accurately observing meaning.

These results emphasise the importance of interactions between genetic and cultural adaptation in models which do not discount one adaptive process as a starting assumption. Specifically, even if a particular learning bias results in the emergence of optimal communication systems through processes of cultural selection we cannot assume that natural selection will reliably find that bias or that the process of natural selection will be unaffected by the process of cultural selection.

References

1. J. Batali. The negotiation and acquisition of recursive grammars as a result of competition among exemplars. In E. Briscoe, editor, *Linguistic Evolution through Language Acquisition: Formal and Computational Models*. Cambridge University Press, Cambridge, 2001.
2. James R. Hurford. Biological evolution of the saussurean sign as a component of the language acquisition device. *Lingua*, 77:187–222, 1989.
3. S. Kirby. Syntax without natural selection: how compositionality emerges from vocabulary in a population of learners. In Chris Knight, Michael Studdert-Kennedy, and James R. Hurford, editors, *The Evolutionary Emergence of Language: Social Function and the Origins of Linguistic Form*, pages 303–323. Cambridge University Press, Cambridge, 2000.
4. M. A. Nowak, J. B. Plotkin, and V. A. A. Jansen. The evolution of syntactic communication. *Nature*, 404:495–498, 2000.
5. M. Oliphant. Rethinking the language bottleneck: Why don't animals learn to communicate? presented at 2nd International Conference on the Evolution of Language, 1998.
6. M. Oliphant. The learning barrier: Moving from innate to learned systems of communication. *Adaptive Behavior*, 7(3/4):371–384, 1999.
7. S. Pinker and P. Bloom. Natural language and natural selection. *Behavioral and Brain Sciences*, 13:707–784, 1990.
8. K. Smith. Key learning biases for the cultural evolution of communication. Presented at the Human Behavior and Evolution Society Conference, 2001.
9. K. Smith. Modelling the evolution of learning rules for associative networks. Technical report, Language Evolution and Computation Research Unit, 2001.

Emergent Syntax: The Unremitting Value of Computational Modeling for Understanding the Origins of Complex Language

Willem H. Zuidema

Artificial Intelligence Laboratory
Vrije Universiteit Brussel
Pleinlaan 2, 1050 Brussels, Belgium
jelle@arti.vub.ac.be
http://arti.vub.ac.be/~jelle

Abstract. In this paper we explore the similarities between a mathematical model of language evolution and several A-life simulations. We argue that the mathematical model makes some problematic simplifications, but that a combination with computational models can help to adapt and extend existing language evolution scenario's.

1 Introduction

The debate on the origins of language has been dominated by "verbal" theories, both in scientific publications (see e.g. [4]) and in popular, best-selling books (e.g. [1]). Recently also mathematical models of the evolution of language, especially those of Martin Nowak et al., have received much attention (e.g. [6]). These models are sometimes seen as a validation of the earlier verbal theories. Steven Pinker, e.g., writes in the accompanying news story of [7] that the paper shows *"the evolvability of [one of] the most striking features of language"*, i.e. its compositionality.

Although we appreciate the major contributions in these books and papers, we still observe many shortcomings in the proposed theories. Both the verbal and the mathematical accounts tend to overlook many crucial details. Verbal theories often underestimate the intricacies of the evolutionary dynamics and take "evolution" too much as a general problem solver. The mathematical models often make crucial simplifications that are linguistically poorly motivated. In particular, both types of theories have shown little appreciation for the importance of the "frequency dependency" of language evolution and the role of selforganization there-in.

A-life models, on the contrary, have shed light on both the dynamics of language evolution and the explanatory role of selforganization. However, A-life models are too often studied as relatively isolated cases, and too seldomly systematically compared with each other and with mathematical models (the review papers [8,3] are exceptions, although they unfortunately do not discuss

J. Kelemen and P. Sosík (Eds.): ECAL 2001, LNAI 2159, pp. 641–644, 2001.

mathematical models). In this paper we explore the similarities between a recently published mathematical model [6], our own A-life simulations [9] and the model of Kirby [5]. We believe that such an approach can eventually both avoid the problematic simplifications of mathematical models, and the *ad hoc-ness* of many A-life models. In the conference presentation we will also discuss some shortcomings of "verbal" theories as revealed by A-life models.

2 The Mathematical Model

Nowak et al. use in [6] an elegant formalism that is in line with our view that one should study both the cultural dynamics of language and the evolutionary dynamics that operate on the parameters of the cultural process. We will discuss here only the model for cultural dynamics.

Nowak et al. assume that there is a finite number of states (grammar types) that an individual can be in. Further, they assume that newcomers (infants) learn their grammar from the population, where more successful grammars have a higher probability to be learned and mistakes are made in learning. The system can now be described in terms of the changes in the relative frequencies x_i of each grammar type i in the population:

$$\dot{x_i} = \sum_{j}^{N} x_j f_j Q_{ji} - \phi x_i \quad (1)$$

In this differential equation, f_i is the *relative fitness* (quality) of grammars of type i and equals $f_i = \sum_j x_j F_{ij}$, where F_{ij} is the expected communicative success from an interaction between an individual of type i and an individual of type j. The relative fitness f of a grammar thus depends on the frequencies of all grammar types, hence it is *frequency dependent*. The proper way to choose F depends on the characteristics of *language use* (production and interpretation).

Q_{ij} is the probability that a child learning from a parent of type i, will end up with grammar of type j. The probability that the child ends up with the same grammar, Q_{ii}, is defined as q, the copying fidelity. The proper way to choose Q depends on the characteristics of *language acquisition* (learning and development). (ϕ is the average fitness in the population and equals $\phi = \sum_i x_i f_i$. This term is needed to keep the sum of all fractions at 1).

The main result that Nowak et al. obtain is a "coherence threshold": they show mathematically that there is a minimum value for q to keep coherence in the population. If q is lower than this value, all possible grammar types are equally frequent in the population and the communicative success in minimal. If q is higher than this value, one grammar type is dominant; the communicative success is much higher than before and reaches 100% if $q = 1$. Further, Nowak et al. derive an upper and a lower bound on the number of sample sentences that a child needs to acquire its parents' language with the required fidelity q.

3 A-Life Models

We argue that computational models that we [9] and others [2] have studied fit the general format of equation 1 well, but differ significantly in the particular choices for the representation of language use and language acquisition, i.e. the functions F and Q. In the limited space that is available here we will only shortly mention two examples of interesting, qualitative differences that these choices bring.

First, for sake of simplicity Nowak et al. assume that all grammars are *equally expressive*, and are all *equally similar* to each other. This has the unrealistic consequence that the benefits of interacting with another individual (F) are either maximal or minimal. We studied a computational model [9] were we used context-free grammars to represent the linguistic abilities of agents. This formalism can represent "languages" of many different types and levels of expressiveness. In that study, we did not model learning explicitly, but in stead assumed (as in equation 1) that children end up with a slightly different grammar than their parents.

One of the surprising findings was that once a certain type of language was established in the population, the language kept changing but remained of the same type. The language types formed "self-enforcing regimes", because the language present at time t determines which agents will be successful and reproduce to the next generation, and therefore indirectly determine the language at time $t+1$. We found three such regimes: (i) idiosyncratic, non-syntactic languages, (ii) compositional languages and (iii) recursive languages. In a population where a rich but idiosyncratic language is established, syntax could not emerge. This phenomenon is important for understanding the consequences of the frequency dependency of language evolution, but is excluded in the simplifications of the mathematical model.

Second, Nowak et al. consider two extreme possibilities for the learning algorithm, and claim to have found a lower and a upper bound on the number of training samples that a learning algorithm needs to reach the coherence threshold. However, in their analysis they have not taken into account that the choice of the grammar that a child has to learn is biased by how well previous generations have been able to learn and maintain it.

In a follow-up of the study above, we have implemented a variant of the "iterated learning model" of Kirby [5], in which agents are endowed with a language-acquisition algorithm to learn the context-free grammars. Kirby found that in the process of iterated cultural transmission the language adapts itself to be better learnable by individual agents. Concretely, this means that the language becomes compositional (syntactic) and that agents are more successful in learning it than would be expected a priori. We replicated this finding, and can show that agents in fact need less training samples than Nowak et al. calculate as a lower bound for maintaining a stable language in the population. The reason is that not only do individuals evolve to be better at language-learning, but also do languages evolve to be better learned [1]. Again, this phenomenon is important for our understanding of the origins of language, but excluded in the simplifications of the mathematical model.

4 Conclusions

Research on the evolution of language faces two aspects of language that are particularly important: (i) it is transmitted, at least in part, culturally, and learned by one individual from the other; (ii) it is a group phenomenon, that occurs only between individuals and has no apparent value for an individual in isolation. These aspects make that the fitness of individual is not a function of its language acquisition system alone, but is dependent on the cultural dynamics and the composition of the group it is in as well. This observation brings *restrictions* and *opportunities* for language evolution scenario's that are deemed to be overlooked in both verbal and mathematical theorizing. We conclude that A-life models can help to evaluate the validity of these scenarios and help to adapt them, while at the same time mathematical models can help to compare computational models and to identify common themes between them.

Acknowledgements. Part of the work reported here has been done in close collaboration with Paulien Hogeweg. I thank her and other members of the Theoretical Biology group in Utrecht, the Netherlands, and members of the AI-Laboratory in Brussels, Belgium, for many helpful discussions.

References

1. Terrence Deacon. *Symbolic species, the co-evolution of language and the human brain.* The Penguin Press, 1997.
2. Takashi Hashimoto and Takashi Ikegami. The emergence of a net-grammar in communicating agents. *BioSystems*, 38:1–14, 1996.
3. James R. Hurford. Expression / induction models of language. In Ted Briscoe, editor, *Linguistic Evolution through Language Acquisition: Formal and Computational Models.* Cambridge University Press, 2000.
4. James R. Hurford, Michael Studdert-Kennedy, and Chris Knight, editors. *Approaches to the evolution of language.* Cambridge University Press, 1998.
5. Simon Kirby. Syntax without natural selection: How compositionality emerges from vocabulary in a population of learners. In C. Knight, J. Hurford, and M. Studdert-Kennedy, editors, *The Evolutionary Emergence of Language: Social function and the origins of linguistic form.* Cambridge University Press, 2000.
6. Martin A. Nowak, Natalia Komarova, and Partha Niyogi. Evolution of universal grammar. *Science*, 291:114–118, 2001.
7. Martin A. Nowak, Joshua B. Plotkin, and Vincent A.A. Jansen. The evolution of syntactic communication. *Nature*, 404:495–498, 2000.
8. Luc Steels. The synthetic modeling of language origins. *Evolution of Communication*, 1:1–35, 1997.
9. Willem H. Zuidema and Paulien Hogeweg. Selective advantages of syntax: a computational model study. In *Proceedings of the 22nd Annual Meeting of the Cognitive Science Society*, pages 577–582. Lawrence Erlbaum Associates, 2000.

Amorphous Geometry

Ellie D'Hondt[1] and Theo D'Hondt[2]

[1] Centrum Leo Apostel (CLEA)
[2] Programming Technology Laboratory
Vrije Universiteit Brussel, Pleinlaan 2, 1050 Brussels, Belgium
{eldhondt, tjdhondt}@vub.ac.be

Abstract. Amorphous computing is a recently introduced paradigm that favours geometrical configurations. The physical layout of an amorphous computer is based on a large number of simple processing components that are well-suited for handling spatial structures. It has come to our attention that the discipline of computational geometry could benefit from this approach and it seemed natural to refer to it by the notion of amorphous geometry. Although at this stage our exploration of this concept is fairly modest, we feel that our experiments are sufficiently convincing and merit further study. We are confident that amorphous geometry can deal with various classes of problems while providing a basis for useful applications.

1 Introduction

The setting for this paper is a newly emerging domain called *Amorphous Computing* or AC [1]. It was developed to support mass production of small and possibly faulty computational units with limited power. A random distribution of these particles, locally interconnected, constitutes the basic model of an amorphous computer. AC is the paradigm providing the programming techniques required for this computer. It addresses the appropriate organising principles and methodologies for obtaining predefined global behaviour through local interaction. See [4] for a more detailed overview of AC.

We propose introducing AC into the field of *Computational Geometry* (see for instance [3]). The fact that AC addresses problems that exist in physical space, led us to the conviction that it might be a useful addition to this discipline. It seemed natural to introduce the notion of *amorphous geometry* (AG). Consequently, we can imagine a complete range of applications involving smart surfaces interacting with their environment.

In order to show that AC can deal with various computational geometry problems, more is required than we can present here. We will limit ourselves to a number of modest experiments that illustrate the point we want to make. The significance of our work is dual: first, it provides an insight in how metaphors from computational geometry are viewed amorphously; next, concrete simulations of AG programs provide us with an idea of its power. Both aspects are discussed using the construction of polygons as a case study.

J. Kelemen and P. Sosík (Eds.): ECAL 2001, LNAI 2159, pp. 645–648, 2001.

2 Metaphors and Experiments in Amorphous Geometry

Ultimately, the objective of AC is to develop languages to manipulate a set of programmable computing particles. The first serious attempt to achieve this is the *Growing Point Language* or GPL [2], a language adopting the construction of complex patterns as a model for global behaviour. Therefore, we use GPL to describe metaphors from AG. While the emphasis lies with the actual notion of geometrical metaphors, concretising them with the help of GPL also sheds light on the architecture of an amorphous solution to a particular geometrical problem.

AG has no meaning without a clear interpretation of both points and lines. We will rely on the one-to-one correspondence between points in space and AC particles; a contiguous set of such particles arranged as a linear array play the role of a line. Consider two AC particles as being the endpoints of a line segment. Through local communication only, one endpoint has to initiate a "search" for the other endpoint, without knowing the global direction in which the second endpoint is to be found. Inspired by [2], we will use the biological metaphor of chemical gradients as follows:

```
(define-growing-point (a-to-b-segment)
  (material a-material)
  (tropism (ortho+ b-pheromone))
  (actions (when ((sensing? b-material) (terminate))
                 (default (propagate)))))
```

The principal concept in GPL is the *growing point*, a locus of activity that describes a path through connected elements in the system. At all times, a growing point resides at a single location in the GPL domain called its *active site*, where it may deposit *material* and secrete *pheromones*. A growing point propagates itself by transferring its activity from a particle to one of its neighbours according to its *tropism*. A growing point's tropism is specified in terms of the neighbouring pheromone concentrations; in this case, `ortho+` directs the growing point towards increasing concentrations of `b-pheromone`.

In order to produce the desired line segment, some cooperation is required from the endpoint B. This is indicated by the presence of `b-pheromone` and `b-material` in the code above. Hence, a growing point is installed at B in the following way

```
(define-growing-point (b-point)
  (material b-material)
  (for-each-step (secrete EDGE-LENGTH b-pheromone)))
```

The above code extracts are to be viewed as examples of concrete amorphous representations of the geometrical metaphors 'point' and 'line segment'. While we return to the GPL-specific code in the discussion of our experiments below, it should be clear that it is important that we can successfully incorporate geometrical concepts into AC; the specific code required in order to do so is at this stage of lesser significance.

After extracting metaphors, the next step in developing AG is developing a concrete implementation in order to carry out experiments. As an illustration, we discuss polygon construction within the GPL framework.

The problem of drawing a polygon can be decomposed into the construction of line segments from one vertex to another. For example, in drawing a quadrangle each vertex is equipped with two growing points similar to those defined in the previous section, a <*vertex*>`-to-`<*next-vertex*>`-segment` and a <*vertex*>`-point` growing point. We say "similar" because we want to avoid interference between pheromones: we have to use a unique pheromone and material for each vertex, and define an associated segment-like growing point that grows towards that particular pheromone. The whole process is presented schematically below:

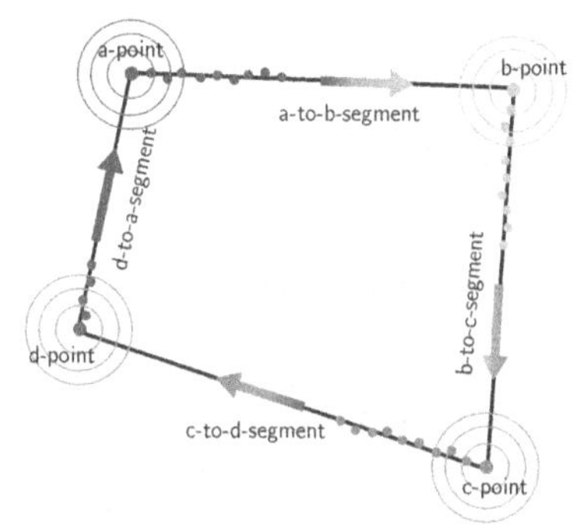

Fig. 1. Drawing a quadrangle.

In this solution for drawing a quadrangle, we are immediately confronted with the inability of GPL to parametrise pheromones. While material parametrisation is already present, we still need a different growing point definition for every vertex to avoid pheromone interference. Since GPL would gain considerably in expressiveness by allowing pheromone parametrisation, we felt it was useful to include this in the GPL framework. Pheromone parameters can occur either in a `secrete`-expression (ensuring that the pheromone parameter is evaluated first) or in a `tropism`-expression. The second case is more complicated since tropisms influence growing point propagation: when tropisms contain pheromone parameters they are no longer part of a growing point's static information, since parameters have to be evaluated dynamically at each active site.

Once these additional GPL-features are implemented, the following code constructs the desired quadrangle through the generic growing points `edge` and `vertex` (defined similarly as in the previous section but *with* pheromone parametrisation):

```
(with-locations (a b c d)
 (at a (start-gp (vertex 'a-pheromone 'a-material))
   (--> (start-gp (edge 'b-pheromone 'b-material))
    (--> (start-gp (edge 'c-pheromone 'c-material))
     (--> (start-gp (edge 'd-pheromone 'd-material))
```

```
          (start-gp (edge 'a-pheromone 'a-material))))))
  (at b (start-gp (vertex 'b-pheromone 'b-material)))
  (at c (start-gp (vertex 'c-pheromone 'c-material)))
  (at d (start-gp (vertex 'd-pheromone 'd-material))))
```

Here `-->` is the `connect`-command, which allows us to connect several growing points. Explicit pheromone and material names are required to be quoted, due to our extensions to GPL. Some simulation results for the construction of polygons are shown below:

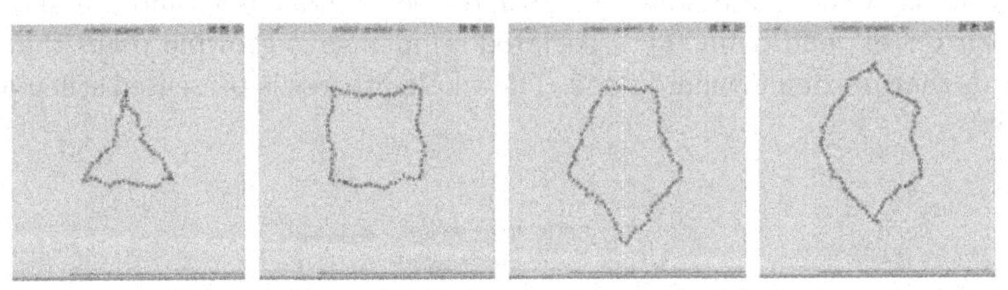

Fig. 2. Several polygons constructed with GPL.

3 Conclusion and Future Work

AC is a fascinating new programming paradigm, which seems an obvious alternative to more conventional computational techniques used to tackle geometrical problems. In this paper, we suggest the notion of AG to explore this idea. We only scratched the surface by considering points, lines and polygons; we showed that they can be simulated by very simple programs, composed in GPL. Although we experienced GPL as too limited even for elementary tasks, the introduction of a particular kind of parametrisation seemed to solve the shortcoming.

Extending GPL into a sufficiently expressive language so as to cover more general problems from computational geometry would seem to be a worthwhile continuation of the work described in this paper. For instance, applying AG to the triangulation of a polygon already delivered some initial results [5]. While still preliminary, it is sufficiently promising to warrant mentioning it as future work.

References

1. Abelson, H. et al.: Amorphous computing. Communications of the ACM, 43(5) (2000)
2. Coore, D.: Botanical Computing: A Developmental Approach to Generating Interconnect Topologies on an Amorphous Computing. PhD thesis, MIT (1999
3. de Berg, M., van Kreveld, M., Overmars, M., Schwarzkopf, O.: Computational Geometry: Algorithms and Applications. Springer, 2nd edition (1998)
4. D'Hondt, E.: Amorphous Computing. Masters thesis, Vrije Universiteit Brussel, `student.vub.ac.be/~eldhondt/PDF directory/thesisAmorphous.pdf` (2000)
5. D'Hondt, E., D'Hondt, T.: Experiments in Amorphous Geometry. Proceedings of the 2001 International Conference on Artificial Intelligence (to appear) (2001)

Artificial Life and Historical Processes

Ezequiel A. Di Paolo

School of Cognitive and Computing Sciences, University of Sussex, UK
ezequiel@cogs.susx.ac.uk

Abstract. Artificial Life is partly aimed at understanding the organisation and complexity of living processes. In this paper the concept of a historical process is discussed with the aim of providing a framework with which to approach diverse phenomena in organismic, ecological, and evolutionary contexts. A historical process is such, not because it is subject to contingencies, nor because it may be explained in historical terms, but because it presents a special relation between its dynamics and changes in its own conditions of realisation. Such processes may lead to durable spontaneous patterns and novelty. It is argued that such patterns can provide powerful explanatory tools and that Artificial Life simulation techniques are well fitted for their exploration.

1 Introduction

To different degrees of explicitness, the central theme of much of the work that currently goes under the rubric of Artificial Life (AL) is the understanding of processes that lead to innovations, transitions, and spontaneous organisations which are difficult to explore using more traditional modelling tools, and which are often associated with biological phenomena. The use in this literature of much worn terms such as 'emergence', 'self-organisation', and 'complexity' bears witness to this aim. And, indeed, evidence supporting the case that AL modelling tools are capable of shedding new light on problems involving the synergies between processes situated at different timescales or 'levels', such as the ecological and evolutionary [1], the behavioural and the social [2], the behavioural and the ecological [3], and others, has not been lacking.

What has been less conspicuous, however, is an attempt to describe such phenomena in a systematic form, equally valid for the different problems areas. Not that similar attempts do not exist, (e.g., [4]). It may be questioned whether this is an useful enterprise. What common theme can be fruitfully sought in the spontaneous formation of social hierarchies, autocatalytic organisations, and wasps' nests? Here, instead of a full justification, we will offer a programmatic bet: Systematization is a key element in the scientific toolkit; it leads to shared knowledge between subdisciplines, to the identification of analogous problem areas and search for analogous solutions – these reasons ought to make the attempt worthwhile, although full systematization might be ultimately an utopia.

This article attempts to describe a central theme of AL research which is a mode of explaining the phenomena of interest that appeals to certain properties

J. Kelemen and P. Sosík (Eds.): ECAL 2001, LNAI 2159, pp. 649–658, 2001.

of the dynamics of the processes involved, namely that these are *historical processes.* The precise meaning of this term will be explored and illustrated by the use of some examples. To this aim, the idea of what constitutes a *constraint* to a process will be examined, as well as how it relates to the dynamics of the process both in operational and explanatory terms. This will permit a specialization of the word 'historical' to processes that are able to introduce some temporal heterogeneity due to the interplay of variations at different timescales. As a corollary, it will be found that any process leading to innovations or transitions (which generate much interest within AL) is, by definition, historical.

Some of the concepts presented here are related to the ideas of scientists who have been influenced by A. N. Whitehead's metaphysics, [4,5,6]. However, the purpose of the article is to make a basic presentation of some central concepts in order to facilitate their subsequent use and not to provide a review and comparative exposition of the philosophical and scientific extent of these ideas.

2 From Homogeneous Time to Historical Time

There are different senses in which the word 'historical' may be applied to a process. For instance, a process may be so called if its unfolding involves a set of contingencies that cannot be predicted until the moment they occur. Such factors could take the form of discrete events (e.g., founder effects or catastrophes in biological evolution) or they could operate with constancy, in which case their effects may become manifested over long periods of time (e.g., random fixation of alleles due to genetic drift).

Another related criterion would consider adequate to apply the name 'historical' to a process if an explanation of how its current state has been attained would be best given in historical terms. Such explanations (see [7, pp. 25 – 26] and [8, pp. 283 – 284]) would account for a state or event in a process in terms of previous *key* states or events. A chain of these events would be understandable if it is possible to understand the connection between one link and the next.

The word 'historical', in the current context, is not intended strictly in any of the above senses. Rather, a historical process would lie roughly at the intersection between the cases just mentioned (i.e., contingent or noisy processes and processes explainable in historical terms) and the set of processes which are sometimes characterized as *self-organising.* Such historical processes are indeed contingent and probably many of them afford historical explanations. However, the key feature to be highlighted is their capability to influence their own constraints and thus to introduce an interplay between dynamics at different timescales which may result in open temporal inhomogeneities. In order to understand this capability the concept of constraint needs to be expanded.

2.1 Constraints

All observable events and processes are underdetermined by the fixed universal laws that are presumably at play in them. The trivial reason for this is that such laws can only be universal because they are disembodied and refer to no concrete system in particular. In order to apply them to the understanding of a specific process a description must be provided of how these laws are constrained by the actual structures and conditions that make up that process.

There are two senses for the word 'constraint'. Consider a physical pendulum. A finite mass is hanged from the ceiling by a piece of string. A description of this system could be offered that would permit the application of universal dynamical laws. Thus, a series of idealisations would allow a description in terms of a zero-dimensional particle hanged from a fixed point by an inelastic string under the exclusive influence of gravity, and so forth. In mechanical terms a constraint describes those relations that place direct limitations to the variation of the variables with which the system is described, (see [9]). For the pendulum, such a constraint is found in the position of the particle which must, at all times, conserve its distance to the point in the ceiling from which it hangs.

In a second, more general sense, a constraint indicates not just these relations but also the set of parameters and other relations that make it possible to embody a universal law into a description of an actual system. If the system remains ideally isolated and such contextual factors remain fixed, it seems that calling these factors 'constraints' would be unnecessary. However, the meaning of the word is recovered when one considers that the system may participate in time-dependent coupling with other systems which, through their effect in such contextual factors, may influence the system's behaviour. Thus, the ceiling may vibrate and the length or the elasticity of the string may change with time – changes that would necessitate a redescription of the system.

It is clear though, that any addition of new boundary conditions or any re-description will end up with a new fixedly defined system and a known relation to its environment. Such a tendency for re-describing actual systems is obviously limited since future changes in the contextual (and internal) conditions need not be predictable either because of random factors or because of unexpected effects of the dynamics on the conditions which granted validity to the initial idealisations. In view of this, it makes sense to associate all these contextual factors and a description of the internal structures of the systems involved in a process under the single name of 'constraint'. In this more general sense, a constraint indicates any factor which may exert some influence on the evolution of a process as described by some generalised dynamical principle.

This usage is a generalization of the meaning favoured by S. J. Gould for the case of evolution. According to him, a constraint is "theory-bound term for causes of change and evolutionary direction by principles and forces outside an explanatory orthodoxy", [10, p. 519]. Thus, any source of change apart from the general explanatory framework for the type of process in question would qualify as a constraint. Readers familiar with the work of H. H. Pattee will also have noticed certain similarity between his idea of constraint as an alternative description of a process and the concept as presented here, (see for instance [11]).

The term thus loses the negative connotation of the more formal notion of constraint as limitation and acquires a more encompassing meaning which may include the senses of direction or canalisation, (see also [10], p. 518). The word will be used in this general sense in what follows.

2.2 The Identity of a Process

Although, as seen above, constraints are not necessarily fixed, one could tentatively distinguish their variations from the actual process by one of the following criteria: a) these variations are independent of the operation of the system or b), if they vary dependently, they do so at a much slower timescale so that, at the scale in which the changes of state of the system occur, constraints may effectively be considered fixed. It can easily be seen that these criteria are qualitative rather than strict. In the first case, influence on the constraints to a process may be exerted through coupling with other processes which operate independently. But such coupling may also reflect how those contextual processes were in turn *previously* influenced by the central process in question – a process may so influence its own constraints indirectly. In the second case, when variations in the constraints depend directly on the dynamics of the process, one could question what is exactly meant by a much slower timescale and why are not such changes included as part of the original process itself.

It is necessary to have a more strict criterion. This issue is a manifestation of a bigger problem. If the dynamics of a process may alter the constraints that define the process, is it not possible that things could change so much that the systems involved would effectively become different systems? In such a case, with what right can one speak of a unique and well-defined process? A fixed set of constraints used to do the job of assuring that the systems remained the same from one moment to the next; in consequence it was possible to speak of a process with a single identity. Such rigidity, however, entailed that no process involving some sort of innovation could be so described. But if the constraints can also change there must be something else that one can point to in order to be able to say that one is referring to a same process. There must be an *organisational invariant* of the process which maintains certain relations fixed.

A process can be *defined* as the dynamics of a set of systems whose actual structures, rules or laws of operation as well as their relationships conserve some global organisational feature unchanged. In the example of the pendulum, one could include the applicability of Newton's second law, the relative positions between hanging mass, string, and ceiling, the very existence of these components, and so on. If the string is chemically unstable it will break at a certain point. When this happens, the process, *as defined by the above invariants*, has ceased. There is clearly certain freedom of choice on the part of the observer regarding what is to be called a process. That freedom is in the distinction of the relevant invariants. Thus, if the only invariant in the case of the pendulum is the mass that hangs and the process is the variation in position of this mass, then it does not matter if the string breaks in two, this is just a change of constraints, the process goes on with the free fall dynamics, the bouncing on the floor, etc. *This* particular process would cease only if the mass disintegrates.

These comments apply to processes in general, but they hold a special significance for historical processes, as these are the only processes in which, besides the basic invariants distinguished by the observer, the interplay between process and constraints may lead to the *spontaneous* formation of *new* invariants.

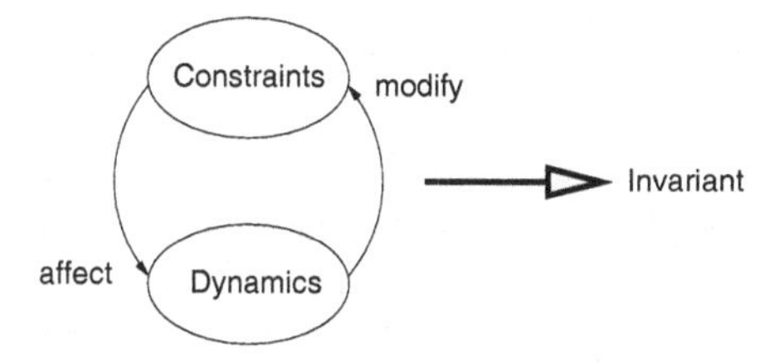

Fig. 1. Circular 'causation' defining a general historical process.

Such spontaneous, durable patterns are constituted by an interplay between the dynamics and the constraints to the process. Due to the amplification of the effects of fluctuations and the breaking of in-built symmetries, complex processes in which many variables interact non-linearly may exhibit transitions to highly ordered dynamics. Such transitions are manifested in a coherent regime which is not pre-specified in the initial definition of the process nor externally imposed. Such processes are often called self-organising[1], [4].

Spontaneous invariants, when they occur and while they last, can be thought of as 'equilibrium' stages in the reciprocal 'causation' showed in Fig. 1. When the accumulated influence of the process on the variation of their own constraints results in little or no extra effect back on its dynamics, constraints will cease to change and the situation will be maintained. This state of order is manifested in the form of durable patterns in the dynamics and its constraints. With a shift of viewpoint these patterns can be seen as affecting the process in ways that tend to their own perpetuation. From this perspective, it is possible to say that a invariant, once established, may be used to 'explain itself'. In addition, these organisational features may also exert an influence over other aspects of the process which need not be directly involved in the conservation of the invariant.

3 Different Manifestations of History

The above considerations give a rough idea of how to differentiate historical processes from processes which are non-historical or merely contingent. A historical process is a process subject to fluctuations whose dynamics affects its constraints either directly or though recurrent coupling with other processes. In order to make the meaning of these concepts clearer it will be helpful to consider some examples of historical processes. Many processes that would qualify as paramount examples, such as stigmergy, cognitive development, cultural change

[1] Historical processes include such instances of self-organisation as a possibility, but describe a wider class. Self-organisation can be a problematic concept (see [12]); especially when dealing with entities that are formed or destroyed in the process. The question of what is the *self* that organises can be better approached from the historical point of view than from the self-organising perspective which would require the identity of a newly formed 'self' to preexist its own formation.

and social norms, structural epigenesis, the economics of increasing returns, etc. will not be discussed due to lack of space.

3.1 Trails on Grass and Pask's Artificial Ear

Consider the trails made naturally by pedestrians on areas that are covered with grass. These trails are made by the action of walking which makes it difficult for grass to grow on zones which are frequently trodden upon. The lack of grass makes walking along the trail easier and people tend to use the trail rather than cutting across the grass, even if this implies a small deviation from the optimal route to their destination. Trail formation has been studied using a very simple and powerful individual-based model, [13]. The process is self-reinforcing and, in the bigger picture, it is also a historical process.

Let the process be the set of individual pedestrian trajectories within a piece of land covered with grass (say a square) with a few preferred entry and exit points. Walkers are driven by two preferences: they want to arrive at their destination cutting across the square and they prefer to walk where the grass is less grown. Initially, no path is marked on the grass and walkers choose a direct route to their destinations. As time passes, and for a certain frequencies of crossings, the effect of the initial trajectories will begin to be manifested in areas where the grass is worn. In the most used trajectories the effect of wear will be so much that the grass will not be able to compensate by growing again before the path is re-used. Thus, trails are formed and maintained in a dynamical equilibrium. The process can be quite complex since the different trails may 'interact' during the process. For instance, it will be common to observe a single exit point halfway between two frequently used and relatively close destinations instead of two exit points corresponding to each one of them, which means that two trails may have converged.

Once a pattern of trails is formed the history of the process has become partially embodied in it and walkers are constrained by its shape to walk along the trails. Thus, the pattern modulates the dynamics of the process but, at the same time, is constantly being constituted by the process as trails can only be maintained if enough people use them.

A similar process was used by cybernetician Gordon Pask for the construction of artificial sensors and effectors out of an initially undifferentiated physical medium, [14]. The system consisted of a network of amplifiers and associated electrodes which were not directly connected but submerged close to one another in a solution of ferrous sulfate. The electrodes acted as sources or sinks of direct electrical current depending of the activity of the system. Crucially, if direct current is passed from a source to a sink, a metallic thread of very low resistance is formed in the ferrous solution which, as the trails on grass, will be much easier to use if current is to pass again between the same electrodes. In contrast, if the thread is not re-used, it will gradually dissolve because of local acidity. After some time, a network of threads may be formed and maintained dynamically.

The system could be 'trained' to respond to different sorts of couplings. The method of training consisted simply in increasing the available energy for forming

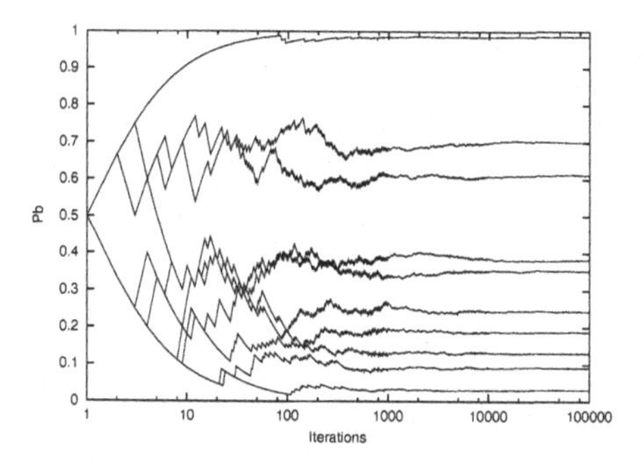

Fig. 2. Illustration of Polya's urn. Probability of drawing a black ball (*Pb*) vs. number of iterations for 10 different runs.

and reinforcing threads if the system's performance was close to the desired one. Such a scheme translated into a growth and pruning dynamics at the level of the network of threads. Interestingly, being a physical system, there were many ways in which the process of thread formation could be affected: mechanical, thermal, chemical, and electrical. Pask was successful in training the system to respond to acoustic vibrations of a specific frequency. The system responded by growing a network of threads around the vibrating regions of the apparatus.

3.2 Polya's Urn Scheme

Consider the following stochastic process known as Polya's urn scheme [15]. Put two balls in an urn, one red and the other one black. Extract one of the balls, observe its colour and then replace it and put another ball of the same colour into the urn. Repeat indefinitely. What is the expected probability for extracting a black ball after a large number of iterations?

This process was originally proposed as a model of epidemics and it has been applied to models of market dominance [16]. Interestingly, in can be shown that the probability of extracting a black ball will converge to a specific value which can be *any* number between 0 and 1. Figure 2 illustrates this convergence for 10 different instantiations.

The process may be understood as historical if its dynamics are taken to be the extraction, observation and double replacement of balls in a repeated manner. At any moment, the probability of extracting a black ball depends on the number and colour of the balls present in the urn. This is taken as the context or constraint of the dynamics. Such context is itself affected by the same process that it constrains. After many iterations, this interplay between dynamics and constraints reaches an equilibrium. This is because the addition of a new ball, whatever its colour, will not affect significantly the existing distribution within the urn and the accumulated set of added balls will tend to reflect this distribution over a number of iterations. The actual equilibrium, however, is strongly dependent upon the history of the process. In particular, much weight is given to the initial steps; figure 2 shows how the variations can be extreme during the initial 10 iterations, then more moderate in the next 100 iterations and from then on less and less significant. This example shows that historical dynamics may be instantiated in processes which are relatively simple.

3.3 Evolution

Evolutionary processes are historical *par excellence*. Their historical character is rarely denied, although there is a tendency to think of evolution as historical only in the sense of being a process subject to contingencies. These may take the form of 'frozen accidents' or they may indeed be the result of the accumulation of small events, as mentioned earlier.

Until recently, the neo-Darwinian perspective has tended to confine the role of historical factors to that of contextual or initial conditions in a process subject to an 'universal law' of fitness maximization. The process in itself has not often been considered historical in the sense given here to that term. Through a process resembling trial-and-error, random changes in the material inherited by an organism will affect differentially the match between organismic and environmental properties so that some variants will be selected as better adapted to the environment than others. It is the assumption that evolution proceeds mainly in this problem-solving fashion that allows (even requires) the historical nature of evolution to be relegated to that of independent contextual factors. In effect, evolution becomes a process of optimising the adaptation to a pre-existing environment which does not depend significantly of the evolving organisms.

The problem with this view, as pointed out on many occasions [17,18, among others], is that the key environmental features that are significant for the reproductive success of an organism are not independent of the organism itself. According to Lewontin, the "world external to a given organism can be partitioned into *a priori* ecological niches in a non-denumerable infinity of ways. Yet only some niches are occupied by the organisms. How can we know, in the absence of the organisms that already occupy them, which of the partitions of the world are niches?", [17, pp. 159 - 160].

In addition to 'choosing', rather than just adapting to, their own niches, organisms also alter their medium, and that of other organisms, in significant ways, [17,18,19]. Birds and social insects build nests and other structures, rabbits and rats dig tunnels, beavers create ponds and alter local water levels, leaves accumulate under high plants, etc. These alterations may have both short and long term effects.

In spite of the mutual inter-dependence between organism and environment, evolution has been approximated as non-historical by sweeping all contingent factors under the carpet of independent environmental variation. This variation is external, i.e., not part of the process itself; this is characteristic of non-historical processes. It is, therefore, not surprising that the problems related to novelty in biological evolution cannot be so easily accounted for from this perspective, [20, 21], since such innovation can only take place in historical processes.

4 Open Issues and Some Consequences for AL

This fairly broad exposition of historical processes, in no way comprehensive, may be enough to suggest that there is some gain in giving expression to the unifying themes implied by grouping together phenomena as diverse as the construction of wasps' nests, the development of a cognitive skill, the maintenance of a social norm, or the evolutionary conservation of a body plan. The main practical consequence of this perspective is a shift in how these phenomena are studied. History implies a subtle dynamical interplay between change and conservation. It cannot be modelled, like the above phenomena have often been modelled, as changes in the external relations between fixed entities themselves not subject to change.

Historical entities are not fixed in the sense that all changes are subordinated to their fixed identity (a point of view giving rise to extreme structuralist thinking), nor are they fully malleable, yielding without inertia to the optimisation of some objective function (a point of view that leads to some forms of functionalism). The historical perspective steers a careful middle course between these extremes by focusing on understanding why certain patterns are *durable* (as opposed to either fixed or unstable) as a consequence of, and not despite, the constant variations that make up the dynamics of the process.

An important notion in this context is that of spontaneous invariants. Once a durable pattern is constituted, understanding the dynamical relations that allow it to persist can provide a powerful frame of reference for addressing specific questions of what goes on in a complex historical process. It allows the researcher to understand why certain things can change while others remain the same. In other words it can provide a *norm* intrinsic to the process. Contained within a spontaneous invariant lies an explanation of its own perpetuation. Even if the properties of the process in need of explanation are not directly related with its maintenance, the invariant sets conditions to how these properties can change usually by limiting a high dimensional space of possibilities into a few ordered modes.

Saying that novelty and qualitative transitions can only occur in historical processes is not the same as having explained how such phenomena happen. This is indeed one of the major areas for development. What causes the disappearance or transformation of an existing durable structure? Does novelty occur when invariants cannot self-maintain any longer? Or does it occur in historical processes that do not lead to new invariants in the first place? These are important open questions that deserve further development, and in which AL simulation models may play an important role. Such models can indeed show more flexibility than strictly analytical models, although their use as scientific tools also brings a whole new set of problems [22]. For the moment, the historical perspective can offer only a negative take on the issue of novelty. If the process cannot be said to be historical, then it is pointless to look for the conditions that will lead to novelty.

Acknowledgements. The author wishes to acknowledge the support of the Nuffield Foundation, (grant no. NAL/00274/G).

References

1. Boerlijst, M.C., Hogeweg, P.: Spiral wave structure in pre-biotic evolution: Hypercycles stable against parasites. Physica D **48** (1991) 17 – 28
2. Hemelrijk, C.K.: An individual-oriented model on the emergence of despotic and egalitarian societies. Proc. Roy. Soc. Lond. **B 266** (1999) 361 – 369
3. Theraulaz, G., Bonabeau, E.: A brief history of stigmergy. Artificial Life **5** (1999) 97 – 116
4. Prigogine, I., Stengers, I.: Order out of chaos: Man's new dialogue with nature. Heinemann, London (1984)
5. Waddington, C.H.: The evolution of an evolutionist. Edinburgh, UP (1975)
6. Woodger, J.H.: Biological principles: A critical study. Routledge, London (1929)
7. Nagel, E.: The structure of science: Problems in the logic of scientific explanations. Routledge, London (1961)
8. Gould, S.J.: Wonderful life: The Burgess Shale and the nature of history. Hutchinson Radius, London (1989)
9. Goldstein, H.: Classical Mechanics. 2nd edn. Addison-Wesley, Reading, MA (1980)
10. Gould, S.J.: A developmental constraint in *Cerion*, with comments on the definition and interpretation of constraint in evolution. Evolution **43(3)** (1989) 516 – 539
11. Pattee, H.H.: Laws and constraints, symbols and languages. In Waddington, C.H., ed.: Towards a Theoretical Biology 4, Essays. Edinburgh UP (1972) 248 – 258
12. Ashby, W.R.: Principles of the self-organizing system. In von Foerster, H., Zopf, G., eds.: Principles of Self-Organization, Pergammon Press, NY (1962) 255 – 278
13. Helbing, D., Keltsch, J., Molnár, P.: Modelling the evolution of human trail systems. Nature **388** (1997) 45 – 50
14. Pask, G.: Physical analogues to the growth of a concept. In: Mechanisation of thought processes: Proceedings of a symposium held at the National Physical Laboratory on 24 - 27 November 1958. Vol. II, HM Stationary Office, London (1959)
15. Blackwell, D., Kendall, D.: The Martin boundary for Polya's urn scheme and an application to stochastic population growth. J. Appl. Prob. **1** (1964) 284
16. Arthur, W.B.: Increasing returns and path dependence in the economy. Univ. of Michigan Press, Ann Arbor (1994)
17. Lewontin, R.: Organism and environment. In Plotkin, H.C., ed.: Learning, development and culture: Essays in evolutionary epistemology. Wiley, Chichester (1982)
18. Laland, K.N., Odling-Smee, J., Feldman, M.W.: Niche construction, biological evolution and cultural change. Behavioral and Brain Sciences **21** (2000) 131–146
19. Jones, C.G., Lawton, J.H., Schachak, M.: Organisms as ecosystem engineers. Oikos **69** (1994) 373 – 386
20. Crutchfield, J.P.: The calculi of emergence: Computation, dynamics and induction. Physica **D 75** (1994) 11 – 54
21. Fontana, W., Wagner, G., Buss, L.W.: Beyond digital naturalism. Artificial Life **1/2** (1994) 211 – 227
22. Di Paolo, E.A., Noble, J., Bullock, S.: Simulation models as opaque thought experiments. In Bedau, M.A., McCaskill, J.S., Packard, N.H., Rasmussen, S., eds.: Artificial Life VII: Proceedings of the Seventh International Conference on Artificial Life, MIT Press, Cambridge, MA (2000) 497–506

Aesthetic Fitness and Artificial Evolution for the Selection of Imagery from the Mythical Infinite Library

Alan Dorin

School of Computer Science and Software Engineering
Monash University, Clayton, Australia 3800
aland@cs.monash.edu.au

Abstract. Aesthetic selection and artificial evolution have been two of the more successful companions introduced to the toolbox of electronic image-makers in recent years. This paper examines the niche in which this technique, often associated with the simulation of biological processes, has positioned itself and some of the reasons for its success. Some remarks concerning the meaningfulness of a user's search for images through a genetic space are made and the relationship of this search to traditional artistic practice is examined. Suggestions on how to link other Artificial Life techniques, especially those involving self-organizing and self-assembling systems, with aesthetic evolution and electronic art are also made.

1 Introduction: A Parable

"This thinker observed that all the books, no matter how diverse they might be, are made up of the same elements: the space, the period, the comma, the twenty-two letters of the alphabet. He also alleged a fact which travellers have confirmed: *In the vast Library there are no two identical books*. From these two incontrovertible facts he deduced that the Library is total and that its shelves register all the possible combinations of the orthographical symbols: in other words, all that it is given to express, in all languages", J.L. Borges, *The Library of Babel*

Borges writes of a Library containing every possible book. Take a minute to imagine instead, a giant catacomb in which is stored every possible image, then read on…

Picasso strolls through the entrance of the image catacombs one morning. Upon encountering a librarian at reception he requests, "Please direct me to an image of a woman." Nodding, the librarian leads Picasso by the light of a sputtering candle through musty corridors, twisted and disused staircases. The countless halls and stairwells are drawer-lined from top to bottom, stretching beyond reach of the feeble candle. Occasionally Picasso observes dusty footprints which invariably vanish into the blackness. Each of the timber drawers has affixed to its surface a square of paper labelled with complex symbols. The librarian approaches a chest of drawers, briefly examines its label, then slides it open. After peering inside for a moment, he pulls from its depths a painted canvas. This he holds unsteadily in the candlelight before the great man.

"No, I'm sorry sir", mutters Picasso to the librarian, "That is the Mona Lisa, Da Vinci has already used that one. Have you anything else?" The librarian shuffles across the hallway and opens a drawer on the other side, removing from it another canvas. "Not bad", says Picasso, "What's in the drawer above that one?" The librarian returns the canvas and opens the next drawer in the chest. He steps backward, humbly gesturing that Picasso may examine the drawer's contents personally. Picasso obliges. Thoughtfully he flicks through the neat array of

J. Kelemen and P. Sosík (Eds.): ECAL 2001, LNAI 2159, pp. 659-668, 2001.

images in the drawer whilst the librarian holds the candle high. "Hmmm, this one *is* different!" exclaims Picasso, "I think I'll call it *The Weeping Woman*. Its blocky, distorted form is full of an anguish which matches my mood today. Fascinating. I'll take it! Do you have any other images in this geometric style?"

Reader, if Picasso, like Da Vinci before him, had wandered through a vast catacomb and, with the assistance of a guide, selected images to claim as his own, would you hold Da Vinci and Picasso in high esteem for the genius they exhibit in their work? Would you label these characters as *artists*? What does the term *artist* mean in this instance? Perhaps the concept of an artist driven by their passion and insight has been replaced by the concept of an *explorer* in search of artefacts to be labelled as his/her "works". These are the main issues which will be discussed in what follows, as will means of employing other A-Life techniques to art-making practice.

2 Background of Aesthetic Evolution

The use of artificial evolution to achieve engineering goals has been discussed at least since Holland published on the topic [1]. Much later, the concept of *aesthetic evolution*, was illustrated by Dawkins' *Blind Watchmaker* software, which accompanied his book of the same name [2]. The concept behind the code was simple–a small population of visual[1] representations (phenotypes) produced from a set of numbers (genotypes) are displayed on-screen. A user selects a pleasing phenotype which gives rise to a new generation whose genotypes are produced by mutation from that of the selected parent. The process of aesthetic selection in this manner continues until a desirable phenotype is produced.

Sims' venture into aesthetic evolution resulted in the construction of images and solid models [3,4]. His system borrowed from Koza's *Genetic Programming* paradigm [5] in that the genotype was a hierarchy of nodes (like those in the parse tree of a mathematical expression) the traversal of which constructs the phenotype. Sims also explored the evolution of cellular automata, dynamical systems and movement of virtual creatures [6,7,8].

Whilst Dawkins' Blind Watchmaker did not include operations for crossover between multiple parents, this has been added by others [7,9,10,11]. Its use is not always required in aesthetic evolution but its inclusion does allow the combination of traits distributed across multiple phenotypes.

3 Advantages of Aesthetic Evolution

The advantages of aesthetic selection have been explored in detail elsewhere [12,13] and are outlined only briefly here. Aesthetic evolution might be considered to have captured interest as strongly as *Boids* [14] and the *Game of Life* [15,16]. Thankfully, the process may produce imagery more diverse than the fractal zooms popular in the

[1] Aesthetic evolution need not be visually guided, however the focus of this paper is on the production of imagery using the technique.

80's! These algorithms nevertheless do share the idea of "complexity from simplicity" which is a theme of A-Life [17] and a goal for much human endeavour.
In the case of image production, the lag between what can be conceived and what can be realized can be partially bridged if a digital assistant is employed. Some have suggested that a computer may even display imagery *beyond* the imagination of the artist [12]. Those who would like to produce images but are not skilled with brush or camera may also enlist the assistance of a computer to produce complex, "interesting" outcomes.

The popularity of aesthetically-driven evolution lies in its simplicity from an implementation standpoint, the broad range of possible outcomes, and the degree of control a user feels over the process. Further, one characteristic of the technique is that the human need not understand the details behind the construction of "their" images or artefacts – all that is required is a critical eye to assess the merits of the current phenotypes. This is a skill all artists and "artists" have, regardless of their level of formal training. Hence, all-comers may be satisfied by the technique, which is also a powerful image-making tool for those with the knowledge to put it to good use.

4 Roles of Programmers and Users

At one level the advantages of aesthetic evolution seem tremendous. The technique certainly provides help with complex design tasks. What are the drawbacks of working in this way as a user with no programming input? This section gives concrete examples of the restrictions imposed on the user by the programmer of the software.
Whenever one uses software to create an image, the finished product speaks partly of the user's input, but more than most would care to admit, of the constraints imposed by the software and the output it is capable of producing. This is true of traditional media also: the limitations of watercolour on paper influence the kind of work a painter may create; the subject matter, film and lenses available to a photographer limit the work so produced.

U&I Software's *Artmatic*, demonstrates procedural production of 2D images *without* aesthetic evolution. Interactively changing an image's parameters nevertheless reveals countless "attractive" relatives. Elements of the equation specifying the image may also be altered to land the user at distant locations within the image catacomb. The designers have thus far resisted the trend to incorporate evolutionary guidance through the vast parameter space.

Exploring massive pre-defined image spaces in such a top-down fashion is a fundamental change from traditional art's bottom-up synthesis. Creating a work becomes a process of eliminating undesirable qualities from an image, or substituting them for other fortuitously appearing properties. The options are not created by the user, but by the software, and are layed down in the form of a complex "choose your own adventure" book.

Within this constraint is some room for creativity, but rather than an approach where an image is synthesized from a blank canvas and a filled mind, this type of image-making involves selecting from filled canvases (with, possibly, a blank mind).

The canvases displayed are selected from a tightly constrained (but possibly infinite[2]) set. Image-making by aesthetic selection of two-dimensional textures or by sliding parameter values is even more tightly constrained than photography since the images themselves are presented ready-made to the user. However when a model or texture is created using these techniques it may become a subject which yet requires creative deployment.

In the sea of imagery or models produced by *Artmatic*, the *Blind Watchmaker*, *Mutator*, Sims' Tokyo ICC installation *Galapagos*, *The Artificial Painter* (offered on-line by Pagliarini et al) or the *Animaland Bauhaus* (the author's own software), there is an abundance of "interesting" forms. One might say "Ooooh, that's a *good* one!", but this could equally be said about millions of such images. Unless there is a reason imposed by the artist for choosing one image over another, then gigabytes can be filled with countless "good" images. Where are the "excellent" images? What distinguishes them from the "good" images? Has artist X who employs aesthetic selection ever produced a *masterpiece* which eclipses X's other works in its sophistication? Are all images produced by such tools equally meaningful? If this were the case, there'd be little point in playing with aesthetic evolution in the first place!

If the images are not equally meaningful, or if an image amongst the works of X may be identified as superior, how so? What are its special qualities? These questions remain unanswered. Where is X's *Mona Lisa*, *David*, *The Scream*, or *Guernica*? How can this masterwork be distinguished from the other works produced by X? Is this traditional view of the artist striving for perfect expression of relevance to those who practice this form of art-making? If they are searching a catacomb to find their masterpiece, what will it look like? These days, can anybody tell a masterpiece by the way it *looks*?

The skill of the artist at capturing a feeling, providing comment, engaging with a process or any one of a myriad of *reasons* people produce art need not be ignored completely. The search for an "attractive" image need not begin (nor end) with computer-selected options, but with a human drive to *create*. In the case of evolutionary design, the non-programming artist searches through the space of interest to the programmer for an image which, however clumsily, reflects their artistic goal. This is not a healthy situation for any art...

Evidence for the dominance of programmer over user lies in the results produced by artists using *different* pieces of software. Try to use the *Blind Watchmaker* to evolve an image like one created by *Mutator* and this is immediately apparent. These are different *media*, not the same medium being employed by different artists. Each program, despite its reliance on procedural or evolutionary mechanisms, hard-codes *different* constraints on the user and the forms which may be created. A work produced by one piece of software will have a trademark *style* imposed by the code before any imposition by the user. Whilst a skilled copy-artist may use oils to imitate another oil painter's style, (demonstrating that each is employing the *same* medium), in the case of the *Blind Watchmaker* and *Mutator*, the *programmers* are employing the same medium, software end-users are not.

Specifically, constraints which ensure a trademark style emerges from aesthetic evolution vary but may include the modelling and rendering methods. For example,

[2] Infinity can be constrained. E.g. Both the sequence of integers and that of floating point numbers are infinite. Yet the integer sequence is more highly constrained than that of floating point numbers.

Mutator begins to constrain Latham by its use of Constructive Solid Geometry. This, combined with the repetition of elements along paths and organized in high-level features like ribs, stacks and horns, gives the images a characteristic segmented appearance reminiscent of the work of Giger [18]. The graphics primitives selected by Latham are usually ellipsoids and torii–these work well to form organic structure–as opposed to cuboids which give a geometric feel to (and simplify the physical simulation of) the virtual creatures of Sims [7,8].

Dawkins' biomorphs retain their segmentary, stick-constructed origins. Dawkins writes "I did allow myself the luxury of using some of my biological knowledge and intuition. Among the most evolutionarily successful animal groups are those that have a *segmented* body plan. And among the most fundamental features of animal body plans are their plans of *symmetry*" [2,p329]. Thereby making clear his understanding of the assumptions he made about the kinds of forms to be produced.

Ventrella's aim was to animate figures by aesthetic evolution and so he predetermined their topology [11]. He writes also, "A *qualitative* physics model is used to constrain motion in these figures. Although it is quaint and home grown, this model does produce most of the salient features of interacting physical bodies". Hence the movements Ventrella's creatures may make are dictated by an ad hoc physical simulation built for simplicity and speed of execution.

The comments above are not criticism, the programmers were *forced* to make decisions/assumptions to write code which met their needs. These have been highlighted to indicate the degree to which the *software* determines the outcome of a user's "art-making" using aesthetic evolution as a guide. Although these catacombs are infinite, they are not so in all directions. Whilst writing software for generating imagery via aesthetic selection might have similar scope to the processes of painting or sculpture, once the program is written, the images implicit in its architecture serve only to bring visual form to the code. Continual application of the software for the generation of multitudes of images reveals only trivial information about the artist and the software. There is nothing more to reveal! It is as if a photographer were obsessively clicking the shutter on a single inanimate subject.

5 Relevance of Software Limitations

A culture develops around those artists who use a particular medium. This culture includes curators who display and collect works. It includes critics who write extensively about artists using a particular medium. It includes communities of artists who explore techniques for working within a medium. Implicit in the culture is the assumption that artists working within a medium have similar constraints within which they work to solve artistic problems. Perhaps nowhere is this more clear than in the battle which raged during the twentieth century over the use of a musical scale of equal temperament [19,20].

Rarely does debate rage *across* media: advances in cinema are seldom compared to architecture (although the two fields may borrow from one another); musical accomplishment is rarely judged by reference to painting. The arts are related to one another in broad terms but it makes little sense to make value judgements between works created under different constraints. Yet *within* a medium, this is exactly the kind of debate which proliferates – people make implicit and explicit qualitative judgements about art and artists. For a historical example see [21, p250-254].

Such a culture exists also within the community of computer programmers. They discuss their ideas for algorithms, swap code and criticize each other's work. Programmers might discuss the merits of including the crossover operation in a program for aesthetic evolution, or the degree and frequency of mutation a genotype undergoes. Similar discussion occurs amongst the users of complex software tools such as Adobe Photoshop to which are devoted discussion lists, text books and magazines. There are of course numerous and broad-ranging discussions and debates amongst those who use specific techniques (such as artificial evolution) on computers for art-making, this paper being one example.

Where is the critical debate about the special tricks for using Latham's *Mutator*? How do different images made with Rooke's [22] software compare to one another? Why are artists not seriously comparing the biomorphs they produced with those constructed by others[3]? The matter that authors claiming artistic merit do not generally distribute their tools for others to use, highlights the answer to this question: there is little to discuss when it comes to *using* the software. There are no subtle tricks, using the software is trivial and reveals no more than did the original output of the tool. All the skill has gone into the programming of the tool, not the use of it. Hence access to the tool is all that is required to make images "in the style of X" where X is the *artist* who is best known for using it.

Any visually-skilled artist, whether they have ten years of experience with aesthetic evolution or one can produce a "quality" image using a properly implemented tool which uses the technique. Unlike painting with a brush and ink, there is no room for improvement. Studying the *use* of these tools for ten years is a laughable pursuit since they may be "mastered" in a matter of hours. There is no means for distinguishing a master from a relatively inexperienced user, in this context the terms are meaningless.

Debate does rage about the outcome of employing aesthetic evolution within the wider context of cinema or image-making. For example Latham's *Bio-Genesis* [23], Sims' *Panspermia* [24], McCormack's *Turbulence* [25], Dorin's *Hydroid Medusae* [26] may be judged as pieces of cinema. It is therefore clear that when combined with talent in the making of cinema (or prints) the raw material produced by aesthetic evolution can be manipulated to produce works of artistic merit. Note however that within the field of cinema there *is* room to improve. A novice film maker and a master may be distinguished from one another. Similarly for a photographer using digital subjects or others. There is always room for even a master to improve and a culture (which includes competitions, critics and reviewers) to comment on this progress.

Aesthetic evolution is a powerful tool, but the imagery a particular implementation may produce is laid down in code–a *creative* process which takes practice to master. The "art" of creating an image using aesthetic selection is indeed mindless. Participation in the process of evolution by clicking on favourite images and playing the role of a garden-weeder or pigeon-breeder is not a process like that traditionally associated with art-making. Nevertheless, the specific tool and the assumptions its code contains,

[3] An online class exercise asked "Try your best to come up with an interesting, attractive (or gross!) result in the final parent frame that will win the immense admiration of your fellow students in your section.... You are encouraged to arrange some sort of contest to judge the 'best' biomorph in each section." It assumed here that value will be attributed by association.

as well as the manner in which its output is used in a wider context, provide ample scope for the fine artist.

6 Future Applications

Complex systems are again in the research spotlight, no longer because of unpredictability, but due to the spontaneous emergence of order from chaos. Fascination with these processes was behind the fame of the *Game of Life* and the explorations of cellular automata it provoked [15,16]. This section looks at how these complex nonlinear processes may be combined with aesthetic evolution for the construction of imagery.

Researchers like Prigogine [27], Maturana & Varela [28] and Kauffman [29] have long championed self-assembly through auto/cross-catalysis as a defining property of life. If technology is to move beyond its current phase in which software/images/music are constructed through an all-knowing controller, then the dynamics of multiple interacting, self-*ordering* primitives must be understood. This is no simple problem, but its impact on art-making will be substantial. It is through the local interactions of countless inanimate components that the universe's most spectacular creation has emerged–the organism. Whilst this is freely acknowledged by researchers in A-Life, little research has investigated the area with any level of sophistication as it applies to the creation of artistic works.

Tolson has created a system where interacting agents leave a trace as virtual brush strokes [12]. The behaviours of these agents may operate under automatic artificial evolution to produce the desired imagery. As mentioned above, Sims used aesthetic evolution to guide the discovery of dynamical systems for the production of two-dimensional imagery. Explorations of cellular automata abound and are naturally displayed in visual form [30,31]. Nevertheless, artists employing cellular automata beyond "the grid of squares" are rare.

McCormack has utilized interacting agents in his installation work *Turbulence* [25] to control pole-mounted pods which signal one another by emitting rings of light. Dorin has created a virtual prism of nodes which initiate musical events as they interact in *Liquiprism* [32]. Brown's mesmerizing tiles rotate to form connected *Sandlines* [33]. In all of these works, the level of signalling is controlled by little more than a creatively implemented cellular automata.

Of course Reynolds' *boids* have also been utilized cinematically [14]. Boids and cellular automata are simple models from which complex behaviour of a discernible style emerges. How may the sophistication of interactions be increased to the point where behaviours which may not be predicted, even in general, arise? Flesicher's *Cellular Texture* generation [34] is an example of how studies of interacting agents for image-making may progress: individual entities follow programs instructing them on movement and behaviour including the release of and response to chemicals in the environment. These units may be used to form complex self-organized pattern. Dorin's *Solid Cellular Automata* operate similarly [35,36]. These virtual solids move through space under attraction and repulsion according to their current state and the state of neighbouring elements. Elements trigger changes of state amongst one another and self-assemble into dynamic or static structures.

In Dorin's and Fleischer's systems, the rules of interaction are laid down by the programmer. This complex task might be surrendered to the computer: the table of floating point parameters which governs the behaviour of Dorin's system could be converted to a Holland-style genotype upon which aesthetic evolution might act; the cell programs of Fleischer might be encoded using Koza's Genetic Programming technique. In either case, the combination of complex dynamical systems and aesthetic evolution seems a promising avenue for further exploration. Unfortunately, the nature of the evolutionary landscape defined by parameters of non-linear systems makes this a difficult task. Attempts such as Sims' to evolve cellular automata rule tables are not as successful as might be hoped for reasons he outlines[6]. Unlike the undulating landscape of "attractive" images, the evolutionary landscape of these systems is not necessarily smooth enough for an evolutionary algorithm to traverse effectively. Instead, the landscapes are characterized by sharp spikes in a bleak plane of parameter combinations producing disorderly behaviour. What is required is an undulating parameter space allowing the search algorithm to find points of local maxima.

For now, it is not clear how this difficulty might be overcome. It is apparent that the physical systems built by natural selection contain a high degree of redundancy, something which may be the key to their evolvability. A system is required in which small genotypic changes produce corresponding small phenotypic variation. Complex dynamical systems are brittle and do not degrade in performance gracefully. Tiny changes within cellular automata rule tables for example, may completely destroy the balance between activity and stasis necessary for complex, orderly dynamic behaviour.

In the case of Sims' virtual creatures, and the work of van de Panne and Fiume [37], huge spaces are scanned automatically for parameters which yield effective locomotion. Such searching is not feasible using a human as sole selector. Ventrella sensibly combines automatic and aesthetic evolution in his software – a somewhat effective means of evolving dynamic systems. Alternatively, an objective means of specifying the "interestingness" of a system is required. Possible criteria include entropy change, the proximity and regularity of elements or behaviour in a model or other related measures of organization.

7 Conclusions

Whilst aesthetic evolution has been applied to the production of various styles of image and model specification, it is yet to make a mark as a control mechanism for the self-assembly/organization of multiple independent elements. The complexity which may arise from carefully orchestrated self-assembly of models and imagery is currently an untapped resource for visual artists.

Aesthetic selection is clearly helpful in the search amongst the possibilities defined by a set of constraints imposed by a particular image-making tool. Once the software has been constructed and a set of images has revealed its form, further rambling through the image space without re-programming is of dubious artistic merit. There is certainly little point in labelling a user of a tool who has no direct or indirect programming input, with the title *artist* due to their lack of control over the image-making process. Yet it is clear that the programmer/software/user combination may produce works as significant as with any other technique. The creative process then lies in the constraints imposed through the software development process on the images which will be produced.

As long as the programmer/user link is strong, aesthetic evolution is a tool with much to recommend it. If this link is severed or weak, the user becomes only a lost soul in the infinite image catacombs. None of the pixel arrays voice the thoughts of the wanderer. Instead, they tirelessly repeat the name of labyrinth.

References

1. Holland, J.H.: Adaption In Natural and Artificial Systems, (reprint MIT Press 1982)
2. Dawkins, R.: The Blind Watchmaker, Penguin Books, (reprint 1991)
3. Sims, K.: Artificial Evolution for Computer Graphics, SIGGRAPH 91, 25(4), ACM Press, 319-328, (1991)
4. Sims, K.: Interactive Evolution of Equations for Procedural Models, *Visual Computer*, Springer-Verlag, 466-476, (1993)
5. Koza, J.R.: Genetic Programming: on the Programming of Computers by Means of Natural Selection, MIT Press, (1992)
6. Sims, K.: Interactive Evolution of Dynamical Systems, *Toward a Practice of Autonomous Systems: Proceedings of the First European Conference of Artificial Life*, Varela & Bourgine (eds), MIT Press, 171-178, (1992)
7. Sims, K.: Evolving Virtual Creatures, SIGGRAPH 94, ACM Press, July, 15-34, (1994)
8. Sims, K.: Evolving 3d Morphology and Behaviour by Competition, *Artificial Life*, 1(4), Langton (ed), MIT Press, 353-372, (1994)
9. Todd, S.: Latham, W., Evolutionary Art and Computers, Academic Press, (1992)
10. Dorin, A.: A Model of Protozoan Movement for Artificial Life, in *Insight Through Computer Graphics, Proceedings of CGI94: Computer Graphics International 1994*, World Scientific Press, 28-38, (1994)
11. Ventrella, J.: Disney Meets Darwin-The Evolution of Funny Animated Figures, in *Proc's of Computer Animation 1995.* Thalmann & Thalmann (eds), IEEE, 35-43, (1995)
12. Joblove, G. (chair): The Applications of Evolutionary and Biological Processes to Computer Art and Animation, panel session, SIGGRAPH 93, 389-390, (1993)
13. Whitelaw, M.: Breeding Aesthetic Objects: Art and Artificial Evolution, in *Proc's of the AISB'99 Symposium on Creative Evolutionary Systems*, Society for the Study of Artificial Intelligence and Simulation of Behaviour, United Kingdom, 1-7. (1999)
14. Reynolds, C.W.: Flocks, Herds and Schools: A Distributed Behavioural Model, SIGGRAPH 87, 21(4), ACM Press, 25-34, (1987)
15. Gardner, M.: Mathematical Games: The Fantastic Combinations of John Conway's New Solitaire Game 'Life', *Scientific American*, 223(4), 120-123, (1970)
16. Gardner, M.: Mathematical Games: On Cellular Automata, Self-Reproduction, the Garden of Eden and the Game 'Life', *Scientific American*, 224(2), 112-117, (1971)
17. Langton, C.G.: Artificial Life, *Artificial Life*, SFI Studies in the Sciences of Complexity, Langton (ed), Addison-Wesley, 1-47, (1989)
18. Falk, G. (ed.): H R GIGER Arh+, Benedikt-Taschen-Verlag, (1993)
19. Ford, A.: Illegal Harmonies, Music in the 20th Century, Hale and Iremonger, (1997)
20. Nyman, M.: Experimental Music, Cage and Beyond, 2nd edn. Cambridge Univ. Press, (reprint 1999)
21. Vasari, G.: The Lives of the Artists, translated by Bull, G., Penguin Books, (reprint 1976)
22. Rooke, S.: Artist's Talk, in, *First Iteration: Proc's of the First International Conference on Generative Systems in the Electronic Arts*, Dorin & McCormack (eds), CEMA, Monash Univ., Melbourne, Australia, 29-30, (1999)

23. Latham, W.: Bio-Genesis, Electronic Theatre, Computer Graphics International 94, Melbourne, Australia, (1994)
24. Sims, K.: Panspermia, Electronic Theatre, SIGGRAPH 92, (1992)
25. McCormack, J.: Turbulence, interactive laser-disk, SIGGRAPH 94 art show, (1994)
26. Dorin, A.: Hydroid Medusae, Comp. Anim. Prog., Digital State, Next Wave Festival, Melbourne, Australia, (1995)
27. Prigogine, I. & Stengers, I.: Order Out Of Chaos - Man's New Dialogue with Nature, Flamingo, (1985)
28. Maturana, H., Varela, F.: Autopoiesis, The Organization of the Living, in *Autopoiesis and Cognition, The Realization of the Living,*, Reidel, 73-140, (1980)
29. Kauffman, S.A.: The Origins of Order, Self Organization and Selection in Evolution, Oxford Univ. Press, (1993)
30. Langton, C.G.: Studying Artificial Life with Cellular Automata, *Physica* 22D, North-Holland, 120-149, (1986)
31. Wolfram, S.: Universality and Complexity in Cellular Automata, in *Physica* 10D, North-Holland, 1-35, (1984)
32. Dorin, A.: Liquiprism, installation, Process Philosophies, Dorin & McCormack (curators) First Iteration Conference, CEMA, Monash Univ., Melbourne, Australia, (1999)
33. Brown, P.: Sandlines, installation, Process Philosophies, Dorin & McCormack (curators) First Iteration Conference, CEMA, Monash Univ., Melbourne, Australia, (1999)
34. Fleischer, K.W., Laidlaw, D.H., Currin, B.L., Barr, A.H.: Cellular Texture Generation, in *Proc's* SIGGRAPH 95, ACM Press, 239-248, (1995)
35. Dorin, A.: Self-Organizing Cellular Automata, in *Proceedings Workshop on Distributed Artificial Intelligence*, Springer Verlag, LNAI, No. 1544, (1998)
36. Dorin, A.: Creating a Physically-based, Virtual-Metabolism with Solid Cellular Automata, in *Proc's Artificial Life 7*, Bedau et al (eds), MIT Press, (2000)
37. van de Panne, M., Fiume, E.: Sensor-Actuator Networks, in *Proc's* SIGGRAPH 93, ACM Press, 335-342, (1993)

Distributing a Mind on the Internet: The World-Wide-Mind

Mark Humphrys

Dublin City University, School of Computer Applications
Glasnevin, Dublin 9, Ireland
humphrys@compapp.dcu.ie
www.compapp.dcu.ie/~humphrys

Abstract. It is proposed that researchers in AI[1] and ALife construct their agent minds and agent worlds as *servers* on the Internet. Under this scheme, not only will 3rd parties be able to re-use agent worlds in their own projects (a long-standing aim of other schemes), but 3rd parties will be able to re-use agent minds as components in larger, multiple-mind, cognitive systems. Under this scheme, any 3rd party user on the Internet may select multiple minds from different remote "mind servers", select a remote "Action Selection server" to resolve the conflicts between them, and run the resulting "society of mind" in the *world* provided on another "world server". Re-use is done not by installing the software, but rather by using a remote service. Hence the term, the "World-Wide-Mind" (WWM), referring to the fact that the mind may be physically distributed across the world. This model addresses the possibility that the AI project may be too big for any single laboratory to complete, so it will be necessary both to decentralise the work and to allow a massive and ongoing experiment with different schemes of decentralisation. We expect that researchers will *not* agree on how to divide up the AI work, so components will overlap and be duplicated and we need multiple-conflicting-minds models [21]. We define the *set of queries and responses* that the servers should implement. Initially we consider schemes of *low-bandwidth* communication, e.g. schemes using numeric weights to resolve competition. This protocol may initially be more suitable to *sub-symbolic* AI. The first prototype implementation is described in [47]. It may be *premature* in some areas of AI to formulate a "mind network protocol", but in the sub-symbolic domain it could be attempted now.

[1] *The use of the term "AI" may cause confusion to an ALife audience. Logically, it should be the case [22] that artificial intelligence is a subfield of artificial animals which is a subfield of artificial life. However, this taxonomy has not caught on, so throughout this paper I use "AI" to refer to all artificial modelling of life, animals and humans. In the sense in which we use it, classic symbolic AI, sub-symbolic AI, Animats, Agents and ALife are all subfields of "AI". To summarise, this paper applies to all types of artificial minds, whether the types popular in ALife, or not. To illustrate this further, see implementation of various ALife models in the section: "**How to implement existing agent architectures as networks of WWM servers**" below.*

J. Kelemen and P. Sosík (Eds.): ECAL 2001, LNAI 2159, pp. 669–680, 2001.

1 Introduction

The starting point for our motivation is the argument that the AI project is too big for any single laboratory to do it all. Many authors have argued along these lines, and a number of different approaches have evolved:

- The traditional AI approach has been to work on *subsections* of the postulated mind. The criticism of this approach [4,5] is that the "whole mind" never actually gets built, and each subsection can avoid hard problems as "someone else's problem", so that *no-one* addresses them. For a symbolic AI call to build whole systems see [30].
- The *Animats* approach [50] is to start with *simple whole creatures* and work up gradually to more complex whole creatures. But as the complexity scales up, it cannot avoid the question of whether one lab can really do it all. Perhaps the Cog project [6,7] is now beginning to hit those limits.
- The *evolutionary* approach is to say that control systems are too hard to design and must be evolved [17]. In practice this has also usually seemed to share with the animat approach an implicit assumption that one lab can do it all.

It seems to me that all these approaches still avoid the basic question: **If the AI project is too big for any single laboratory to do it all, then as we scale up, how will we link the work of multiple laboratories?** Who decides who works on which piece? What if people can't agree on how to divide up the work, or indeed what the pieces are? [5] Will this scheme force everyone to use a common programming language? Will it enforce a common AI methodology and exclude others?

This paper proposes a scheme for decentralising the work in AI *without* having to agree on any of the above issues. Briefly (for a fuller discussion see [24]):

1. Researchers will *never* agree on how to divide up the work, so we need models in which this is not a problem, i.e. *models of multiple overlapping, conflicting and duplicated minds* [28,29], and conflict-resolution mechanisms to generate winning actions from such a collection [21].
2. Researchers do not re-use each others' work for the same reasons that software re-use in general [20] has not been as easy as promised – complex installation, and incompatibility of libraries, versions, languages and operating systems. Therefore it is suggested that we look to the most successful recent model of re-use – the using of other people's documents and programs off remote Web servers. We suggest a model where the agent minds and agent worlds *stay* at the remote server and are used from there, instead of being installed.
3. We need *total language and platform independence*, so that researchers can concentrate on AI and not on networks.
4. This will be easier for virtual agents, but is not impossible with robotic agents, as the field of *Internet tele-robotics* demonstrates [44].

2 The World-Wide-Mind

The proposed scheme to address these issues is called the "World-Wide-Mind" (WWM). In this scheme, it is proposed that researchers construct their agent minds and worlds as *servers* on the Internet.

2.1 Types of Servers

In the basic scheme, there are the following types of server:

1. A **World and Body server** together. This server can be queried for the current state of the world: x as detected by the body, and can be sent actions: a for the body to perform in the world.
2. A **Mind server**, which is a behavior-producing system, capable of suggesting an action: a given a particular input state: x. A **Mind$_{\text{M}}$ server** is a Mind server that calls other Mind servers. For example:
   1 An **Action Selection or AS or Mind$_{\text{AS}}$ server**, which resolves competition among multiple Mind servers. Each Mind server i suggests an action a_i to execute. The AS server queries them and somehow produces a winning action a_k. To the outside world, the AS server looks like just another Mind server producing a given x.

2.2 Types of Societies

By allowing Mind servers call each other we can incrementally build up more and more complex hierarchies, networks or societies of mind. We will call any collection of more than one Mind server acting together a *Society*. A Society is built up in stages. At each stage, there is a single Mind server that serves as the interface to the whole Society, to which we send the state and receive back an action.

1. A Mind$_{\text{M}}$ server calls other Mind servers. To run this Society you talk to the Mind$_{\text{M}}$ server.
2. A Mind$_{\text{AS}}$ server adjudicates among multiple Mind servers. To run this Society you talk to the Mind$_{\text{AS}}$ server.

2.3 Types of Users

1. **A non-technical client user** – essentially any user on the Internet. Basically, the client user will run *other people's minds in other people's worlds*. Without needing any technical ability, the client should be able to do the following:
   1 Pick one Mind server to run in one World. Even this apparently simple choice may be the product of a lot of hard work – in picking 2 suitable servers that work together. So it is suggested that the client can present the results of this work for others to use at some URL. No new server is created, but rather a "link" to 2 existing servers with particular arguments.

2 Even a non-technical client may be able to construct a Society. For instance: Select a combination of remote Mind servers, a remote AS server to resolve the competition between these, and a World server to run this Society in. To be precise: Pick a Mind_{AS} server, pass it a list of Mind servers as a startup argument, and then just pick a World to run the Mind_{AS} server in. Again the client can present the results of his work for others to use. Again, no new server is created, but rather a "link" to 2 existing servers with particular arguments.

2. **A technically-proficient server author** – again any user on the Internet, if they have the ability. They will need to understand how to construct a server, but their understanding of AI does not *necessarily* have to be profound. For example:
   1 Write a *wrapper* around an existing, working Mind server, i.e. Write a new Mind_{M} server. The most simple type of wrapper would not provide any actions itself, but just selectively call other servers: "*If input is x then do whatever Mind server M1 does – otherwise do whatever Mind server M2 does.*"
   2 An AI-proficient server author might try writing a Mind_{M} server that attempts to provide some actions itself: "*If input is x then take this action a – otherwise do whatever the old server does.*" The author may need little understanding of how the existing Mind server works. If overriding it in one area of the input space doesn't work (doesn't perform better) he may try overriding it in a different area.
   3 At the most advanced level, AI researchers would write their own servers from scratch. But it is envisaged that even AI researchers will make use of the techniques above.

2.4 Low-Bandwidth Communication

If Mind servers come from diverse sources, written according to different AI methodologies in different languages, and do not understand each other's goals, there is a limit to how much information they can usefully communicate to each other. A central question is: **What information does the AS server need from the Mind servers to resolve the competition?** For example, if it just gets a simple list of their suggested actions: a_i it seems it could do little more than just pick the most popular one. For more sophisticated Action Selection, the Mind server needs to provide *more information*. We first consider schemes where competition is resolved using numeric weights rather than symbolic reasoning. For example, Mind server i tells the AS server what action a_i it wants to take, plus a weight W_i expressing how much it is willing to "pay" to win this competition. We define the following weights:

1. The "Q-value" defines how good this action is in pursuit of the Mind server's goal, Mind server i might build up a table $Q_i(x, a)$ showing the expected value for each action in each state.
2. The "W-value" defines how *bad* it is for the Mind server to lose this competition. This depends on what action will be taken if it loses. Mind server i may

maintain a table $W_i(x)$ defining how bad it is to lose in each state. It may judge the badness of a *specific* action a by the quantity: $Q_i(x, ai) - Q_i(x, a)$.

The usage of Q and W comes from [21]. For the differences between Q and W see [21, §5.5, §6.1, §16.2]. Higher-bandwidth communications than numeric weights would seem difficult if we are not to impose some structure on what is inside each Mind server. So I begin the WWM implementation with a *sub-symbolic Society of Mind*, rather than a symbolic one.

2.5 What Is the Definition of State and Action?

We have so far avoided the question of what is the exact data structure that is being passed back and forth as the state or action. It seems that this definition will be different in different domains. This scheme proposes that we allow different definitions to co-exist. Each server will explain the format of the state and action they generate and expect (most of the time this will just involve linking to another server's definition).

2.6 The Name "The World-Wide-Mind"

The name "The World-Wide-Mind" makes a number of points:

1. **The mind stays at the server:** The mind will be *literally* decentralised across the world, with parts of the mind at different remote servers.
2. **Parts of the mind are separate from each other:** The important thing is not the separation of *mind from world*, but the separation of different parts of the mind from each other, so they can be maintained by different authors.
3. **This is separate from the Web:** This is a different thing to the *World-Wide-Web*. During the recent rise of the Internet, many people have talked about seeing some sort of "global intelligence" emerge [3]. But these writers are talking about the intelligence being embodied in the *humans* using the network, plus the pages they create [32], or at most intelligence being embodied implicitly in the hyperlinks from page to page [18,14]. Claims that the network *itself* might be intelligent are at best vague and unconvincing analogies between the network and the brain [37]. For a *real* society of mind or network mind, we need a network of *AI programs* rather than a network of pages and links.
4. **This may not even interact with the Web:** This is separate from existing work that might go under the name of "AI on the Web", namely, AI systems learning from the Web. A WWM system is not *necessarily* interested in learning from or interacting with the current Web or its users.

3 How the WWM Will Be Used in AI

We imagine that a scheme such as this is inevitable in AI – that the days of isolated experiments are numbered. For a detailed discussion see [24]. Briefly, it addresses these issues:

1. **Duplication of Effort** – Until now, sharing work has been so difficult that researchers tend to build their own agent minds and worlds from scratch, duplicating work that has been done elsewhere. There have been a number of attempts to re-use agent minds [40,43] and worlds [11], but the model of re-use often requires installation, or even a particular programming language. Here we propose a language-independent model of remote re-use. [31] is probably the closest previous work to this philosophy.
2. **Unused agents and worlds** – Having invented a robotic or agent testbed, few experiments are often done with it. For example, I was in fact one of the first people to put an AI mind online [51,52], an "Eliza"-type chat program in 1989 [19]. Many people talked to it, but soon (by 1990), it had ceased to interact with the world. A *brief*, finite interaction with the world, seen by only a few people, is the *norm* in autonomous agents research. In this field it has become acceptable not to have direct access to many of the major systems under discussion. How many action selection researchers have ever *seen* Tyrrell's world running, for example? [45] How many robotics researchers have ever *seen* Cog move (not in a movie)? [7] Due to incompatibilities of software and expense of hardware, we accept that we will never experiment with, or even *see*, many of these things ourselves, but only read papers on them.
3. **Making AI Science – 3rd party experimentation** – Building your own agent world also means your new algorithms are not tested in the same world as previous algorithms. How to *prove* one agent architecture is better than another has become an important issue. [8] points out that, essentially, no one uses each other's architectures, because they are not convinced by each other's tests. In any branch of AI, the existence of objective tests that cannot be argued with tends to provide a major impetus to research. This is one of the reasons for the popularity of *rule-based games* in AI, and, more recently, *robotic soccer* [31]. [9] suggests the setting up of a website repository of standard tests for adaptive behavior. The WWM goes further than that, where the standard test worlds need not even be installed, but are run remotely. And the WWM goes further to support testing. By its emphasis on *3rd party* experimentation, algorithms will be subjected to constant examination by populations of testers with no vested interest in any outcome. 3rd parties will test servers in environments their authors never thought of, and combine them with other servers that their authors did not write. The servers should get a much more thorough testing than their own authors could ever have given them.
4. **Minds will be too complex to be fully understood** – Finally, there is definitely something in the evolutionary criticism that advanced artificial minds may be too complex to be *understood*. In the system we propose, of a vast network of servers calling other servers, each individual *link* in the network will make sense to the person who set it up, but there is no need for the system as a whole to be grasped by any one individual.

4 Objections to the Model

For a discussion of possible objections see [24]. Here we mention a few points:

1. **Models of Broken links and Brain Damage** – The WWM will need well-under-stood models of *fault-tolerant* artificial minds. e.g. In [21] I explicitly addressed the issue of brain damage in a large society of mind [21, §17.2.2, §18]. The reader might have wondered what is the point of a model of AI that can survive brain damage. Here is the point – a model of AI that can survive *broken links*.
2. **It is premature at symbolic level to attempt to define mind network protocols.** – Probably true. Researchers have long debated symbolic-AI knowledge-sharing protocols, with [13] arguing that it is premature to define such protocols. Recently this debate has continued in the Agents community as the debate over *agent communication languages* [27] and, to some extent, XML [2]. Agreement is weak, and it may be that the whole endeavour is still premature. For example [35] attempts to implement a Society of Mind on the Internet, but they insist on a symbolic model, with which they make limited progress. We argue, though, that it is not premature to start defining mind network protocols at the sub-symbolic level.
3. **The network is not up to this yet.** – Possibly true. But that will change.
4. **"Agents" researchers have already done this.** – Apparently not. There are some major differences between this and the Agents approach:
   1 **Agents researchers imagine that agents should be *installed*.** I disagree. Agents should be *servers*.
   2 **Agents have 1 localised (installed) mind per body. I have multiple remote minds (servers) per body.** Consider that *Distributed AI* (DAI) [42,33] has split into two camps:
      1 Distributed Problem Solving (DPS) – where the Minds are cooperating to solve the *same problem* in one Body.
      2 Multi-Agent Systems (MAS) – where the Minds are in different Bodies. We have 1 mind – 1 body actors, and coordination of multiple actors. This is what the field of "*Agents*" has come to mean. Indeed [33, §4.3] makes clear that our servers here are not Agents.
      This is neither of these, but is multiple minds solving multiple problems in one body. It is closer to *Adaptive Behavior* and its interest in whole, multi-goal creatures whose goals may simply *conflict*. Previous work in ALife has been more towards the MAS approach, e.g. in [36] it is a *society of agents* that is distributed across the network, not a *single agent mind*.
   3 In short, Agents researchers simply *aren't* trying to solve the problem of **how to divide up the work in AI, and link the work of multiple laboratories**, that we addressed at the start of this paper.

5 Implementation

We now define a way of implementing this idea on today's network. We suggest the standard client-server model of short, limited-length transactions. The server responds to a short query with a response within a limited time. The server does not know when, if ever, it will receive the next query. The client software, which is driving a single top-level Mind and World, implements a program like:

1. For each server:
   - Connect to server – Send "New run" command – Receive unique run ID
2. Repeat:
   1 Connect to World – Send ID – Query state – Get state x
   2 Connect to Mind – Send ID – Send state x – Get action a
   3 Connect to World – Send ID – Send action a – Get new state y
   4 Connect to Mind – Send ID – Tell it new state y – Receive confirm
3. For each server:
   - Connect to server – Send ID – Send "End run" – Receive confirm

The unique run ID is because the server may be simultaneously involved in many other runs with other clients. We will not lay down what the client algorithm should be (for example, should it implement time-outs). It may implement any general-purpose algorithm using the server queries. Similarly, a server may implement any algorithm it likes provided it responds to the set of queries expected of it (e.g. it may *itself* be a client to yet another server). So the definition of the WWM really comes down to just the definition of the possible queries and responses of WWM servers. For a detailed list of queries see [23].

What technology should we use to implement these queries? I suggest one overriding objective: *That the WWM server authors be required to know as little as possible to get their servers on the network.* The server authors are interested in AI, not necessarily in networks. They may only know AI programming languages. They may have never written a network application, and may not want to learn. As a result, it is proposed that the basic implementation of the WWM be done using CGI across HTTP. Every AI programmer has access to a HTTP server with CGI, and every AI programmer can write a program that receives stdin and writes to stdout. For further justification see [24].

What format should the data transmitted be? We suggest plain text XML [2]. We need some format, and XML provides an *extensible* format, where the invention of new queries won't crash old servers. For examples of XML encoding of the server queries see [23]. To summarise, all requests to a WWM server are requests to a CGI program on a Web server: `http://site/path/program`. All arguments (including the type of WWM query being sent) are passed as XML in stdin. The server writes the response as XML to stdout.

6 How to Implement Some Existing Agent Architectures as Networks of WWM Servers

6.1 Sub-symbolic Mind Servers

[24] shows in more detail than is possible here how many existing agent architectures can be implemented as networks of WWM servers, including:

1. Internet tele-robotics [53,44]. One issue is if multiple clients can connect to the World at the same time [41,34,15,38,16].
2. The Subsumption Architecture [4,5], using a hierarchy of Mind_{M} servers.
3. Serial models [39], using a master Mind_{M} server.
4. Any state-space learner [10], including Reinforcement Learning [25,48].
5. Hierarchical Q-Learning [26], using a master Mind_{AS} server.
6. Static measures of W-values (e.g. W=Q) [21, §5.3].
7. Dynamic measures of W-values [21, §5.5], including W-learning [21, §5, §6], including where Minds do not share the same suite of actions [21, §11, §13], e.g. Minds from different authors.
8. Strong and Weak Mind servers [21, §8, §C, §D], passing "mind strength" as an argument to the server.
9. Mind servers with different senses in the same Society [21, §6.6, §7, §8, §10]. The top-level Mind server sees the full state.
10. Global Action Selection decisions [21, §14], including Minimize the Worst Unhappiness and others [21, §F]. All such schemes in [45] etc. Some require a Mind_{AS} server that makes *multiple* queries of each Mind server.
11. Nested Q-learning [12], and Nested W-learning [21, §18.1], where the action returned is: "do whatever server k wants to do".
12. Feudal Q-learning [48] and Feudal W-learning [21, §18.2], where the Mind server accepts commands: "Take me to state c"
13. Economy of Mind [1], where the Mind_{AS} server will redistribute payments.

6.2 Symbolic Mind Servers – Single

There are a vast number of models of agent mind, whether hand-coded, learnt or evolved, symbolic or non-symbolic, that will repeatedly produce an action given a state. Most of these could be implemented as WWM servers without raising any particular issues apart from having to agree on the format of state and action with the World server. For example, "Eliza"-type chatbots [19,49].

6.3 Symbolic Mind Servers – Multiple

The difficulty arises when we consider *competition* between multiple symbolic Minds. So far we only defined a protocol for conflict resolution using numeric weights. Higher-bandwidth communication leads us into the field of *Agents* and its problems with defining agent *communication languages* (formerly symbolic AI knowledge-sharing protocols) that we discussed above.

A lot could be done, though, without having to define symbolic queries. Master Mind_{M} servers can switch from server to server. The drawback is the Mind_{M} server needs a lot of intelligence. This relates to the "homunculus" problem, or the need for an intelligent headquarters [21, §5]. Another possibility is the subsidiary Mind servers can be symbolic, while the master Mind_{AS} server is sub-symbolic – e.g. a Hierarchical Q-learner which just *learns* which subsidiary symbolic Mind server to let through.

7 Future Work

This is clearly the start of an open-ended program of implementation and testing. The first prototype implementation is now described in [47]. Immediate issues are:

1. Define the server queries – Define the full list of server queries, arguments, responses and error conditions [23] and encode them in XML. Is this list sufficient to implement *all* current sub-symbolic agent minds and worlds?
2. Define the client user view – The basic question is whether the client user software can be provided through *existing Web browsers* (perhaps through a *WWM portal site*), or whether a *separate client application* needs to be installed.

8 Conclusion

There are two issues here – first, that we need a system of decentralised network AI minds, and second a proposed protocol for it. Even if the protocol here is not adopted, the first part of this paper (the need to decentralise AI) stands on its own. For further reading see [24]. For the first prototype implementation see [47]. The first test of this system may be in the domain of language evolution [46].

8.1 Endnote – Showing the World What a Mind Looks Like

If a scheme like the WWM becomes successful, much of the user population of the Internet will gradually become familiar with minds made up of hundreds or even thousands of distributed components; minds that have little identifiable headquarters, but contain crowded collections of sub-minds, duplicating, competing, overlapping, communicating and learning, with "*alternative strategies constantly bubbling up, seeking attention, wanting to be given control of the body*" [21, §18.3]. Such models may be long familiar to AI researchers, but they are not much understood outside AI. The WWM scheme may help large numbers of people expand their imagination to think about what a mind *could* be.

References

1. Baum, E.B. (1996), Toward a Model of Mind as a Laissez-Faire Economy of Idiots, *13th Int. Conf. on Machine Learning.*
2. Bosak, J. and Bray, T. (1999), XML and the Second-Generation Web, *Scientific American*, May 1999.
3. Brooks, M. (2000), Global Brain, *New Scientist*, 24th June 2000.
4. Brooks, R.A. (1986), A robust layered control system for a mobile robot, *IEEE Journal of Robotics and Automation* 2:14–23.
5. Brooks, R.A. (1991), Intelligence without Representation, *Artificial Intelligence* 47:139–160.
6. Brooks, R.A. (1997), From Earwigs to Humans, *Robotics and Autonomous Systems*, Vol. 20, Nos. 2-4, pp. 291–304.

7. Brooks, R.A. et al. (1998), The Cog Project: Building a Humanoid Robot, *Computation for Metaphors, Analogy and Agents*, Springer-Verlag.
8. Bryson, J. (2000), Cross-Paradigm Analysis of Autonomous Agent Architecture, *JETAI* 12(2):165–89.
9. Bryson, J.; Lowe, W. and Stein, L.A. (2000), Hypothesis Testing for Complex Agents, *NIST Workshop on Performance Metrics for Intelligent Systems.*
10. Clocksin, W.F. and Moore, A.W. (1989), Experiments in Adaptive State-Space Robotics, *AISB-89.*
11. Daniels, M. (1999), Integrating Simulation Technologies With Swarm, *Workshop on Agent Simulation: Applications, Models, and Tools*, Univ. Chicago, Oct 1999.
12. Digney, B.L. (1996), Emergent Hierarchical Control Structures: Learning Reactive/Hierarchical Relationships in Reinforcement Environments, *SAB-96.*
13. Ginsberg, M.L. (1991), Knowledge Interchange Format: The KIF of Death, *AI Magazine*, Vol.5, No.63, 1991.
14. Goertzel, B. (1996), *The WorldWideBrain: Using the WorldWideWeb to Implement Globally Distributed Cognition*, `http://www.goertzel.org/papers/wwb.html`
15. Goldberg, K. et al. (1996), A Tele-Robotic Garden on the World Wide Web, *SPIE Robotics and Machine Perception Newsletter*, 5(1), March 1996.
16. Goldberg, K. et al. (2000), Collaborative Teleoperation via the Internet, *IEEE Int. Conf. on Robotics and Automation (ICRA-00).*
17. Harvey, I.; Husbands, P. and Cliff, D. (1992), Issues in Evolutionary Robotics, *SAB-92.*
18. Heylighen, F. (1997), Towards a Global Brain, in *Der Sinn der Sinne*, Steidl Verlag, Göttingen, `http://www.pespmc1.vub.ac.be/papers/GBrain-Bonn.html`
19. Humphrys, M. (1989), The MGonz program, `http://www.compapp.dcu.ie/~humphrys/eliza.html`
20. Humphrys, M. (1991), *The Objective evidence: A real-life comparison of Procedural and Object-Oriented Programming*, technical report, IBM Ireland.
21. Humphrys, M. (1997), *Action Selection methods using Reinforcement Learning*, PhD thesis, University of Cambridge, `http://www.compapp.dcu.ie/~humphrys/PhD`
22. Humphrys, M. (1997a), *AI is possible .. but AI won't happen: The future of Artificial Intelligence*, "Next Generation" symposium, Jesus College, Cambridge, Aug 1997.
23. Humphrys, M. (2001), *The World-Wide-Mind: Draft Proposal*, Dublin City University, School of Computer Applications, Technical Report CA-0301.
24. Humphrys, M. (2001a), The World-Wide-Mind: A protocol for building a client-server sub-symbolic Society of Mind distributed over the network, submitted to *Adaptive Behavior.*
25. Kaelbling, L.P.; Littman, M.L. and Moore, A.W. (1996), Reinforcement Learning: A Survey, *JAIR* 4:237–285.
26. Lin, L.-J. (1993), Scaling up Reinforcement Learning for robot control, *10th Int. Conf. on Machine Learning.*
27. Martin, F.J.; Plaza, E. and Rodriguez-Aguilar, J.A. (2000), An Infrastructure for Agent-Based Systems: an Interagent Approach, *Int. Journal of Intelligent Systems* 15(3):217–240.
28. Minsky, M. (1986), *The Society of Mind.*
29. Minsky, M. (1991), Society of Mind: a response to four reviews, *Artificial Intelligence* 48:371–96.

30. Nilsson, N.J. (1995), Eye on the Prize, *AI Magazine* 16(2):9–17, Summer 1995.
31. Noda, I. et al. (1998), Soccer Server: A Tool for Research on Multiagent Systems, *Applied Artificial Intelligence* 12:233–50.
32. Numao, M. (2000), Long-term learning in Global Intelligence, *17th Workshop on Machine Intelligence* (MI-17).
33. Nwana, H.S. (1996), Software agents: an overview, *Knowledge Engineering Review*, 11(3).
34. Paulos, E. and Canny, J. (1996), Delivering Real Reality to the World Wide Web via Telerobotics, *IEEE Int. Conf. on Robotics and Automation (ICRA-96).*
35. Porter, B.; Rangaswamy, S. and Shalabi, S. (undated), *Collaborative Intelligence – Agents over the Internet*, Undergraduate final year project, MIT Laboratory of Computer Science.
36. Ray, T.S. (1995), *A proposal to create a network-wide biodiversity reserve for digital organisms*, Technical Report TR-H-133, ATR Research Laboratories, Japan.
37. Russell, P. (2000), *The Global Brain Awakens*, Element Books.
38. Simmons, R.G. et al. (1997), Xavier: Experience with a Layered Robot Architecture, ACM SIGART *Intelligence* magazine.
39. Singh, S.P. (1992), Transfer of Learning by Composing Solutions of Elemental Sequential Tasks, *Machine Learning* 8:323–339.
40. Sloman, A. and Logan, B. (1999), Building cognitively rich agents using the SIM_AGENT toolkit, *Communications of the ACM*, 43(2):71–7, March 1999.
41. Stein, M.R. (1998), Painting on the World Wide Web: The PumaPaint Project, IEEE / RSJ Int. *Conf. on Intelligent Robotic Systems* (IROS-98).
42. Stone, P. and Veloso, M. (2000), Multiagent Systems: A Survey from a Machine Learning Perspective, *Autonomous Robots*, 8(3), July 2000.
43. Sutton, R.S. and Santamaria, J.C. (undated), A Standard Interface for Reinforcement Learning Software,
`http://www-anw.cs.umass.edu/~rich/RLinterface/RLinterface.html`
44. Taylor, K. and Dalton, B. (1997), Issues in Internet Telerobotics, *Int. Conf. on Field and Service Robotics* (FSR-97).
45. Tyrrell, T. (1993), *Computational Mechanisms for Action Selection*, PhD thesis, University of Edinburgh.
46. Walshe, R. (2001), The Origin of the Speeches: language evolution through collaborative reinforcement learning, submitted to *3rd Int. Workshop on Intelligent Virtual Agents* (IVA-2001).
47. Walshe, R. and Humphrys, M. (2001), First Implementation of the World-Wide-Mind, poster to appear in *ECAL-01*.
48. Watkins, C.J.C.H. (1989), *Learning from delayed rewards*, PhD thesis, University of Cambridge.
49. Weizenbaum, J. (1966), ELIZA – A computer program for the study of natural language communication between man and machine, *Communications of the ACM* 9:36–45.
50. Wilson, S.W. (1990), The animat path to AI, SAB-90.
51. Yahoo list of AI programs online, yahoo.com/.../Web_Games/Artificial_Intelligence
52. Yahoo list of ALife programs online, yahoo.com/.../Artificial_Life/Online_Examples
53. Yahoo list of robots online, yahoo.com/.../Devices_Connected_to_the_Internet/Robots

The Dimensions of the Cyber Universe

Ninad Jog

iPlanet, Sun Microsystems. Inc.
6905 Rockledge Drive Suite 820, Bethesda, Maryland 20817, USA.
Ninad.Jog@sun.com

Abstract. Digital organisms are the focus of much research and development, but the nature of their cyber universe habitat has received lesser scientific scrutiny. Here the cyber universe dimensions are described. A third dimension called Hop is proposed apart from cyberspace and cyber time. The cyber and the physical universes are compared and a number of questions about the two are raised.

1 Introduction

There is debate about whether network computing digital organism habitats such as Network Tierra [5] and others constitute a distinct universe or merely form a subset of the physical universe. Here I assume that they do form a domain distinct from the physical universe, which I term the *cyber universe*. Its dimensions are then described. Comparing the cyber universe dimensions with the physical universe dimensions should yield clues about whether the cyber universe is truly a distinct universe.

A computer's memory space, referred to here as *cyberspace*, differs fundamentally from physical space. Physical-world objects such as a chair cannot exist in cyberspace except as representations; conversely, digital organisms can exist in physical space only as representations. Cyberspace can therefore be considered a cyber universe dimension.

Digital organisms consist of charges in electronic computer circuits and are built primarily of energy rather than matter. They do not experience forces acting on physical-world biochemical organisms such as gravity, inertia, etc. [5]. Organisms living in cyberspace can perceive each other but cannot directly perceive physical-world entities, as they lack a physical manifestation and are alive only in the computing environment. They are also blind to the devices implementing their habitat such as memory and processing chips, input-output devices, etc.

2 Cyberspace

Cyberspace consists of a computer's RAM, disk, cache and registers. One measure of a space's dimensionality is the number of numbers needed to specify a point's location relative to another point in the space. Computer memory is organized and implemented as a linear array and a location's address is ultimately a single number. Cyberspace therefore appears one-dimensional.

J. Kelemen and P. Sosík (Eds.): ECAL 2001, LNAI 2159, pp. 681–684, 2001.

In large memories oganized by segments and pages, location addresses are specified by segment, page and offset numbers. Cyberspace can appear to have three dimensions when addresses are viewed as triplets. However, each triplet maps to a single unique number. Besides, each page has a fixed number of offsets and each segment a fixed number of pages, implying bounded page and offset dimensions but an unbounded segment dimension. Cyberspace therefore has one unbounded dimension.

Can digital organisms have a multi-dimensional cyberspace embodiment in one-dimensional memory? Memory locations contain charge and voltage patterns labeled 1's and 0's, that can be logically considered *cyber mass* particles. Organisms are composed of such particles, which are electrons in electronic computers and photons in optical computers. Operating systems map organisms linearly in memory in multiple linked blocks that may or may not be contiguous. An organism's shape is thus a set of one-dimensional line fragments rather than rectangles, implying a one-dimensional organism embodiment.

3 Cyber Time

In a computer environment many time-sliced processes execute round-robin on a small number of CPUs [6]. An organism is a single or multi-threaded process that alternates between CPU execution and queued for execution. A clock carried by an organism will move forward only during execution and will stay still when the organism awaits execution. An organism would therefore experience time's forward flow only during execution by the CPU. Relative to physical time, the experienced time flows in a stop-and-go manner due to time-shared processing, but the organism itself would perceive smooth and continuous time flow.

Organisms living on high-speed or multi-processor computers would take less physical time to execute, yet would not experience any speedups relative to the time they keep. Since the organism's time flows forward differently from the time experienced by physical-world entities, this time can be thought of as *cyber time*. Cyber time is built from physical time but its flow rate is governed by computing environment specifics. Cyber time's passage rate variation among computers with differing architectures, clock speeds and execution loads is analogous to the differing rates of time passage on physical-world celestial bodies. By the general theory of relativity, the time passage rate on a celestial body is determined by its mass, spin rate and distance from rotational axis. Each computer in a network is analogous to a planet with regards to cyber time. Like physical time, cyber time is one-dimensional.

4 The Hop Dimension

How can cyber universe organisms have a single spatial dimension when physical-world organisms cannot be two-dimensional Flatlanders [1] or one-dimensional Linelanders? The answer may lie in the existence of another dimension, the *Hop dimension*. Hop is best understood in the context of organism motion through cyberspace and cyber time. An organism advances in time when the CPU executes its next instruction, and moves in cyberspace when the CPU copies it to a new location. In essence it teleports to the new location without passing through the intervening cyberspace. A teleporting entity necessarily travels through a different dimension as it teleports, and Hop facilitates teleportation in the cyber universe. Figure 1 shows two teleportation examples: one of a physical-world Flatland object and the other of a software entity in cyberspace.

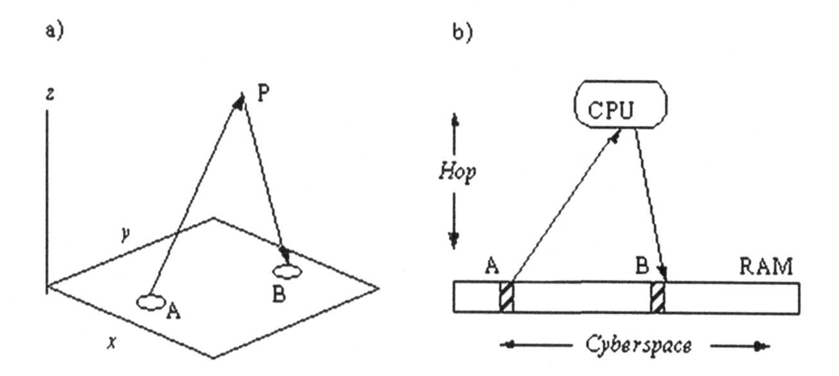

Fig. 1. Teleportation in two media. a) Physical-world Flatland object uses dimension z to teleport from A to B in the xy plane. b) Digital organism uses Hop dimension to teleport from A to B in cyberspace RAM

Organisms reside in memory and move forward in time when executed by the CPU. Cyberspace is therefore implemented by memory and cyber time by the CPU. The CPU-memory connecting bus facilitates transfers through cyberspace and cyber time, and hence implements Hop. The bus' finite length makes Hop a bounded dimension. An organism's clock stays still during Hop traversal, but the traversal time manifests as elapsed physical-world time. Hop connects cyberspace and cyber time and may also have a qualitatively intermediate nature. Lacking memory locations, Hop's distances cannot be measured in cyberspace address units. Physical-world objects can displace into any of three spatial dimensions to make way for objects moving into their path, but cyberspace's single dimension prevents organisms from entering other cyberspace dimensions. Motion in cyberspace is thus necessarily mediated by Hop.

5 Implications and Conclusions

Ray states that cyberspace is non-Euclidean since all cyberspace locations are equidistant from each other [5]. If we disregard cyberspace motion and define cyberspace distance in memory location units, cyberspace is Euclidean as all locations have an identical storage capacity. But the topology of the cyberspace-Hop combine appears to be non-Euclidean. Since secondary storage access is much slower than RAM access, secondary storage may appear to organisms as a cyberspace area with a higher 'viscosity' or 'density' than RAM. Computer network links appear to be Hop instances since organisms cannot be operational as they pass along network links, and link distances cannot be measured in memory location units. Hop's dimensionality equals the number of CPUs in shared-memory multiprocessors. We deduced Hop's existence by studying cyberspace entity teleports. In hindsight it could have been deduced from a computer block diagram. The CPU implements cyber time and RAM implements cyberspace, so the bus — the other major block — forms a third dimension of an intermediate nature.

Does Hop have physical-universe analogues? String theory [3] points to a physical universe with four space-time and seven curled dimensions. Are the additional dimensions Hop-like, facilitating object motion through physical space-time? Are the cyber and physical universe's particularities too disparate to develop a unified theory of the two? To paraphrase Langton [4], viewing computer networks as Artificial Universes would allow studies of universes-as-they-could-be than just the universe-as-it-is. The three-dimensional cyber universe can be thought to be embedded in the ambient four-dimensional physical universe, a view that parallels Arkani-Hamed et. al.'s suggestion that the physical universe itself is embedded in a higher dimensional universe [2].

In conclusion, the cyber universe has the cyberspace, cyber time and Hop dimensions. Hop facilitates motion and is intermediate between cyberspace and cyber time. The cyber universe structure may hold clues to physical universe structure and vice versa, and its characterization will offer insights into digital organisms' adaptability to their habitat.

Acknowledgments. Thanks to Thomas Baby, Sanjeev Khudanpur, Sridhar Srinivasan, Amar Karvir, Eric Ellington, John Werry, Prashant Nagaraddi and others for discussions.

References

1. Abbot, M. A.: Flatland: A Romance of Many Dimensions. Princeton University Press, Princeton (1991)
2. Arkani-Hamed, N., Dimopoulos, S., Dvali, G.: The Universe's Unseen Dimensions. Scientific American, 62-69 (August 2000)
3. Greene, B. R.: The Elegant Universe. Random House, Inc., New York (1999)
4. Langton, C. G.: Artificial Life. In: Nadel, L. and Stein, D. (eds.): Lectures in Complex Systems. Addison-Wesley, Reading, Mass (1992) 189-241
5. Ray, T. S.: An Evolutionary Approach to Synthetic Biology: Zen and the Art of Creating Life. In: Langton, C. (ed.): Artificial Life. MIT Press, Mass (1997)
6. Silberschatz A. and Galvin, P.: Operating System Concepts. John Wiley and Sons, New York (1999)

Taxonomy in Alife. Measures of Similarity for Complex Artificial Organisms

Maciej Komosiński and Marek Kubiak

Institute of Computing Science,
Poznan University of Technology
Piotrowo 3A, 60-965 Poznan, Poland
Maciej.Komosinski@cs.put.poznan.pl

Abstract. In this paper a formal approach to construction of a similarity measure for complex creatures is presented. The simulation model is described, and a Framsticks agent is expressed in a formal way. This helps in defining a dissimilarity measure. Two main ideas are discussed with reference to biology, namely genotypic and phenotypic methods. The holistic phenotypic measure is then proposed, where a fast, heuristic algorithm is used. Examples of its application are shown, including mutation and crossing over analysis, and a clustering tree based on distances between pairs of seven artificial individuals.

1 Introduction

Many research works in the field of artificial life concern studies of evolutionary processes, their dynamics and efficiency. Various measures and methods have been developed in order to be able to analyze evolution, complexity, and interaction in the observed systems. Other works try to understand behaviors of artificial creatures, regarding them as subjects of survey rather than black boxes with assigned fitness and performance. There are some works which employ formal views for these purposes [2,5,14].

Artificial life systems, especially those applied to evolutionary robotics and design [4,6,11], are quite complex and it is difficult to understand the behavior of existing agents in detail. The only way is to observe them carefully and use human intelligence to draw conclusions. Usually, the behavior of such agents is non-deterministic, and their control systems are sophisticated, often coupled with morphology and very strongly connected functionally [12].

Thus for the purposes of studying behaviors and populations of individuals, one needs high-level, intelligent support tools [10]. It is not likely that automatic tools will soon be able to produce understandable, non-trivial explanations of sophisticated artificial agents. However, it is possible to devise automatic measures of similarity which will help in observation of regularities, groups of related individuals, etc.

Similarity can be identified in many ways, including aspects of morphology (body), brain, size, function, behavior, performance, fitness, etc. Whatever definition is used, automatic measure of similarity can be useful for

J. Kelemen and P. Sosík (Eds.): ECAL 2001, LNAI 2159, pp. 685-694, 2001.

- optimization to introduce artificial niches by modification of fitness values [7,13],
- studies of evolutionary processes and the structure of populations of individuals,
- studies of function/behavior of agents,
- reduction of the number of agents to a small subset of interesting, diverse, unique individuals,
- inferring dendrograms (and hopefully, phylogenetic trees) based on distances between organisms.

For such reasons we developed a measure of dissimilarity on the realistic agents model of *Framsticks*. Although the model is simpler than biological creatures, it is general enough to observe various properties and difficulties. Devising a good measure of similarity on agents of such complexity is not an easy task, and was not studied so far. The results and conclusions of this long-term work are meaningful for the fields of artificial life, optimization, biology, and robotics. In this paper we report the first step towards such measure.

The paper is organized as follows: section 2 describes the model of simulation and evolution, as well as the formal definition of an agent. In section 3 we focus on the dissimilarity measure, discussing our motivations and biological background. In section 4 a few simple experiments are presented. Section 5 summarizes the work and outlines future goals.

2 The Model

2.1 Simulation

Agents in *Framsticks* are built of body and brain. Body is composed of material points (called *parts* in this paper) connected by elastic *joints*. Brain is made from *neurons* and their elements called *neuroitems* (these are receptors, effectors, and neural connections). For more detailed description of the model mechanics and neural network refer to [9,8,1].

It is possible to use the system for simulation of various processes, including local optimization, evolutionary optimization, coevolution of populations, spontaneous evolution, multi-criteria optimization, etc. Such universality is obtained by the use of a scripting language, which controls the overall architecture of the system as well as more specific issues.

Framsticks supports a few genetic representations of agents, including direct and developmental [10]. It is possible to easily "plug in" a new encoding and genetic operators. The fact of diverse genotype languages influences the characteristics of evolutionary process in each case, but finally, each encoding has to produce an agent. Thus it is possible to compare phenotypes which are "compatible" because they are build of a set of standardized components (parts, joints, neurons, and neuroitems).

2.2 Formal View

Elements of a framstick agent are characterized by many properties. Morphological elements have mass, friction, stiffness, etc., and control elements – weights, sigmoid coefficients, etc. Body and brain are connected. All these aspects make the definition of a similarity measure difficult; to devise a formal measure it is helpful to mathematically describe an agent.

Let P denote a non-empty set of parts (material points) of a creature's body. We define several functions which describe physical and biological attributes of parts ($\mathbf{R}$ denotes the set of real numbers):

- *position:* $P \to \mathbf{R}^3$
- *orientation:* $P \to [0, 2\pi)^3$
- *mass:* $P \to \mathbf{R}_{>0}$
- *volume:* $P \to \mathbf{R}_{>0}$
- *friction:* $P \to \mathbf{R}_{>0}$
- *ingestion:* $P \to \mathbf{R}_{>0}$
- *assimilation:* $P \to \mathbf{R}_{>0}$

Furthermore, we introduce J as a non-empty set of joints, and a function:

- *joint_connects:* $J \to P^2$

which represents connections (arcs) between pairs of parts (vertices). The graph corresponding to morphology must be connected, which means that the creature's body cannot have isolated parts. No arcs joining the same pair of vertices may exist. The properties of joints include

- *stiffness:* $J \to \mathbf{R}_{\geq 0}$
- *rot_stiffness:* $J \to \mathbf{R}_{\geq 0}$
- *stamina:* $J \to \mathbf{R}_{\geq 0}$

For brain, we define a set of neurons N, which may be empty. Properties of a neuron can be defined as functions:

- *force:* $N \to [0, 1]$
- *inertia:* $N \to [0, 1]$
- *sigmoid:* $N \to \mathbf{R}$
- *initial_state:* $N \to \mathbf{R}$
- *placement:* $N \to P$

The set C (which may be empty) concerns neuroitems: elements related to connections between neurons and interfacing parts of the body. This set is divided into pair wise distinct subsets, so that $C = Const \cup RecG \cup RecT \cup RecS \cup RotMuscle \cup BendMuscle \cup Conn$. These subsets describe constant signal inputs, receptors (equilibrium, touch, smell), muscles with neural control and connections between neurons.

There are also some properties that these items have (some functions are typical to some specific types of neuroitems):

- *parent_neuron:* $C \to N$
- *connected_to:* $Conn \to N$
- *weight:* $(Const \cup RecG \cup RecT \cup RecS \cup Conn) \to \mathbf{R}$
- *on_part_placement:* $(RecT \cup RecS) \to P$

- *on_joint_placement:* $(RecG \cup RotMuscle \cup BendMuscle) \rightarrow J$
- *strength:* $(RotMuscle \cup BendMuscle) \rightarrow [0, 1]$
- *rot_range:* $RotMuscle \rightarrow [0, 1]$

The two top functions describe the structure of a neural network of a creature. The third one defines weights of connections in the network.

Thus a creature can be described by four sets: *P*, *J*, *N*, and *C*, and relations between their elements. It can be clearly seen that the structure of an agent is quite complicated – it has different aspects: structural (one graph defining body, another one defining neural network which is located on the body's graph), geometrical (position and orientation in a 3-D space), and also several parameters describing elements of body and brain.

3 Similarity Measure

3.1 Biological Reference

The measures of similarity (or dissimilarity) are widely adopted in practice by biologists for classification and constructing taxonomies of organisms. Such measures are built based on two kinds of information: genotypic (DNA sequences) [15] and phenotypic (construction and behavior of individuals) [16]. These two streams (apart from phenetics/cladistics difference) can be characterized as molecular and morphological approach.

The first emphasizes the role of a DNA sequence as the true indicator of on-going evolutionary processes. The analyzed DNA fragments (or RNA transcripts/protein sequences) are selected based on knowledge of their relative position in the genome and their function (or the lack thereof). The analyzed sequences are often introns – the non-coding regions of genes, which are assumed to accumulate random mutations non-selectively, according to the defined probabilistic model [3]. However, using this approach with artificial systems may raise a few problem issues. With *Framsticks* system, these are:

- organisms' genomes are very short compared to living organisms, their length varies from a few characters to a few thousands,
- all formats encode organisms on a very high level: the building blocks are entire elements of an ALifeform body structure,
- difficulties arise in the process of selecting common properties, on basis of which the analysis could be performed,
- there are almost no non-coding, coevolving sequences in the artificial genomes, for which the assumption of a probabilistic model could be made.

Translation of the coding format to the two-state binary model (akin to DNA sequences) would not eliminate these problems.

The latter approach (the morphological one), which entails defining the similarity measure using phenetic resemblance, is more promising. In biology, the overall phenetic resemblance is inferred from many different characters' states – which, if possible, should utilize clearly distinguished and independent properties. This serves to minimize the randomness effect in the (dis)similarity measure. Additional operations in this approach cope with the problems of affine transformations (organism size, scale, and rotations of body parts). The end results are recalculated into an overall measure value, most often using a weighted sum model.

The important question is how to choose the properties of organisms for analysis. Answering this requires delving into the ALife model itself. The numerical similarity analysis and clustering of living organisms require such discrimination because biological perspective is inherently incomplete; an all-inclusive model cannot be built mainly because the principles of the entire system are not known. As we know all the details of ALife individuals, such a holistic approach is possible. Thus such approach in construction of a dissimilarity measure is presented in this work.

3.2 Preliminary Considerations

The problem of similarity estimation in *Framsticks* is closely related to the problem of isomorphism of graphs. Taking only morphology into consideration, the task would be to find the matching between parts of the two agents. Let $G_1=(P_1, J_1, joint\_connects_1)$ and $G_2=(P_2, J_2, joint\_connects_2)$ be the graphs representing bodies of two agents. We assume that $|P_1| \leq |P_2|$ without loss of generality.

In the common situation when $|P_1| \neq |P_2|$, it is not possible to find a mutually single-valued function *matching:* $P_1 \rightarrow P_2$. However, one can add a number of artificial points to P_1 forming P_1' set, so that $|P_1'| = |P_2|$, and connect these points with artificial joints so that the graph is consistent. Thus the sets P_1', J_1' and the function $joint\_connects_1'$ would be obtained. The desired function *matching:* $P_1' \rightarrow P_2$ should then maximize the covering of edges in both sets.

However, such approach is not feasible. It requires a few steps where it is unknown how to act (adding points-vertices and joints-edges), and the final similarity value will most probably be sensitive to these arbitrary choices. Furthermore, an exact algorithm working on such sophisticated graph representations would have an unacceptably high computational complexity.

3.3 Heuristic Algorithm

Following these considerations we constructed a heuristic method. This method tries to match the body structure of two individuals based on the degrees of parts as the main piece of information. The problematic step of adding parts and joints to the smaller organism was abandoned, so the matching function has the form *matching:* $P_1 \rightarrow M$, where $M \subseteq P_2$. In the overall measure, information about the number of neurons and neuroitems located on point p was also used. The corresponding functions are:

$$neuron\_count(p) = |\{n \in N: placement(n)=p\}|$$
$$nitem\_count(p) = neuron\_count(p) + |\{c \in C: placement(parent\_neuron(c))=p\}|$$

To match parts in two organisms, the degree of vertices was used:

$$dDeg(matching) = \sum_{p1\in P1,\, p2\in M:\, p2=matching(p1)} |degree(p_1) - degree(p_2)| + \sum_{p2\notin M} |degree(p_2)|$$

The second criterion of similarity considers the numbers of neurons on already matched parts:

$$dNeu(matching) = \sum_{p1\in P1,\, p2\in M:\, p2=matching(p1)} |neuron\_count(p_1) - neuron\_count(p_2)| + \sum_{p2\notin M} |neuron\_count(p_2)|$$

Sets of parts are sorted by *degree*, then by *nitem_count* and *neuron_count*. Then the algorithm is matching parts starting from the highest degree groups and trying to find parts that have similar *nitem_count* and *neuron_count*. Starting from the highest degree is likely to preserve the most important parts from being unmatched. In case of ambiguity (when points have the same *degree*, *nitem_count* and *neuron_count*), the remaining properties of parts are used as discriminating features (except of *position*).

Based on the function *matching* constructed in the algorithm presented, two parameters *dDeg* and *dNeu* are estimated. Finally, the dissimilarity between two organisms is evaluated using the weighted sum model:

$$dissimilarity(O_1, O_2) = w_{Deg}\cdot dDeg + w_{Neu}\cdot dNeu$$

where w_{Deg} and w_{Neu} are weights of these two criteria. The proposed measure has the following properties:

- *dissimilarity*(i, i) = 0,
- *dissimilarity*(i, j) = *dissmilarity*(j, i),
- *dDeg* component holds the triangular inequality.

4 Computational Experiments

For all experiments, the weights of similarity criteria were w_{Deg}=1 and w_{Neu}=0.5. These values were adjusted experimentally, constituting an acceptable tradeoff between importance of differences in body and brain.

4.1 Mutation

The measure was tested on two random mutations in the "recurrent language" genetic encoding [1,18] (figure 1).

In both cases, the measure of dissimilarity of ancestors and descendants (which were in genetic terms distant by a single mutation) indicated *dissimilarity* = 2. This small value characterizes well the minor disturbances of the mutation operation. Considering weights used, dissimilarity value of 2 corresponds to the difference of 2 unmatched parts of degree 1, or 4 unmatched neurons.

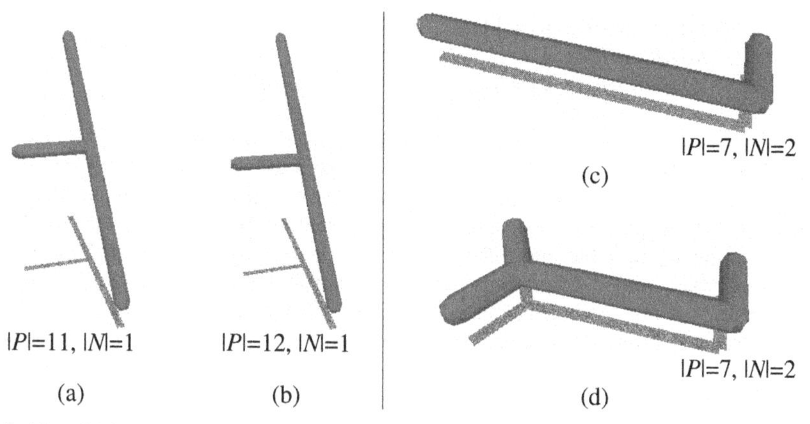

Fig. 1. (a) and (c) ancestors, (b) and (d) descendants. Only morphology is shown. In (a) and (b), the branching nodes are in different places.

4.2 Crossover

In the crossover experiment, we combined two agents described by the direct representation, which simply lists all elements of body and brain. The first object was a simple table, which was a human-designed simple object with no neural network. The second one was an individual with a neural network. This individual was the result of a long-term evolution oriented for speed in a land environment (figure 2).

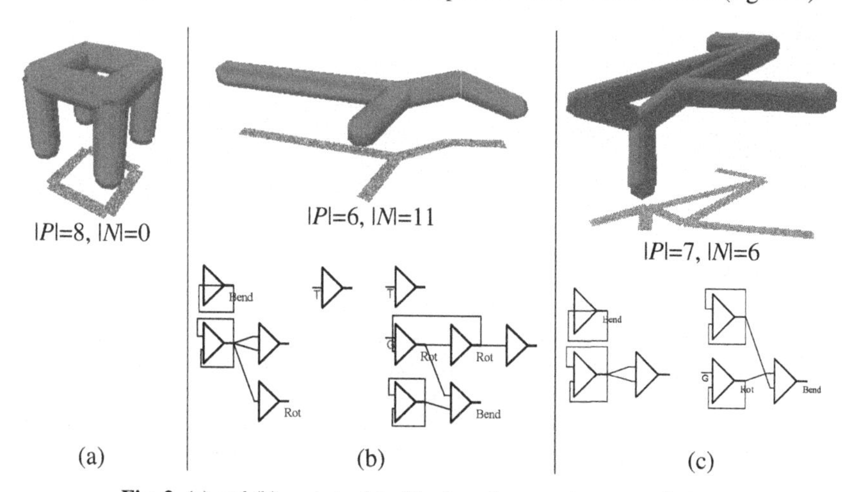

Fig. 2. (a) and (b) parents, (c) offspring after crossing over of parents.

The crossing over operation working on direct representation consists in cutting the parent agents by a 3-D plane and constructing an offspring from two halves of the two parents. It can be seen that the child has some structure parts of parents in its body and brain. The values for the measure are as follows:

dissimilarity(a, b) = 11.5
dissimilarity(a, c) = 2
dissimilarity(b, c) = 9.5

It is interesting that *dissimilarity*(a, b) = *dissimilarity*(a, c) + *dissimilarity*(b, c), so the child is exactly "between" parents, which reflects the averaging role of the crossing over operator used. This relation holds in the above example, but it is not the general property of the proposed measure. According to the above values, the offspring is very similar to the table and different from the speed-evolved parent. Note that this measure takes into account only structural morphological and neural quantitative properties, but not geometrical aspects (position of parts and length of joints). Thus it can be seen that the child (fig. 2c) has similar structure of the body graph to the parent (fig. 2a).

4.3 Clustering

The similarity measure was finally applied to a small example of a clustering problem. Seven diverse creatures were chosen (figure 3).

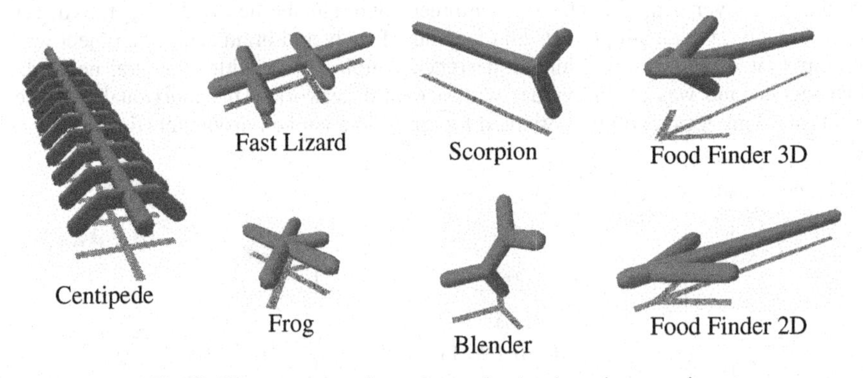

Fig. 3. The seven organisms chosen for the clustering experiment.

The two creatures out of seven can seek food in water. These are food finders; their bodies were designed, and neural networks were evolved. The Centipede and Blender were constructed entirely by a human; Frog's and Fast Lizard's morphologies were constructed and brains were evolved; the remaining ones were evolved from scratch.

For each pair of these creatures, the dissimilarity was computed (table 1).

Table 1. The symmetric matrix of dissimilarity values of selected creatures.

	Centipede	Fast Lizard	Food 2D	Food 3D	Frog	Blender	Scorpion
Centipede	0	116	109.5	99.5	128.5	124.5	123
Fast Lizard	116	0	11.5	19.5	13.5	12.5	10
Food 2D	109.5	11.5	0	11	20	18	14.5
Food 3D	99.5	19.5	11	0	31	27	25.5
Frog	128.5	13.5	20	31	0	8	9.5
Blender	124.5	12.5	18	27	8	0	4.5
Scorpion	123	10	14.5	25.5	9.5	4.5	0

Based on this matrix, it is possible to construct a clustering tree. UPGMA method was used (unweighted pair group method with arithmetic averages) [16]. It is a simple and intuitively appealing example of the phenetic approach to data summary. UPGMA groups objects that differ least according to the similarity measure without other points of consideration. The obtained hierarchical tree (with distances respected roughly) is shown on figure 4. Considering structures of creatures shown on figure 3, it can be seen that the hierarchy is sound.

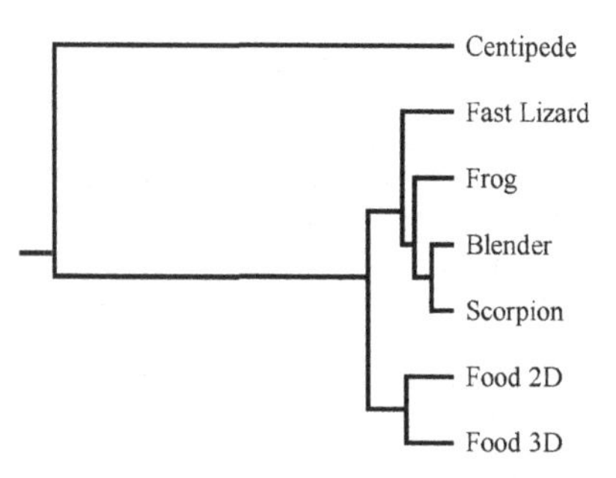

Fig. 4. The resulting clustering tree.

5 Conclusions and Future Work

In this paper, a formal approach to agent analysis was presented. Based on this approach, a structural dissimilarity measure was proposed, and its prototypic applications were shown on some examples.

The measure will be further developed to take into account all the relevant features of an agent, including geometrical and biological properties. Considering geometry creates additional problems, which are also present in biological analyses, namely affine transformations (scaling, rotation, etc.). However, geometry has a great impact on the behavior of agent, functionality, and dynamics of its movement, so it cannot be neglected. Even though, the structural properties considered in this paper are an important holistic property of a creature and cannot be easily modified by single mutations, so it is a good basis for the matching algorithm.

There are strong complexity constraints imposed on such measure, because it is to be used many times. For artificial speciation, the measure has to be computed for all pairs of individuals in each generation. In analysis of structure of populations, it has to be computed once, but the population size is usually much greater than in the case of a typical evolutionary algorithm. Thus only the heuristic approach is possible; exact measures would be too slow to compute for all pairs of individuals. It is required for the measure to have some good properties, like robustness, simplicity, clarity, etc. Another properties are required if the measure is to be used in a biological, evolutionary context. Our future work will proceed in these directions. Application of this measure to diversify populations of individuals will also be tested.

Acknowledgements. This work has been supported by the State Committee for Scientific Research, from KBN research grant no. 8T11F 006 19, and by the Foundation for Polish Science, from subsidy no. 11/2001.

References

1. Adamatzky A., Komosiński M. and Ulatowski S. (2000) Software review: Framsticks, *Kybernetes: The International Journal of Systems & Cybernetics*, **29** (9/10), 1344-1351.
2. Agre P. and Horswill I. (1997) Lifeworld analysis, *Journal of Artificial Intelligence Research*, **6**, 111-145.
3. Avise J.C. (1994) *Molecular Markers, Natural History and Evolution*. Chapman & Hall, New York.
4. Bentley P. (1999) *Evolutionary design by computers*. Morgan Kaufmann.
5. Bongard J.C. and Paul C. (2000) Investigating morphological symmetry and locomotive efficiency using virtual embodied evolution. In: *Proceedings of the Sixth International Conference on Simulation of Adaptive Behaviour* (ed. by J.-A. Meyer), pp. 420-429. MIT Press.
6. Funes P. and Pollack J.B. (1998) Evolutionary body building: adaptive physical designs for robots, *Artificial Life*, **4** (4, Autumn), 337-357.
7. Goldberg D.E. (1989) *Genetic Algorithms in Search, Optimization and Machine Learning*. Addison-Wesley Publishing Co.
8. Komosiński, M. and Ulatowski, Sz. Framsticks Internet site, http://www.frams.poznan.pl/
9. Komosiński M. (2000) The World of Framsticks: Simulation, Evolution, Interaction. In: *Virtual Worlds. Lecture Notes in Artificial Intelligence 1834* (ed. by J.-C. Heudin), pp. 214-224. Springer-Verlag.
10. Komosiński M. and Rotaru-Varga A. (2000) From Directed to Open-Ended Evolution in a Complex Simulation Model. In: *Artificial Life VII* (ed. by M.A. Bedau, J.S. McCaskill, N.H. Packard and S. Rasmussen), pp. 293-299. MIT Press.
11. Lipson H. and Pollack J.B. (2000) Automatic design and manufacture of robotic lifeforms, *Nature*, **406** (6799), 974-978.
12. Lund H.H., Hallam J. and Lee W.-P. (1997) Evolving Robot Morphology. In: *Proceedings of IEEE 4th International Conference on Evolutionary Computation*. NJ. IEEE Press. Invited paper.
13. Michalewicz Z. (1996) *Genetic Algorithms + Data Structures = Evolution Programs*. Springer-Verlag.
14. Rasmussen S. and Barrett C.L. (1995) Elements of a theory of simulation. In: *Proceedings of the European Conference on Artificial Life (ECAL 95), Lecture Notes in Computer Science*. Springer-Verlag, Berlin.
15. Setubal J. and Meidanis J. (1997) *Introduction to Computational Molecular Biology*. PWS Publishing Company.
16. Sneath P.H. and Sokal R.R. (1973) *Numerical Taxonomy*. Freeman & Co, San Francisco.

The λ-Game System: An Approach to a Meta-game

Gen Masumoto and Takashi Ikegami

General Systems Sciences, The Graduate School of Arts and Sciences,
The University of Tokyo 3-8-1 Komaba Meguro-ku, Tokyo, 153-8902, Japan
{gen,ikeg}@sacral.c.u-tokyo.ac.jp

Abstract. A new game system formalism is presented. The system describes both strategies and a game master (who computes scores in a given game system) in terms of the λ-calculus. A prisoners' dilemma game is revisited with this formalism, to discuss how meta-strategies emerge in this classical game, even without repetition.

1 Introduction

A relation between a game and its meta-game is the main concern of this paper. While game theory is based on Nash equilibrium with rational players, we often observe irrational behaviour and sometimes unexpected behaviour in practical situations (e.g., changing the rules of a game during play). We consider this discrepancy as a difference between a game and "play". Infants do play, but seldom show game behaviours; a rule of a game may be violated so that a game theoretical solution such as Nash equilibrium is only an artifact in play situations.

In artificial life studies (e.g., Artificial Life II meeting (1992)), an open-ended evolution concept was proposed and has been taken as a key concept to understand evolutionary dynamics. Rather than studying a closed static system, open-ended evolution treats an open system. Good examples can be found in Lindgren (1992), Ikegami (1994), and Ikegami and Taiji (1999). They studied the (noisy) iterated prisoners' dilemma game without fixing the number of strategies in advance. On the other hand, Howard (1966, 1971) pondered the paradox of rationality and developed meta-game theories. Hofstadter (1985) introduced "Nomic" games, where players can update the rules of the game that provide how to change themselves. These studies gave us new perspectives of open-endedness in game situations.

We propose here a new approach based on the λ-calculus. Fontana (1994) introduced a λ-calculus formalism into studies of chemical evolution. An advantage of λ formalism is that each λ-term can be either a function or a variable. Here we use the formalism in terms of game strategies. Instead of maintaining game theoretical concepts (i.e., rational players and Nash equilibrium), we hope to provide a model that bridges between game and 'play' behaviours. Key concepts are meta-strategies and interventions in game rules.

J. Kelemen and P. Sosík (Eds.): ECAL 2001, LNAI 2159, pp. 695–699, 2001.

2 λ-Game System

In the λ-calculus, Boolean truth values and a conditional can be represented as $\mathbf{true} \equiv \lambda xy.x$, $\mathbf{false} \equiv \lambda xy.y$. If B only takes Boolean terms (i.e., either **true** or **false**) and P, Q take arbitrary λ-terms, then a conditional "if x then P else Q" is represented by a specific λ-term $\lambda x.xPQ$. They actually function as follows: $(\lambda x.xPQ)\,\mathbf{true} = P$, $(\lambda x.xPQ)\,\mathbf{false} = Q$. One can define natural numbers in a variety of ways in the λ-calculus. Here we will use Barendregt (1981) numerals and henceforth refer to them simply as numerals **n**. Barendregt numerals are defined inductively in the following way: $\mathbf{0} \equiv \lambda x.x$, $\mathbf{n+1} \equiv \lambda x.(x\,\mathbf{false}\,\mathbf{n})$.

The prisoners' dilemma game consists of two elementary actions, cooperation (**C**) and defection (**D**). The rewards of the game for each strategy are as follows: $(\mathbf{C},\mathbf{C}) = (3,3)$, $(\mathbf{C},\mathbf{D}) = (0,5)$, $(\mathbf{D},\mathbf{C}) = (5,0)$, $(\mathbf{D},\mathbf{D}) = (1,1)$.

We express **C** by $\mathbf{true}(= \lambda xy.x)$ and **D** by $\mathbf{false}(= \lambda xy.y)$. We will now construct a game master to enable us to interpret the true and false values as **C** and **D** actions. A requirement of a game master is to simulate the prisoners' dilemma game with the payoff matrix described above. Therefore we define the "game master" (**G**) as:

$$\mathbf{G} \equiv \lambda x.x(\lambda y.y\,\mathbf{3}\,\mathbf{5})(\lambda y.y\,\mathbf{0}\,\mathbf{1}). \tag{1}$$

Given two strategies S_i and S_j in λ-terms, the game master computes the corresponding reward of S_j by entering the two strategies successively into the game master: $\mathbf{G}S_iS_j \to \mathbf{G}'S_j \to Reward_j$. A basic requirement for the game master is to form the correct PD rewards. That is, a match between elementary strategies **C** and **D** should be computed as, for example, $\mathbf{GCC} \to \mathbf{G'C} \to \mathbf{3}$. A simple construction of such an intermediate $\mathbf{G}'(= \mathbf{G}S_i)$ is to have a conditional as follows:

$$\mathbf{G\,C} = \lambda y.y\mathbf{35}, \qquad \mathbf{G\,D} = \lambda y.y\mathbf{01} \tag{2}$$

It is worth noting that there is no distinction between strategy, score and the game master in the λ-calculus. Therefore we can apply **G** to a λ-term that represents a function giving a strategy instead of a mere Boolean value. In this case, a Barendregt numeral is sometimes obtained via an unexpected reduction process.

3 Numerical Experiments and Analysis

Since the application of the game master on non-closed λ-terms never yields Barendregt numerals, we here only consider closed terms as the strategies.

A strategic closed λ-term is represented by $\lambda x_1 \ldots x_m.M$, where $0 \le m \le 3$ and M is composed of n (here: $1 \le n \le 20$) applications of the atomic elements $X_i \in \{x_1, \ldots, x_m, \mathbf{true}, \mathbf{false}, \mathbf{I}, \mathbf{S}, \mathbf{V}\}$ in an arbitrary order, where $\mathbf{I} \equiv \lambda x.x$, $\mathbf{S} \equiv \lambda xyz.xz(yz)$ and $\mathbf{V} \equiv \lambda xyz.zxy$. For example, $\lambda x_1x_2.(x_2(\mathbf{I}(\mathbf{true}\,x_1)))$, and $(\mathbf{S}((\mathbf{false}(\mathbf{I}\,\mathbf{true}))\mathbf{V}))$ are allowable strategies in our λ-game model.

Generating random λ-terms L of the above form, we first transform them into normal forms before we use them as game strategies. Then, the strategy's rewards against **C**, **D** and itself are computed, i.e., $(\mathbf{GLC})$, $(\mathbf{GC}L)$, $(\mathbf{G}L\mathbf{D})$, $(\mathbf{GD}L)$ and $(\mathbf{G}LL)$. Some matches between two strategies cannot return any numerical values. When that occurs, those strategies are eliminated, and we term those strategies syntactically incorrect strategies. In addition, λ-terms that do not acquire a normal form within a preset limit (here: 1000) are removed from meta-game strategies. Therefore, our meta-game system has no halting problem.

While examining randomly-generated strategies, we noticed that many syntactically correct strategies exist. A certain set of strategies can give and take rewards that are not expected in the original payoff matrix. Examples of such strategies are depicted in Table 1.

We have good reason to call these strategies *meta-strategies*, as they do not merely follow the PD rule but interfere with the game master. All the meta-strategies obtained in our simulation can be categorized with respect to their intermediate $\mathbf{G}'$ forms. There are roughly three types as follows.

Type I (constant function): $\mathbf{G}'$ is computed as $\lambda x.\mathbf{n}$, where $\mathbf{n}$ is a Barendregt numeral. $\mathbf{G}'$ always outputs $\mathbf{n}$ against any other strategies. Examples of such strategies are (1) and (2) in Table 1, and they correspond to all D and all C respectively.

Type II (non-conditional form): Examples of this type $\mathbf{G}'$ are $\lambda x.(x\mathbf{01})(x\mathbf{35})$ and $\lambda z.(z\,\mathbf{false}\,\mathbf{2})(z\,\mathbf{0}\,\mathbf{1})$. A type II strategy is a reactive strategy sensitive to the given opponent. It returns rewards of a Barendregt numeral against many strategies through interference reduction processes with the game master. A known good strategy tit-for-tat falls into this group ((3) in Table 1).

Type III (conditional form): $\mathbf{G}'$ is computed as $\mathbf{G}' = \lambda x.x\,\mathbf{m}\,\mathbf{n}$, where $\mathbf{m}$ and $\mathbf{n}$ are Barendregt numerals. This strategy has the conditional structure shown in Eq.(2). Some strategies of type III that construct $\lambda x.x\mathbf{01}$ behave as $\mathbf{D}$ players to other strategies and exploit $\mathbf{C}$ strategies. ((5) in Table 1) Moreover, some of the strategies with $\mathbf{G}' = \lambda x.x\mathbf{01}$ ((6) in Table 1) can give two points reward when playing against themselves. These two points are obtained through interference processes with the conditional $\mathbf{G}'$, because the strategies of this type can deceive the conditional $\lambda x.x\mathbf{01}$. This means that strategies such as (6) in Table 1 can exploit $\mathbf{D}$ strategies, because both strategies generate the conditional $\mathbf{G}'(= \lambda x.x\mathbf{01})$, and strategies such as (6) obtain two points from $\mathbf{G}'$ whereas $\mathbf{D}$ can obtain only one point. On the other hand, the strategies of (6) are generally self-cooperative because their $\mathbf{G}'$s induce cooperative outputs when played against themselves.

We have discussed outcomes against $\mathbf{C}$, $\mathbf{D}$ and the same strategy. Random matches of meta-strategies are now computed to examine the above analysis. It should be noted that each type shows some characteristic features described as follows. Strategy (1) of type I avoids being exploited as it gives a constant reward of zero to the opponent, but this type of strategy never cooperates with a similar strategy. Strategy (3) of type II has the sophisticated protocol for mutual cooperation described above, but it cannot work correctly for randomly selected strategies and sometimes this strategy is exploited. Strategy (6) of type III is a robust strategy, as it outputs conditional $\mathbf{G}'$ against other strategies. Fortunately, this $\mathbf{G}'$ can be interpreted by itself to obtain two points reward.

Table 1. Examples of meta-strategies

type	vs. **C**	vs. **D**	vs. self	examples of λ-terms	
C	3, 3	0, 5	3	$\lambda xy.x$ (= **true**)	
D	5, 0	1, 1	1	$\lambda xy.y$ (= **false**)	
type I	5, 0	1, 0	0	$\lambda xy.(((x(\lambda z.(z\mathbf{IS})))\mathbf{S})y)y$	(1)
	3, 5	0, 5	5	$\lambda xyz.x((((((x\mathbf{I})\mathbf{I})y)\mathbf{I})\mathbf{true})z)\mathbf{false})$	(2)
type II	3, 3	0, 0	3	$\lambda xy.(((((y\mathbf{I})\mathbf{S})\mathbf{I})\mathbf{S})y)x$	(3)
	5, 0	2, 1	0	$\lambda x.x(\lambda yz.(z(\lambda y.(((((x(x\mathbf{true}))\mathbf{S})\mathbf{I})\mathbf{true})\mathbf{S}))y))$	(4)
type III	4, 0	1, 1	1	$\lambda xy.y(\lambda z.((x\mathbf{S})z(\lambda yw.(wzy))))$	(5)
	6, 0	2, 1	2	$\lambda xy.((((y(x\mathbf{false}))\mathbf{V})\mathbf{false})y)$	(6)

4 Discussion

An abstract game system has been described that makes use of a λ-calculus formalism. A great advantage of λ-calculus formalism compared with dynamical systems formalism is that it can take both variables and functions as its input values. This advantage has allowed us to study a meta-game structure that provided a way to formalize play behaviours.

For a λ-game version of the PD game, we found that there were many meta-strategies that violated the rules of the game. With respect to λ-terms, there were three common structures for syntactically correct strategies. The well-known tit-for-tat strategy emerges as a type II strategy. On the other hand, some good strategies found in type III exploit both **C** and **D**, but cooperate against each other.

An apparent disadvantage of the λ-game comparing with the dynamical systems approach is that it lacks "dynamics". To compensate for the lack of dynamics, adaptation and/or evolution are possible ways to incorporate a time structure into the λ-game system. It is interesting to see the coevolution of strategy and the game master. A marriage between a λ-game and a dynamical systems approach is our next challenge.

Acknowledgments. This work is partially supported by Grant-in-aid (No. 09640454) from the Ministry of Education, Science, Sports and Culture.

References

[1981]Barendregt, H.G.: *The Lambda Calculus: Its Syntax and Semantics* Amsterdam, North Holland (1981).

[1994]Fontana, W., Buss L.W.: The Arrival of the Fittest: Toward a Theory of Biological Organization. *Bulletin of Mathematical Biology* 56 (1994) 1–64.

[1985]Hofstadter, D.R.: Metamagical Themas. BasicBooks (1985).

[1966]Howard, N.: The Theory of Meta-games. *General Systems* 11 (1966) 167–186.

[1971]Howard, N.: *Paradoxes of Rationality: Theory of Metagames and Political Behavior.* MIT Press, Cambridge MA. (1971).

[1994]Ikegami, T.: From Genetic Evolution to Emergence of Game Strategies. *Physica D* 75 (1994) 310–327.

[1999]Ikegami, T., Taiji, M.: Imitation and Cooperation in Coupled Dynamical Recognizers. In: Floreano, D. et al. (eds.): *Advances in Artificial Life*, Springer-Verlag (1999) 545–554.
[1992]Langton, C.G. et al. (eds): *Artificial Life II.* Addison-Wesley, CA. (1992).
[1992]Lindgren, K.: Evolutionary Phenomena in Simple Dynamics. In: Langton, C.G. et al. (eds): *Artificial Life II.* Addison-Wesley, CA. (1992) 295–311.

Formal Description of Autopoiesis Based on the Theory of Category

Tatsuya Nomura

Faculty of Management Information, Hannan University,
5–4–33, Amamihigashi, Matsubara, Osaka 580–8502, Japan

Abstract. Since the concept of autopoiesis was proposed as a model of minimal living systems by Maturana and Varela, there has been still few mathematically strict models to represent the characteristics of it because of its difficulty for interpretation. This paper proposes a formal description of autopoiesis based on the theory of category and Rosen's perspective of "closure under efficient cause".

1 Introduction

Autopoiesis gives a framework in which a system exists as an organism through physical and chemical processes, based on the assumption that organisms are machinary [3]. According to the original definition of it by Maturana and Varela, an autopoietic system is one that continuously produces the components that specify it, while at the same time realizing itself to be a concrete unity in space and time; this makes the network of production of components possible. An autopoietic system is organized as a network of processes of production of components, where these components:

1. continuously regenerate and realize the network that produces them, and
2. constitute the system as a distinguishable unity in the domain in which they exist.

The characteristics of autopoietic systems Maturana gives are as follows:

1. Autonomy by integration of various changes into the maintenance of their organization,
2. Individuality independent of mutual actions between them and external observers by repeatedly reproducing and maintaining the organization,
3. Self–determination of the boundary of the system through the self–reproduction processes,
4. Absence of input and output in the system by the fact that changes by any stimulus are subordinate to the maintenance of the organization which specifies the machine.

However, there has been still few mathematically strict models that represent autopoiesis. In [4,5], we discussed the difficulty of interpreting autopoiesis within

J. Kelemen and P. Sosík (Eds.): ECAL 2001, LNAI 2159, pp. 700–703, 2001.

system theories using state spaces and problems of some models proposed for representing autopoiesis. The aim in this paper is to clarify whether autopoiesis can really be represented within more abstract mathematical frameworks by introducing the theory of category [9], one of the most abstract algebraic structure representing relations between components. The focus is the concept of "closure under efficient cause" in "relational biology" by Rosen [6].

In relational analysis, a system is regarded as a network that consists of components having functions. Rosen compared machine systems with living systems to clarify the difference between them, based on the relationship among components through entailment [6]. In other words, he focused his attention on where the function of each component results from in the sense of Aristotle's four causal categories, that is, material cause, efficient cause, formal cause, and final cause. As a result, Rosen claimed that a material system is an organism if and only if it is closed to efficient causation. In this paper, we consider that closure under entailment or production is a necessary condition for a system to be autopoietic because the components reproduce themselves in the system. Then, we give a system closed under entailment in a category theoretic framework.

2 Systems Closed under Entailment in a Category Theoretic Framework

In this paper, we assume that a category $\mathcal{C}$ has a final object 1 and product object $A \times B$ for any pair of objects A and B. The category of all sets is an example of this category. Moreover, we describe the set of morphisms from A to B as $H_{\mathcal{C}}(A, B)$ for any pair of objects A and B. An element of $H_{\mathcal{C}}(1, X)$ is called a morphic point on X. For a morphism $f \in H_{\mathcal{C}}(X, X)$ and a morphic point x on X, x is called a fixed point of f iff $f \circ x = x$ ($\circ$ means composition of morphisms) [8]. Morphic points and fixed points are respectively abstraction of elements of a set and fixed points of maps in the category of sets.

When there exists the power object Y^X for objects X and Y (that is, the functor $\cdot \times X$ on $\mathcal{C}$ has the right adjoint functor $\cdot^X$ for X), note that there is a natural one–to–one correspondence between $H_{\mathcal{C}}(Z \times X, Y)$ and $H_{\mathcal{C}}(Z, Y^X)$ for any objects X, Y, Z satisfying the diagram in the left figure of Fig. 1. Thus, there is a natural one–to–one corrspondence between morphic points on Y^X and morphisms from X to Y satisfying the diagram in the right figure of Fig. 1.

One of the easist methods for representing the self–reproductive aspect of autopoiesis is considered to assume that components in a system are not only operands but also operators [2]. Thus, we assume that there is an isomorphism from the space of operands to the space of operators, that is, an object X with powers and an isomorphism $f : X \simeq X^X$ in $\mathcal{C}$. Then, there uniquely exists a morphic point p on $(X^X)^X$ corresponding to f in the above sense. Since the morphism from X^X to $(X^X)^X$ entailed by the functor $\cdot^X$, f^X, is also isomorphic, there uniquely exists a morphic point q on X^X such that $f^X \circ q = p$. We can consider that p and q entail each other by f^X. Furthermore, there uniquely exists a morphic point x on X such that $f \circ x = q$. Since we can consider

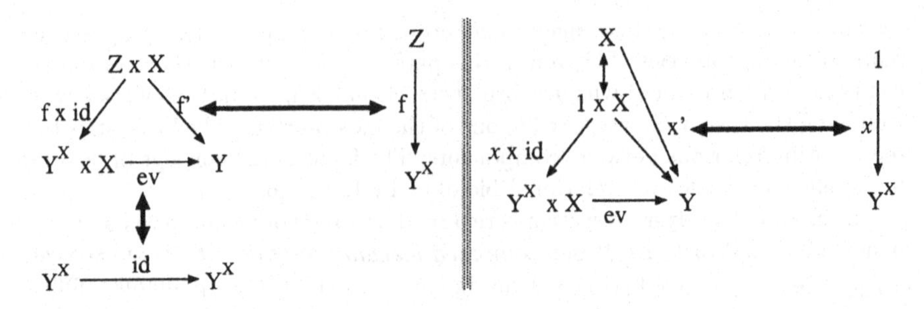

Fig. 1. Natural One–To–One Correspondence between $Hc(Z \times X, Y)$ and $Hc(Z, Y^X)$

that x and q entail each other by f, and f and p entail each other by the natural correspondence, the system consisting of x, q, p, f, and f^X is completely closed under entailment. Moreover, if a set S of morphic points on X is fixed by $q' : X \to X$ naturally corresponding to q, that is, $\forall x' \in S\ q' \circ x' \in S$ and $\exists x'' \in S\ s.t.,\ q' \circ x'' = x'$, we can consider that S entails itself by f (the existence of these sets is guaranteed by Theorem 1 in [8], that is, the fact that q' has fixed points by f as a labelling of X^X by X).

Fig. 2 shows the diagrams of this completely closed system and its hyper-digraph [1] representing the relationship on entailment between components (a thick line starting from a thin line means that the components connected by the thin line entails the component at which the thick line ends). Thus, one isomorphism from X to X^X generates one completely closed system.

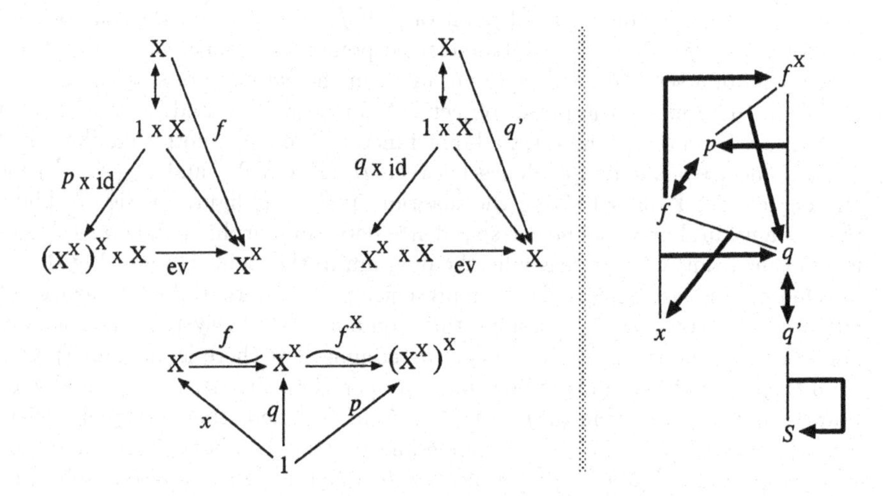

Fig. 2. Diagrams of a Completely Closed System and Its Hyperdigraph on Entailment

3 Conclusion and Discussion

We proposed completely closed systems under entailment in Sec. 2 by assuming the existence of an isomorphism between an object and its power object. Although we cannot do in this paper due to the page limit, we can provide another type of closed systems by assuming similar conditions and abstracting Rosen's (M,R) systems [7].

Although we need to consider some future problems such as coupling of these closed systems, the most important problem is the condition of the category used for constructing closed system. Although we required that operands coincide with operators ($X \simeq X^X$), this condition is difficult to be satisfied in the naive set theory. Although Soto–Andrade and Varela provided a category satisfying this condition (the category of partially ordered sets and continuous monotone maps with special conditions)[8], this category is very special. Furthermore, Rosen showed that systems closed under efficient cause cannot be described with their states because they lead to infinite regress [6]. We have still not clarified whether the existence of an isomorphism between an object and its power object is a sufficient condition for a system to be closed under entailment in the category theoretic framework. If these closed systems can exist only in special categories not observable in the conventional sense, however, autopoiesis may be hard to be a general theory of a variety of systems.

References

1. Higuchi, A., Matsuno, K., and Tsujishita, T.: Deductive Hyperdigraphs, A Method of Describing Diversity of Coherence. preprint (1997).
2. Kampis, G: Self–Modifying Systems in Biology and Cognitive Science: A New Framework for Synamics, Information, and Complexity. Pergamon Press (1991).
3. Maturana, H. R., and Varela, F. J.: Autopoiesis and Cognition: The Realization of the Living. D. Reidel Publishing (1980). (Japanese Edition: Kawamoto, H. Kokubun–sha Publishers. (1991).)
4. Nomura, T: An Attempt for Description of Quasi–Autopoietic Systems Using Metabolism–Repair Systems. Proc. the Fourth European Conference on Artificial Life (1997) 48–56.
5. Nomura, T: A Computational Aspect of Autopoiesis. Proc. the Fourth Asian Fuzzy Systems Symposium (2000) 1–6.
6. Rosen, R: LIFE ITSELF. Columbia University Press (1991).
7. Rosen, R: Some Relational Cell Models: The Metabolism–Repair Systems. In FOUNDATIONS OF MATHEMATICAL BIOLOGY. Academic Press (1972) 217–253.
8. Soto–Andrade, J., and Varela, F. J.: Self–Reference and Fixed Points: A Discussion and an Extension of Lawvere's Theorem. Acta Applicandae Mathematicae **2** (1984) 1–19.
9. Takeuchi, G: Sheaf, Category, and Topos. Nihon Hyoron–sha (1978) (Japanese).

An Information-Theoretic Approach for the Quantification of Relevance

Daniel Polani, Thomas Martinetz, and Jan Kim

Institute for Neuro- and Bioinformatics, University of Lübeck, Germany
{polani,martinetz,kim}@inb.mu-luebeck.de

Abstract. We propose a concept for a Shannon-type quantification of information relevant to a decision unit or agent. The proposed measure is operational, can – at least in principle – be calculated for a given system and has an immediate interpretation as an information quantity. Its use as a natural framework for the study of sensor evolution is discussed.

1 Introduction

1.1 Shannon Information and Semantics

Shannon's seminal study of the transmission capacity of a (possibly noisy) channel introduced one of the most important notions of modern science, namely a quantification of information and of information transmission capability. Shannon's notion of *entropy* of a memory-less source has deep connections to the notion of entropy known from physics and to a generalized principle of insufficient reason which can be extended to a Bayesian framework (Jaynes 1957a,b; Reichl 1980).

One of the important steps taken by Shannon was to do away with the notion of semantics when dealing with transmitted data. In the Shannon view, when a sequence of symbols is transmitted, every symbol carries the same importance; in particular, there is no additional structure on the set of symbols that can be transmitted. Only their probability of occurrence plays a role in the Shannon model.

The non-interpretation of symbols is one of the strengths of the Shannon model and allows it to introduce a powerful and universal notion of information. However, it has long been felt that, for the purposes of understanding the more complex levels of information processing in natural and artificial computation, and of better understanding natural and artificial language, a variant of Shannon's information would be in place that would take into account the semantic content of the symbol sequences transmitted.

As Shannon's approach is not straightforward to generalize in this direction, in the last decades approaches for a formal characterization or quantification of semantics and relevance have been the topic of many discussions. It has been attempted to treat this problem e.g. on the level of philosophy (for a concept of relevance based upon a philosophy of human communication, see e.g.

J. Kelemen and P. Sosík (Eds.): ECAL 2001, LNAI 2159, pp. 704–713, 2001.

Sperber and Wilson 1995), or on the level of conceptual foundations of mathematics (e.g. predicate logic, see Bar-Hillel 1964). However, these approaches lack the interpretability, the conceptual crispness and the universality of the original Shannon approach. What is frequently sought is just a simple operational quantification of "semantic" information, conceptually similar to Shannon's information, but differing in its treatment of semantics and, in particular, being able to quantify *relevance*.

1.2 Adaptive Sensors, POMDPs, and Relevant Information

A new incentive to develop a quantification of *semantic*, or as we will prefer to say here, *relevant* information has been brought forward due to the efforts to study adaptive and evolving sensors. The study of sensor evolution is a branch of Artificial Life that lies at the focal point between models to understand biological sensors (Shaaban et al. 1998; Liese et al. 2000), to construct evolving hardware (Lee et al. 1996; Lund et al. 1997) and between fundamental questions concerning the capabilities of evolution regarding the adaptation to the environment (Pask 1959; Cariani 1993; Menczer and Belew 1994; Mark et al. 1998) or the development of selective perception of environmental stimuli acting as triggers for certain behavior patterns (von Uexküll 1956a,b). It is no surprise that this topic has obtained increased interest in recent work (Dautenhahn, Polani, Uthmann 2001).

In the field of sensor evolution, one is interested in agents that adapt or evolve to solve a certain task. This task can be e.g. maximizing survival time or the number of offspring. For this purpose, the agents are expected to evolve appropriate sensors. Such sensor optimization procedures can be driven by explicit sensor quality measures (Nilsson and Pelger 1994). Often, however, the task is implicitly formulated and there is no a priori knowledge about which sensor properties are going to be relevant for the agents to be able to solve it.

The question important in this context is now when sensors will evolve that attempt to capture as much information about the agent environment as possible and when the sensors will be limited to capture only a very restricted bandwidth of information channels. This intuitively depends, among other factors, on what amount of information in the environment is *relevant* for the agent. What we are interested in is to give a precise quantification of this amount of information.

This view is also closely related to the study of partially observable Markovian decision processes (POMDPs, Kaelbling et al. 1996), as also in those it is of particular interest to quantify how much information from the environment is needed to form an adequate decision.

These considerations show that quantifying *relevant* information is not an abstract or exotic requirement, but promises to provide a central measure for the quality of information processing in a system, which is an essential aspect in the studies of Artificial Life systems.

2 Quantification Approaches

To quantify the relevance of information or, alternatively, the amount of semantics or relevance in a given data channel, it has become clear that the semantics has to be brought in by recipient of the data (Wittgenstein 1958; Nehaniv 1999). Nehaniv considers the meaning to arise from the "decoding" of information by given agents. An agent is here seen as an automaton where equivalent states with respect to the behavior of the agent or with respect to to its goals are collapsed into a single state (see e.g. Crutchfield 1994). In this model, the external data influence the agent automaton (for instance, to attain certain goal states).

A framework for the modeling of strategies to achieve goals is provided by decision theory (Bertsekas 1976) which also underlies the theory of Markovian Decision Problems (Sutton and Barto 1998). In the context of decision theory, a theory of information value has been developed (Howard 1966; Poh and Horvitz 1996). A utility function is a suitably normalized measure that quantifies the reward obtained by taking a certain action in a given state. Solving the decision problem means finding that action that gives the highest expected reward. If the exact state is not perfectly known to the decision system, the attained reward will typically drop below that of the fully informed decision system. In general, given a piece of information (e.g. in form of a random variable that is correlated with the true state variable of the system), it will be possible to attain a higher expected utility than without it. The difference between the two attained values is the information value of the variable according to information value theory. Note that this quantifies the *value* of the information which, in turn, is measured in the units the utility function is given in, be it monetary value, fitness, score or other.

For a quantified notion of *relevant* information in the vein of the Shannon sense, however, different properties would be desirable. In particular, one would like to have is a measure that is able to be interpreted in a sense similar to the classical Shannon information; loosely speaking, its value should be expressed in *bits*, the "universal unit of information" and not in the arbitrary units of some utility measure.

Nehaniv (1999) grounds the meaning of Shannon-type information in the usefulness of signals detected by sensors and of actuators manipulating the world and producing signals. This usefulness is there put into the context of fitness (e.g. reproductive success) or also to more short-term goals like homeostasis. In particular, Nehaniv considers this concept in relation to sensor and actuator evolution; as example, he discusses the trade-off of cost and benefit versus the signal length for information transmission evolved for the signaling behavior by squids. In addition, different meanings and, thus, different degrees of informativity can arise for different observers in the same signal channel, even in the same context, depending on the respective goals perceived (von Uexküll 1956b; Nehaniv et al. 1999).

Like Nehaniv, we will relate the amount of relevant information present in a channel to the usefulness to the agent that can be derived from its knowledge. To do so, we here formalize the concept of usefulness via the framework of decision theory. In addition, we develop a formalism that allows to model the actuator channel as a random variable and to compute the mutual information between sensor and actuator channel which then will serve as measure for the amount of relevant information.

3 A Measure for Relevant Information

First, we introduce some notation. We will denote random variables by uppercase letters (e.g. X), the values they can assume by lowercase letters (e.g. x), the set of these values by calligraphic letters (e.g. $\mathcal{X}$). The probability that a random variable X assumes a value $x \in \mathcal{X}$ will be denoted by $P(X = x)$ or, for simplicity by $p(x)$ by abuse of notation. Likewise, we will write $p(y|x)$ instead of $P(Y = y|X = x)$ for the conditional probability for Y assuming the value y, given that X assumes the value x.

To keep the mathematics simple, we will always assume the random variables assume a finite number of (therefore discrete) values. The *entropy* of a random variable X is defined as

$$H(X) = -\sum_{x\in\mathcal{X}} p(x) \log p(x) , \tag{1}$$

where for convenience we assume the binary logarithm. The conditional entropy $H(Y|X)$ of a random variable Y given X is defined as

$$H(Y|X) := \sum_{x\in\mathcal{X}} p(x) H(Y|X = x) \tag{2}$$

$$:= -\sum_{x\in\mathcal{X}} p(x) \sum_{y\in\mathcal{Y}} p(y|x) \log p(y|x) . \tag{3}$$

To construct a measure for the relevant information present in a random variable X, Tishby et al. (1999); Kaski and Sinkkonen (2000) assume the existence of some relevance indicator variable Y that is jointly distributed with X. Then they define the relevant information in X via the *mutual information*

$$I(X;Y) = H(Y) - H(Y|X) = H(X) - H(X|Y) \tag{4}$$

between the variables X and Y. The mutual information quantifies (in a Shannon sense) how many bits of uncertainty about Y can be resolved by knowing X. In their work, the relevance indicator variable has to be given a priori, e.g. by appropriate labeling of training samples for X or by human intervention. The relevance is determined by Y. In other words, the model is to determine how many relevant information (about Y) can be extracted from knowing X.

We introduce now our notion of *relevant information* by instantiating the relevance indicator variable Y from the models of Tishby et al.; Kaski and Sinkkonen with a specific, decision-theoretic model (Polani et al. 2001), grounding, as Nehaniv (1999), the relevance of an action in its utility to the agent.

We assume a decision system with (for simplicity of exposition) finite state and action spaces. Our key assumption is that for a given state x, there exists a set $\mathcal{Y}^*(x)$ of equivalently optimal actions $\mathcal{Y}^*(x) := \{y_{i_1^*} \dots y_{i_l^*}\} \subseteq \mathcal{Y} = \{y_1 \dots y_k\}$, where $\mathcal{Y}$ is the set of possible actions. At this point we make no assumptions about the model that selects the optimal actions; one could, for instance, imagine the optimal actions to be maximizing the expectation value of some suitable utility function. Given that state x, we define a conditional probability $p(y|x)$ of selecting an action y by $p(y|x) := 1/|\mathcal{Y}^*(x)|$ if $y \in \mathcal{Y}^*(x)$, else setting $p(y|x) := 0$, where $|\,.\,|$ denotes the number of elements in a set. In other words, given a state x, then the conditional probability $p(y|x)$ that an optimal action y will be selected arises from the principle of insufficient reason (or *maximum entropy*, Jaynes 1957a,b) if several optimal actions exist. We call $p(y|x)$ the *action selection model.*

Let now $p(x)$ be the probability that the system is in state x; denote the respective random variable by X. Define now the random variable Y^* in such a way that it is jointly distributed with X via $p(x,y) = p(y|x)p(x)$, where the $p(y|x)$ is the action selection model given above. Then Y^* denotes a random variable modeling the selection of an optimal action. Then, using Y^* as relevance indicator variable, we define in the sense of Tishby et al.; Kaski and Sinkkonen the relevant information in X for the decision system by the mutual information $I(X;Y^*)$ between the state variable X and the action selection variable Y^*.

4 Some Simple Examples

In this section, some very simple prototypical examples are given that illustrate how the notion of relevant information operates. For this purpose, a couple of instructive constellations are investigated. We generally consider a simple scenario where the decision maker can assume 8 possible states x and can select between 8 possible actions y in each state x. Given a set $\mathcal{Y}^*(x)$ of optimal actions for state x, the action selection model from Sec. 3 becomes

$$p(y|x) = P(Y^* = y \mid X = x) = \frac{1}{|\mathcal{Y}^*(x)|} \mathbb{I}_{\mathcal{Y}^*(x)}(y) , \tag{5}$$

where $\mathbb{I}_{\mathcal{Y}^*(x)}$ is the characteristic function of $\mathcal{Y}^*(x)$ which is 1 for $y \in \mathcal{Y}^*(x)$ and 0 else.

Consider first the situation in Fig. 1(a). The vertical axis denotes the state of the (decision making) agent, the horizontal axis the possible actions. A given state x defines a row, in which darkened squares indicate the action selection model. In the example of Fig. 1(a), for each distinct state there is a unique distinct optimal action to take. One obtains $p(y|x) = \delta_{x,f(x)}$ with a suitably chosen function f. It is easy to see that, in this case, $H(Y^*|X) = 0$ since the optimal action variable Y^* is perfectly determined by knowing the value of X. The relevant information becomes $I(X;Y^*) = H(Y^*) - H(Y^*|X) = H(Y^*) = H(X)$ (the last equation holds because f is injective, i.e. $f(x') \neq f(x'')$ for

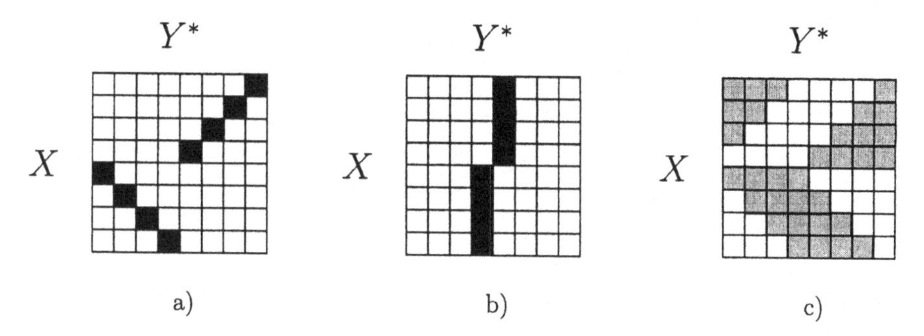

Fig. 1. Illustration of the operation of relevant information. The darkened squares indicate the optimal actions y for a state x. The depth of the shading represents the value $p(y|x)$ of the action selection model, where black denotes a value of 1.

$x' \neq x''$). Thus, full information about the state X is relevant since each state x requires its individual optimal response.

In Figure 1(b), this is different. It is still possible to consider the (unique) optimal action as function $y = f(x)$ of the present state x, but f is no longer injective. The relevant information is still given by $H(Y^*)$, but now $H(Y^*) < H(X)$. For equiprobabilistic X, one obtains $I(X;Y^*) = H(Y^*) - 0$ bit $= 1$ bit as relevant information. The full uncertainty of the present state is $H(X) = 3$ bit. This means that to find a relevant action, the current state has to be determined only to the precision of 1 bit and not to its full precision of 3 bit.

Fig. 1(c) somewhat reminds of Fig. 1(a), but now for each state x, there is a selection of four different actions that are equivalently optimal. In this case, the action selection model becomes $p(y|x) = 1/4$ for an action y that is optimal for x and $p(y|x) = 0$ else. Easy calculation shows that the relevant information becomes $I(X;Y^*) = H(Y^*) - H(Y^*|X) = 3$ bit $-$ 2 bit $= 1$ bit, assuming X is equiprobabilistic. This fits with intuition, as one has more freedom to select an action and less specific state information is required to decide which actions to take. Thus, less relevant information is therefore attributed to the system.

5 Discussion

Above definition of relevant information is natural in many senses. It has an immediate interpretation as an information quantity of Shannon type. Via the decision mechanism it addresses and incorporates the subjectivity of relevance (i.e. it takes into account the question "*relevant for what purpose?*"). When there is no available choice in actions, i.e. always the same unique action is optimal, the relevant information vanishes. In this case, it is of no value knowing the state of the system.

The same holds if there is no correlation between the state and the possible actions that can be chosen, e.g. if the action selection model $p(y|x)$ is identical for

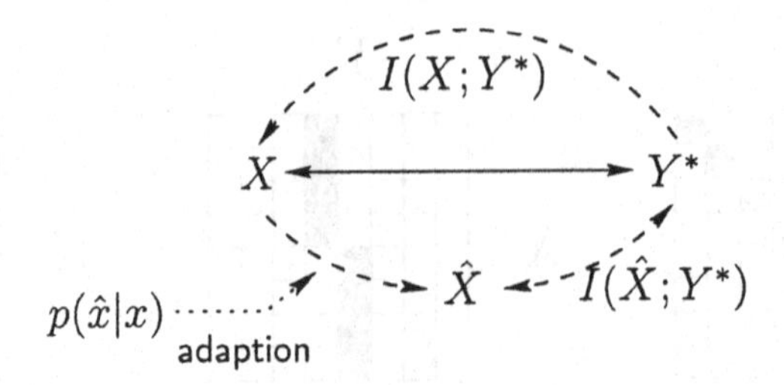

Fig. 2. Relevant information and information acquisition via a sensoric channel. See text for details.

all x. This fits intuition that, if, by knowing X, the choice of Y^* is not affected, there is no relevant information in X with respect to Y^*.

The potential of the approach lies in the fact that, given an action selection model, it does not only provide a very natural quantification of the amount of relevant information given a certain state, but it allows also the study of the effect of incomplete state information. For illustration, assume the (true) state variable X of the agent to be perceived by the agent not directly, but through a sensorial channel which is modeled by a random variable $\hat{X}$, obtained from X via some probabilistic mapping $p(\hat{x}|x)$, analogously to the scenario of Tishby et al.'s information bottleneck method (Fig. 2). For the selection of an action, the total amount of relevant information is given by $I(X;Y^*)$. However, the "useful" information available to the agent via the sensoric state variable $\hat{X}$ is only $I(\hat{X};Y^*)$. Elementary information theory allows to show that $I(\hat{X};Y^*) \leq I(X;Y^*)$. The relevant information $I(X;Y^*)$ therefore serves to gauge the maximum amount of useful information that the sensorial channel can possibly extract from the present state of the agent.

A paradigmatic instance of such a scenario is the adaption or evolution of sensors. In this context, $I(\hat{X};Y^*)$ as compared to $I(X;Y^*)$ provides a measure for the usefulness of a sensory variable for an agent to solve the task at hand. If $I(X;Y^*)$ is already small (which can e.g. be the case even for large state spaces if the set of optimal actions is small) then even a simple and thus cheap sensor may capture most of the relevant information in the environment and this amout can instantly be quantified by $I(\hat{X};Y^*)$. The trade-off between the bandwidth of the sensor channel and the overall relevant information can be used to fine-tune the necessary resources. Here, the information bottleneck framework by Tishby et al. (1999) provides powerful tools for theoretical studies of the setting. In addition, we emphasize that the concept of relevant information developed here is immediately applicable to the study of POMDPs (Kaelbling et al. 1996). POMDPs, in turn, we believe to have deep and intricate connections to the field of sensor evolution whose exploration is still to be undertaken.

6 Summary and Future Work

A framework for the quantification of relevant information has been presented. Combining decision- and information-theoretic aspects, the concept incorporates many properties that are intuitively expected from such a quantity. It quantifies relevance with respect to a given agent or decision system, it yields a measure for the usefulness of sensors or about an agent's state of knowledge with respect to a POMDP. The concept introduces an operational notion of a Shannon-type quantification of relevant information that can, in principle, always be computed.

We intend to apply the notion of relevant information that has been introduced here on framework level to a couple of sensor evolution scenarios in the near future. Abstracted sensor evolution models can be easily interpreted in the terms developed in the present paper. They provides a suitable framework to better understand the mechanisms that govern sensor evolution and to pursue methods of improving it that may be generalizable to hardware. Jung et al. (2001b) developed a simple but instructive sensor evolution model in whose analysis our present concept of relevant information was used (Jung et al. 2001a).

On a conceptual level, there are several interesting aspects by which we plan to extend our model. First, the notion of relevant information can be strengthened to quantify the *minimum* required information about the system state X to choose a right action Y^*. This view slightly changes the method how the action selection model is constructed and the consequences of this (slightly more complicated) modification are to be explored.

A second important aspect is the fact that often there may be few exactly optimal actions, but a large selection of suboptimal actions which lead almost to the same reward as the optimal one. Thirdly, it is in general not easy to identify precisely the optimal action to take in a given state. The decision system will usually have only a limited model about which action may be the best.

Interestingly, the last two problems can be solved at once by attacking them simultaneously. One can regard the decision system or agent as an estimator for the optimal action to take. The estimator, e.g. being based on some learning or evolution system, will be trained by experimental data. It will typically give imperfect estimations about which actions are optimal, depending on the history of its training. The action selection model $p(y|x)$ then reflects this imperfection. As optimal and suboptimal rewards converge, by virtue of the continuity of the estimator will cause also the action model to converge in a way that is consistent with the specialized definition from Sec. 3. The imperfect model $\hat{Y}^*$ for the action to be taken is then be regarded as an imperfect "sensor" for the best possible action, in analogy to the sensor part. A forthcoming paper will develop this generalization of the model in detail.

Acknowledgements. We wish to express our gratitude to the anonymous reviewers for their detailed and helpful comments.

References

Bar-Hillel, Y., (1964). *Language and Information.* Addison-Wesley.

Bertsekas, D. P., (1976). *Dynamic Progamming and Stochastic Control*, vol. 125 of *Mathematics in Science and Engineering.* New York: Academic Press, Inc. First edition edition.

Cariani, P., (1993). To evolve an ear: epistemological implications of Gordon Pask's electrochemical devices. *Systems Research*, 10(3):19–33.

Crutchfield, J. P., (1994). Observing Complexity and the Complexity of Observation. In Atmanspacher, H., editor, *Inside versus Outside*, 234–272. Springer.

Dautenhahn, Polani, Uthmann, (2001). Artificial Life Journal — Special Issue on Sensor Evolution. *Artificial Life Journal — Special Issue on Sensor Evolution*, 7(2). (in press).

Howard, R. A., (1966). Information value theory. *IEEE Transactions on Systems Science and Cybernetics*, SSC-2:22–26.

Jaynes, E. T., (1957a). Information theory and statistical mechanics. *Phys. Rev.*, 106(4):620–630.

Jaynes, E. T., (1957b). Information theory and statistical mechanics II. *Phys. Rev.*, 108(2).

Jung, T., Dauscher, P., and Uthmann, T., (2001a). On Individual Learning, Evolution of Sensors and Relevant Information. In Polani, D., Uthmann, T., and Dautenhahn, K., editors, *Proc. GECCO-2001 Workshop on Evolution of Sensors in Nature, Hardware and Simulation.*

Jung, T., Dauscher, P., and Uthmann, T., (2001b). Some Effects of Individual Learning on the Evolution of Sensors. In *Proc. European Conference on Artificial Life (ECAL 2001, Prague).*

Kaelbling, L. P., Littman, M. L., and Moore, A. W., (1996). Reinforcement learning: a survey. *Journal of Artificial Intelligence Research*, 4:237–285.

Kaski, S., and Sinkkonen, J., (2000). Metrics that learn relevance. In *Proc. IJCNN-2000*, vol. V, 547–552. IEEE Service Center, Piscataway, NJ.

Lee, W.-P., Hallam, J., and Lund, H. H., (1996). A Hybrid GP/GA Approach for Co-evolving Controllers and Robot Bodies to Achieve Fitness-Specified Tasks. In *Proc. IEEE 3rd Int. Conference on Evolutionary Computation.* NJ: IEEE Pr.

Liese, A., Polani, D., and Uthmann, T., (2000). On the development of spectral properties of visual agent receptors through evolution. In Whitley, D., Goldberg, D., Cantú-Paz, E., Spector, L., Parmee, I., and Beyer, H.-G., editors, *Proc. Genetic and Evolutionary Computation Conference (GECCO), Las Vegas, Nevada*, 857–864. Morgan Kaufmann.

Lund, H. H., Hallam, J., and Lee, W.-P., (1997). Evolving Robot Morphology. In *Proc. IEEE 4th Int. Conference on Evolutionary Computation.* NJ: IEEE Press.

Mark, A., Polani, D., and Uthmann, T., (1998). A Framework for Sensor Evolution in a Population of Braitenberg Vehicle-like Agents. In Adami, C., Belew, R., Kitano, H., and Taylor, C., editors, *Proc. of Artificial Life VI, Los Angeles, June 26-29*, 428–432. MIT Press.

Menczer, F., and Belew, R. K., (1994). Evolving Sensors in Environments of Controlled Complexity. In Rodney A. Brooks, P. M., editor, *Artificial Life IV: Proceedings of the Fourth International Workshop on the Synthesis and Simulation of Living Systems*, 210–221. MIT Press.

Nehaniv, C. L., (1999). Meaning for Observers and Agents. In *Proc. IEEE International Symposium on Intelligent Control / Intelligent Systems and Semiotics, ISIC/ISAS'99*, 435–440. Cambridge, Massachusetts, USA.

Nehaniv, C. L., Dautenhahn, K., and Loomes, M. J., (1999). Constructive Biology and Approaches to Temporal Grounding in Post-Reactive Robotics. In McKee, G. T., and Schenker, P., editors, *Sensor Fusion and Decentralized Control in Robotics Systems II, Boston, Mass. (Proc. SPIE Vol. 3839)*, 156–167.

Nilsson, D.-E., and Pelger, S., (1994). A pessimistic estimate of the time required for an eye to evolve. *Proc. Roy. Soc. Lond.*, 256:53–58.

Pask, G., (1959). Physical Analogues to the Growth of a Concept. In *Mechanisation of Thought Processes: Proceedings of a Symposium held at the National Physical Laboratory (No. 10)*, vol. II, 877–928. London: Her Majesty's Stationary Office.

Poh, K. L., and Horvitz, E., (1996). A Graph-Theoretic Analysis of Information Value. In Horvitz, E., and Jensen, F., editors, *Proc. Twelfth Conf. on Uncertainty in Artificial Intelligence*, 427–435. San Francisco, CA: Morgan Kaufman.

Polani, D., Martinetz, T., and Kim, J., (2001). On the Quantification of Relevant Information. Presented at SCAI'01 (Scandinavian Conference on Artificial Intelligence), Feb. 19-21, 2001.

Reichl, L., (1980). *A Modern Course in Statistical Physics.* Austin: University of Texas Press.

Shaaban, S. A., Crognale, M. A., Calderone, J. B., Huang, J., Jacobs, G. H., and Deeb, S. S., (1998). Transgenic Mice Expressing a Functional Human Photopigment. *Investigative Ophtalmology & Visual Science*, 39(6):1036–1043.

Sperber, D., and Wilson, D., (1995). *Relevance: Communication and cognition.* Oxford: Blackwell. Second edition.

Sutton, R. S., and Barto, A. G., (1998). *Reinforcement Learning.* Bradford.

Tishby, N., Pereira, F. C., and Bialek, W., (1999). The Information Bottleneck Method. In *Proc. 37th Annual Allerton Conference on Communication, Control and Computing, Illinois.*

von Uexküll, J., (1956a). *Bedeutungslehre.* Hamburg: Rowohlt.

von Uexküll, J., (1956b). *Streifzüge durch die Umwelten von Tieren und Menschen.* Hamburg: Rowohlt.

Wittgenstein, L., (1958). *The Blue and Brown Books.* Harper & Brothers.

First Implementation of the World-Wide-Mind

Ray Walshe and Mark Humphrys

Dublin City University, School of Computer Applications
Glasnevin, Dublin 9, Ireland
{ray,humphrys}@compapp.dcu.ie
www.compapp.dcu.ie/~ray
www.compapp.dcu.ie/~humphrys

Abstract. The "World-Wide-Mind" (WWM) is introduced in [1]. For a short introduction see [2]. Under this scheme, it is proposed that autonomous agents researchers in AI and ALife construct their agent minds and agent worlds as *servers* on the Internet. Users will be able to run remote 3rd party Minds in other remote 3rd party Worlds. And users will be able to construct complex "Societies of Mind" out of many different remote Mind servers, and run this Society as a single Mind in some World. The motivation is: (a) to re-use other people's Worlds, (b) to re-use other people's Minds as components in larger, multiple-mind cognitive systems, and (c) to divide up the work in AI, so people can specialise on different parts.
This poster details the first working implementation of this idea. The key principle behind this implementation is to make it *as trivially easy as possible* for any author of a World or Mind to put it online without having to learn any network programming or any particular language. All technology involving a particular programming language (e.g. Java) or requiring the learning of particular programming skills (e.g. sockets) has therefore been rejected. A solution is proposed where the Mind or World author need only know how to repeatedly read plain text from and write plain text to a local file.

1 Introduction

Using the terminology in [1], a "World" server is a general environment that can be queried for the current "state", and that receives "actions" to be executed in that World. A "Mind" server receives state and returns an action. [1, §9] attempts to define a full set of queries that WWM servers should respond to. The lowest common denominator approach explained in [1, §11] argues that all queries to a server and all response data should be plain text in a standard, extensible XML format.

2 Putting a Mind or World Online

Given that a researcher in ALife (or related fields) has constructed a Mind or World, how can this be put online? We propose that the Mind or World be a

J. Kelemen and P. Sosík (Eds.): ECAL 2001, LNAI 2159, pp. 714–718, 2001.

program running on a Web server, repeatedly reading and writing local files. A separate program will then provide online access to these files. For both Minds and Worlds, the Web server needs installed:

1. **Program 1 – The WWM server – A persistent program, representing the Mind or World.** – This will repeatedly read plaintext queries from and write plaintext responses to local files on the server. It will have its own persistent data structures in memory (global variables, gridworld, state-space, neural network, etc.). It doesn't have to be repeatedly launched. It launches once, and then repeatedly reads queries from the file until (if ever) the query "End run" is read.
2. **Program 2 – The CGI script – A non-persistent program through which the outside world talks to the persistent program.** – This accepts a WWM command across CGI, reads or writes a local file on the server, and exits. The "Start run" command should start a persistent instance of Program 1.

Now, to put his Mind or World online, the server author only has to write Program 1 (i.e. rewrite his program so it repeatedly reads/writes a local file). He may download a standard Program 2 from other WWM researchers. He then installs Program 2 in `cgi-bin` (or wherever) on his Web server, and then installs his Program 1 in some directory where the CGI script can run it from.

3 Program 1

Pseudo-Code examples are now provided. Here, the XML queries are not yet implemented. The World simply repeatedly reads actions and outputs state, and the Mind vice-versa. An initial system like this is running on our first node: `wwm.compapp.dcu.ie` with CGI scripts in Perl and WWM servers in Java. Obviously it must be shown that WWM servers can be written in any language.

3.1 The Original AI Program

The original AI program is probably an offline, combined World and Mind, program something like the following:

```
repeat
{
 State = World.GetState();
 Action = Mind.GetAction(State);
 World.ExecuteAction(Action);
 State = World.GetState();
}
```

3.2 The Re-written World Server

To put his system online, the AI author will break it into two programs, a World server and a Mind server. The World server will look something like this:

```
// Set up initial state, so as to provoke initial action:
State = GetState();
WriteStateFile(State);

repeat
{
 Action = ReadActionFile();

 if (Action != null)
 {
  ExecuteAction(Action);
  State = GetState();
  WriteStateFile(State);
 }
}
```

There is a StateFile and an ActionFile on the World server.

3.3 The Re-written Mind Server

```
repeat
{
 State = ReadStateFile();

 if (State != null)
 {
  Action = GetAction(State);
  WriteActionFile(Action);
 }
}
```

There is also a StateFile and an ActionFile on the Mind server. Remember this is a different machine, so these are different files to the ones the World server sees.

4 Program 2

Remember, Program 2 is simply downloaded from a standard source.

4.1 World CGI Script

```
if (Query == StartRun)
        run_command(StartWorld);
if (Query == EndRun)
        run_command(EndWorld);
if (Query == GetState)
{
        State = ReadStateFile();
        echo State
}
if (Query == ExecuteAction)
{
        get Action from QUERY_STRING
        WriteActionFile(Action);
}
```

There are two basic commands – `GetState()` is how the states that the World is writing to the StateFile are read by the outside world – and `ExecuteAc-tion(Action)` is how Actions from the outside world get into the ActionFile for the World to read.

4.2 Mind CGI Script

```
if (Query == StartRun)
        run_command(StartMind);
if (Query == EndRun)
        run_command(EndMind);
if (Query == HereIsState)
{
        get State from QUERY_STRING
        WriteStateFile(State);
}
if (Query == GetAction)
{
        Action = ReadActionFile();
        echo Action
}
```

The single basic command `GetAction(State)` could be broken into *two commands* – one that tells the Mind the state, and another that collects the Action from the ActionFile later.

5 Conclusion

We have demonstrated a lowest-common-denominator approach to implementing the WWM concept. We believe considerable work needs to be done on making installation as simple as possible in any language on any OS for this scheme to become widely used. We are working on defining rudimentary "template" Minds and Worlds in all major programming languages to show mind and world authors how to get them online. Funding has also been secured to work on client user software.

References

1. Humphrys, M. (2001), *The World-Wide-Mind: Draft Proposal*, Dublin City University, School of Computer Applications, Technical Report CA-0301, `http://www.compapp.dcu.ie/~humphrys/WWM`
2. Humphrys, M. (2001a), Distributing a Mind on the Internet: The World-Wide-Mind, to appear in *ECAL-01*.

Evolving Lives: The Individual Historical Dimension in Evolution

Rachel Wood

Centre for Computational Neuroscience and Robotics (CCNR)
School of Cognitive and Computing Sciences
University of Sussex, Brighton, UK
rachelwo@cogs.susx.ac.uk

Abstract. Some benefits of a dialogue between evolutionary robotics and developmental ethology are presented with discussion of how developmental models might inform approaches to evolution. Notions of the importance of historical processes in adaptation are outlined and parallels between evolution and ontogeny as sources of change examined.

1 Introduction

Evolutionary robotics encompasses an extraordinary range of ideas and approaches. This paper is intended as an addition to the significant body of existing work which calls for consideration of ontogeny in evolution (e.g. [11,7,13,12,10, 9,4].) The perspective presented here has two parts, the first is epistemological and concerns the descriptive dichotomies built into models of adaptive behaviour. The seond argues for explicit investigation of morphology and the role of changes in physical scale and proportion in adaptation. Although evolution is invoked as an explanation for developmental phenomena, the converse perspective is rarely explored. Dobzhansky's claim "[n]othing in biology makes sense except in the light of evolution."[5] has been taken so much to heart that evolutionary models often make no reference to ontogeny. In development behavioural change is demonstrably embedded in, and scaffolded by, systematic changes both within and without the agent. On this view the development of adaptation should be understood as a history comprising multi-layered feedback between form and function. Development changes what an agent *is*, thus altering what it *does* and what constitutes its world. This characterisation is equally applicable to evolution both consist in historical processes which produce an adaptive fit between organism and environment. Thus the most useful question in relating macro and micro processes concerns how the products of evolution are enacted through development.

There are straightforward reasons for using artificial systems to study developmental processes. Natural populations are non-homogeneous, there are constraints on the manipulations that can be done and it is often impossible to control for the effects of extraneous variables. The advantages of a developmental perspective for robotics are less often noted. Evolutionary robotics has its

J. Kelemen and P. Sosík (Eds.): ECAL 2001, LNAI 2159, pp. 719–722, 2001.

roots in the failure of AI systems and is founded on ideas of adaptive behaviour as shaped by dynamic temporal interactions between elements of embodied, situated systems. In approaches to ontogeny these concerns are echoed most clearly by Developmental Systems Theory [11]. According to Oyama, developmental systems comprise "networks of mutually dependent controls and effects evolving over time."(ibid, pp. 129-130) This view emphasises control - the problem of explaining how change is constrained and adaptation conserved is non-trivial at both evolutionary and onotgenetic scales. There is a strong tendency (not just in A.I.) to divide developmental processes into classes based on causation, where the cause is located in relation to an agent and the behavioural effects produced. These epistemological points would be of restricted theoretical interest except that they reflect deeper dualisms inherent in many models of adaptation. Morphological development is associated with innateness, genetic transmission and mechanisms internal to the agent. the development of behaviour invokes acquisition, environmental constraint and external causation. The information metaphor crops up everywhere. Adaptive systems research has already demonstrated scaffolding of action by embodiment (eg. [12]), but the broader picture of interaction between physical and behavioural change has had little attention.

2 Boundary Problems

The boundary problem begins at the divide between observer and the thing observed, it includes categorisation of causes (internal vs external), and extends over notions of what develops (learning vs innateness). An observer necessarily stands outside the system in question and defines it in terms of her own perspective, it is necessary to have a meaningful definition of the entity under observation and to be able to locate sources of variance. But, is it fair to assume that these are fixed boundaries or that phenomena may be neatly categorised by which side of the agent's skin they originated from. Questioning these dichotomies enables examination of developmental variables as components of a system, without prior categorisation. This approach draws heavily on developmental systems theory [11,7] and the concept of enaction -

> "cognitive capacities are inextricably linked to a history that is lived, much like a path that does not exist but is laid down in walking. Consequently the view of cognition is not that of solvingproblems...but as a creative bringing forth of a world where the only required condition is that it is effective action: it permits the continued integrity of the system involved." (Varela, 1998 in [14, p.8]).

An enactive perspective rejects notions of behavioural adaptation as a series of encapsulated acquisitions by modular systems and emphasises constructive feedback between processes which induce change. Development is not best understood by 'telling tales' - taking an adaptation and working backward to reconstruct the 'problem' it solves. These attempts inevitably entail non-trivial assumptions at all levels. Gould and Lewontin criticise the 'adaptationist programme' [6, p.141] for seeking to decompose organisms into unitary traits forged by natural selection. This functionalism may be seen in developmental models with behaviour treated as discrete units serving domain specific ends. Such approaches tacitly assume optimality of outcome (either developmental or evolutionary), and explicitly assume the validity of hypothetical system boundaries. Just as the concept of a trait defies absolute definition, (ibid) so surely, do behavioural units and systems when divorced from a temporal and dynamic context.

3 Developing Embodiment

The epistemological problems inherent in classical models of cognition are from resolved in modern behavioural AI - there is still a strong tendency to abstract functional modules or units of behaviour as a basis for investigating adaptation. The Cog project (see [3] has the stated aim of developing human-like intelligence through "human-like interaction with the world" (ibid p.1). Cog is embodied, (to the extent of having an humanoid torso), and physically situated, yet its 'engineered' ontogeny overlooks cardinal features of development - *concurrency* and *scaling*. In real development the whole organism shares an ontogeny and experiences incremental, physical changes in its own scale and relative proportions. By contrast, Cog's physical ontogeny proceeds via the addition of functional modules developed and tested on other platforms. Similarly, Cog's body is of fixed proportions and incapable of growth. These omissions seem curious in a project which is founded on ideas of how physical embodiment scaffolds development. Rutkowska highlights another pitfall of ontogeny by design, pre-selection of behavioural units entails significant assumptions about how function is constructed [13]. A striking example of this principle can be seen in Held's studies of plasticity in sensory-motor systems [8]. Held found development of visio-motor coordination in kittens to be dependent on correlated activation in sensory and motor systems. Most importantly the results show that normal perceptual development requires experience of changes in visual stimulation through self-initiated motor-activity. This evidence, (that kittens need to initiate motor activity in order to develop perceptual skill), presents clear difficulties for a functional engineering approach - unless we are prepared to view feet as perceptual organs.

A first step in unravelling the contribution of morpholgical development to adaptation would be to explore issues raised by previous robotics work showing strong effects of form on function. For example Pfeifer and Scheier [12], describe experiments in which sensor morphology apparently determines agent behaviour. With one arrangement of sensors the robot avoids obstacles, with another it 'tidies' them into heaps. If the obstacles are labelled as a resource (e.g. food) then the piling behaviour is interesting (though limited, the robot has no forward vison). It would be useful to extend this line of research to look at non-linear interactions between physical and functional development. Beer uses minimal cognition, to denote examples of simple cognitively, interesting behaviour [1,

2]. Evolving a minimal developmental system would mean finding the simplest level at which it is possible to see developmental self-organisation produced out of a history of structural coupling. On this view evolution can be employed to design systems of mutual developmental feedback. Experimentally this requires an approach in which qualitative behavioural change results from changes in the embodiment of an agent coupled with a history of interaction.

References

1. R.D. Beer. Toward the evolution of dynamical neural networks for minimally cognitive behaviour. In P. Maes, M. Mataric, J. Meyer, J. Pollack, and S. Wilson, editors, *From Animals to Animats 4; Proceedings of the Fourth International Conference on SimulationofAdaptive Behaviour*, pages 421–429. MIT Press, 1996.
2. R.D. Beer. Dynamical approaches to cognitive science. *Trends in Cognitive Science*, 4(3):91–99, 2000.
3. R.A. Brooks, C. Breazeal, M. Marjanovic, B. Scassellati, and M.M Williamson. The cog project: Building a humanoid robot. `http://www.ai.mit.edu/projects/cog`, 1999.
4. K Dautenhahn and C. Nehaniv. Artificial life real stories. *In Proceedings of the Third International Symposium on Artificial Life and Robotics*, volume 2, pages 435–439, 1998.
5. T. Dobzhansky. Nothing in biology makes sense except in the light ofevolution. In M. Ridley, editor, *Evolution*. Oxford University Press, 1997.
6. S.J Gould and R.C. Lewontin. The spandrels of san·marco and the panglossian paradigm: A critique of the adaptationist program. In M. Ridley, editor, *Evolution*. Oxford University Press, 1997.
7. P.E. Griffiths and R.D. Gray. Darwinism and developmental system. In S. Oyama, P.E. Griffiths, and R.D. Gray, editors, *Cycles of Contingency: Developmental Systems and Evolution*. MIT Pres, 2001.
8. R. Held. Plasticity in sensory-motor systems. In *Perception, Mechanisms and Models; Readings from Scientific American*. Scientific American, 1965.
9. C. Nehaniv and K. Dautenhahn. Embodiment and memories - algebras of time and history for autobiographic agents. In R. Trappl, editor, *Proceedings of the 14th European meeting on Cybernetics and Systems Research*, pages 651–656. Austrian Society for Cybernetics Studies, 1998.
10. S Nolfi and D. Parisi. Evolving artificial neural networks that develop in time. In P. Husbands and I. Harvey, editors, *European Conference on Artificial Life*, pages 353–367. MIT Press, 1995.
11. S. Oyama. *The Ontogeny of Information*. Duke University Press, second edition edition, 2000.
12. R. Pfeifer and C. Scheier. *Understanding Intelligence*. MIT Press, 1999.
13. J.C. Rutkowska. Can development be designed? what we may learn from the cog project. *In Procedings of the third European Conference on Artificial Life*. MIT Press, 1995.
14. J.C. Rutkowska. Reassessing Piaget's theory of sensorimotor intelligence: A view from cognitive science. CSRP 369, University of Sussex, 1995.

Author Index

Lecture Notes in Artificial Intelligence (LNAI)

Vol. 1986: C. Castelfranchi, Y. Lespérance (Eds.), Intelligent Agents VII. Proceedings, 2000. XVII, 357 pages. 2001.

Vol. 1991: F. Dignum, C. Sierra (Eds.), Agent Mediated Electronic Commerce. VIII, 241 pages. 2001.

Vol. 1994: J. Lind, Iterative Software Engineering for Multiagent Systems. XVII, 286 pages. 2001.

Vol. 1996: P.L. Lanzi, W. Stolzmann, S.W. Wilson (Eds.), Advances in Learning Classifier Systems. Proceedings, 2000. VIII, 273 pages. 2001. (Subseries LNAI).

Vol. 2003: F. Dignum, U. Cortés (Eds.), Agent-Mediated Electronic Commerce III. XII, 193 pages. 2001.

Vol. 2007: J.F. Roddick, K. Hornsby (Eds.), Temporal, Spatial, and Spatio-Temporal Data Mining. Proceedings, 2000. VII, 165 pages. 2001.

Vol. 2014: M. Moortgat (Ed.), Logical Aspects of Computational Linguistics. Proceedings, 1998. X, 287 pages. 2001.

Vol. 2019: P. Stone, T. Balch, G. Kraetzschmar (Eds.), RoboCup 2000: Robot Soccer World Cup IV. XVII, 658 pages. 2001.

Vol. 2033: J. Liu, Y. Ye (Eds.), E-Commerce Agents. VI, 347 pages. 2001.

Vol. 2035: D. Cheung, G.J. Williams, Q. Li (Eds.), Advances in Knowledge Discovery and Data Mining – PAKDD 2001. Proceedings, 2001. XVIII, 596 pages. 2001.

Vol. 2036: S. Wermter, J. Austin, D. Willshaw (Eds.), Emergent Neural Computational Architectures Based on Neuroscience. X, 577 pages. 2001.

Vol. 2039: M. Schumacher, Objective Coordination in Multi-Agent System Engineering. XIV, 149 pages. 2001.

Vol. 2049: G. Paliouras, V. Karkaletsis, C.D. Spyropoulos (Eds.), Machine Learning and Its Applications. VIII, 325 pages. 2001.

Vol. 2056: E. Stroulia, S. Matwin (Eds.), Advances in Artificial Intelligence. Proceedings, 2001. XII, 366 pages. 2001.

Vol. 2062: A. Nareyek, Constraint-Based Agents. XIV, 178 pages. 2001.

Vol. 2070: L. Monostori, J. Váncza, M. Ali (Eds.), Engineering of Intelligent Systems. Proceedings, 2001. XVIII, 951 pages. 2001.

Vol. 2080: D.W. Aha, I. Watson (Eds.), Case-Based Reasoning Research and Development. Proceedings, 2001. XII, 758 pages. 2001.

Vol. 2083: R. Goré, A. Leitsch, T. Nipkow (Eds.), Automated Reasoning. Proceedings, 2001. XV, 708 pages. 2001.

Vol. 2086: M. Luck, V. Mařík, O. Štěpánková, R. Trappl (Eds.), Multi-Agent Systems and Applications. Proceedings, 2001. X, 437 pages. 2001.

Vol. 2087: G. Kern-Isberner, Conditionals in Nonmonotonic Reasoning and Belief Revision. X, 190 pages. 2001.

Vol. 2099: P. de Groote, G. Morrill, C. Retoré (Eds.), Logical Aspects of Computational Linguistics. Proceedings, 2001. VIII, 311 pages. 2001.

Vol. 2100: R. Küsters, Non-Standard Inferences in Description Logics. X, 250 pages. 2001.

Vol. 2101: S. Quaglini, P. Barahona, S. Andreassen (Eds.), Artificial Intelligence in Medicine. Proceedings, 2001. XIV, 469 pages. 2001.

Vol. 2103: M. Hannebauer, J. Wendler, E. Pagello (Eds.), Balancing Reactivity and Social Deliberation in Multi-Agent Systems. VIII, 237 pages. 2001.

Vol. 2109: M. Bauer, P.J. Gmytrasiewicz, J. Vassileva (Eds.), User Modeling 2001. Proceedings, 2001. XIII, 318 pages. 2001.

Vol. 2111: D. Helmbold, B. Williamson (Eds.), Computational Learning Theory. Proceedings, 2001. IX, 631 pages. 2001.

Vol. 2116: V. Akman, P. Bouquet, R. Thomason, R.A. Young (Eds.), Modeling and Using Context. Proceedings, 2001. XII, 472 pages. 2001.

Vol. 2117: M. Beynon, C.L. Nehaniv, K. Dautenhahn (Eds.), Cognitive Technology: Instruments of Mind. Proceedings, 2001. XV, 522 pages. 2001.

Vol. 2120: H.S. Delugach, G. Stumme (Eds.), Conceptual Structures: Broadening the Base. Proceedings, 2001. X, 377 pages. 2001.

Vol. 2123: P. Perner (Ed.), Machine Learning and Data Mining in Pattern Recognition. Proceedings, 2001. XI, 363 pages. 2001.

Vol. 2132: S.-T. Yuan, M. Yokoo (Eds.), Intelligent Agents: Specification, Modeling, and Application. Proceedings, 2001. X, 237 pages. 2001.

Vol. 2143: S. Benferhat, P. Besnard (Eds.), Symbolic and Quantitative Approaches to Reasoning with Uncertainty. Proceedings, 2001. XIV, 818 pages. 2001.

Vol. 2157: C. Rouveirol, M. Sebag (Eds.), Inductive Logic Programming. Proceedings, 2001. X, 261 pages. 2001.

Vol. 2159: J. Kelemen, P. Sosík (Eds.), Advances in Artificial Life. Proceedings, 2001. XIX, 724 pages. 2001.

Vol. 2166: V. Matoušek, P. Mautner, R. Mouček, K. Taušer (Eds.), Text, Speech and Dialogue. Proceedings, 2001. XIII, 542 pages. 2001.

Vol. 2190: A. de Antonio, R. Aylett, D. Ballin (Eds.), Intelligent Virtual Agents. Proceedings, 2001. VIII, 245 pages. 2001.

Lecture Notes in Computer Science